Student Solutions Manual
to accompany

Chemistry
The Molecular Nature of Matter and Change

Fifth Edition

Prepared by
Patricia Amateis
Virginia Tech
and
Martin S. Silberberg

 **McGraw-Hill**
Higher Education

Boston   Burr Ridge, IL   Dubuque, IA   New York   San Francisco   St. Louis
Bangkok   Bogotá   Caracas   Kuala Lumpur   Lisbon   London   Madrid   Mexico City
Milan   Montreal   New Delhi   Santiago   Seoul   Singapore   Sydney   Taipei   Toronto

The McGraw-Hill Companies

McGraw-Hill
Higher Education

Student Solutions Manual to accompany
CHEMISTRY: THE MOLECULAR NATURE OF MATTER AND CHANGE, FIFTH EDITION
MARTIN S. SILBERBERG

Published by McGraw-Hill Higher Education, an imprint of The McGraw-Hill Companies, Inc., 1221 Avenue of the Americas, New York, NY 10020. Copyright © 2009 by The McGraw-Hill Companies, Inc. All rights reserved. Previous editions © 2006, 2003, and 2000. No part of this publication may be reproduced or distributed in any form or by any means, or stored in a database or retrieval system, without the prior written consent of The McGraw-Hill Companies, Inc., including, but not limited to, network or other electronic storage or transmission, or broadcast for distance learning.

This book is printed on recycled, acid-free paper containing 10% postconsumer waste.

1 2 3 4 5 6 7 8 9 0 QPD/QPD 0 9 8

ISBN: 978-0-07-304860-4
MHID: 0-07-304860-7

www.mhhe.com

CONTENTS

PREFACE

WELCOME TO YOUR STUDENT SOLUTIONS MANUAL

Your Student Solutions Manual (SSM) includes detailed solutions for the Follow-up Problems and highlighted End-of-Chapter Problems in the fifth edition of Martin Silberberg's *Chemistry: The Molecular Nature of Matter and Change.*

You should use the SSM in your study of chemistry as a study tool:
- to better understand the reasoning behind problem solutions. The plan-solution-check format illustrates the problem-solving thought process for the Follow-up Problems and for selected End-of-Chapter Problems.
- to better understand the concepts through their applications in problems. Explanations and hints for problems are in response to questions from students in general chemistry courses.
- to check your problem solutions. Solutions provide comments on the solution process as well as the answer.

To succeed in general chemistry you must develop skills in problem solving. Not only does this mean being able to follow a solution path and reproduce it on your own, but also to analyze problems you have never seen before and develop a solution strategy. Chemistry problems are story problems that bring together chemistry concepts and mathematical reasoning. The analysis of new problems is the most challenging step for general chemistry students. You may face this in the initial few chapters or not until later in the year when the material is less familiar. When you find that you are having difficulty starting problems, do not be discouraged. This is an opportunity to learn new skills that will benefit you in future courses and your future career. The following two strategies, tested and found successful by many students, may help you develop the skills you need.

The first strategy is to become aware of your own thought processes as you solve problems. As you solve a problem, make notes in the margin concerning your thoughts. Why are you doing each step and what questions do you ask yourself during the solution? After you complete the problem, review your notes and make an outline of the process you used in the solution while reviewing the reasoning behind the solution. It may be useful to write a paragraph describing the solution process you used.

The second strategy helps you develop the ability to transfer a solution process to a new problem. After solving a problem, rewrite the problem to ask a different question. One way to rewrite the question is to ask the question backwards—find what has been given in the problem from the answer to the problem. Another approach is change the conditions—for instance, ask yourself what if the temperature is higher or there is twice as much carbon dioxide present? A third method is to change the reaction or process taking place—what if the substance is melting instead of boiling?

At times, you may find slight differences between your answer and the one in the SSM. Two reasons may account for the differences. First, SSM calculations do not round answers until the final step so this may impact the exact numerical answer. Note that in preliminary calculations extra significant figures are retained and shown in the intermediate answers. The second reason for discrepancies may be that your solution route was different from the one given in the SSM. Valid alternate paths exist for many problems, but the SSM does not have space to show all alternate solutions. So, trust your solution as long as the discrepancy with the SSM answer is small, and use the different solution route to understand the concepts used in the problem.

Some problems have more than one correct answer. For example, if you are asked to name a metal, there are over 80 correct answers.

I would like to thank Donna Nemmers and the staff at McGraw-Hill, for their assistance in completing this project.

Patricia Amateis
Virginia Tech

CHAPTER 1 KEYS TO THE STUDY OF CHEMISTRY

Most numerical problems will have two answers, a "calculator" answer and a "true" answer. The calculator answer, as seen on a calculator, will have one or more additional significant figures. These extra digits are retained in all subsequent calculations to avoid intermediate rounding error. Rounding of a calculator answer to the correct number of significant figures gives the true (final) answer.

FOLLOW–UP PROBLEMS

1.1   Plan: The real question is "Does the substance change composition or just change form?"
Solution: The figure on the left shows red atoms and molecules composed of one red atom and one blue atom. The figure on the right shows a change to blue atoms and molecules containing two red atoms. The change is **chemical** since the substances themselves have changed in composition.

1.2   Plan: The real question is "Does the substance change composition or just change form?"
Solution:
a) Both the solid and the vapor are iodine, so this must be a physical change.
b) The burning of the gasoline fumes produces energy and products that are different gases. This is a chemical change.
c) The scab forms due to a chemical change.

1.3   Plan: We need to know the total area supplied by the bolts of fabric, and the area required for each chair. These two areas need to be in the same units. The area units can be either ft^2 or m^2. The conversion of one to the other will use the conversion given in the problem: 1 m = 3.281 ft.
Solution:

$$(3 \text{ bolts})\left(\frac{200 \text{ m}^2}{1 \text{ bolt}}\right)\left(\frac{(3.281 \text{ ft})^2}{(1 \text{ m})^2}\right)\left(\frac{1 \text{ chair}}{31.5 \text{ ft}^2}\right) = 205.047 = \textbf{205 chairs}$$

Check: 200 chairs require about 200 x 32 / 10 = 640 m^2 of fabric. (Round 31.5 to 32, and 1 / $(3.281)^2$ to 1 / 10) Three bolts contain 3 x 200. = 600. m^2.

1.4   Plan: The volume of the ribosome must be determined using the equation given in the problem. This volume may then be converted to the other units requested in the problem.
Solution:

$$V = \frac{4}{3}\pi r^3 = \frac{4}{3}(3.14159)\left(\frac{21.4 \text{ nm}}{2}\right)^3 = 5131.45 \text{ nm}^3$$

$$(5131.45 \text{ nm}^3)\left(\frac{(10^{-9} \text{ m})^3}{(1 \text{ nm})^3}\right)\left(\frac{(1 \text{ dm})^3}{(0.1 \text{ m})^3}\right) = 5.13145 \times 10^{-21} = \textbf{5.13 x 10}^{-21} \textbf{ dm}^3$$

$$(5.13145 \times 10^{-21} \text{ dm}^3)\left(\frac{1 \text{ L}}{(1 \text{ dm})^3}\right)\left(\frac{1 \text{ } \mu\text{L}}{10^{-6} \text{ L}}\right) = 5.13145 \times 10^{-15} = \textbf{5.13 x 10}^{-15} \textbf{ } \mu\textbf{L}$$

Check: The magnitudes of the answers are reasonable. The units also agree.

1.5   Plan: The time is given in hours and the rate of delivery is in drops per second. A conversion(s) relating seconds to hours is needed. This will give the total number of drops, which may be combined with their mass to get the total mass. The mg of drops will then be changed to kilograms. The other conversions are given in the inside back cover of the book.

Solution:

$$8.0 \text{ h} \left(\frac{60 \text{ min}}{1 \text{ h}} \right) \left(\frac{60 \text{ s}}{1 \text{ min}} \right) \left(\frac{1.5 \text{ drops}}{1 \text{ s}} \right) \left(\frac{65 \text{ mg}}{1 \text{ drop}} \right) \left(\frac{10^{-3} \text{ g}}{1 \text{ mg}} \right) \left(\frac{1 \text{ kg}}{10^3 \text{ g}} \right) = 2.808 = \textbf{2.8 kg}$$

Check: Estimating the answer — (8 h) (3600 s / h) (2 drops / s) (0.070 kg/10^3 drop) = 4 kg. The final units in the calculation are the desired units.

1.6   Plan: The volume unit may be factored away by multiplying by the density. Then it is simply a matter of changing grams to kilograms.
Solution:

$$4.6 \text{ cm}^3 \left(\frac{7.5 \text{ g}}{\text{cm}^3} \right) \left(\frac{1 \text{ kg}}{1000 \text{ g}} \right) = 0.0345 = \textbf{0.034 kg}$$

Check: 5 x 7 / 1000 = 0.035, and the calculated units are correct.

1.7   Plan: Using the relationship between the Kelvin and Celsius scales, change the Kelvin temperature to the Celsius temperature. Then convert the Celsius temperature to the Fahrenheit value using the relationship between these two scales.

$$T \text{ (in} ^\circ\text{C)} = T \text{ (in K)} - 273.15 \qquad\qquad T \text{ (in} ^\circ\text{F)} = \frac{9}{5} T \text{ (in} ^\circ\text{C)} + 32$$

Solution:
$T \text{ (in} ^\circ\text{C)} = 234 \text{ K} - 273.15 = -39.15 = \textbf{--39} ^\circ\textbf{C}$

$T \text{ (in} ^\circ\text{F)} = \frac{9}{5} (-39.15 ^\circ\text{C}) + 32 = -38.47 = \textbf{--38} ^\circ\textbf{F}$

Check: Since the Kelvin temperature is below 273, the Celsius temperature must be negative. The low Celsius value gives a negative Fahrenheit value.

1.8   Plan: Determine the significant figures by determining the digits present, and accounting for the zeros. Only zeros between non-zero digits and zeros on the right, if there is a decimal point, are significant. The units make no difference.
Solution:
a) 31.070 mg; **five** significant figures
b) 0.06060 g; **four** significant figures
c) 850.°C; **three** significant figures — note the decimal point that makes the zero significant.
d) 2.000 x 10^2 mL; **four** significant figures
e) 3.9 x 10^{-6} m; **two** significant figures — note that none of the zeros are significant.
f) 4.01 x 10^{-4} L; **three** significant figures
Check: All significant zeros must come after a significant digit.

1.9   Plan: Use the rules presented in the text. Add the two values in the numerator before dividing. The time conversion is an exact conversion, and, therefore, does not affect the significant figures in the answer.
The addition of 25.65 and 37.4 gives an answer where the last significant figure is the one after the decimal point (giving three significant figures total). When a four significant figure number divides a three significant figure number, the answer must round to three significant figures. An exact number (1 min / 60 s) will have no bearing on the number of significant figures.
Solution:

$$\frac{25.65 \text{ mL} + 37.4 \text{ mL}}{73.55 \text{ s} \left(\dfrac{1 \text{ min}}{60 \text{ s}} \right)} = 51.434 = \textbf{51.4 mL/min}$$

END–OF–CHAPTER PROBLEMS

1.2   Plan: Apply the definitions of the states of matter to a container. Next, apply these definitions to the examples.
       <u>Solution:</u>
       Gas molecules fill the entire container; the volume of a gas is the volume of the container. Solids and liquids have
       a definite volume. The volume of the container does not affect the volume of a solid or liquid.
       a) The helium fills the volume of the entire balloon. The addition or removal of helium will change the volume of
       a balloon. Helium is a **gas**.
       b) At room temperature, the mercury does not completely fill the thermometer. The surface of the **liquid** mercury
       indicates the temperature.
       c) The soup completely fills the bottom of the bowl, and it has a definite surface. The soup is a **liquid**, though it is
       possible that solid particles of food will be present.

1.4   Plan: Define the terms and apply these definitions to the examples.
       <u>Solution:</u>
       Physical property – A characteristic shown by a substance itself, without interacting with or changing into other
       substances.
       Chemical property – A characteristic of a substance that appears as it interacts with, or transforms into, other
       substances.
       a) The change in color (yellow–green and silvery to white), and the change in physical state (gas and metal to
       crystals) are examples of **physical properties**. The change in the physical properties indicates that a chemical
       change occurred. Thus, the interaction between chlorine gas and sodium metal producing sodium chloride is an
       example of a **chemical property**.
       b) The sand and the iron are still present. Neither sand nor iron became something else. Colors along with
       magnetism are **physical properties**. No chemical changes took place, so there are no chemical properties to
       observe.

1.6   Plan: Apply the definitions of chemical and physical changes to the examples.
       <u>Solution:</u>
       a) Not a chemical change, but a **physical change** — simply cooling returns the soup to its original form.
       b) There is a **chemical change** — cooling the toast will not "un–toast" the bread.
       c) Even though the wood is now in smaller pieces, it is still wood. There has been no change in composition, thus
       this is a **physical change**, and not a chemical change.
       d) This is a **chemical change** converting the wood (and air) into different substances with different compositions.
       The wood cannot be "unburned."

1.8   Plan: A system has a higher potential energy before the energy is released (used).
       <u>Solution:</u>
       a) The exhaust is lower in energy than the fuel by an amount of energy equal to that released as the fuel burns.
       The **fuel** has a higher potential energy.
       b) **Wood**, like the fuel, is higher in energy by the amount released as the wood burns.

1.13   Plan: Re-read the material in the Chapter about Lavoisier.
       <u>Solution:</u>
       Lavoisier measured the total mass of the reactants and products, not just the mass of the solids. The total mass of
       the reactants and products remained constant. His measurements showed that a gas was involved in the reaction.
       He called this gas oxygen (one of his key discoveries).

1.16   Plan: Re-read the section in the Chapter on experimental design.
       <u>Solution:</u>
       A well-designed experiment must have the following essential features:
       1) There must be two variables that are expected to be related.
       2) There must be a way to control all the variables, so that only one at a time may be changed.
       3) The results must be reproducible.

1.19   <u>Plan:</u> Review the table of conversions in the Chapter or inside the back cover of the book.
        <u>Solution:</u>

a) To convert from in^2 to cm^2, use $\dfrac{(2.54\,\text{cm})^2}{(1\,\text{in})^2}$; to convert from cm^2 to m^2, use $\dfrac{(1\,\text{m})^2}{(100\,\text{cm})^2}$

b) To convert from km^2 to m^2, use $\dfrac{(1000\,\text{m})^2}{(1\,\text{km})^2}$; to convert from m^2 to cm^2, use $\dfrac{(100\,\text{cm})^2}{(1\,\text{m})^2}$

c) This problem requires two conversion factors: one for distance and one for time. It does not matter which conversion is done first. Alternate methods may be used.
To convert distance, mi to m, use:

$$\left(\dfrac{1.609\,\text{km}}{1\,\text{mi}}\right)\left(\dfrac{1000\,\text{m}}{1\,\text{km}}\right) = \textbf{1.609 x 10}^3\textbf{ m/mi}$$

To convert time, h to s, use:

$$\left(\dfrac{1\,\text{h}}{60\,\text{min}}\right)\left(\dfrac{1\,\text{min}}{60\,\text{s}}\right) = \textbf{1 h / 3600 s}$$

Therefore, the complete conversion factor is $\left(\dfrac{1.609 \times 10^3\,\text{m}}{1\,\text{mi}}\right)\left(\dfrac{1\,\text{h}}{3600\,\text{s}}\right) = \dfrac{\textbf{0.4469 m h}}{\textbf{mi s}}$. Do the units cancel

when you start with a measurement of mi/h?

d) To convert from pounds (lb) to grams (g), use $\dfrac{1000\,\text{g}}{2.205\,\text{lb}}$;

to convert volume from ft^3 to cm^3 use: $\left(\dfrac{(1\,\text{ft})^3}{(12\,\text{in})^3}\right)\left(\dfrac{(1\,\text{in})^3}{(2.54\,\text{cm})^3}\right) = \textbf{3.531 x 10}^{-5}\textbf{ ft}^3\textbf{/cm}^3$

1.21   <u>Plan:</u> Review the definitions of extensive and intensive properties.
        <u>Solution:</u>
An extensive property depends on the amount of material present. An intensive property is the same regardless of how much material is present.
a) Mass is an **extensive property**. Changing the amount of material will change the mass.
b) Density is an **intensive property**. Changing the amount of material changes both the mass and the volume, but the ratio (density) remains fixed.
c) Volume is an **extensive property**. Changing the amount of material will change the size (volume).
d) The melting point is an **intensive property**. The melting point depends on the substance not on the amount of substance.

1.23   <u>Plan:</u> Anything that increases the mass or decreases the volume will increase the density $\left(\text{density} = \dfrac{\text{mass}}{\text{volume}}\right)$.
        <u>Solution:</u>
a) Density **increases**. The mass of the chlorine gas is not changed, but its volume is smaller.
b) Density **remains the same**. Neither the mass nor the volume of the solid has changed.
c) Density **decreases**. Water is one of the few substances that expands on freezing. The mass is constant, but the volume increases.
d) Density **increases**. Iron, like most materials, contracts on cooling, thus the volume decreases while the mass does not change.
e) Density **remains the same**. The water does not alter either the mass or the volume of the diamond.

1.26   <u>Plan:</u> Use conversion factors from the inside back cover: 10^{-12} m = 1 pm; 10^{-9} m = 1 nm
        <u>Solution:</u>

$$\text{Radius} = 1430\,\text{pm}\left(\dfrac{10^{-12}\,\text{m}}{1\,\text{pm}}\right)\left(\dfrac{1\,\text{nm}}{10^{-9}\,\text{m}}\right) = \textbf{1.43 nm}$$

1.28   <u>Plan:</u> Use conversion factors: 0.01 m = 1 cm, 1 inch = 2.54 cm.
       <u>Solution:</u>

$$\text{Length} = 100.\ \text{m}\left(\frac{1\ \text{cm}}{0.01\ \text{m}}\right)\left(\frac{1\ \text{inch}}{2.54\ \text{cm}}\right) = 3.9370 \times 10^3 = \mathbf{3.94 \times 10^3\ in}$$

1.30   a) <u>Plan:</u> Use conversion factors: $(0.01\ \text{m})^2 = (1\ \text{cm})^2$; $(1\ \text{km})^2 = (1000\ \text{m})^2$
       <u>Solution:</u>

$$\mathbf{20.7\ cm^2}\left(\frac{(0.01\ \text{m})^2}{(1\ \text{cm})^2}\right)\left(\frac{(1\ \text{km})^2}{(1000\ \text{m})^2}\right) = \mathbf{2.07 \times 10^{-9}\ km^2}$$

       b) <u>Plan:</u> Use conversion factor: $(1\ \text{inch})^2 = (2.54\ \text{cm})^2$
       <u>Solution:</u>

$$\mathbf{20.7\ cm^2}\left(\frac{(1\ \text{in})^2}{(2.54\ \text{cm})^2}\right)\left(\frac{\$3.25}{1\ \text{in}^2}\right) = 10.4276 = \mathbf{\$10.43}$$

1.32   <u>Plan:</u> Use conversion factor: 1 kg = 2.205 lb. The following assumes a body weight of 155 lbs. Use your own body weight in place of the 155 lbs.
       <u>Solution:</u>

$$155\ \text{lb}\left(\frac{1\ \text{kg}}{2.205\ \text{lb}}\right) = \mathbf{70.3\ kg}$$

       Answers will vary, depending on the person's mass.

1.34   <u>Plan:</u> Use the relationships from the inside back cover of the book. The conversions may be performed in any order.
       <u>Solution:</u>

$$\text{a)}\ \left(\frac{5.52\ \text{g}}{\text{cm}^3}\right)\left(\frac{(1\ \text{cm})^3}{(0.01\ \text{m})^3}\right)\left(\frac{1\ \text{kg}}{1000\ \text{g}}\right) = \mathbf{5.52 \times 10^3\ kg/m^3}$$

$$\text{b)}\ \left(\frac{5.52\ \text{g}}{\text{cm}^3}\right)\left(\frac{(2.54\ \text{cm})^3}{(1\ \text{in})^3}\right)\left(\frac{(12\ \text{in})^3}{(1\ \text{ft})^3}\right)\left(\frac{1\ \text{kg}}{1000\ \text{g}}\right)\left(\frac{2.205\ \text{lb}}{1\ \text{kg}}\right) = 344.661 = \mathbf{345\ lb/ft^3}$$

1.36   <u>Plan:</u> The conversions may be done in any order. Use the definitions of the various SI prefixes.
       <u>Solution:</u>

$$\text{a) Volume} = \left(\frac{2.56\ \mu\text{m}^3}{\text{cell}}\right)\left(\frac{(1 \times 10^{-6}\ \text{m})^3}{(1\ \mu\text{m})^3}\right)\left(\frac{(1\ \text{mm})^3}{(1 \times 10^{-3}\ \text{m})^3}\right) = \mathbf{2.56 \times 10^{-9}\ mm^3/cell}$$

$$\text{b) Volume} = (10^5\ \text{cells})\left(\frac{2.56\ \mu\text{m}^3}{\text{cell}}\right)\left(\frac{(1 \times 10^{-6}\ \text{m})^3}{(1\ \mu\text{m})^3}\right)\left(\frac{1\ \text{L}}{1 \times 10^{-3}\ \text{m}^3}\right) = 2.56 \times 10^{-10} = \mathbf{10^{-10}\ L}$$

1.38   <u>Plan:</u> The mass of the contents is the mass of the full container minus the mass of the empty container. The volume comes from the mass density relationship. Reversing this process gives the answer to part b.
       <u>Solution:</u>
       a) Mass of mercury = 185.56 g – 55.32 g = 130.24 g

$$\text{Volume of mercury} = \text{volume of vial} = (130.24\ \text{g})\left(\frac{1\ \text{cm}^3}{13.53\ \text{g}}\right) = 9.626016 = \mathbf{9.626\ cm^3}$$

$$\text{b) Mass of water} = (9.626016\ \text{cm}^3)\left(\frac{0.997\ \text{g}}{1\ \text{cm}^3}\right) = 9.59714\ \text{g}\ \text{water}$$

       Mass of vial filled with water = 55.32 g + 9.59714 g = 64.91714 = **64.92 g**

1.40   Plan: Volume of a cube = (length of side)^3
Solution:
The value 15.6 means this is a three significant figure problem.
The final answer, and no other, is rounded to the correct number of significant figures.

$$15.6 \text{ mm} \left(\frac{10^{-3} \text{ m}}{1 \text{ mm}} \right) \left(\frac{1 \text{ cm}}{10^{-2} \text{ m}} \right) = 1.56 \text{ cm} \quad \text{(convert to cm to match density unit)}$$

Al cube volume = $(1.56 \text{ cm})^3 = 3.7964 \text{ cm}^3$

$$\text{density} = \frac{\text{mass}}{\text{volume}} = \frac{10.25 \text{ g}}{3.7964 \text{ cm}^3} = 2.69993 = \textbf{2.70 g/cm}^3$$

1.42   Plan: Use the equations given in the text for converting between the three temperature scales.
Solution:

a) T (in °C) = [T (in °F) – 32]$\frac{5}{9}$ = [68°F – 32]$\frac{5}{9}$ = **20. °C**

   T (in K) = T (in °C) + 273.15 = 20. °C + 273.15 = 293.15 = **293 K**
b) T (in K) = T (in °C) + 273.15 = –164°C + 273.15 = 109.15 = **109 K**

   T (in °F) = $\frac{9}{5}$ T (in °C) + 32 = $\frac{9}{5}$ (–164°C) + 32 = –263.2 = **–263°F**

c) T (in °C) = T (in K) –273.15 = 0 K – 273.15 = –273.15 = **–273°C**

   T (in °F) = $\frac{9}{5}$ T (in °C) + 32 = $\frac{9}{5}$ (–273.15°C) + 32 = –459.67 = **–460. °F**

1.45   Plan: Use 10^{-9} m = 1 nm to convert wavelength. Use 0.01 Å = 1 pm, 10^{-12} m = 1 pm, and 10^{-9} m = 1 nm
Solution:

a) 247 nm $\left(\dfrac{10^{-9} \text{ m}}{1 \text{ nm}} \right)$ = **2.47 x 10^{-7} m**

b) 6760 pm $\left(\dfrac{0.01 \text{ Å}}{1 \text{ pm}} \right)$ = **67.6 Å**

1.52   Plan: Review the rules for significant zeros.
Solution:
a) No significant zeros (leading zeros are not significant)
b) No significant zeros (leading zeros are not significant)
c) 0.041<u>0</u> (terminal zeros to the right of the decimal point are significant)
d) 4.<u>0100</u> x 10^4 (zeros between nonzero digits are significant; terminal zeros to the right of the decimal point are significant)

1.54   Plan: Review the rules for rounding.
Solution: (significant figures are underlined)
a) 0.000<u>3554</u>: the extra digits are 54 at the end of the number. When the digit to be removed is 5 and that 5 is followed by nonzero numbers, the last digit kept is increased by 1: **0.00036**
b) <u>35.83</u>48: the extra digits are 48. Since the digit to be removed (4) is less than 5, the last digit kept is unchanged: **35.83**
c) <u>22.4</u>555: the extra digits are 555. When the digit to be removed is 5 and that 5 is followed by nonzero numbers, the last digit kept is increased by 1: **22.5**

1.56   Plan: Review the rules for rounding.
Solution: 19 rounds to 20; 155 rounds to 160; 8.3 rounds to 8; 3.2 rounds to 3; 2.9 rounds to 3; 4.7 rounds to 5.

$$\left(\frac{20 \times 160 \times 8}{3 \times 3 \times 5} \right) = 568.89 = \textbf{6 x 10}^2 \text{ (vs. 560 with original number of significant digits)}$$

1.58   Plan: Use a calculator to obtain an initial value. Use the rules for significant figures and rounding to get the final answer.
        Solution:

a) $\dfrac{(2.795 \text{ m})(3.10 \text{ m})}{6.48 \text{ m}} = 1.3371 = \textbf{1.34 m}$ (maximum of 3 significant figures allowed since two of the original numbers in the calculation have only 3 significant figures)

b) $V = \left(\dfrac{4}{3}\right)\pi(17.282 \text{ m})^3 = 72.3907 = \textbf{72.391 m}^3$ (maximum of 5 significant figures allowed)

c) $1.110 \text{ cm} + 17.3 \text{ cm} + 108.2 \text{ cm} + 316 \text{ cm} = 442.61 = \textbf{443 cm}$ (no digits allowed to the right of the decimal since 316 has no digits to the right of the decimal point)

1.60   Plan: Review the procedure for changing a number to scientific notation.
        Solution:
        a) $\textbf{1.310000 x 10}^5$ (Note that all zeros are significant.)
        b) $\textbf{4.7 x 10}^{-4}$ (No zeros are significant.)
        c) $\textbf{2.10006 x 10}^5$
        d) $\textbf{2.1605 x 10}^3$

1.62   Plan: Review the examples for changing a number from scientific notation to standard notation.
        Solution:
        a) **5550**   (Do not use terminal decimal point since the zero is not significant)
        b) **10070.**   (Use terminal decimal point since final zero is significant)
        c) **0.000000885**
        d) **0.003004**

1.64   Plan: In most cases, this involves a simple addition or subtraction of values from the exponents. There can be only 1 nonzero digit to the left of the decimal point in correct scientific notation.
        Solution:
        a) $\textbf{8.025 x 10}^4$ (The decimal point must be moved an additional 2 places to the left: $10^2 + 10^2 = 10^4$)
        b) $\textbf{1.0098 x 10}^{-3}$ (The decimal point must be moved an additional 3 places to the left: $10^3 + 10^{-6} = 10^{-3}$)
        c) $\textbf{7.7 x 10}^{-11}$ (The decimal point must be moved an additional 2 places to the right: $10^{-2} + 10^{-9} = 10^{-11}$)

1.66   Plan: Calculate a temporary answer by simply entering the numbers into a calculator. Then you will need to round the value to the appropriate number of significant figures. Cancel units as you would cancel numbers, and place the remaining units after your numerical answer.
        Solution:

a) $\dfrac{\left(6.626 \text{ x } 10^{-34} \text{ Js}\right)\left(2.9979 \text{ x } 10^8 \text{ m/s}\right)}{489 \text{ x } 10^{-9} \text{ m}} = 4.06218 \text{ x } 10^{-19}$

$\textbf{4.06 x 10}^{-19}$ **J** (489 x 10^{-9} m limits the answer to 3 significant figures; units of m and s cancel)

b) $\dfrac{\left(6.022 \text{ x } 10^{23} \text{ molecules/mol}\right)\left(1.23 \text{ x } 10^2 \text{ g}\right)}{46.07 \text{ g/mol}} = 1.6078 \text{ x } 10^{24}$

$\textbf{1.61 x 10}^{24}$ **molecules** (1.23 x 10^2 g limits answer to 3 significant figures; units of mol and g cancel)

c) $\left(6.022 \text{ x } 10^{23} \text{ atoms/mol}\right)\left(2.18 \text{ x } 10^{-18} \text{ J/atom}\right)\left(\dfrac{1}{2^2} - \dfrac{1}{3^2}\right) = 1.82333 \text{ x } 10^5$

$\textbf{1.82 x 10}^5$ **J/mol** (2.18 x 10^{-18} J/atom limits answer to 3 significant figures; unit of atoms cancels)

1.68   Plan: Exact numbers are those that have no uncertainty. Unit definitions and number counts of items in a group are examples of exact numbers.
Solution:
a) The height of Angel Falls is a measured quantity. This is <u>not</u> an exact number.
b) The number of planets in the solar system is a number count. This <u>is</u> an exact number.
c) The number of grams in a pound is not a unit definition. This is <u>not</u> an exact number.
d) The number of millimeters in a meter is a definition of the prefix "milli–." This <u>is</u> an exact number.

1.70   Plan: Observe the figure, and estimate a reading the best you can.
Solution:
The scale markings are 0.2 cm apart. The end of the metal strip falls between the mark for 7.4 cm and 7.6 cm. If we assume that one can divide the space between markings into fourths, the uncertainty is one-fourth the separation between the marks. Thus, since the end of the metal strip falls between 7.45 and 7.55 we can report its length as **7.50 ± 0.05 cm**. (Note: If the assumption is that one can divide the space between markings into halves only, then the result is 7.5 ± 0.1 cm.)

1.72   Plan: Calculate the average of each data set. Remember that accuracy refers to how close a measurement is to the actual or true value while precision refers to how close multiple measurements are to each other.
Solution:
a)
$$I_{avg} = \frac{8.72 \text{ g} + 8.74 \text{ g} + 8.70 \text{ g}}{3} = 8.7200 = \textbf{8.72 g}$$

$$II_{avg} = \frac{8.56 \text{ g} + 8.77 \text{ g} + 8.83 \text{ g}}{3} = 8.7200 = \textbf{8.72 g}$$

$$III_{avg} = \frac{8.50 \text{ g} + 8.48 \text{ g} + 8.51 \text{ g}}{3} = 8.4967 = \textbf{8.50 g}$$

$$IV_{avg} = \frac{8.41 \text{ g} + 8.72 \text{ g} + 8.55 \text{ g}}{3} = 8.5600 = \textbf{8.56 g}$$

Sets **I** and **II** are most accurate since their average value, 8.72 g, is closest to the true value, 8.72 g.
b)   To get an idea of precision, calculate the range of each set of values: largest value – smallest value. A small range is an indication of good precision since the values are close to each other.
I_{range} = 8.74 g – 8.70 g = 0.04 g
II_{range} = 8.83 g – 8.56 g = 0.27 g
III_{range} = 8.51 g – 8.48 g = 0.03 g
IV_{range} = 8.72 g – 8.41 g = 0.31 g
Set **III** is the most precise (smallest range), but is the least accurate (the average is the farthest from the actual value).
c)   Set **I** has the best combination of high accuracy (average value = actual value) and high precision (relatively small range).
d)   Set **IV** has both low accuracy (average value differs from actual value) and low precision (has the largest range).

1.74   Plan: If it is necessary to force something to happen, the potential energy will be higher.
Solution:
a)

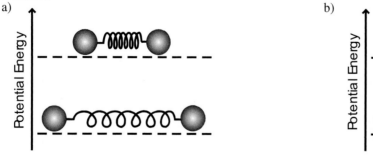

b)

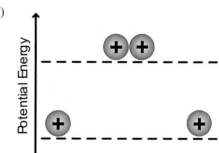

8

a) The balls on the relaxed spring have a lower potential energy and are more stable. The balls on the compressed spring have a higher potential energy, because the balls will move once the spring is released. This configuration is less stable.
b) The two + charges apart from each other have a lower potential energy and are more stable. The two + charges near each other have a higher potential energy, because they repel one another. This arrangement is less stable.

1.76   Plan: Use the concentrations of bromide given.
        Solution:

$$\frac{\text{mass bromine in Dead Sea}}{\text{mass bromine in sea water}} = \frac{0.50 \text{ g/L}}{0.065 \text{ g/L}} = 7.7{:}1$$

1.78   Plan: In each case, calculate the overall density of the sphere and contents.
        Solution:

a)   Density of evacuated ball: $d = \dfrac{\text{mass}}{\text{volume}} = \dfrac{0.12 \text{ g}}{560 \text{ cm}^3} = 2.1429 \times 10^{-4} \text{ g/cm}^3$

   Convert density to units of g/L: $\dfrac{2.1429 \times 10^{-4} \text{ g}}{\text{cm}^3} \left(\dfrac{1 \text{ cm}^3}{1 \text{ mL}}\right)\left(\dfrac{1 \text{ mL}}{10^{-3} \text{ L}}\right) = 0.21429 = 0.21 \text{ g/L}$

The evacuated ball will **float** because its density is less than that of air.
b) Because the density of CO_2 is greater than that of air, a ball filled with CO_2 will **sink**.

c) $560 \text{ cm}^3 \left(\dfrac{1 \text{ mL}}{1 \text{ cm}^3}\right)\left(\dfrac{10^{-3} \text{ L}}{1 \text{ mL}}\right) = 0.560 = \textbf{0.56 L}$

Mass of hydrogen: $(0.56 \text{ L})\left(\dfrac{0.0899 \text{ g}}{1 \text{ L}}\right) = 0.0503 \text{ g}$

The H_2 filled ball will have a total mass of 0.0503 + 0.12 g = 0.17 g, and a resulting density of

$d = \dfrac{0.17 \text{ g}}{0.56 \text{ L}} = 0.30 \text{ g/L}$ The ball will **float** because density of the ball filled with hydrogen is less than density

of air. Note: *The densities are additive because the volume of the ball and the volume of the H_2 gas are the same: 0.0899 + 0.21 g/L = 0.30 g/L.*
d) Because the density of O_2 is greater than that of air, a ball filled with O_2 will **sink**.
e) Density of ball filled with nitrogen: 0.21 g/L + 1.165 g/L = 1.38 g/L. Ball will **sink** because density of ball filled with nitrogen is greater than density of air.

f) To sink, the total mass of the ball and gas must weigh $\left(\dfrac{1.189 \text{ g}}{1 \text{ L}}\right)\left(\dfrac{0.560 \text{ L}}{}\right) = 0.66584 \text{ g}$

For the evacuated ball: 0.66584 – 0.12 g = 0.54585 = **0.55 g.** More than 0.55 g would have to be added to make the ball sink.
For ball filled with hydrogen: 0.66584 – 0.17 g = 0.4958 = **0.50 g.** More than 0.50 g would have to be added to make the ball sink.

1.81   Plan: Use the surface area and the depth to determine the volume of the oceans (area x depth = volume). The volume may then be converted to liters, and finally to the mass of gold using the density of gold in g/L. Once the mass of the gold is known, other conversions stem from it.
        Solution:

a) $(3800 \text{ m})(3.63 \times 10^8 \text{ km}^2)\left(\dfrac{(1000 \text{ m})^2}{(1 \text{ km})^2}\right)\left(\dfrac{1 \text{ L}}{10^{-3} \text{ m}^3}\right)\left(\dfrac{5.8 \times 10^{-9} \text{ g}}{\text{L}}\right) = 8.00052 \times 10^{12} = \textbf{8.0} \times \textbf{10}^{\textbf{12}} \textbf{ g}$

b) Use the density of gold to convert mass of gold to volume:

$(8.00052 \times 10^{12} \text{ g})\left(\dfrac{1 \text{ cm}^3}{19.3 \text{ g}}\right)\left(\dfrac{(0.01 \text{ m})^3}{(1 \text{ cm})^3}\right) = 4.14535 \times 10^5 = \textbf{4.1} \times \textbf{10}^{\textbf{5}} \textbf{ m}^3$

c) $\left(8.00052 \times 10^{12}\ g\right)\left(\dfrac{1\ tr.\ oz.}{31.1\ g}\right)\left(\dfrac{\$370.00}{1\ tr.\ oz.}\right) = 9.51830 \times 10^{13} = \mathbf{\$9.5 \times 10^{13}}$

1.83   Plan: Use the equations for temperature conversion given in the chapter. The mass and density will then give the volume.
Solution:
a) T (in °C) = T (in K) – 273.15 = 77.36 K – 273.15 = **–195.79°C**

b) T (in °F) = $\dfrac{9}{5}$ T (in °C) + 32 = $\dfrac{9}{5}$ (–195.79°C) + 32 = –320.422 = **–320.42°F**

c) The mass of nitrogen is conserved, meaning that the mass of the nitrogen gas equals the mass of the liquid nitrogen. We first find the mass of liquid nitrogen and then convert that quantity to volume using the density of the liquid.

Mass of liquid nitrogen = mass of gaseous nitrogen = $(895.0\ L)\left(\dfrac{4.566\ g}{1\ L}\right) = 4086.57\ g\ N_2$

Volume of liquid N_2 = $(4086.57\ g)\left(\dfrac{1\ L}{809\ g}\right) = 5.0514 = \mathbf{5.05\ L}$

1.86   Plan: Each step involves one or more conversions from the inside back cover of the book.
Solution:

a) $\left(\dfrac{5.9\ mi}{h}\right)\left(\dfrac{1000\ m}{0.62\ mi}\right)\left(\dfrac{1\ h}{3600\ s}\right) = 2.643 = \mathbf{2.6\ m/s}$

b) $(98\ min)\left(\dfrac{1\ h}{60\ min}\right)\left(\dfrac{5.9\ mi}{h}\right)\left(\dfrac{1\ km}{0.62\ mi}\right) = 15.543 = \mathbf{16\ km}$

c) $4.75 \times 10^4\ ft\left(\dfrac{1\ mi}{5280\ ft}\right)\left(\dfrac{1\ h}{5.9\ mi}\right) = 1.5248 = 1.5\ h$

If she starts running at 11:15 am, 1.5 hours later the time is **12:45 pm.**

1.90   Plan: In visualizing the problem, the two scales can be set next to each other.
Solution:
There are 50 divisions between the freezing point and boiling point of benzene on the °X scale and 74.6 divisions (80.1°C – 5.5°C) on the °C scale. So °X = $\left(\dfrac{50°\ X}{74.6°\ C}\right)°C$

This does not account for the offset of 5.5 divisions in the °C scale from the zero point on the °X scale.

So °X = $\left(\dfrac{50°\ X}{74.6°\ C}\right)(°C - 5.5°C)$

Check: Plug in 80.1°C and see if result agrees with expected value of 50°X.

So °X = $\left(\dfrac{50°\ X}{74.6°\ C}\right)(80.1°C - 5.5°C) = 50°X$

Use this formula to find the freezing and boiling points of water on the °X scale.

FP_{water} °X = $\left(\dfrac{50°\ X}{74.6°\ C}\right)(0.00°C - 5.5°C) = \mathbf{-3.7°X}$

BP_{water} °X = $\left(\dfrac{50°\ X}{74.6°\ C}\right)(100.0°C - 5.5°C) = \mathbf{63.3°X}$

1.91   Plan: Determine the total mass of the Earth's crust in metric tons (t) by finding the volume of crust (surface area x depth) and then using the density to find the mass of this volume, using conversions from the inside back cover. The mass of each individual element comes from the concentration of that element multiplied by the mass of the crust.
Solution:

$$\text{Mass of crust} = (35\,\text{km})(5.10 \times 10^8\,\text{km}^2)\left(\frac{(1000\,\text{m})^3}{(1\,\text{km})^3}\right)\left(\frac{(1\,\text{cm})^3}{(0.01\,\text{m})^3}\right)\left(\frac{2.8\,\text{g}}{1\,\text{cm}^3}\right)\left(\frac{1\,\text{kg}}{1000\,\text{g}}\right)\left(\frac{1\,\text{t}}{1000\,\text{kg}}\right) = 4.998 \times 10^{19}\,\text{t}$$

$$\left(4.998 \times 10^{19}\,\text{t}\right)\left(\frac{4.55 \times 10^5\,\text{g oxygen}}{1\,\text{t}}\right) = 2.2741 \times 10^{25} = \mathbf{2.3 \times 10^{25}\,\text{g Oxygen}}$$

$$\left(4.998 \times 10^{19}\,\text{t}\right)\left(\frac{2.72 \times 10^5\,\text{g silicon}}{1\,\text{t}}\right) = 1.3595 \times 10^{25} = \mathbf{1.4 \times 10^{25}\,\text{g Silicon}}$$

$$\left(4.998 \times 10^{19}\,\text{t}\right)\left(\frac{1 \times 10^{-4}\,\text{g ruthenium}}{1\,\text{t}}\right) = 4.998 \times 10^{15} = \mathbf{5 \times 10^{15}\,\text{g Ruthenium (and Rhodium)}}$$

CHAPTER 2 THE COMPONENTS OF MATTER

FOLLOW–UP PROBLEMS

2.1 Plan: An element has only one kind of atom; a compound is composed to at least two kinds of atoms.
Solution:
The circle on the left contains molecules with either only orange atoms or only blue atoms. This is a mixture of two different elements. In the circle on the right, the molecules are composed of one orange atom and one blue atom so this is a compound.

2.2 Plan: Use the mass fraction of uranium in pitchblende to find the mass of pitchblende that contains 2.3 t of uranium. Find the mass fraction of oxygen in pitchblende. The total mass of pitchblende and the mass fraction of oxygen may then be used to determine the mass of oxygen.
Solution:

$$\text{Mass of pitchblende} = 2.3 \text{ t uranium} \left(\frac{84.2 \text{ t pitchblende}}{71.4 \text{ t uranium}} \right) = 2.7123 = \textbf{2.7 t pitchblende}$$

$$\text{Mass of oxygen in 84.2 t of pitchblende} = 84.2 \text{ t pitchblende} - 71.4 \text{ t uranium} = \textbf{12.8 t oxygen}$$

$$\text{Mass of oxygen} = 2.7123 \text{ t pitchblende} \left(\frac{12.8 \text{ t oxygen}}{84.2 \text{ t pitchblende}} \right) = 0.4123 = \textbf{0.41 t oxygen}$$

Check: Adding the amount of oxygen calculated to the mass of uranium gives the calculated 2.7 t of pitchblende.

2.3 Plan: The Law of Multiple Proportions states that when two elements react to form two compounds, the different masses of element B that react with a fixed mass of element A is a ratio of small whole numbers.
Solution:
Only **Sample B** shows two different bromine-fluorine compounds. In one compound there are three fluorine atoms for every one bromine atom; in the other compound, there is one fluorine atom for every bromine atom. Therefore, in the two compounds, the ratio of fluorine atoms combining with one bromine atom is 3/1.

2.4 Plan: The subscript (Atomic Number = Z) gives the number of protons, and for an atom, the number of electrons. The Atomic Number identifies the element. The superscript gives the mass number (A) which is the total of the protons plus neutrons. The number of neutrons is simply the Mass Number minus the Atomic Number ($A - Z$).
Solution:
a) $Z = 5$ and $A = 11$, there are 5 p^+ and 5 e^- and $11 - 5 = 6$ n^0 Atomic number $= 5 = \textbf{B}$.
b) $Z = 20$ and $A = 41$, there are 20 p^+ and 20 e^- and $41 - 20 = 21$ n^0 Atomic number $= 20 = \textbf{Ca}$.
c) $Z = 53$ and $A = 131$, there are 53 p^+ and 53 e^- and $131 - 53 = 78$ n^0 Atomic number $= 53 = \textbf{I}$.

2.5 Plan: To find the percent abundance of each B isotope, let x equal the fractional abundance of ^{10}B and $(1 - x)$ equal the fractional abundance of ^{11}B. Remember that atomic mass = isotopic mass of ^{10}B x fractional abundance) + (isotopic mass of ^{11}B x fractional abundance).
Solution:
Atomic Mass = (^{10}B mass) (fractional abundance of ^{10}B) + (^{11}B mass) (fractional abundance of ^{11}B)
Amount of ^{10}B + Amount ^{11}B = 1 (setting ^{10}B = x gives ^{11}B = 1 – x)
  10.81 amu = (10.0129 amu)(x) + (11.0093 amu) (1 – x)
  10.81 amu = 11.0093 – 11.0093x + 10.0129 x
  10.81 amu = 11.0093 – 0.9964 x
  –0.1993 = – 0.9964x
  x = 0.20;   1 – x = 0.80   (10.81 – 11.0093 limits the answer to 2 significant figures)
  Fraction x 100% = percent abundance.
  % abundance of ^{10}B = **20.%**; % abundance of ^{11}B = **80.%**
Check: The atomic mass of B is 10.81 (close to the mass of ^{11}B, thus the sample must be mostly this isotope.)

2.6   Plan: Locate these elements on the periodic table and predict what ions they will form. For A-group cations (metals), ion charge = group number; for anions (nonmetals), ion charge = group number – 8. Or, relate the element's position to the nearest noble gas. Elements after a noble gas lose electrons to become positive ions, while those before a noble gas gain electrons to become negative ions.
Solution:
a) $_{16}S^{2-}$ [Group 6A(16); $6 – 8 = –2$]; sulfur needs to gain 2 electrons to match the number of electrons in $_{18}Ar$.
b) $_{37}Rb^+$ [Group 1A(1)]; rubidium needs to lose 1 electron to match the number of electrons in $_{36}Kr$.
c) $_{56}Ba^{2+}$ [Group 2A(2)]; barium needs to lose 2 electrons to match the number of electrons in $_{54}Xe$.

2.7   Plan: When dealing with ionic binary compounds, the first name belongs to the metal and the second name belongs to the nonmetal. If there is any doubt, refer to the periodic table. The metal name is unchanged, while the nonmetal has an -ide suffix added to the nonmetal root.
Solution:
a) **Zinc** is in **Group 2B(12)** and **oxygen**, from oxide, is in **Group 6A(16)**.
b) **Silver** is in **Group 1B(11)** and **bromine**, from bromide, is in **Group 7A(17)**.
c) **Lithium** is in **Group 1A(1)** and **chlorine**, from chloride, is in **Group 7A(17)**.
d) **Aluminum** is in **Group 3A(13)** and **sulfur**, from sulfide, is in **Group 6A(16)**.

2.8   Plan: Use the charges to predict the lowest ratio leading to a neutral compound.
Solution:
a) Zinc should form Zn^{2+} and oxygen should form O^{2-}, these will combine to give **ZnO**. The charges cancel $(+2 – 2 = 0)$, so this is an acceptable formula.
b) Silver should form Ag^+ and bromine should form Br^-, these will combine to give **AgBr**. The charges cancel $(+1 – 1 = 0)$, so this is an acceptable formula.
c) Lithium should form Li^+ and chlorine should form Cl^-, these will combine to give **LiCl**. The charges cancel $(+1 – 1 = 0)$, so this is an acceptable formula.
d) Aluminum should form Al^{3+} and sulfur should form S^{2-}, to produce a neutral combination the formula is **Al$_2$S$_3$**. This way the charges will cancel $[2(+3) + 3(–2) = 0]$, so this is an acceptable formula.

2.9   Plan: Determine the names or symbols of each of the species present. Then combine the pieces to produce a name or formula. The metal or positive ions always go first. Review the rules for nomenclature covered in the chapter. For metals like many transition metals, that can form more than one ion each with a different charge, the ionic charge of the metal ion is indicated by a Roman numeral within parentheses immediately following the metal's name.
Solution:
a) The Roman numerals mean that the lead is Pb^{4+} and oxygen produces the usual O^{2-}. The neutral combination is $[+4 + 2(–2) = 0]$, so the formula is **PbO$_2$**.
b) Sulfide, like oxide, is –2. This is split between two copper ions, each of which must be +1. This is one of the two common charges for copper ions. The +1 charge on the copper is indicated with a Roman numeral. This gives the name **copper(I) sulfide** (common name = cuprous sulfide).
c) Bromide, like other elements in the same column of the periodic table, forms a –1 ion. Two of these ions requires a total of +2 to cancel them out. Thus, the iron must be +2 (indicated with a Roman numeral). This is one of the two common charges on iron ions. This gives the name **iron(II) bromide** (or ferrous bromide).
d) The mercuric ion is Hg^{2+}, and two –1 ions (Cl^-) are needed to cancel the charge. This gives the formula **HgCl$_2$**.

2.10   Plan: Determine the names or symbols of each of the species present. Then combine the pieces to produce a name or formula. The metal or positive ions always go first.
Solution:
a) The cupric ion, Cu^{2+}, requires two nitrate ions, NO_3^-, to cancel the charges. Trihydrate means three water molecules. These combine to give: **Cu(NO$_3$)$_2$·3H$_2$O**.
b) The zinc ion, Zn^{2+}, requires two hydroxide ions, OH^-, to cancel the charges. These combine to give: **Zn(OH)$_2$**.
c) Lithium only forms the Li^+ ion, so Roman numerals are unnecessary. The cyanide ion, CN^-, has the appropriate charge. These combine to give **lithium cyanide**.

2.11   Plan: Determine the names or symbols of each of the species present. Then combine the pieces to produce a name
        or formula. The metal or positive ions always go first. Make corrections accordingly.
        Solution:
        a) The ammonium ion is NH_4^+ and the phosphate ion is PO_4^{3-}. To give a neutral compound they should combine
        [3(+1) + (–3) = 0] to give the correct formula $(NH_4)_3PO_4$.
        b) Aluminum gives Al^{3+} and the hydroxide ion is OH^-. To give a neutral compound they should combine
        [+3 + 3(–1) =0] to give the correct formula $Al(OH)_3$. Parentheses are required around the polyatomic ion.
        c) Manganese is Mn, and Mg, in the formula, is magnesium. Magnesium only forms the Mg^{2+} ion, so Roman
        numerals are unnecessary. The other ion is HCO_3^-, which is called the hydrogen carbonate (or bicarbonate) ion.
        The correct name is **magnesium hydrogen carbonate** or **magnesium bicarbonate**.
        d) Either use the "-ic" suffix or the "(III)" but not both. Nit*ride* is N^{3-}, and nit*rate* is NO_3^-. This gives the correct
        name: **chromium(III) nitrate** (the common name is chromic nitrate).
        e) Cadmium is Cd, and Ca, in the formula, is calcium. Nit*rate* is NO_3^-, and nit*rite* is NO_2^-. The correct name is
        calcium nitrite.

2.12   Plan: Determine the names or symbols of each of the species present. The number of hydrogen atoms equals
        the charge on the anion. Then combine the pieces to produce a name or formula. The hydrogen always goes first.
        For the oxoanions, the -ate suffix changes to -ic acid and the -ite suffix changes to -ous acid.
        Solution:
        a) Chlor*ic acid* is derived from the chlor*ate* ion, ClO_3^-. The –1 charge on the ion requires one hydrogen. These
        combine to give the formula: $HClO_3$.
        b) As a binary acid, HF requires a "hydro-" prefix and an "ic" suffix on the "fluor" root. These combine to give
        the name: **hydrofluoric acid**.
        c) Acet*ic acid* is derived from the acet*ate* ion, which may be written as CH_3COO^- or as $C_2H_3O_2^-$. The –1 charge
        means that one H is needed. These combine to give the formula: CH_3COOH or $HC_2H_3O_2$.
        d) Sulfur*ous acid* is derived from the sulf*ite* ion, SO_3^{2-}. The –2 charge on the ion requires two hydrogen atoms.
        These combine to give the formula: H_2SO_3.
        e) HBrO is an oxoacid containing the BrO^- ion (hypobromite ion). To name the acid, the "-ite" must be
        replaced with "-ous." This gives the name: **hypobromous acid**.

2.13   Plan: Determine the names or symbols of each of the species present. The number of atoms leads to prefixes, and
        the prefixes lead to the number of that type of atom.
        Solution:
        a) **Sulfur trioxide** — one sulfur and three (tri) oxygens, as oxide, are present.
        b) **Silicon dioxide** — one silicon and two (di) oxygens, as oxide, are present.
        c) N_2O Nitrogen has the prefix "di" = 2, and oxygen has the prefix "mono" = 1 (understood in the formula).
        d) SeF_6 Selenium has no prefix (understood as = 1), and the fluoride has the prefix "hexa" = 6.

2.14   Plan: Determine the names or symbols of each of the species present. For compounds between nonmetals, the
        number of atoms of each type is indicated by a prefix.
        Solution:
        a) Suffixes are not used in the common names of the nonmetal listed first in the formula. Sulfur does not
        qualify for the use of a suffix. Chlorine correctly has an "ide" suffix. There are two of each nonmetal atom, so
        both names require a "di" prefix. This gives the name: **disulfur dichloride**.
        b) Both elements are nonmetals, and there is just one nitrogen and one oxygen. These combine to give the formula
        NO.
        c) The heavier Br should be named first. The three chlorides are correctly named. The correct name is **bromine
        trichloride**.

2.15   Plan: First, write a formula to match the name. Next, multiply the number of each type of atom by the atomic
        mass of that atom. Sum all the masses to get an overall mass.
        Solution:
        a) The peroxide ion is O_2^{2-}, which requires two hydrogen atoms to cancel the charge: H_2O_2.
        Molecular mass = (2 x 1.008 amu) + (2 x 16.00 amu) = 34.016 = **34.02 amu**.
        b) A Cs^{+1} ion requires one Cl^{-1}, ion to give: **CsCl**; formula mass = 132.9 amu +35.45 amu = 168.35 = **168.4 amu**.

c) Sulfuric acid contains the sulfate ion, SO_4^{2-}, which requires two hydrogen atoms to cancel the charge: H_2SO_4; molecular mass = (2 x 1.008 amu) + 32.07 amu + (4 x 16.00 amu) = 98.086 = **98.09 amu**.
d) The sulfate ion, SO_4^{2-}, requires two +1 potassium ions, K^+, to give K_2SO_4; formula mass = (2 x 39.10 amu) + 32.07 amu + (4 x 16.00 amu) = **174.27 amu**.

2.16   Plan: Since the compounds only contain two elements, finding the formulas involve simply counting each type of atom and developing a ratio.
Solution:
a)There are two brown atoms (sodium) for every red (oxygen). The compound contains a metal with a nonmetal. Thus, the compound is **sodium oxide**, with the formula Na_2O. The formula mass is twice the mass of sodium plus the mass of oxygen: 2 (22.99 amu) + (16.00 amu) = **61.98 amu**.
b) There is one blue (nitrogen) and two reds (oxygen) in each molecule. The compound only contains nonmetals. Thus, the compound is **nitrogen dioxide**, with the formula NO_2. The molecular mass is the mass of nitrogen plus twice the mass of oxygen: (14.01 amu) + 2 (16.00 amu) = **46.01 amu**.

END–OF–CHAPTER PROBLEMS

2.1   Plan: Refer to the definitions of an element and a compound.
Solution:
Unlike compounds, elements cannot be broken down by chemical changes into simpler materials. Compounds contain different types of atoms; there is only one type of atom in an element.

2.4   Plan: Review the definitions of elements, compounds and mixtures.
Solution:
a) The presence of more than one element (calcium and chlorine) makes this pure substance a **compound**.
b) There are only atoms from one element, sulfur, so this pure substance in an **element**.
c) The presence of more than one compound makes this a **mixture**.
d) The presence of more than one type of atom means it cannot be an element. The specific, not variable, arrangement means it is a **compound**.

2.12   Plan: Restate the three laws in your own words.
Solution:
a) The law of mass conservation applies to all substances — **elements, compounds, and mixtures**. Matter can neither be created nor destroyed, whether it is an element, compound, or mixture.
b) The law of definite composition applies to **compounds** only, because it refers to a constant, or definite, composition of elements within a compound.
c) The law of multiple proportions applies to **compounds** only, because it refers to the combination of elements to form compounds.

2.14   Plan: Review the three laws: Law of Mass Conservation, Law of Definite Composition, and Law of Multiple Proportions.
Solution:
a) Law of Definite Composition — The compound potassium chloride, KCl, is composed of the same elements and same fraction by mass, regardless of its source (Chile or Poland).
b) Law of Mass Conservation — The mass of the substance inside the flashbulb did not change during the chemical reaction (formation of magnesium oxide from magnesium and oxygen).
c) Law of Multiple Proportions — Two elements, O and As, can combine to form two different compounds that have different proportions of As present.

2.16   Plan: Review the definition of percent by mass.
Solution:
a) **No**, the mass percent of each element in a compound is fixed. The percentage of Na in the compound NaCl is 39.34% (22.99 amu / 58.44 amu), whether the sample is 0.5000 g or 50.00 g.
b) **Yes**, the mass of each element in a compound depends on the mass of the compound. A 0.5000 g sample of NaCl contains 0.1967 g of Na (39.34% of 0.5000 g), whereas a 50.00 g sample of NaCl contains 19.67 g of Na (39.34% of 50.00 g).

2.18   Plan: Review the mass laws: Law of Mass Conservation, Law of Definite Composition, and Law of Multiple Proportions.
Solution:
Experiments 1 and 2 together demonstrate the **Law of Definite Composition**. When 3.25 times the amount of blue compound in experiment 1 is used in experiment 2, then 3.25 times the amount of products were made and the relative amounts of each product are the same in both experiments. In experiment 1, the ratio of white compound to colorless gas is 0.64:0.36 or 1.78:1 and in experiment 2, the ratio is 2.08:1.17 or 1.78:1. The two experiments also demonstrate the **Law of Conservation of Mass** since the total mass before reaction equals the total mass after reaction.

2.20   Plan: The difference between the mass of fluorite and the mass of calcium gives the mass of fluorine. The masses of calcium, fluorine, and fluorite combine to give the other values.
Solution:
Fluorite is a mineral containing only calcium and fluorine.
a) Mass of fluorine = mass of fluorite – mass of calcium = 2.76 g – 1.42 g = **1.34 g fluorine**

b) To find the mass fraction of each element, divide the mass of each element by the mass of fluorite:

$$\text{Mass fraction of Ca} = \frac{\text{mass Ca}}{\text{mass fluorite}} = \frac{1.42 \text{ g Ca}}{2.76 \text{ g fluorite}} = 0.51449 = \textbf{0.514}$$

$$\text{Mass fraction of F} = \frac{\text{mass F}}{\text{mass fluorite}} = \frac{1.34 \text{ g F}}{2.76 \text{ g fluorite}} = 0.48551 = \textbf{0.486}$$

c) To find the mass percent of each element, multiply the mass fraction by 100:
Mass % Ca = (0.514) (100) = 51.449 = **51.4%**
Mass % F = (0.486) (100) = 48.551 = **48.6%**

2.22   Plan: Dividing the mass of magnesium by the mass of the oxide gives the ratio. Use this ratio, and the mass of the second sample to determine the mass of magnesium.
Solution:
a) If 1.25 g of MgO contains 0.754 g of Mg, then the mass ratio (or fraction) of magnesium in the oxide

$$\text{compound is } \frac{\text{mass Mg}}{\text{mass MgO}} = \frac{0.754 \text{ g Mg}}{1.25 \text{ g MgO}} = 0.6032 = \textbf{0.603.}$$

$$\text{b) Mass of magnesium} = \left(534 \text{ g MgO}\right)\left(\frac{0.6032 \text{ g Mg}}{1 \text{ g MgO}}\right) = 322.109 = \textbf{322 g magnesium}$$

2.24   Plan: Since copper is a metal and sulfur is a nonmetal, the sample contains 88.39 g Cu and 44.61 g S. Calculate the mass fraction of each element in the sample.
Solution:
Mass of compound = 88.39 g copper + 44.61 g sulfur = 133.00 g compound

$$\text{Mass of copper} = \left(5264 \text{ kg compound}\right)\left(\frac{10^3 \text{ g compound}}{1 \text{ kg compound}}\right)\left(\frac{88.39 \text{ g copper}}{133.00 \text{ g compound}}\right) = 3.49838 \times 10^6$$

$$= \textbf{3.498} \times \textbf{10}^6 \textbf{ g copper}$$

$$\text{Mass of sulfur} = \left(5264 \text{ kg compound}\right)\left(\frac{10^3 \text{ g compound}}{1 \text{ kg compound}}\right)\left(\frac{44.61 \text{ g sulfur}}{133.00 \text{ g compound}}\right) = 1.76562 \times 10^6$$

$$= \textbf{1.766} \times \textbf{10}^6 \textbf{ g sulfur}$$

2.26   Plan: The law of multiple proportions states that if two elements form two different compounds, the relative amounts of the elements in the two compounds form a whole number ratio. To illustrate the law we must calculate the mass of one element to one gram of the other element for each compound and then compare this mass for the two compounds. The law states that the ratio of the two masses should be a small whole number ratio such as 1:2, 3:2, 4:3, etc.

Solution:

Compound 1: $\dfrac{47.5 \text{ mass } \% \text{ S}}{52.5 \text{ mass } \% \text{ Cl}} = 0.90476 = 0.905$

Compound 2: $\dfrac{31.1 \text{ mass } \% \text{ S}}{68.9 \text{ mass } \% \text{ Cl}} = 0.451379 = 0.451$

Ratio: $\dfrac{0.905}{0.451} = 2.0067 = 2.00 / 1.00$

Thus, the ratio of the mass of sulfur per gram of chlorine in the two compounds is a small whole number ratio of 2 to 1, which agrees with the law of multiple proportions.

2.29   Plan: Determine the mass percent of sulfur in each sample by dividing the grams of sulfur in the sample by the total mass of the sample. The coal type with the smallest mass percent of sulfur has the smallest environmental impact.
Solution:

$$\text{Mass } \% \text{ in Coal A} = \left(\dfrac{11.3 \text{ g sulfur}}{378 \text{ g sample}} \right)(100\%) = 2.9894 = \mathbf{2.99\% \text{ S (by mass)}}$$

$$\text{Mass } \% \text{ in Coal B} = \left(\dfrac{19.0 \text{ g sulfur}}{495 \text{ g sample}} \right)(100\%) = 3.8384 = \mathbf{3.84\% \text{ S (by mass)}}$$

$$\text{Mass } \% \text{ in Coal C} = \left(\dfrac{20.6 \text{ g sulfur}}{675 \text{ g sample}} \right)(100\%) = 3.0519 = \mathbf{3.05\% \text{ S (by mass)}}$$

Coal A has the smallest environmental impact.

2.31   Plan: This question is based on the Law of Definite Composition. If the compound contains the same types of atoms, they should combine in the same way to give the same mass percentages of each of the elements.
Solution:
Potassium nitrate is a compound composed of three elements — potassium, nitrogen, and oxygen — in a specific ratio. If the ratio of these elements changed, then the compound would be changed to a different compound, for example, to potassium nitrite with different physical and chemical properties. Dalton postulated that atoms of an element are identical, regardless of whether that element is found in India or Italy. Dalton also postulated that compounds result from the chemical combination of specific ratios of different elements. Thus, Dalton's theory explains why potassium nitrate, a compound comprised of three different elements in a specific ratio, has the same chemical composition regardless of where it is mined or how it is synthesized.

2.32   Plan: Review the discussion of the experiments in this chapter.
Solution:
Millikan determined the minimum *charge* on an oil drop and that the minimum charge was equal to the charge on one electron. Using Thomson's value for the *mass-to-charge ratio* of the electron and the determined value for the charge on one electron, Millikan calculated the mass of an electron (charge/(charge/mass)) to be 9.109×10^{-28} g.

2.36   Plan: Re-examine the definitions of atomic number and the mass number.
Solution:
The atomic number is the number of protons in the nucleus of an atom. When the atomic number changes, the identity of the element also changes. The mass number is the total number of protons and neutrons in the nuclues of an atom. Since the identity of an element is based on the number of protons and not the number of neutrons, the mass number can vary (by a change in number of neutrons) without changing the identity of the element.

2.39   Plan: The superscript is the mass number. Consult the periodic table to get the atomic number (the number of protons). The mass number – the number of protons = the number of neutrons. For atoms, the number of protons and electrons are equal.

Solution:

Isotope	Mass Number	# of Protons	# of Neutrons	# of Electrons
^{36}Ar	36	18	18	18
^{38}Ar	38	18	20	18
^{40}Ar	40	18	22	18

2.41   Plan: The superscript is the mass number (A) and the subscript is the atomic number (Z, number of protons). The mass number – the number of protons = the number of neutrons. For atoms, the number of protons = the number of electrons.
Solution:

a) $^{16}_{8}$O and $^{17}_{8}$O have the same number of protons and electrons (8), but different numbers of neutrons.

$^{16}_{8}$O and $^{17}_{8}$O are isotopes of oxygen, and $^{16}_{8}$O has $16 - 8 = 8$ neutrons whereas $^{17}_{8}$O has $17 - 8 = 9$ neutrons. **Same Z value**

b) $^{40}_{18}$Ar and $^{41}_{19}$K have the same number of neutrons (Ar: $40 - 18 = 22$; K: $41 - 19 = 22$) but different numbers of protons and electrons (Ar = 18 protons and 18 electrons; K = 19 protons and 19 electrons). **Same N value**

c) $^{60}_{27}$Co and $^{60}_{28}$Ni have different numbers of protons, neutrons, and electrons. Co: 27 protons, 27 electrons and $60 - 27 = 33$ neutrons; Ni: 28 protons, 28 electrons and $60 - 28 = 32$ neutrons. However, both have a mass number of 60. **Same A value**

2.43   Plan: Combine the particles in the nucleus (protons + neutrons) to give the mass number (superscript, A). The number of protons gives the atomic number (subscript, Z) and identifies the element.
Solution:

a) $^{38}_{18}$Ar   b) $^{55}_{25}$Mn   c) $^{109}_{47}$Ag

2.45   Plan: Determine the number of each type of particle. The superscript is the mass number (A) and the subscript is the atomic number (Z, number of protons). The mass number – the number of protons = the number of neutrons. For atoms, the number of protons = the number of electrons.
Solution:

a) $^{48}_{22}$Ti   b) $^{79}_{34}$Se   c) $^{11}_{5}$B

22 protons   34 protons   5 protons
22 electrons   34 electrons   5 electrons
$48 - 22 = 26$ neutrons $79 - 34 = 45$ neutrons $11 - 5 = 6$ neutrons

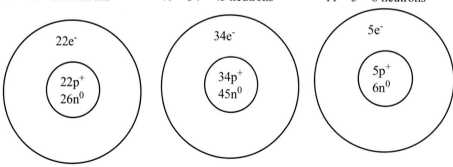

2.47   Plan: To calculate the atomic mass of an element, take a weighted average based on the natural abundance of the isotopes: (isotopic mass of isotope 1 x fractional abundance) + (isotopic mass of isotope 2 x fractional abundance).
Solution:

$$\text{Atomic mass of gallium} = (68.9256 \text{ amu})\left(\frac{60.11\%}{100\%}\right) + (70.9247 \text{ amu})\left(\frac{39.89\%}{100\%}\right) = 69.7230 = \textbf{69.72 amu}$$

2.49   Plan: To find the percent abundance of each Cl isotope, let x equal the fractional abundance of ^{35}Cl and $(1 - x)$ equal the fractional abundance of ^{37}Cl. Remember that atomic mass =(isotopic mass of ^{35}Cl x fractional abundance) + (isotopic mass of ^{37}Cl x fractional abundance)
Solution:
Atomic mass of Cl = 35.4527 amu
$$35.4527 = 34.9689x + 36.9659(1 - x)$$
$$35.4527 = 34.9689x + 36.9659 - 36.9659x$$
$$35.4527 = 36.9659 - 1.9970x$$
$$1.9970x = 1.5132$$
$$x = 0.75774 \text{ and } 1 - x = 0.24226$$
% abundance ^{35}**Cl = 75.774%**   % abundance ^{37}**Cl = 24.226%**

2.52   Plan: Review the section in the Chapter on the periodic table.
Solution:
a) In the modern periodic table, the elements are arranged in order of increasing atomic **number.**
b) Elements in a **column or group** (or family) have similar chemical properties, not those in the same period or row.
c) Elements can be classified as **metals,** metalloids, or nonmetals.

2.55   Plan: Review the properties of these two columns in the periodic table.
Solution:
The alkali metals (Group 1A(1)) are metals and readily lose one electron to form cations whereas the halogens (Group 7A(17)) are nonmetals and readily gain one electron to form anions.

2.56   Plan: Locate each element on the periodic table. The Z value is the atomic number of the element. Metals are to the left of the "staircase", nonmetals are to the right of the "staircase" and the metalloids are the elements that lie along the "staircase" line.
Solution:
a) Germanium   Ge   4A(14)   metalloid
b) Phosphorus   P   5A(15)   nonmetal
c) Helium   He   8A(18)   nonmetal
d) Lithium   Li   1A(1)   metal
e) Molybdenum   Mo   6B(6)   metal

2.58   Plan: Review the section in the chapter on the periodic table. Remember that alkaline earth metals are in Group 2A(2) and the halogens are in Group 7A(17); periods are horizontal rows.
Solution:
a) The symbol and atomic number of the heaviest alkaline earth metal are **Ra** and **88**.
b) The symbol and atomic number of the lightest metalloid in Group 4A(14) are **Si** and **14**.
c) The symbol and atomic mass of the coinage metal whose atoms have the fewest electrons are **Cu** and **63.55 amu**.
d) The symbol and atomic mass of the halogen in Period 4 are **Br** and **79.90 amu**.

2.60   Plan: Review the section of the chapter on the formation of ionic compounds.
Solution:
These atoms will form **ionic** bonds, in which one or more electrons are transferred from the metal atom to the nonmetal atom to form a cation and an anion, respectively. The oppositely charged ions attract, forming the ionic bond.

2.63   Plan: Assign charges to each of the ions. Since the sizes are similar, there are no differences due to the sizes.
Solution:
Coulomb's law states the energy of attraction in an ionic bond is directly proportional to the *product of charges* and inversely proportional to the *distance between charges*. The *product of charges* in MgO (+2 x –2 = –4) is greater than the *product of charges* in LiF (+1 x –1 = –1). Thus, MgO has stronger ionic bonding.

2.66   <u>Plan:</u> Locate these groups on the periodic table and assign charges to the ions that would form.
        <u>Solution:</u>
        The monatomic ions of Group 1A(1) have a +1 charge (e.g., Li^+, Na^+, and K^+) whereas the monatomic ions of Group 7A(17) have a –1 charge (e.g., F^-, Cl^-, and Br^-). Elements gain or lose electrons to form ions with the same number of electrons as the nearest noble gas. For example, Na loses one electron to form a cation with the same number of electrons as Ne. The halogen Cl gains one electron to form an anion with the same number of electrons as Ar.

2.68   <u>Plan:</u> A metal and a nonmetal will form an ionic compound. Locate these elements on the periodic table and predict their charges.
        <u>Solution:</u>
        Potassium sulfide (K_2S) is an ionic compound formed from a metal (potassium) and a nonmetal (sulfur). Potassium atoms transfer electrons to sulfur atoms. Each potassium atom loses one electron to form an ion with +1 charge and the same number of electrons (18) as the noble gas argon. Each sulfur atom gains two electrons to form an ion with a –2 charge and the same number of electrons (18) as the noble gas argon. The oppositely charged ions, K^+ and S^{2-}, attract each other to form an ionic compound with the ratio of two K^+ ions to one S^{2-} ion. The total number of electrons lost by the potassium atoms equals the total number of electrons gained by the sulfur atoms.

2.70   <u>Plan:</u> Locate these elements on the periodic table and predict what ions they will form. For A group cations (metals), ion charge = group number; for anions (nonmetals), ion charge = group number minus 8.
        <u>Solution:</u>
        Potassium (K) is in Group 1A(1) and forms the $\mathbf{K^+}$ ion. Iodine (I) is in Group 7A(17) and forms the $\mathbf{I^-}$ ion $(7 - 8 = -1)$.

2.72   <u>Plan:</u> Use the number of protons (atomic number) to identify the element. Add the number of protons and neutrons together to get the mass number. Locate the element on the periodic table and assign its group and period number.
        <u>Solution:</u>
        a) Oxygen (atomic number = 8)   mass number = 8p + 9n = 17   Group 6A(16)   Period 2
        b) Fluorine (atomic number = 9)   mass number = 9p + 10 n = 19   Group 7A(17)   Period 2
        c) Calcium (atomic number = 20)   mass number = 20p + 20n = 40   Group 2A(2)   Period 4

2.74   <u>Plan:</u> Determine the charges of the ions based on their position on the periodic table. Next, determine the ratio of the charges to get the ratio of the ions.
        <u>Solution:</u>
        Lithium [Group 1A(1)] forms the Li^+ ion; oxygen [Group 6A(16)] forms the O^{2-} ion $(6 - 8 = -2)$. The ionic compound that forms from the combination of these two ions must be electrically neutral, so two Li^+ ions combine with one O^{2-} ion to form the compound Li_2O. There are twice as many Li^+ ions as O^{2-} ions in a sample of Li_2O.

$$\text{Number of } O^{2-} \text{ ions} = (8.4 \times 10^{21} \text{ } Li^+ \text{ ions}) \left(\frac{1 \text{ } O^{2-} \text{ ion}}{2 \text{ } Li^+ \text{ ions}} \right) = \mathbf{4.2 \times 10^{21} \text{ } O^{2-} \text{ ions}}$$

2.76   <u>Plan:</u> The key is the size of the two alkali ions. The charges on the sodium and potassium ions are the same as both are in Group 1A(1), so there will be no difference due to the charge. The chloride ions are the same, so there will be no difference due to the chloride.
        <u>Solution:</u>
        Coulomb's law states that the energy of attraction in an ionic bond is directly proportional to the *product of charges* and inversely proportional to the *distance between charges*. The *product of the charges* is the same in both compounds because both sodium and potassium ions have a +1 charge. Attraction increases as distance decreases, so the ion with the smaller radius, Na^+, will form a stronger ionic interaction (**NaCl**).

2.78   <u>Plan:</u> Review the definitions of empirical and molecular formulas.
        <u>Solution:</u>
        An empirical formula describes the type and **simplest ratio** of the atoms of each element present in a compound whereas a molecular formula describes the type and **actual number** of atoms of each element in a molecule of the compound. The empirical formula and the molecular formula can be the same. For example, the compound formaldehyde has the molecular formula, CH_2O. The carbon, hydrogen, and oxygen atoms are present in the ratio of 1:2:1. The ratio of elements cannot be further reduced, so formaldehyde's empirical formula and molecular formula are the same. Acetic acid has the molecular formula, $C_2H_4O_2$. The carbon, hydrogen, and oxygen atoms are present in the ratio of 2:4:2, which can be reduced to 1:2:1. Therefore, acetic acid's empirical formula is CH_2O, which is different from its molecular formula. Note that the empirical formula does not *uniquely* identify a compound, because acetic acid and formaldehyde share the same empirical formula but are not the same compound.

2.80   <u>Plan:</u> Review the concepts of atoms and molecules.
        <u>Solution:</u>
        The mixture is similar to the sample of hydrogen peroxide in that both contain 20 billion oxygen atoms and 20 billion hydrogen atoms since both O_2 and H_2O_2 contain 2 oxygen atoms per molecule and both H_2 and H_2O_2 contain 2 hydrogen atoms per molecule. They differ in that they contain different types of molecules: H_2O_2 molecules in the hydrogen peroxide sample and H_2 and O_2 molecules in the mixture. In addition, the mixture contains 20 billion molecules (10 billion H_2 molecules + 10 billion O_2 molecules) while the hydrogen peroxide sample contains 10 billion molecules.

2.84   <u>Plan:</u> The empirical formula describes the **simplest ratio** of the atoms of each element present in a compound. Examine the subscripts and see if there is a common divisor. If one exists, divide all subscripts by this value.
        <u>Solution:</u>
        a) To find the empirical formula for N_2H_4, divide the subscripts by the highest common divisor, 2: N_2H_4 becomes **NH_2**.

        b) To find the empirical formula for $C_6H_{12}O_6$, divide the subscripts by the highest common divisor, 6: $C_6H_{12}O_6$ becomes **CH_2O**.

2.86   <u>Plan:</u> Locate each of the individual elements on the periodic table, and assign charges to each of the ions. For A group cations (metals), ion charge = group number; for anions (nonmetals), ion charge = group number minus 8. Find the smallest number of each ion that gives a neutral compound.
        <u>Solution:</u>
        a) Sodium is a metal that forms a +1 (Group **1**A) ion and nitrogen is a nonmetal that forms a –3 ion (Group **5**A, $5 - 8 = -3$).

        $\begin{array}{cc} +1 \ \ -3 & +1 \\ Na \ \ N & Na_3N \end{array}$ $\begin{array}{c} +3 \ \ -3 \end{array}$ The compound is **Na_3N, sodium nitride**.

        b) Oxygen is a nonmetal that forms a –2 ion (Group **6**A, $6 - 8 = -2$) and strontium is a metal that forms a +2 ion (Group **2**A).

        $\begin{array}{c} +2 \ -2 \\ Sr \ \ O \end{array}$ The compound is **SrO, strontium oxide**.

        c) Aluminum is a metal that forms a +3 ion (Group **3**A) and chlorine is a nonmetal that forms a –1 ion (Group **7**A, $7 - 8 = -1$).

        $\begin{array}{cc} +3 \ \ -1 & +3 \ \ -1 \\ Al \ \ Cl & AlCl_3 \end{array}$ $\begin{array}{c} +3 \ \ -3 \end{array}$ The compound is **$AlCl_3$, aluminum chloride**.

2.88   <u>Plan:</u> Based on the atomic numbers (the subscripts) locate the elements on the periodic table. Once the atomic numbers are located, identify the element and based on its position, assign a charge. For A group cations (metals), ion charge = group number; for anions (nonmetals), ion charge = group number minus 8. Find the smallest number of each ion that gives a neutral compound.
        <u>Solution:</u>
        a) $_{12}L$ is the element Mg ($Z = 12$). Magnesium [Group **2**A(2)] forms the Mg^{2+} ion. $_9M$ is the element F ($Z = 9$). Fluorine [Group **7**A(17)] forms the F^- ion ($7–8 = -1$). The compound formed by the combination of these two elements is **MgF_2, magnesium fluoride**.

b) $_{30}$L is the element Zn ($Z = 30$). Zinc forms the Zn^{2+} ion (see Table 2.3). $_{16}$M is the element S ($Z = 16$). Sulfur[Group **6**A(16)] will form the S^{2-} ion ($6-8 = -2$). The compound formed by the combination of these two elements is **ZnS, zinc sulfide**.

c)$_{17}$L is the element Cl ($Z = 17$). Chlorine [Group 7A(17)] forms the Cl^- ion ($7-8 = -1$). $_{38}$M is the element Sr ($Z = 38$). Strontium [Group **2**A(2)] forms the Sr^{2+} ion. The compound formed by the combination of these two elements is **SrCl$_2$, strontium chloride**.

2.90   Plan: Review the rules for nomenclature covered in the chapter. For metals, like many transition metals, that can form more than one ion each with a different charge, the ionic charge of the metal ion is indicated by a Roman numeral within parentheses immediately following the metal's name.
Solution:
a) tin (IV) chloride = **SnCl$_4$**   The (IV) indicates that the metal ion is Sn^{4+} which requires 4 Cl^- ions for a neutral compound.
b) FeBr$_3$ = **iron(III) bromide** (common name is ferric bromide); the charge on the iron ion is +3 to match the –3 charge of 3 Br^- ions. The +3 charge of the Fe is indicated by (III).   +6 –6
c) cuprous bromide = **CuBr** (cuprous is +1 copper ion, cupric is +2 copper ion)   +3 –2
d) Mn$_2$O$_3$ = **manganese(III) oxide**   Use (III) to indicate the +3 ionic charge of Mn: Mn$_2$O$_3$

2.92   Plan: Review the rules for nomenclature covered in this chapter.
Solution: +**2** –2
a) **cobalt (II) oxide** (cobalt forms more than one monatomic ion and the ion charge must be specified): CoO
b) **Hg$_2$Cl$_2$** (mercury is an unusual case in which the +1 ion form is Hg_2^{2+}, not Hg^+, see Table 2.4).
c) **lead(II) acetate trihydrate**   The $C_2H_3O_2^-$ ion has a –1 charge (see Table 2.5); since there are two of these ions, the lead ion has a +2 charge which must be indicated with the Roman numeral II. The · 3H$_2$O indicates a hydrate in which the number of H$_2$O molecules is indicated by the prefix –tri.   +3 –2   +6 –6
d) **Cr$_2$O$_3$** "chromic" denotes a +3 charge (see Table 2.4), oxygen has a –2 charge: CrO → Cr$_2$O$_3$

2.94   Plan: Review the rules for nomenclature covered in the chapter. Compounds must be neutral.
Solution:
a) Barium [Group **2**A(2)] forms Ba^{2+} and oxygen [Group 6A(16)] forms O^{2-} ($6 – 8 = -2$) so the neutral compound forms from one Ba^{2+} ion and one O^{2-} ion. Correct formula is **BaO**.
b) Iron(II) indicates Fe^{2+} and nitrate is NO_3^- so the neutral compound forms from one iron(II) ion and two nitrate ions. Correct formula is **Fe(NO$_3$)$_2$**.
c) Mn is the symbol for manganese. Mg is the correct symbol for magnesium. Correct formula is **MgS**. Sulfide is the S^{2-} ion and sulfite is the SO_3^{2-} ion.

2.96   Plan: Acids donate H^+ ion to solution, so the acid is a combination of H^+ and a negatively charged ion. Binary acids (H plus one other nonmetal) are named hydro + nonmetal root + ic acid Oxoacids (H + an oxoanion) are named by changing the suffix of the oxoanion: -ate becomes -ic acid and -ite becomes -ous acid.
Solution:
a) Hydrogen sulfate is HSO_4^-, so its source acid is **H$_2$SO$_4$**. Name of acid is **sulfuric acid** (-ate becomes -ic acid).
b) **HIO$_3$, iodic acid** (IO_3^- is the iodate ion: -ate becomes -ic acid)
c) Cyanide is CN^-; its source acid is **HCN hydrocyanic acid** (binary acid).
d) **H$_2$S, hydrosulfuric acid** (binary acid)

2.98   Plan: Consult Table 2.5 for the names and formulas of polyatomic ions.
Solution:
a) ammonium ion = **NH$_4^+$**   ammonia = **NH$_3$**
b) magnesium sulfide = **MgS**   magnesium sulfite = **MgSO$_3$**   magnesium sulfate = **MgSO$_4$**
c) hydrochloric acid = **HCl**   chloric acid = **HClO$_3$**   chlorous acid = **HClO$_2$**
d) cuprous bromide = **CuBr**   cupric bromide = **CuBr$_2$**

2.100   Plan: This compound is composed of two nonmetals. Rule 1 ("Names and Formulas of Binary Covalent Compounds") indicates that the element with the lower group number is named first. Greek numerical prefixes are used to indicate the number of atoms of each element in the compound.
Solution:
Disulfur tetrafluoride S_2F_4 di- indicates two S atoms and tetra- indicates four F atoms.

2.102   Plan: Review the nomenclature rules in the chapter.
Solution:
a) Calcium(II) dichloride, $CaCl_2$
The name becomes **calcium chloride** because calcium does not require "(II)" since it only forms +2 ions. Prefixes like di are only used in naming covalent compounds between nonmetal elements.
b) Copper(II) oxide, Cu_2O
The charge on the oxide ion is O^{2-}, which makes each copper a Cu^+. The name becomes **copper(I) oxide** to match the charge on the copper.
c) Stannous fluoride, SnF_4
Stannous refers to Sn^{2+}, but the tin in this compound is Sn^{4+} due to the charge on the fluoride ion. The tin(IV) ion is the stannic ion, this gives the name **stannic fluoride or tin (IV) fluoride**.
d) Hydrogen chloride acid, HCl
Binary acids consist of the root name of the nonmetal (chlor in this case) with a hydro- prefix and an - ic suffix. The word acid is also needed. This gives the name **hydrochloric acid**.

2.104   Plan: Break down each formula to the individual elements and count the number of each. The molecular (formula) mass is the sum of the atomic masses of all of the atoms.
Solution:
a) There are **12 atoms of oxygen** in $Al_2(SO_4)_3$. The molecular mass is:

Al	=	2(26.98 amu)	=	53.96 amu
S	=	3(32.07 amu)	=	96.21 amu
O	=	12(16.00 amu)	=	192.0 amu
				342.2 amu

b) There are **9 atoms of hydrogen** in $(NH_4)_2HPO_4$. The molecular mass is:

N	=	2(14.01 amu)	=	28.02 amu
H	=	9(1.008 amu)	=	9.072 amu
P	=	1(30.97 amu)	=	30.97 amu
O	=	4(16.00 amu)	=	64.00 amu
				132.06 amu

c) There are **8 atoms of oxygen** in $Cu_3(OH)_2(CO_3)_2$. The molecular mass is:

Cu	=	3(63.55 amu)	=	190.6 amu
O	=	8(16.00 amu)	=	128.0 amu
H	=	2(1.008 amu)	=	2.016 amu
C	=	2(12.01 amu)	=	24.02 amu
				344.6 amu

2.106   Plan: Review the rules of nomenclature and then assign a name. The molecular (formula) mass is the sum of the atomic masses of all of the atoms.
Solution:
a) $(NH_4)_2SO_4$ ammonium is NH_4^+ and sulfate is SO_4^{2-}

N	=	2(14.01 amu)	=	28.02 amu
H	=	8(1.008 amu)	=	8.064 amu
S	=	1(32.07 amu)	=	32.07 amu
O	=	4(16.00 amu)	=	64.00 amu
				132.15 amu

b) **NaH₂PO₄**   sodium is Na^+ and dihydrogen phosphate is $H_2PO_4^-$

Na	=	1(22.99 amu)	=	22.99 amu
H	=	2(1.008 amu)	=	2.016 amu
P	=	1(30.97 amu)	=	30.97 amu
O	=	4(16.00 amu)	=	64.00 amu
				119.98 amu

c) **KHCO₃**   potassium is K^+ and bicarbonate is HCO_3^-

K	=	1(39.10 amu)	=	39.10 amu
H	=	1(1.008 amu)	=	1.008 amu
C	=	1(12.01 amu)	=	12.01 amu
O	=	3(16.00 amu)	=	48.00 amu
				100.12 amu

2.108   <u>Plan:</u> Convert the names to the appropriate chemical formulas. The molecular (formula) mass is the sum of the masses of each atom times its atomic mass.
<u>Solution:</u>
a) dinitrogen pentaoxide N_2O_5 (di = 2 and penta = 5)

N	=	2(14.01 amu)	=	28.02 amu
O	=	5(16.00 amu)	=	80.00 amu
				108.02 amu

b) lead (II) nitrate $Pb(NO_3)_2$ (lead(II) is Pb^{2+} and nitrate is NO_3^-)

Pb	=	1(207.2 amu)	=	207.2 amu
N	=	2(14.01 amu)	=	28.02 amu
O	=	6(16.00 amu)	=	96.00 amu
				331.2 amu

c) calcium peroxide CaO_2 (calcium is Ca^{2+} and peroxide is O_2^{2-})

Ca	=	1(40.08 amu)	=	40.08 amu
O	=	2(16.00 amu)	=	32.00 amu
				72.08 amu

2.110   <u>Plan:</u> Use the chemical symbols and count the atoms of each type to give a molecular formula. Use the nomenclature rules in the chapter to derive the name. The molecular (formula) mass is the sum of the masses of each atom times its atomic mass.
<u>Solution:</u>
a)   Formula is **SO₃**. Name is **sulfur trioxide** (the prefix tri indicates 3 oxygen atoms).

S	=	1(32.07 amu)	=	32.07 amu
O	=	3(16.00 amu)	=	48.00 amu
				80.07 amu

b) Formula is **C₃H₈**. Since it contains only carbon and hydrogen it is a hydrocarbon and with three carbons its name is **propane**.

C	=	3(12.01 amu)	=	36.03 amu
H	=	8(1.008 amu)	=	8.064 amu
				44.09 amu

2.112   <u>Plan:</u> Use the chemical symbols and count the atoms of each type to give a molecular formula. Divide the molecular formula by the largest factor to give the empirical formula. Use the nomenclature rules in the chapter to derive the name. This compound is composed of two nonmetals. Rule 1 ("Names and Formulas of Binary Covalent Compounds") indicates that the element with the lower group number is named first. Greek numerical prefixes are used to indicate the number of atoms of each element in the compound. The molecular (formula) mass is the sum of the atomic masses of all of the atoms.
<u>Solution:</u>
The compound's name is **disulfur dichloride**. (Note: Are you unsure when and when not to use a prefix? If you leave off a prefix, can you definitively identify the compound? The name *sulfur* dichloride would not exclusively identify the molecule in the diagram because sulfur dichloride could be any combination of sulfur atoms with two chlorine atoms. Use prefixes when the name may not uniquely identify a compound.) The empirical formula is $S_{2/2}Cl_{2/2}$ or **SCl**.

The molecular mass is:

S	=	2(32.07 amu)	=	64.14 amu
Cl	=	2(35.45 amu)	=	70.90 amu
				135.04 amu

2.116   Plan: Review the discussion on separations.
Solution:
Separating the components of a mixture requires physical methods only; that is, no chemical changes (no changes in composition) take place and the components maintain their chemical identities and properties throughout. Separating the components of a compound requires a chemical change (change in composition).

2.119   Plan: Review the definitions in the chapter.
Solution:
a) Distilled water is a **compound** that consists of H_2O molecules only. How would you classify tap water?
b) Gasoline is a **homogeneous mixture** of hydrocarbon compounds of uniform composition that can be separated by physical means (distillation).
c) Beach sand is a **heterogeneous mixture** of different size particles of minerals and broken bits of shells.
d) Wine is a **homogeneous mixture** of water, alcohol, and other compounds that can be separated by physical means (distillation).
e) Air is a **homogeneous mixture** of different gases, mainly N_2, O_2, and Ar.

2.121   Plan: Review the discussion on separations.
Solution:
a) Salt dissolves in water and pepper does not. Procedure: add water to mixture, filter to remove solid pepper. Evaporate water to recover solid salt.
b) Sugar dissolves in water and sand does not. Procedure: add water to mixture, filter to remove sand. Evaporate water to recover sugar.
c) Oil is less volatile (lower tendency to vaporize to a gas) than water. Procedure: separate the components by distillation (boiling), and condense the water from vapor thus leaving the oil in the distillation flask.
d) Vegetable oil and vinegar (solution of acetic acid in water) are not soluble in each other. Procedure: allow mixture to separate into two layers and drain off the lower (denser) layer.

2.123   Plan: Review the discussion on separations.
Solution:
a)   **Filtration** — separating the mixture on the basis of differences in particle size. The water moves through the holes in the colander but the larger pasta cannot.
b) **Extraction** — The colored impurities are extracted into a solvent that is rinsed away from the raw sugar (or **chromatography**). A sugar solution is passed through a column in which the impurities stick to the stationary phase and the sugar moves through the column in the mobile phase.
c) **Extraction and filtration** — The hot water extracts the soluble tannins, flavoring and caffeine from the coffee bean. The grounds are excluded from the coffee pot by filtering the mixture through a coffee filter.

2.125   Plan: Use the equation for the volume of a sphere in Part a) to find the volume of the nucleus and the volume of the atom. Calculate the fraction of the atom volume that is occupied by the nucleus. For Part b), calculate the total mass of the two electrons; subtract the electron mass from the mass of the atom to find the mass of the nucleus. Then calculate the fraction of the atom's mass contributed by the mass of the nucleus.
Solution:

a) Fraction of volume $= \dfrac{\text{Volume of Nucleus}}{\text{Volume of Atom}} = \dfrac{\left(\dfrac{4}{3}\right)\pi\left(2.5 \times 10^{-15}\text{ m}\right)^3}{\left(\dfrac{4}{3}\right)\pi\left(3.1 \times 10^{-11}\text{ m}\right)^3} = 5.2449 \times 10^{-13} = \mathbf{5.2 \times 10^{-13}}$

b) Mass of nucleus = mass of atom – mass of electrons

$$= 6.64648 \times 10^{-24} \text{ g} - 2(9.10939 \times 10^{-28} \text{ g}) = 6.64466 \times 10^{-24} \text{ g}$$

$$\text{Fraction of mass} = \frac{\text{Mass of Nucleus}}{\text{Mass of Atom}} = \frac{\left(6.64466 \times 10^{-24} \text{ g}\right)}{\left(6.64648 \times 10^{-24} \text{ g}\right)} = 0.99972617 = \mathbf{0.999726}$$

As expected, the volume of the nucleus relative to the volume of the atom is small while its relative mass is large.

2.126 Plan: Use Coulomb's law. Choose the largest ionic charges and smallest radii for the strongest ionic bonding and the smallest ionic charges and largest radii for the weakest ionic bonding.
Solution:
Coulomb's law states that the energy of attraction in an ionic bond is directly proportional to the *product of charges* and inversely proportional to the *distance between charges*.
Strongest ionic bonding: **MgO**. Mg^{2+}, Ba^{2+} and O^{2-} have the largest charges. Attraction increases as distance decreases, so the positive ion with the smaller radius, Mg^{2+}, will form a stronger ionic bond than the larger ion Ba^{2+}.
Weakest ionic bonding: **RbI**. K^+, Rb^+, Cl^- and I^- have the smallest charges. Attraction decreases as distance increases, so the ions with the larger radius, Rb^+ and I^- will form the weakest ionic bond.

2.130 Plan: The rock is composed of 5.0% Fe_2SiO_4, 7.0% Mg_2SiO_4, and 88.0% SiO_2. It is comprised of four elements — iron, magnesium, silicon, and oxygen. At first glance, one would expect that silicon and oxygen would have the highest mass percentages because the rock is mostly silicon dioxide.
Solution:
1) First, determine the fraction of each element in each mineral.
Molecular mass of Fe_2SiO_4 = 2(55.85 amu) + 28.09 amu + 4(16.00 amu) = 203.79 amu

$$\text{Mass fraction of Fe} = \frac{\text{mass Fe}}{\text{molecular mass}} = \frac{2 \times 55.85 \text{ amu}}{203.79 \text{ amu}} = 0.5481$$

$$\text{Mass fraction of Si} = \frac{\text{mass Si}}{\text{molecular mass}} = \frac{1 \times 28.09 \text{ amu}}{203.79 \text{ amu}} = 0.1378$$

$$\text{Mass fraction of O} = \frac{\text{mass O}}{\text{molecular mass}} = \frac{4 \times 16.00 \text{ amu}}{203.79 \text{ amu}} = 0.3140$$

Molecular mass of Mg_2SiO_4 = 2(24.31 amu) + 28.09 amu + 4(16.00 amu) = 140.71 amu

$$\text{Mass fraction of Mg} = \frac{\text{mass Mg}}{\text{molecular mass}} = \frac{2 \times 24.31 \text{ amu}}{140.71 \text{ amu}} = 0.3455$$

$$\text{Mass fraction of Si} = \frac{\text{mass Si}}{\text{molecular mass}} = \frac{1 \times 28.09 \text{ amu}}{140.71 \text{ amu}} = 0.1996$$

$$\text{Mass fraction of O} = \frac{\text{mass O}}{\text{molecular mass}} = \frac{4 \times 16.00 \text{ amu}}{140.71 \text{ amu}} = 0.4548$$

Molecular mass of SiO_2 = 28.09 amu + 2(16.00 amu) = 60.09 amu

$$\text{Mass fraction of Si} = \frac{\text{mass Si}}{\text{molecular mass}} = \frac{1 \times 28.09 \text{ amu}}{60.09 \text{ amu}} = 0.4675$$

$$\text{Mass fraction of O} = \frac{\text{mass O}}{\text{molecular mass}} = \frac{2 \times 16.00 \text{ amu}}{60.09 \text{ amu}} = 0.5325$$

2) The percent mass of each element in the rock can be found by multiplying the percent of each element in each mineral by the percent of that mineral in the rock.
% Fe = (0.050) (0.5481) (100%) = **2.7% Fe**
% Mg = (0.070) (0.3455) (100%) = **2.4% Mg**
% Si = [(0.050) (0.1378) + (0.070) (0.1996) + (0.880) (0.4675)] (100%) = **43.2% Si**
% O = [(0.050) (0.3140) + (0.070) (0.4548) + (0.880) (0.5325)] (100%) = **51.6% O**
3) The percentages add up to ~100% (rounding accounts for the small error) and the majority of the rock is silicon and oxygen as expected.

2.133   <u>Plan:</u> Determine the percent oxygen in each oxide by subtracting the percent nitrogen from 100%. Express the percentage in grams and divide by the atomic mass of the appropriate elements. Then divide by the smaller ratio and convert to a whole number.
<u>Solution:</u>

a)   I $(100.00 - 46.69\ N)\% = 53.31\%\ O$

$$\left(46.69\ g\ N\right)\left(\frac{1\ mol\ N}{14.01\ g\ N}\right) = 3.3326\ mol\ N$$

$$\left(53.31\ g\ O\right)\left(\frac{1\ mol\ O}{16.00\ g\ O}\right) = 3.3319\ mol\ O$$

$$\frac{3.3326\ mol\ N}{3.3319} = 1.0002\ mol\ N \qquad \frac{3.3319\ mol\ O}{3.3319} = 1.0000\ mol\ O$$

1:1 N:O gives the empirical formula: **NO**

II $(100.00 - 36.85\ N)\% = 63.15\%\ O$

$$\left(36.85\ g\ N\right)\left(\frac{1\ mol\ N}{14.01\ g\ N}\right) = 2.6303\ mol\ N$$

$$\left(63.15\ g\ O\right)\left(\frac{1\ mol\ O}{16.00\ g\ O}\right) = 3.9469\ mol\ O$$

$$\frac{2.6303\ mol\ N}{2.6303} = 1.0000\ mol\ N$$

$$\frac{3.9469\ mol\ O}{2.6303} = 1.5001\ mol\ O$$

1:1.5 N:O gives the empirical formula: **N$_2$O$_3$**

III $(100.00 - 25.94\ N)\% = 74.06\%\ O$

$$\left(25.94\ g\ N\right)\left(\frac{1\ mol\ N}{14.01\ g\ N}\right) = 1.8515\ mol\ N$$

$$\left(74.06\ g\ O\right)\left(\frac{1\ mol\ O}{16.00\ g\ O}\right) = 4.6288\ mol\ O$$

$$\frac{1.8515\ mol\ N}{1.8515} = 1.0000\ mol\ N$$

$$\frac{4.6288\ mol\ O}{1.8515} = 2.5000\ mol\ O$$

1:2.5 N:O gives the empirical formula: **N$_2$O$_5$**

b)   I $\left(1.00\ g\ N\right)\left(\dfrac{53.31\ g\ O}{46.69\ g\ N}\right) = 1.1418 = \textbf{1.14 g O}$

II $\left(1.00\ g\ N\right)\left(\dfrac{63.15\ g\ O}{36.85\ g\ N}\right) = 1.7137 = \textbf{1.71 g O}$

III $\left(1.00\ g\ N\right)\left(\dfrac{74.06\ g\ O}{25.94\ g\ N}\right) = 2.8550 = \textbf{2.86 g O}$

2.139 Plan: The mass percent comes from determining the kilograms of each substance in a kilogram of seawater. The percent of an ion is simply the mass of that ion divided by the total mass of ions.
Solution:
a) For chloride ions:

$$\text{Mass \% Cl}^- = \left(\frac{18980.\ \text{mg Cl}^-}{1\ \text{kg seawater}}\right)\left(\frac{0.001\ \text{g}}{1\ \text{mg}}\right)\left(\frac{1\ \text{kg}}{1000\ \text{g}}\right)(100\%) = 1.898\%\ \text{Cl}^-$$

Cl⁻:   **1.898%**
Na⁺:   **1.056%**
SO₄²⁻: **0.265%**
Mg²⁺:   **0.127%**
Ca²⁺:   **0.04%**
K⁺:   **0.038%**
HCO₃⁻: **0.014%**   The mass percents do not add to 100% since the majority of seawater is H_2O.

b) Total mass of ions in 1 kg of seawater
= 18980 mg + 10560 mg + 2650 mg + 1270 mg + 400 mg + 380 mg + 140 mg = **34380 mg**

$$\%\ \text{Na}^+ = \left(\frac{10560\ \text{mg Na}^+}{34380\ \text{mg total ions}}\right)(100) = 30.71553 = \mathbf{30.72\%}$$

c) Alkaline earth metal ions are Mg^{2+} and Ca^{2+}. Total mass % = 0.127 + 0.04 = 0.167 = **0.17%**
Alkali metal ions are K^+ and Na^+. Total mass % = 1.056 + 0.038 = **1.094%**
Total mass percent for alkali metal ions is 6.6 times greater than the total mass percent for alkaline earth metal ions. Sodium ions (alkali metal ions) are dominant in seawater.
d) Anions are Cl^-, SO_4^{2-}, and HCO_3^-. Total mass % = 1.898 + 0.265 + 0.014 = **2.177%**
Cations are Na^+, Mg^{2+}, Ca^{2+}, and K^+. Total mass % = 1.056 + 0.127 + 0.04 + 0.038 = 1.2610 = **1.26%**
The mass fraction of **anions** is larger than the mass fraction of cations. Is the solution neutral since the mass of anions exceeds the mass of cations? Yes, although the mass is larger, the number of positive charges equals the number of negative charges.

2.142 Plan: First, count each type of atom present to produce a molecular formula. Divide the molecular formula by the largest divisor to produce the empirical formula. The molecular mass comes from the sum of each of the atomic masses times the number of each atom. The atomic mass times the number of each type of atom divided by the molecular mass times 100 percent gives the mass percent of each element.
Solution:
The molecular formula of succinic acid is $\mathbf{C_4H_6O_4}$.
Dividing the subscripts by 2 yields the empirical formula $\mathbf{C_2H_3O_2}$.
The molecular mass of succinic acid is 4(12.01 amu) + 6(1.008 amu) + 4(16.00 amu) = 118.088 = **118.09 amu**.

$$\%\ C = \left(\frac{4\,(12.01\ \text{amu C})}{118.088\ \text{amu}}\right)100\% = 40.6815 = \mathbf{40.68\%\ C}$$

$$\%\ H = \left(\frac{6\,(1.008\ \text{amu H})}{118.088\ \text{amu}}\right)100\% = 5.1216 = \mathbf{5.122\%\ H}$$

$$\%\ O = \left(\frac{4\,(16.00\ \text{amu O})}{118.088\ \text{amu}}\right)100\% = 54.1969 = \mathbf{54.20\%\ O}$$

Check: Total = (40.68 + 5.122 + 54.20)% = 100.00% The answer checks.

2.144 Plan: Write the formulas in the form $C_xH_yO_z$. Reduce the formulas to obtain the empirical formulas. Add the atomic masses in that empirical formula to obtain the molecular mass.
Solution:
Compound A: $C_4H_{10}O_2 = C_2H_5O$ Compound B: C_2H_4O
Compound C: $C_4H_8O_2 = C_2H_4O$ Compound D: $C_6H_{12}O_3 = C_2H_4O$
Compound E: $C_5H_8O_2$
Compounds **B, C and D** all have the same empirical formula, C_2H_4O. The molecular mass of this formula is (2 x 12.01 amu) + (4 x 1.008 amu) + (1 x 16.00 amu) = **44.05 amu**.

2.146   Plan: List all possible combinations of the isotopes. Determine the masses of each isotopic composition. The molecule consisting of the lower abundance isotopes is the least common, and the one containing only the more abundant isotopes will be the most common.
Solution:
a)   b)

Formula	Mass (amu)	
$^{15}N_2^{18}O$	48	least common
$^{15}N_2^{16}O$	46	
$^{14}N_2^{18}O$	46	
$^{14}N_2^{16}O$	44	most common
$^{15}N^{14}N^{18}O$	47	
$^{15}N^{14}N^{16}O$	45	

2.149   Plan: To find the formula mass of potassium fluoride, add the atomic masses of potassium and fluorine. Fluorine has only one naturally occurring isotope, so the mass of this isotope equals the atomic mass of fluorine. The atomic mass of potassium is the weighted average of the two isotopic masses: (isotopic mass of isotope 1 x fractional abundance) + (isotopic mass of isotope 2 x fractional abundance)
Solution:

$$\text{Average atomic mass of K} = (38.9637\ \text{amu})\left(\frac{93.258\%}{100\%}\right) + (40.9618\ \text{amu})\left(\frac{6.730\%}{100\%}\right) = 39.093\ \text{amu}$$

The formula for potassium fluoride is KF, so its molecular mass is $(39.093 + 18.9984)$ = **58.091 amu**

2.151   Plan: One molecule of NO is released per atom of N in the medicine. Divide the total mass of NO released by the molecular mass of the medicine for mass percent.
Solution:
NO = $(14.01 + 16.00)$ amu = 30.01 amu
Nitroglycerin = $C_3H_5N_3O_9$ = $[3(12.01) + 5(1.008) + 3(14.01) + 9(16.00)]$ amu = 227.10 amu
Isoamyl nitrate = $C_5H_{11}NO_3$ = $[5(12.01) + 11(1.008) + (14.01) + 3(16.00)]$ amu = 133.15 amu
In $C_3H_5N_3O_9$ (molecular mass = 227.10 amu), there are 3 atoms of N; since 1 molecule of NO is released per atom of N, this medicine would release 3 molecules of NO. The molecular mass of NO = 30.01 amu.

$$\text{mass percent of NO} = \frac{\text{total mass of NO}}{\text{mass of compound}}(100) = \frac{3(30.01\ \text{amu})}{227.10\ \text{amu}}(100) = 39.6433 = \textbf{39.64\%}$$

In $(CH_3)_2CHCH_2CH_2ONO_2$ (molecular mass = 133.15 amu), there is one atom of N; since 1 molecule of NO is released per atom of N, this medicine would release 1 molecule of NO.

$$\text{mass percent of NO} = \frac{\text{total mass of NO}}{\text{mass of compound}}(100) = \frac{1(30.01\ \text{amu})}{133.15\ \text{amu}}(100) = 22.5385 = \textbf{22.54\%}$$

2.152   Plan: First, count each type of atom present to produce a molecular formula. Determine the mass percent of each element. Mass percent = $\left(\dfrac{\text{total mass of the element}}{\text{molecular mass of TNT}}\right)100$. The mass of TNT times the mass percent of each element gives the mass of that element.
Solution:
The molecular formula for TNT is $C_7H_5O_6N_3$. (What is its empirical formula?) The molecular mass of TNT is:

C	=	7(12.01 amu)	=	84.07 amu
H	=	5(1.008 amu)	=	5.040 amu
O	=	6(16.00 amu)	=	96.00 amu
N	=	3(14.01 amu)	=	42.03 amu
				227.14 amu

The mass percent of each element is:

$$C = \left(\frac{84.07\ \text{amu}}{227.14\ \text{aum}}\right)100 = 37.01\%\ C \qquad\qquad H = \left(\frac{5.040\ \text{amu}}{227.14\ \text{aum}}\right)100 = 2.219\%\ H$$

$$O = \left(\frac{96.00 \text{ amu}}{227.14 \text{ aum}} \right) 100 = 42.26\% \text{ O} \qquad\qquad N = \left(\frac{42.03 \text{ amu}}{227.14 \text{ aum}} \right) 100 = 18.50\% \text{ N}$$

Masses:

$$\text{mass C} = \left(\frac{37.01\% \text{ C}}{100\% \text{ TNT}} \right) (1.00 \text{ lbs}) = \mathbf{0.370 \text{ lb C}}$$

$$\text{mass H} = \left(\frac{2.219\% \text{ H}}{100\% \text{ TNT}} \right) (1.00 \text{ lbs}) = \mathbf{0.0222 \text{ lb H}}$$

$$\text{mass O} = \left(\frac{42.26\% \text{ O}}{100\% \text{ TNT}} \right) (1.00 \text{ lbs}) = \mathbf{0.423 \text{ lb O}}$$

$$\text{mass N} = \left(\frac{18.50\% \text{ N}}{100\% \text{ TNT}} \right) (1.00 \text{ lbs}) = \mathbf{0.185 \text{ lb N}}$$

Note: The percent ratio yields the mass of a substance in the compound.

2.156   Plan: Mass percent = $\left(\dfrac{\text{total mass of Mg}}{\text{molecular mass of compound}} \right) 100$.

Solution:

a) If x = 0, the formula is: CH_3MgBr and the formula mass is:
   $[12.01 + 3(1.008) + 24.31 + 79.90]$ amu = 119.24 amu

$$\text{Mass \% Mg} = \frac{24.31 \text{ amu}}{119.24 \text{ amu}} \text{ x } 100\% = 20.3875 = \mathbf{20.39\% \text{ Mg}}$$

b) If x = 5, the formula is: $CH_3(CH_2)_5MgBr$ and the formula mass is:
   $[6(12.01) + 13(1.008) + 24.31 + 79.90]$ amu = 189.37 amu

$$\text{Mass \% Mg} = \frac{24.31 \text{ amu}}{189.37 \text{ amu}} \text{ x } 100\% = 12.8373 = \mathbf{12.84\% \text{ Mg}}$$

c) Mg = 16.5%

$$(24.31 \text{ amu Mg}) \left(\frac{100\% \text{ compound}}{16.5 \% \text{ Mg}} \right) = 147.33 \text{ amu} = \text{mass of compound}$$

147.33 amu - mass (CH_3MgBr) = (147.33 - 119.24) amu = 28.09 amu in CH_2 groups
(28.09 amu / 14.03 amu CH_2) = 2, thus **x = 2**.

2.158   Plan: No new substances result in a physical change; a chemical change results in new substances.
Solution:
1) Initially, all the molecules are present in blue-blue or red-red pairs. After the change, there are no red-red pairs, and there are now red-blue pairs. Changing some of the pairs means there has been a **chemical change**.
2) There are two blue-blue pairs and four red-blue pairs both before and after the change, thus no chemical change occurred. The different types of molecules are separated into different boxes. This is a **physical change**.
3) The identity of the box contents has changed from pairs to individuals. This requires a **chemical change**.
4) The contents have changed from all pairs to all triplets. This is a change in the identity of the particles, thus, this is a **chemical change**.
5) There are four red-blue pairs both before and after, thus there has been no change in the identity of the individual units. There has been a **physical change**.

CHAPTER 3 STOICHIOMETRY OF FORMULAS AND EQUATIONS

FOLLOW–UP PROBLEMS

3.1   Plan: The mass of carbon must be changed from mg to g. The molar mass of carbon can then be used to determine the number of moles.
Solution:

$$315 \text{ mg C}\left(\frac{10^{-3} \text{ g}}{1 \text{ mg}}\right)\left(\frac{1 \text{ mol C}}{12.01 \text{ g C}}\right) = 2.6228 \times 10^{-2} = \mathbf{2.62 \times 10^{-2} \text{ mol C}}$$

Check: 315 mg is less than a gram, which is less than 1/12 mol of C.

3.2   Plan: Avogadro's number is needed to convert the number of atoms to moles. The molar mass of manganese can then be used to determine the number of grams.
Solution:

$$3.22 \times 10^{20} \text{ Mn Atoms}\left(\frac{1 \text{ mol Mn}}{6.022 \times 10^{23} \text{ Mn Atoms}}\right)\left(\frac{54.94 \text{ g Mn}}{1 \text{ mol Mn}}\right) = 2.9377 \times 10^{-2}$$
$$= \mathbf{2.94 \times 10^{-2} \text{ g Mn}}$$

Check: The exponent, 20, from the Mn atoms is much smaller than that for Avogadro's number, 23, thus the mass is much smaller than the molar mass of Mn.

3.3   a) Plan: Avogadro's number is used to change the number of molecules to moles. Moles may be changed to mass using the molar mass. The molar mass of tetraphosphorus decaoxide requires the chemical formula.
Solution:
Tetra = 4, and deca = 10 to give P_4O_{10}.
The molar mass, $\mathscr{M}$, is the sum of the atomic weights, expressed in g/mol:

$$\begin{aligned} P &= 4(30.97) & &= 123.88 \text{ g/mol} \\ O &= 10(16.00) & &= \underline{160.00 \text{ g/mol}} \\ & & &= 283.88 \text{ g/mol of } P_4O_{10} \end{aligned}$$

$$4.65 \times 10^{22} \text{ molecules } P_4O_{10}\left(\frac{1 \text{ mol}}{6.022 \times 10^{23} \text{ molecules}}\right)\left(\frac{283.88 \text{ g}}{1 \text{ mol}}\right) = 21.9203 = \mathbf{21.9 \text{ g } P_4O_{10}}$$

Check: The exponents indicate there is less than 1/10 mole of P_4O_{10}. Thus the grams must be less than 1/10 the molar mass.
b) Plan: Each molecule has four phosphorus atoms, so the total number of atoms is four times the number of molecules.
Solution:

$$4.65 \times 10^{22} \text{ molecules } P_4O_{10}\left(\frac{4 \text{ atoms P}}{1 \text{ } P_4O_{10} \text{ molecule}}\right) = \mathbf{1.86 \times 10^{23} \text{ P atoms}}$$

Check: There are more phosphorus atoms than molecules so the answer should be larger than the original number.

3.4   Plan: The formula of ammonium nitrate and the molar mass are needed. The total mass of nitrogen over the molar mass times 100% gives the answer.
Solution:
The formula for ammonium nitrate is NH_4NO_3. There are 2 atoms of N per each formula.
a) Molar mass $NH_4NO_3 = (2 \times 14.01) + (4 \times 1.008) + (3 \times 16.00) = 80.05 \text{ g/mol}$

$$\frac{\text{total mass of N}}{\text{molar mass of } NH_4NO_3}(100) = \frac{2 \times 14.01 \text{ g/mol}}{80.05 \text{ g/mol}}(100) = 35.0031 = \mathbf{35.00\% \text{ N}}$$

b) <u>Plan:</u> Convert kg to grams. Use the mass percent found in (a) to find the mass of N in the sample.
<u>Solution:</u>

$$(35.8 \text{ kg})\left(\frac{10^3 \text{ g}}{1 \text{ kg}}\right)\left(\frac{35.00 \text{ g N}}{100 \text{ g NH}_4\text{NO}_3}\right) = 1.2530 \times 10^4 = \textbf{1.25} \times \textbf{10}^4 \textbf{ g Nitrogen}$$

Note: The percent ratio yields the mass of nitrogen in the compound.
<u>Check:</u> Nitrogen and oxygen are about the same mass. The difference is very slight and is neglected. Now 40% of the atoms remaining are N, so the answer should be about 40%.

3.5   <u>Plan:</u> The moles of sulfur may be calculated from the mass of sulfur and the molar mass of sulfur. The moles of sulfur and the chemical formula will give the moles of M. The mass of M divided by the moles of M will give the molar mass of M.
<u>Solution:</u>

$$(2.88 \text{ g S})\left(\frac{1 \text{ mol S}}{32.07 \text{ g S}}\right)\left(\frac{2 \text{ mol M}}{3 \text{ mol S}}\right) = 0.0599 \text{ mol M}$$

$(3.12 \text{ g M}) / (0.0599 \text{ mol M}) = 52.1 = 52.1 \text{ g M} / \text{mol M}$
The element is Cr (52.00 g/mol); M is **Chromium** and M_2S_3 is **chromium(III) sulfide**.
<u>Check:</u> Given that, the starting masses are similar, but the final formula has 1.5 times as many S ions as M ions. M should have a molar mass near 1.5 times the molar mass of sulfur: 32 x 1.5 = 48

3.6   <u>Plan:</u> Two calculations are needed — one for carbon and one for hydrogen. This is because these are the only elements present in benzo[a]pyrene. If we assume there are 100 grams of this compound, then the masses of carbon and hydrogen, in grams, are numerically equivalent to the percentages. Using the atomic masses of these two elements, the moles of each may be calculated. Dividing each of the moles by the smaller value gives the simplest ratio of C and H. The smallest multiplier to convert the ratios to whole numbers gives the empirical formula. Comparing the molar mass of the empirical formula to the molar mass given in the problem allows the molecular formula to be determined.
<u>Solution:</u>
Assuming 100 g of compound gives 95.21 g C and 4.79 g H. Then find the moles of each element using their molar masses.

$$95.21 \text{ g C}\left(\frac{1 \text{ mol C}}{12.01 \text{ g C}}\right) = 7.92756 \text{ mol C}$$

$$4.79 \text{ g H}\left(\frac{1 \text{ mol H}}{1.008 \text{ g H}}\right) = 4.75198 \text{ mol H}$$

Divide each of the moles by 4.75198, the smaller value.

$$\frac{7.92756 \text{ mol C}}{4.75198} = 1.6683 \text{ mol C}; \quad \frac{4.75198 \text{ mol H}}{4.75198} = 1.000 \text{ mol H}$$

The value 1.668 is 5/3, so the moles of C and H must each be multiplied by 3. If it is not obvious that the value is near 5/3, use a trial and error procedure whereby the value is multiplied by the successively larger integer until a value near an integer results. This gives C_5H_3 as the empirical formula. The molar mass of this formula is:
      $(5 \times 12.01 \text{ g/mol}) + (3 \times 1.008 \text{ g/mol}) = 63.074 \text{ g/mol}$
Dividing 252.30 g/mol by 63.074 g/mol gives 4.000. Thus, the empirical formula must be multiplied by 4 to give $4(C_5H_3) = \textbf{C}_{20}\textbf{H}_{12}$ as the molecular formula of benzo[a]pyrene.
<u>Check:</u> Determine the molar mass of the formula given and compare it to the value given in the problem:
$\mathcal{M} = (20 \times 12.01 \text{ g/mol}) + (12 \times 1.008 \text{ g/mol}) = 252.296 \text{ g/mol}$. This is very close to the given value.

3.7   <u>Plan:</u> The carbon in the sample is converted to carbon dioxide, the hydrogen is converted to water, and the remaining material is chlorine. The grams of carbon dioxide and the grams of water are both converted to moles. One mole of carbon dioxide gives one mole of carbon, while one mole of water gives two moles of hydrogen. Using the molar masses of carbon and hydrogen, the grams of each of these elements in the original sample may be determined. The original mass of sample minus the masses of carbon and hydrogen gives the mass of chlorine. The mass of chlorine and the molar mass of chlorine will give the moles of chlorine. Once the moles of each of the elements have been calculated, divide by the smallest value, and, if necessary, multiply by the smallest number required to give a set of whole numbers for the empirical formula. Compare the molar mass of the empirical formula to the molar mass given in the problem to find the molecular formula.

Solution:

Determine the moles and the masses of carbon and hydrogen produced by combustion of the sample.

$$0.451 \text{ g CO}_2 \left(\frac{1 \text{ mol CO}_2}{44.01 \text{ g CO}_2} \right) \left(\frac{1 \text{ mol C}}{1 \text{ mol CO}_2} \right) = 0.010248 \text{ mol C} \left(\frac{12.01 \text{ g C}}{1 \text{ mol C}} \right) = 0.12307 \text{ g C}$$

$$0.0617 \text{ g H}_2\text{O} \left(\frac{1 \text{ mol H}_2\text{O}}{18.016 \text{ g H}_2\text{O}} \right) \left(\frac{2 \text{ mol H}}{1 \text{ mol H}_2\text{O}} \right) = 0.0068495 \text{ mol H} \left(\frac{1.008 \text{ g H}}{1 \text{ mol H}} \right) = 0.006904 \text{ g H}$$

The mass of chlorine is given by: 0.250 g sample – (0.12307 g C + 0.006904 g H) = 0.120 g Cl
The moles of chlorine are:

$$0.120 \text{ g Cl} \left(\frac{1 \text{ mol Cl}}{35.45 \text{ g Cl}} \right) = 0.0033850 \text{ mol Cl}. \text{ This is the smallest number of moles.}$$

Dividing each mole value by the lowest value, 0.0033850:

$$\frac{0.010248 \text{ mol C}}{0.0033850} = 3.03 \text{ mol C}; \quad \frac{0.0068495 \text{ mol H}}{0.0033850} = 2.02 \text{ mol H}; \quad \frac{0.0033850 \text{ mol Cl}}{0.0033850} = 1.00 \text{ mol Cl}$$

These values are all close to whole numbers, thus the empirical formula is C_3H_2Cl.
The empirical formula has the following molar mass:
(3 x 12.01 g/mol) + (2 x 1.008 g/mol) + (35.45 g/mol) = 73.496 g/mol C_3H_2Cl
Dividing the given molar mass by the empirical formula mass: (146.99 g/mol) / (73.496 g/mol) = 2.00 Thus, the molecular formula is two times the empirical formula, $2(C_3H_2Cl) = \mathbf{C_6H_4Cl_2}$.
Check: Carbon is about one-fourth the mass of carbon dioxide, or 0.11 grams of C in the compound. The mass of hydrogen is very small, thus, most of the remaining mass of the compound (0.250 – 0.11 = 0.14) must be chlorine. Chlorine is about three times as heavy as carbon, thus, there must be three carbons for each chlorine. Also, check that the molar mass of the molecular formula agrees with the given molar mass.
(6 x 12.01 g/mol) + (4 x 1.008 g/mol) + (2 x 35.45 g/mol) = 146.992 g/mol

3.8 Plan: In each part it is necessary to determine the chemical formulas, including the physical states, for both the reactants and products. The formulas are then placed on the appropriate sides of the reaction arrow. The equation is then balanced.

a) Solution:
Sodium is a metal (solid) that reacts with water (liquid) to produce hydrogen (gas) and a solution of sodium hydroxide (aqueous). Sodium is Na; water is H_2O; hydrogen is H_2; and sodium hydroxide is NaOH.

$$Na(s) + H_2O(l) \rightarrow H_2(g) + NaOH(aq) \text{ is the equation.}$$

Balancing will precede one element at a time. One way to balance hydrogen gives:

$$Na(s) + 2 H_2O(l) \rightarrow H_2(g) + 2 NaOH(aq)$$

Next, the sodium will be balanced:

$$2 Na(s) + 2 H_2O(l) \rightarrow H_2(g) + 2 NaOH(aq)$$

On inspection, we see that the oxygens are already balanced.
Check: Reactants (2 Na, 4 H, 2 O) = Products (2 Na, 4 H, 2 O)

b) Solution:
Aqueous nitric acid reacts with calcium carbonate (solid) to produce carbon dioxide (gas), water (liquid), and aqueous calcium nitrate. Nitric acid is HNO_3; calcium carbonate is $CaCO_3$; carbon dioxide is CO_2; water is H_2O; and calcium nitrate is $Ca(NO_3)_2$. The starting equation is

$$HNO_3(aq) + CaCO_3(s) \rightarrow CO_2(g) + H_2O(l) + Ca(NO_3)_2(aq)$$

Initially, Ca and C are balanced. Proceeding to another element, such as N, or better yet the group of elements in NO_3^- gives the following partially balanced equation:

$$2 HNO_3(aq) + CaCO_3(s) \rightarrow CO_2(g) + H_2O(l) + Ca(NO_3)_2(aq)$$

Now, all the elements are balanced.
Check: Reactants (2 H, 2 N, 9 O, 1 Ca, 1 C) = Products (2 H, 2 N, 9 O, 1 Ca, 1 C)

c) Solution:

We are told all the substances involved are gases. The reactants are phosphorus trichloride and hydrogen chloride, while the products are phosphorus trifluoride and hydrogen chloride. Phosphorus trifluoride is PF_3; phosphorus trichloride is PCl_3; hydrogen fluoride is HF; and hydrogen chloride is HCl. The initial equation is:

$$PCl_3(g) + HF(g) \rightarrow PF_3(g) + HCl(g)$$

Initially, P and H are balanced. Proceed to another element (either F or Cl); if we will choose Cl, it balances as:

$$PCl_3(g) + HF(g) \rightarrow PF_3(g) + 3HCl(g)$$

The balancing of the Cl unbalances the H, this should be corrected by balancing the H as:

$$PCl_3(g) + 3HF(g) \rightarrow PF_3(g) + 3HCl(g)$$

Now, all the elements are balanced.

Check: Reactants (1 P, 3 Cl, 3 H, 3 F) = Products (1 P, 3 Cl, 3 H, 3 F)

d) Solution:

We are told that nitroglycerine is a liquid reactant, and that all the products are gases. The formula for nitroglycerine is given. Carbon dioxide is CO_2; water is H_2O; nitrogen is N_2; and oxygen is O_2. The initial equation is:

$$C_3H_5N_3O_9(l) \rightarrow CO_2(g) + H_2O(g) + N_2(g) + O_2(g)$$

Counting the atoms shows no atoms are balanced.

One element should be picked and balanced. Any element except oxygen will work. Oxygen will not work in this case because it appears more than once on one side of the reaction arrow. We will start with carbon. Balancing C gives:

$$C_3H_5N_3O_9(l) \rightarrow 3\ CO_2(g) + H_2O(g) + N_2(g) + O_2(g)$$

Now balancing the hydrogen gives:

$$C_3H_5N_3O_9(l) \rightarrow 3\ CO_2(g) + 5/2\ H_2O(g) + N_2(g) + O_2(g)$$

Similarly, if we balance N we get:

$$C_3H_5N_3O_9(l) \rightarrow 3\ CO_2(g) + 5/2\ H_2O(g) + 3/2\ N_2(g) + O_2(g)$$

Clearing the fractions by multiplying everything except the unbalanced oxygen by 2 gives:

$$2\ C_3H_5N_3O_9(l) \rightarrow 6\ CO_2(g) + 5\ H_2O(g) + 3\ N_2(g) + O_2(g)$$

This leaves oxygen to balance. Balancing oxygen gives:

$$2\ C_3H_5N_3O_9(l) \rightarrow 6\ CO_2(g) + 5\ H_2O(g) + 3\ N_2(g) + 1/2\ O_2(g)$$

Again clearing fractions by multiplying everything by 2 gives:

$$4\ C_3H_5N_3O_9(l) \rightarrow 12\ CO_2(g) + 10\ H_2O(g) + 6\ N_2(g) + O_2(g)$$

Now all the elements are balanced.

Check: Reactants (12 C 20 H 12 N 36 O) = Products (12 C 20 H 12 N 36 O)

3.9   Plan: Count the number of each type of atom in each molecule to write the formulas of the reactants and products.

Solution:

$$6\ CO(g) + 3\ O_2(g) \rightarrow 6\ CO_2(g)$$
$$\text{or,} \qquad 2\ CO(g) + O_2(g) \rightarrow 2\ CO_2(g)$$

Check: Reactants (2 C   4 O) = Products (2 C   4 O)

3.10   a) Plan: The reaction, like all reactions, needs a balanced chemical equation. The atomic mass of aluminum is used to determine the moles of aluminum. The mole ratio, from the balanced chemical equation, converts the moles of aluminum to moles of iron. Finally, the atomic mass of iron is used to change the moles of iron to the grams of iron.

Solution:

The names and formulas of the substances involved are: iron(III) oxide, Fe_2O_3, and aluminum, Al, as reactants, and aluminum oxide, Al_2O_3, and iron, Fe. The iron is formed as a liquid; all other substances are solids. The equation begins as:

$$Fe_2O_3(s) + Al(s) \rightarrow Al_2O_3(s) + Fe(l)$$

There are 2 Fe, 3 O, and 1 Al on the reactant side and 1 Fe, 3 O, and 2 Al on the product side.

Balancing aluminum: $Fe_2O_3(s) + 2\ Al(s) \rightarrow Al_2O_3(s) + Fe(l)$

Balancing iron: $Fe_2O_3(s) + 2\ Al(s) \rightarrow Al_2O_3(s) + 2\ Fe(l)$

Check: reactants: (2 Fe, 3 O, 2 Al) = products: (2 Fe, 3 O, 2 Al)

Using the balanced equation and the atomic masses, calculate the grams of iron:

$$\left(135 \text{ g Al}\right)\left(\frac{1 \text{ mol Al}}{26.98 \text{ g Al}}\right)\left(\frac{2 \text{ mol Fe}}{2 \text{ mol Al}}\right)\left(\frac{55.85 \text{ g Fe}}{1 \text{ mol Fe}}\right) = 279.457 = \textbf{279 g Fe}$$

Check: The number of moles of aluminum produces an equal number of moles of iron. Iron has about double the mass of Al (2 x 27 = 54). Thus, the initial mass of aluminum should give approximately 2 x 135 g (= 270) of iron.

b) Plan: The grams of aluminum oxide must be converted to moles. The formula shows there are two moles of aluminum for every mole of aluminum oxide. Avogadro's number will then convert the moles of aluminum to the number of atoms.

Solution:

$$\left(1.00 \text{ g Al}_2\text{O}_3\right)\left(\frac{1 \text{ mol Al}_2\text{O}_3}{101.96 \text{ g Al}_2\text{O}_3}\right)\left(\frac{2 \text{ mol Al}}{1 \text{ mol Al}_2\text{O}_3}\right)\left(\frac{6.022 \times 10^{23} \text{ atoms Al}}{1 \text{ mol Al}}\right) = 1.18125 \times 10^{22} = \textbf{1.18 x 10}^{22} \textbf{ atoms Al}$$

Check: The molar mass of aluminum oxide is about 100, so one mole of aluminum oxide is about 1/100 the mass of aluminum oxide. Doubling the moles of aluminum oxide gives the moles of aluminum atoms. Multiplying Avogadro's number by 2/100 gives about 1.2×10^{22} atoms.

3.11   Plan: Write the balanced chemical equation for each step. Add the equations, canceling common substances.
        Solution:
            Step 1 $2 \text{ SO}_2(g) + \text{O}_2(g) \rightarrow 2 \text{ SO}_3(g)$
            Step 2 $\text{SO}_3(g) + \text{H}_2\text{O}(l) \rightarrow \text{H}_2\text{SO}_4(aq)$
        Adjust the coefficients since 2 moles of SO_3 are produced in Step 1 but only 1 mole of SO_3 is consumed in Step 2. We have to double all of the coefficients in Step 2 so that the amount of SO_3 formed in Step 1 is used in Step 2.
            Step 1 $2 \text{ SO}_2(g) + \text{O}_2(g) \rightarrow 2 \text{ SO}_3(g)$
            Step 2 $2 \text{ SO}_3(g) + 2 \text{ H}_2\text{O}(l) \rightarrow 2 \text{ H}_2\text{SO}_4(aq)$
        Add the two equations and cancel common substances.
            Step 1 $2 \text{ SO}_2(g) + \text{O}_2(g) \rightarrow 2 \text{ SO}_3(g)$
            Step 2 $\underline{2 \text{ SO}_3(g) + 2 \text{ H}_2\text{O}(l) \rightarrow 2 \text{ H}_2\text{SO}_4(aq)}$
            $2 \text{ SO}_2(g) + \text{O}_2(g) + 2\text{ SO}_3(g) + 2 \text{ H}_2\text{O}(l) \rightarrow 2\text{ SO}_3(g) + 2 \text{ H}_2\text{SO}_4(aq)$
        Or $2 \text{ SO}_2(g) + \text{O}_2(g) + 2 \text{ H}_2\text{O}(l) \rightarrow 2 \text{ H}_2\text{SO}_4(aq)$

3.12   Plan: (a) Count the molecules of each type, and find the simplest ratio. The simplest ratio leads to a balanced chemical equation. The substance with no remaining particles is the limiting reagent. (b) Use the balanced chemical equation to determine the mole ratio for the reaction.
        Solution:
        (a) The balanced chemical equation is
            $\text{B}_2(g) + 2 \text{ AB}(g) \rightarrow 2 \text{ AB}_2(g)$
            The limiting reagent is **AB** since there is a B_2 molecule left over (excess).
        (b) Moles of AB_2 from AB: $\left(1.5 \text{ mol AB}\right)\left(\frac{2 \text{ mol AB}_2}{2 \text{ mol AB}}\right) = 1.5 \text{ mol AB}_2$

            Moles of AB_2 from B_2: $\left(1.5 \text{ mol B}_2\right)\left(\frac{2 \text{ mol AB}_2}{1 \text{ mol B}_2}\right) = 3.0 \text{ mol AB}_2$

        Thus AB is the Limiting reagent and only **1.5 mol of AB$_2$** will form.
        Check: The balanced chemical equation says twice as many moles of AB are needed for every mole of B_2. Equal numbers of moles means AB is not twice as great as B_2. Thus, AB should be limiting.

3.13   Plan: First, determine the formulas of the materials in the reaction and begin a chemical equation. Balance the equation. Using the molar mass of each reactant, determine the moles of each reactant. Use mole ratios from the balanced equation to determine the moles of aluminum sulfide that may be produced from each reactant. The reactant that generates the smaller number of moles is limiting. Change the moles of aluminum sulfide from the limiting reactant to the grams of product using the molar mass of aluminum sulfide. To find the excess reactant amount, find the amount of excess reactant required to react with the limiting reagent and subtract that amount from the amount given in the problem.

Solution:
The balanced equation is: $2Al(s) + 3S(s) \rightarrow Al_2S_3(s)$
Determining the moles of product from each reactant:

$$(10.0 \text{ g Al})\left(\frac{1 \text{ mol Al}}{26.98 \text{ g Al}}\right)\left(\frac{1 \text{ mol Al}_2S_3}{2 \text{ mol Al}}\right) = 0.18532 \text{ mol Al}_2S_3$$

$$(15.0 \text{ g S})\left(\frac{1 \text{ mol S}}{32.07 \text{ g S}}\right)\left(\frac{1 \text{ mol Al}_2S_3}{3 \text{ mol Al}}\right) = 0.155909 \text{ mol Al}_2S_3$$

Sulfur produces less product, thus it is limiting. The moles of product formed are calculated from the moles and the molar mass.

$$(0.155909 \text{ mol Al}_2S_3)\left(\frac{150.17 \text{ g Al}_2S_3}{1 \text{ mol Al}_2S_3}\right) = 23.413 = \textbf{23.4 g Al}_2\textbf{S}_3$$

The mass of aluminum used in the reaction is now determined:

$$(15.0 \text{ g S})\left(\frac{1 \text{ mol S}}{32.07 \text{ g S}}\right)\left(\frac{2 \text{ mol Al}}{3 \text{ mol Al}}\right)\left(\frac{26.98 \text{ g Al}}{1 \text{ mol Al}}\right) = 8.41258 \text{ g Al used}$$

Subtracting the mass of aluminum used from the initial aluminum gives the mass remaining.
Un–reacted aluminum = 10.0 g – 8.41285 g = 1.5872 = **1.6 g Al excess**
Check: The reactant ratios in the balanced chemical equation are the same as the mass ratio given in the problem. Thus, the substance with the larger molar mass should be limiting. This is the sulfur. An additional check is based on the aluminum calculation. If the sulfur were not the limiting reactant, the "remaining" aluminum would have been a negative number.

3.14   Plan: Determine the formulas, and then balance the chemical equation. The grams of marble are converted to moles, a mole ratio (from the balanced equation) gives the moles of CO_2, and finally the theoretical yield of CO_2 is determined from the moles of CO_2 and its molar mass.
Solution:
The balanced equation: $CaCO_3(s) + 2 \text{ HCl}(aq) \rightarrow CaCl_2(aq) + H_2O(l) + CO_2(g)$
The theoretical yield of carbon dioxide:

$$(10.0 \text{ g CaCO}_3)\left(\frac{1 \text{ mol CaCO}_3}{100.09 \text{ g CaCO}_3}\right)\left(\frac{1 \text{ mol CO}_2}{1 \text{ mol CaCO}_3}\right)\left(\frac{44.01 \text{ g CO}_2}{1 \text{ mol CO}_2}\right) = 4.39704 \text{ g CO}_2$$

The percent yield:

$$\left(\frac{\text{actual yield}}{\text{theoretical yield}}\right)(100\%) = \left(\frac{3.65 \text{ g CO}_2}{4.39704 \text{ g CO}_2}\right)(100\%) = 83.0104 = \textbf{83.0\%}$$

Check: Initially, there are about 10/100 moles of marble (mass of marble/molar mass). Theoretically, about 1/10 the molar mass of carbon dioxide should form.

3.15   Plan: Multiply the molarity by the liters to determine the moles.
Solution:

$$\left(\frac{0.50 \text{ mol KI}}{L}\right)\left(\frac{10^{-3} \text{ L}}{1 \text{ mL}}\right)(84 \text{ mL}) = \textbf{0.042 mol KI}$$

Check: The molarity is 0.5, this ratio implies the moles should be 0.5 (84 mL) = 42 mmole (0.042)

3.16   Plan: Use the molar mass to change the grams to moles. The molarity may then be used to convert to volume.
Solution:

$$(135 \text{ g C}_{12}\text{H}_{22}\text{O}_{11})\left(\frac{1 \text{ mol C}_{12}\text{H}_{22}\text{O}_{11}}{342.30 \text{ g C}_{12}\text{H}_{22}\text{O}_{11}}\right)\left(\frac{1 \text{ L}}{3.30 \text{ mol C}_{12}\text{H}_{22}\text{O}_{11}}\right) = 0.11951 = \textbf{0.120 L}$$

Check: All units cancel except the volume units requested. The mass given is about 1/3 mole and the molarity term gives another 1/3 ratio, thus the answer should be about (1/3) (1/3) = 1/9 L.

3.17   <u>Plan:</u> Determine the new concentration from the dilution equation $(M_{conc})(V_{conc}) = (M_{dil})(V_{dil})$. Convert the molarity (mol/L) to g/mL in two steps (one–step is moles to grams, and the other step is L to mL.)
<u>Solution:</u>

$$M_{dil} = \frac{M_{conc}V_{conc}}{V_{dil}} = \frac{(7.50\ M)(25.0\ m^3)}{500.\ m^3} = 0.375\ M$$

$$\left(\frac{0.375\ mol\ H_2SO_4}{1\ L}\right)\left(\frac{98.09\ g\ H_2SO_4}{1\ mol\ H_2SO_4}\right)\left(\frac{10^{-3}\ L}{1\ mL}\right) = 0.036784 = \mathbf{3.68\ x\ 10^{-2}}\ g/mL\ solution$$

<u>Check:</u> The dilution will give a concentration lower by 25/500 = 1/20. Checking the calculation by doing a rough calculation gives (8) (1/20) (100) (0.001) = 0.04.

3.18   <u>Plan:</u> Count the number of particles in each solution per unit volume.
<u>Solution:</u>
Solution A has 6 particles per unit volume while Solution B has 12 particles per unit volume. Solution B is more concentrated than Solution A. To obtain B:
the total volume of Solution A was reduced by half:

$$V_{conc} = \frac{N_{dil}V_{dil}}{N_{conc}} = \frac{(6\ particles)(1.0\ mL)}{(12\ particles)} = 0.50\ mL$$

Solution C has 4 particles and is thus more dilute than Solution A. To obtain Solution C, one-half the volume of solvent must be added for every volume of Solution A:

$$V_{dil} = \frac{N_{conc}V_{conc}}{N_{dil}} = \frac{(6\ particles)(1.0\ mL)}{(4\ particles)} = 1.5\ mL$$

3.19   <u>Plan:</u> The problem should be solved like sample problem 3.19 to facilitate the comparison. A new balanced chemical equation is needed. A 0.10 g sample of aluminum hydroxide, for direct comparison, is converted to moles. The mole ratio from the balanced equation converts from moles of $Al(OH)_3$ to moles of HCl. Finally, the moles of acid reacting with the $Al(OH)_3$ are compared to the $3.4\ x\ 10^{-3}$ mol of HCl that was shown in the Sample Problem to react with 0.10 g of $Mg(OH)_2$.
<u>Solution:</u>
Balanced equation: $3\ HCl(aq) + Al(OH)_3(s) \rightarrow AlCl_3(aq) + 3\ H_2O(l)$

$$(0.10\ g\ Al(OH)_3)\left(\frac{1\ mol\ Al(OH)_3}{78.00\ g\ Al(OH)_3}\right)\left(\frac{3\ mol\ HCl}{1\ mol\ Al(OH)_3}\right) = 3.84615\ x\ 10^{-3} = 3.8\ x\ 10^{-3}\ mol\ HCl$$

Aluminum hydroxide is more effective than magnesium hydroxide because it reacts with more ($3.8\ x\ 10^{-3}$) moles of HCl.
<u>Check:</u> The greater molar mass of $Al(OH)_3$ is about 6/8, and this compound has a 3/2 larger mole ratio from the balanced equations. This gives $(3.4\ x\ 10^{-3})$ (6/8) (1.5) = $3.8\ x\ 10^{-3}$.

3.20   a) <u>Plan:</u> The formula of lead(II) acetate is needed. The moles of compound are converted to moles of lead; this value times the inverse of the molarity gives the volume of the solution.
<u>Solution:</u>
The formula is $Pb(C_2H_3O_2)_2$.

$$(0.400\ mol\ Pb^{2+})\left(\frac{1\ mol\ Pb(C_2H_3O_2)_2}{1\ mol\ Pb^{2+}}\right)\left(\frac{1\ L}{1.50\ mol\ Pb(C_2H_3O_2)_2}\right) = 0.266667 = \mathbf{0.267\ L}$$

<u>Check:</u> A reverse calculation may serve as a check (complete units are unnecessary):

$$(0.267\ L)\left(\frac{1.5\ mol}{L}\right)\left(\frac{1\ mol}{1\ mol}\right) = 0.4005 = 0.400\ mol$$

b) <u>Plan:</u> The formulas are used for the balanced chemical equation. The molarity and the volume of the sodium chloride solution are used to determine the moles of NaCl. Either the lead or the chloride ion is limiting — the one producing the fewer moles of product is limiting. Convert the lower number of moles of product to grams of product.

Solution:
Balanced reaction:

$$Pb(C_2H_3O_2)_2(aq) + 2\ NaCl(aq) \rightarrow PbCl_2(s) + 2\ NaC_2H_3O_2(aq)$$

Because of the 1:1 ratio, there will be 0.400 mol of $PbCl_2$ produced from 0.400 mol of Pb^{2+}.
The moles of $PbCl_2$ produced from the NaCl are:

$$\left(\frac{3.40\ mol\ NaCl}{1\ L}\right)\left(\frac{10^{-3}\ L}{1\ mL}\right)(125\ mL)\left(\frac{1\ mol\ PbCl_2}{2\ mol\ NaCl}\right) = 0.2125\ mol\ PbCl_2$$

The NaCl is limiting. The mass of $PbCl_2$ may now be determined using the molar mass.

$$(0.2125\ mol\ PbCl_2)\left(\frac{278.1\ g\ PbCl_2}{1\ mol\ PbCl_2}\right) = 59.0962 = \textbf{59.1 g PbCl}_2$$

Check: There are 0.400 mol of lead ions; they will require 2 x 0.400 = 0.800 mol of chloride ion. The volume of NaCl solution needed to supply the chloride ions is (0.800/3.4) = 0.235 L (235 mL). The NaCl must be limiting because less than 235 mL are supplied.

END–OF–CHAPTER PROBLEMS

3.2   Plan: The molecular formula of sucrose gives the mole amount of each element in 1 mole of sucrose. Avogadro's number allows the change from moles to atoms.
Solution:

a) Moles of C atoms = $(1\ mol\ Sucrose)\left(\dfrac{12\ mol\ C}{1\ mol\ C_{12}H_{22}O_{11}}\right) = \textbf{12 mol C}$

b) C atoms = $(2\ mol\ C_{12}H_{22}O_{11})\left(\dfrac{12\ mol\ C}{1\ mol\ C_{12}H_{22}O_{11}}\right)\left(\dfrac{6.022\ x\ 10^{23}\ C\ atoms}{1\ mol\ C}\right) = \textbf{1.445 x 10}^{25}\ \textbf{C atoms}$

3.7   Plan: It is possible to relate the relative atomic masses by counting the number of atoms.
Solution:
a) The element on the **left** (green) has the higher molar mass because only 5 green balls are necessary to counterbalance the mass of 6 yellow balls. Since the green ball is heavier, its atomic mass is larger, and therefore its molar mass is larger.
b) The element on the **left** (red) has more atoms per gram. This figure requires more thought because the number of red and blue balls is unequal and their masses are unequal. If each pan contained 3 balls, then the red balls would be lighter. The presence of six red balls means that they are that much lighter. Because the red ball is lighter, more red atoms are required to make 1 gram.
c) The element on the **left** (orange) has fewer atoms per gram. The orange balls are heavier, and it takes fewer orange balls to make 1 gram.
d) **Neither** element has more atoms per mole. Both the left and right elements have the same number of atoms per mole. The number of atoms per mole ($6.022\ x\ 10^{23}$) is constant and so is the same for every element.

3.8   Plan: Locate each of the elements on the periodic table and record its atomic mass. The atomic mass of the element times the number of atoms present in the formula gives the mass of that element in one mole of the substance. The molar mass is the sum of the masses of the elements in the substance.
Solution:
a) The molar mass, $\mathcal{M}$, is the sum of the atomic weights, expressed in g/mol:

Sr = (1 mol Sr) (87.62 g Sr/mol Sr) = 87.62 g Sr/mol Sr(OH)₂
O = (2 mol O) (16.00 g O/mol O)   = 32.00 g O/mol Sr(OH)₂
H = (2 mol H) (1.008 g H/mol H)   = 2.016 g H/mol Sr(OH)₂
                                  = **121.64 g/mol of Sr(OH)₂**

b) $\mathcal{M}$ = (2 mol N) (14.01 g N/mol N) + (3 mol O) (16.00 g O/mol O) = **76.02 g/mol of N₂O₃**
c) $\mathcal{M}$ = (1 mol Na) (22.99 g Na/mol Na) + (1 mol Cl) (35.45 g Cl/mol Cl)
      + (3 mol O) (16.00 g O/mol O) = **106.44 g/mol of NaClO₃**
d) $\mathcal{M}$ = (2 mol Cr) (52.00 g Cr/mol Cr) + (3 mol O) (16.00 g O/mol O) = **152.00 g/mol of Cr₂O₃**

3.10   Plan: Locate each of the elements on the periodic table and record its atomic mass. The atomic mass of the element times the number of atoms present in the formula gives the mass of that element in one mole of the substance. The molar mass is the sum of the masses of the elements in the substance.
Solution:
a) $\mathcal{M}$ = (1 mol Sn) (118.7 g Sn/mol Sn) + (1 mol O) (16.00 g O/mol O) = **134.7 g/mol of SnO_2**
b) $\mathcal{M}$ = (1 mol Ba) (137.3 g Ba/mol Ba) + (2 mol F) (19.00 g F/mol F) = **175.3 g/mol of BaF_2**
c) $\mathcal{M}$ = (2 mol Al) (26.98 g Al/mol Al) + (3 mol S) (32.07 g S/mol S) + (12 mol O) (16.00 g O/mol O)
       = **342.17 g/mol of $Al_2(SO_4)_3$**
d) $\mathcal{M}$ = (1 mol Mn) (54.94 g Mn/mol Mn) + (2 mol Cl) (35.45 g Cl/mol Cl) = **125.84 g/mol of $MnCl_2$**

3.12   Plan: The mass of a substance and its number of moles are related through the conversion factor of $\mathcal{M}$, the molar mass expressed in g/mol. The moles of a substance and the number of entities per mole are related by the conversion factor, Avogadro's number.
Solution:
a) $\mathcal{M}$ of $KMnO_4$ = 39.10 + 54.94 + (4 x 16.00) = 158.04 g/mol of $KMnO_4$

$$\text{Mass of } KMnO_4 = \left(0.68 \text{ mol } KMnO_4\right)\left(\frac{158.04 \text{ g } KMnO_4}{1 \text{ mol } KMnO_4}\right) = 107.47 = \textbf{1.1 x } \mathbf{10^2} \textbf{ g } \mathbf{KMnO_4}$$

b) $\mathcal{M}$ of $Ba(NO_3)_2$ = 137.3 + (2 x 14.01) + (6 x 16.00) = 261.3 g/mol $Ba(NO_3)_2$

$$\text{Moles of O atoms} = \left(8.18 \text{ g } Ba(NO_3)_2\right)\left(\frac{1 \text{ mol } Ba(NO_3)_2}{261.3 \text{ g } Ba(NO_3)_2}\right)\left(\frac{6 \text{ mol O atoms}}{1 \text{ mol } Ba(NO_3)_2}\right)$$

= 0.18783 = **0.188 mol O atoms**
c) $\mathcal{M}$ of $CaSO_4\cdot 2H_2O$ = 40.08 + 32.07 + (6 x 16.00) + (4 x 1.008) = 172.18 g/mol
(Note that the waters of hydration are included in the molar mass.)

$$\text{O atoms} = \left(7.3 \text{ x } 10^{-3} \text{ g Ca Cmpd}\right)\left(\frac{1 \text{ mol Ca Cmpd}}{172.18 \text{ g Ca Cmpd}}\right)\left(\frac{6 \text{ mol O atoms}}{1 \text{ mol Ca Cmpd}}\right)\left(\frac{6.022 \text{ x } 10^{23} \text{ O atoms}}{1 \text{ mol O atoms}}\right)$$

= 1.5319 x 10^{20} = **1.5 x 10^{20} O atoms**

3.14   Plan: Determine the molar mass of each substance, then perform the appropriate molar conversions.
Solution:
a) $\mathcal{M}$ of $MnSO_4$ = (54.94 g Mn/mol Mn) + (32.07 g S/mol S) + [(4 mol O) (16.00 g O/mol O)]
            = 151.01 g/mol of $MnSO_4$

$$\text{Mass of } MnSO_4 = \left(6.44 \text{ x } 10^{-2} \text{ mol } MnSO_4\right)\left(\frac{151.01 \text{ g } MnSO_4}{1 \text{ mol } MnSO_4}\right) = 9.7250 = \textbf{9.73 g } \mathbf{MnSO_4}$$

b) $\mathcal{M}$ of $Fe(ClO_4)_3$ = (55.85 g Fe/mol Fe) + [(3 mol Cl) (35.45 g Cl/mol Cl)] + [(12 mol O) (16.00 g O/mol O)]
            = 354.20 g/mol of $Fe(ClO_4)_3$

$$\text{Moles } Fe(ClO_4)_3 = \left(15.8 \text{ kg } Fe(ClO_4)_3\right)\left(\frac{10^3 \text{ g}}{1 \text{ kg}}\right)\left(\frac{1 \text{ mol } Fe(ClO_4)_3}{354.20 \text{ g } Fe(ClO_4)_3}\right)$$

= 44.6076 = **44.6 mol $Fe(ClO_4)_3$**
c) $\mathcal{M}$ of NH_4NO_2 =[(2 mol N) (14.01 g N/mol N)] + [(4 mol H) (1.008 g H/mol H)]
        + [(2 mol O) (16.00 g O/mol O)] = 64.05 g/mol of NH_4NO_2
        N atoms =

$$\left(92.6 \text{ mg } NH_4NO_2\right)\left(\frac{10^{-3} \text{ g}}{1 \text{ mg}}\right)\left(\frac{1 \text{ mol } NH_4NO_2}{64.05 \text{ g } NH_4NO_2}\right)\left(\frac{2 \text{ mol N atoms}}{1 \text{ mol } NH_4NO_2}\right)\left(\frac{6.022 \text{ x } 10^{23} \text{ N atoms}}{1 \text{ mol N atoms}}\right)$$

= 1.74126 x 10^{21} = **1.74 x 10^{21} N atoms**

3.16   Plan: The formula of each compound must be determined from its name. The molar mass for each formula comes from the formula and atomic masses from the periodic table. Avogadro's number is also necessary to find the number of particles.
Solution:
a) Carbonate is a polyatomic anion with the formula, CO_3^{2-}. Copper (I) indicates Cu^+. The correct formula for this ionic compound is Cu_2CO_3.

$$\mathcal{M} \text{ of } Cu_2CO_3 = (2 \times 63.55) + 12.01 + (3 \times 16.00) = 187.11 \text{ g/mol}$$

$$\text{Mass } Cu_2CO_3 = \left(8.35 \text{ mol } Cu_2CO_3\right)\left(\frac{187.11 \text{ g } Cu_2CO_3}{1 \text{ mol } Cu_2CO_3}\right) = 1562.4 = \mathbf{1.56 \times 10^3 \text{ g } Cu_2CO_3}$$

b) Dinitrogen pentaoxide has the formula N_2O_5. Di- indicates 2 N atoms and penta- indicates 5 O atoms.

$$\mathcal{M} \text{ of } N_2O_5 = (2 \times 14.01) + (5 \times 16.00) = 108.02 \text{ g/mol}$$

$$\text{Mass } N_2O_5 = \left(4.04 \times 10^{20} \, N_2O_5 \text{ molecules}\right)\left(\frac{1 \text{ mol } N_2O_5}{6.022 \times 10^{23} \, N_2O_5 \text{ molecules}}\right)\left(\frac{108.02 \text{ g } N_2O_5}{1 \text{ mol } N_2O_5}\right)$$

$$= 0.072468 = \mathbf{0.0725 \text{ g } N_2O_5}$$

c) The correct formula for this ionic compound is $NaClO_4$; Na has a charge of +1 (Group 1 ion) and the perchlorate ion is ClO_4^-. There are Avogadro's number of entities (in this case, formula units) in a mole of this compound.

$$\mathcal{M} \text{ of } NaClO_4 = 22.99 + 35.45 + (4 \times 16.00) = 122.44 \text{ g/mol}$$

$$\text{Moles } NaClO_4 = \left(78.9 \text{ g } NaClO_4\right)\left(\frac{1 \text{ mol } NaClO_4}{122.44 \text{ g } NaClO_4}\right) = 0.644397 = \mathbf{0.644 \text{ mol } NaClO_4}$$

FU = formula units

$$\text{FU } NaClO_4 = \left(78.9 \text{ g } NaClO_4\right)\left(\frac{1 \text{ mol } NaClO_4}{122.44 \text{ g } NaClO_4}\right)\left(\frac{6.022 \times 10^{23} \text{ FU } NaClO_4}{1 \text{ mol } NaClO_4}\right) =$$

$$3.88056 \times 10^{23} = \mathbf{3.88 \times 10^{23} \text{ FU } NaClO_4}$$

d) The number of ions or atoms is calculated from the formula units given in part c. Note the unrounded initially calculated value is used to avoid intermediate rounding.

$$3.88056 \times 10^{23} \text{ mol } NaClO_4 \left(\frac{1 \, Na^+ \text{ ion}}{1 \text{ FU } NaClO_4}\right) = \mathbf{3.88 \times 10^{23} \, Na^+ \text{ ions}}$$

$$3.88056 \times 10^{23} \text{ mol } NaClO_4 \left(\frac{1 \, ClO_4^- \text{ ion}}{1 \text{ FU } NaClO_4}\right) = \mathbf{3.88 \times 10^{23} \, ClO_4^- \text{ ions}}$$

$$3.88056 \times 10^{23} \text{ mol } NaClO_4 \left(\frac{1 \text{ Cl atom}}{1 \text{ FU } NaClO_4}\right) = \mathbf{3.88 \times 10^{23} \text{ Cl atoms}}$$

$$3.88056 \times 10^{23} \text{ mol } NaClO_4 \left(\frac{4 \text{ O atoms}}{1 \text{ FU } NaClO_4}\right) = \mathbf{1.55 \times 10^{24} \text{ O atoms}}$$

3.18   Plan: Determine the formula and the molar mass of each compound. The formula gives the number of atoms of each type of element present. Masses come from the periodic table. Mass percent = (total mass of element in the substance/molar mass of substance) x 100.
Solution:
a) Ammonium bicarbonate is an ionic compound consisting of ammonium ions, NH_4^+ and bicarbonate ions, HCO_3^-. The formula of the compound is NH_4HCO_3.

$$\mathcal{M} \text{ of } NH_4HCO_3 = (14.01 \text{ g/mol}) + (5 \times 1.008 \text{ g/mol}) + (12.01 \text{ g/mol}) + (3 \times 16.00 \text{ g/mol})$$
$$= 79.06 \text{ g/mol } NH_4HCO_3$$

In 1 mole of ammonium bicarbonate, with a mass of 79.06 g, there are 5 moles of H atoms with a mass of 5.040 g.

$$\frac{(5 \text{ mol H}) \, (1.008 \text{ g/mol H})}{79.06 \text{ g/mol } NH_4HCO_3} \times 100\% = 6.374905 = \mathbf{6.375\% \text{ H}}$$

b) Sodium dihydrogen phosphate heptahydrate is a salt that consists of sodium ions, Na^+, dihydrogen phosphate ions, $H_2PO_4^-$, and seven waters of hydration. The formula is $NaH_2PO_4 \cdot 7H_2O$. Note that the waters of hydration are included in the molar mass.

$\mathcal{M}$ of $NaH_2PO_4 \cdot 7H_2O$ = (22.99 g/mol) + (16 x 1.008 g/mol) + (30.97 g/mol) + (11 x 16.00 g/mol)
= 246.09 g/mol $NaH_2PO_4 \cdot 7H_2O$

In each mole of $NaH_2PO_4 \cdot 7H_2O$ (with mass of 246.09 g), there are 11 x 16.00 g/mol or 176.00 g of oxygen.

$$\frac{(11 \text{ mol O})(16.00 \text{ g/mol O})}{246.09 \text{ g/mol}} \text{ x } 100\% = 71.51855 = \textbf{71.52\% O}$$

3.20   Plan: Mass fraction is related to the mass percentage, however the mass fraction is expressed in decimal rather than percentage form.
Solution:
a) Cesium acetate is an ionic compound consisting of Cs^+ cations and $C_2H_3O_2^-$ anions. (Note that the formula for acetate ions can be written as either $C_2H_3O_2^-$ or CH_3COO^-.) The formula of the compound is $CsC_2H_3O_2$. One mole of $CsC_2H_3O_2$ weighs 191.9 g:

$\mathcal{M}$ of $CsC_2H_3O_2$ = 132.9 g/mol + (2 x 12.01 g/mol) + (3 x 1.008 g/mol) + (2 x 16.00 g/mol) = 191.9 g/mol

$$\text{Mass fraction of C} = \frac{(2 \text{ mol C})(12.01 \text{ g/mol C})}{191.9 \text{ g/mol}} = 0.125169 = \textbf{0.1252 mass fraction C}$$

b) The formula for this compound is $UO_2SO_4 \cdot 3H_2O$.

$\mathcal{M}$ of $UO_2SO_4 \cdot 3H_2O$ = 238.0 g/mol + (9 x 16.00 g/mol) + 32.07 g/mol + (6 x 1.008 g/mol) = 420.1 g/mol

$$\text{Mass fraction of O} = \frac{(9 \text{ mol O})(16.00 \text{ g/mol O})}{420.1 \text{ g/mol}} = 0.3427755 = \textbf{0.3428 mass fraction O}$$

3.23   Plan: Determine the formula of cisplatin from the figure, and then calculate the molar mass from the formula. The molar mass is necessary for the subsequent calculations.
Solution:
The formula for cisplatin is $Pt(Cl)_2(NH_3)_2$

$\mathcal{M}$ of $Pt(Cl)_2(NH_3)_2$ = 195.1 + (2 x 35.45) + (2 x 14.01) + (6 x 1.008) = 300.1 g/mol

a) Moles cisplatin = $(285.3 \text{ g cisplatin})\left(\dfrac{1 \text{ mol cisplatin}}{300.1 \text{ g cisplatin}}\right)$ = 0.9506831 = **0.9507 mol cisplatin**

b) H atoms = $(0.98 \text{ mol cisplatin})\left(\dfrac{6 \text{ mol H}}{1 \text{ mol cisplatin}}\right)\left(\dfrac{6.022 \text{ x } 10^{23} \text{ H atoms}}{1 \text{ mol H}}\right)$

= 3.540936 x 10^{24} = **3.5 x 10^{24} H atoms**

3.25   Plan: Determine the molar mass of rust. Use the molar mass to find the moles of rust. The moles of rust may be related to the grams of iron through the mole ratio.
Solution:
a) $\mathcal{M}$ of $Fe_2O_3 \cdot 4H_2O$ = (2 x 55.85 g/mol) + (7 x 16.00 g/mol) + (8 x 1.008 g/mol) = 231.76 g/mol

Moles of compound = $(45.2 \text{ kg rust})\left(\dfrac{10^3 \text{ g}}{1 \text{ kg}}\right)\left(\dfrac{1 \text{ mol rust}}{231.76 \text{ g rust}}\right)$ = 195.029 = **195 mol rust**

b) The formula shows that there is one mole of Fe_2O_3 for every mole of rust, so there are also **195 mol of Fe_2O_3.**
c) Calculate grams of iron by determining the mole ratio and converting to grams.

Grams of iron = $(45.2 \text{ kg rust})\left(\dfrac{10^3 \text{ g}}{1 \text{ kg}}\right)\left(\dfrac{1 \text{ mol rust}}{231.76 \text{ g rust}}\right)\left(\dfrac{1 \text{ mol Fe}_2\text{O}_3}{1 \text{ mol rust}}\right)\left(\dfrac{2 \text{ mol Fe}}{1 \text{ mol Fe}_2\text{O}_3}\right)\left(\dfrac{55.85 \text{ g Fe}}{1 \text{ mol Fe}}\right)$

= 21784.78 = **2.18 x 10^4 g Fe**

3.27   <u>Plan:</u> Determine the formulas for the compounds where needed. Determine the molar mass of each formula. Calculate the percent nitrogen by dividing the mass of nitrogen in a mole of compound by the molar mass of the compound, and multiply the result by 100%. Then rank the values.
<u>Solution:</u>

Name	Formula	Molar Mass (g/mol)
Potassium nitrate	KNO_3	101.11
Ammonium nitrate	NH_4NO_3	80.05
Ammonium sulfate	$(NH_4)_2SO_4$	132.15
Urea	$CO(NH_2)_2$	60.06

Calculating the nitrogen percentages:

Potassium nitrate
$$\frac{(1 \text{ mol N})(14.01 \text{ g/mol N})}{101.11 \text{ g/mol}} \times 100 = 13.856196 = \textbf{13.86\% N}$$

Ammonium nitrate
$$\frac{(2 \text{ mol N})(14.01 \text{ g/mol N})}{80.05 \text{ g/mol}} \times 100 = 35.003123 = \textbf{35.00\% N}$$

Ammonium sulfate
$$\frac{(2 \text{ mol N})(14.01 \text{ g/mol N})}{132.15 \text{ g/mol}} \times 100 = 21.20318 = \textbf{21.20\% N}$$

Urea
$$\frac{(2 \text{ mol N})(14.01 \text{ g/mol N})}{60.06 \text{ g/mol}} \times 100 = 46.6533 = \textbf{46.65\% N}$$

Rank is $CO(NH_2)_2 > NH_4NO_3 > (NH_4)_2SO_4 > KNO_3$

3.28   <u>Plan:</u> The volume must be converted from cubic feet to cubic centimeters (or vice versa). The volume and the density will give you mass, and the mass with the molar mass gives you moles. Part (b) requires a conversion from cubic decimeters to cubic centimeters. The density allows you to change these cubic centimeters to mass, the molar mass allows you to find moles, and finally Avogadro's number allows you to make the last step.
<u>Solution:</u>
The molar mass of galena (PbS) is 239.3 g/mol.

a) Moles PbS $= \left(1.00 \text{ ft}^3 \text{ PbS}\right)\left(\frac{(12 \text{ in})^3}{(1 \text{ ft})^3}\right)\left(\frac{(2.54 \text{ cm})^3}{(1 \text{ in})^3}\right)\left(\frac{7.46 \text{ g PbS}}{1 \text{ cm}^3}\right)\left(\frac{1 \text{ mol PbS}}{239.3 \text{ g PbS}}\right) = 882.7566886 = \textbf{883 mol PbS}$

b) Lead atoms =

$\left(1.00 \text{ dm}^3 \text{ PbS}\right)\left(\frac{(0.1 \text{ m})^3}{(1 \text{ dm})^3}\right)\left(\frac{(1 \text{ cm})^3}{(10^{-2} \text{ m})^3}\right)\left(\frac{7.46 \text{ g PbS}}{1 \text{ cm}^3}\right)\left(\frac{1 \text{ mol PbS}}{239.3 \text{ g PbS}}\right)\left(\frac{1 \text{ mol Pb}}{1 \text{ mol PbS}}\right)\left(\frac{6.022 \times 10^{23} \text{ Pb atoms}}{1 \text{ mol Pb}}\right)$

$= 1.87731 \times 10^{25} = \textbf{1.88} \times \textbf{10}^{\textbf{25}} \textbf{ Pb atoms}$

3.31   <u>Plan:</u> Remember that the molecular formula tells the *actual* number of moles of each element in one mole of compound.
<u>Solution:</u>
a) No, this information does not allow you to obtain the molecular formula. You can obtain the empirical formula from the number of moles of each type of atom in a compound, but not the molecular formula.
b) Yes, you can obtain the molecular formula from the mass percentages and the total number of atoms.
Solution plan:
    1) Assume a 100.0 g sample and convert masses (from the mass % of each element) to moles using molar mass.
    2) Identify the element with the lowest number of moles and use this number to divide into the number of moles for each element. You now have at least one elemental mole ratio (the one with the smallest number of moles) equal to 1.00 and the remaining mole ratios that are larger than one.
    3) Examine the numbers to determine if they are whole numbers. If not, multiply each number by a whole number factor to get whole numbers for each element. You will have to use some judgment to decide when to round.
    4) Write the empirical formula using the whole numbers from step 3.

5) Check the total number of atoms in the empirical formula. If it equals the total number of atoms given then the empirical formula is also the molecular formula. If not, then divide the total number of atoms given by the total number of atoms in the empirical formula. This should give a whole number. Multiply the number of atoms of each element in the empirical formula by this whole number to get the molecular formula. If you do not get a whole number when you divide, return to step 3 and revise how you multiplied and rounded to get whole numbers for each element.

c) Yes, you can determine the molecular formula from the mass percent and the number of atoms of one element in a compound. Solution plan:
1) Follow steps 1-4 in part b.
2) Compare the number of atoms given for the one element to the number in the empirical formula. Determine the factor the number in the empirical formula must be multiplied by to obtain the given number of atoms for that element. Multiply the empirical formula by this number to get the molecular formula.

d) No, the mass % will only lead to the empirical formula.

e) Yes, a structural formula shows all the atoms in the compound. Solution plan: Count the number of atoms of each type of element and record as the number for the molecular formula.

3.33   Plan: Examine the number of atoms of each type in the compound. Divide all atom numbers by any common factor. The final answers must be the lowest whole-number values.
Solution:
a) C_2H_4 has a ratio of 2 carbon atoms to 4 hydrogen atoms, or 2:4. This ratio can be reduced to 1:2, so that the empirical formula is **CH_2**. The empirical formula mass is $12.01 + 2(1.008) = $ **14.03 g/mol**.
b) The ratio of atoms is 2:6:2, or 1:3:1. The empirical formula is **CH_3O** and its empirical formula mass is $12.01 + 3(1.008) + 16.00 = $ **31.03 g/mol**.
c) Since, the ratio of elements cannot be further reduced, the molecular formula and empirical formula are the same, **N_2O_5**. The formula mass is $2(14.01) + 5(16.00) = $ **108.02 g/mol**.
d) The ratio of elements is 3 atoms of barium to 2 atoms of phosphorus to 8 atoms of oxygen, or 3:2:8. This ratio cannot be further reduced, so the empirical formula is also **$Ba_3(PO_4)_2$**, with a formula mass of $3(137.3) + 2(30.97) + 8(16.00) = $ **601.8 g/mol**.
e) The ratio of atoms is 4:16, or 1:4. The empirical formula is **TeI_4**, and the formula mass is $127.6 + 4(126.9) = $ **635.2 g/mol**.

3.35   Plan: Determine the molar mass of each empirical formula. The molar mass of each compound divided by its empirical formula mass gives the number of times the empirical formula is within the molecule. Multiply the empirical formula by the number of times the empirical formula appears to get the molecular formula.
Solution:
Only approximate whole number values are needed.
a) CH_2 has empirical mass equal to 14.03 g/mol

$$\left(\frac{42.08 \text{ g/mol}}{14.03 \text{ g/mol}} \right) = 3$$

Multiplying the subscripts in CH_2 by 3 gives **C_3H_6**
b) NH_2 has empirical mass equal to 16.03 g/mol

$$\left(\frac{32.05 \text{ g/mol}}{16.03 \text{ g/mol}} \right) = 2$$

Multiplying the subscripts in NH_2 by 2 gives **N_2H_4**
c) NO_2 has empirical mass equal to 46.01 g/mol

$$\left(\frac{92.02 \text{ g/mol}}{46.01 \text{ g/mol}} \right) = 2$$

Multiplying the subscripts in NO_2 by 2 gives **N_2O_4**

d) CHN has empirical mass equal to 27.03 g/mol

$$\left(\frac{135.14 \text{ g/mol}}{27.03 \text{ g/mol}}\right) = 5$$

Multiplying the subscripts in CHN by 5 gives $C_5H_5N_5$

3.37   Plan: The empirical formula is the smallest whole-number ratio of the atoms or moles in a formula. All data must be converted to moles of an element. Divide each mole number by the smallest mole number to convert the mole ratios to whole numbers.
Solution:

a) $\left(\dfrac{0.063 \text{ mol Cl}}{0.063 \text{ mol Cl}}\right) = 1$ $\left(\dfrac{0.22 \text{ mol O}}{0.063 \text{ mol Cl}}\right) = 3.5$

The formula is $Cl_1O_{3.5}$, which in whole numbers (x 2) is Cl_2O_7

b) $\left(2.45 \text{ g Si}\right)\left(\dfrac{1 \text{ mol Si}}{28.09 \text{ g Si}}\right) = 0.08722 \text{ mol Si}$ $\left(12.4 \text{ g Cl}\right)\left(\dfrac{1 \text{ mol Cl}}{35.45 \text{ g Cl}}\right) = 0.349788 \text{ mol Cl}$

$\left(\dfrac{0.08722 \text{ mol Si}}{0.08722 \text{ mol Si}}\right) = 1$ $\left(\dfrac{0.349788 \text{ mol Cl}}{0.08722 \text{ mol Si}}\right) = 4$

The empirical formula is $SiCl_4$.

c) Assume a 100 g sample and convert the masses to moles.

$\left(100 \text{ g}\right)\left(\dfrac{27.3\% \text{ C}}{100\%}\right)\left(\dfrac{1 \text{ mol C}}{12.01 \text{ g C}}\right) = 2.2731 \text{ mol C}$ $\left(100 \text{ g}\right)\left(\dfrac{72.7\% \text{ O}}{100\%}\right)\left(\dfrac{1 \text{ mol O}}{16.00 \text{ g O}}\right) = 4.5438 \text{ mol O}$

$\left(\dfrac{2.2731 \text{ mol C}}{2.2731 \text{ mol C}}\right) = 1$ $\left(\dfrac{4.5438 \text{ mol O}}{2.2731 \text{ mol C}}\right) = 2$

The empirical formula is CO_2.

3.39   Plan: The percent oxygen is 100% minus the percent nitrogen. Assume 100 grams of sample, and then the moles of each element may be found. Divide each of the moles by the smaller value, and convert to whole numbers to get the empirical formula. The empirical formula mass and the given mass lead to the molecular formula.
Solution:

a) % O = 100% − % N = 100% − 30.45% N = 69.55% O
For a 100-gram sample, the mass, in grams, is numerically identical to the mass percent.

Moles N $= \left(30.45 \text{ g N}\right)\left(\dfrac{1 \text{ mol N}}{14.01 \text{ g N}}\right) = 2.1734 \text{ mol N}$

Moles O $= \left(69.55 \text{ g O}\right)\left(\dfrac{1 \text{ mol O}}{16.00 \text{ g O}}\right) = 4.3469 \text{ mol O}$

$\left(\dfrac{2.1734 \text{ mol N}}{2.1734 \text{ mol N}}\right) = 1$ $\left(\dfrac{4.3469 \text{ mol O}}{2.1734 \text{ mol N}}\right) = 2$

The empirical formula is NO_2.

b) Formula mass of empirical formula = 14.01 g N/mol + 2(16.00 g O/mol) = 46.01 g/mol

$$\frac{\text{molar mass}}{\text{empirical formula mass}} = \frac{90 \text{ g/mol}}{46 \text{ g/mol}} = 2$$

Thus, the molecular formula is twice the empirical formula. Note: only an approximate value is needed.
The molecular formula is $2(NO_2) = N_2O_4$.

3.41   <u>Plan:</u> The balanced equation for this reaction is: $M(s) + F_2(g) \rightarrow MF_2(s)$ since fluorine, like other halogens, exists as a diatomic molecule. The moles of the metal are known, and the moles of everything else may be found from these moles using the balanced chemical equation.
<u>Solution:</u>
a) Determine the moles of fluorine.

$$\text{Moles F} = (0.600 \text{ mol M})\left(\frac{2 \text{ mol F}}{1 \text{ mol M}}\right) = \textbf{1.20 mol F}$$

b) The grams of M are the grams of MF_2 minus the grams of F present.

$$\text{Grams M} = 46.8 \text{ g (MF}_2) - (1.20 \text{ mol F})\left(\frac{19.00 \text{ g F}}{1 \text{ mol F}}\right) = \textbf{24.0 g M}$$

c)The molar mass is needed to identify the element.

$$\text{Molar mass of M} = \frac{24.0 \text{ g M}}{0.600 \text{ mol M}} = 40.0 \text{ g/mol}$$

The metal with the closest molar mass to 40.0 g/mol is **calcium**.

3.44   <u>Plan:</u> Assume 100 grams of cortisol so the percentages are numerically equivalent to the masses of each element. Convert each of the masses to moles by using the molar mass of each element involved. Divide all moles by the lowest number of moles and convert to whole numbers to determine the empirical formula. The empirical formula mass and the given molar mass will then relate the empirical formula to the molecular formula.
<u>Solution:</u>

$$\text{Moles C} = (69.6 \text{ g C})\left(\frac{1 \text{ mol C}}{12.01 \text{ g C}}\right) = 5.7952 \text{ mol C}$$

$$\text{Moles H} = (8.34 \text{ g H})\left(\frac{1 \text{ mol H}}{1.008 \text{ g H}}\right) = 8.2738 \text{ mol H}$$

$$\text{Moles O} = (22.1 \text{ g O})\left(\frac{1 \text{ mol O}}{16.00 \text{ g O}}\right) = 1.38125 \text{ mol O}$$

$$\left(\frac{5.7952 \text{ mol C}}{1.38125 \text{ mol O}}\right) = 4.20 \qquad \left(\frac{8.2738 \text{ mol H}}{1.38125 \text{ mol O}}\right) = 6.00 \qquad \left(\frac{1.38125 \text{ mol O}}{1.38125 \text{ mol O}}\right) = 1.00$$

The carbon value is not close enough to a whole number to round the value. The smallest number that 4.20 may be multiplied by to get close to a whole number is 5. (You may wish to prove this to yourself.) All three ratios need to be multiplied by five to get the empirical formula of $C_{21}H_{30}O_5$.
The empirical formula mass is:
        21 (12.01 g C/mol) + 30 (1.008 g H/mol) + 5 (16.00 g O/mol) = 362.45 g/mol
The empirical formula mass and the molar mass given are the same, so the empirical and the molecular formulas are the same. The molecular formula is $\textbf{C}_{21}\textbf{H}_{30}\textbf{O}_5$.

3.46   <u>Plan:</u> In combustion analysis, finding the moles of carbon and hydrogen is relatively simple because all of the carbon present in the sample is found in the carbon of CO_2, and all of the hydrogen present in the sample is found in the hydrogen of H_2O. The moles of oxygen are more difficult to find, because additional O_2 was added to cause the combustion reaction. The masses of CO_2 and H_2O are used to find both the mass of C and H and the moles of C and H. Subtracting the masses of C and H from the mass of the sample gives the mass of O. Convert the mass of O to moles of O. Take the moles of C, H, and O and divide by the smallest value, and convert to a whole number to get the empirical formula. Determine the empirical formula mass and compare it to the molar mass given in the problem to see how the empirical and molecular formulas are related. Finally, determine the molecular formula.
<u>Solution:</u>   (There is no intermediate rounding.)
Initial mole determination:

$$\text{Moles C} = (0.449 \text{ g CO}_2)\left(\frac{1 \text{ mol CO}_2}{44.01 \text{ g CO}_2}\right)\left(\frac{1 \text{ mol C}}{1 \text{ mol CO}_2}\right) = 0.010202 \text{ mol C}$$

$$\text{Moles H} = \left(0.184 \text{ g H}_2\text{O}\right)\left(\frac{1 \text{ mol H}_2\text{O}}{18.02 \text{ g H}_2\text{O}}\right)\left(\frac{2 \text{ mol H}}{1 \text{ mol H}_2\text{O}}\right) = 0.020422 \text{ mol H}$$

Now determine the masses of C and H:

$$\text{Grams C} = \left(0.010202 \text{ mol C}\right)\left(\frac{12.01 \text{ g C}}{1 \text{ mol C}}\right) = 0.122526 \text{ g C}$$

$$\text{Grams H} = \left(0.020422 \text{ mol H}\right)\left(\frac{1.008 \text{ g H}}{1 \text{ mol H}}\right) = 0.020585 \text{ g H}$$

Determine the mass and then the moles of O:

$$0.1595 \text{ g (C, H, and O)} - (0.122526 \text{ g C} + 0.020585 \text{ g H}) = 0.016389 \text{ g O}$$

$$\text{Moles O} = \left(0.016389 \text{ g O}\right)\left(\frac{1 \text{ mol O}}{16.00 \text{ g O}}\right) = 0.0010243 \text{ mol O}$$

Divide by the smallest number of moles: (Rounding is acceptable for these answers.)

$$\left(\frac{0.010202 \text{ mol C}}{0.0010243 \text{ mol O}}\right) = 9.96 = 10 \qquad \left(\frac{0.020422 \text{ mol H}}{0.0010243 \text{ mol O}}\right) = 19.9 = 20 \qquad \left(\frac{0.0010243 \text{ mol O}}{0.0010243 \text{ mol O}}\right) = 1$$

Empirical formula = $C_{10}H_{20}O$
Empirical formula mass = 10 (12.01 g C/mol) + 20 (1.008 g H/mol) + 1 (16.00 g O/mol) = 156.26 g/mol
The empirical formula mass matches the given molar mass so the empirical and molecular formulas are the same.
The molecular formula is **$C_{10}H_{20}O$**.

3.47   A balanced chemical equation describes:
1) The identities of the reactants and products.
2) The molar (and molecular) ratios by which reactants form products.
3) The physical states of all substances in the reaction.

3.50   Plan: Examine the diagram and label each formula. We will use A for red atoms and B for green atoms.
Solution:
The reaction shows A_2 and B_2 molecules forming AB molecules. Equal numbers of A_2 and B_2 combine to give twice as many molecules of AB. Thus, the reaction is $A_2 + B_2 \rightarrow 2$ AB. This is the answer to **part b**.

3.51   Plan: Balancing is a trial and error procedure. Do one blank/one element at a time.
Solution:
a) $\underline{16}$ Cu(s) + ___ S_8(s) $\rightarrow$ $\underline{8}$ Cu_2S(s)
Hint: Balance the S first, because there is an obvious deficiency of S on the right side of the equation. Then balance the Cu.
b)   ___ P_4O_{10}(s) + $\underline{6}$ H_2O(l) $\rightarrow$ $\underline{4}$ H_3PO_4(l)
Hint: Balance the P first, because there is an obvious deficiency of P on the right side of the equation. Balance the H next, because H is present in only one reactant and only one product. Balance the O last, because it appears in both reactants and is harder to balance.
c)   ___B_2O_3(s) + $\underline{6}$ NaOH(aq) $\rightarrow$ $\underline{2}$ Na_3BO_3(aq) + $\underline{3}$ H_2O(l)
Hint: Oxygen is again the hardest element to balance because it is present in more than one place on each side of the reaction. If you balance the easier elements first (B, Na, H), the oxygen will automatically be balanced.
d) $\underline{2}$ CH_3NH_2(g) + $\underline{9/2}$ O_2(g) $\rightarrow$ $\underline{2}$ CO_2(g) + $\underline{5}$ H_2O(g) + ___N_2(g)
         $\underline{4}$ CH_3NH_2(g) + $\underline{9}$ O_2(g) $\rightarrow$ $\underline{4}$ CO_2(g) + $\underline{10}$ H_2O(g) + $\underline{2}$ N_2(g)
Hint: Balance odd/even numbers of oxygen using the "half" method, and then multiply all coefficients by two.

3.53   Plan: The balancing is a trial and error procedure. Do one blank/one element at a time.
Solution:
a) $\underline{2}$ SO_2(g) + ___O_2(g) $\rightarrow$ $\underline{2}$ SO_3(g)
b) ___Sc_2O_3(s) + $\underline{3}$ H_2O(l) $\rightarrow$ $\underline{2}$ Sc(OH)$_3$(s)
c) ___H_3PO_4(aq) + $\underline{2}$ NaOH(aq) $\rightarrow$ ___Na_2HPO_4(aq) + $\underline{2}$ H_2O(l)
d) $C_6H_{10}O_5$(s) + $\underline{6}$ O_2(g) $\rightarrow$ $\underline{6}$ CO_2(g) + $\underline{5}$ H_2O(g)

3.55 <u>Plan:</u> The names must first be converted to chemical formulas. The balancing is a trial and error procedure. Do one blank/one element at a time. Remember that oxygen is diatomic.
<u>Solution:</u>
a) $\underline{4}\ Ga(s) + \underline{3}\ O_2(g) \rightarrow \underline{2}\ Ga_2O_3(s)$
b) $\underline{2}\ C_6H_{14}(l) + \underline{19}\ O_2(g) \rightarrow \underline{12}\ CO_2(g) + \underline{14}\ H_2O(g)$
c) $\underline{3}\ CaCl_2(aq) + \underline{2}\ Na_3PO_4(aq) \rightarrow Ca_3(PO_4)_2(s) + \underline{6}\ NaCl(aq)$

3.61 <u>Plan:</u> First, write a balanced chemical equation. Since A is the limiting reagent (B is in excess), A is used to determine the amount of C formed, using the mole ratio between reactant A and product C.
<u>Solution:</u>
First, write a balanced equation: $a\ A + b\ B \rightarrow c\ C$

$$\text{Mass C} = \left(25\ \text{g A}\right)\left(\frac{1\ \text{mol A}}{\text{Molar Mass A}}\right)\left(\frac{c\ \text{Mol C}}{a\ \text{Mol A}}\right)\left(\frac{\text{Molar Mass C}}{1\ \text{mol C}}\right) = \text{g C}$$

3.63 <u>Plan:</u> Use the mole ratio from the balanced chemical equation to determine the moles produced. Use the moles with the molar mass to determine the grams produced.
<u>Solution:</u>

a) Moles $Cl_2 = \left(1.82\ \text{mol HCl}\right)\left(\dfrac{1\ \text{mol Cl}_2}{4\ \text{mol HCl}}\right) = \mathbf{0.455\ mol\ Cl_2}$

b) Grams $Cl_2 = \left(1.82\ \text{mol HCl}\right)\left(\dfrac{1\ \text{mol Cl}_2}{4\ \text{mol HCl}}\right)\left(\dfrac{70.90\ \text{g Cl}_2}{1\ \text{mol Cl}_2}\right) = 32.2595 = \mathbf{32.3\ g\ Cl_2}$

The beginning of the calculation is repeated to emphasize that the second part of the problem is simply an extension of the first part. There is no need to repeat the entire calculation, as only the final step times the answer of the first part will give the final answer to this part.
Hint: Always check to see if the initial equation is balanced. If the equation is not balanced, it should be balanced before proceeding.

3.65 <u>Plan:</u> Convert the kilograms of oxygen to the moles of oxygen. Use the moles of oxygen and the mole ratios from the balanced chemical equation to determine the moles of KNO_3. The moles of KNO_3 and its molar mass will give the grams.
<u>Solution:</u>

a) Moles $KNO_3 = \left(56.6\ \text{kg O}_2\right)\left(\dfrac{10^3\ \text{g}}{1\ \text{kg}}\right)\left(\dfrac{1\ \text{mol O}_2}{32.00\ \text{g O}_2}\right)\left(\dfrac{4\ \text{mol KNO}_3}{5\ \text{mol O}_2}\right) = 1415 = \mathbf{1.42 \times 10^3\ mol\ KNO_3}$

b) Grams $KNO_3 = \left(56.6\ \text{kg O}_2\right)\left(\dfrac{10^3\ \text{g}}{1\ \text{kg}}\right)\left(\dfrac{1\ \text{mol O}_2}{32.00\ \text{g O}_2}\right)\left(\dfrac{4\ \text{mol KNO}_3}{5\ \text{mol O}_2}\right)\left(\dfrac{101.11\ \text{g KNO}_3}{1\ \text{mol KNO}_3}\right)$

$= 143070.65 = \mathbf{1.43 \times 10^5\ g\ KNO_3}$

The beginning of the calculation is repeated to emphasize that the second part of the problem is simply an extension of the first part. There is no need to repeat the entire calculation, as only the final step times the answer of the first part will give the final answer to this part.

3.67 <u>Plan:</u> First, balance the equation. Convert the grams of diborane to moles of diborane using its molar mass. Use mole ratios from the balanced chemical equation to determine the moles of the products. Use the moles and molar mass of each product to determine the mass formed.
<u>Solution:</u>
The balanced equation is: $B_2H_6(g) + 6\ H_2O(l) \rightarrow 2\ H_3BO_3(s) + 6\ H_2(g)$.

$$\text{Mass H}_3BO_3 = \left(43.82\ \text{g B}_2H_6\right)\left(\frac{1\ \text{mol B}_2H_6}{27.67\ \text{g B}_2H_6}\right)\left(\frac{2\ \text{mol H}_3BO_3}{1\ \text{mol B}_2H_6}\right)\left(\frac{61.83\ \text{g H}_3BO_3}{1\ \text{mol H}_3BO_3}\right)$$

$= 195.83597 = \mathbf{195.8\ g\ H_3BO_3}$

$$\text{Mass H}_2 = \left(43.82 \text{ g B}_2\text{H}_6\right)\left(\frac{1 \text{ mol B}_2\text{H}_6}{27.67 \text{ g B}_2\text{H}_6}\right)\left(\frac{6 \text{ mol H}_2}{1 \text{ mol B}_2\text{H}_6}\right)\left(\frac{2.016 \text{ g H}_2}{1 \text{ mol H}_2}\right)$$
$$= 19.156007 = \textbf{19.16 g H}_2$$

3.69   Plan: Write the balanced equation by first writing the formulas for the reactants and products. Reactants: formula for phosphorus is given as P_4 and formula for chlorine gas is Cl_2 (chlorine occurs as a diatomic molecule). Products: formula for phosphorus pentachloride — the name indicates one phosphorus atom and five chlorine atoms to give the formula PCl_5. Convert the mass of phosphorus to moles, use the mole ratio from the balanced chemical equation, and finally use the molar mass of chlorine to get the mass of chlorine.
Solution:
Formulas give the equation: $P_4 + Cl_2 \rightarrow PCl_5$
Balancing the equation: $P_4 + 10 \text{ Cl}_2 \rightarrow 4 \text{ PCl}_5$
$$\text{Grams Cl}_2 = \left(455 \text{ g P}_4\right)\left(\frac{1 \text{ mol P}_4}{123.88 \text{ g P}_4}\right)\left(\frac{10 \text{ mol Cl}_2}{1 \text{ mol P}_4}\right)\left(\frac{70.90 \text{ g Cl}_2}{1 \text{ mol Cl}_2}\right) = 2604.09267 = \textbf{2.60 x 10}^3 \textbf{ g Cl}_2$$

3.71   Plan: Begin by writing the chemical formulas of the reactants and products in each step. Next, balance each of the equations. Combine the equations for the separate steps by adjusting the equations so the intermediate (iodine monochloride) cancels. Finally, change the mass of product to mole and use the mole ratio and molar mass of iodine to determine the mass of iodine.
Solution:
a) First step: $I_2(s) + Cl_2(g) \rightarrow 2 \text{ ICl}(s)$
   Second step: $ICl(s) + Cl_2(g) \rightarrow ICl_3(s)$
b) Multiply the coefficients of the second equation by 2, so that $ICl(s)$, an intermediate product, can be eliminated from the overall equation.
        $I_2(s) + Cl_2(g) \rightarrow 2 \text{ ICl}(s)$
        $\underline{2 \text{ ICl}(s) + 2 \text{ Cl}_2(g) \rightarrow 2 \text{ ICl}_3(s)}$
        $I_2(s) + Cl_2(g) + \cancel{2 \text{ ICl}(s)} + 2 \text{ Cl}_2(g) \rightarrow \cancel{2 \text{ ICl}(s)} + 2 \text{ ICl}_3(s)$
        Overall equation: $\textbf{I}_2(s) + \textbf{3 Cl}_2(g) \rightarrow \textbf{2 ICl}_3(s)$
$$\text{c) Grams I}_2 = \left(2.45 \text{ kg ICl}_3\right)\left(\frac{10^3 \text{ g}}{1 \text{ kg}}\right)\left(\frac{1 \text{ mol ICl}_3}{233.2 \text{ g ICl}_3}\right)\left(\frac{1 \text{ mol I}_2}{2 \text{ mol ICl}_3}\right)\left(\frac{253.8 \text{ g I}_2}{1 \text{ mol I}_2}\right)$$
$$= 1333.2118 = \textbf{1.33 x 10}^3 \textbf{ g I}_2$$

3.73   Plan: Convert the given masses to moles and use the mole ratio from the balanced chemical equation to find the moles of CaO that will form. The reactant that produces the least moles of CaO is the limiting reactant. Convert the moles of CaO from the limiting reactant to grams using the molar mass.
Solution:
$$\text{a) Moles CaO from Ca} = \left(4.20 \text{ g Ca}\right)\left(\frac{1 \text{ mol Ca}}{40.08 \text{ g Ca}}\right)\left(\frac{2 \text{ mol CaO}}{2 \text{ mol Ca}}\right) = 0.104790 = \textbf{0.105 mol CaO}$$

$$\text{b) Moles CaO from O}_2 = \left(2.80 \text{ g O}_2\right)\left(\frac{1 \text{ mol O}_2}{32.00 \text{ g O}_2}\right)\left(\frac{2 \text{ mol CaO}}{1 \text{ mol O}_2}\right) = 0.17500 = \textbf{0.175 mol CaO}$$

c) **Calcium** is the limiting reactant since it will form less calcium oxide.
d) The mass of CaO formed is determined by the limiting reactant, Ca.
$$\text{Grams CaO} = \left(4.20 \text{ g Ca}\right)\left(\frac{1 \text{ mol Ca}}{40.08 \text{ g Ca}}\right)\left(\frac{2 \text{ mol CaO}}{2 \text{ mol Ca}}\right)\left(\frac{56.08 \text{ g CaO}}{1 \text{ mol CaO}}\right) = 5.8766 = \textbf{5.88 g CaO}$$

3.75   Plan: First, balance the chemical equation. Determine which of the reactants is the limiting reagent. Use the limiting reagent and the mole ratio from the balanced chemical equation to determine the amount of material formed and the amount of the other reactant used. The difference between the amount of reactant used and the initial reactant supplied gives the amount of excess reactant remaining.

Solution:
The balanced chemical equation for this reaction is:

$$2\ ICl_3 + 3\ H_2O \rightarrow ICl + HIO_3 + 5\ HCl$$

Hint: Balance the equation by starting with oxygen. The other elements are in multiple reactants and/or products and are harder to balance initially.

Next, find the limiting reactant by using the molar ratio to find the smaller number of moles of HIO_3 that can be produced from each reactant given and excess of the other:

$$\text{Moles } HIO_3 \text{ from } ICl_3 = \left(635\ g\ ICl_3\right)\left(\frac{1\ mol\ ICl_3}{233.2\ g\ ICl_3}\right)\left(\frac{1\ mol\ HIO_3}{2\ mol\ ICl_3}\right) = 1.361492 = 1.36\ mol\ HIO_3$$

$$\text{Moles } HIO_3 \text{ from } H_2O = \left(118.5\ g\ H_2O\right)\left(\frac{1\ mol\ H_2O}{18.02\ g\ H_2O}\right)\left(\frac{1\ mol\ HIO_3}{3\ mol\ H_2O}\right) = 2.19201 = 2.19\ mol\ HIO_3$$

ICl_3 is the limiting reagent and will produce **1.36 mol HIO_3**. Use the limiting reagent to find the grams of HIO_3 formed.

$$\text{Grams } HIO_3 = \left(635\ g\ ICl_3\right)\left(\frac{1\ mol\ ICl_3}{233.2\ g\ ICl_3}\right)\left(\frac{1\ mol\ HIO_3}{2\ mol\ ICl_3}\right)\left(\frac{175.9\ g\ HIO_3}{1\ mol\ HIO_3}\right) = 239.4865 = \textbf{239 g } HIO_3$$

The remaining mass of the excess reagent can be calculated from the amount of H_2O combining with the limiting reagent.

$$\text{Grams } H_2O \text{ required to react with 635 g } ICl_3: \left(635\ g\ ICl_3\right)\left(\frac{1\ mol\ ICl_3}{233.2\ g\ ICl_3}\right)\left(\frac{3\ mol\ H_2O}{2\ mol\ ICl_3}\right)\left(\frac{18.02\ g\ H_2O}{1\ mol\ H_2O}\right)$$

$$= 73.6023 = 73.6\ g\ H_2O \text{ reacted}$$
Remaining $H_2O = 118.5\ g - 73.6\ g = \textbf{44.9 g } H_2O$

3.77 Plan: Write the balanced equation: formula for carbon is C, formula for oxygen is O_2 and formula for carbon dioxide is CO_2. Determine the limiting reagent by seeing which reactant will yield the smaller amount of product. The limiting reactant is used for all subsequent calculations.
Solution:
$$C(s) + O_2(g) \rightarrow CO_2(g)$$

$$\text{Moles } CO_2 \text{ from } C = \left(0.100\ mol\ C\right)\left(\frac{1\ mol\ CO_2}{1\ mol\ C}\right) = 0.100\ mol\ CO_2$$

$$\text{Moles } CO_2 \text{ from } O_2 = \left(8.00\ g\ O_2\right)\left(\frac{1\ mol\ O_2}{32.00\ g\ O_2}\right)\left(\frac{1\ mol\ CO_2}{1\ mol\ O_2}\right) = 0.25000 = 0.250\ mol\ CO_2$$

The C is the limiting reactant and will be used to determine the amount of CO_2 that will form.

$$\text{Grams } CO_2 = \left(0.100\ mol\ C\right)\left(\frac{1\ mol\ CO_2}{1\ mol\ C}\right)\left(\frac{44.01\ g\ CO_2}{1\ mol\ CO_2}\right) = 4.401 = \textbf{4.40 g } CO_2$$

Since the C is limiting, the **O_2 is in excess**. The amount remaining depends on how much combines with the limiting reagent.

$$O_2 \text{ required to react with 0.100 mol of } C = \left(0.100\ mol\ C\right)\left(\frac{1\ mol\ O_2}{1\ mol\ C}\right)\left(\frac{32.00\ g\ O_2}{1\ mol\ O_2}\right) = 3.20\ g\ O_2$$

Remaining $O_2 = 8.00\ g - 3.20\ g = \textbf{4.80 g } O_2$

3.79 Plan: The question asks for the mass of each substance present at the end of the reaction. "Substance" refers to both reactants and products. Solve this problem using multiple steps. Recognizing that this is a limiting reactant problem, first write a balanced chemical equation. Using the molar relationships from the balanced equation, determine which reactant is limiting. Any product can be used to predict the limiting reactant; in this case, $AlCl_3$ is used. Additional significant figures are retained until the last step.

Solution:
The balanced chemical equation is:

$$Al(NO_2)_3(aq) + 3\ NH_4Cl(aq) \rightarrow AlCl_3(aq) + 3\ N_2(g) + 6\ H_2O(l)$$

Now determine the limiting reagent. We will use the moles of $AlCl_3$ produced to determine which is limiting.

$$\text{Mole } AlCl_3 \text{ from } Al(NO_2)_3 = \left(72.5 \text{ g } Al(NO_2)_3\right)\left(\frac{1 \text{ mol } Al(NO_2)_3}{165.01 \text{ g } Al(NO_2)_3}\right)\left(\frac{1 \text{ mol } AlCl_3}{1 \text{ mol } Al(NO_2)_3}\right)$$

$$= 0.439367 = 0.439 \text{ mol } AlCl_3$$

$$\text{Mole } AlCl_3 \text{ from } NH_4Cl = \left(58.6 \text{ g } NH_4Cl\right)\left(\frac{1 \text{ mol } NH_4Cl}{53.49 \text{ g } NH_4Cl}\right)\left(\frac{1 \text{ mol } AlCl_3}{3 \text{ mol } NH_4Cl}\right) = 0.365177 = 0.365 \text{ mol } AlCl_3$$

Ammonium chloride is the limiting reactant, and it is used for all subsequent calculations.

Mass of substances after the reaction:
$Al(NO_2)_3$:
$Al(NO_2)_3$ required to react with 58.6 g of NH_4Cl:

$$\left(58.6 \text{ g } NH_4Cl\right)\left(\frac{1 \text{ mol } NH_4Cl}{53.49 \text{ g } NH_4Cl}\right)\left(\frac{1 \text{ mol } Al(NO_2)_3}{3 \text{ mol } NH_4Cl}\right)\left(\frac{165.01 \text{ g } Al(NO_2)_3}{1 \text{ mol } Al(NO_2)_3}\right)$$

$$= 60.2579 = 60.3 \text{ g } Al(NO_2)_3$$

$Al(NO_2)_3$ remaining: $72.5 \text{ g} - 60.3 \text{ g} = \mathbf{12.2\ g\ Al(NO_2)_3}$
NH_4Cl: **None left** since it is the limiting reagent.
$AlCl_3$:

$$\left(58.6 \text{ g } NH_4Cl\right)\left(\frac{1 \text{ mol } NH_4Cl}{53.49 \text{ g } NH_4Cl}\right)\left(\frac{1 \text{ mol } AlCl_3}{3 \text{ mol } NH_4Cl}\right)\left(\frac{133.33 \text{ g } AlCl_3}{1 \text{ mol } AlCl_3}\right) = 48.6891 = \mathbf{48.7\ g\ AlCl_3}$$

N_2:

$$\left(58.6 \text{ g } NH_4Cl\right)\left(\frac{1 \text{ mol } NH_4Cl}{53.49 \text{ g } NH_4Cl}\right)\left(\frac{3 \text{ mol } N_2}{3 \text{ mol } NH_4Cl}\right)\left(\frac{28.02 \text{ g } N_2}{1 \text{ mol } N_2}\right) = 30.697 = \mathbf{30.7\ g\ N_2}$$

H_2O:

$$\left(58.6 \text{ g } NH_4Cl\right)\left(\frac{1 \text{ mol } NH_4Cl}{53.49 \text{ g } NH_4Cl}\right)\left(\frac{6 \text{ mol } H_2O}{3 \text{ mol } NH_4Cl}\right)\left(\frac{18.02 \text{ g } H_2O}{1 \text{ mol } H_2O}\right) = 39.48297 = \mathbf{39.5\ g\ H_2O}$$

3.81   Plan: Multiply the yield of the first step by that of the second step to get the overall yield.
Solution:
It is simpler to use the decimal equivalents of the percent yields, and then convert to percent using 100%.
$(0.73)(0.68)(100\%) = 49.64 = \mathbf{50.\%}$

3.83   Plan: Balance the chemical equation using the formulas of the substances. Determine the yield (theoretical yield) for the reaction from the mass of tungsten(VI) oxide. Use the density of water to determine the actual yield of water in grams. The actual yield divided by the yield just calculated (with the result multiplied by 100%) gives the percent yield.
Solution: (Rounding to the correct number of significant figures will be postponed until the final result.)
The balanced chemical equation is:
$WO_3(s) + 3\ H_2(g) \rightarrow W(s) + 3\ H_2O(l)$
Theoretical yield of H_2O:

$$\left(45.5 \text{ g } WO_3\right)\left(\frac{1 \text{ mol } WO_3}{231.9 \text{ g } WO_3}\right)\left(\frac{3 \text{ mol } H_2O}{1 \text{ mol } WO_3}\right)\left(\frac{18.02 \text{ g } H_2O}{1 \text{ mol } H_2O}\right) = 10.60686 \text{ g } H_2O$$

Actual yield, in grams, of H_2O:

$$\left(9.60 \text{ mL } H_2O\right)\left(\frac{1.00 \text{ g } H_2O}{1 \text{ mL } H_2O}\right) = 9.60 \text{ g } H_2O$$

Calculate the percent yield:

$$\left(\dfrac{\text{Actual Yield}}{\text{Theoretical Yield}}\right) \times 100\% = \left(\dfrac{9.60 \text{ g } H_2O}{10.60686 \text{ g } H_2O}\right) \times 100\% = 90.5075 = \mathbf{90.5\%}$$

3.85   Plan: Write the balanced chemical equation. Since quantities of reactants are present, we must determine which is limiting. We will use the moles of CH_3Cl produced to determine which is limiting. Only 80.0% of the calculated amounts of products will form. (Rounding to the correct number of significant figures will be postponed until the final result.)
Solution:
$$CH_4(g) + Cl_2(g) \rightarrow CH_3Cl(g) + HCl(g)$$

$$\text{Mole } CH_3Cl \text{ from } CH_4 = \left(20.5 \text{ g } CH_4\right)\left(\dfrac{1 \text{ mol } CH_4}{16.04 \text{ g } CH_4}\right)\left(\dfrac{1 \text{ mol } CH_3Cl}{1 \text{ mol } CH_4}\right) = 1.278055 \text{ mol } CH_3Cl$$

$$\text{Mole } CH_3Cl \text{ from } Cl_2 = \left(45.0 \text{ g } Cl_2\right)\left(\dfrac{1 \text{ mol } Cl_2}{70.90 \text{ g } Cl_2}\right)\left(\dfrac{1 \text{ mol } CH_3Cl}{1 \text{ mol } Cl_2}\right) = 0.634697 \text{ mol } CH_3Cl$$

Chlorine is the limiting reactant.

$$\text{Grams } CH_3Cl = \left(45.0 \text{ g } Cl_2\right)\left(\dfrac{1 \text{ mol } Cl_2}{70.90 \text{ g } Cl_2}\right)\left(\dfrac{1 \text{ mol } CH_3Cl}{1 \text{ mol } Cl_2}\right)\left(\dfrac{50.48 \text{ g } CH_3Cl}{1 \text{ mol } CH_3Cl}\right)\left(\dfrac{80.0\%}{100\%}\right)$$

$$= 25.6316 = \mathbf{25.6 \text{ g } CH_3Cl}$$

The beginning of the calculation is repeated to emphasize that the second part of the problem is simply an extension of the first part. There is no need to repeat the entire calculation as only the final step(s) times the answer of the first part will give the final answer to this part.

3.87   Plan: The first step is to determine the chemical formulas so a balanced chemical equation can be written. The limiting reactant must be determined. Finally, the mass of CF_4 is determined from the limiting reactant.
Solution: (Rounding to the correct number of significant figures will be postponed until the final result.)
The balanced chemical equation is:
$$(CN)_2(g) + 7 F_2(g) \rightarrow 2 CF_4(g) + 2 NF_3(g)$$

$$\text{Mole } CF_4 \text{ from } (CN)_2 = \left(60.0 \text{ g } (CN)_2\right)\left(\dfrac{1 \text{ mol } (CN)_2}{52.04 \text{ g } (CN)_2}\right)\left(\dfrac{2 \text{ mol } CF_4}{1 \text{ mol } (CN)_2}\right) = 2.30592 \text{ mol } CF_4$$

$$\text{Mole } CF_4 \text{ from } F_2 = \left(60.0 \text{ g } F_2\right)\left(\dfrac{1 \text{ mol } F_2}{38.00 \text{ g } F_2}\right)\left(\dfrac{2 \text{ mol } CF_4}{7 \text{ mol } F_2}\right) = 0.4511278 \text{ mol } CF_4$$

F_2 is the limiting reactant, and will be used to calculate the yield.

$$\text{Grams } CF_4 = \left(60.0 \text{ g } F_2\right)\left(\dfrac{1 \text{ mol } F_2}{38.00 \text{ g } F_2}\right)\left(\dfrac{2 \text{ mol } CF_4}{7 \text{ mol } F_2}\right)\left(\dfrac{88.01 \text{ g } CF_4}{1 \text{ mol } CF_4}\right) = 39.70376 = \mathbf{39.7 \text{ g } CF_4}$$

3.88   Plan: Write and balance the chemical reaction. Use the mole ratio in the balanced reaction to choose the correct scene.
Solution:
The reaction is $2 Cl_2O(g) \rightarrow 2 Cl_2(g) + O_2(g)$
Both oxygen and chlorine are diatomic and the ratio between Cl_2 and O_2 is 2:1. **Scene A** best represents the product mixture as there are three O_2 molecules and twice as many, or six Cl_2 molecules in Scene A. This is the proper mole ratio. Scene B shows oxygen and chlorine atoms and Scene C shows oxygen atoms. Oxygen and chlorine atoms are NOT products of this reaction.

3.92   Plan: The spheres represent particles of solute and the amount of *solute* per given volume of *solution* determines its concentration. Molarity = moles of solute/volume (L) of solution.
Solution:
a) **Box C** has more solute added because it contains 2 more spheres than Box A contains.
b) **Box B** has more solvent because solvent molecules have displaced two solute molecules.
c) **Box C** has a higher molarity, because it has more moles of solute per volume of solution.

d) **Box B** has a lower concentration (and molarity), because it has fewer moles of solute per volume of solution.

3.95   Plan: Remember that molarity is moles of solute in volume of solution.
Solution: Volumes may **not** be additive when two different solutions are mixed, so the final volume may be slightly different from 1000.0 mL. The correct method would state, "Take 100.0 mL of the 10.0 M solution and add water until the total volume is 1000 mL."

3.96   Plan: In all cases, the definition of molarity ($M = \dfrac{\text{moles solute}}{\text{L of solution}}$) will be important. Volume must be expressed in liters. The molar mass is used to convert moles to grams. The chemical formulas must be written to determine the molar mass.
Solution:

a) Grams $Ca(C_2H_3O_2)_2$ = $(185.8 \text{ mL})\left(\dfrac{10^{-3} \text{ L}}{1 \text{ mL}}\right)\left(\dfrac{0.267 \text{ mol } Ca(C_2H_3O_2)_2}{1 \text{ L}}\right)\left(\dfrac{158.17 \text{ g } Ca(C_2H_3O_2)_2}{1 \text{ mol } Ca(C_2H_3O_2)_2}\right)$

   $= 7.84659 = $ **7.85 g $Ca(C_2H_3O_2)_2$**

b) Moles KI = $(21.1 \text{ g KI})\left(\dfrac{1 \text{ mol KI}}{166.0 \text{ g KI}}\right) = 0.127108$ moles KI

   Volume = $(500. \text{ mL})\left(\dfrac{10^{-3} \text{ L}}{1 \text{ mL}}\right) = 0.500$ L

   Molarity KI = $\dfrac{0.127108 \text{ mol KI}}{0.500 \text{ L}} = 0.254216 = $ **0.254 M KI**

c) Moles NaCN = $\left(\dfrac{0.850 \text{ mol NaCN}}{1 \text{ L}}\right)(145.6 \text{ L}) = 123.76 = $ **124 mol NaCN**

3.98   Plan: It will help to rewrite M as its definition (mole/L). (a) You will need to convert milliliters to liters, and determine the molar mass of potassium sulfate. (b) The simplest way will be to convert the milligrams to millimoles. Molarity may not only be expressed as moles/L, but also as mmoles/mL. (c) Convert the milliliters to liters, and find the moles of solute. It will be necessary to use Avogadro's number to determine the number of ions present.
Solution:

a) Grams potassium sulfate = $(475 \text{ mL})\left(\dfrac{10^{-3} \text{ L}}{1 \text{ mL}}\right)\left(\dfrac{5.62 \text{ x } 10^{-2} \text{ mol } K_2SO_4}{\text{L}}\right)\left(\dfrac{174.27 \text{ g } K_2SO_4}{1 \text{ mol } K_2SO_4}\right)$

   $= 4.6521 = $ **4.65 g K_2SO_4**

b) Molarity calcium chloride = $\left(\dfrac{7.25 \text{ mg } CaCl_2}{1 \text{ mL}}\right)\left(\dfrac{1 \text{ mmol } CaCl_2}{110.98 \text{ mg } CaCl_2}\right) = 0.065327 = $ **0.0653 M $CaCl_2$**

   If you believe that molarity must be moles/liters then the calculation becomes:

   $\left(\dfrac{7.25 \text{ mg } CaCl_2}{1 \text{ mL}}\right)\left(\dfrac{10^{-3} \text{ g}}{1 \text{ mg}}\right)\left(\dfrac{1 \text{ mL}}{10^{-3} \text{ L}}\right)\left(\dfrac{1 \text{ mol } CaCl_2}{110.98 \text{ g } CaCl_2}\right) = 0.065327 = $ **0.0653 M $CaCl_2$**

   Notice that the two central terms cancel each other.

c) Number of Mg^{2+} ions = $(1 \text{ mL})\left(\dfrac{10^{-3} \text{ L}}{1 \text{ mL}}\right)\left(\dfrac{0.184 \text{ mol } MgBr_2}{1 \text{ L}}\right)\left(\dfrac{1 \text{ mol } Mg^{2+}}{1 \text{ mol } MgBr_2}\right)\left(\dfrac{6.022 \text{ x } 10^{23} \text{ } Mg^{2+} \text{ ions}}{1 \text{ mol } Mg^{2+}}\right)$

   $= 1.1080 \text{ x } 10^{20} = $ **1.11 x 10^{20} Mg^{2+} ions**

3.100   <u>Plan:</u> These are dilution problems. Dilution problems can be solved by converting to moles and using the new volume, however, it is much easier to use $M_1V_1 = M_2V_2$. Part (c) may be done as two dilution problems or as a mole problem. The dilution equation does not require a volume in liters; it only requires that the volume units match.
<u>Solution:</u>
a) $M_1 = 0.250\ M$ KCl $V_1 = 37.00$ mL $M_2 = ?$ $V_2 = 150.00$ mL

$M_1V_1 = M_2V_2$ $(0.250\ \text{M})(37.00\ \text{mL}) = (M_2)(150.0\ \text{mL})$

$$\frac{(0.250\ \text{M})(37.00\ \text{mL})}{150.0\ \text{mL}} = M_2 \qquad M_2 = 0.061667 = \textbf{0.0617 M KCl}$$

b) $M_1 = 0.0706\ M$ $(NH_4)_2SO_4$ $V_1 = 25.71$ mL $M_2 = ?$ $V_2 = 500.00$ mL

$M_1V_1 = M_2V_2$ $(0.0706\ \text{M})(25.71\ \text{mL}) = (M_2)(1500.0\ \text{mL})$

$$\frac{(0.0706\ \text{M})(25.71\ \text{mL})}{500.0\ \text{mL}} = M_2 \qquad M_2 = 0.003630 = \textbf{0.00363 M (NH}_4\textbf{)}_2\textbf{SO}_4$$

c) When working this as a mole problem it is necessary to find the individual number of moles of sodium ions in each separate solution. (Rounding to the proper number of significant figures will only be done for the final answer.)

$$\text{Moles Na}^+ \text{ from NaCl solution} = \left(3.58\ \text{mL}\right)\left(\frac{10^{-3}\ \text{L}}{1\ \text{mL}}\right)\left(\frac{0.348\ \text{mol NaCl}}{1\ \text{L}}\right)\left(\frac{1\ \text{mol Na}^+}{1\ \text{mol NaCl}}\right)$$

$$= 0.00124584\ \text{mol Na}^+$$

$$\text{Moles Na}^+ \text{ from Na}_2\text{SO}_4 \text{ solution} = \left(500.\ \text{mL}\right)\left(\frac{10^{-3}\ \text{L}}{1\ \text{mL}}\right)\left(\frac{6.81 \times 10^{-2}\ \text{mol Na}_2\text{SO}_4}{1\ \text{L}}\right)\left(\frac{2\ \text{mol Na}^+}{1\ \text{mol Na}_2\text{SO}_4}\right)$$

$$= 0.0681\ \text{mol Na}^+$$

$$\text{Molarity of Na}^+ = \left(\frac{(0.00124584 + 0.0681)\ \text{mol Na}^+}{(3.58 + 500.)\ \text{mL}}\right)\left(\frac{1\ \text{mL}}{10^{-3}\ \text{L}}\right) = 0.1377058 = \textbf{0.138 } M \textbf{ Na}^+ \textbf{ ions}$$

3.102   <u>Plan:</u> Use the density of the solution to find the mass of 1 L of solution. The 70.0% by mass translates to 70.0 g solute/100 g solution and is used to find the mass of HNO_3 in 1 L of solution. Convert mass of HNO_3 to moles to obtain moles/L, molarity.
<u>Solution:</u>

a) Mass HNO_3 per liter $= \left(\frac{1.41\ \text{g Solution}}{1\ \text{mL}}\right)\left(\frac{1\ \text{mL}}{10^{-3}\ \text{L}}\right)\left(\frac{70.0\ \text{g HNO}_3}{100\ \text{g Solution}}\right) = \textbf{987 g HNO}_3\textbf{ / L}$

b) Molarity of $HNO_3 = \left(\frac{1.41\ \text{g Solution}}{1\ \text{mL}}\right)\left(\frac{1\ \text{mL}}{10^{-3}\ \text{L}}\right)\left(\frac{70.0\ \text{g HNO}_3}{100\ \text{g Solution}}\right)\left(\frac{1\ \text{mol HNO}_3}{63.02\ \text{g HNO}_3}\right)$

$= 15.6617 = \textbf{15.7 } M \textbf{ HNO}_3$

3.104   <u>Plan:</u> Convert the mass of calcium carbonate to moles, and use the balanced chemical equation to find the moles of hydrochloric acid required. The moles of acid along with the molarity of the acid will give the volume required. The molarity of the solution is given in the calculation as mol/L.
<u>Solution:</u>
$2\ HCl(aq) + CaCO_3(s) \rightarrow CaCl_2(aq) + CO_2(g) + H_2O(l)$

Volume required $= \left(16.2\ \text{g CaCO}_3\right)\left(\frac{1\ \text{mol CaCO}_3}{100.09\ \text{g CaCO}_3}\right)\left(\frac{2\ \text{mol HCl}}{1\ \text{mol CaCO}_3}\right)\left(\frac{1\ \text{L}}{0.383\ \text{mol HCl}}\right)\left(\frac{1\ \text{mL}}{10^{-3}\ \text{L}}\right)$

$= 845.1923 = \textbf{845 mL HCl solution}$

3.106   Plan: The first step is to write and balance the chemical equation for the reaction. Use the molarity and volume of each of the reactants to determine the moles of each as a prelude to determining which is the limiting reactant. Use the limiting reactant to determine the mass of barium sulfate that will form.

Solution:

The balanced chemical equation is:

$BaCl_2(aq) + Na_2SO_4(aq) \rightarrow BaSO_4(s) + 2\ NaCl(aq)$

The mole and limiting reactant calculations are:

$$\text{Moles BaSO}_4 \text{ from BaCl}_2 = \left(35.0\ \text{mL}\right)\left(\frac{10^{-3}\ \text{L}}{1\ \text{mL}}\right)\left(\frac{0.160\ \text{mol BaCl}_2}{1\ \text{L}}\right)\left(\frac{1\ \text{mol BaSO}_4}{1\ \text{mol BaCl}_2}\right) = 0.00560\ \text{mol BaSO}_4$$

$$\text{Moles BaSO}_4 \text{ from Na}_2\text{SO}_4 = \left(58.0\ \text{mL}\right)\left(\frac{10^{-3}\ \text{L}}{1\ \text{mL}}\right)\left(\frac{0.065\ \text{mol Na}_2\text{SO}_4}{1\ \text{L}}\right)\left(\frac{1\ \text{mol BaSO}_4}{1\ \text{mol Na}_2\text{SO}_4}\right) = 0.00377\ \text{mol BaSO}_4$$

Sodium sulfate is the limiting reactant.

$$\text{Grams BaSO}_4 = \left(58.0\ \text{mL}\right)\left(\frac{10^{-3}\ \text{L}}{1\ \text{mL}}\right)\left(\frac{0.065\ \text{mol Na}_2\text{SO}_4}{1\ \text{L}}\right)\left(\frac{1\ \text{mol BaSO}_4}{1\ \text{mol Na}_2\text{SO}_4}\right)\left(\frac{233.4\ \text{g BaSO}_4}{1\ \text{mol BaSO}_4}\right)$$

$$= 0.879918 = \mathbf{0.88\ g\ BaSO_4}$$

3.109   Plan: The first part of the problem is a simple dilution problem ($M_1V_1 = M_2V_2$). The second part requires the molar mass of the HCl along with the molarity.

Solution:

a) $M_1 = 11.7\ M$ $V_1 = ?$ $M_2 = 3.5\ M$ $V_2 = 3.0\ \text{gal}$

      $M_1V_1 = M_2V_2$ $(11.7\ M)(V_1) = (3.5\ M)(3.0\ \text{gal})$

$$\frac{(3.5\ M)(3.0\ \text{gal})}{11.7\ M} = V_1 \qquad\qquad V_1 = 0.897436\ \text{gallons (unrounded)}$$

Instructions: Be sure to wear goggles to protect your eyes! Pour approximately 2.0 gallons of water into the container. Add slowly and with mixing 0.90 gallons of 11.7 M HCl into the water. Dilute to 3.0 gallons with water.

$$\text{b) Volume needed} = \left(9.66\ \text{g HCl}\right)\left(\frac{1\ \text{mol HCl}}{36.46\ \text{g HCl}}\right)\left(\frac{1\ \text{L}}{11.7\ \text{mol HCl}}\right)\left(\frac{1\ \text{mL}}{10^{-3}\ \text{L}}\right)$$

$$= 22.64512 = \mathbf{22.6\ mL\ muriatic\ acid\ solution}$$

3.114   Plan: The moles of narceine and the moles of water are required. We can assume any mass of narceine hydrate (we will use 100 g), and use this mass to determine the mass of water present and convert the mass to moles of the hydrate. The mass of water will be converted to moles. Finally, the ratio of the moles of hydrate to moles of water will give the amount of water present.

Solution:

$$\text{Moles narceine hydrate} = \left(100\ \text{g narceine hydrate}\right)\left(\frac{1\ \text{mol narceine hydrate}}{499.52\ \text{g narceine hydrate}}\right)$$

$$= 0.20019\ \text{mol narceine hydrate}$$

$$\text{Moles H}_2\text{O} = \left(100\ \text{g narceine hydrate}\right)\left(\frac{10.8\%\ \text{H}_2\text{O}}{100\%\ \text{narceine hydrate}}\right)\left(\frac{1\ \text{mol H}_2\text{O}}{18.02\ \text{g H}_2\text{O}}\right)$$

$$= 0.59933\ \text{mol H}_2\text{O}$$

The ratio of water to hydrate is: $(0.59933\ \text{mol}) / (0.20019\ \text{mol}) = 3$

Thus, there are three water molecules per mole of hydrate. The formula for narceine hydrate is **narceine•3H$_2$O**.

3.115 Plan: Determine the formula, then the molar mass of each compound. Determine the mass of hydrogen in each formula. The mass of hydrogen divided by the molar mass of the compound (with the result multiplied by 100%) will give the mass percent hydrogen. Ranking, based on the percents, is easy.
Solution:

Name	Chemical formula	Molar mass (g/mol)	Mass percent H [(mass H) / (molar mass)] x 100%
Ethane	C_2H_6	30.07	[(6 x 1.008) / (30.07)] x 100% = 20.11% H
Propane	C_3H_8	44.09	[(8 x 1.008) / (44.09)] x 100% = 18.29% H
Cetyl palmitate	$C_{32}H_{64}O_2$	480.83	[(64 x 1.008) / (480.83)] x 100% = 13.42% H
Ethanol	C_2H_5OH	46.07	[(6 x 1.008) / (46.07)] x 100% = 13.13% H
Benzene	C_6H_6	78.11	[(6 x 1.008) / (78.11)] x 100% = 7.743% H

The hydrogen percentage decreases in the following order:

Ethane > Propane > Cetyl palmitate > Ethanol > Benzene

3.120 Plan: The key to solving this problem is determining the overall balanced equation. Each individual step must be set up and balanced first. The separate equations can then be combined to get the overall equation. The mass of iron may be converted to moles of iron, and the mole ratio and molar mass of carbon monoxide used to determine the mass of carbon monoxide.
Solution:
a) In the first step, ferric oxide (ferric denotes Fe^{3+}) reacts with carbon monoxide to form Fe_3O_4 and carbon dioxide:
$$3\ Fe_2O_3(s) + CO(g) \rightarrow 2\ Fe_3O_4(s) + CO_2(g)\ (1)$$
In the second step, Fe_3O_4 reacts with more carbon monoxide to form ferrous oxide:
$$Fe_3O_4(s) + CO(g) \rightarrow 3\ FeO(s) + CO_2(g)\ (2)$$
In the third step, ferrous oxide reacts with more carbon monoxide to form molten iron:
$$FeO(s) + CO(g) \rightarrow Fe(l) + CO_2(g)\ (3)$$
Common factors are needed to allow these equations to be combined. The intermediate products are Fe_3O_4 and FeO, so multiply equation (2) by 2 to cancel Fe_3O_4 and equation (3) by 6 to cancel FeO:
$$3\ Fe_2O_3(s) + CO(g) \rightarrow 2\ \cancel{Fe_3O_4(s)} + CO_2(g)$$
$$2\ \cancel{Fe_3O_4(s)} + 2\ CO(g) \rightarrow 6\ \cancel{FeO(s)} + 2\ CO_2(g)$$
$$6\ \cancel{FeO(s)} + 6\ CO(g) \rightarrow 6\ Fe(l) + 6\ CO_2(g)$$
$$3\ Fe_2O_3(s) + 9\ CO(g) \rightarrow 6\ Fe(s) + 9\ CO_2(g)$$

Then divide by 3 to obtain the smallest integer coefficients:
$$\mathbf{Fe_2O_3(s) + 3\ CO(g) \rightarrow 2\ Fe(s) + 3\ CO_2(g)}$$
b) A metric ton is equal to 1000 kg.

$$\text{Grams carbon monoxide} = (45.0\ \text{ton Fe})\left(\frac{10^3\ \text{kg}}{1\ \text{ton}}\right)\left(\frac{10^3\ \text{g}}{1\ \text{kg}}\right)\left(\frac{1\ \text{mol Fe}}{55.85\ \text{g Fe}}\right)\left(\frac{3\ \text{mol CO}}{2\ \text{mol Fe}}\right)\left(\frac{28.01\ \text{g CO}}{1\ \text{mol CO}}\right)$$

$$= 3.385273\ \text{x}\ 10^7 = \mathbf{3.39\ x\ 10^7\ g\ CO}$$

3.122 Plan: If 100.0 g of dinitrogen tetroxide reacts with 100.0 g of hydrazine (N_2H_4), what is the theoretical yield of nitrogen if no side reaction takes place? First, we need to identify the limiting reactant. The limiting reactant can be used to calculate the theoretical yield. Determine the amount of limiting reactant required to produce 10.0 grams of NO. Reduce the amount of limiting reactant by the amount used to produce NO. The reduced amount of limiting reactant is then used to calculate an "actual yield." The "actual" and theoretical yields will give the maximum percent yield.
Solution:
Determining the limiting reactant:

$$N_2\ \text{from}\ N_2O_4 = (100.0\ \text{g}\ N_2O_4)\left(\frac{1\ \text{mol}\ N_2O_4}{92.02\ \text{g}\ N_2O_4}\right)\left(\frac{3\ \text{mol}\ N_2}{1\ \text{mol}\ N_2O_4}\right) = 3.26016\ \text{mol}\ N_2$$

$$N_2\ \text{from}\ N_2H_4 = (100.0\ \text{g}\ N_2H_4)\left(\frac{1\ \text{mol}\ N_2H_4}{32.05\ \text{g}\ N_2H_4}\right)\left(\frac{3\ \text{mol}\ N_2}{2\ \text{mol}\ N_2H_4}\right) = 4.68019\ \text{mol}\ N_2$$

N_2O_4 is the limiting reactant.

$$\text{Theoretical yield of N}_2 = \left(100.0 \text{ g N}_2\text{O}_4\right)\left(\frac{1 \text{ mol N}_2\text{O}_4}{92.02 \text{ g N}_2\text{O}_4}\right)\left(\frac{3 \text{ mol N}_2}{1 \text{ mol N}_2\text{O}_4}\right)\left(\frac{28.02 \text{ g N}_2}{1 \text{ mol N}_2}\right)$$

$$= 91.3497 \text{ g N}_2 \text{ (unrounded)}$$

How much limiting reactant is used to produce 10.0 g NO?

$$\text{Grams N}_2\text{O}_4 \text{ used} = \left(10.0 \text{ g NO}\right)\left(\frac{1 \text{ mol NO}}{30.01 \text{ g NO}}\right)\left(\frac{2 \text{ mol N}_2\text{O}_4}{6 \text{ mol NO}}\right)\left(\frac{92.02 \text{ g N}_2\text{O}_4}{1 \text{ mol N}_2\text{O}_4}\right) = 10.221 \text{ g N}_2\text{O}_4 \text{ (unrounded)}$$

Determine the "actual yield."
Amount of N_2O_4 available to produce N_2 = 100.0 g N_2O_4 – mass of N_2O_4 required to produce 10.0 g NO
100.0 g – 10.221 g = 89.779 g N_2O_4 (unrounded)

$$\text{"Actual yield" of N}_2 = \left(89.779 \text{ g N}_2\text{O}_4\right)\left(\frac{1 \text{ mol N}_2\text{O}_4}{92.02 \text{ g N}_2\text{O}_4}\right)\left(\frac{3 \text{ mol N}_2}{1 \text{ mol N}_2\text{O}_4}\right)\left(\frac{28.02 \text{ g N}_2}{1 \text{ mol N}_2}\right)$$

$$= 82.01285 \text{ g N}_2 \text{ (unrounded)}$$

$$\text{Theoretical yield} = \left(\frac{\text{actual yield}}{\text{theoretical yield}}\right)(100) = \left(\frac{82.01285}{91.3497}\right)(100) = 89.7790 = \textbf{89.8\%}$$

3.124   Plan: Count the number of each type of molecule in the reactant box and in the product box. Subtract any molecules of excess reagent (molecules appearing in both boxes). The remaining material is the overall equation. This will need to be simplified if there is a common factor among the substances in the equation. The balanced chemical equation is necessary for the remainder of the problem.
Solution:
a) The contents of the boxes give:
$$6 \text{ AB}_2 + 5 \text{ B}_2 \rightarrow 6 \text{ AB}_3 + 2 \text{ B}_2$$
B_2 is in excess, so two molecules need to be removed from each side. This gives:
$$6 \text{ AB}_2 + 3 \text{ B}_2 \rightarrow 6 \text{ AB}_3$$
Three is a common factor among the coefficients, and all coefficients need to be divided by this value to give the final balanced equation:
$$\textbf{2 AB}_2 + \textbf{B}_2 \rightarrow \textbf{2 AB}_3$$
b) B_2 was in excess, thus $\textbf{AB}_2$ is the limiting reactant.

c) Moles of AB_3 from AB_2 = $\left(5.0 \text{ mol AB}_2\right)\left(\dfrac{2 \text{ mol AB}_3}{2 \text{ mol AB}_2}\right) = 5.0 \text{ mol AB}_3$

Moles of AB_3 from B_2 = $\left(3.0 \text{ mol B}_2\right)\left(\dfrac{2 \text{ mol AB}_3}{1 \text{ mol B}_2}\right) = 6.0 \text{ mol AB}_3$

AB_2 is the limiting reagent and maximum is $\textbf{5.0 mol AB}_3$.

d) Moles of B_2 that reacts with 5.0 mol AB_2 = $\left(5.0 \text{ mol AB}_2\right)\left(\dfrac{1 \text{ mol B}_2}{2 \text{ mol AB}_2}\right) = 2.5 \text{ mol B}_2$

The un-reacted B_2 is 3.0 mol – 2.5 mol = $\textbf{0.5 mol B}_2$.

3.127   Plan: The balanced chemical equation is needed. From the balanced chemical equation and the masses we can calculate the limiting reagent. The limiting reagent will be used to calculate the theoretical yield, and finally the percent yield.
Solution:
Determine the balanced chemical equation:
$$\text{ZrOCl}_2 \cdot 8\text{H}_2\text{O}(s) + 4 \text{ H}_2\text{C}_2\text{O}_4 \cdot 2\text{H}_2\text{O}(s) + 4 \text{ KOH}(aq) \rightarrow \text{K}_2\text{Zr}(\text{C}_2\text{O}_4)_3(\text{H}_2\text{C}_2\text{O}_4) \cdot \text{H}_2\text{O}(s) + 2 \text{ KCl}(aq) + 20 \text{ H}_2\text{O}(l)$$

Determine the limiting reactant:

$$\text{Moles product from } ZrOCl_2 \cdot 8H_2O = \left(1.68 \text{ gZr cmpd}\right)\left(\frac{1 \text{ mol Zr cmpd}}{322.25 \text{ g Zr cmpd}}\right)\left(\frac{1 \text{ mol Product}}{1 \text{ mol Zr cmpd}}\right)$$

$$= 0.00521334 \text{ mol product (unrounded)}$$

Moles product from $H_2C_2O_4 \cdot 2H_2O =$

$$\left(5.20 \text{ g } H_2C_2O_4 \cdot 2H_2O\right)\left(\frac{1 \text{ mol } H_2C_2O_4 \cdot 2H_2O}{126.07 \text{ g } H_2C_2O_4 \cdot 2H_2O}\right)\left(\frac{1 \text{ mol Product}}{4 \text{ mol } H_2C_2O_4 \cdot 2H_2O}\right)$$

$$= 0.0103117 \text{ mol product (unrounded)}$$

Moles product from KOH = irrelevant because KOH is stated to be in excess.

The $ZrOCl_2 \cdot 8H_2O$ is limiting, and will be used to calculate the theoretical yield:

$$\text{Grams product} = \left(1.68 \text{ g Zr cmpd}\right)\left(\frac{1 \text{ mol Zr cmpd}}{322.25 \text{ g Zr cmpd}}\right)\left(\frac{1 \text{ mol Product}}{1 \text{ mol Zr cmpd}}\right)\left(\frac{541.53 \text{ g Product}}{1 \text{ mol Product}}\right)$$

$$= 2.82318 \text{ g product (unrounded)}$$

Finally, calculating the percent yield:

$$\text{Percent yield} = \left(\frac{\text{Actual Yield}}{\text{Theoretical Yield}}\right) \times 100\% = \left(\frac{1.25 \text{ g}}{2.82318 \text{ g}}\right) \times 100\% = 44.276 = \textbf{44.3\% yield}$$

3.130 **Plan:** Count the total number of spheres in each box. The number in box A divided by the volume change in each part will give the number we are looking for and allow us to match boxes.
Solution:
The number in each box is: A = 12, B = 6, C = 4, and D = 3.
a) When the volume is tripled, there should be 12/3 = 4 spheres in a box. This is box **C**.
b) When the volume is doubled, there should be 12/2 = 6 spheres in a box. This is box **B**.
c) When the volume is quadrupled, there should be 12/4 = 3 spheres in a box. This is box **D**.

3.134 **Plan:** This problem may be done as two dilution problems with the two final molarities added, or, as done here, it may be done by calculating, then adding the moles and dividing by the total volume.
Solution:

$$M \text{ KBr} = \frac{\text{Total Moles KBr}}{\text{Total Volume}} = \frac{\text{Moles KBr from Solution 1} + \text{Moles KBr from Solution 2}}{\text{Volume Solution 1} + \text{Volume Solution 2}}$$

$$M \text{ KBr} = \frac{\left(\frac{0.053 \text{ mol KBr}}{1 \text{ L}}\right)\left(0.200 \text{ L}\right) + \left(\frac{0.078 \text{ mol KBr}}{1 \text{ L}}\right)\left(0.550 \text{ L}\right)}{0.200 \text{ L} + 0.550 \text{ L}} = 0.071333 = \textbf{0.071 } M \textbf{ KBr}$$

3.138 **Plan:** Deal with the methane and propane separately, and combine the results. Balanced equations are needed for each hydrocarbon. The total mass and the percentages will give the mass of each hydrocarbon. The mass of each hydrocarbon is changed to moles, and through the balanced chemical equation the amount of CO_2 produced by each gas may be found. Summing the amounts of CO_2 gives the total from the mixture. For Part b), let x and 252 – x represent the masses of CH_4 and C_3H_8, respectively.
Solution:
a) The balanced chemical equations are:

Methane: $CH_4(g) + 2 O_2(g) \rightarrow CO_2(g) + 2 H_2O(l)$

Propane: $C_3H_8(g) + 5 O_2(g) \rightarrow 3 CO_2(g) + 4 H_2O(l)$

Mass of CO_2 from each:

$$\text{Methane: } \left(200. \text{ g Mixture}\right)\left(\frac{25.0\%}{100\%}\right)\left(\frac{1 \text{ mol } CH_4}{16.04 \text{ g } CH_4}\right)\left(\frac{1 \text{ mol } CO_2}{1 \text{ mol } CH_4}\right)\left(\frac{44.01 \text{ g } CO_2}{1 \text{ mol } CO_2}\right) = 137.188 \text{ g } CO_2$$

$$\text{Propane: } \left(200. \text{ g Mixture}\right)\left(\frac{75.0\%}{100\%}\right)\left(\frac{1 \text{ mol } C_3H_8}{44.09 \text{ g } C_3H_8}\right)\left(\frac{3 \text{ mol } CO_2}{1 \text{ mol } C_3H_8}\right)\left(\frac{44.01 \text{ g } CO_2}{1 \text{ mol } CO_2}\right) = 449.183 \text{ g } CO_2$$

Total $CO_2 = 137.188 \text{ g} + 449.183 \text{ g} = 586.371 = \textbf{586 g } CO_2$

b) Since the mass of CH_4 + the mass of C_3H_8 = 252 g, let x = mass of CH_4 in the mixture and 252 – x = mass of C_3H_8 in the mixture. Use mole ratios to calculate the amount of CO_2 formed from x amount of CH_4 and the amount of CO_2 formed from 252 – x amount of C_3H_8.
The total mass of CO_2 produced = 748 g.

The total moles of CO_2 produced = $\left(748 \text{ g CO}_2\right)\left(\dfrac{1 \text{ mol CO}_2}{44.01 \text{ g CO}_2}\right)$ = 16.996 mol CO_2

$$16.996 \text{ mol CO}_2 = \left(x \text{ g CH}_4\right)\left(\dfrac{1 \text{ mol CH}_4}{16.04 \text{ g CH}_4}\right)\left(\dfrac{1 \text{ mol CO}_2}{1 \text{ mol CH}_4}\right) + \left(252 - x \text{ g C}_3\text{H}_8\right)\left(\dfrac{1 \text{ mol C}_3\text{H}_8}{44.09 \text{ g C}_3\text{H}_8}\right)\left(\dfrac{3 \text{ mol CO}_2}{1 \text{ mol C}_3\text{H}_8}\right)$$

$$16.996 \text{ mol CO}_2 = \dfrac{x}{16.04} \text{ mol CO}_2 + \dfrac{3(252 - x)}{44.09} \text{ mol CO}_2$$

$$16.996 \text{ mol CO}_2 = \dfrac{x}{16.04} \text{ mol CO}_2 + \dfrac{756 - 3x}{44.09} \text{ mol CO}_2$$

$16.996 \text{ mol CO}_2 = 0.06234x \text{ mol CO}_2 + (17.147 - 0.06804x \text{ mol CO}_2)$
$16.996 = 17.147 – 0.0057 \, x$
$x = 26.49 \text{ g CH}_4$ $252 – x = 252 \text{ g} – 26.49 \text{ g} = 225.51 \text{ g C}_3\text{H}_8$

$$\text{mass \% CH}_4 = \left(\dfrac{26.49 \text{ g CH}_4}{252 \text{ g mixture}}\right)100 = \textbf{10.5\% CH}_4$$

$$\text{mass \% C}_3\text{H}_8 = \left(\dfrac{225.51 \text{ g C}_3\text{H}_8}{252 \text{ g mixture}}\right)100 = \textbf{89.5\% C}_3\textbf{H}_8$$

3.140   Plan: If we assume a 100-gram sample of fertilizer, then the 30:10:10 percentages become the masses, in grams, of N, P_2O_5, and K_2O. These masses may be changed to moles of substance, and then to moles of each element. To get the desired x:y:1.0 ratio, divide the moles of each element by the moles of potassium.
Solution:
A 100-gram sample of 30:10:10 fertilizer contains 30 g N, 10 g P_2O_5, and 10 g K_2O.

$$\text{Moles N} = \left(30\text{ g N}\right)\left(\dfrac{1 \text{ mol N}}{14.01 \text{ g N}}\right) = 2.1413 \text{ mol N (unrounded)}$$

$$\text{Moles P} = \left(10 \text{ g P}_2\text{O}_5\right)\left(\dfrac{1 \text{ mol P}_2\text{O}_5}{141.94 \text{ g P}_2\text{O}_5}\right)\left(\dfrac{2 \text{ mol P}}{1 \text{ mol P}_2\text{O}_5}\right) = 0.14090 \text{ mol P (unrounded)}$$

$$\text{Moles K} = \left(10 \text{ g K}_2\text{O}\right)\left(\dfrac{1 \text{ mol K}_2\text{O}}{94.20 \text{ g K}_2\text{O}}\right)\left(\dfrac{2 \text{ mol K}}{1 \text{ mol K}_2\text{O}}\right) = 0.21231 \text{ mol K (unrounded)}$$

This gives a N:P:K ratio of 2.1413:0.14090:0.21231
The ratio must be divided by the moles of K and rounded.

$\dfrac{2.1413 \text{ mol N}}{0.21231} = 10.086$ $\dfrac{0.14090 \text{ mol P}}{0.21231} = 0.66365$ $\dfrac{0.21231 \text{ mol K}}{0.21231} = 1$

        10.086 : 0.66365 : 1.000   or   **10:0.66:1.0**

3.146   Plan: Determine the molecular formula from the figure. Once the molecular formula is known, use the periodic table to determine the molar mass. Convert the volume in (b) from quarts to mL and use the density to convert from mL to mass in grams. Take 6.82% of that mass and use the molar mass to convert to moles.
Solution:
a) The formula of citric acid obtained by counting the number of carbon atoms, oxygen atoms, and hydrogen atoms is $\textbf{C}_6\textbf{H}_8\textbf{O}_7$.
        Molar mass = (6 x 12.01) + (8 x 1.008) + (7 x 16.00) = **192.12 g/mol**

b) Determine the mass of citric acid in the lemon juice, and then use the molar mass to find the moles.

$$\text{Moles } C_6H_8O_7 = (1.50 \text{ qt})\left(\frac{1 \text{ L}}{1.057 \text{ qt}}\right)\left(\frac{1 \text{ mL}}{10^{-3} \text{ L}}\right)\left(\frac{1.09 \text{ g}}{\text{mL}}\right)\left(\frac{6.82\%}{100\%}\right)\left(\frac{1 \text{ mol } C_6H_8O_7}{192.12 \text{ g acid}}\right)$$

$$= 0.549104 = \textbf{0.549 mol } \mathbf{C_6H_8O_7}$$

3.147   Plan: Determine the formulas of each reactant and product, then balance the individual equations. Combine the three smaller equations to give the overall equation, where some substances serve as intermediates and will cancel.   The amount of nitrogen plus the balanced overall chemical equation will give the amount of nitric acid formed.
Solution:
a) Derive the formulas and balance the equations.
Nitrogen and oxygen combine to form nitrogen monoxide:

$$N_2(g) + O_2(g) \rightarrow 2 \text{ NO}(g)$$

Nitrogen monoxide reacts with oxygen to form nitrogen dioxide:

$$2 \text{ NO}(g) + O_2(g) \rightarrow 2 \text{ NO}_2(g)$$

Nitrogen dioxide combines with water to form nitric acid and nitrogen monoxide:

$$3 \text{ NO}_2(g) + H_2O(l) \rightarrow 2 \text{ HNO}_3(g) + \text{NO}(g)$$

b) Combining the reactions may involve adjusting the equations in various ways to cancel out as many materials as possible other than the reactants added and the desired products.

$$2 \text{ x } (N_2(g) + O_2(g) \rightarrow 2 \text{ NO}(g)) = \qquad\qquad 2 N_2(g) + 2 O_2(g) \rightarrow \cancel{4 \text{ NO}(g)}$$
$$3 \text{ x } (2 \text{ NO}(g) + O_2(g) \rightarrow 2 \text{ NO}_2(g)) = \qquad \cancel{6 \text{ NO}(g)} + 3 O_2(g) \rightarrow \cancel{6 \text{ NO}_2(g)}$$
$$2 \text{ x } (3 \text{ NO}_2(g) + H_2O(l) \rightarrow 2 \text{ HNO}_3(g) + \text{NO}(g)) = \quad \cancel{6 \text{ NO}_2(g)} + 2 H_2O(l) \rightarrow 4 \text{ HNO}_3(g) + \cancel{2 \text{ NO}(g)}$$

Multiplying the above equations, summing the results, and canceling any substance appearing on both sides gives:

$$2 N_2(g) + 5 O_2(g) + 2 H_2O(l) \rightarrow 4 \text{ HNO}_3(g)$$

c) Metric tons HNO_3 =

$$(1350 \text{ t } N_2)\left(\frac{10^3 \text{ kg}}{1 \text{ t}}\right)\left(\frac{10^3 \text{ g}}{1 \text{ kg}}\right)\left(\frac{1 \text{ mol } N_2}{28.02 \text{ g } N_2}\right)\left(\frac{4 \text{ mol } HNO_3}{2 \text{ mol } N_2}\right)\left(\frac{63.02 \text{ g } HNO_3}{1 \text{ mol } HNO_3}\right)\left(\frac{1 \text{ kg}}{10^3 \text{ g}}\right)\left(\frac{1 \text{ t}}{10^3 \text{ kg}}\right)$$

$$= 6.0726 \text{ x } 10^3 = \textbf{6.07 x } 10^3 \textbf{ metric tons } \mathbf{HNO_3}$$

3.149   Plan: Write and balance the chemical reaction. Use the mole ratio to find the amount of product that should be produced and take 66% of that amount to obtain the actual yield.
Solution:

$$2 \text{ NO}(g) + O_2(g) \rightarrow 2 \text{ NO}_2(g)$$

With 6 molecules of NO and 3 molecules of O_2 reacting, 6 molecules of NO_2 can be produced.
If the reaction only has a 66% yield, then $(0.66)(6) = 4$ molecules of NO_2 will be produced. **Circle A** shows the formation of 4 molecules of NO_2. Circle B also shows the formation of 4 molecules of NO_2 but also has 2 unreacted molecules of NO and 1 unreacted molecule of O_2. Since neither reactant is limiting, there will be no unreacted reactant left after reaction.

3.152   Plan: Use the mass percent to find the mass of heme in the sample; use the molar mass to convert the mass of heme to moles. Then find the mass of Fe in the sample.
Solution:

a) Grams of heme = $(0.65 \text{ g hemoglobin})\left(\dfrac{6.0\% \text{ heme}}{100\% \text{ hemoglobin}}\right) = \textbf{0.039 g heme}$

b) Mole of heme = $(0.039 \text{ g heme})\left(\dfrac{1 \text{ mol heme}}{616.49 \text{ g heme}}\right) = 6.32614 \text{ x } 10^{-5} = \textbf{6.3 x } 10^{-5} \textbf{ mol heme}$

c) Grams of Fe = $(6.32614 \text{ x } 10^{-5} \text{ mol heme})\left(\dfrac{1 \text{ mol Fe}}{1 \text{ mol heme}}\right)\left(\dfrac{55.85 \text{ g Fe}}{1 \text{ mol Fe}}\right)$

$$= 3.5331 \text{ x } 10^{-3} = \textbf{3.5 x } 10^{-3} \textbf{ g Fe}$$

d) Grams of hemin = $\left(6.32614 \times 10^{-5} \text{ mol heme}\right)\left(\dfrac{1 \text{ mol hemin}}{1 \text{ mol heme}}\right)\left(\dfrac{651.94 \text{ g hemin}}{1 \text{ mol hemin}}\right)$

$= 4.1243 \times 10^{-2} = \mathbf{4.1 \times 10^{-2} \text{ g hemin}}$

3.155   <u>Plan:</u> Determine the molecular formula and the molar mass of each of the compounds. From the amount of nitrogen present and the molar mass, the percent nitrogen may be determined. The moles need to be determined for part (b).
<u>Solution:</u>
a) To find mass percent of nitrogen, first determine molecular formula, then the molar mass of each compound. Mass percent is then calculated from the mass of nitrogen in the compound divided by the molar mass of the compound, and multiply by 100%.

Urea: CH_4N_2O, $\mathcal{M} = 60.06$ g/mol

$$\% \text{ N} = \left(\frac{2(14.01 \text{ g/mol N})}{60.06 \text{ g/mol } CH_4N_2O}\right)100\% = 46.6533 = \mathbf{46.65\% \text{ N in urea}}$$

Arginine: $C_6H_{15}N_4O_2$, $\mathcal{M} = 175.22$ g/mol

$$\% \text{ N} = \left(\frac{4(14.01 \text{ g/mol N})}{175.22 \text{ g/mol } C_6H_{15}N_4O_2}\right)100\% = 31.98265 = \mathbf{31.98\% \text{ N in arginine}}$$

Ornithine: $C_5H_{13}N_2O_2$, $\mathcal{M} = 133.17$ g/mol

$$\% \text{ N} = \left(\frac{2(14.01 \text{ g/mol N})}{133.17 \text{ g/mol } C_5H_{13}N_2O_2}\right)100\% = 21.04077 = \mathbf{21.04\% \text{ N in ornithine}}$$

b) Grams of N =

$\left(135.2 \text{ g } C_5H_{13}N_2O_2\right)\left(\dfrac{1 \text{ mol } C_5H_{13}N_2O_2}{133.17 \text{ g } C_5H_{13}N_2O_2}\right)\left(\dfrac{1 \text{ mol } CH_4N_2O}{1 \text{ mol } C_5H_{13}N_2O_2}\right)\left(\dfrac{2 \text{ mol N}}{1 \text{ mol } CH_4N_2O}\right)\left(\dfrac{14.01 \text{ g N}}{1 \text{ mol N}}\right)$

$= 28.447 = \mathbf{28.45 \text{ g N}}$

3.157   <u>Plan:</u> Determine the molar mass of each product and plug into atom percent equation.
<u>Solution:</u>
Molar masses of product: N_2H_4: 32.05 g/mol   NaCl: 58.44 g/mol H_2O: 18.02 g/mol

$$\% \text{ atom economy} = \frac{\left(\text{Number of Moles}\right)\left(\text{Molar Mass of Desired Product}\right)}{\left(\text{Sum of Number of Moles}\right)\left(\text{Molar Mass for all Products}\right)} \times 100\%$$

% atom economy =

$$\frac{\left(1 \text{ mol}\right)\left(\text{Molar Mass of } N_2H_4\right)}{\left(1 \text{ mol}\right)\left(\text{Molar Mass of } N_2H_4\right) + \left(1 \text{ mol}\right)\left(\text{Molar Mass of NaCl}\right) + \left(1 \text{ mol}\right)\left(\text{Molar Mass of } H_2O\right)} \times 100\%$$

$$\frac{\left(1 \text{ mol}\right)\left(32.05 \text{ g/mol}\right)}{\left(1 \text{ mol}\right)\left(32.05 \text{ g/mol}\right) + \left(1 \text{ mol}\right)\left(58.44 \text{ g/mol}\right) + \left(1 \text{ mol}\right)\left(18.02 \text{ g/mol}\right)} \times 100\% = 29.5364 = \mathbf{29.54\% \text{ atom economy}}$$

3.159   <u>Plan:</u> Convert the mass of ethanol to moles, and use moles to determine the theoretical yield of diethyl ether. This theoretical yield, along with the given (actual) yield, is used to determine the percent yield. The difference between the actual and theoretical yields is related to the quantity of ethanol that did not produce diethyl ether. Forty-five percent of the mass difference is converted to moles of diethyl ether, then to moles of ethanol and ethylene, and finally to the mass of ethylene.
<u>Solution:</u>
a) The determination of the theoretical yield:
Grams diethyl ether =

$\left(50.0 \text{ g } CH_3CH_2OH\right)\left(\dfrac{1 \text{ mol } CH_3CH_2OH}{46.07 \text{ g } CH_3CH_2OH}\right)\left(\dfrac{1 \text{ mol } CH_3CH_2OCH_2CH_3}{2 \text{ mol } CH_3CH_2OH}\right)\left(\dfrac{74.12 \text{ g } CH_3CH_2OCH_2CH_3}{1 \text{ mol } CH_3CH_2OCH_2CH_3}\right)$

$= 40.2214 \text{ g diethyl ether (unrounded)}$

Determining the percent yield:

$$\text{Percent yield} = \left(\frac{\text{Actual Yield}}{\text{Theoretical Yield}}\right) \times 100\% = \left(\frac{35.9\ \text{g}}{40.2214\ \text{g}}\right) \times 100\% = 89.2560 = \mathbf{89.3\%\ yield}$$

b) To determine the amount of ethanol not producing diethyl ether, we will use the difference between the theoretical yield and actual yield (unrounded) to determine the amount of diethyl ether that did not form and hence, the amount of ethanol that did not produce the desired product. Forty-five percent of this amount will be used to determine the amount of ethylene formed.

Mass difference = 40.2214 g – 35.9 g = 4.3214 g diethyl ether that did not form

Grams ethylene =

$$\left(4.3214\ \text{g}\ (C_2H_5)_2O\right)\left(\frac{1\ \text{mol}\ (C_2H_5)_2O}{74.12\ \text{g}\ (C_2H_5)_2O}\right)\left(\frac{2\ \text{mol}\ C_2H_6O}{1\ \text{mol}\ (C_2H_5)_2O}\right)\left(\frac{1\ \text{mol}\ C_2H_4}{1\ \text{mol}\ C_2H_6O}\right)\left(\frac{28.05\ \text{g}\ C_2H_4}{1\ \text{mol}\ C_2H_4}\right)\left(\frac{45.0\%}{100\%}\right)$$

$$= 1.47185 = \mathbf{1.47\ g\ ethylene}$$

3.161 <u>Plan:</u> Careful tracking of the units will carry you through this problem.
<u>Solution:</u>

a) Dissolved salt = $\left(50.0\ \text{mL}\ H_2O\right)\left(\frac{10^{-3}\ \text{L}\ H_2O}{1\ \text{mL}\ H_2O}\right)\left(\frac{2.50\ \text{kg Salt}}{1\ \text{L}\ H_2O}\right)\left(\frac{10^3\ \text{g}}{1\ \text{kg}}\right) = \mathbf{125\ g\ salt}$

b) Convert the grams from part (a) to moles of salt, then to moles of cocaine, and finally to grams of cocaine. A calculation similar to that in part (a) will give the volume required.

Mass cocaine = $\left(125\ \text{g Salt}\right)\left(\frac{1\ \text{mol Salt}}{339.81\ \text{g Salt}}\right)\left(\frac{1\ \text{mol Cocaine}}{1\ \text{mol Salt}}\right)\left(\frac{303.35\ \text{g Cocaine}}{1\ \text{mol Cocaine}}\right)$

$$= 111.588\ \text{g cocaine (unrounded)}$$

Additional water needed = $\left(111.588\ \text{g Cocaine}\right)\left(\frac{1\ \text{L}}{1.70\ \text{g Cocaine}}\right) - \left(50.0\ \text{mL}\right)\left(\frac{10^{-3}\ \text{L}}{1\ \text{mL}}\right)$

$$= 65.59 = \mathbf{65.6\ L\ H_2O}$$

CHAPTER 4 THREE MAJOR CLASSES OF CHEMICAL REACTIONS

FOLLOW–UP PROBLEMS

4.1   Plan: We must write an equation showing the dissociation of one mole of compound into its ions. The number of moles of compound times the total number of ions formed gives the moles of ions in solution. If the moles of compound are not given directly, they must be calculated from the information given.
Solution:
a) One mole of $KClO_4$ dissociates to form one mole of potassium ions and one mole of perchlorate ions.
$$KClO_4(s) \rightarrow K^+(aq) + ClO_4^-(aq)$$
Therefore, 2 moles of solid $KClO_4$ produce **2 mol of K^+** ions and **2 mol of ClO_4^-** ions.
b) $Mg(C_2H_3O_2)_2(s) \rightarrow Mg^{2+}(aq) + 2\ C_2H_3O_2^-(aq)$
First convert grams of $Mg(C_2H_3O_2)_2$ to moles of $Mg(C_2H_3O_2)_2$ and then use molar ratios to determine the moles of each ion produced.
$$\left(354\ g\ Mg(C_2H_3O_2)_2\right)\left(\frac{1\ mol\ Mg(C_2H_3O_2)_2}{142.40\ g\ Mg(C_2H_3O_2)_2}\right) = 2.48596\ mol\ (unrounded)$$
The dissolution of 2.48596 mol $Mg(C_2H_3O_2)_2(s)$ produces **2.49 mol Mg^{2+}** and (2 x 2.48596) = **4.97 mol $C_2H_3O_2^-$**
c) $(NH_4)_2CrO_4(s) \rightarrow 2\ NH_4^+(aq) + CrO_4^{2-}(aq)$
First convert formula units to moles and then use molar ratios as in part b).
$$\left(1.88\ x\ 10^{24}\ Formula\ Units\right)\left(\frac{1\ mol\ (NH_4)_2CrO_4}{6.022\ x\ 10^{23}\ Formula\ Units}\right) = 3.121886\ mol\ (unrounded)$$
The dissolution of 3.121886 mol $(NH_4)_2CrO_4(s)$ produces (2 x 3.121886) = **6.24 mol NH_4^+** and **3.12 mol CrO_4^{2-}**.
d) $NaHSO_4(s) \rightarrow Na^+(aq) + HSO_4^-(aq)$
The solution contains (1.32 L solution) (0.55 mol $NaHSO_4$/L solution) = 0.726 mol of $NaHSO_4$ and therefore **0.73 mol of Na^+** and **0.73 mol HSO_4^-**.
Check: The ratio of the moles of each ion in the solution should equal the ratio in the original formula.

4.2   Plan: Determine the ions present in each substance on the reactant side. Use Table 4.1 to determine if any combination of ions is not soluble. If a precipitate forms there will be a reaction and chemical equations may be written. The molecular equation simply includes the formulas of the substances and balancing. In the total ionic equation, all soluble substances are written as separate ions. The net ionic equation comes from the total ionic equation by eliminating all substances appearing in identical form (spectator ions) on each side of the reaction arrow.
Solution:
a) The resulting ion combinations that are possible are iron(III) phosphate and cesium chloride. According to Table 4.1, iron(III) phosphate is a common phosphate and insoluble, so a reaction occurs. We see that cesium chloride is soluble.
Total ionic equation:
$$Fe^{3+}(aq) + 3\ Cl^-(aq) + 3\ Cs^+(aq) + PO_4^{3-}(aq) \rightarrow FePO_4(s) + 3\ Cl^-(aq) + 3\ Cs^+(aq)$$
Net ionic equation:
$$Fe^{3+}(aq) + PO_4^{3-}(aq) \rightarrow FePO_4(s)$$
b) The resulting ion combinations that are possible are sodium nitrate (soluble) and cadmium hydroxide (insoluble). A reaction occurs.
Total ionic equation:
$$2\ Na^+(aq) + 2\ OH^-(aq) + Cd^{2+}(aq) + 2\ NO_3^-(aq) \rightarrow Cd(OH)_2(s) + 2\ Na^+(aq) + 2\ NO_3^-(aq)$$
Note: The coefficients for Na^+ and OH^- are necessary to balance the reaction and must be included.
Net ionic equation:
$$Cd^{2+}(aq) + 2\ OH^-(aq) \rightarrow Cd(OH)_2(s)$$

c) The resulting ion combinations that are possible are magnesium acetate (soluble) and potassium bromide (soluble). No reaction occurs.

d) The resulting ion combinations that are possible are silver chloride (insoluble, an exception) and barium nitrate (soluble). A reaction occurs.

Total ionic equation:

$$2\,Ag^+(aq) + 2\,NO_3^-(aq) + Ba^{2+}(aq) + 2\,Cl^-(aq) \rightarrow 2\,AgCl(s) + Ba^{2+}(aq) + 2\,NO_3^-(aq)$$

Net ionic equation:

$$Ag^+(aq) + Cl^-(aq) \rightarrow AgCl(s)$$

4.3   Plan: Determine the ions present in each substance on the reactant side. Use Table 4.1 to determine if any combination of ions is not soluble. If a precipitate forms there will be a reaction and chemical equations may be written. The molecular equation simply includes the formulas of the substances, and balancing. In the total ionic equation, all soluble substances are written as separate ions. The net ionic equation comes from the total ionic equation by eliminating all substances appearing in identical form (spectator ions) on each side of the reaction arrow.

Solution:

a) Beaker A has four ions with a +2 charge and eight ions with a –1 charge. The beaker contains dissolved **$Zn(NO_3)_2$** which dissolves to produce Zn^{2+} and NO_3^- ions in a 1:2 ratio. The compound $PbCl_2$ also has a +2 ion and –1 ion in a 1:2 ratio but $PbCl_2$ is insoluble so ions would not result from this compound.

b) Beaker B has three ions with a +2 charge and six ions with a –1 charge.   The beaker contains dissolved **$Ba(OH)_2$** which dissolves to produce Ba^{2+} and OH^- ions in a 1:2 ratio. $Cd(OH)_2$ also has a +2 ion and a –1 ion in a 1:2 ratio but $Cd(OH)_2$ is insoluble so ions would not result from this compound.

c) The resulting ion combinations that are possible are zinc hydroxide (insoluble) and barium nitrate (soluble). The precipitate formed is **$Zn(OH)_2$**. The spectator ions are **Ba^{2+} and NO_3^-**.

Balanced molecular equation: $Zn(NO_3)_2(aq) + Ba(OH)_2(aq) \rightarrow Zn(OH)_2(s) + Ba(NO_3)_2(aq)$

Total ionic equation:

$$Zn^{2+}(aq) + 2\,NO_3^-(aq) + Ba^{2+}(aq) + 2\,OH^-(aq) \rightarrow Zn(OH)_2(s) + Ba^{2+}(aq) + 2\,NO_3^-(aq)$$

Net ionic equation:

$$Zn^{2+}(aq) + 2\,OH^-(aq) \rightarrow Zn(OH)_2(s)$$

d) Since there are only six OH- ions and four Zn2+ ions, the OH- is the limiting reactant.

$$\left(6\ OH^-\ ions\right)\left(\frac{0.050\ \text{mol } OH^-}{1\ OH^-\ \text{ion}}\right)\left(\frac{1\ \text{mol } Zn(OH)_2}{2\ \text{mol } OH^-}\right)\left(\frac{99.43\ \text{g } Zn(OH)_2}{1\ \text{mol } Zn(OH)_2}\right) = 14.9145 = \textbf{15 g } Zn(OH)_2$$

4.4   Plan: Each mole of strong base will produce one mole of hydroxide ions. It is convenient to express molarity as its definition (moles/L).

Solution:

$$KOH(s) \rightarrow K^+(aq) + OH^-(aq)$$

$$\text{Moles } OH^- = \left(451\ \text{mL}\right)\left(\frac{10^{-3}\,\text{L}}{1\ \text{mL}}\right)\left(\frac{1.20\ \text{mol } KOH}{L}\right)\left(\frac{1\ \text{mol } OH^-}{1\ \text{mol } KOH}\right) = 0.5412 = \textbf{0.541 mol } OH^-$$

4.5   Plan: According to Table 4.2, both reactants are strong. Thus, the key reaction is the formation of water. The other product of the reaction is soluble.

Solution:

Molecular equation: $2\ HNO_3(aq) + Ca(OH)_2(aq) \rightarrow Ca(NO_3)_2(aq) + 2\ H_2O(l)$

Total ionic equation:

$$2\,H^+(aq) + 2\,NO_3^-(aq) + Ca^{2+}(aq) + 2\,OH^-(aq) \rightarrow Ca^{2+}(aq) + 2\,NO_3^-(aq) + 2\,H_2O(l)$$

Net ionic equation: $2\,H^+(aq) + 2\,OH^-(aq) \rightarrow 2\,H_2O(l)$ which simplifies to: **$H^+(aq) + OH^-(aq) \rightarrow H_2O(l)$**

4.6   Plan: A balanced chemical equation is necessary. Determine the moles of HCl, and, through the balanced chemical equation, determine the moles of $Ba(OH)_2$ required for the reaction. The moles of base and the molarity may be used to determine the volume necessary.
Solution:
The molarity of the HCl solution is 0.1016 M. However, the molar ratio is not 1:1 as in the example problem. According to the balanced equation, the ratio is 2 moles of acid per 1 mole of base:
$2\ HCl(aq) + Ba(OH)_2(aq) \rightarrow BaCl_2(aq) + 2\ H_2O(l)$

$$\text{Volume} = \left(50.00\ \text{mL}\right)\left(\frac{10^{-3}\,L}{1\ \text{mL}}\right)\left(\frac{0.1016\ \text{mol HCl}}{L}\right)\left(\frac{1\ \text{mol Ba(OH)}_2}{2\ \text{mol HCl}}\right)\left(\frac{L}{0.1292\ \text{mol Ba(OH)}_2}\right)$$

$= 0.0196594 = \textbf{0.01966 L Ba(OH)}_2\ \textbf{solution}$

4.7   Plan: Apply Table 4.3 to the compounds. Do not forget that the sum of the O.N.'s (oxidation numbers) for a compound must sum to zero, and for a polyatomic ion, the sum must equal the charge on the ion.
Solution:
a) **Sc = +3 O = –2** In most compounds, oxygen has a –2 O.N., so oxygen is often a good starting point. If each oxygen atom has a –2 O.N., then each scandium must have a +3 oxidation state so that the sum of O.N.'s equals zero: 2(+3) + 3(–2) = 0.
b) **Ga = +3 Cl = –1** In most compounds, chlorine has a –1 O.N., so chlorine is a good starting point. If each chlorine atom has a –1 O.N., then the gallium must have a +3 oxidation state so that the sum of O.N.'s equals zero: 1(+3) + 3(–1) = 0.
c) **H = +1 P = +5 O = –2** The hydrogen phosphate ion is HPO_4^{2-}. Again, oxygen has a –2 O.N. Hydrogen has a +1 O.N. because it is combined with nonmetals. The sum of the O.N.'s must equal the ionic charge, so the following algebraic equation can be solved for P: 1(+1) + 1(P) + 4(–2) = –2; O.N. for P = +5.
d) **I = +3 F = –1** The formula of iodine trifluoride is IF_3. In all compounds, fluorine has a –1 O.N., so fluorine is often a good starting point. If each fluorine atom has a –1 O.N., then the iodine must have a +3 oxidation state so that the sum of O.N.'s equal zero: 1(+3) + 3(–1) = 0.

4.8   Plan: Apply Table 4.3 to assign oxidation numbers to all of the atoms. If the oxidation numbers of any of the atoms do not change, the reaction is not a redox reaction. If the oxidation numbers of some atoms do change, the reaction is a redox reaction.
Solution:
a) +3 –3   +2 –2   –3 +3
     –1   +1   +1   +1 –2 +1
   $NCl_3(l) + 3\ H_2O(l) \rightarrow NH_3(aq) + 3\ HOCl(aq)$
N in NCl_3 reduces from +3 to –3 in NH_3; Cl in NCl_3 oxidizes from –1 to +1 in HOCl. This is a redox reaction.
b) +1 +5 –6   –3 +4 –1   –3 +4 +5 –6
        –2   +1   +1 –1   +1   –2
   $AgNO_3(aq) + NH_4I(aq) \rightarrow AgI(s) + NH_4NO_3(aq)$
This is not a redox reaction. None of the atoms have a change in oxidation number. This is a precipitation reaction.
c)   +2 –2   +4 –4   +2 –2
       +1   0   –2   +1
   $2\ H_2S(g) + 3\ O_2(g) \rightarrow 2\ SO_2(g) + 2\ H_2O(g)$
S in H_2S oxidizes from –2 to +4 in SO_2; O in O_2 reduces from 0 to –2 in SO_2 and H_2O. This is a redox reaction.

4.9   Plan: Use Table 4.3 and assign O.N.'s to each element. A reactant containing an element that is oxidized (increasing O.N.) is the reducing agent, and a reactant containing an element that is reduced (decreasing O.N.) is the oxidizing agent. (The O.N.'s are placed under the symbols of the appropriate atoms.)
Solution:
a) $2\ Fe(s) + 3\ Cl_2(g) \rightarrow 2\ FeCl_3(s)$
   O.N.:   Fe = 0   Cl = 0   Fe = +3
                                              Cl = –1
         Fe is oxidized from 0 to +3; **Fe is the reducing agent.**
         Cl is reduced from 0 to –1; **Cl_2 is the oxidizing agent.**

b) $2 C_2H_6(g) + 7 O_2(g) \rightarrow 4 CO_2(g) + 6 H_2O(l)$
   O.N.:   C = –3   O = 0   C = +4   H = +1
              H = +1   O = –2   O = –2
              C is oxidized from –3 to +4; **C_2H_6 is the reducing agent.**
              O is reduced from 0 to –2; **O_2 is the oxidizing agent.**
              H remains +1
c) $5 CO(g) + I_2O_5(s) \rightarrow I_2(s) + 5 CO_2(g)$
   O.N.:   C = +2   I = +5   I = 0   C = +4
              O = –2   O = –2   O = –2
              C is oxidized from +2 to +4; **CO is the reducing agent.**
              I is reduced from +5 to 0; **I_2O_5 is the oxidizing agent.**
              O remains –2

4.10   Plan: Use Table 4.3 to assign oxidation numbers to all atoms. Identify the oxidized and reduces species and compare the electrons lost with the electrons gained. Multiply by factors if necessary so that the electron loss equals the electron gain. Complete the balancing by inspection.
   Solution:

$$
\begin{array}{cccc}
+2\ +12\ -14 & & +3\ -3 & +2\ -2 \\
+1\ +6\ \ -2 & +1\ -1 \quad +1\ -1 & -1 \qquad 0 & +1
\end{array}
$$

$K_2Cr_2O_7(aq) + HI(aq) \rightarrow KI(aq) + CrI_3(aq) + I_2(s) + H_2O(l)$

Cr in $K_2Cr_2O_7$ decreased in oxidation number of +6 to +3 in CrI_3. Two Cr atoms have been reduced and each Cr has lost 3 electrons for a total electron loss of 6 electrons. Note: a coefficient of 2 will be needed in front of CrI_3 to balance the Cr atoms.
I in HI has increased in oxidation number from –1 to 0 in I_2 as each of the 2 I atoms gains one electron for a total gain in 2 electrons. Note: a coefficient of 2 will be needed in front of HI to balance the I atoms.
Cr lost 6 electrons and I gained 2 electrons so the 2 electrons gained by I should be multiplied by 2. The coefficients in front of HI and I_2 are doubled.

$K_2Cr_2O_7(aq) + 2 HI(aq) \rightarrow KI(aq) + 2 CrI_3(aq) + I_2(s) + H_2O(l)$
$K_2Cr_2O_7(aq) + 4 HI(aq) \rightarrow KI(aq) + 2 CrI_3(aq) + 2 I_2(s) + H_2O(l)$

Finish balancing by inspection. A coefficient of 2 will be needed in front of KI to balance K:

$K_2Cr_2O_7(aq) + 4 HI(aq) \rightarrow 2 KI(aq) + 2 CrI_3(aq) + 2 I_2(s) + H_2O(l)$

A coefficient of 7 will be needed in front of H_2O to balance oxygen:

$K_2Cr_2O_7(aq) + 4 HI(aq) \rightarrow 2 KI(aq) + 2 CrI_3(aq) + 2 I_2(s) + 7 H_2O(l)$

The coefficient in front of HI will need to be changed to 14 to balance hydrogen:

$K_2Cr_2O_7(aq) + 14 HI(aq) \rightarrow 2 KI(aq) + 2 CrI_3(aq) + 2 I_2(s) + 7 H_2O(l)$

The coefficient in front of I2 must be changed to 3 to balance iodine:

$K_2Cr_2O_7(aq) + 14 HI(aq) \rightarrow 2 KI(aq) + 2 CrI_3(aq) + 3 I_2(s) + 7 H_2O(l)$

4.11   Plan: Write a balanced reaction. Use the volume and concentration of the $KMnO_4$ solution to determine the moles of Ca^{2+} in the sample using the molar ratio from the reaction. In Part b), moles of Ca will be converted to grams.
   Solution:
$2 KMnO_4(aq) + 5 CaC_2O_4(s) + H_2SO_4(aq) \rightarrow$
$$2 MnSO_4(aq) + K_2SO_4(aq) + 5 CaSO_4(s) + 10 CO_2(g) + 8 H_2O(l)$$

$$\text{Moles of } KMnO_4 = (6.53 \text{ mL})\left(\frac{10^{-3} \text{ L}}{1 \text{ mL}}\right)\left(\frac{4.56 \times 10^{-3} \text{ mol } KMnO_4}{1 \text{ L}}\right) = 2.97768 \times 10^{-5} \text{ mol } KMnO_4$$

$$\text{Molarity of } Ca^{2+} = (2.97768 \times 10^{-5} \text{ mol } KMnO_4)\left(\frac{5 \text{ mol } CaC_2O_4}{2 \text{ mol } KMnO_4}\right)\left(\frac{1 \text{ mol } Ca^{2+}}{1 \text{ mol } CaC_2O_4}\right)\left(\frac{1}{2.50 \text{ mL}}\right)\left(\frac{1 \text{ mL}}{10^{-3} \text{ L}}\right)$$

$$= 2.97768 \times 10^{-2} = 2.98 \times 10^{-2} \text{ mol/L} = \textbf{2.98} \times \textbf{10}^{-2} \textbf{ M } \textbf{Ca}^{2+}$$

b) $\left(\dfrac{2.97768 \times 10^{-2} \text{ mol } Ca^{2+}}{L}\right)\left(\dfrac{40.08 \text{ g } Ca^{2+}}{1 \text{ mol } Ca^{2+}}\right) = 1.193454 = \textbf{1.19 g/L}$

This concentration is consistent with the typical value in milk.

4.12   Plan: To classify a reaction, compare the number of reactants used versus the number of products formed. Also examine the changes, if any, in the oxidation numbers. Recall the definitions of each type of reaction:

Combination: $X + Y \rightarrow Z$;   decomposition: $Z \rightarrow X + Y$

Single displacement: $X + YZ \rightarrow XZ + Y$ double displacement: $WX + YZ \rightarrow WZ + YX$

Solution:

a) **Combination**; $S_8(s) + 16\ F_2(g) \rightarrow 8\ SF_4(g)$

O.N.: $S = 0$ $F = 0$ $S = +4$
                                            $F = -1$

Sulfur changes from 0 to +4 oxidation state; it is oxidized and **S_8 is the reducing agent**.
Fluorine changes from 0 to -1 oxidation state; it is reduced and **F_2 is the oxidizing agent**.

b) **Displacement**; $2\ CsI(aq) + Cl_2(aq) \rightarrow 2\ CsCl(aq) + I_2(aq)$

O.N.: $Cs = +1$ $Cl = 0$ $Cs = +1$ $I = 0$
            $I = -1$ $Cl = -1$

Total ionic eqn: $2\ Cs^+(aq) + 2\ I^-(aq) + Cl_2(aq) \rightarrow 2\ Cs^+(aq) + 2\ Cl^-(aq) + I_2(aq)$
Net ionic eqn: $2\ I^-(aq) + Cl_2(aq) \rightarrow 2\ Cl^-(aq) + I_2(aq)$
Iodine changes from -1 to 0 oxidation state; it is oxidized and **CsI is the reducing agent**. Chlorine changes from 0 to -1 oxidation state; it is reduced and **Cl_2 is the oxidizing agent**.

c) **Displacement**; $3\ Ni(NO_3)_2 + 2\ Cr(s) \rightarrow 2\ Cr(NO_3)_3(aq) + 3\ Ni(s)$

O.N.: $Ni = +2$ $Cr = 0$ $Cr = +3$ $Ni = 0$
            $N = +5$ $N = +5$
            $O = -2$ $O = -2$

Total ionic eqn: $3\ Ni^{+2}(aq) + 6\ NO_3^-(aq) + 2\ Cr(s) \rightarrow 2\ Cr^{+3}(aq) + 6\ NO_3^-(aq) + 3\ Ni(s)$
Net ionic eqn: $3\ Ni^{+2}(aq) + 2\ Cr(s) \rightarrow 2\ Cr^{+3}(aq) + 3\ Ni(s)$
Nickel changes from +2 to 0 oxidation state; it is reduced and **$Ni(NO_3)_2$ is the oxidizing agent**.
Chromium changes from 0 to +3 oxidation state; it is oxidized and **Cr is the reducing agent**.

END–OF–CHAPTER PROBLEMS

4.2   Plan: Review the discussion on water soluble compounds.
       Solution:
       Ionic and polar covalent compounds are most likely to be soluble in water. Because water is polar, the partial charges in water molecules are able to interact with the charges, either ionic or dipole-induced, in other substances.

4.3   Plan: Review the definition of electrolytes.
       Solution:
       Ions must be present in an aqueous solution for it to conduct an electric current. Ions come from ionic compounds or from other electrolytes such as acids and bases.

4.6   Plan: Write the formula for magnesium nitrate and note the ratio of magnesium ions to nitrate ions.
       Solution:
       The box in (**2**) best represents a volume of magnesium nitrate solution. Upon dissolving the salt in water, magnesium nitrate, $Mg(NO_3)_2$, would dissociate to form one Mg^{2+} ion for every two NO_3^- ions, thus forming twice as many nitrate ions. Only box (2) has twice as many nitrate ions (red circles) as magnesium ions (blue circles).

4.10   Plan: Compounds that are soluble in water tend to be ionic compounds or covalent compounds that have polar bonds.
       Solution:
       a) Benzene is likely to be insoluble in water because it is non-polar and water is polar.
       b) Sodium hydroxide (NaOH), an ionic compound, is likely to be soluble in water since the ions from sodium hydroxide will be held in solution through ion-dipole attractions with water.
       c) Ethanol (CH_3CH_2OH) will likely be soluble in water because the alcohol group ($-OH$) will hydrogen bond with the water.

d) Potassium acetate ($KC_2H_3O_2$), an ionic compound, will likely dissolve in water to form sodium ions and acetate ions that are held in solution through ion-dipole attractions to water.

4.12   Plan: Substances whose aqueous solutions conduct an electric current are electrolytes such as ionic compounds and acids and bases.
Solution:
a) An aqueous solution that contains ions conducts electricity. CsBr is a soluble ionic compound, and a solution of this salt in water contains Cs^+ and Br^- ions. Its solution conducts electricity.
b) HI is a strong acid that dissociates completely in water. Its aqueous solution contains H^+ and I^- ions, so it conducts electricity.

4.14   Plan: To determine the total moles of ions released, write a dissolution equation showing the correct molar ratios, and convert the given amounts to moles if necessary.
Solution:
a) Each mole of NH_4Cl dissolves in water to form 1 mole of NH_4^+ ions and 1 mole of Cl^- ions, or a total of 2 moles of ions.

$$\left(0.32 \text{ mol } NH_4Cl\right)\left(\frac{2 \text{ mol ions}}{1 \text{ mol } NH_4Cl}\right) = \textbf{0.64 mol of ions}$$

b) Each mole of $Ba(OH)_2 \cdot 8H_2O$ forms 1 mole of barium ions (Ba^{2+}) and 2 mol of hydroxide ions (OH^-), or a total of 3 mol of ions.

$$\left(25.4 \text{ g } Ba(OH)_2 \cdot 8H_2O\right)\left(\frac{1 \text{ mol } Ba(OH)_2 \cdot 8H_2O}{315.4 \text{ g } Ba(OH)_2 \cdot 8H_2O}\right)\left(\frac{3 \text{ mol ions}}{1 \text{ mol } Ba(OH)_2 \cdot 8H_2O}\right)$$
$$= 0.2415980 = \textbf{0.242 mol of ions}$$

c) Each mole of LiCl produces 2 moles of ions (1 mole of Li^+ ions and 1 mole of Cl^- ions).

$$\left(3.55 \times 10^{19} \text{ FU LiCl}\right)\left(\frac{1 \text{ mol LiCl}}{6.022 \times 10^{23} \text{ FU LiCl}}\right)\left(\frac{2 \text{ mol ions}}{1 \text{ mol LiCl}}\right)$$
$$= 1.17901 \times 10^{-4} = \textbf{1.18} \times \textbf{10}^{-4} \textbf{ mol of ions}$$

4.16   Plan: To determine the total moles of ions released, write a dissolution equation showing the correct molar ratios, and convert the given amounts to moles if necessary.
Solution:
a) Recall that phosphate, PO_4^{3-}, is a polyatomic anion and does not dissociate further in water.
$K_3PO_4(s) \rightarrow 3\ K^+(aq) + PO_4^{3-}(aq)$
4 moles of ions are released when one mole of K_3PO_4 dissolves, so the total number of moles released is

$$\left(0.75 \text{ mol } K_3PO_4\right)\left(\frac{4 \text{ mol ions}}{1 \text{ mol } K_3PO_4}\right) = \textbf{3.0 mol of ions}$$

b) $NiBr_2 \cdot 3H_2O(s) \rightarrow Ni^{2+}(aq) + 2\ Br^-(aq)$
Three moles of ions are released when 1 mole of $NiBr_2 \cdot 3H_2O$ dissolves.
The waters of hydration become part of the larger bulk of water. Convert the grams of $NiBr_2 \cdot 3H_2O$ to moles using the molar mass (be sure to include the mass of the water):

$$\left(6.88 \times 10^{-3} \text{ g } NiBr_2 \cdot 3H_2O\right)\left(\frac{1 \text{ mol } NiBr_2 \cdot 3H_2O}{272.54 \text{ g } NiBr_2 \cdot 3H_2O}\right)\left(\frac{3 \text{ mol ions}}{1 \text{ mol } NiBr_2 \cdot 3H_2O}\right)$$
$$= 7.5732 \times 10^{-5} = \textbf{7.57} \times \textbf{10}^{-5} \textbf{ mol of ions}$$

c) $FeCl_3(s) \rightarrow Fe^{3+}(aq) + 3\ Cl^-(aq)$

Recall that a mole contains 6.022×10^{23} entities, so a mole of $FeCl_3$ contains 6.022×10^{23} units of $FeCl_3$, (more easily expressed as formula units). Since the problem specifies only 2.23×10^{22} formula units, we know that the amount is some fraction of a mole. 4 moles of ions are released when one mole of $FeCl_3$ dissolves.

$$\left(2.23 \times 10^{22}\ FU\ FeCl_3\right)\left(\frac{1\ mol\ FeCl_3}{6.022 \times 10^{23}\ FU\ FeCl_3}\right)\left(\frac{4\ mol\ ions}{1\ mol\ FeCl_3}\right)$$

$$= 0.148124 = \textbf{0.148 mol of ions}$$

4.18   Plan: To determine the moles of each type of ion released, write a dissolution equation showing the correct molar ratios, and convert the given amounts to moles if necessary. Use Avogadro's number to convert moles of ions to number of ions.
Solution:
a) $AlCl_3(s) \rightarrow Al^{3+}(aq) + 3\ Cl^-(aq)$ 1 mol of Al^{3+} and 3 mol of Cl^- per mole of $AlCl_3$ dissolved

$$\text{Moles } Al^{3+} = \left(130.\ mL\right)\left(\frac{10^{-3}\ L}{1\ mL}\right)\left(\frac{0.45\ mol\ AlCl_3}{L}\right)\left(\frac{1\ mol\ Al^{3+}}{1\ mol\ AlCl_3}\right) = 0.0585 = \textbf{0.058 mol } Al^{3+}$$

$$Al^{3+}\text{ ions} = \left(0.0585\ mol\ Al^{3+}\right)\left(\frac{6.022 \times 10^{23}\ Al^{3+}}{1\ mol\ Al^{3+}}\right) = 3.52287 \times 10^{22} = \textbf{3.5} \times \textbf{10}^{\textbf{22}}\ Al^{3+}\textbf{ ions}$$

$$\text{Moles } Cl^- = \left(130.\ mL\right)\left(\frac{10^{-3}\ L}{1\ mL}\right)\left(\frac{0.45\ mol\ AlCl_3}{L}\right)\left(\frac{3\ mol\ Cl^-}{1\ mol\ AlCl_3}\right) = 0.1755 = \textbf{0.18 mol } Cl^-$$

$$Cl^-\text{ ions} = \left(0.1755\ mol\ Cl^-\right)\left(\frac{6.022 \times 10^{23}\ Cl^-}{1\ mol\ Cl^-}\right) = 1.05686 \times 10^{23} = \textbf{1.1} \times \textbf{10}^{\textbf{23}}\ Cl^-\textbf{ ions}$$

b) $Li_2SO_4(s) \rightarrow 2\ Li^+(aq) + SO_4^{2-}(aq)$ 2 mol of Li^+ and 1 mol of SO_4^{2-} per mol of Li_2SO_4 dissolved

$$\text{Moles } Li^+ = \left(9.80\ mL\right)\left(\frac{10^{-3}\ L}{1\ mL}\right)\left(\frac{2.59\ g\ Li_2SO_4}{1\ L}\right)\left(\frac{1\ mol\ Li_2SO_4}{109.95\ g\ Li_2SO_4}\right)\left(\frac{2\ mol\ Li^+}{1\ mol\ Li_2SO_4}\right)$$

$$= 4.6170 \times 10^{-4} = \textbf{4.62} \times \textbf{10}^{-\textbf{4}}\textbf{ mol } Li^+$$

$$Li^+\text{ ions} = \left(4.6170 \times 10^{-4}\ mol\ Li^+\right)\left(\frac{6.022 \times 10^{23}\ Li^+}{1\ mol\ Li^+}\right) = 2.7804 \times 10^{20} = \textbf{2.78} \times \textbf{10}^{\textbf{20}}\ Li^+\textbf{ ions}$$

$$\text{Moles } SO_4^{2-} = \left(9.80\ mL\right)\left(\frac{10^{-3}\ L}{1\ mL}\right)\left(\frac{2.59\ g\ Li_2SO_4}{1\ L}\right)\left(\frac{1\ mol\ Li_2SO_4}{109.95\ g\ Li_2SO_4}\right)\left(\frac{1\ mol\ SO_4^{2-}}{1\ mol\ Li_2SO_4}\right)$$

$$= 2.3085 \times 10^{-4} = \textbf{2.31} \times \textbf{10}^{-\textbf{4}}\textbf{ mol } SO_4^{2-}$$

$$SO_4^{2-}\text{ ions} = \left(2.3085 \times 10^{-4}\ mol\ SO_4^{2-}\right)\left(\frac{6.022 \times 10^{23}\ SO_4^{2-}}{1\ mol\ SO_4^{2-}}\right) = 1.39018 \times 10^{20} = \textbf{1.39} \times \textbf{10}^{\textbf{20}}\ SO_4^{2-}\textbf{ ions}$$

c) $KBr(s) \rightarrow K^+(aq) + Br^-(aq)$ 1 mol of Li^+ and 1 mol of Br^- per mole of KBr dissolved

$$K^+\text{ ions} = \left(245\ mL\right)\left(\frac{10^{-3}\ L}{1\ mL}\right)\left(\frac{3.68 \times 10^{22}\ FU\ KBr}{L}\right)\left(\frac{1\ K^+}{1\ FU\ KBr}\right) = 9.016 \times 10^{21} = \textbf{9.02} \times \textbf{10}^{\textbf{21}}\ K^+\textbf{ ions}$$

$$\text{Moles } K^+ = \left(9.016 \times 10^{21}\ K^+\right)\left(\frac{1\ mol\ K^+}{6.022 \times 10^{23}\ K^+}\right) = 1.49718 \times 10^{-2} = \textbf{1.50} \times \textbf{10}^{-\textbf{2}}\textbf{ mol } K^+$$

$$Br^-\text{ ions} = \left(245\ mL\right)\left(\frac{10^{-3}\ L}{1\ mL}\right)\left(\frac{3.68 \times 10^{22}\ FU\ KBr}{L}\right)\left(\frac{1\ Br^-}{1\ FU\ KBr}\right) = 9.016 \times 10^{21} = \textbf{9.02} \times \textbf{10}^{\textbf{21}}\ Br^-\textbf{ ions}$$

$$\text{Moles Br}^- = \left(9.016 \times 10^{21} \text{ Br}^-\right)\left(\frac{1 \text{ mol Br}^-}{6.022 \times 10^{23} \text{ Br}^-}\right) = 1.49718 \times 10^{-2} = \mathbf{1.50 \times 10^{-2} \text{ mol Br}^-}$$

4.20   Plan: The acids in this problem are all strong acids, so you can assume that all acid molecules dissociate completely to yield H^+ ions and associated anions. One mole of $HClO_4$, HNO_3 and HCl each produce one mole of H^+ upon dissociation, so moles H^+ = moles acid. Molarity is expressed as moles/L instead of as M.
Solution:
a) $HClO_4(aq) \rightarrow H^+(aq) + ClO_4^-(aq)$

$$\text{Moles } H^+ = \text{mol } HClO_4 = \left(1.40 \text{ L}\right)\left(\frac{0.25 \text{ mol}}{1 \text{ L}}\right) = \mathbf{0.35 \text{ mol } H^+}$$

b) $HNO_3(aq) \rightarrow H^+(aq) + NO_3^-(aq)$

$$\text{Moles } H^+ = \text{mol } HNO_3 = \left(6.8 \text{ mL}\right)\left(\frac{10^{-3} \text{ L}}{1 \text{ mL}}\right)\left(\frac{0.92 \text{ mol}}{1 \text{ L}}\right) = 6.256 \times 10^{-3} = \mathbf{6.3 \times 10^{-3} \text{ mol } H^+}$$

c) $HCl(aq) \rightarrow H^+(aq) + Cl^-(aq)$

$$\text{Moles } H^+ = \text{mol } HCl = \left(2.6 \text{ L}\right)\left(\frac{0.085 \text{ mol}}{1 \text{ L}}\right) = 0.221 = \mathbf{0.22 \text{ mol } H^+}$$

4.24   Plan: Review the definition of spectator ions.
Solution:
Ions in solution that do not participate in the reaction do not appear in a net ionic equation. These spectator ions remain as dissolved ions throughout the reaction. These ions are only present to balance charge.

4.28   Plan: Use Table 4.1 to predict the products of this reaction. Ions not involved in the precipitate are spectator ions and are not included in the net ionic equation.
Solution:
Assuming that the left beaker is $AgNO_3$ (because it has a gray Ag^+ ion) and the right must be $NaCl$, then the NO_3^- is blue, the Na^+ is brown, and the Cl^- is green. (Cl^- must be green since it is present with Ag+ in the precipitate in the beaker on the right.)
Molecular equation: $AgNO_3(aq) + NaCl(aq) \rightarrow AgCl(s) + NaNO_3(aq)$
Total ionic equation: $Ag^+(aq) + NO_3^-(aq) + Na^+(aq) + Cl^-(aq) \rightarrow AgCl(s) + Na^+(aq) + NO_3^-(aq)$
Net ionic equation: $Ag^+(aq) + Cl^-(aq) \rightarrow AgCl(s)$

4.29   Plan: Check to see if any of the ion pairs are not soluble according to the solubility rules in Table 4.1 Ions not involved in the precipitate are spectator ions and are omitted from the net ionic equation.
a) Molecular: $Hg_2(NO_3)_2(aq) + 2 KI(aq) \rightarrow Hg_2I_2(s) + 2 KNO_3(aq)$
      Total ionic: $Hg_2^{2+}(aq) + 2 NO_3^-(aq) + 2 K^+(aq) + 2 I^-(aq) \rightarrow Hg_2I_2(s) + 2 K^+(aq) + 2 NO_3^-(aq)$
      Net ionic: $Hg_2^{2+}(aq) + 2 I^-(aq) \rightarrow Hg_2I_2(s)$
      Spectator ions are K^+ and NO_3^-.
b) Molecular: $FeSO_4(aq) + Sr(OH)_2(aq) \rightarrow Fe(OH)_2(s) + SrSO_4(s)$
      Total ionic: $Fe^{2+}(aq) + SO_4^{2-}(aq) + Sr^{2+}(aq) + 2 OH^-(aq) \rightarrow Fe(OH)_2(s) + SrSO_4(s)$
      Net ionic: This is the same as the total ionic equation, because there are no spectator ions.

4.31   Plan: A precipitate forms if reactant ions can form combinations that are insoluble, as determined by the solubility rules in Table 4.1. Create cation-anion combinations other than the original reactants and determine if they are insoluble. Any ions not involved in a precipitate are spectator ions and are omitted from the net ionic equation.
Solution:
a) $NaNO_3(aq) + CuSO_4(aq) \rightarrow Na_2SO_4(aq) + Cu(NO_3)_2(aq)$
No precipitate will form. The ions Na^+ and SO_4^{2-} will not form an insoluble salt according to the first solubility rule which states that *all common compounds of Group 1A ions are soluble*. The ions Cu^{2+} and NO_3^- will not form an insoluble salt according to the solubility rule #2: *All common nitrates are soluble*. There is no reaction.

b) A precipitate will form because silver ions, Ag^+, and bromide ions, Br^-, will combine to form a solid salt, silver bromide, AgBr. The ammonium and nitrate ions do not form a precipitate.

Molecular: $NH_4Br(aq) + AgNO_3(aq) \rightarrow AgBr(s) + NH_4NO_3(aq)$

Total ionic: $NH_4^+(aq) + Br^-(aq) + Ag^+(aq) + NO_3^-(aq) \rightarrow AgBr(s) + NH_4^+(aq) + NO_3^-(aq)$

Net ionic: $Ag^+(aq) + Br^-(aq) \rightarrow AgBr(s)$

4.33   <u>Plan:</u> A precipitate forms if reactant ions can form combinations that are insoluble, as determined by the solubility rules in Table 4.1. Create cation-anion combinations other than the original reactants and determine if they are insoluble.
<u>Solution:</u>
a) New cation-anion combinations are potassium nitrate (KNO_3) and iron(III) chloride ($FeCl_3$). The solubility rules state that all common nitrates and chlorides (with some exceptions) are soluble, so no precipitate forms.

$3 KCl(aq) + Fe(NO_3)_3(aq) \rightarrow 3 KNO_3(aq) + FeCl_3(aq)$

b) New cation-anion combinations are ammonium chloride and barium sulfate. Again, the rules state that most chlorides are soluble; however, another rule states that sulfate compounds containing barium are insoluble. Barium sulfate is a precipitate and its formula is $BaSO_4$.

Molecular: $(NH_4)_2SO_4(aq) + BaCl_2(aq) \rightarrow BaSO_4(s) + 2 NH_4Cl(aq)$

Total ionic: $2 NH_4^+(aq) + SO_4^{2-}(aq) + Ba^{2+}(aq) + 2 Cl^-(aq) \rightarrow BaSO_4(s) + 2 NH_4^+(aq) + Cl^-(aq)$

Net ionic: $Ba^{2+}(aq) + SO_4^{2-}(aq) \rightarrow BaSO_4(s)$

4.35   <u>Plan:</u> Write a balanced equation for the chemical reaction described in the problem. By applying the solubility rules to the two possible products ($NaNO_3$ and PbI_2), determine that PbI_2 is the precipitate. By using molar relationships, determine how many moles of $Pb(NO_3)_2$ are required to produce 0.628 g of PbI_2.
<u>Solution:</u>
The reaction is: $Pb(NO_3)_2(aq) + 2 NaI(aq) \rightarrow PbI_2(s) + 2 NaNO_3(aq)$.

$$M\ Pb^{2+} = \left(0.628\ g\ PbI_2\right)\left(\frac{1\ mol\ PbI_2}{461.0\ g\ PbI_2}\right)\left(\frac{1\ mol\ Pb(NO_3)_2}{1\ mol\ PbI_2}\right)\left(\frac{1}{38.5\ mL}\right)\left(\frac{1\ mL}{10^{-3}\ L}\right)$$

$$= 0.035383 = \mathbf{0.0354\ \textit{M}\ Pb^{2+}}$$

4.37   a) The blue spheres are Pb^{2+} and the yellow spheres are SO_4^{2-}. The precipitate is thus $PbSO_4$.
b) $Pb^{2+}(aq) + SO_4^{2-}(aq) \rightarrow PbSO_4(s)$

$$c)\ Mass\ PbSO_4 = \left(10\ Pb^{2+}\ spheres\right)\left(\frac{5.0\ x\ 10^{-4}\ mol\ Pb^{2+}}{1\ Pb^{2+}\ sphere}\right)\left(\frac{1\ mol\ PbSO_4}{1\ mol\ Pb^{2+}}\right)\left(\frac{303.3\ g\ PbSO_4}{1\ mol\ PbSO_4}\right)$$

$$= 1.5165 = \mathbf{1.5\ g\ PbSO_4}$$

4.39   <u>Plan:</u> The balanced equation for this reaction is $AgNO_3(aq) + Cl^-(aq) \rightarrow AgCl(s) + NO_3^-(aq)$.
First, find the moles of $AgNO_3$ and thus the moles of Cl^- present in the 25.00 mL sample by using the molar ratio in the balanced equation. Second, convert moles of Cl^- into grams, and convert the sample volume into grams using the given density. The mass percent of Cl^- is found by dividing the mass of Cl^- by the mass of the sample volume and multiplying by 100.
<u>Solution:</u>

$$\left(53.63\ mL\right)\left(\frac{10^{-3}\ L}{1\ mL}\right)\left(\frac{0.2970\ mol\ AgNO_3}{L}\right)\left(\frac{1\ mol\ Cl^-}{1\ mol\ AgNO_3}\right)\left(\frac{35.45\ g\ Cl}{1\ mol\ Cl^-}\right) = 0.56465\ g\ Cl\ (unrounded)$$

$$Mass\ of\ sample = \left(25.00\ mL\right)\left(\frac{1.024\ g}{mL}\right) = 25.60\ g\ sample$$

$$Mass\ \%\ Cl = \frac{Mass\ Cl}{Mass\ Sample}\ x\ 100\% = \frac{0.56465\ g\ Cl}{25.6\ g\ Sample}\ x\ 100\% = 2.20566 = \mathbf{2.206\%\ Cl}$$

4.45   a) The formation of a gas, $SO_2(g)$, and formation of water drive this reaction to completion, because both products remove reactants from solution.
b) The formation of a precipitate, $Ba_3(PO_4)_2(s)$, will cause this reaction to go to completion. This reaction is one between an acid and a base, so the formation of water molecules through the combination of H^+ and OH^- ions also drives the reaction.

4.47   Plan: Remember that strong acids and bases can be written as ions in the total ionic equation but weak acids and bases cannot be written as ions. Omit spectator ions from the net ionic equation.
Solution:
a) KOH is a strong base and HBr is a strong acid; both may be written in dissociated form. KBr is a soluble compound since all Group 1A(1) compounds are soluble.
Molecular equation: $KOH(aq) + HBr(aq) \rightarrow KBr(aq) + H_2O(l)$
Total ionic equation: $K^+(aq) + OH^-(aq) + H^+(aq) + Br^-(aq) \rightarrow K^+(aq) + Br^-(aq) + H_2O(l)$
Net ionic equation: $OH^-(aq) + H^+(aq) \rightarrow H_2O(l)$
The spectator ions are $K^+(aq)$ and $Br^-(aq)$
b) NH_3 is a weak base and is written in the molecular form. HCl is a strong acid and is written in the dissociated form (as ions). NH_4Cl is a soluble compound, because all ammonium compounds are soluble.
Molecular equation: $NH_3(aq) + HCl(aq) \rightarrow NH_4Cl(aq)$
Total ionic equation: $NH_3(aq) + H^+(aq) + Cl^-(aq) \rightarrow NH_4^+(aq) + Cl^-(aq)$
Net ionic equation: $NH_3(aq) + H^+(aq) \rightarrow NH_4^+(aq)$
Cl^- is the only spectator ion.

4.49   Plan: Write an acid-base reaction between $CaCO_3$ and HCl.
Solution:
Calcium carbonate dissolves in $HCl(aq)$ because the carbonate ion, a base, reacts with the acid to form H_2CO_3 which decomposes into $CO_2(g)$ and $H_2O(l)$.
$$CaCO_3(s) + 2HCl(aq) \rightarrow CaCl_2(aq) + H_2CO_3(aq)$$
Total ionic equation:
$$CaCO_3(s) + 2H^+(aq) + 2\ Cl^-(aq) \rightarrow Ca^{2+}(aq) + 2\ Cl^-(aq) + H_2O(l) + CO_2(g)$$
Net ionic equation:
$$CaCO_3(s) + 2H^+(aq) \rightarrow Ca^{2+}(aq) + H_2O(l) + CO_2(g)$$

4.51   Plan: Write a balanced equation and use the molar ratios to convert the amount of KOH to the amount of CH_3COOH.
Solution:
The reaction is: $KOH(aq) + CH_3COOH(aq) \rightarrow KCH_3COO(aq) + H_2O(l)$
$$M = \left(\frac{0.1180 \text{ mol KOH}}{L}\right)\left(\frac{10^{-3} \text{ L}}{1 \text{ mL}}\right)(25.98 \text{ mL})\left(\frac{1 \text{ mol } CH_3COOH}{1 \text{ mol KOH}}\right)\left(\frac{1}{52.50 \text{ mL}}\right)\left(\frac{1 \text{ mL}}{10^{-3} \text{ L}}\right)$$
$$= 0.05839314 = \textbf{0.05839 } \textbf{\textit{M}} \textbf{ CH}_3\textbf{COOH}$$

4.62   Plan: An oxidizing agent has an atom whose oxidation number decreases during the reaction.
Solution:
a) The S in SO_4^{2-} (i.e., H_2SO_4) has O.N. = +6, and in SO_2, O.N. (s) = +4, so the S has been reduced (and the I^- oxidized), so the H_2SO_4 is an oxidizing agent.
b) The oxidation numbers remain constant throughout; H_2SO_4 transfers a proton to F^- to produce HF, so it acts as an acid.

4.64   Refer to the rules in Table 4.3. Remember that oxidation number (O.N.) is not the same thing as ionic charge. The O.N. is the charge an atom would have **if** electrons were transferred completely. The compounds in this problem contain covalent bonds, or a combination of ionic and covalent bonds ($Na_2C_2O_4$). Covalent bonds do not involve the complete transfer of electrons.
a) CF_2Cl_2. The rules dictate that F and Cl each have an O.N. of –1; two F and two Cl yield a sum of –4, so C must have a +4 O.N. **C = +4**

b) $Na_2C_2O_4$. The rule dictates that Na has a +1 O.N.; rule 5 dictates that O has a –2 O.N.; two Na and four O yield a sum of –6 [(+2) + (–8)]. Therefore, the total of the O.N.'s on the two C atoms is +6 and each C is +3. **C = +3**

c) HCO_3^-. H is combined with nonmetals and has an O.N. of +1; O has an O.N. of –2 and three O have a sum of –6. To have an overall oxidation state equal to –1, C must be +4 because (+1) + (+4) + (–6) = –1. **C = +4**

d) C_2H_6. Each H has an O.N. of +1; six H gives +6. The sum of O.N.'s for the two C atoms must be –6, so each C is –3. **C = –3**

4.66   <u>Plan:</u> Consult Table 4.3 for the rules for assigning oxidation numbers.
    <u>Solution:</u>

a) NH_2OH:	(O.N. for N) + 3(+1 for H) + 1(–2 for O) = 0	O.N. for N = **–1**
b) N_2F_4:	2(O.N. for N) + 4(–1 for F) = 0	O.N. for N = **+2**
c) NH_4^+:	(O.N. for N) + 4(+1 for H) = +1	O.N. for N = **–3**
d) HNO_2:	(O.N. for N) + 1(+1 for H) + 2(–2 for O) = 0	O.N. for N = **+3**

4.68   <u>Plan:</u> Consult Table 4.3 for the rules for assigning oxidation numbers.
    <u>Solution:</u>
a) AsH_3. H is combined with a nonmetal, so its O.N. is +1 (Rule 3). Three H have a sum of +3. To have a sum of 0 for the molecule, the O.N. for As is **–3**.
b) $H_2AsO_4^-$. The O.N. of H in this compound is +1, or +2 for 2 H's. Oxygen's O.N. is –2, with total O.N. of –8 (4 times –2), so As needs to have an O.N. of **+5**: +2 + (+5) + (–8) = –1.
c) $AsCl_3$. Cl has an O.N. of –1, total of –3, so As must have an O.N. of **+3**.

4.70   <u>Plan:</u> Consult Table 4.3 for the rules for assigning oxidation numbers.
    <u>Solution:</u>

a) MnO_4^{2-}:	(O.N. for Mn) + 4(–2 for O) = –2	O.N. for Mn = **+6**
b) Mn_2O_3:	{2(O.N. for Mn)} + 3(–2 for O) = 0	O.N. for Mn = **+3**
c) $KMnO_4$:	1(+1 for K) + (O.N. for Mn) + 4(–2 for O) = 0	O.N. for Mn = **+7**

4.72   <u>Plan:</u> Oxidizing agent: substance that accepts the electrons released by the substance that is oxidized; the oxidation number of the atom accepting electrons decreases. The oxidizing agent undergoes reduction. Reducing agent: substance that provides the electrons accepted by the substance that is reduced; the oxidation number of the atom providing the electrons increases. The reducing agent undergoes oxidation. First, assign oxidation numbers to all atoms. Second, recognize that the agent is the compound that contains the atom that is gaining or losing electrons, not just the atom itself.
    <u>Solution:</u>
a) $5 H_2C_2O_4(aq) + 2 MnO_4^-(aq) + 6 H^+(aq) \rightarrow 2 Mn^{2+}(aq) + 10 CO_2(g) + 8 H_2O(l)$

H = +1	Mn = +7	H = +1	Mn = +2	C = +4	H = +1
C = +3	O = –2			O = –2	O = –2
O = –2					

Hydrogen and oxygen do not change oxidation state. The Mn changes from +7 to +2 (reduction). Therefore, **MnO_4^- is the oxidizing agent**. C changes from +3 to +4 (oxidation), so **$H_2C_2O_4$ is the reducing agent**.
b) $3 Cu(s) + 8 H^+(aq) + 2 NO_3^-(aq) \rightarrow 3 Cu^{2+}(aq) + 2 NO(g) + 4 H_2O(l)$

Cu = 0	H = +1	N = +5	Cu = +2	N = +2	H = +1
		O = –2		O = –2	O = –2

Cu changes from 0 to +2 (is oxidized) and **Cu is the reducing agent**. N changes from +5 (in NO_3^-) to +2 (in NO) and is reduced, so **NO_3^- is the oxidizing agent**.

4.74   <u>Plan:</u> Oxidizing agent: substance that accepts the electrons released by the substance that is oxidized; the oxidation number of the atom accepting electrons decreases. The oxidizing agent undergoes reduction. Reducing agent: substance that provides the electrons accepted by the substance that is reduced; the oxidation number of the atom providing the electrons increases. The reducing agent undergoes oxidation. First, assign oxidation numbers to all atoms. Second, recognize that the agent is the compound that contains the atom that is gaining or losing electrons, not just the atom itself.

Solution:

a) $8H^+(aq) + 6Cl^-(aq) + Sn(s) + 4NO_3^-(aq) \rightarrow SnCl_6^{2-}(aq) + 4NO_2(g) + 4H_2O(l)$

   $H^+ = +1$ $Cl^- = -1$ $Sn = 0$ $N = +5$ $Sn = +4$ $N = +4$ $H = +1$

                                 $O = -2$ $Cl = -1$ $O = -2$ $O = -2$

Oxidizing agent is NO_3^- because nitrogen changes from +5 O.N. in NO_3^- to +4 O.N. in NO_2. **Reducing agent is Sn** because its O.N. changes from 0 as the element to +4 in $SnCl_6^{2-}$.

b) $2MnO_4^-(aq) + 10Cl^-(aq) + 16H^+(aq) \rightarrow 5Cl_2(g) + 2Mn^{2+}(aq) + 8H_2O(l)$

   $Mn = +7$ $Cl^- = -1$ $H^+ = +1$ $Cl = 0$ $Mn^{2+} = +2$ $H = +1$

   $O = -2$ $O = -2$

Oxidizing agent is MnO_4^- because manganese changes from +7 O.N. in MnO_4^- to +2 O.N. in Mn^{2+}. **Reducing agent is Cl^-** because its O.N. changes from –1 in Cl^- to 0 as the element to Cl_2.

4.76   Plan: S is in Group 6A (16), so its highest possible O.N. is +6 and its lowest possible O.N. is $6 - 8 = -2$. Remember that a reducing agent has an atom whose O.N. increases while an oxidizing agent has an atom whose O.N. decreases.

Solution:

a) The lowest O.N. for S [(Group 6A(16)] is $6 - 8 = -2$, which occurs in S^{2-}. Therefore, when S^{2-} reacts in an oxidation-reduction reaction, S can only increase its O.N. (oxidize), so S^{2-} can only function as a reducing agent.

b) The highest O.N. for S [(Group 6A(16)] is +6, which occurs in SO_4^{2-}. Therefore, when SO_4^{2-} reacts in an oxidation-reduction reaction, the S can only decrease its O.N. (reduce), so SO_4^{2-} can only function as an oxidizing agent.

c) The O.N. of S in SO_2 is +4, so it can increase or decrease its O.N. Therefore, SO_2 can function as either an oxidizing or reducing agent.

4.78   Plan: Follow these steps:

Step: 1: Assign oxidation numbers to all elements.
Step: 2: Identify oxidized and reduced species.
Step: 3: Compute number of electrons lost in oxidation and gained in reduction.
Step: 4: Multiply by factors to make e^- lost equal e^- gained.
Step: 5: Finish balancing by inspection.

Solution:

a)

     –6   +2   –8   +10 –12   –6   +15 –18   +15 –18   +2

+1 –5 –2   +1 +6 –2   +2 +5 –2   +1 +5 –2   +3 +5 –2   +3 +5 –2   +1 –2

   $HNO_3(aq) + K_2CrO_4(aq) + Fe(NO_3)_2(aq) \rightarrow KNO_3(aq) + Fe(NO_3)_3(aq) + Cr(NO_3)_3(aq) + H_2O(l)$

The O.N. of Cr decreases from +6 in K_2CrO_4 to +3 in $Cr(NO_3)_3$, so chromium is gaining 3 electrons and is reduced. The O.N. of Fe increases from +2 in $Fe(NO_3)_2$ to +3 in $Fe(NO_3)_3$, so iron is losing 1 electron and is oxidized. Multiply Fe by 3 for a loss of 3 electrons to match the 3 electron gain of Cr:

   $HNO_3(aq) + K_2CrO_4(aq) + 3\ Fe(NO_3)_2(aq) \rightarrow KNO_3(aq) + 3\ Fe(NO_3)_3(aq) + Cr(NO_3)_3(aq) + H_2O(l)$

Finish balancing by inspection:

8 $HNO_3(aq) + K_2CrO_4(aq) + 3\ Fe(NO_3)_2(aq) \rightarrow 2\ KNO_3(aq) + 3\ Fe(NO_3)_3(aq) + Cr(NO_3)_3(aq) + 4\ H_2O(l)$

Oxidizing agent: K_2CrO_4   **Reducing agent: $Fe(NO_3)_2$**

b)

     –6   –4 +6   +2 +12 –14   –6   –2 +4   +2   +15 –18

+1 +5 –2   –2 +1 –2   +1 +6 –2   +1 +5 –2   –1 +1 –2   +1 –2   +3 +5 –2

   $HNO_3(aq) + C_2H_6O(l) + K_2Cr_2O_7(aq) \rightarrow KNO_3(aq) + C_2H_4O(l) + H_2O(l) + Cr(NO_3)_3(aq)$

The O.N. of each Cr decreases from +6 in $K_2Cr_2O_7$ to +3 in $Cr(NO_3)_3$, so each chromium is gaining 3 electrons for a total of 6 electrons and is reduced. The O.N. of each C increases from –2 in C_2H_6O to –1 in C_2H_4O, so each carbon is losing 1 electron for a total of 2 electrons and is oxidized. Multiply C by 2 for a loss of 6 electrons to match the 6 electron gain of Cr:

   $HNO_3(aq) + 3\ C_2H_6O(l) + K_2Cr_2O_7(aq) \rightarrow KNO_3(aq) + 3\ C_2H_4O(l) + H_2O(l) + Cr(NO_3)_3(aq)$

Finish balancing by inspection:

8 $HNO_3(aq) + 3\ C_2H_6O(l) + K_2Cr_2O_7(aq) \rightarrow 2\ KNO_3(aq) + 3\ C_2H_4O(l) + 7\ H_2O(l) + 2\ Cr(NO_3)_3(aq)$

Oxidizing agent: $K_2Cr_2O_7$ **Reducing agent: C_2H_6O**

c)
$$\overset{+1\ -1}{HCl}(aq) + \overset{-3\ +1\ -1}{NH_4Cl}(aq) + \overset{+1\ +6\ -2}{K_2Cr_2O_7}(aq) \rightarrow \overset{+1\ -1}{KCl}(aq) + \overset{+3\ -1}{CrCl_3}(aq) + \overset{0}{N_2}(g) + \overset{+1\ -2}{H_2O}(l)$$

(with overall labels +4, +2 +12 –14, –3, +2)

The O.N. of each Cr decreases from +6 in $K_2Cr_2O_7$ to +3 in $CrCl_3$, so each chromium is gaining 3 electrons for a total of 6 electrons and is reduced. The O.N. of each N increases from –3 in NH_4Cl to 0 in N_2, so each nitrogen is losing 3 electrons for a total of 6 electrons and is oxidized. The electron gain equals the electron loss. Finish balancing by inspection:

$$6\ HCl(aq) + 2\ NH_4Cl(aq) + K_2Cr_2O_7(aq) \rightarrow 2\ KCl(aq) + 2\ CrCl_3(aq) + N_2(g) + 7\ H_2O(l)$$

Oxidizing agent: $K_2Cr_2O_7$ **Reducing agent: NH_4Cl**

d)
$$\overset{+1\ +5\ -2}{KClO_3}(aq) + \overset{+1\ -1}{HBr}(aq) \rightarrow \overset{0}{Br_2}(l) + \overset{+1\ -2}{H_2O}(l) + \overset{+1\ -1}{KCl}(aq)$$

(with overall labels –6, +2)

The O.N. of Cl decreases from +5 in $KClO_3$ to –1 in KCl, so chlorine is gaining 6 electrons and is reduced. The O.N. of each Br increases from –1 in HBr to 0 in Br_2, so each bromine is losing 1 electron for a total of 2 electrons and is oxidized. Multiply Br by 3 for a total loss of 6 electrons to match the 6 electron gain of Cl:

$$KClO_3(aq) + 6\ HBr(aq) \rightarrow 3\ Br_2(l) + H_2O(l) + KCl(aq)$$

Finish balancing by inspection:

$$KClO_3(aq) + 6\ HBr(aq) \rightarrow 3\ Br_2(l) + 3\ H_2O(l) + KCl(aq)$$

Oxidizing agent: $KClO_3$ **Reducing agent: HBr**

4.80   Plan: Determine the moles of permanganate ion present. The moles of permanganate ion along with the balanced chemical equation give the moles of hydrogen peroxide. The molar mass of hydrogen peroxide is required to finish the problem. The unrounded numbers from each step are carried to the next step to minimize rounding errors.
Solution:

a) Moles $MnO_4^- = \left(\dfrac{0.105\ mol\ MnO_4^-}{L} \right) \left(\dfrac{10^{-3}\ L}{1\ mL} \right) (43.2\ mL) = 4.536 \times 10^{-3} = \mathbf{4.54 \times 10^{-3}\ mol\ MnO_4^-}$

b) Moles $H_2O_2 = \left(4.536 \times 10^{-3}\ MnO_4^- \right) \left(\dfrac{5\ mol\ H_2O_2}{2\ mol\ MnO_4^-} \right) = 0.01134 = \mathbf{0.0113\ mol\ H_2O_2}$

c) Mass $H_2O_2 = \left(0.01134\ mol\ H_2O_2 \right) \left(\dfrac{34.02\ g\ H_2O_2}{1\ mol\ H_2O_2} \right) = 0.3857868 = \mathbf{0.386\ g\ H_2O_2}$

d) Mass percent $H_2O_2 = \left(\dfrac{0.3857868\ g\ H_2O_2}{14.8\ g\ Sample} \right) \times 100\% = 2.606668 = \mathbf{2.61\%\ H_2O_2}$

e) $\mathbf{H_2O_2}$ (The hydrogen peroxide reduces the manganese from +7 to +2.)

4.86   A common example of a combination/redox reaction is the combination of a metal and nonmetal to form an ionic salt, such as $2\ Mg(s) + O_2(g) \rightarrow 2\ MgO(s)$. A common example of a combination/non-redox reaction is the combination of a metal oxide and water to form an acid, such as: $CaO(s) + H_2O(l) \rightarrow Ca(OH)_2(aq)$.

4.87   Plan: In a combination reaction, two or more reactants form one product. In a decomposition reaction, one reactant forms two or more products. In a displacement reaction, atoms or ions exchange places. Balance the reactions by inspection.
Solution:
a) $Ca(s) + 2\ H_2O(l) \rightarrow Ca(OH)_2(aq) + H_2(g)$
     Displacement: one Ca atom displaces 2 H atoms.
b) $2\ NaNO_3(s) \rightarrow 2\ NaNO_2(s) + O_2(g)$
     Decomposition: one reactant breaks into two products.
c) $C_2H_2(g) + 2\ H_2(g) \rightarrow C_2H_6(g)$
     Combination: reactants combine to form one product.

4.89   Plan: Recall the definitions of each type of reaction:
        Combination: $X + Y \rightarrow Z$;   decomposition: $Z \rightarrow X + Y$
        Single displacement: $X + YZ \rightarrow XZ + Y$ double displacement: $WX + YZ \rightarrow WZ + YX$
        Solution:
        a) $2\ Sb(s) + 3\ Cl_2(g) \rightarrow 2\ SbCl_3(s)$ **combination**
        b) $2\ AsH_3(g) \rightarrow 2\ As(s) + 3\ H_2(g)$ **decomposition**
        c) $Zn(s) + Fe(NO_3)_2(aq) \rightarrow Zn(NO_3)_2(aq) + Fe(s)$ **displacement**

4.91   Plan: Recall the definitions of each type of reaction:
        Combination: $X + Y \rightarrow Z$;   decomposition: $Z \rightarrow X + Y$
        Single displacement: $X + YZ \rightarrow XZ + Y$ double displacement: $WX + YZ \rightarrow WZ + YX$
        Balance the reactions by inspection.
        Solution:
        a) The combination between a metal and a nonmetal gives a binary ionic compound.
                $Sr(s) + Br_2(l) \rightarrow SrBr_2(s)$
        b) Many metal oxides release oxygen gas upon thermal decomposition.
                $2\ Ag_2O(s) \xrightarrow{\ \Delta\ } 4\ Ag(s) + O_2(g)$
        c) This is a single displacement reaction. Mn is a more reactive metal and displaces Cu^{2+} from solution.
                $Mn(s) + Cu(NO_3)_2(aq) \rightarrow Mn(NO_3)_2(aq) + Cu(s)$

4.93   Plan: Two elements as reactants often results in a combination reaction while one reactant only often indicates
        a decomposition reaction. Review the types of reactions in Section 4.6.
        Solution:
        a) $N_2(g) + 3\ H_2(g) \rightarrow 2\ NH_3(g)$
        b) $2\ NaClO_3(s) \xrightarrow{\ \Delta\ } 2\ NaCl(s) + 3\ O_2(g)$
        c) $Ba(s) + 2\ H_2O(l) \rightarrow Ba(OH)_2(aq) + H_2(g)$

4.95   Plan: Review the types of reactions in Section 4.6.
        Solution:
        a) Cs, a metal, and I_2, a nonmetal, react to form the binary ionic compound, CsI.
                $2\ Cs(s) + I_2(s) \rightarrow 2\ CsI(s)$
        b) Al is a stronger reducing agent than Mn and is able to displace Mn from solution, i.e., cause the reduction from
        $Mn^{2+}(aq)$ to $Mn^0(s)$.
                $2\ Al(s) + 3\ MnSO_4(aq) \rightarrow Al_2(SO_4)_3(aq) + 3\ Mn(s)$
        c) Sulfur dioxide, SO_2, is a nonmetal oxide that reacts with oxygen, O_2, to form the higher oxide, SO_3.
                $2\ SO_2(g) + O_2(g) \xrightarrow{\ \Delta\ } 2\ SO_3(g)$
        It is not clear from the problem, but energy must be added to force this reaction to proceed.
        d) Butane is a four-carbon hydrocarbon with the formula C_4H_{10}. It burns in the presence of oxygen, O_2, to form
        carbon dioxide gas and water vapor. Although this is a redox reaction that could be balanced using the oxidation
        number method, it is easier to balance by considering only atoms on either side of the equation. First, balance
        carbon and hydrogen (because they only appear in one species on each side of the equation), and then balance
        oxygen.
                $2C_4H_{10}(g) + 13O_2(g) \rightarrow 8CO_2(g) + 10H_2O(g)$
        e) Total ionic equation:
                $2\ Al(s) + 3\ Mn^{2+}(aq) + 3\ SO_4^{2-}(aq) \rightarrow 2\ Al^{3+}(aq) + 3\ SO_4^{2-}(aq) + 3\ Mn(s)$
        Net ionic equation:
                $2\ Al(s) + 3\ Mn^{2+}(aq) \rightarrow 2\ Al^{3+}(aq) + 3\ Mn(s)$
        Note that the molar coefficients are not simplified because the number of electrons lost (6 e⁻) must equal the
        electrons gained (6 e⁻).

4.97   Plan: Write a balanced equation; convert the mass of HgO to moles and use the molar ratio from the balanced equation to find the moles and then the mass of O_2. Perform the same calculation to find the mass of the other product.
Solution:

The balanced chemical equation is $2\ HgO(s) \xrightarrow{\Delta} 2\ Hg(l) + O_2(g)$.

$$\text{Mass } O_2 = \left(4.27\text{ kg HgO}\right)\left(\frac{10^3\text{ g}}{1\text{ kg}}\right)\left(\frac{1\text{ mol HgO}}{216.6\text{ g HgO}}\right)\left(\frac{1\text{ mol }O_2}{2\text{ mol HgO}}\right)\left(\frac{32.00\text{ g }O_2}{1\text{ mol }O_2}\right) = 315.420 = \mathbf{315\ g\ O_2}$$

The other product is **mercury**.

$$\text{Mass Hg} = \left(4.27\text{ kg HgO}\right)\left(\frac{10^3\text{ g}}{1\text{ kg}}\right)\left(\frac{1\text{ mol HgO}}{216.6\text{ g HgO}}\right)\left(\frac{2\text{ mol Hg}}{2\text{ mol HgO}}\right)\left(\frac{200.6\text{ g Hg}}{1\text{ mol Hg}}\right)\left(\frac{1\text{ kg}}{10^3\text{ g}}\right)$$

$$= 3.95458 = \mathbf{3.95\ kg\ Hg}$$

4.99   Plan: To determine the reactant in excess, write the balanced equation (metal + O_2 → metal oxide), convert reactant masses to moles, and use molar ratios to see which reactant makes the smaller ("limiting") amount of product. Use the limiting reactant to calculate the amount of product formed.
Solution:

The balanced equation is $4\ Li(s) + O_2(g) \rightarrow 2\ Li_2O(s)$.

a) Moles Li_2O if Li limiting $= \left(1.62\text{ g Li}\right)\left(\dfrac{1\text{ mol Li}}{6.941\text{ g Li}}\right)\left(\dfrac{2\text{ mol }Li_2O}{4\text{ mol Li}}\right) = 0.1166979\text{ mol }Li_2O$ (unrounded)

Moles Li_2O if O_2 limiting $= \left(6.50\text{ g }O_2\right)\left(\dfrac{1\text{ mol }O_2}{32.00\text{ g }O_2}\right)\left(\dfrac{2\text{ mol }Li_2O}{1\text{ mol }O_2}\right) = 0.40625\text{ mol }Li_2O$ (unrounded)

Li is the limiting reactant; **O_2 is in excess**.
b) $0.1166979 = \mathbf{0.117\ mol\ Li_2O}$
c) Li is limiting, thus there will be none remaining (**0 g Li**).

$$\text{Grams } Li_2O = \left(1.62\text{ g Li}\right)\left(\frac{1\text{ mol Li}}{6.941\text{ g Li}}\right)\left(\frac{2\text{ mol }Li_2O}{4\text{ mol Li}}\right)\left(\frac{29.88\text{ g }Li_2O}{1\text{ mol }Li_2O}\right) = 3.4869 = \mathbf{3.49\ g\ Li_2O}$$

$$\text{Grams } O_2 \text{ used} = \left(1.62\text{ g Li}\right)\left(\frac{1\text{ mol Li}}{6.941\text{ g Li}}\right)\left(\frac{1\text{ mol }O_2}{4\text{ mol Li}}\right)\left(\frac{32.00\text{ g }O_2}{1\text{ mol }O_2}\right) = 1.867166\text{ g }O_2 \text{ (unrounded)}$$

Remaining $O_2 = 6.50\text{ g }O_2 - 1.867166\text{ g }O_2 = 4.632834 = \mathbf{4.63\ g\ O_2}$

4.101   Plan: Since mass must be conserved, the original amount of mixture – amount of residue = mass of oxygen produced. Write a balanced equation and use mole ratios to convert from the mass of oxygen produced to the amount of $KClO_3$ reacted. Use the mass percent definition.
Solution:

$$2\ KClO_3(s) \xrightarrow{\Delta} 2\ KCl(s) + 3\ O_2(g)$$

$$\text{Mass } KClO_3 = \left((0.950 - 0.700)\text{g }O_2\right)\left(\frac{1\text{ mol }O_2}{32.00\text{ g }O_2}\right)\left(\frac{2\text{ mol }KClO_3}{3\text{ mol }O_2}\right)\left(\frac{122.55\text{ g }KClO_3}{1\text{ mol }KClO_3}\right)$$

$$= 0.63828125\text{ g }KClO_3 \text{ (unrounded)}$$

$$\text{Mass \% } KClO_3 = \left(\frac{0.63828125\text{ g }KClO_3}{0.950\text{ g Sample}}\right) \times 100\% = 67.1875 = \mathbf{67.2\%\ KClO_3}$$

4.103   Plan: To find the mass of Fe, write a balanced equation for the reaction, determine whether Al or Fe_2O_3 is the limiting reactant, and convert to mass.
Solution:
$$2\ Al(s) + Fe_2O_3(s) \rightarrow 2\ Fe(s) + Al_2O_3(s)$$
When the masses of both reactants are given, you must determine which reactant is limiting.

Mole Fe (from Al) = $(1.50\ kg\ Al)\left(\dfrac{10^3\ g}{1\ kg}\right)\left(\dfrac{1\ mol\ Al}{26.98\ g\ Al}\right)\left(\dfrac{2\ mol\ Fe}{2\ mol\ Al}\right) = 55.59674\ mol\ Fe$

Mole Fe (from Fe_2O_3) = $(25.0\ mol\ Fe_2O_3)\left(\dfrac{2\ mol\ Fe}{1\ mol\ Fe_2O_3}\right) = 50.0\ mol\ Fe$

Fe_2O_3 is limiting, so 50.0 moles of Fe forms.

Mass = $(50.0\ mol\ Fe)\left(\dfrac{55.85\ g\ Fe}{1\ mol\ Fe}\right) = 2792.5 = 2790\ g\ Fe$

Though not required by the problem, this could be converted to **2.79 kg**.

4.104   Plan: In ionic compounds, iron has two common oxidation states, +2 and +3. First, write the balanced equations for the formation and decomposition of compound A. Then, determine which reactant is limiting and from the amount of the limiting reactant calculate how much compound B will form.
Solution:
Compound A is iron chloride with iron in the higher, +3, oxidation state. Thus, the formula for compound A is $FeCl_3$ and the correct name is iron(III) chloride. The balanced equation for formation of $FeCl_3$ is:
$$2\ Fe(s) + 3\ Cl_2(g) \rightarrow 2\ FeCl_3(s)$$
The $FeCl_3$ decomposes to $FeCl_2$ with iron in the +2 oxidation state. The balanced equation for the decomposition of $FeCl_3$ is:
$$2\ FeCl_3(s) \rightarrow 2\ FeCl_2(s) + Cl_2(g)$$
To find out whether 50.6 g Fe or 83.8 g Cl_2 limits the amount of product, calculate the number of moles of iron(III) chloride that could form based on each reactant.

Mole $FeCl_3$ (from Fe) = $(50.6\ g\ Fe)\left(\dfrac{1\ mol\ Fe}{55.85\ g\ Fe}\right)\left(\dfrac{2\ mol\ FeCl_3}{2\ mol\ Fe}\right) = 0.905998\ FeCl_3$ (unrounded)

Mole $FeCl_3$ (from Cl_2) = $(83.8\ g\ Cl_2)\left(\dfrac{1\ mol\ Cl_2}{70.90\ g\ Cl_2}\right)\left(\dfrac{2\ mol\ FeCl_3}{3\ mol\ Cl_2}\right) = 0.787964\ FeCl_3$ (unrounded)

Since fewer moles of $FeCl_3$ are produced from the available amount of chlorine, the chlorine is limiting and we calculate the amount of compound B from the amount of chlorine gas.

Mass $FeCl_2$ = $(83.8\ g\ Cl_2)\left(\dfrac{1\ mol\ Cl_2}{70.90\ g\ Cl_2}\right)\left(\dfrac{2\ mol\ FeCl_3}{3\ mol\ Cl_2}\right)\left(\dfrac{2\ mol\ FeCl_2}{2\ mol\ FeCl_3}\right)\left(\dfrac{126.75\ g\ FeCl_2}{1\ mol\ FeCl_2}\right)$
$= 99.874 = $ **99.9 g $FeCl_2$**

4.108   Plan: Review the concept of dynamic equilibrium.
Solution:
$$NO(g) + Br_2(g) \leftrightarrow 2\ NOBr(g)$$
On a molecular scale, chemical reactions are dynamic. If NO and Br_2 are placed in a container, molecules of NO and molecules of Br_2 will react to form NOBr. Some of the NOBr molecules will decompose and the resulting NO and Br_2 molecules will recombine with different NO and Br_2 molecules to form more NOBr. In this sense, the reaction is dynamic because the original NO and Br_2 pairings do not remain permanently attached to each other. Eventually, the rate of the forward reaction (combination of NO and Br_2) will equal the rate of the reverse reaction (decomposition of NOBr) at which point the reaction is said to have reached equilibrium. If you could take "snapshot" pictures of the molecules at equilibrium, you would see a constant number of reactant (NO, Br_2) and product molecules (NOBr), but the pairings would not stay the same.

4.110   <u>Plan:</u> Ferrous ion is Fe^{2+}. Write a reaction to show the conversion of Fe to Fe^{2+}. Convert the mass of Fe in a 125-g serving to the mass of Fe in a 737-g sample. Use molar mass to convert mass to moles and use Avogadro's number to convert moles of Fe to moles of ions.
<u>Solution:</u>

a) $Fe(s) + 2 H^+(aq) \rightarrow Fe^{2+}(aq) + H_2(g)$
O.N.:   0   +1   +2   0

b) Fe^{2+} ions $= \left(737 \text{ g Sauce}\right)\left(\dfrac{49 \text{ mg Fe}}{125 \text{ g Sauce}}\right)\left(\dfrac{10^{-3} \text{ g}}{1 \text{ mg}}\right)\left(\dfrac{1 \text{ mol Fe}}{55.85 \text{ g Fe}}\right)\left(\dfrac{1 \text{ mol Fe}^{2+}}{1 \text{ mol Fe}}\right)\left(\dfrac{6.022 \times 10^{23} \text{ Fe}^{2+}\text{ions}}{1 \text{ mol Fe}^{2+}}\right)$

$= 3.11509 \times 10^{21} = \textbf{3.1} \times \textbf{10}^{\textbf{21}} \textbf{ Fe}^{\textbf{2+}}$ **ions** per jar of sauce

4.112   <u>Plan:</u> Convert the mass of glucose to moles and use the molar ratios from the balanced equation to find the moles of ethanol and CO_2. The amount of ethanol is converted from moles to grams using its molar mass.
<u>Solution:</u>

Mass of $C_2H_5OH = \left(10.0 \text{ g } C_6H_{12}O_6\right)\left(\dfrac{1 \text{ mol } C_6H_{12}O_6}{180.16 \text{ g } C_6H_{12}O_6}\right)\left(\dfrac{2 \text{ mol } C_2H_5OH}{1 \text{ mol } C_6H_{12}O_6}\right)\left(\dfrac{46.07 \text{ g } C_2H_5OH}{1 \text{ mol } C_2H_5OH}\right)$

$= 5.1143 = \textbf{5.11 g } \textbf{C}_{\textbf{2}}\textbf{H}_{\textbf{5}}\textbf{OH}$

Volume $CO_2 = \left(10.0 \text{ g } C_6H_{12}O_6\right)\left(\dfrac{1 \text{ mol } C_6H_{12}O_6}{180.16 \text{ g } C_6H_{12}O_6}\right)\left(\dfrac{2 \text{ mol } CO_2}{1 \text{ mol } C_6H_{12}O_6}\right)\left(\dfrac{22.4 \text{ L } CO_2}{1 \text{ mol } CO_2}\right)$

$= 2.4866785 = \textbf{2.49 L } \textbf{CO}_{\textbf{2}}$

4.114   <u>Plan:</u> Remember that spectator ions are omitted from net ionic equations. Find the number of moles of $KMnO_4$ and use molar ratios to find the moles of $CaCl_2$. The mass of $CaCl_2$ is found using the molar mass.
<u>Solution:</u>
a) A solution of $Na_2C_2O_4$ (soluble salt) is added to a solution containing 1.9348 g of road salt (NaCl and $CaCl_2$, both soluble salts) and an insoluble precipitate is formed (CaC_2O_4). The Na^+ and Cl^- ions are spectators in this reaction because all Na^+ salts are soluble and most Cl^- salts are soluble.
   $Ca^{2+}(aq) + C_2O_4{}^{2-}(aq) \rightarrow CaC_2O_4(s)$
b) You may recognize this reaction as a redox titration, because the permanganate ion, $MnO_4{}^-$, is a common oxidizing agent. The $MnO_4{}^-$ oxidizes the oxalate ion, $C_2O_4{}^{2-}$ to CO_2. Mn changes from +7 to +2 (reduction) and C changes from +3 to +4 (oxidation). The equation that describes this process is:
   $H_2C_2O_4(aq) + MnO_4{}^-(aq) \rightarrow Mn^{2+}(aq) + CO_2(g)$
$H_2C_2O_4$ is a weak acid, so it cannot be written in a fully dissociated form ($2 H^+(aq) + C_2O_4{}^{2-}(aq)$). $KMnO_4$ is a soluble salt, so it can be written in its dissociated form. $K^+(aq)$ is omitted because it is a spectator ion. We will balance the equation using the oxidation number method (you can verify by doing the half reaction method) and first assigning oxidation numbers to all elements in the reaction.

Oxidation numbers: +1 +3 –2   +7 –2   +2   +4 –2
                 $H_2C_2O_4(aq) + MnO_4{}^-(aq) \rightarrow Mn^{2+}(aq) + 2 CO_2(g)$
Identify the oxidized and reduced species and multiply one or both species by the appropriate factors to make the electrons lost equal the electrons gained. Each C in $H_2C_2O_4$ increases in O.N. from +3 to +4 for a total gain of 2 electrons. Mn in $MnO_4{}^-$ decreases in O.N. from +7 to +2 for a loss of 5 electrons. Multiply C by 5 and Mn by 2 to get an electron loss = electron gain of 10 electrons.
            $5 H_2C_2O_4(aq) + 2 MnO_4{}^-(aq) \rightarrow 2 Mn^{2+}(aq) + 10 CO_2(g)$
Adding water and $H^+(aq)$ to finish balancing the equation is appropriate since the reaction takes place in acidic medium. Add 8 $H_2O(l)$ to right side of equation to balance the oxygen and then add 6 $H^+(aq)$ to the left to balance hydrogen.
    $5 H_2C_2O_4(aq) + 2 MnO_4{}^-(aq) + 6 H^+(aq) \rightarrow 10 CO_2(g) + 2 Mn^{2+}(aq) + 8 H_2O(l)$
c) Oxidizing agent = **$KMnO_4$**
d) Reducing agent = **$H_2C_2O_4$** (Remember that the *agent* refers to the whole compound, not a particular element within a compound.)

e) The balanced equations provide the accurate molar ratios between species. Working backwards, we know that
(Moles of MnO_4^-) is related to moles of $C_2O_4^{2-}$ at the endpoint (5:2)
(Moles of $C_2O_4^{2-}$) is related to moles of CaC_2O_4 (1:1)
(Moles of CaC_2O_4) is related to moles of $CaCl_2$ (1:1)

$$\text{Mass } CaCl_2 = (37.68 \text{ mL})\left(\frac{10^{-3} \text{ L}}{1 \text{ mL}}\right)\left(\frac{0.1019 \text{ mol KMnO}_4}{1 \text{ L}}\right)\left(\frac{5 \text{ mol H}_2\text{C}_2\text{O}_4}{2 \text{ mol KMnO}_4}\right)\left(\frac{1 \text{ mol CaCl}_2}{1 \text{ mol H}_2\text{C}_2\text{O}_4}\right)\left(\frac{110.98 \text{ g CaCl}_2}{1 \text{ mol CaCl}_2}\right)$$

$= 1.06529$ g $CaCl_2$ (unrounded)

$$\text{Mass percent } CaCl_2 = \frac{1.06529 \text{ g CaCl}_2}{1.9348 \text{ g mixture}}(100) = 55.0594 = \mathbf{55.06\% \ CaCl_2}$$

4.117   Plan: Write a balanced equation for the reaction. This is an acid-base reaction between HCl and the base CO_3^{2-} to form H_2CO_3 which then decomposes into CO_2 and H_2O. To find mass %, convert the solution to moles of HCl, use the molar ratio to find moles of dolomite, convert to mass using molar mass and divide by the mass of soil.
Solution:

The balanced equation for this reaction is:

$$CaMg(CO_3)_2(s) + 4 \text{ HCl}(aq) \rightarrow Ca^{2+}(aq) + Mg^{2+}(aq) + 2 \text{ H}_2\text{O}(l) + 2 \text{ CO}_2(g) + 4 \text{ Cl}^-(aq)$$

$$\text{Mass } CaMg(CO_3)_2 = (33.56 \text{ mL})\left(\frac{10^{-3} \text{ L}}{1 \text{ mL}}\right)\left(\frac{0.2516 \text{ mol HCl}}{1 \text{ L}}\right)\left(\frac{1 \text{ mol CaMg(CO}_3)_2}{4 \text{ mol HCl}}\right)\left(\frac{184.41 \text{ g CaMg(CO}_3)_2}{1 \text{ mol CaMg(CO}_3)_2}\right)$$

$= 0.389275$ g $CaMg(CO_3)_2$ (unrounded)

$$\text{Mass percent } CaMg(CO_3)_2 = \frac{0.389275 \text{ g CaMg(CO}_3)_2}{13.86 \text{ g soil}}(100) = 2.80863 = \mathbf{2.809\% \ CaMg(CO_3)_2}$$

4.119   Plan: For part (a), assign oxidation numbers to each element; the oxidizing agent has an atom whose oxidation number decreases while the reducing agent has an atom whose oxidation number increases. For part (b), use the molar ratios, beginning with Step 3, to find the moles of NO_2, then moles of NO, then moles of NH_3 required to produce the given mass of HNO_3.
Solution:
a) Step 1. $4NH_3(g) + 5O_2(g) \rightarrow 4NO(g) + 6H_2O(l)$
        O.N.: $N = -3$ $O = 0$ $N = +2$ $H = +1$
                     $H = +1$ $O = -2$ $O = -2$
        N oxidized from -3 to $+2$ by O_2, and O is reduced from 0 to -2.
        Oxidizing agent = O_2 Reducing agent = NH_3
    Step 2. $2NO(g) + O_2(g) \rightarrow 2NO_2(g)$
        O.N.: $N = +2$ $O = 0$ $N = +4$
                     $O = -2$ $O = -2$
        N oxidized from $+2$ to $+4$ by O_2, and O is reduced from 0 to -2.
        Oxidizing agent = O_2 Reducing agent = NO
    Step 3. $3NO_2(g) + H_2O(l) \rightarrow 2 \text{ HNO}_3(l) + NO(g)$
        O.N.: $N = +4$ $H = +1$ $H = +1$ $N = +2$
                     $O = -2$ $O = -2$ $N = +5$ $O = -2$
                                                $O = -2$
        N oxidized from $+4$ to $+5$ by NO_2, and N is reduced from $+4$ to $+2$.
        Oxidizing agent = NO_2 Reducing agent = NO_2
b) Mass of NH_3 =

$$(3.0 \times 10^4 \text{ kg HNO}_3)\left(\frac{10^3 \text{ g}}{1 \text{ kg}}\right)\left(\frac{1 \text{ mol HNO}_3}{63.02 \text{ g HNO}_3}\right)\left(\frac{3 \text{ mol NO}_2}{2 \text{ mol HNO}_3}\right)\left(\frac{2 \text{ mol NO}}{2 \text{ mol NO}_2}\right)\left(\frac{4 \text{ mol NH}_3}{4 \text{ mol NO}}\right)\left(\frac{17.03 \text{ g NH}_3}{1 \text{ mol NH}_3}\right)\left(\frac{1 \text{ kg}}{10^3 \text{ g}}\right)$$

$= 1.21604 \times 10^4 = \mathbf{1.2 \times 10^4 \ kg \ NH_3}$

4.124   <u>Plan:</u> Write a balanced equation and use the molar ratio between Na_2O_2 and CO_2 to convert the amount of Na_2O_2 given to an amount of CO_2.
   <u>Solution:</u>
   The reaction is: $2\,Na_2O_2(s) + 2\,CO_2(g) \rightarrow 2\,Na_2CO_3(s) + O_2(g)$.

$$\text{Volume} = \left(80.0\ g\ Na_2O_2\right)\left(\frac{1\ mol\ Na_2O_2}{77.98\ g\ Na_2O_2}\right)\left(\frac{2\ mol\ CO_2}{2\ mol\ Na_2O_2}\right)\left(\frac{44.01\ g\ CO_2}{1\ mol\ CO_2}\right)\left(\frac{L\ Air}{0.0720\ g\ CO_2}\right)$$

$$= 627.08 = \textbf{627 L Air}$$

4.127   A 1.00 kg piece of glass of composition 75% SiO_2, 15% Na_2O, and 10% CaO would contain:
   (1.00 kg) (75.0% / 100%) = **0.750 kg SiO_2**
   (1.00 kg) (15.0% / 100%) = **0.150 kg Na_2O**
   (1.00 kg) (10.0% / 100%) = **0.100 kg CaO**
   In this example, the SiO_2 is added directly while the sodium oxide comes from decomposition of sodium carbonate and the calcium oxide from decomposition of calcium carbonate:
   $$Na_2CO_3(s) \rightarrow Na_2O(s) + CO_2(g)$$
   $$CaCO_3(s) \rightarrow CaO(s) + CO_2(g)$$

$$\text{Mass } Na_2CO_3 = \left(0.150\ kg\ Na_2O\right)\left(\frac{10^3\ g}{1\ kg}\right)\left(\frac{1\ mol\ Na_2O}{61.98\ g\ Na_2O}\right)\left(\frac{1\ mol\ Na_2CO_3}{1\ mol\ Na_2O}\right)\left(\frac{105.99\ g\ Na_2CO_3}{1\ mol\ Na_2CO_3}\right)\left(\frac{1\ kg}{10^3\ g}\right)$$

$$= 0.25651 = \textbf{0.257 kg } Na_2CO_3$$

$$\text{Mass } CaCO_3 = \left(0.100\ kg\ CaO\right)\left(\frac{10^3\ g}{1\ kg}\right)\left(\frac{1\ mol\ CaO}{56.08\ g\ CaO}\right)\left(\frac{1\ mol\ CaCO_3}{1\ mol\ CaO}\right)\left(\frac{100.09\ g\ CaCO_3}{1\ mol\ CaCO_3}\right)\left(\frac{1\ kg}{10^3\ g}\right)$$

$$= 0.178477 = \textbf{0.178 kg } CaCO_3$$

   Combine 0.750 kg SiO_2, 0.257 kg Na_2CO_3, and 0.178 kg $CaCO_3$ and heat to make 1.00 kg glass.

4.130   a) There is not enough information to write complete chemical equations, but the following equations can be written:
   $$C_5H_{11}I_4NO_4(s) + Na_2CO_3(s) \rightarrow 4\ I^-(aq) + \text{other products}$$
   $$I^-(aq) + Br_2(l) + HCl(aq) \rightarrow IO_3^-(aq) + \text{other products}$$
   For every mole of thyroxine reacted, **4 moles IO_3^- are produced**.
   b) $IO_3^-(aq) + H^+(aq) + I^-(aq) \rightarrow I_2(aq) + H_2O(l)$
   Oxidation numbers: +5 –2   +1   –1   0   +1 –2
   This is a difficult equation to balance because the iodine species are both reducing and oxidizing. I in IO_3^- gains 5 electrons as the O.N changes from +5 to 0 and I in I^- loses 1 electron as the O.N changes from –1 to 0. Start balancing the equation by placing a coefficient of 5 in front of $I^-(aq)$, so the electrons lost equal the electrons gained. Do <u>not</u> place a 5 in front of $I_2(aq)$, because not all of the $I_2(aq)$ comes from oxidation of $I^-(aq)$. Some of the $I_2(aq)$ comes from the reduction of $IO_3^-(aq)$. Place a coefficient of 3 in front of $I_2(aq)$ to correctly balance iodine. The reaction is now balanced from a redox standpoint, so finish balancing the reaction by balancing the oxygen and hydrogen.
   $$IO_3^-(aq) + 6\ H^+(aq) + 5\ I^-(aq) \rightarrow 3\ I_2(aq) + 3\ H_2O(l)$$
   IO_3^- is the oxidizing agent, and **I^- is the reducing agent**.
   If 3 moles of I_2 are produced per mole of IO_3^-, and 4 moles of IO_3^- are produced per mole of thyroxine, then **12 moles of I_2 are produced per mole of thyroxine**.
   c) Using Thy to represent thyroxine:
   The balanced equation for this reaction is $I_2(aq) + 2\,S_2O_3^{2-}(aq) \rightarrow 2\,I^-(aq) + S_4O_6^{2-}(aq)$.

$$\text{Mass Thy} = \left(17.23\ mL\right)\left(\frac{10^{-3}\ L}{1\ mL}\right)\left(\frac{0.1000\ mol\ S_2O_3^{2-}}{L}\right)\left(\frac{1\ mol\ I_2}{2\ mol\ S_2O_3^{2-}}\right)\left(\frac{1\ mol\ Thy}{12\ mol\ I_2}\right)\left(\frac{776.8\ g\ Thy}{1\ mol\ Thy}\right)$$

$$= 0.055767766\ g\ Thy\ \text{(unrounded)}$$

$$\text{Mass \% Thy} = \frac{\text{mass of Thy}}{\text{mass of extract}}(100) = \frac{0.055767766\ g\ Thy}{0.4332\ g}(100) = 12.8734 = \textbf{12.87\% thyroxine}$$

4.134   <u>Plan:</u> Balance the equation to obtain the correct molar ratios. Convert the mass of each reactant to moles and use the molar ratios to find the limiting reactant and the amount of CO_2 produced.
<u>Solution:</u>
a) Here is a suggested method for approaching balancing the equation.
— Since PO_4^{2-} remains as a unit on both sides of the equation, treat it as a unit when balancing.
— On first inspection, one can see that Na needs to be balanced by adding a "2" in front of $NaHCO_3$. This then affects the balance of C, so add a "2" in front of CO_2.
— Hydrogen is not balanced, so change the coefficient of water to "2," as this will have the least impact on the other species.
— Verify that the other species are balanced.

$$Ca(H_2PO_4)_2(s) + 2\ NaHCO_3(s) \xrightarrow{\Delta} 2\ CO_2(g) + 2\ H_2O(g) + CaHPO_4(s) + Na_2HPO_4(s)$$

Determine whether $Ca(H_2PO_4)_2$ or $NaHCO_3$ limits the production of CO_2. In each case calculate the moles of CO_2 that might form.

$$\text{Mole } CO_2\ (NaHCO_3) = \left(1.00\ g\right)\left(\frac{31.0\%}{100\%}\right)\left(\frac{1\ mol\ NaHCO_3}{84.01\ g\ NaHCO_3}\right)\left(\frac{2\ mol\ CO_2}{2\ mol\ NaHCO_3}\right)$$

$$= 3.690 \times 10^{-3}\ mol\ CO_2\ \text{(unrounded)}$$

$$\text{Mole } CO_2\ (Ca(H_2PO_4)_2) = \left(1.00\,g\right)\left(\frac{35.0\%}{100\%}\right)\left(\frac{1\ mol\ Ca(H_2PO_4)_2}{234.05\ g\ Ca(H_2PO_4)_2}\right)\left(\frac{2\ mol\ CO_2}{1\ mol\ Ca(H_2PO_4)_2}\right)$$

$$= 2.9908 \times 10^{-3}\ mol\ CO_2\ \text{(unrounded)} = 3.0 \times 10^{-3}\ mol\ CO_2$$

Since $Ca(H_2PO_4)_2$ is limiting, **2.99×10^{-3} mol CO_2** will be produced.

b) $\text{Volume } CO_2 = \left(2.9908 \times 10^{-3}\ mol\ CO_2\right)\left(\dfrac{37.0\ L}{1\ mol\ CO_2}\right) = 0.1106596 = \mathbf{0.111\ L\ CO_2}$

4.136   <u>Plan:</u> To determine the empirical formula, find the moles of each element present and divide by the smallest number of moles to get the smallest ratio of atoms. To find the molecular formula, divide the molar mass by the mass of the empirical formula to find the factor by which to multiply the empirical formula.
<u>Solution:</u>
a) Determine the moles of each element present. The sample was burned in an unknown amount of O_2, therefore, the moles of oxygen must be found by a different method.

$$\text{Moles } C = \left(0.1880\ g\ CO_2\right)\left(\frac{1\ mol\ CO_2}{44.01\ g\ CO_2}\right)\left(\frac{1\ mol\ C}{1\ mol\ CO_2}\right) = 4.271756 \times 10^{-3}\ mol\ C\ \text{(unrounded)}$$

$$\text{Moles } H = \left(0.02750\ g\ H_2O\right)\left(\frac{1\ mol\ H_2O}{18.02\ g\ H_2O}\right)\left(\frac{2\ mol\ H}{1\ mol\ H_2O}\right) = 3.052164 \times 10^{-3}\ mol\ H\ \text{(unrounded)}$$

$$\text{Moles } Bi = \left(0.1422\ g\ Bi_2O_3\right)\left(\frac{1\ mol\ Bi_2O_3}{466.0\ g\ Bi_2O_3}\right)\left(\frac{2\ mol\ Bi}{1\ mol\ Bi_2O_3}\right) = 6.103004 \times 10^{-4}\ mol\ Bi\ \text{(unrounded)}$$

Subtracting the mass of each element present from the mass of the sample will give the mass of oxygen originally present in the sample. This mass is used to find the moles of oxygen.

$$\text{Mass } C = \left(4.271756 \times 10^{-3}\ mol\ C\right)\left(\frac{12.01\,g\ C}{1\,mol\ C}\right) = 0.0513038\ g\ C$$

$$\text{Mass } H = \left(3.052164 \times 10^{-3}\ mol\ H\right)\left(\frac{1.008\ g\ H}{1\,mol\ H}\right) = 0.0030766\ g\ H$$

$$\text{Mass } Bi = \left(6.103004 \times 10^{-4}\ mol\ Bi\right)\left(\frac{209.0\ g\ Bi}{1\,mol\ Bi}\right) = 0.127553\ g\ Bi$$

$$\text{Mass } O = 0.22105\ g\ \text{sample} - (0.0513038\ g\ C + 0.0030766\ g\ H + 0.127553\ g\ Bi) = 0.0391166\ g\ O$$

$$\text{Moles } O = \left(0.0391166\,g\ O\right)\left(\frac{1\,mol\ O}{16.00\,g\ O}\right) = 0.0024448\ mol\ O$$

Divide each of the moles by the smallest value (moles Bi).

$C = (4.271756 \times 10^{-3} \text{ mol}) / (6.103004 \times 10^{-4} \text{ mol}) = 7$

$H = (3.052164 \times 10^{-3} \text{ mol}) / (6.103004 \times 10^{-4} \text{ mol}) = 5$

$O = (2.4448 \times 10^{-3} \text{ mol}) / (6.103004 \times 10^{-4} \text{ mol}) = 4$

$Bi = (6.103004 \times 10^{-4} \text{ mol}) / (6.103004 \times 10^{-4} \text{ mol}) = 1$

Empirical formula = **$C_7H_5O_4Bi$**

b) The empirical formula mass is 362 g/mol. Therefore, there are 1086 / 362 = 3 empirical formula units per molecular formula making the molecular formula = 3 x $C_7H_5O_4Bi$ = **$C_{21}H_{15}O_{12}Bi_3$**.

c) $Bi(OH)_3(s) + 3\ HC_7H_5O_3(aq) \rightarrow Bi(C_7H_5O_3)_3(s) + 3\ H_2O(l)$

d) $(0.600 \text{ mg})\left(\dfrac{10^{-3}\text{ g}}{1\text{ mg}}\right)\left(\dfrac{1\text{ mol Active}}{1086\text{ g}}\right)\left(\dfrac{3\text{ mol Bi}}{1\text{ mol Active}}\right)\left(\dfrac{1\text{ mol Bi(OH)}_3}{1\text{ mol Bi}}\right)\left(\dfrac{260.0\text{ g Bi(OH)}_3}{1\text{ mol Bi(OH)}_3}\right)\left(\dfrac{1\text{ mg}}{10^{-3}\text{ g}}\right)\left(\dfrac{100\%}{88.0\%}\right)$

$= 0.48970 = $ **0.490 mg $Bi(OH)_3$**

4.138   Plan: Write balanced equations and use the molar ratios to convert mass of each fuel to the mass of oxygen required for the reaction.

Solution:

a) Complete combustion of hydrocarbons involves heating the hydrocarbon in the presence of oxygen to produce carbon dioxide and water.

Ethanol: $C_2H_5OH(l) + 3\ O_2(g) \rightarrow 2\ CO_2(g) + 3\ H_2O(l)$

Gasoline: $2\ C_8H_{18}(l) + 25\ O_2(g) \rightarrow 16\ CO_2(g) + 18\ H_2O(g)$

b) The amounts of each fuel must be found:

Gasoline $= (1.00 \text{ L})\left(\dfrac{90\%}{100\%}\right)\left(\dfrac{1\text{ mL}}{10^{-3}\text{ L}}\right)\left(\dfrac{0.742\text{ g}}{1\text{ mL}}\right) = 667.8 \text{ g gasoline (unrounded)}$

Ethanol $= (1.00 \text{ L})\left(\dfrac{10\%}{100\%}\right)\left(\dfrac{1\text{ mL}}{10^{-3}\text{ L}}\right)\left(\dfrac{0.789\text{ g}}{1\text{ mL}}\right) = 78.9 \text{ g ethanol}$

Mass O_2 (gasoline) $= (667.8 \text{ g } C_8H_{18})\left(\dfrac{1\text{ mol } C_8H_{18}}{114.22\text{ g } C_8H_{18}}\right)\left(\dfrac{25\text{ mol } O_2}{2\text{ mol } C_8H_{18}}\right)\left(\dfrac{32.00\text{ g } O_2}{1\text{ mol } O_2}\right)$

$= 2338.64 \text{ g } O_2 \text{ (unrounded)}$

Mass O_2 (ethanol) $= (78.9 \text{ g } C_2H_5OH)\left(\dfrac{1\text{ mol } C_2H_5OH}{46.07\text{ g } C_2H_5OH}\right)\left(\dfrac{3\text{ mol } O_2}{1\text{ mol } C_2H_5OH}\right)\left(\dfrac{32.00\text{ g } O_2}{1\text{ mol } O_2}\right)$

$= 164.41 \text{ g } O_2 \text{ (unrounded)}$

Total $O_2 = 2338.64 \text{ g } O_2 + 164.41 \text{ g } O_2 = 2503.05 = $ **2.50 x 10³ g O_2**

c) $(2503.05 \text{ g } O_2)\left(\dfrac{1\text{ mol } O_2}{32.00\text{ g } O_2}\right)\left(\dfrac{22.4\text{ L}}{1\text{ mol } O_2}\right) = 1752.135 = $ **1.75 x 10³ L O_2**

d) $(1752.135 \text{ L } O_2)\left(\dfrac{100\%}{20.9\%}\right) = 8383.42 = $ **8.38 x 10³ L air**

4.140 Plan: From the molarity and volume of the base NaOH, find the moles of NaOH and use the molar ratios from the two balanced equations to convert the moles of NaOH to moles of HBr to moles of vitamin C. Use the molar mass of vitamin C to convert moles to grams.
Solution:
Mass vitamin C:

$$\left(43.20 \text{ mL NaOH}\right)\left(\frac{10^{-3}\text{ L}}{1\text{ mL}}\right)\left(\frac{0.1350\text{ mol NaOH}}{1\text{ L}}\right)\left(\frac{1\text{ mol HBr}}{1\text{ mol NaOH}}\right)\left(\frac{1\text{ mol C}_6\text{H}_8\text{O}_6}{2\text{ mol HBr}}\right)\left(\frac{176.12\text{ g C}_6\text{H}_8\text{O}_6}{1\text{ mol C}_6\text{H}_8\text{O}_6}\right)\left(\frac{1\text{ mg}}{10^{-3}\text{ g}}\right)$$

$$= 513.5659 = 513.6 \text{ mg C}_6\text{H}_8\text{O}_6$$

Yes, the tablets have the quantity advertised.

4.142 Plan: Remember that oxidation numbers change in a redox reaction. For the calculations, use the molarity and volume of HCl to find the moles of HCl and use the molar ratios from the balanced equation to convert moles of HCl to moles and then grams of the desired substance.
Solution:
a) The **second reaction is a redox process** because the O.N. of iron changes from 0 to +2 (it oxidizes) while the O.N. of hydrogen changes from +1 to 0 (it reduces).
b) Determine the moles of HCl present and use the balanced chemical equation to determine the appropriate quantities.

$$\text{Mass Fe}_2\text{O}_3 = \left(2.50 \times 10^3\text{ L}\right)\left(\frac{3.00\text{ mol HCl}}{\text{L}}\right)\left(\frac{1\text{ mol Fe}_2\text{O}_3}{6\text{ mol HCl}}\right)\left(\frac{159.70\text{ g Fe}_2\text{O}_3}{1\text{ mol Fe}_2\text{O}_3}\right)$$

$$= 199625 = \textbf{2.00} \times \textbf{10}^5 \textbf{ g Fe}_2\textbf{O}_3$$

$$\text{Mass FeCl}_3 = \left(2.50 \times 10^3\text{ L}\right)\left(\frac{3.00\text{ mol HCl}}{\text{L}}\right)\left(\frac{2\text{ mol FeCl}_3}{6\text{ mol HCl}}\right)\left(\frac{162.20\text{ g FeCl}_3}{1\text{ mol FeCl}_3}\right)$$

$$= 405500 = \textbf{4.06} \times \textbf{10}^5 \textbf{ g FeCl}_3$$

c) Use reaction 2 like reaction 1 was used in part b.

$$\text{Mass Fe} = \left(2.50 \times 10^3\text{ L}\right)\left(\frac{3.00\text{ mol HCl}}{\text{L}}\right)\left(\frac{1\text{ mol Fe}}{2\text{ mol HCl}}\right)\left(\frac{55.85\text{ g Fe}}{1\text{ mol Fe}}\right)$$

$$= 209437.5 = \textbf{2.09} \times \textbf{10}^5 \textbf{ g Fe}$$

$$\text{Mass FeCl}_2 = \left(2.50 \times 10^3\text{ L}\right)\left(\frac{3.00\text{ mol HCl}}{\text{L}}\right)\left(\frac{1\text{ mol FeCl}_2}{2\text{ mol HCl}}\right)\left(\frac{126.75\text{ g FeCl}_2}{1\text{ mol FeCl}_2}\right)$$

$$= 475312.5 = \textbf{4.75} \times \textbf{10}^5 \textbf{ g FeCl}_2$$

d) Use 1.00 g Fe$_2$O$_3$ to determine the mass of FeCl$_3$ formed (reaction 1), and 0.280 g Fe to determine the mass of FeCl$_2$ formed (reaction 2).

$$\text{Mass FeCl}_3 = \left(1.00\text{ g Fe}_2\text{O}_3\right)\left(\frac{1\text{ mol Fe}_2\text{O}_3}{159.70\text{ g Fe}_2\text{O}_3}\right)\left(\frac{2\text{ mol FeCl}_3}{1\text{ mol Fe}_2\text{O}_3}\right)\left(\frac{162.20\text{ g FeCl}_3}{1\text{ mol FeCl}_3}\right)$$

$$= 2.0313 \text{ g FeCl}_3 \text{ (unrounded)}$$

$$\text{Mass FeCl}_2 = \left(0.280\text{ g Fe}\right)\left(\frac{1\text{ mol Fe}}{55.85\text{ g Fe}}\right)\left(\frac{1\text{ mol FeCl}_2}{1\text{ mol Fe}}\right)\left(\frac{126.75\text{ g FeCl}_2}{1\text{ mol FeCl}_2}\right)$$

$$= 0.635452 \text{ g FeCl}_2 \text{ (unrounded)}$$

Ratio $= (0.635452 \text{ g FeCl}_2) / (2.0313 \text{ g FeCl}_3) = 0.312830 = \textbf{0.313}$

CHAPTER 5 GASES AND THE KINETIC-MOLECULAR THEORY

FOLLOW–UP PROBLEMS

5.1 Plan: Use the equation for gas pressure in an open-end manometer. Use conversion factors to convert pressure in mmHg to units of torr, pascals and lb/in^2.
Solution:
$P_{gas} = P_{atm} - \Delta h$
$P_{gas} = 753.6$ mm Hg $- 174.0$ mm Hg $= 579.6$ mm Hg

Converting from mm Hg to torr: $P = (579.6$ mm Hg$)\left(\dfrac{1 \text{ torr}}{1 \text{ mm Hg}}\right) = \textbf{579.6 torr}$

Converting from mm Hg to pascals: $P = (579.6$ mm Hg$)\left(\dfrac{1 \text{ atm}}{760 \text{ mm Hg}}\right)\left(\dfrac{1.01325 \times 10^5 \text{Pa}}{1 \text{atm}}\right)$

$= 7.727364 \times 10^4 = \textbf{7.727} \times \textbf{10}^4 \textbf{ Pa}$

Converting from mm Hg to lb/in^2: $P = (579.6$ mm Hg$)\left(\dfrac{1 \text{ atm}}{760 \text{ mm Hg}}\right)\left(\dfrac{14.7 \text{ } lb/in^2}{1 \text{ atm}}\right) = 11.21068 = \textbf{11.2 } \textbf{lb/in}^2$

Check: For conversion to torr, the value of the pressure should stay the same. The order of magnitude for each conversion corresponds to the calculated answer. For conversion to pascals the order of magnitude calculation is $10^2 \times 10^5/10^2 = 10^5$. For conversion to lb/in^2 the order of magnitude calculation is $10^2 \times 10^1/10^2 = 10^1$.

5.2 Plan: Given in the problem is an initial volume, initial pressure, and final pressure for the argon gas. The final volume can be calculated from the relationship $P_1V_1 = P_2V_2$. Unit conversions for mL to L and atm to kPa must be included.
Solution:

$P_1 = 0.871$ atm; $V_1 = (105$ mL$)\left(\dfrac{10^{-3} L}{1 \text{ mL}}\right) = 0.105$ L; $P_2 = (26.3$ kPa$)\left(\dfrac{10^3 \text{ Pa}}{1 \text{ kPa}}\right)\left(\dfrac{1 \text{ atm}}{1.01325 \times 10^5 \text{ Pa}}\right)$

$= 0.25956$ atm

$V_2 = (V_1)\left(\dfrac{P_1}{P_2}\right) = \dfrac{(0.871 \text{ atm})(0.105 \text{ L})}{(0.25956 \text{ atm})} = 0.352346 = \textbf{0.352 L}$

Check: As the pressure goes from 0.871 atm to 0.25956 atm, it is decreasing by a factor of about 3 and the volume should increase by the same factor. The calculated volume of 0.352 L is approximately 3 times greater than the initial volume of 0.105 L.

5.3 Plan: The problem asked for a temperature with given initial temperature, initial volume, and final volume. The relationship to use is $V_1/T_1 = V_2/T_2$. The units of volume in both cases are the same, but the initial temperature must be converted to Kelvin.
Solution:
$V_1 = 6.83$ cm^3; $T_1 = 0°C + 273 = 273$ K; $V_2 = 9.75$ cm^3; $T_2 = ?$
$\dfrac{V_1}{T_1} = \dfrac{V_2}{T_2}$

$T_2 = T_1\left(\dfrac{V_2}{V_1}\right) = (273 \text{ K})\dfrac{(9.75 \text{ cm}^3)}{(6.83 \text{ cm}^3)} = 389.714 = \textbf{390. K}$

Check: The volume increases by a factor of about 1.4 so the temperature must have increased by the same factor. The initial temperature of 273 K times 1.4 is 380 K, so the answer 390 K appears to be correct.

5.4   Plan: In this problem, the amount of gas is decreasing. Since the container is rigid, the volume of the gas will not change with the decrease in moles of gas. The temperature is also constant. So, the only change will be that the pressure of the gas will decrease since fewer moles of gas will be present after removal of the 5.0 g of ethylene. To calculate the final pressure, use the relationship $n_1/P_1 = n_2/P_2$. Since the ratio of moles of ethylene is equal to the ratio of grams of ethylene, there is no need to convert the grams to moles. (This is illustrated in the solution by listing the molar mass conversion twice.)
Solution:

$$\frac{n_1}{P_1} = \frac{n_2}{P_2}$$

$$P_2 = (P_1)\left(\frac{n_2}{n_1}\right) = (793 \text{ torr})\frac{\left((35.0-5.0)\text{ g } C_2H_4\right)\left(\dfrac{1 \text{ mol } C_2H_4}{28.05 \text{ g } C_2H_4}\right)}{(35.0 \text{ g } C_2H_4)\left(\dfrac{1 \text{ mol } C_2H_4}{28.05 \text{ g } C_2H_4}\right)} = 679.714 = \textbf{680. torr}$$

Check: The amount of gas decreases by a factor of 6/7. Since pressure is proportional to the amount of gas, the pressure should decrease by the same factor, $6/7 \times 793 = 680$.

5.5   Plan: From Sample Problem 5.5 the temperature of 21°C and volume of 438 L are given. The pressure is 1.37 atm and the unknown is the moles of oxygen gas. Use the ideal gas equation $PV = nRT$ to calculate the number of moles of gas.
Solution:

$$n = PV / RT = \frac{(1.37 \text{ atm})(438 \text{ L})}{\left(\dfrac{0.0821 \text{ atm} \cdot L}{\text{mol} \cdot K}\right)\left((273.15 + 21)K\right)} = 24.847 \text{ mol } O_2$$

Mass of O_2 = (24.847 mol O_2) x (32.00 g/mol) = 795.104 = **795 g O_2**

5.6   Plan: The pressure is constant and, according to the picture, the volume approximately doubles. The volume change may be due to the temperature and/or a change in moles.
Solution:
The balanced chemical equation must be $2 \text{ CD} \rightarrow C_2 + D_2$.
Thus, the number of mole of gas does not change (2 moles both before and after the reaction). Only the temperature remains as a variable to cause the volume change. Using $V_1 / T_1 = V_2 / T_2$ with $V_2 = 2 V_1$, and $T_1 = (-73 + 273.15)$ K gives:

$$T_2 = T_1\left(\frac{V_2}{V_1}\right) = \frac{(2V_1)(-73 + 273.15)K}{(V_1)} = 400.30 \text{ K} - 273.15 = 127.15 = \textbf{127°C}$$

5.7   Plan: Density of a gas can be calculated using a version of the ideal gas equation, $d = \dfrac{\mathcal{M}P}{RT}$

Two calculations are required, one with T = 0°C = 273 K and P = 380 torr and the other at STP which is defined as T = 273 K and P = 1 atm.
Solution: Density at T = 273 K and P = 380 torr

$$d = \frac{(44.01 \text{ g/mol})(380 \text{ torr})}{\left(\dfrac{0.0821 \text{ atm} \cdot L}{\text{mol} \cdot K}\right)(273 \text{ K})}\left(\frac{1 \text{ atm}}{760 \text{ torr}}\right) = 0.981783 = \textbf{0.982 g/L}$$

Density at T = 273 K and P = 1 atm. (Note: The 1 atm is an exact number and does not affect the significant figures in the answer.)

$$d = \frac{(44.01 \text{ g/mol})(1 \text{ atm})}{\left(\dfrac{0.0821 \text{ atm} \cdot L}{\text{mol} \cdot K}\right)(273 \text{ K})} = 1.9638566 = \textbf{1.96 g/L}$$

The density of a gas increases proportionally to the increase in pressure.

Check: In the two density calculations, the temperature of the gas is the same, but the pressure differs by a factor of two. The calculation of density shows that it is proportional to pressure. The pressure in the first case is half the pressure at STP, so the density at 380 torr should be half the density at STP and it is.

5.8   Plan: Use the ideal gas equation for density, $d = \dfrac{\mathcal{M}P}{RT}$, and solve for molar mass.

Solution:

$$\text{Molar mass} = \frac{dRT}{P} = \frac{(1.26 \text{ g/L}) \left(\dfrac{0.0821 \text{ atm} \bullet \text{L}}{\text{mol} \bullet \text{K}} \right) ((10.0 + 273.15) \text{K})}{(102.5 \text{ kPa})} \left(\frac{101.325 \text{ kPa}}{1 \text{ atm}} \right)$$

$= 28.95496 = $ **29.0 g/mol**

Check: Dry air would consist of about 80% N_2 and 20% O_2. Estimating a molar mass for this mixture gives (0.80 x 28) + (0.20 x 32) = 28.8 g/mol, which is close to the calculated value.

5.9   Plan: Calculate the number of moles of each gas present and then the mole fraction of each gas. The partial pressure of each gas equals the mole fraction times the total pressure. Total pressure equals 1 atm since the problem specifies STP. This pressure is an exact number, and will not affect the significant figures in the answer. No intermediate values, such as the moles of each gas, will be rounded. This will avoid intermediate rounding.

Solution:

$$n_{He} = (5.50 \text{ g He}) \left(\frac{1 \text{ mol He}}{4.003 \text{ g He}} \right) = 1.37397 \text{ mol He}$$

$$n_{Ne} = (15.0 \text{ g Ne}) \left(\frac{1 \text{ mol Ne}}{20.18 \text{ g Ne}} \right) = 0.74331 \text{ mol Ne}$$

$$n_{Kr} = (35.0 \text{ g Kr}) \left(\frac{1 \text{ mol Kr}}{83.80 \text{ g Kr}} \right) = 0.41766 \text{ mol Ke}$$

Total number of moles of gas = 1.37397 + 0.74331 + 0.41766 = 2.53494 mol

$P_A = X_A \times P_{total}$

$$P_{He} = \left(\frac{1.37397 \text{ mol He}}{2.53494 \text{ mol}} \right) (1 \text{ atm}) = 0.54201 = \textbf{0.542 atm He}$$

$$P_{Ne} = \left(\frac{0.74331 \text{ mol Ne}}{2.53494 \text{ mol}} \right) (1 \text{ atm}) = 0.29322 = \textbf{0.293 atm Ne}$$

$$P_{Kr} = \left(\frac{0.41766 \text{ mol Kr}}{2.53494 \text{ mol}} \right) (1 \text{ atm}) = 0.16476 = \textbf{0.165 atm Kr}$$

Check: One way to check is that the partial pressures add to the total pressure,
0.542 + 0.293 + 0.165 = 1.000 atm, which agrees with the total pressure of 1 atm at STP.

5.10   Plan: The gas collected over the water will consist of H_2 and H_2O gas molecules. The partial pressure of the water can be found from the vapor pressure of water at the given temperature given in the text. Subtracting this partial pressure of water from total pressure gives the partial pressure of hydrogen gas collected over the water. Calculate the moles of hydrogen gas using the ideal gas equation. The mass of hydrogen can then be calculated by converting the moles of hydrogen from the ideal gas equation to grams.

Solution: From the table in the text, the partial pressure of water is 13.6 torr at 16°C.

P = 752 torr − 13.6 torr = 738.4 = **738 torr H_2**

The unrounded partial pressure (738.4 torr) will be used to avoid rounding error.

$$\text{Moles of hydrogen} = n = PV/RT = \frac{(738.4 \text{ torr})(1495 \text{ mL})}{\left(\dfrac{0.0821 \text{ atm} \bullet \text{L}}{\text{mol} \bullet \text{K}} \right) ((273.15 + 16) \text{K})} \left(\frac{1 \text{ atm}}{760 \text{ torr}} \right) \left(\frac{10^{-3} \text{ L}}{1 \text{ mL}} \right)$$

$= 0.061186 \text{ mol } H_2$ (unrounded)

Mass of hydrogen = $(0.061186 \text{ mol H}_2)\left(\dfrac{2.016 \text{ g H}_2}{1 \text{ mol H}_2}\right) = 0.123351 = \mathbf{0.123 \text{ g H}_2}$

Check: Since the pressure and temperature are close to STP, and the volume is near 1.5 L, the molar volume at STP (22.4 L/mol) may be used to estimate the mass of hydrogen.
(1 mol/22.4 L) (1.5 L) (2 g/mol) = 0.13 g, which is close to the calculated value.

5.11   Plan: Write a balanced equation for the reaction. Calculate the moles of HCl(g) from the starting amount of sodium chloride using the stoichiometric ratio from the balanced equation. Find the volume of the HCl(g) from the molar volume at STP.
Solution:
The balanced equation is $H_2SO_4(aq) + 2 \text{ NaCl}(aq) \rightarrow Na_2SO_4(aq) + 2 \text{ HCl}(g)$.

$(0.117 \text{ kg NaCl})\left(\dfrac{10^3 \text{ g}}{1 \text{ kg}}\right)\left(\dfrac{1 \text{ mol NaCl}}{58.44 \text{ g NaCl}}\right)\left(\dfrac{2 \text{ mol HCl}}{2 \text{ mol NaCl}}\right) = 2.00205 \text{ mol HCl}$

At STP: $(2.00205 \text{ mol HCl})\left(\dfrac{22.4 \text{ L}}{1 \text{ mol HCl}}\right)\left(\dfrac{1 \text{ mL}}{10^{-3} \text{ L}}\right) = 4.4846 \times 10^4 = \mathbf{4.48 \times 10^4 \text{ mL HCl}}$

Check: 117 g NaCl is about 2 moles, which would form 2 moles of HCl(g). Twice the molar volume is 44.8 L, which is the answer as calculated.

5.12   Plan: Balance the equation for the reaction. Determine the limiting reactant by finding the moles of each reactant from the ideal gas equation, and comparing the values. Calculate the moles of remaining excess reactant. This is the only gas left in the flask, so it is used to calculate the pressure inside the flask. There will be no intermediate rounding.
Solution:
The balanced equation is $NH_3(g) + HCl(g) \rightarrow NH_4Cl(s)$.
The stoichiometric ratio of NH_3 to HCl is 1:1, so the reactant present in the lower quantity of moles is the limiting reactant.

Moles ammonia = $\dfrac{PV}{RT} = \dfrac{(0.452 \text{ atm})(10.0 \text{ L})}{\left(\dfrac{0.0821 \text{ atm} \bullet \text{L}}{\text{mol} \bullet \text{K}}\right)((273.15 + 22)\text{K})} = 0.18653 \text{ mol NH}_3$

Moles hydrogen chloride = $\dfrac{(7.50 \text{ atm})(155 \text{ mL})}{\left(\dfrac{0.0821 \text{ atm} \bullet \text{L}}{\text{mol} \bullet \text{K}}\right)(271 \text{ K})}\left(\dfrac{10^{-3} \text{ L}}{1 \text{ mL}}\right) = 0.052249 \text{ mol HCl}$

The HCl is limiting so the moles of ammonia gas left after the reaction would be
$0.18653 - 0.052249 = 0.134281 \text{ mol NH}_3$

Pressure ammonia = $\dfrac{nRT}{V} = \dfrac{(0.134281 \text{ mol})\left(\dfrac{0.0821 \text{ atm} \bullet \text{L}}{\text{mol} \bullet \text{K}}\right)((273.15 + 22)\text{K})}{(10.0 \text{ L})}$

$= 0.325387 = \mathbf{0.325 \text{ atm NH}_3}$

Check: Doing a rough calculation of moles gives for NH_3
(0.5 x 10) / (0.1 x 300) = 0.17 mol and for HCl (8 x 0.1)/(0.1 x 300) = 0.027 mol which means 0.14 mol NH_3 is left. Plugging this value into a rough calculation of the pressure gives (0.14 x 0.1 x 300)/10 = 0.4 atmospheres. This is close to the calculated answer.

5.13   Plan: Graham's Law can be used to solve for the effusion rate of the ethane since the rate and molar mass of helium is known, along with the molar mass of ethane. In the same way that running slower increases the time to go from one point to another, so the rate of effusion decreases as the time increases. The rate can be expressed as 1/time.
Solution:

$$\frac{\text{Rate He}}{\text{Rate C}_2\text{H}_6} = \sqrt{\frac{M_{C_2H_6}}{M_{He}}}$$

$$\frac{\left(\dfrac{0.010 \text{ mol He}}{1.25 \text{ min}}\right)}{\left(\dfrac{0.010 \text{ mol C}_2\text{H}_6}{t_{C_2H_6}}\right)} = \sqrt{\frac{(30.07 \text{ g/mol})}{(4.003 \text{ g/mol})}}$$

Time for C_2H_6 = 3.42597 = **3.43 min**
Check: The ethane should move slower than the helium since ethane has a larger molar mass. This is consistent with the calculation that the ethane molecule takes longer to effuse. The second check is an estimate. The square root of 30 is estimated as 5 and the square root of 4 is 2. The time 1.25 min x 5/2 = 3.1 min, which validates the calculated answer of 3.42 min.

END–OF–CHAPTER PROBLEMS

5.1   Plan: Review the behavior of the gas phase vs. the liquid phase.
Solution:
a) The volume of the liquid remains constant, but the volume of the gas increases to the volume of the larger container.
b) The volume of the container holding the gas sample increases when heated, but the volume of the container holding the liquid sample remains essentially constant when heated.
c) The volume of the liquid remains essentially constant, but the volume of the gas is reduced.

5.6   Plan: The ratio of the heights of columns of mercury and water are inversely proportional to the ratio of the densities of the two liquids.
Solution:

$$\frac{h_{H_2O}}{h_{Hg}} = \frac{d_{Hg}}{d_{H_2O}}$$

$$h_{H_2O} = \frac{d_{Hg}}{d_{H_2O}} x h_{Hg} = \left(\frac{13.5 \text{ g/mL}}{1.00 \text{ g/mL}}\right)(730 \text{ mmHg})\left(\frac{10^{-3} \text{ m}}{1 \text{ mm}}\right)\left(\frac{1 \text{ cm}}{10^{-2} \text{ m}}\right) = 985.5 = \textbf{990 cmH}_2\textbf{O}$$

5.8   Plan: Use the conversion factors between pressure units:
1 atm = 760 mmHg = 760 torr = 101.325 kPa = 1.01325 bar
Solution:

a) $(0.745 \text{ atm})\left(\dfrac{760 \text{ mmHg}}{1 \text{ atm}}\right) = 566.2 = \textbf{566 mmHg}$

b) $(992 \text{ torr})\left(\dfrac{1.01325 \text{ bar}}{760 \text{ torr}}\right) = 1.32256 = \textbf{1.32 bar}$

c) $(365 \text{ kPa})\left(\dfrac{1 \text{ atm}}{101.325 \text{ kPa}}\right) = 3.60227 = \textbf{3.60 atm}$

d) $(804 \text{ mmHg})\left(\dfrac{101.325 \text{ kPa}}{760 \text{ mmHg}}\right) = 107.191 = \textbf{107 kPa}$

5.10   Plan: Since the height of the mercury column in contact with the gas is higher than the column in contact with the air, the gas is exerting less pressure on the mercury than the air. Therefore the pressure corresponding to the height difference between the two arms is subtracted from the atmospheric pressure.
Solution:

$$(2.35 \text{ cm})\left(\frac{10^{-2} \text{ m}}{1 \text{ cm}}\right)\left(\frac{1 \text{ mm}}{10^{-3} \text{ m}}\right)\left(\frac{1 \text{ torr}}{1 \text{ mmHg}}\right) = 23.5 \text{ torr}$$

738.5 torr - 23.5 torr = 715.0 torr

$$P(\text{atm}) = (715.0 \text{ torr})\left(\frac{1 \text{ atm}}{760 \text{ torr}}\right) = 0.940789 = \textbf{0.9408 atm}$$

5.12   The difference in the height of the Hg is directly related to pressure in atmospheres.

$$P(\text{atm}) = (0.734 \text{ mHg})\left(\frac{1 \text{ mmHg}}{10^{-3} \text{ mHg}}\right)\left(\frac{1 \text{ atm}}{760 \text{ mmHg}}\right) = 0.965789 = \textbf{0.966 atm}$$

5.18   Plan: Review Boyle's and Amonton's Laws.
Solution:
At constant temperature and volume, the pressure of the gas is directly proportional to the number of moles of the gas. Verify this by examining the ideal gas equation. At constant T and V, the ideal gas equation becomes $P = n(RT/V)$ or $P = n \times$ constant.

5.20   Plan: Use the relationship $\dfrac{P_1 V_1}{T_1} = \dfrac{P_2 V_2}{T_2}$ or $V_2 = \dfrac{V_1 P_1 T_2}{P_2 T_1}$

Solution:
a) As the pressure on a gas increases, the molecules move closer together, decreasing the volume. When the pressure is tripled, the **volume decreases to one third of the original volume** at constant temperature (Boyle's

Law). $V_2 = \dfrac{V_1 P_1 T_2}{P_2 T_1} = \dfrac{(V_1)(P_1)(1)}{(3P_1)(1)}$ $V_2 = \frac{1}{3}V_1$

b) As the temperature of a gas increases, the gas molecules gain kinetic energy. With higher energy, the gas molecules collide with the walls of the container with greater force, which increases the size (volume) of the container. If the temperature is increased by a factor of 3.0 (at constant pressure) then the **volume will increase by a factor of 3.0** (Charles's Law). $V_2 = \dfrac{V_1 P_1 T_2}{P_2 T_1} = \dfrac{(V_1)(1)(3T_1)}{(1)(T_1)}$ $V_2 = 3V_1$

c) As the number of molecules of gas increases, the force they exert on the container increases. This results in an increase in the volume of the container. Adding three moles of gas to one mole increases the number of moles by a factor of four, thus the **volume increases by a factor of four** (Avogadro's Law).

5.22   Plan: Use the relationship $\dfrac{P_1 V_1}{T_1} = \dfrac{P_2 V_2}{T_2}$ or $V_2 = \dfrac{V_1 P_1 T_2}{P_2 T_1}$

Solution:
a)   The temperature is decreased by a factor of 2, so **the volume is decreased by a factor of 2** (Charles's Law).

$V_2 = \dfrac{V_1 P_1 T_2}{P_2 T_1} = \dfrac{(V_1)(1)(\frac{1}{2}T_1)}{(1)(T_1)}$ $V_2 = \frac{1}{2} V_1$

b) The temperature increases by a factor of $[(500 + 273) / (250 + 273)] = 1.48$, so **the volume is increased by a factor of 1.48** (Charles's Law). $V_2 = \dfrac{V_1 P_1 T_2}{P_2 T_1} = \dfrac{(V_1)(1)(1.48T_1)}{(1)(T_1)}$ $V_2 = 1.48V_1$

c) The pressure is increased by a factor of 3, so **the volume decreases by a factor of 3** (Boyle's Law).

$V_2 = \dfrac{V_1 P_1 T_2}{P_2 T_1} = \dfrac{(V_1)(P_1)(1)}{(3P_1)(1)}$ $V_2 = \frac{1}{3}V_1$

5.24   Plan: This is Charles's Law. Charles's Law states that at constant pressure and with a fixed amount of gas, the volume of a gas is directly proportional to the absolute temperature of the gas. The temperature must be lowered to reduce the volume of a gas.

Solution:

$$\frac{V_1}{T_1} = \frac{V_2}{T_2} \qquad \text{at constant n and P} \qquad T_2 = T_1 \frac{V_2}{V_1}$$

$V_1 = 9.10 \text{ L};\quad T_1 = 198^{\circ}C + 273 = 471K;\quad V_2 = 2.50 \text{ L}\quad T_2 = ?$

$$T_2 = 471K \left(\frac{2.50 \text{ L}}{9.10 \text{ L}} \right) = 129.396 \text{ K} - 273 = -143.604 = \mathbf{-144^{\circ}C}$$

5.26   Plan: Since the volume, temperature, and pressure of the gas are changing, use the combined gas law.

Solution:

$$\frac{P_1 V_1}{T_1} = \frac{P_2 V_2}{T_2} \qquad P_1 = 153.3 \text{ kPa};\quad V_1 = 25.5 \text{ L};\quad T_1 = 298 \text{ K};\quad P_2 = 101.325 \text{ kPa};\quad T_2 = 273 \text{ K};\quad V_2 = ?$$

$$V_2 = V_1 \left(\frac{T_2}{T_1} \right) \left(\frac{P_1}{P_2} \right) = (25.5 \text{ L}) \left(\frac{273 \text{ K}}{298 \text{ K}} \right) \left(\frac{153.3 \text{ kPa}}{101.325 \text{ kPa}} \right) = 35.3437 = \mathbf{35.3 \ L}$$

5.28   Plan: Given the volume, pressure, and temperature of a gas, the number of moles of the gas can be calculated using the ideal gas equation, $PV = nRT$. The gas constant, $R = 0.0821$ L•atm/mol•K, gives pressure in atmospheres and temperature in Kelvin. The given pressure in torr must be converted to atmospheres and the temperature converted to Kelvin.

Solution:

$$PV = nRT \quad \text{or} \quad n = PV/RT$$

$$P = (328 \text{ torr}) \left(\frac{1 \text{ atm}}{760 \text{ torr}} \right) = 0.43158 \text{ atm}; \qquad V = 5.0 \text{ L}; \qquad T = 37^{\circ}C + 273 = 310 \text{ K}$$

$$n = \frac{PV}{RT} = \frac{(0.43158 \text{ atm})(5.0 \text{ L})}{\left(\dfrac{0.0821 \text{ atm} \bullet \text{L}}{\text{mol} \bullet \text{K}} \right)(310 \text{ K})} = 0.08479 = \mathbf{0.085 \ mol \ chlorine}$$

5.30   Plan: Solve the ideal gas equation for moles and convert to mass using the molar mass of ClF_3. Volume must be converted to L, pressure to atm and temperature to K.

Solution:

$$PV = nRT \quad \text{or } n = PV/RT \quad P = (699 \text{ mmHg}) \left(\frac{1 \text{ atm}}{760 \text{ mmHg}} \right) = 0.91974 \text{ atm}; \qquad V = 0.357 \text{ L}$$

$$T = 45^{\circ}C + 273 = 318 \text{ K}$$

$$n = \frac{PV}{RT} = \frac{(0.91974 \text{ atm})(0.357 \text{ L})}{\left(\dfrac{0.0821 \text{ atm} \bullet \text{L}}{\text{mol} \bullet \text{K}} \right)(318 \text{ K})} = 0.01258 \ \text{mol } ClF_3$$

$$\text{mass } ClF_3 = (0.01258 \text{ mol } ClF_3) \left(\frac{92.45 \text{ g } ClF_3}{1 \text{ mol } ClF_3} \right) = 1.163021 = \mathbf{1.16 \ g \ ClF_3}$$

5.33   Plan: Assuming that while rising in the atmosphere the balloon will neither gain nor lose gas molecules, the number of moles of gas calculated at sea level will be the same as the number of moles of gas at the higher altitude. Using the ideal gas equation, (n x R) is a constant equal to (PV/T). Given the sea-level conditions of volume, pressure and temperature, and the temperature and pressure at the higher altitude for the gas in the balloon, we can set up an equation to solve for the volume at the higher altitude. Comparing the calculated volume to the given maximum volume of 835 L will tell us if the balloon has reached its maximum volume at this altitude.

Solution:

$$\frac{P_1 V_1}{T_1} = \frac{P_2 V_2}{T_2}$$

$$V_2 = \frac{P_1 V_1 T_2}{P_2 T_1}$$

$$V_2 = \frac{(745 \text{ torr})(65.0 \text{ L})((273 - 5)\text{K})}{((273 + 25)\text{K})(0.066 \text{ atm})} \left(\frac{1 \text{ atm}}{760 \text{ torr}}\right) = 868.2217 = \mathbf{870 \text{ L}}$$

The calculated volume of the gas at the higher altitude is more than the maximum volume of the balloon. **Yes**, the balloon will reach its maximum volume.

Check: Should we expect that the volume of the gas in the balloon should increase? At the higher altitude, the pressure decreases; this increases the volume of the gas. At the higher altitude, the temperature decreases, this decreases the volume of the gas. Which of these will dominate? The pressure decreases by a factor of 0.98/0.066 = 15. If we label the initial volume V_1, then the resulting volume is 15 V_1. The temperature decreases by a factor of 296/268 = 1.1, so the resulting volume is V_1/1.1 or 0.91 V_1. The increase in volume due to the change in pressure is greater than the decrease in volume due to change in temperature, so the volume of gas at the higher altitude should be greater than the volume at sea level.

5.35   The molar mass of H_2 is less than the average molar mass of air (mostly N_2, O_2, and Ar), so air is denser. To collect a beaker of $H_2(g)$, **invert** the beaker so that the air will be replaced by the lighter H_2. The molar mass of CO_2 is greater than the average molar mass of air, so $CO_2(g)$ is more dense. Collect the CO_2 holding the beaker **upright**, so the lighter air will be displaced out the top of the beaker.

5.39   Plan: Using the ideal gas equation and the molar mass of xenon, 131.3 g/mol, we can find the density of xenon gas at STP. Standard temperature is 0°C and standard pressure is 1 atm. Do not forget that the pressure at STP is exact and will not affect the significant figures.
Solution:

$$d = \mathcal{M} P / RT = \frac{(131.3 \text{ g/mol})(1 \text{ atm})}{\left(\dfrac{0.0821 \text{ atm} \cdot \text{L}}{\text{mol} \cdot \text{K}}\right)(273 \text{ K})} = 5.8581 = \mathbf{5.86 \text{ g/L}}$$

5.41   Plan: Apply the ideal gas equation to determine the number of moles. Convert moles to mass and divide by the volume to obtain density in g/L. Do not forget that the pressure at STP is exact and will not affect the significant figures.
Solution:

$$n = \frac{PV}{RT} = \frac{(1 \text{ atm})(0.0400 \text{ L})}{\left(\dfrac{0.0821 \text{ atm} \cdot \text{L}}{\text{mol} \cdot \text{K}}\right)(273 \text{ K})} = 1.78465 \times 10^{-3} = \mathbf{1.78 \times 10^{-3} \text{ mol AsH}_3}$$

$$d = \frac{\text{mass}}{\text{volume}} = \frac{(1.78465 \times 10^{-3} \text{ mol})(77.94 \text{ g/mol})}{(0.0400 \text{ L})} = 3.47740 = \mathbf{3.48 \text{ g/L}}$$

5.43   Plan: Rearrange the formula PV = (m / $\mathcal{M}$)RT to solve for molar mass: $\mathcal{M}$ = mRT / PV. Convert the mass in ng to grams and volume in μL to L. Temperature must be in Kelvin and pressure in torr.
Solution:

$$P = (388 \text{ torr})\left(\frac{1 \text{ atm}}{760 \text{ torr}}\right) = 0.510526 \text{ atm} \qquad T = 45^{\circ}\text{C} + 273 = 318 \text{ K}$$

$$V = (0.206 \text{ } \mu\text{L})\left(\frac{10^{-6} \text{ L}}{1 \text{ } \mu\text{L}}\right) = 2.06 \times 10^{-7} \text{ L} \qquad \text{mass} = (206 \text{ ng})\left(\frac{10^{-9} \text{ g}}{1 \text{ ng}}\right) = 2.06 \times 10^{-7} \text{ g}$$

$$\mathcal{M} = \frac{mRT}{PV} = \left(\frac{(2.06 \times 10^{-7} \text{ g})\left(\frac{0.0821 \text{ atm} \cdot \text{L}}{\text{mol} \cdot \text{K}} \right)(318 \text{ K})}{(0.510526 \text{ atm})(2.06 \times 10^{-7} \text{ L})} \right) = 51.1390 = \mathbf{51.1 \text{ g/mol}}$$

5.45   Plan: Use the ideal gas equation to determine the number of moles of Ar and O_2. The gases are combined ($n_{TOT} = n_{Ar} + n_O$) into a 400 mL flask (V) at 27°C (T). Determine the total pressure from n_{TOT}, V, and T. Pressure must be in units of atm, volume in units of L and temperature in K.
Solution:

$$PV = nRT \qquad\qquad n = \frac{PV}{RT}$$

$$\text{Moles Ar} = \frac{PV}{RT} = \frac{(1.20 \text{ atm})(0.600 \text{ L})}{\left(\frac{0.0821 \text{ atm} \cdot \text{L}}{\text{mol} \cdot \text{K}} \right)\big((273 + 227)\text{K}\big)} = 0.017539585 \text{ mol Ar (unrounded)}$$

$$\text{Moles O}_2 = \frac{PV}{RT} = \frac{(501 \text{ torr})(0.200 \text{ L})}{\left(\frac{0.0821 \text{ atm} \cdot \text{L}}{\text{mol} \cdot \text{K}} \right)\big((273 + 127)\text{K}\big)}\left(\frac{1 \text{ atm}}{760 \text{ torr}} \right) = 0.004014680 \text{ mol O}_2 \text{ (unrounded)}$$

$$n_{TOT} = n_{Ar} + n_O = 0.017539585 \text{ mol} + 0.004014680 \text{ mol} = 0.021554265 \text{ mol}$$

$$\text{P mixture} = \frac{nRT}{V} = \frac{(0.021554265 \text{ mol})\left(\frac{0.0821 \text{ atm} \cdot \text{L}}{\text{mol} \cdot \text{K}} \right)\big((273 + 27)\text{K}\big)}{400 \text{ mL}}\left(\frac{1 \text{ mL}}{10^{-3} \text{ L}} \right) = 1.32720 = \mathbf{1.33 \text{ atm}}$$

5.49   Plan: The problem gives the mass, volume, temperature and pressure of a gas, so we can solve for molar mass using $\mathcal{M} = mRT/PV$. The problem also states that the gas is a hydrocarbon, which by, definition, contains only carbon and hydrogen atoms. We are also told that each molecule of the gas contains five carbon atoms so we can use this information and the calculated molar mass to find out how many hydrogen atoms are present and the formula of the compound. Convert pressure to atm and temperature to K.
Solution:

$$\mathcal{M} = \frac{mRT}{PV} = \frac{(0.482 \text{ g})\left(\frac{0.0821 \text{ atm} \cdot \text{L}}{\text{mol} \cdot \text{K}} \right)\big((273 + 101)\text{K}\big)}{(767 \text{ torr})(0.204 \text{ L})}\left(\frac{760 \text{ torr}}{1 \text{ atm}} \right) = 71.8869 \text{ g/mol (unrounded)}$$

The carbon accounts for [5 (12 g/mol)] = 60 g/mol, thus, the hydrogen must make up the difference (72 – 60) = 12 g/mol. A value of 12 g/mol corresponds to 12 H atoms. (Since fractional atoms are not possible, rounding is acceptable.)
Therefore, the molecular formula is $\mathbf{C_5H_{12}}$.

5.51   Plan: Since you have the pressure, volume and temperature, use the ideal gas equation to solve for the total moles of gas.
Solution:
a) $PV = nRT$

$$n = PV / RT = \frac{(850. \text{ torr})(21 \text{ L})}{\left(\frac{0.0821 \text{ atm} \cdot \text{L}}{\text{mol} \cdot \text{K}} \right)\big((273 + 45)\text{K}\big)}\left(\frac{1 \text{ atm}}{760 \text{ torr}} \right) = 0.89961 = \mathbf{0.90 \text{ mol gas}}$$

b) The information given in ppm is a way of expressing the proportion, or fraction, of SO_2 present in the mixture. Since n is directly proportional to V, the *volume* fraction can be used in place of the *mole* fraction used in equation 5.12. There are 7.95×10^3 parts SO_2 in a million parts of mixture, so volume fraction = (7.95×10^3 / 1×10^6) = 7.95×10^{-3}.
Therefore, P_{SO_2} = volume fraction x P_{TOT} = (7.95×10^{-3}) (850. torr) = 6.7575 = **6.76 torr**.

5.52   Plan: We can find the moles of oxygen from the standard molar volume of gases (1 L of gas occupies 22.4 L at STP) and use the stoichiometric ratio from the balanced equation to determine the moles of phosphorus that will react with the oxygen.
Solution:
$$P_4(s) + 5\ O_2(g) \rightarrow P_4O_{10}(s)$$
$$\text{Mass } P_4 = \left(35.5\ L\ O_2\right)\left(\frac{1\ mol\ O_2}{22.4\ L\ O_2}\right)\left(\frac{1\ mol\ P_4}{5\ mol\ O_2}\right)\left(\frac{123.88\ g\ P_4}{1\ mol\ P_4}\right) = 39.2655 = \mathbf{39.3\ g\ P_4}$$

5.54   Plan: To find the mass of PH_3, write the balanced equation and find the number of moles of PH_3 produced by each reactant. The smaller number of moles of product indicates the limiting reagent. Solve for moles of H_2 using the standard molar volume (or use ideal gas equation).
Solution:
$$P_4(s) + 6\ H_2(g) \rightarrow 4\ PH_3(g)$$
$$\text{Moles hydrogen} = \left(83.0\ L\right)\left(\frac{1\ mol}{22.4\ L}\right) = 3.705357\ mol\ H_2\ \text{(unrounded)}$$
$$PH_3 \text{ from } P_4 = \left(37.5\ g\ P_4\right)\left(\frac{1\ mol\ P_4}{123.88\ g\ P_4}\right)\left(\frac{4\ mol\ PH_3}{1\ mol\ P_4}\right) = 1.21085\ mol\ PH_3$$
$$PH_3 \text{ from } H_2 = \left(3.705357\ mol\ H_2\right)\left(\frac{4\ mol\ PH_3}{6\ mol\ H_2}\right) = 2.470238\ mol\ PH_3$$

P_4 is the limiting reactant.
$$\text{Mass } PH_3 = \left(37.5\ g\ P_4\right)\left(\frac{1\ mol\ P_4}{123.88\ g\ P_4}\right)\left(\frac{4\ mol\ PH_3}{1\ mol\ P_4}\right)\left(\frac{33.99\ g\ PH_3}{1\ mol\ PH_3}\right) = 41.15676 = \mathbf{41.2\ g\ PH_3}$$

5.56   Plan: First, write the balanced equation. The moles of hydrogen produced can be calculated from the ideal gas equation and then the stoichiometric ratio from the balanced equation is used to determine the moles of aluminum that reacted. The problem specifies "hydrogen gas collected over water," so the partial pressure of water must first be subtracted. Table 5.3 reports pressure at 26°C (25.2 torr) and 28°C (28.3 torr), so take the average of the two values to obtain the partial pressure of water at 27°C.
Solution:
$$2\ Al(s) + 6\ HCl(aq) \rightarrow 2\ AlCl_3(aq) + 3\ H_2(g)$$
Hydrogen pressure = total pressure – pressure of water vapor =
$$(751\ mmHg) - [(28.3 + 25.2)\ torr\ /\ 2] = 724.25\ torr\ \text{(unrounded)}$$
Moles of hydrogen:

$$PV = nRT \qquad n = \frac{PV}{RT} = \frac{(724.25\ torr)(35.8\ mL)}{\left(\dfrac{0.0821\ atm\cdot L}{mol\cdot K}\right)(273+27)K}\left(\frac{1\ atm}{760\ torr}\right)\left(\frac{10^{-3}\ L}{1\ mL}\right) = 0.00138514\ mol\ H_2$$

$$\text{Mass of Al} = \left(0.00138514\ mol\ H_2\right)\left(\frac{2\ mol\ Al}{3\ mol\ H_2}\right)\left(\frac{26.98\ g\ Al}{1\ mol\ Al}\right) = 0.024914 = \mathbf{0.0249\ g\ Al}$$

5.58   Plan: To find mL of SO_2, write the balanced equation, convert the given mass of P_4S_3 to moles, use the molar ratio to find moles of SO_2, and use the ideal gas equation to find volume.
Solution:
$$P_4S_3(s) + 8\ O_2(g) \rightarrow P_4O_{10}(s) + 3\ SO_2(g)$$
$$\text{Moles } SO_2 = \left(0.800\ g\ P_4S_3\right)\left(\frac{1\ mol\ P_4S_3}{220.09\ g\ P_4S_3}\right)\left(\frac{3\ mol\ SO_2}{1\ mol\ P_4S_3}\right) = 0.010905\ mol\ SO_2$$

$$\text{Volume SO}_2 = \frac{nRT}{P} = \frac{(0.010905 \text{ mol SO}_2)\left(\dfrac{0.0821 \text{ atm} \cdot \text{L}}{\text{mol} \cdot \text{K}}\right)((273 + 32)\text{K})}{725 \text{ torr}}\left(\frac{760 \text{ torr}}{1 \text{ atm}}\right)\left(\frac{1 \text{ mL}}{10^{-3} \text{ L}}\right)$$

$$= 286.249 = \textbf{286 mL SO}_2$$

5.60   Plan: First, write the balanced equation. Given the amount of xenon hexafluoride that reacts, we can find the number of moles of silicon tetrafluoride gas formed. Then, using the ideal gas equation with the moles of gas, the temperature and the volume, we can calculate the pressure of the silicon tetrafluoride gas.
Solution:

$$2 \text{ XeF}_6(s) + \text{SiO}_2(s) \rightarrow 2 \text{ XeOF}_4(l) + \text{SiF}_4(g)$$

$$\text{Mole SiF}_4 = n = (2.00 \text{ g XeF}_6)\left(\frac{1 \text{ mol XeF}_6}{245.3 \text{ g XeF}_6}\right)\left(\frac{1 \text{ mol SiF}_4}{2 \text{ mol XeF}_6}\right) = 0.0040766 \text{ mol SiF}_4$$

$$\text{Pressure SiF}_4 = P = \frac{nRT}{V} = \frac{(0.0040766 \text{ mol SiF}_4)\left(\dfrac{0.0821 \text{ atm} \cdot \text{L}}{\text{mol} \cdot \text{K}}\right)((273 + 25)\text{K})}{1.00 \text{ L}}$$

$$= 0.099737 = \textbf{0.0997 atm SiF}_4$$

5.65   At STP (or any identical temperature and pressure), the volume occupied by a mole of any gas will be identical. This is because at the same temperature, all gases have the same average kinetic energy, resulting in the same pressure.

5.68   Plan: The molar masses of the three gases are 2.016 for H_2 (Flask A), 4.003 for He (Flask B), and 16.04 for CH_4 (Flask C). Since hydrogen has the smallest molar mass of the three gases, 4 g of H_2 will contain more gas molecules (about 2 mole's worth) than 4 g of He or 4 g of CH_4. Since helium has a smaller molar mass than methane, 4 g of He will contain more gas molecules (about 1 mole's worth) than 4 g of CH_4 (about 0.25 mole's worth).
Solution:
a) $\textbf{P}_A > \textbf{P}_B > \textbf{P}_C$ The pressure of a gas is proportional to the number of gas molecules. So, the gas sample with more gas molecules will have a greater pressure.
b) $E_A = E_B = E_C$ Average kinetic energy depends only on temperature. The temperature of each gas sample is 273 K, so they all have the same average kinetic energy.
c) $\textbf{rate}_A > \textbf{rate}_B > \textbf{rate}_C$ When comparing the speed of two gas molecules, the one with the lower mass travels faster.
d) $\textbf{total } E_A > \textbf{total } E_B > \textbf{total } E_C$ Since the average kinetic energy for each gas is the same (part b of this problem) then the total kinetic energy would equal the average times the number of molecules. Since the hydrogen flask contains the most molecules, its total kinetic energy will be the greatest.
e) $d_A = d_B = d_C$ Under the conditions stated in this problem, each sample has the same volume, 5 L, and the same mass, 4 g. Thus, the density of each is 4 g/5 L = 0.8 g/L.
f) $\textbf{Collision frequency (A)} > \textbf{collision frequency (B)} > \textbf{collision frequency (C)}$ The number of collisions depends on both the speed and the distance between gas molecules. Since hydrogen is the lightest molecule it has the greatest speed and the 5 L flask of hydrogen also contains the most molecules, so collisions will occur more frequently between hydrogen molecules than between helium molecules. By the same reasoning, collisions will occur more frequently between helium molecules than between methane molecules.

5.69   Plan: To find the ratio of effusion rates, calculate the inverse of the ratio of the square roots of the molar masses (Graham's Law).
Solution:

$$\frac{\text{Rate H}_2}{\text{Rate UF}_6} = \sqrt{\frac{\text{Molar Mass UF}_6}{\text{Molar Mass H}_2}} = \sqrt{\frac{352.0 \text{ g/mol}}{2.016 \text{ g/mol}}} = 13.2137 = \textbf{13.21}$$

5.71   Plan: Recall that the heavier the gas, the slower the molecular speed. The molar mass of Ar is 39.95 g/mol while the molar mass of He is 4.003 g/mol.
Solution:
a) The gases have the same average kinetic energy because they are at the same temperature. The heavier Ar atoms are moving slower than the lighter He atoms to maintain the same average kinetic energy. Therefore, **Curve 1** better represents the behavior of Ar.
b) A gas that has a slower molecular speed would effuse more slowly, so **Curve 1** is the better choice.
c) Fluorine gas exists as a diatomic molecule, F_2, with $\mathcal{M} = 38.00$ g/mol. Therefore, F_2 is much closer in mass to Ar (39.95 g/mol) than He (4.003 g/mol), so **Curve 1** more closely represents F_2's behavior.

5.73   Plan: To find the ratio of effusion rates, calculate the inverse of the ratio of the square roots of the molar masses (Graham's Law). Then use the ratio of effusion rates to find the time for the F_2 effusion.
Solution:

$$\frac{\text{Rate He}}{\text{Rate } F_2} = \sqrt{\frac{\text{Molar Mass } F_2}{\text{Molar Mass He}}} = \sqrt{\frac{38.00 \text{ g/mol}}{4.003 \text{ g/mol}}} = 3.08105 \text{ (unrounded)}$$

$$\frac{\text{Rate He}}{\text{Rate } F_2} = \frac{\text{Time } F_2}{\text{Time He}} \qquad \frac{3.08105}{1.00} = \frac{\text{Time } F_2}{4.85 \text{ min He}} \qquad \text{Time } F_2 = 14.94309 = \textbf{14.9 min}$$

5.75   Plan: White phosphorus is a molecular form of the element phosphorus consisting of some number, x, of phosphorus atoms. Use the relative rates of effusion of white phosphorus and neon to determine the molar mass of white phosphorous. From the molar mass of white phosphorus, determine the number of phosphorus atoms, x, in one molecule of white phosphorus.
Solution:

$$\frac{\text{Rate } P_x}{\text{Rate Ne}} = 0.404 = \sqrt{\frac{\text{Molar Mass Ne}}{\text{Molar Mass } P_x}}$$

$$0.404 = \sqrt{\frac{20.18 \text{ g/mol}}{\text{Molar Mass } P_x}}$$

$$(0.404)^2 = \frac{20.18 \text{ g/mol}}{\text{Molar Mass } P_x}$$

$$0.163216 = \frac{20.18 \text{ g/mol}}{\text{Molar Mass } P_x}$$

Molar Mass P_x = 123.6398 g/mol

$$\left(\frac{123.6398 \text{ g}}{\text{mol } P_x}\right)\left(\frac{1 \text{ mol P}}{30.97 \text{ g P}}\right) = 3.992244 = 4 \text{ mol P/mol } P_x \quad \text{or} \quad 4 \text{ atoms P/molecule } P_x$$

Thus, **4 atoms per molecule**, so $P_x = P_4$.

5.78   Intermolecular attractions cause the real pressure to be *less than* ideal pressure, so it causes a *negative* deviation. The size of the intermolecular attraction is related to the constant *a*. According to Table 5.5, $a_{N_2} = 1.39$, $a_{Kr} = 2.32$ and $a_{CO_2} = 3.59$. Therefore, CO_2 experiences a greater negative deviation in pressure than the other two gases: **$N_2 < Kr < CO_2$.**

5.80   Nitrogen gas behaves more ideally at **1 atm** than at 500 atm because at lower pressures the gas molecules are farther apart. An ideal gas is defined as consisting of gas molecules that act independently of the other gas molecules. When gas molecules are far apart, they act ideally, because intermolecular attractions are less important and the volume of the molecules is a smaller fraction of the container volume.

5.83   Plan: Use the Ideal Gas Equation to find the number of moles of O_2. Moles of O_2 combine with Hb in a 4:1 ratio. Divide the mass of Hb by the number of moles to obtain molar mass, g/mol.
Solution:
$PV = nRT$

$$\text{Moles } O_2 = n = \frac{PV}{RT} = \frac{(743 \text{ torr})(1.53 \text{ mL})}{\left(\dfrac{0.0821 \text{ atm} \cdot L}{\text{mol} \cdot K}\right)((273 + 37)K)}\left(\frac{1 \text{ atm}}{760 \text{ torr}}\right)\left(\frac{10^{-3} L}{1 \text{ mL}}\right)$$

$$= 5.87708 \times 10^{-5} \text{ mol } O_2 \text{ (unrounded)}$$

$$\text{Moles Hb} = \left(5.87708 \times 10^{-5} \text{ mol } O_2\right)\left(\frac{1 \text{ mol Hb}}{4 \text{ mol } O_2}\right) = 1.46927 \times 10^{-5} \text{ mol Hb (unrounded)}$$

Molar mass hemoglobin = (1.00 g Hb) / (1.46927×10^{-5} mol Hb) = 6.806098×10^4 = **6.81×10^4 g/mol**

5.86   Plan: Convert the mass of Cl_2 to moles and use the ideal gas law and van der Waals equation to find the pressure of the gas.
Solution:

a) Moles Cl_2: $\left(0.5950 \text{ kg } Cl_2\right)\left(\dfrac{10^3 \text{ g}}{1 \text{ kg}}\right)\left(\dfrac{1 \text{ mol } Cl_2}{70.90 \text{ g } Cl_2}\right) = 8.3921016 \text{ mol}$

Ideal gas equation: $PV = nRT$

$$P_{IGL} = nRT/V = \frac{8.3921016 \text{ mol}\left(\dfrac{0.0821 \text{ atm} \cdot L}{\text{mol} \cdot K}\right)((273 + 225)K)}{15.50 \text{ L}}$$

$$= 22.1366 = \textbf{22.1 atm}$$

b) $\left(P + \dfrac{n^2 a}{V^2}\right)(V - nb) = nRT$

$$P_{VDW} = \frac{nRT}{V - nb} - \frac{n^2 a}{V^2}$$ From Table 5.5: a = $6.49 \dfrac{\text{atm} \cdot L^2}{\text{mol}^2}$;   b = $0.0562 \dfrac{L}{\text{mol}}$

n = 8.3921016 mol from Part (a)

$$P_{VDW} = \frac{(8.3921016 \text{ mol } Cl_2)\left(\dfrac{0.0821 \text{ atm} \cdot L}{\text{mol} \cdot K}\right)((273 + 225)K)}{15.50 \text{ L} - (8.3921016 \text{ mol } Cl_2)\left(0.0562 \dfrac{L}{\text{mol}}\right)} - \frac{(8.3921016 \text{ mol } Cl_2)^2\left(6.49 \dfrac{\text{atm} \cdot L^2}{\text{mol}^2}\right)}{(15.50 \text{ L})^2}$$

$$= 20.91773 = \textbf{20.9 atm}$$

5.90   Plan: Partial pressures and mole fractions are calculated from Dalton's Law of Partial Pressures: $P_A = X_A(P_{\text{total}})$ Remember that 1 atm = 760 torr.
Solution:
a) Convert each mole percent to a mole fraction by dividing by 100%.

$$P_{\text{Nitrogen}} = X_{\text{Nitrogen}} P_{\text{Total}} = 0.786(1.00 \text{ atm})\left(\frac{760 \text{ torr}}{1 \text{ atm}}\right) = 597.36 = \textbf{597 torr } N_2$$

$$P_{\text{Oxygen}} = X_{\text{Oxygen}} P_{\text{Total}} = 0.209(1.00 \text{ atm})\left(\frac{760 \text{ torr}}{1 \text{ atm}}\right) = 158.84 = \textbf{159 torr } O_2$$

$$P_{\text{Carbon Dioxide}} = X_{\text{Carbon Dioxide}} P_{\text{Total}} = 4.00 \times 10^{-4}(1.00 \text{ atm})\left(\frac{760 \text{ torr}}{1 \text{ atm}}\right) = 0.304 = \textbf{0.3 torr } CO_2$$

$$P_{\text{Water}} = X_{\text{Water}} P_{\text{Total}} = 0.0046(1.00 \text{ atm})\left(\frac{760 \text{ torr}}{1 \text{ atm}}\right) = 3.496 = \textbf{3.5 torr } O_2$$

b) Mole fractions can be calculated by rearranging Dalton's Law of Partial Pressures:

$X_A = P_A/P_{total}$ and multiply by 100 to express mole fraction as percent.

$P_{Total} = (569 + 104 + 40 + 47)$ torr = 760 torr

N_2:   [(569 torr) / (760 torr)] x 100% = 74.8684 = **74.9 mol% N$_2$**

O_2:   [(104 torr) / (760 torr)] x 100% = 13.6842 = **13.7 mol% O$_2$**

CO_2:   [(40 torr) / (760 torr)] x 100% = 5.263 = **5.3 mol% CO$_2$**

H_2O: [(47 torr) / (760 torr)] x 100% = 6.1842 = **6.2 mol% CO$_2$**

c) Number of molecules of O_2 can be calculated using the Ideal Gas Equation and Avogadro's number.

$PV = nRT$

$$\text{Moles O}_2 = \frac{PV}{RT} = \frac{(104 \text{ torr})(0.50 \text{ L})}{\left(\dfrac{0.0821 \text{ atm} \cdot \text{L}}{\text{mol} \cdot \text{K}}\right)((273+37)\text{K})}\left(\frac{1 \text{ atm}}{760 \text{ torr}}\right) = 0.0026883 \text{ mol O}_2$$

$$\text{Molecules O}_2 = (0.0026883 \text{ mol O}_2)\left(\frac{6.022 \times 10^{23} \text{ molecules O}_2}{1 \text{ mol O}_2}\right)$$

$$= 1.6189 \times 10^{21} = \mathbf{1.6 \times 10^{21} \text{ molecules O}_2}$$

5.92   <u>Plan:</u> For part a, since the volume, temperature, and pressure of the gas are changing, use the combined gas law. For part b, use the ideal gas equation to solve for moles of air and then moles of N_2.

<u>Solution:</u>

a) $\dfrac{P_1 V_1}{T_1} = \dfrac{P_2 V_2}{T_2}$ $P_1 = (1450. \text{ mmHg})\left(\dfrac{1 \text{ atm}}{760 \text{ mmHg}}\right) = 1.9079$ atm; $V_1 = 208$ mL;

$T_1 = 286$ K; $P_2 = 1$ atm; $T_2 = 298$ K; $V_2 = ?$

$$V_2 = V_1\left(\frac{T_2}{T_1}\right)\left(\frac{P_1}{P_2}\right) = (208 \text{ mL})\left(\frac{298 \text{ K}}{286 \text{ K}}\right)\left(\frac{1.9079 \text{ atm}}{1 \text{ atm}}\right) = 4.13494 \text{ mL} = \mathbf{4 \times 10^2 \text{ mL}}$$

b) Mole air $= n = \dfrac{PV}{RT} = \dfrac{(1.9079 \text{ atm})(0.208 \text{ L})}{\left(\dfrac{0.0821 \text{ atm} \cdot \text{L}}{\text{mol} \cdot \text{K}}\right)(286 \text{ K})} = 0.016901$ mol air

Mole $N_2 = (0.016901 \text{ mol})\left(\dfrac{77\% \text{ N}_2}{100\%}\right) = 0.01301 = \mathbf{0.013 \text{ mol N}_2}$

5.94   <u>Plan:</u> The balanced equation and reactant amounts are given, so the first step is to identify the limiting reactant. Then use the limiting reactant and molar ratios to find the moles of NO_2 produced. The volume of NO_2 is found from the Ideal Gas Equation.

<u>Solution:</u>

$Cu(s) + 4 \text{ HNO}_3(aq) \rightarrow Cu(NO_3)_2(aq) + 2 \text{ NO}_2(g) + 2 \text{ H}_2O(l)$

$$\text{Moles NO}_2 \text{ from Cu} = (4.95 \text{ cm}^3)\left(\frac{8.95 \text{ g Cu}}{\text{cm}^3}\right)\left(\frac{1 \text{ mol Cu}}{63.55 \text{ g Cu}}\right)\left(\frac{2 \text{ mol NO}_2}{1 \text{ mol Cu}}\right) = 1.394256 \text{ mol NO}_2 \text{ (unrounded)}$$

$$\text{Moles NO}_2 \text{ from HNO}_3 = (230.0 \text{ mL})\left(\frac{68.0\% \text{ HNO}_3}{100\%}\right)\left(\frac{1 \text{ cm}^3}{1 \text{ mL}}\right)\left(\frac{1.42 \text{ g}}{\text{cm}^3}\right)\left(\frac{1 \text{ mol HNO}_3}{63.02 \text{ g}}\right)\left(\frac{2 \text{ mol NO}_2}{4 \text{ mol HNO}_3}\right)$$

$$= 1.7620 \text{ mol NO}_2 \text{ (unrounded)}$$

Since less product can be made from the copper, it is the limiting reactant and excess nitric acid will be left after the reaction goes to completion. Use the calculated number of moles of NO_2 and the given temperature and pressure in the ideal gas equation to find the volume of nitrogen dioxide produced. Note that nitrogen dioxide is the only gas involved in the reaction.

$$V = \frac{nRT}{P} = \frac{\left(1.394256 \text{ mol NO}_2\right)\left(\dfrac{0.0821 \text{ atm} \cdot \text{L}}{\text{mol} \cdot \text{K}}\right)\left((273.2 + 28.2)\text{K}\right)}{\left(735 \text{ torr}\right)}\left(\frac{760 \text{ torr}}{1 \text{ atm}}\right)$$

$$= 35.67428 = \textbf{35.7 L NO}_2$$

5.99   Plan: The empirical formula for aluminum chloride is $AlCl_3$ (Al^{3+} and Cl^-). The empirical formula mass is (133.33 g/mol). Calculate the molar mass of the gaseous species from the ratio of effusion rates. This molar mass, divided by the empirical weight, should give a whole number multiple that will yield the molecular formula.
Solution:

$$\frac{\text{Rate Unk}}{\text{Rate He}} = 0.122 = \sqrt{\frac{\text{Molar Mass He}}{\text{Molar Mass Unk}}}$$

$$0.122 = \sqrt{\frac{4.003 \text{ g/mol}}{\text{Molar Mass Unk}}}$$

Molar mass Unknown = 268.9465 g/mol
The whole number multiple is 268.9465/133.33, which is about 2. Therefore, the molecular formula of the gaseous species is 2 x ($AlCl_3$) = $\textbf{Al}_2\textbf{Cl}_6$.

5.102   Plan: First, write the balanced equation for the reaction: $2 SO_2 + O_2 \rightarrow 2 SO_3$. The total number of moles of gas will change as the reaction occurs since 3 moles of reactant gas forms 2 moles of product gas. From the volume, temperature, and pressures given, we can calculate the number of moles of gas before and after the reaction using ideal gas equation. For each mole of SO_3 formed, the total number of moles of gas decreases by 1/2 mole. Thus, twice the decrease in moles of gas equals the moles of SO_3 formed.
Solution:
Moles of gas before and after reaction:

$$\text{Initial moles} = \frac{PV}{RT} = \frac{\left(1.90 \text{ atm}\right)\left(2.00 \text{ L}\right)}{\left(\dfrac{0.0821 \text{ atm} \cdot \text{L}}{\text{mol} \cdot \text{K}}\right)\left(800. \text{ K}\right)} = 0.05785627 \text{ mol (unrounded)}$$

$$\text{Final moles} = \frac{PV}{RT} = \frac{\left(1.65 \text{ atm}\right)\left(2.00 \text{ L}\right)}{\left(\dfrac{0.0821 \text{ atm} \cdot \text{L}}{\text{mol} \cdot \text{K}}\right)\left(800. \text{ K}\right)} = 0.050243605 \text{ mol (unrounded)}$$

Moles of SO_3 produced = 2 x decrease in the total number of moles
        = 2 x (0.05785627 mol – 0.050243605 mol)
        = 0.01522533 = $\textbf{1.52 x 10}^{-2}$ **mol**
Check: If the starting amount is 0.0578 total moles of SO_2 and O_2, then x + y = 0.0578 mol, where x = mol of SO_2 and y = mol of O_2. After the reaction:
(x – z) + (y – 0.5z) + z = 0.0502 mol
Where z = mol of SO_3 formed = mol of SO_2 reacted = 2(mol of O_2 reacted).
Subtracting the two equations gives:
        x – (x – z) + y – (y – 0.5z) – z = 0.0578 – 0.0502
        z = 0.0152 mol SO_3
The approach of setting up two equations and solving them gives the same result as above.

5.107   Plan: The balanced equation for the reaction is $Ni(s) + 4 CO(g) \rightarrow Ni(CO)_4(g)$. Although the problem states that Ni is impure, you can assume the impurity is unreactive and thus not include it in the reaction. The mass of Ni can be calculated from the stoichiometric relationship between moles of CO (using the ideal gas equation) and Ni.
Mass Ni = mol Ni x molar mass

Solution:

a) Mass Ni = $\dfrac{(100.7 \text{ kPa})(3.55 \text{ m}^3)}{\left(\dfrac{0.0821 \text{ atm} \cdot \text{L}}{\text{mol} \cdot \text{K}}\right)((273 + 50)\text{K})}\left(\dfrac{1 \text{ atm}}{101.325 \text{ kPa}}\right)\left(\dfrac{1 \text{ L}}{10^{-3} \text{ m}^3}\right)\left(\dfrac{1 \text{ mol Ni}}{4 \text{ mol CO}}\right)\left(\dfrac{58.69 \text{ g Ni}}{1 \text{ mol Ni}}\right)$

= 1952.089 = **1.95 x 10^3 g Ni**

b) Assume the volume is 1 m^3. Use the ideal gas equation to solve for moles of Ni(CO)$_4$, which equals the moles of Ni, and convert moles to grams using the molar mass.

Mass Ni = $\dfrac{(21 \text{ atm})(\text{m}^3)}{\left(\dfrac{0.0821 \text{ atm} \cdot \text{L}}{\text{mol} \cdot \text{K}}\right)((273 + 155)\text{K})}\left(\dfrac{1 \text{ L}}{10^{-3} \text{ m}^3}\right)\left(\dfrac{1 \text{ mol Ni}}{1 \text{ mol Ni(CO)}_4}\right)\left(\dfrac{58.69 \text{ g Ni}}{1 \text{ mol Ni}}\right)$

= 3.50749 x 10^4 = **3.5 x 10^4 g Ni**

The pressure limits the significant figures.

c) The amount of CO needed to form Ni(CO)$_4$ is the same amount that is released on decomposition. The vapor pressure of water at 35°C (42.2 torr) must be accounted for (see Table 5.3). Use the ideal gas equation to calculate the volume of CO, and compare this to the volume of Ni(CO)$_4$. The mass of Ni from a m^3 (part b) can be used to calculate the amount of CO released.

$P_{CO} = P_{total} - P_{water}$ = 769 torr - 42.2 torr = 726.8 torr (unrounded)

V = nRT / P =

$\dfrac{\left[3.50749 \times 10^4 \text{ g Ni}\left(\dfrac{1 \text{ mol Ni}}{58.69 \text{ g Ni}}\right)\left(\dfrac{4 \text{ mol CO}}{1 \text{ mol Ni}}\right)\right]\left(\dfrac{0.0821 \text{ atm} \cdot \text{L}}{\text{mol} \cdot \text{K}}\right)((273 + 35)\text{K})}{726.8 \text{ torr}}\left(\dfrac{760 \text{ torr}}{1 \text{ atm}}\right)\left(\dfrac{10^{-3} \text{ m}^3}{1 \text{ L}}\right)$

= 63.209866 = **63 m^3 CO**

The answer is limited to two significant figures because the mass of Ni comes from part b.

5.109 a) A preliminary equation for this reaction is __C$_x$H$_y$N$_z$ + n O$_2$ → 4 CO$_2$ + 2 N$_2$ + 10 H$_2$O.
Since the organic compound does not contain oxygen, the only source of oxygen as a reactant is oxygen gas. To form 4 volumes of CO$_2$ would require 4 volumes of O$_2$ and to form 10 volumes of H$_2$O would require 5 volumes of O$_2$. Thus, 9 **volumes of O$_2$** was required.
b) Since the volume of a gas is proportional to the number of moles of the gas we can equate volume and moles. From a volume ratio of 4 CO$_2$:2 N$_2$:10 H$_2$O we deduce a mole ratio of 4 C:4 N:20 H or 1 C:1 N:5 H for an empirical formula of **CH$_5$N**.

5.114 Plan: To find the factor by which a diver's lungs would expand, find the factor by which P changes from 125 ft to the surface, and apply Boyle's Law. To find that factor, calculate $P_{seawater}$ at 125 ft by converting the given depth from ft-seawater to mmHg to atm and adding the surface pressure (1.00 atm).
Solution:

$P (H_2O) = (125 \text{ ft})\left(\dfrac{12 \text{ in}}{1 \text{ ft}}\right)\left(\dfrac{2.54 \text{ cm}}{1 \text{ in}}\right)\left(\dfrac{10^{-2} \text{ m}}{1 \text{ cm}}\right)\left(\dfrac{1 \text{ mm}}{10^{-3} \text{ m}}\right) = 3.81 \times 10^4 \text{ mmH}_2\text{O}$

$P (Hg)$: $\dfrac{h_{H_2O}}{h_{Hg}} = \dfrac{d_{Hg}}{d_{H_2O}}$ $\dfrac{3.81 \times 10^4 \text{ mmH}_2\text{O}}{h_{Hg}} = \dfrac{13.5 \text{ g/mL}}{1.04 \text{ g/mL}}$ $h_{Hg} = 2935.1111 \text{ mmHg}$

$P (Hg) = (2935.11111 \text{ mmHg})\left(\dfrac{1 \text{ atm}}{760 \text{ mm Hg}}\right) = 3.861988 \text{ atm (unrounded)}$

$P_{total} = (1.00 \text{ atm}) + (3.861988 \text{ atm}) = 4.861988 \text{ atm (unrounded)}$
Use Boyle's Law to find the volume change of the diver's lungs:
$P_1V_1 = P_2V_2$

$\dfrac{V_2}{V_1} = \dfrac{P_1}{P_2}$ $\dfrac{V_2}{V_1} = \dfrac{4.861988 \text{ atm}}{1 \text{ atm}}$ = **4.86**

To find the depth to which the diver could ascend safely, use the given safe expansion factor (1.5) and the pressure at 125 ft, P_{125}, to find the safest ascended pressure, P_{safe}.

$P_{125} / P_{safe} = 1.5$

$P_{safe} = P_{125} / 1.5 = (4.861988 \text{ atm}) / 1.5 = 3.241325 \text{ atm (unrounded)}$

Convert the pressure in atm to pressure in ft of seawater using the conversion factors above. Subtract this distance from the initial depth to find how far the diver could ascend.

h (Hg): $(4.861988 - 3.241325 \text{ atm})\left(\dfrac{760 \text{ mmHg}}{1 \text{ atm}}\right) = 1231.7039 \text{ mmHg}$

$\dfrac{h_{H_2O}}{h_{Hg}} = \dfrac{d_{Hg}}{d_{H_2O}}$ $\dfrac{h_{H_2O}}{1231.7039 \text{ mmHg}} = \dfrac{13.5 \text{ g/mL}}{1.04 \text{ g/mL}}$ $h_{H_2O} = 15988.464 \text{ mm}$

$(15988.464 \text{ mmH}_2\text{O})\left(\dfrac{10^{-3} \text{ m}}{1 \text{ mm}}\right)\left(\dfrac{1.094 \text{ yd}}{1 \text{ m}}\right)\left(\dfrac{3 \text{ ft}}{1 \text{ yd}}\right) = 52.4741 \text{ ft}$

Therefore, the diver can safely ascend 52.5 ft to a depth of $(125 - 52.4741) = 72.5259 = \textbf{73 ft}$.

5.116   First, write the balanced equation: $2 \text{ H}_2\text{O}_2(aq) \rightarrow 2 \text{ H}_2\text{O}(l) + \text{O}_2(g)$

According to the description in the problem, a given volume of peroxide solution (0.100 L) will release a certain number of "volumes of oxygen gas" (20). Assume that 20 is exact.

0.100 L solution will produce $(20 \times 0.100 \text{ L}) = 2.00 \text{ L O}_2$ gas

You can convert volume of $O_2(g)$ to moles of gas using the ideal gas equation.

Moles $O_2 = PV / RT = \dfrac{(1.00 \text{ atm})(2.00 \text{ L})}{\left(\dfrac{0.0821 \text{ atm} \cdot \text{L}}{\text{mol} \cdot \text{K}}\right)(273 \text{ K})} = 8.92327 \times 10^{-2} \text{ mol O}_2 \text{ (unrounded)}$

The grams of hydrogen peroxide can now be solved from the stoichiometry using the balanced chemical equation.

Mass $H_2O_2 = (8.92327 \times 10^{-2} \text{ mol O}_2) (2 \text{ mol H}_2\text{O}_2 / 1 \text{ mol O}_2) (34.02 \text{ g H}_2\text{O}_2 / 1 \text{ mol H}_2\text{O}_2)$

    $= 6.071395 = \textbf{6.07 g H}_2\textbf{O}_2$

5.121   Plan: The diagram below describes the two Hg height levels within the barometer. To find the mass of N_2, find P and V (T given) and use the ideal gas equation. The P_{N_2} is directly related to the change in column height of Hg. The volume of the space occupied by the $N_2(g)$ is calculated from the dimensions of the barometer.

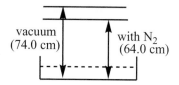

vacuum
(74.0 cm)

with N_2
(64.0 cm)

Solution:

Pressure of the nitrogen $= (74.0 \text{ cm} - 64.0 \text{ cm})\left(\dfrac{10^{-2} \text{ m}}{1 \text{ cm}}\right)\left(\dfrac{1 \text{ mm}}{10^{-3} \text{ m}}\right)\left(\dfrac{1 \text{ atm}}{760 \text{ atm}}\right) = 0.1315789 \text{ atm}$

Volume of the nitrogen $= (1.00 \times 10^2 \text{ cm} - 64.0 \text{ cm})(1.20 \text{ cm}^2)\left(\dfrac{1 \text{ mL}}{1 \text{ cm}^3}\right)\left(\dfrac{10^{-3} \text{ L}}{1 \text{ mL}}\right) = 0.0432 \text{ L}$

The moles of nitrogen may be found using the ideal gas equations, and the moles times the molar mass of nitrogen will give the mass of nitrogen.

Mass of nitrogen $= \dfrac{(0.1315789 \text{ atm})(0.0432 \text{ L})}{\left(\dfrac{0.0821 \text{ atm} \cdot \text{L}}{\text{mol} \cdot \text{K}}\right)((273 + 24)\text{K})}\left(\dfrac{28.02 \text{ g N}_2}{1 \text{ mol}}\right)$

    $= 6.5319 \times 10^{-3} = \textbf{6.53} \times \textbf{10}^{-3} \textbf{ g N}_2$

5.125   **Plan:** Deviations from ideal gas behavior are due to attractive forces between particles which reduce the pressure of the real gas and due to the size of the particle which affects the volume. Compare the size and/or attractive forces between the substances.
   Solution:
   a) **Xenon** would show greater deviation from ideal behavior than argon since xenon is a larger atom than argon. The electron cloud of Xe is more easily distorted so intermolecular attractions are greater. Xe's larger size also means that the volume the gas occupies becomes a greater proportion of the container's volume at high pressures.
   b) **Water** vapor would show greater deviation from ideal behavior than neon gas since the attractive forces between water molecules are greater than the attractive forces between neon atoms. We know the attractive forces are greater for water molecules because it remains a liquid at a higher temperature than neon (water is a liquid at room temperature while neon is a gas at room temperature).
   c) **Mercury** vapor would show greater deviation from ideal behavior than radon gas since the attractive forces between mercury atoms is greater than that between radon atoms. We know that the attractive forces for mercury are greater because it is a liquid at room temperature while radon is a gas.
   d) **Water** is a liquid at room temperature; methane is a gas at room temperature (think about where you have heard of methane gas before - Bunsen burners in lab, cows' digestive system). Therefore, water molecules have stronger attractive forces than methane molecules and should deviate from ideal behavior to a greater extent than methane molecules.

5.129   **Plan:** V and T are not given, so the ideal gas equation cannot be used. The total pressure of the mixture is given. Use $P_A = X_A \times P_{total}$ to find the mole fraction of each gas and then the mass fraction. The total mass of the two gases is 35.0 g.
   Solution:
$$P_{total} = P_{krypton} + P_{carbon\ dioxide} = 0.708\ atm$$
   The NaOH absorbed the CO_2 leaving the Kr, thus $P_{krypton} = 0.250$ atm.
$$P_{carbon\ dioxide} = P_{total} - P_{krypton} = 0.708\ atm - 0.250\ atm = 0.458\ atm$$
   Determining mole fractions: $P_A = X_A \times P_{total}$

   Carbon dioxide: $X = \dfrac{P_{CO_2}}{P_{total}} = \dfrac{0.458\ atm}{0.708\ atm} = 0.64689$ (unrounded)

   Krypton: $X = \dfrac{P_{Kr}}{P_{total}} = \dfrac{0.250\ atm}{0.708\ atm} = 0.353107$ (unrounded)

   Relative mass fraction $= \left[\dfrac{(0.353107)\left(\dfrac{83.80\ g\ Kr}{mol}\right)}{(0.64689)\left(\dfrac{44.01\ g\ CO_2}{mol}\right)} \right] = 1.039366$ (unrounded)

   35.0 g = x g CO_2 + (1.039366 x) g Kr
   35.0 g = 2.039366 x
   Grams CO_2 = x = (35.0 g) / (2.039366) = 17.16219581 = **17.2 g CO_2**
   Grams Kr = 35.0 g – 17.162 g CO_2 = 17.83780419 = **17.8 g Kr**

5.134   The root mean speed is related to temperature and molar mass as shown in equation:
$$u_{rms} = \sqrt{\dfrac{3\ RT}{M}}$$

$$u_{rms}\ Ne = \sqrt{\dfrac{3(8.314\ J/mol\cdot K)(370\ K)}{(20.18\ g/mol)}\left(\dfrac{10^3\ g}{kg}\right)\left(\dfrac{kg\cdot m^2/s^2}{J}\right)} = 676.24788 = \textbf{676 m/s Ne}$$

$$u_{rms}\ Ar = \sqrt{\dfrac{3(8.314\ J/mol\cdot K)(370\ K)}{(39.95\ g/mol)}\left(\dfrac{10^3\ g}{kg}\right)\left(\dfrac{kg\cdot m^2/s^2}{J}\right)} = 480.6269 = \textbf{481 m/s Ar}$$

$$u_{rms} \text{ He} = \sqrt{\frac{3(8.314 \text{ J/mol} \cdot K)(370 \text{ K})}{(4.003 \text{ g/mol})}\left(\frac{10^3 \text{ g}}{kg}\right)\left(\frac{kg \cdot m^2/s^2}{J}\right)} = 1518.356 = \textbf{1.52 x 10}^3 \textbf{ m/s He}$$

5.136   a) The number of moles of water gas can be found using the ideal gas equation, and moles times molar mass gives the mass of water.

$$\text{Moles } H_2O = \frac{PV}{nT} = \frac{(9.0 \text{ atm})(0.25 \text{ mL})}{\left(\dfrac{0.0821 \text{ atm} \cdot L}{mol \cdot K}\right)((273 + 170)K)}\left(\frac{10^{-3} \text{ L}}{1 \text{ mL}}\right)\left(\frac{75\%}{100\%}\right)$$

$$\text{Mass} = (4.639775 \text{ x } 10^{-5} \text{ mol } H_2O)\left(\frac{18.02 \text{ g } H_2O}{1 \text{ mol } H_2O}\right)\left(\frac{100\%}{1.6\%}\right) = 0.052255 = \textbf{0.052 g}$$

The last percent adjustment compensates for the water content being 1.6% of the mass.

$$\text{b) } V = nRT / P = \frac{(4.639775 \text{ x } 10^{-5} \text{ mol } H_2O)\left(\dfrac{0.0821 \text{ atm} \cdot L}{mol \cdot K}\right)((273 + 25)K)}{1.00 \text{ atm}}\left(\frac{1 \text{ mL}}{10^{-3} \text{ L}}\right)$$

$$= 1.13516 = \textbf{1.1 mL}$$

5.143   Plan: Find the number of moles of carbon dioxide produced by converting the mass of glucose in grams to moles and using the stoichiometric ratio from the balanced equation. Then use the T and P given to calculate volume from the ideal gas equation.
Solution:
a) $C_6H_{12}O_6(s) + 6 O_2(g) \rightarrow 6 CO_2(g) + 6 H_2O(g)$

$$\text{Moles } CO_2: \quad (20.0 \text{ g } C_6H_{12}O_6)\left(\frac{1 \text{ mol } C_6H_{12}O_6}{180.16 \text{ g } C_6H_{12}O_6}\right)\left(\frac{6 \text{ mol } CO_2}{1 \text{ mol } C_6H_{12}O_6}\right) = 0.666075 \text{ mol } CO_2$$

$$V = \frac{nRT}{P} = \frac{(0.666075 \text{ mol})\left(\dfrac{0.0821 \text{ atm} \cdot L}{mol \cdot K}\right)((273 + 37)K)}{(780. \text{ torr})}\left(\frac{760 \text{ torr}}{1 \text{ atm}}\right)$$

$$= 16.5176 = \textbf{16.5 Liters CO}_2$$

This solution assumes that partial pressure of O_2 does not interfere with the reaction conditions.
b) Plan: From the stoichiometric ratios in the balanced equation, calculate the moles of each gas and then use Dalton's law of partial pressures to determine the pressure of each gas.
Solution:

$$\text{Moles } CO_2 = \text{Moles } O_2 = (10.0 \text{ g } C_6H_{12}O_6)\left(\frac{1 \text{ mol } C_6H_{12}O_6}{180.16 \text{ g } C_6H_{12}O_6}\right)\left(\frac{6 \text{ mol}}{1 \text{ mol } C_6H_{12}O_6}\right)$$

$$= 0.333037 \text{ mol } CO_2 = \text{ mol } O_2$$

At 37°C, the vapor pressure of water is 48.8 torr. No matter how much water is produced, the partial pressure of H_2O will still be 48.8 torr. The remaining pressure, 780 torr – 48.8 torr = 731.2 torr is the sum of partial pressures for O_2 and CO_2. Since the mole fractions of O_2 and CO_2 are equal, their pressures must be equal, and be one–half of sum of the partial pressures just found.
$\textbf{P}_{\textbf{water}} = \textbf{48.8 torr}$
(731.2 torr) / 2 = 365.6 = $\textbf{3.7 x 10}^2 \textbf{ torr P}_{\textbf{oxygen}} = \textbf{P}_{\textbf{carbon dioxide}}$

5.148   Plan: To find the number of steps through the membrane, calculate the molar masses to find the ratio of effusion rates. This ratio is the enrichment factor for each step.
Solution:

$$\frac{\text{Rate}_{^{235}UF_6}}{\text{Rate}_{^{238}UF_6}} = \sqrt{\frac{\text{Molar Mass } ^{238}UF_6}{\text{Molar Mass } ^{235}UF_6}} = \sqrt{\frac{352.04 \text{ g/mol}}{349.03 \text{ g/mol}}}$$

$$= 1.004302694 \text{ enrichment factor (unrounded)}$$

Therefore, the abundance of $^{235}UF_6$ after one membrane is 0.72% x 1.004302694;
Abundance of $^{235}UF_6$ after "N" membranes = 0.72% * $(1.004302694)^N$
Desired abundance of $^{235}UF_6$ = 3.0% = 0.72% * $(1.004302694)^N$
Solving for N:
3.0% = 0.72% * $(1.004302694)^N$
4.16667 = $(1.004302694)^N$
ln 4.16667 = ln $(1.004302694)^N$
ln 4.16667 = N * ln (1.004302694)
N = (ln 4.16667) / (ln 1.004302694) = 332.392957 = **332 steps**

5.150   Plan: The amount of each gas that leaks from the balloon is proportional to its effusion rate. Using 35% as the rate for H_2, the rate for O_2 can be determined from Graham's Law.
    Solution:

$$\frac{Rate\ O_2}{Rate\ H_2} = \sqrt{\frac{Molar\ Mass\ H_2}{Molar\ Mass\ O_2}} = \sqrt{\frac{2.016\ g/mol}{32.00\ g/mol}} = \frac{Rate\ O_2}{35}$$

Rate O_2 = 0.250998(35) = 8.78493 (unrounded)
Amount of H_2 that leaks = 35%; 100–35 = 65% H_2 remains
Amount of O_2 that leaks = 8.78493%; 100–8.78493 = 91.21507% O_2 remains

$$\frac{O_2}{H_2} = \frac{91.21507}{65} = 1.40331 = \textbf{1.4}$$

5.154   The balanced chemical equation is: $7\ F_2(g) + I_2(s) \rightarrow 2\ IF_7(g)$.
    It will be necessary to determine the limiting reactant.

$$Moles\ IF_7\ from\ F_2 = \frac{(350.\ torr)(2.50\ L)}{\left(\frac{0.0821\ atm \bullet L}{mol \bullet K}\right)(250.\ K)}\left(\frac{1\ atm}{760\ torr}\right)\left(\frac{2\ mol\ IF_7}{7\ mol\ F_2}\right)$$

    = 0.016026668 mol IF_7 (unrounded)

$$Moles\ IF_7\ from\ I_2 = (2.50\ g\ I_2)\left(\frac{1\ mol\ I_2}{253.8\ g\ I_2}\right)\left(\frac{2\ mol\ IF_7}{1\ mol\ I_2}\right) = 0.019700551\ mol\ IF_7\ (unrounded)$$

    F_2 is limiting.
    Mole I_2 remaining = original amount of moles of I_2 – number of I_2 moles reacting with F_2

$$Mole\ I_2\ remaining = (2.50\ g\ I_2)\left(\frac{1\ mol\ I_2}{253.8\ g\ I_2}\right) - \frac{(350.\ torr)(2.50\ L)}{\left(\frac{0.0821\ atm \bullet L}{mol \bullet K}\right)(250.\ K)}\left(\frac{1\ atm}{760\ torr}\right)\left(\frac{1\ mol\ I_2}{7\ mol\ F_2}\right)$$

    = 1.83694×10^{-3} mol I_2 (unrounded)
Total moles of gas = (0 mol F_2) + (0.016026668 mol IF_7) + (1.83694×10^{-3} mol I_2)
    = 0.0178636 mol gas (unrounded)

$$P_{total} = nRT\ /\ V = \frac{(0.0178636\ mol)\left(\frac{0.0821\ atm \bullet L}{mol \bullet K}\right)(550.\ K)}{(2.50\ L)} = 0.322652 = \textbf{0.323 atm}$$

$P_{iodine} = X_{iodine}\ P_{total} = [(1.83694 \times 10^{-3}\ mol\ I_2)\ /\ (0.0178636\ mol)]\ (0.322652\ atm) = 0.03317877 = \textbf{0.0332 atm}$

CHAPTER 6 THERMOCHEMISTRY: ENERGY FLOW AND CHEMICAL CHANGE

FOLLOW–UP PROBLEMS

6.1   Plan: The system is the reactant and products of the reaction. Since heat is absorbed by the surroundings, the system releases heat and q is negative. Because work is done on the system, w is positive. Use the equation $\Delta E = q + w$ to calculate ΔE.
Solution:

$$\Delta E = q + w = \left(-26.0 \text{ kcal}\right)\left(\frac{4.184 \text{ kJ}}{1 \text{ kcal}}\right) + \left(+15.0 \text{ Btu}\right)\left(\frac{1.055 \text{ kJ}}{1 \text{ Btu}}\right) = -92.959 = \textbf{– 93 kJ}$$

Check: A negative ΔE seems reasonable since energy must be removed from the system to condense gaseous reactants into a liquid product.

6.2   Plan: Since heat is a "product" in this reaction, the reaction is **exothermic** ($\Delta H < 0$) and the reactants are above the products in an enthalpy diagram.
Solution:

6.3   Plan: Heat is transferred away from the ethylene glycol as it cools. Table 6.2 lists the specific heat of ethylene glycol as 2.42 J/gK. This value becomes negative because heat is lost. The heat released is calculated using the equation $q = c \times \text{mass} \times \Delta T$
Solution:
$\Delta T = 37.0°C - 25.0°C = 12°C = 12.0 \text{ K}$

$$q = c \times \text{mass} \times \Delta T = \left(2.42 \text{ J/gK}\right)\left[\left(5.50 \text{ L}\right)\left(\frac{1 \text{ mL}}{10^{-3} \text{ L}}\right)\left(\frac{1.11 \text{ g}}{\text{mL}}\right)\right]\left(12.0 \text{ K}\right)\left(\frac{1 \text{ kJ}}{10^3 \text{ J}}\right)$$

$$= -177.289 = \textbf{–177 kJ}$$

6.4   Plan: To find ΔT, find the T_{final} by applying the equation $q = c \times \text{mass} \times \Delta T$. The diamond loses heat ($-q_{diamond}$) whereas the water gains heat (q_{water}). Although the Celsius degree and Kelvin degree are the same size, be careful when interchanging the units. To be safe, always convert the Celsius temperature to Kelvin temperatures. Use the initial temperatures, masses, and specific heat capacities to solve the expression below.
Solution:

$-q_{diamond} = q_{water}$
$-(m \times c_{diamond} \times \Delta T) = m \times c_{water} \times \Delta T$

$$\text{mass of diamond} = \left(10.25 \text{ carat}\right)\left(\frac{0.2000 \text{ g}}{1 \text{ carat}}\right) = 2.050 \text{ g}$$

T_{init} (diamond) $= 74.21 + 273.15 = 347.36 \text{ K}$
T_{init} (water) $= 27.20 + 273.15 = 300.35 \text{ K}$
$- (2.050 \text{ g}) (0.519 \text{ J/g–K}) (T_{final} - 347.36) = (26.05 \text{ g}) (4.184 \text{ J/g–K}) (T_{final} - 300.35)$
$(-1.06395 \text{ J/K}) (T_{final} - 347.36) = (108.9932 \text{ J/K}) (T_{final} - 300.35)$
$- 1.06395 T_{final} + 369.5737 = 108.9932 T_{final} - 32736.108$
$33105.6817 = 110.05715 T_{final}$

$$T_{final} = 300.80446 \text{ K}$$
$$\Delta T_{diamond} = 300.80446 - 347.36 = -46.55554 = \textbf{-46.56 K}$$
$$\Delta T_{water} = 300.80446 - 300.35 = 0.45446 \text{ K} = \textbf{0.45 K}$$

Check: Rounding errors account for the slight differences in the final answer when compared to the identical calculation using Celsius temperatures. The temperature of the diamond decreases as the temperature of the water increases. The temperature of the water does not increase significantly because its specific heat capacity is large in comparison to the diamond, so it takes more heat to raise 1 g of the substance by 1°C.

6.5   Plan: The bomb calorimeter gains heat from the combustion of graphite, so $-q_{graphite} = q_{calorimeter}$. Convert the mass of graphite from grams to moles and use the given kJ/mol to find $q_{graphite}$. The heat lost by graphite equals the heat gained from the calorimeter, or ΔT multiplied by $C_{calorimeter}$.
Solution:
$$-(\text{mol graphite x kJ/mol graphite}) = C_{calorimeter}\Delta T_{calorimeter}$$
$$-(0.8650 \text{ g C})\left(\frac{1 \text{ mol C}}{12.01 \text{ g C}}\right)\left(\frac{-393.5 \text{ kJ}}{1 \text{ mol C}}\right) = C_{calorimeter}(2.613 \text{ K})$$
$$-28.34117 = C_{calorimeter}(2.613 \text{ K})$$
$$C_{calorimeter} = 10.84622 = \textbf{10.85 kJ/K}$$

6.6   Plan: To find the heat required, write a balanced thermochemical equation and use appropriate molar ratios to solve for required heat.
Solution:
$$C_2H_4(g) + H_2(g) \rightarrow C_2H_6(g) \qquad \Delta H = -137 \text{ kJ}$$
$$\text{Heat} = (15.0 \text{ kg } C_2H_6)\left(\frac{10^3 \text{ g}}{1 \text{ kg}}\right)\left(\frac{1 \text{ mol } C_2H_6}{30.07 \text{ g } C_2H_6}\right)\left(\frac{-137 \text{ kJ}}{1 \text{ mol } C_2H_6}\right) = -6.83405 \times 10^4 = \textbf{-6.83} \times \textbf{10}^4 \textbf{ kJ}$$

6.7   Plan: Manipulate the two equations so that their sum will result in the overall equation. Reverse the first equation (and change the sign of ΔH); reverse the second equation and multiply the coefficients (and ΔH) by two.
Solution:

2 ~~NO(g)~~ + 3/2 $O_2(g) \rightarrow N_2O_5(s)$	$\Delta H = -(223.7 \text{ kJ}) = -223.7 \text{ kJ}$
2 $NO_2(g) \rightarrow$ 2 ~~NO(g)~~ + $O_2(g)$	$\Delta H = -2(-57.1 \text{ kJ}) = 114.2 \text{ kJ}$
Total: 2 $NO_2(g)$ + 1/2 $O_2(g) \rightarrow N_2O_5(s)$	$\Delta H = \textbf{-109.5 kJ}$

6.8   Plan: Write the elements as reactants (each in its standard state), and place one mole of the substance formed on the product side. Balance the equation with the following differences from "normal" balancing — only one mole of the desired product can be on the right hand side of the arrow (and nothing else), and fractional coefficients are allowed on the reactant side. The values for the standard heats of formation (ΔH_f°) may be found in the appendix.
Solution:
a) $C(\text{graphite}) + 2 H_2(g) + 1/2 O_2(g) \rightarrow CH_3OH(l) \qquad \Delta H_f^\circ = \textbf{-238.6 kJ}$
b) $Ca(s) + 1/2 O_2(g) \rightarrow CaO(s) \qquad \Delta H_f^\circ = \textbf{-635.1 kJ}$
c) $C(\text{graphite}) + 1/4 S_8(\text{rhombic}) \rightarrow CS_2(l) \qquad \Delta H_f^\circ = \textbf{87.9 kJ}$

6.9   Plan: Apply the ΔH_{rxn}° to this reaction, substitute given values, and solve for the ΔH_f° (CH_3OH).
Solution:
$$\Delta H_{rxn}^\circ = \Sigma \Delta H_f^\circ \text{ (products)} - \Sigma \Delta H_f^\circ \text{ (reactants)}$$
$$\Delta H_{rxn}^\circ = [\Delta H_f^\circ (CO_2(g)) + 2 \Delta H_f^\circ (H_2O(g))] - [\Delta H_f^\circ (CH_3OH(l)) + 3/2 \Delta H_f^\circ (O_2(g))]$$
$$-638.5 \text{ kJ} = [1 \text{ mol}(-393.5 \text{ kJ/mol}) + 2 \text{ mol}(-241.826 \text{ kJ/mol})] - [\Delta H_f^\circ (CH_3OH(l)) + 3/2(0)]$$
$$-638.5 \text{ kJ} = (-877.152 \text{ kJ}) - \Delta H_f^\circ (CH_3OH(l))$$
$$\Delta H_f^\circ (CH_3OH(l)) = -238.652 = \textbf{-238.6 kJ}$$

END–OF–CHAPTER PROBLEMS

6.4 <u>Plan:</u> Remember that an increase in internal energy is a result of the system (body) gaining heat or having work done on it and a decrease in internal energy is a result of the system (body) losing heat or doing work.
<u>Solution:</u>
The internal energy of the body is the sum of the cellular and molecular activities occurring from skin level inward. The body's internal energy can be increased by adding food, which adds energy to the body through the breaking of bonds in the food. The body's internal energy can also be increased through addition of work and heat, like the rubbing of another person's warm hands on the body's cold hands. The body can lose energy if it performs work, like pushing a lawnmower, and can lose energy by losing heat to a cold room.

6.6 <u>Plan:</u> Use the Law of Conservation of Energy.
<u>Solution:</u>
The amount of the change in internal energy in the two cases is the same. By the law of energy conservation, the change in energy of the universe is zero. This requires that the change in energy of the system (heater or air conditioner) equals an opposite change in energy of the surroundings (room air). Since both systems consume the same amount of electrical energy, the change in energy of the heater equals that of the air conditioner.

6.8 The change in a system's energy is $\Delta E = q + w$. If the system <u>receives</u> heat, then its q_{final} is greater than $q_{initial}$ so q is positive. Since the system <u>performs</u> work, its $w_{final} < w_{initial}$ so w is negative.
$\Delta E = q + w$
$\Delta E = (+425 \text{ J}) + (-425 \text{ J}) = \mathbf{0\ J}$.

6.10 <u>Plan:</u> A system that releases thermal energy has a negative value for q and a system that has work done on it has a positive value for work.
<u>Solution:</u>
$\Delta E = -675 \text{ J} + (530 \text{ cal} \times 4.184 \text{ J/cal}) = 1542.52 = \mathbf{1.54 \times 10^3\ J}$

6.12 <u>Plan:</u> Convert 6.6×10^{10} J to the other units using conversion factors.
<u>Solution:</u>
$C(s) + O_2(g) \rightarrow CO_2(g) + 6.6 \times 10^{10} \text{ J}$
(2.0 ton)

a) $\Delta E(\text{kJ}) = (6.6 \times 10^{10} \text{ J})\left(\dfrac{1 \text{ kJ}}{10^3 \text{ J}}\right) = \mathbf{6.6 \times 10^7\ kJ}$

b) $\Delta E(\text{kcal}) = (6.6 \times 10^{10} \text{ J})\left(\dfrac{1 \text{ cal}}{4.184 \text{ J}}\right)\left(\dfrac{1 \text{ kcal}}{10^3 \text{ cal}}\right) = 1.577 \times 10^7 = \mathbf{1.6 \times 10^7\ kcal}$

c) $\Delta E(\text{Btu}) = (6.6 \times 10^{10} \text{ J})\left(\dfrac{1 \text{ Btu}}{1055 \text{ J}}\right) = 6.256 \times 10^7 = \mathbf{6.3 \times 10^7\ Btu}$

6.15 $\text{Time} = (1.0 \text{ lb})\left(\dfrac{4.1 \times 10^3 \text{ Cal}}{1.0 \text{ lb}}\right)\left(\dfrac{10^3 \text{ cal}}{1 \text{ Cal}}\right)\left(\dfrac{4.184 \text{ J}}{1 \text{ cal}}\right)\left(\dfrac{1 \text{ kJ}}{10^3 \text{ J}}\right)\left(\dfrac{\text{h}}{1950 \text{ kJ}}\right) = 8.79713 = \mathbf{8.8\ hour}$

6.17 Since many reactions are performed in an open flask, the reaction proceeds at constant pressure. The determination of ΔH (constant pressure conditions) requires a measurement of heat only, whereas ΔE requires measurement of heat and PV work.

6.19   Plan: An exothermic process releases heat and an endothermic process absorbs heat.
        Solution:
        a) **Exothermic,** the system (water) is releasing heat in changing from liquid to solid.
        b) **Endothermic,** the system (water) is absorbing heat in changing from liquid to gas.
        c) **Exothermic,** the process of digestion breaks down food and releases energy.
        d) **Exothermic,** heat is released as a person runs and muscles perform work.
        e) **Endothermic,** heat is absorbed as food calories are converted to body tissue.
        f) **Endothermic,** the wood being chopped absorbs heat (and work).
        g) **Exothermic,** the furnace releases heat from fuel combustion. Alternatively, if the system is defined as the air in the house, the change is endothermic since the air's temperature is increasing by the input of heat energy from the furnace.

6.22   Plan: An exothermic reaction releases heat, so the reactants have greater H ($H_{initial}$) than the products (H_{final}). $\Delta H = H_{final} - H_{initial} < 0$.
        Solution:

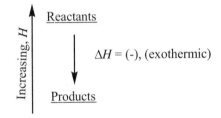

6.24   a) Combustion of ethane: $2\ C_2H_6(g) + 7\ O_2(g) \rightarrow 4\ CO_2(g) + 6\ H_2O(g) + heat$

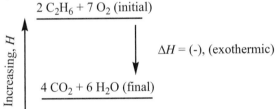

        b) Freezing of water: $H_2O(l) \rightarrow H_2O(s) + heat$

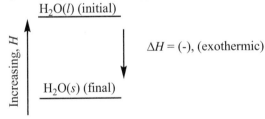

6.26   a) Combustion of hydrocarbons and related compounds require oxygen (and a heat catalyst) to yield carbon dioxide gas, water vapor, and heat.
        $2\ CH_3OH(l) + + 3\ O_2(g) \rightarrow\ 2\ CO_2(g)\ +\ 4\ H_2O(g)\ + heat$

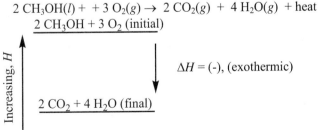

b) Nitrogen dioxide, NO_2, forms from N_2 and O_2.
$1/2\ N_2(g) + O_2(g) + heat \rightarrow NO_2(g)$

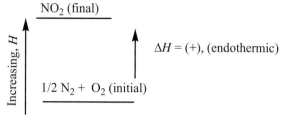

$\Delta H = (+)$, (endothermic)

6.28   a) This is a phase change from the solid phase to the gas phase. Heat is absorbed by the system so q_{sys} is positive (+).
b) The system is expanding in volume as more moles of gas exist after the phase change than were present before the phase change. So the system has done work of expansion and w is negative. $\Delta E_{sys} = q + w$. Since q is positive and w is negative, the sign of ΔE_{sys} cannot be predicted. It will be positive if $q > w$ and negative if $q < w$.
c) $\Delta E_{univ} = 0$. If the system loses energy, the surroundings gain an equal amount of energy. The sum of the energy of the system and the energy of the surroundings remains constant.

6.31   To determine the specific heat capacity of a substance, you need its mass, the heat added (or lost) and the change in temperature.

6.33   Specific heat capacity is the quantity of heat required to raise one gram of a substance by one kelvin. Heat capacity is also the quantity of heat required for a one kelvin temperature change, but it applies to an object instead of a specified amount of a substance. Thus, specific heat capacity is used when talking about an element or compound while heat capacity is used for a calorimeter or other object.
a) Use heat capacity because the fixture is a combination of substances.
b) Use specific heat capacity because the copper wire is a pure substance.
c) Use specific heat capacity because the water is a pure substance.

6.35   Plan: The heat required to raise the temperature of water is found by using the equation
$q = c \times mass \times \Delta T$. The specific heat capacity, c_{water}, is found in Table 6.2. Because the Celsius degree is the same size as the kelvin degree, $\Delta T = 100.^{\circ}C - 20.^{\circ}C = 80.^{\circ}C = 80.\ K$.
Solution:

$$q(J) = c \times mass \times\ \Delta T = \left(4.184\frac{J}{g\ ^{\circ}C}\right)(22.0\ g)\left((100. - 25)^{\circ}C\right) = 6903.6 = \mathbf{6.9 \times 10^3\ J}$$

6.37 $q(J) = c \times mass \times\ \Delta T \qquad T_i = 13.00^{\circ}C \qquad T_f = ? \qquad mass = 295\ g \qquad c = 0.900\ J/g^{\circ}C$

$$q = \left(75.0\ kJ\right)\left(\frac{10^3\ J}{1\ kJ}\right) = 7.50 \times 10^4\ J$$

$7.50 \times 10^4\ J = (0.900\ J/g^{\circ}C)\ (295\ g)\ (T_f - 13.00)^{\circ}C$

$$(T_f - 13.00)^{\circ}C = \frac{\left(7.50 \times 10^4\ J\right)}{(295\ g)\left(\dfrac{0.900\ J}{g\ ^{\circ}C}\right)}$$

$(T_f - 13.00)^{\circ}C = 282.49^{\circ}C$ (unrounded)
$T_f = 295.49 = \mathbf{295^{\circ}C}$

6.39   Plan: Since the bolts have the same mass, and one must cool as the other heats, the intuitive answer is:
       Solution:

$$\left[\frac{(T_1 + T_2)}{2}\right] = \left[\frac{(100.°C + 55°C)}{2}\right] = \textbf{77.5°C}.$$

6.41   Plan: The heat lost by the water originally at 85°C is gained by the water that is originally at 26°C. Therefore $-q_{lost} = q_{gained}$. Both volumes are converted to mass using the density.
       Solution:

Mass of 75 mL $= (75 \text{ mL})\left(\dfrac{1.00 \text{ g}}{1 \text{ mL}}\right) = 75 \text{ g}$ Mass of 155 mL $= (155 \text{ mL})\left(\dfrac{1.00 \text{ g}}{1 \text{ mL}}\right) = 155 g$

$-q_{lost} = q_{gained}$
c x mass x ΔT (85°C water)= c x mass x ΔT (26°C water)
$-$ (4.184 J/g°C) (75 g) $(T_f - 85)°C =$(4.184 J/g°C) (155 g) $(T_f - 26)°C$
$-$ [75]$(T_f - 85)$ = [155] $(T_f - 26)$
$6375 - 75 \, T_f = 155 \, T_f - 4030$
$6375 + 4030 = 155 \, T_f + 75 \, T_f$
$10405 = 230. \, T_f$
$T_f = (10405 / 230.) = 45.24 = \textbf{45°C}$

6.43   Plan: Heat gained by water plus heat lost by copper tubing must equal zero, so $q_{water} = -q_{copper}$. However, some heat will be lost to the insulated container.
       Solution:
$- q_{lost} = q_{gained} = q_{water} + q_{calorimeter}$
$-$ (455 g Cu) (0.387 J/g°C) $(T_f - 89.5)°C =$ (159 g H_2O) (4.184 J/g°C) $(T_f - 22.8)°C +$ (10.0 J/°C) $(T_f - 22.8)°C$
$-$ (176.085) $(T_f - 89.5) =$ (665.256) $(T_f - 22.8) +$ (10.0) $(T_f - 22.8)$
$15759.6075 - 176.085 \, T_f = 665.256 \, T_f - 15167.8368 + 10.0 \, T_f - 228$
$15759.6075 + 15167.8368 + 228 = 176.085 \, T_f + 665.256 \, T_f + 10.0 \, T_f$
$31155.4443 = (851.341) \, T_f$
$T_f = 31155.4443 / (851.341) = 36.59573 = \textbf{36.6°C}$

6.50   The reaction has a **positive ΔH_{rxn}**, because this reaction requires the input of energy to break the oxygen-oxygen bond in O_2:
$$O_2(g) + \text{energy} \rightarrow 2O(g)$$

6.51   As a substance changes from the gaseous state to the liquid state, energy is released so ΔH would be negative for the condensation of 1 mol of water. The value of ΔH for the vaporization of 2 mol of water would be twice the value of ΔH for the condensation of 1 mol of water vapor but would have an opposite sign $(+\Delta H)$.

$H_2O(g) \rightarrow H_2O(l) + \text{Energy}$ $2 H_2O(l) + \text{Energy} \rightarrow 2 H_2O(g)$
    $\Delta H_{cond} = (-)$ $\Delta H_{vap} = (+)2[\Delta H_{cond}]$

The enthalpy for 1 mole of water condensing would be opposite in sign to and one-half the value for the conversion of 2 moles of liquid H_2O to H_2O vapor.

6.52   a) This reaction is **exothermic** because ΔH is negative.
       b) Because ΔH is a state function, the total energy required for the reverse reaction, regardless of how the change occurs, is the same magnitude but different sign of the forward reaction. Therefore, $\Delta H = \textbf{+20.2 kJ}$.
       c) The ΔH_{rxn} is specific for the reaction as written, meaning that 20.2 kJ is released when 1/8 of a mole of sulfur reacts. In this case, 3.2 moles of sulfur react and we therefore expect that much more energy will be released.

$$\Delta H_{rxn} = (2.6 \text{ mol } S_8)\left(\frac{-20.2 \text{ kJ}}{(1/8) \text{ mol } S_8}\right) = -420.16 = \textbf{-4.2 x 10}^2 \textbf{ kJ}$$

d) The mass of S_8 requires conversion to moles and then a calculation identical to c) can be performed.

$$\Delta H_{rxn} = \left(25.0 \text{ g } S_8\right)\left(\frac{1 \text{ mol } S_8}{256.56 \text{ g } S_8}\right)\left(\frac{-20.2 \text{ kJ}}{(1/8) \text{ mol } S_8}\right) = -15.7468 = \textbf{-15.7 kJ}$$

6.54   a) $1/2 \text{ N}_2(g) + 1/2 \text{ O}_2(g) \rightarrow NO(g)$ $\Delta H = 90.29$ kJ

b) $\Delta H_{rxn} = \left(3.50 \text{ g NO}\right)\left(\dfrac{1 \text{ mol NO}}{30.01 \text{ g NO}}\right)\left(\dfrac{-90.29 \text{ kJ}}{1 \text{ mol NO}}\right) = -10.5303 = \textbf{-10.5 kJ}$

6.56   For the reaction written, 2 moles of H_2O_2 release 196.1 kJ of energy upon decomposition.
$$2 \text{ H}_2O_2(l) \rightarrow 2 \text{ H}_2O(l) + O_2(g) \qquad\qquad \Delta H_{rxn} = -196.1 \text{ kJ}$$

$$\text{Heat} = q = \left(652 \text{ kg H}_2O_2\right)\left(\frac{10^3 \text{ g}}{1 \text{ kg}}\right)\left(\frac{1 \text{ mol H}_2O_2}{34.02 \text{ g H}_2O_2}\right)\left(\frac{-196.1 \text{ kJ}}{2 \text{ mol H}_2O_2}\right) = -1.87915 \times 10^6 = \textbf{-1.88} \times \textbf{10}^6 \textbf{ kJ}$$

6.60   a) $C_2H_4(g) + 3 \text{ O}_2(g) \rightarrow 2 \text{ CO}_2(g) + 2 \text{ H}_2O(l)$ $\Delta H_{rxn} = -1411$ kJ

b) Mass $C_2H_4 = \left(-70.0 \text{ kJ}\right)\left(\dfrac{1 \text{ mol C}_2H_4}{-1411 \text{ kJ}}\right)\left(\dfrac{28.05 \text{ g C}_2H_4}{1 \text{ mol C}_2H_4}\right) = 1.39157 = \textbf{1.39 g C}_2\textbf{H}_4$

6.64   Plan: Two chemical equations can be written based on the description given:

$$C(s) + O_2(g) \rightarrow CO_2(g) \qquad\qquad \Delta H_1 \quad (1)$$
$$CO(g) + 1/2 \text{ O}_2(g) \rightarrow CO_2(g) \qquad\qquad \Delta H_2 \quad (2)$$

The second reaction can be reversed, its ΔH sign changed, and appropriate coefficients can be multiplied to one or both of the equations to allow addition of the two reactions. In this case, no coefficients are necessary because the CO_2 cancels.
Solution:

$$C(s) + O_2(g) \rightarrow \cancel{CO_2(g)} \qquad\qquad \Delta H_1$$
$$\cancel{CO_2(g)} \rightarrow CO(g) + 1/2 \text{ O}_2(g) \qquad -\Delta H_2$$
Total $C(s) + 1/2 \text{ O}_2(g) \rightarrow CO(g) \qquad \Delta H_{rxn} = \Delta H_1 + -(\Delta H_2)$

How are the ΔH's for each reaction determined? The ΔH_1 can be found by using the heats of formation in Appendix B: $\Delta H_1 = [\Delta H_f(CO_2)] - [\Delta H_f(C) + \Delta H_f(O_2)]$ $= [-393.5 \text{ kJ/mol}] - [0 + 0] = -393.5 \text{ kJ/mol}$. The ΔH_2 can be found by using the heats of formation in Appendix B: $\Delta H_2 = [\Delta H_f(CO) + 1/2\Delta H_f(O_2)] - [\Delta H_f(CO_2)] = [-110.5$ kJ/mol $+ 0)] - [-393.5] = -283$ kJ/mol. $\Delta H_{rxn} = \Delta H_1 + -(\Delta H_2) = -393.5 \text{ kJ} + -(-283.0 \text{ kJ}) = \textbf{-110.5 kJ}$.

6.65   Plan: To obtain the overall reaction, add the first reaction to the reverse of the second. When the second reaction is reversed, the sign of its enthalpy change is reversed from positive to negative.
Solution:

$Ca(s) + 1/2 \text{ O}_2(g) \rightarrow \cancel{CaO(s)}$	$\Delta H = \quad -635.1 \text{ kJ}$
$\cancel{CaO(s)} + CO_2(g) \rightarrow CaCO_3(s)$	$\Delta H = \quad -178.3 \text{ kJ}$
$Ca(s) + 1/2 \text{ O}_2(g) + CO_2(g) \rightarrow CaCO_3(s)$	$\Delta H = \quad \textbf{-813.4 kJ}$

6.67   1) $N_2(g) + O_2(g) \rightarrow \cancel{2NO(g)}$ $\Delta H = 180.6$ kJ
2) $\cancel{2NO(g)} + O_2(g) \rightarrow 2NO_2(g)$ $\Delta H = -114.2$ kJ
3) $N_2(g) + 2 \text{ O}_2(g) \rightarrow 2 \text{ NO}_2(g)$ $\Delta H_{rxn} = +66.4$ kJ

In Figure P6.67, **A represents reaction 1** with a larger amount of energy absorbed, **B represents reaction 2** with a smaller amount of energy released and **C represents reaction 3** as the sum of A and B.

6.69   Plan: Vaporization is the change in state from a liquid to a gas: $H_2O(l) \rightarrow H_2O(g)$. The two equations describing the chemical reactions for the formation of gaseous and liquid water can be combined to yield the equation for vaporization.
Solution:

1) Formation of $H_2O(g)$: $H_2(g) + 1/2 \text{ O}_2(g) \rightarrow H_2O(g)$ $\Delta H = -241.8$ kJ
2) Formation of $H_2O(l)$: $H_2(g) + 1/2 \text{ O}_2(g) \rightarrow H_2O(l)$ $\Delta H = -285.8$ kJ
Reverse Reaction 2 (change the sign of ΔH) and add the two reactions:

$$H_2(g) + \cancel{1/2\, O_2(g)} \rightarrow H_2O(g) \qquad \Delta H = -241.8\ \text{kJ}$$
$$\cancel{H_2O(l) \rightarrow H_2(g) + 1/2\, O_2(g)} \qquad \Delta H = +285.8\ \text{kJ}$$
$$H_2O(l) \rightarrow H_2O(g) \qquad \Delta H_{vap} = \mathbf{44.0\ kJ}$$

6.72   The standard heat of reaction, ΔH°_{rxn}, is the enthalpy change for any reaction where all substances are in their standard states. The standard heat of formation, ΔH°_{f}, is the enthalpy change that accompanies the formation of one mole of a compound in its standard state from elements in their standard states. Standard state is 1 atm for gases, 1 M for solutes, and pure state for liquids and solids. Standard state does not include a specific temperature, but a temperature must be specified in a table of standard values.

6.74   <u>Plan:</u> ΔH°_{f} is for the reaction that shows the formation of one mole of compound from its elements in their standard states.

<u>Solution:</u>
a) **1/2** $Cl_2(g) + Na(s) \rightarrow NaCl(s)$ The element chlorine occurs as Cl_2, not Cl.
b) $H_2(g) + 1/2\, O_2(g) \rightarrow H_2O(g)$ The element hydrogen exists as H_2, not H, and the formation of water is written with water as the product.
c) No changes

6.75   <u>Plan:</u> Formation equations show the formation of <u>one</u> mole of compound from its elements. The elements must be in their most stable states ($\Delta H^{\circ}_{f} = 0$)

<u>Solution:</u>
a) $Ca(s) + Cl_2(g) \rightarrow CaCl_2(s)$
b) $Na(s) + 1/2\, H_2(g) + C(graphite) + 3/2\, O_2(g) \rightarrow NaHCO_3(s)$
c) $C(graphite) + 2\, Cl_2(g) \rightarrow CCl_4(l)$
d) $1/2\, H_2(g) + 1/2\, N_2(g) + 3/2\, O_2(g) \rightarrow HNO_3(l)$

6.77   <u>Plan:</u> The enthalpy change of a reaction is the sum of the ΔH_{f} of the products minus the sum of the ΔH_{f} of the reactants. Since the ΔH_{f} values (Appendix B) are reported as energy per one mole, use the appropriate stoichiometric coefficient to reflect the higher number of moles.

<u>Solution:</u>
$$\Delta H^{\circ}_{rxn} = \Sigma m[\Delta H^{\circ}_{f}\ (\text{products})] - \Sigma n[\Delta H^{\circ}_{f}\ (\text{reactants})]$$

a) $\Delta H^{\circ}_{rxn} = [2\ \Delta H^{\circ}_{f}\ (SO_2(g)) + 2\ \Delta H^{\circ}_{f}\ (H_2O(g))] - [2\ \Delta H^{\circ}_{f}\ (H_2S(g)) + 3\ \Delta H^{\circ}_{f}\ (O_2(g))]$

      $= [2\ \text{mol}(-296.8\ \text{kJ/mol}) + 2\ \text{mol}(-241.826\ \text{kJ/mol})] - [2\ \text{mol}(-20.2\ \text{kJ/mol}) + 3(0.0)] = \mathbf{-1036.8\ kJ}$

b) The balanced equation is $CH_4(g) + 4\, Cl_2(g) \rightarrow CCl_4(l) + 4\, HCl(g)$

$\Delta H^{\circ}_{rxn} = [1\ \Delta H^{\circ}_{f}\ (CCl_4(l)) + 4\ \Delta H^{\circ}_{f}\ (HCl(g))] - [1\ \Delta H^{\circ}_{f}\ (CH_4(g)) + 4\ \Delta H^{\circ}_{f}\ (Cl_2(g))]$

$\Delta H^{\circ}_{rxn} = [1\ \text{mol}(-139\ \text{kJ/mol}) + 4\ \text{mol}(-92.31\ \text{kJ/mol})] - [1\ \text{mol}(-74.87\ \text{kJ/mol}) + 4\ \text{mol}(0)]$

      $= \mathbf{-433\ kJ}$

6.79 $\Delta H^{\circ}_{rxn} = \Sigma m[\Delta H^{\circ}_{f}\ (\text{products})] - \Sigma n[\Delta H^{\circ}_{f}\ (\text{reactants})]$

$Cu_2O(s) + 1/2\, O_2(g) \rightarrow 2\, CuO(s) \qquad\qquad \Delta H^{\circ}_{rxn} = -146.0\ \text{kJ}$

$\Delta H^{\circ}_{rxn} = [2\ \text{mol}\ (\Delta H^{\circ}_{f}, CuO(s))] - [1\ \text{mol}\ (\Delta H^{\circ}_{f}, Cu_2O(s)) + 1/2\ \text{mol}\ (\Delta H^{\circ}_{f}, O_2(g))]$

    $-146.0\ \text{kJ} = [2\ \text{mol}\ (\Delta H^{\circ}_{f}, CuO(s))] - [1\ \text{mol}\ (-168.6\ \dfrac{\text{kJ}}{\text{mol}}) + 1/2\ \text{mol}\ (0)]$

    $-146.0\ \text{kJ} = 2\ \text{mol}\ (\Delta H^{\circ}_{f}, CuO(s)) + 168.6\ \text{kJ}$

$\Delta H^{\circ}_{f}, CuO(s) = -\dfrac{314.6\ \text{kJ}}{2\ \text{mol}} = \mathbf{-157.3\ kJ/mol}$

6.81 $2 \text{ PbSO}_4(s) + 2 \text{ H}_2\text{O}(l) \rightarrow \text{Pb}(s) + \text{PbO}_2(s) + 2 \text{ H}_2\text{SO}_4(l)$

      a) $\Delta H^{\circ}_{rxn} = [1 \text{ mol } (\Delta H^{\circ}_{f}, \text{Pb}(s)) + 1 \text{ mol } (\Delta H^{\circ}_{f}, \text{PbO}_2(s)) + 2 \text{ mol } (\Delta H^{\circ}_{f}, \text{H}_2\text{SO}_4(l))] - [2 \text{ mol } (\Delta H^{\circ}_{f}, \text{PbSO}_4(s))$

           $+ 2 \text{ mol } (\Delta H^{\circ}_{f}, \text{H}_2\text{O}(l))]$

      $= [1 \text{ mol } (0 \frac{\text{kJ}}{\text{mol}}) + 1 \text{ mol } (-276.6 \frac{\text{kJ}}{\text{mol}}) + 2 \text{ mol } (-813.989 \frac{\text{kJ}}{\text{mol}})]$

                $- [2 \text{ mol } (-918.39 \frac{\text{kJ}}{\text{mol}}) + 2 \text{ mol } (-285.840 \frac{\text{kJ}}{\text{mol}})] = \textbf{503.9 kJ}$

      b) Use Hess's Law to rearrange equations (1) and (2) to give the equation wanted.

      Reverse the first equation (changing the sign of ΔH°_{rxn}) and multiply the coefficients (and ΔH°_{rxn}) of the second reaction by 2.

          $2 \text{ PbSO}_4(s) \rightarrow \text{Pb}(s) + \text{PbO}_2(s) + \cancel{2 \text{ SO}_3(g)}$ $\Delta H^{\circ} = -(-768 \text{ kJ})$

          $\underline{\cancel{2 \text{ SO}_3(g)} + 2 \text{ H}_2\text{O}(l) \rightarrow 2 \text{ H}_2\text{SO}_4(l)}$ $\underline{\Delta H^{\circ} = 2(-132 \text{ kJ})}$

      Reaction: $2 \text{ PbSO}_4(s) + 2 \text{ H}_2\text{O}(l) \rightarrow \text{Pb}(s) + \text{PbO}_2(s) + 2 \text{ H}_2\text{SO}_4(l)$ $\Delta H^{\circ}_{rxn} = \textbf{504 kJ}$

6.82   a) $C_{18}H_{36}O_2(s) + 26 \text{ O}_2(g) \rightarrow 18 \text{ CO}_2(g) + 18 \text{ H}_2\text{O}(g)$

      b) $\Delta H^{\circ}_{comb} = [18 \text{ mol } (\Delta H^{\circ}_{f} \text{ CO}_2(g)) + 18 \text{ mol } (\Delta H^{\circ}_{f} \text{ H}_2\text{O}(g))] - [1 \text{ mol } (\Delta H^{\circ}_{f} \text{ C}_{18}H_{36}O_2(s)) +$

            $26 \text{ mol } (\Delta H^{\circ}_{f} \text{ O}_2(g))]$

            $= [18 \text{ mol } (-393.5 \text{ kJ/mol}) + 18 \text{ mol } (-241.826 \text{ kJ/mol})] - [1 \text{ mol } (-948 \text{ kJ/mol}) + 26 \text{ mol } (0 \text{ kJ/mol})]$

            $= \textbf{-10,488 kJ}$

      c) $q(\text{kJ}) = (1.00 \text{ g C}_{18}H_{36}O_2) \left(\dfrac{1 \text{ mol C}_{18}H_{36}O_2}{284.47 \text{ C}_{18}H_{36}O_2} \right) \left(\dfrac{-10,488 \text{ kJ}}{1 \text{ mol C}_{18}H_{36}O_2} \right)$

          $= \textbf{-36.9 kJ}$

          $q(\text{kcal}) = (-36.9 \text{ kJ}) \left(\dfrac{1 \text{ kcal}}{4.184 \text{ kJ}} \right) = \textbf{-8.81 kcal}$

      d) $q(\text{kcal}) = (11.0 \text{ g fat}) \left[\dfrac{-8.81 \text{ kcal}}{1.0 \text{ g fat}} \right] = \textbf{96.9 kcal}$

      The calculated calorie content is consistent with the package information.

6.84   a) A first read of this problem suggests there is insufficient information to solve the problem. Upon more careful reading, you find that the question asks volumes for each mole of helium.

           $T = 273 + 15 = 288 \text{ K}$ or $T = 273 + 30 = 303 \text{ K}$

      $\dfrac{V_{15}}{n} = \dfrac{RT}{P} = \dfrac{\left(0.0821 \frac{\text{L} \cdot \text{atm}}{\text{mol} \cdot \text{K}} \right)(288 \text{ K})}{(1.00 \text{ atm})} = 23.6448 = \textbf{23.6 L/mol}$

      $\dfrac{V_{30}}{n} = \dfrac{RT}{P} = \dfrac{\left(0.0821 \frac{\text{L} \cdot \text{atm}}{\text{mol} \cdot \text{K}} \right)(303 \text{ K})}{(1.00 \text{ atm})} = 24.8763 = \textbf{24.9 L/mol}$

      b) Internal energy is the sum of the potential and kinetic energies of each He atom in the system (the balloon). The energy of one mole of helium atoms can be described as a function of temperature, $E = 3/2 \text{ nRT}$, where n = 1mole. Therefore, the internal energy at 15°C and 30°C can be calculated. The inside back cover lists values of R with different units.

      $E = 3/2 \text{ nRT} = (3/2) (1.00 \text{ mol}) (8.314 \text{ J/mol}\cdot\text{K}) (303 - 288)\text{K} = 187.065 = \textbf{187 J}$

      c) When the balloon expands as temperature rises, the balloon performs PV work. However, the problem specifies that pressure remains constant, so work done on the surroundings by the balloon is defined by the equation: $w = -P\Delta V$. When pressure and volume are multiplied together, the unit is L·atm. Since we would like to express work in Joules, we can create a conversion factor between L·atm and J.

$$w = -P\Delta V = -(1.00 \text{ atm})((24.8763 - 23.6448) \text{ L})\left[\left(\frac{101325 \text{ Pa}}{1 \text{ atm}}\right)\left(\frac{\text{kg}/\text{ms}^2}{\text{Pa}}\right)\left(\frac{10^{-3} \text{ m}^3}{1 \text{ L}}\right)\right]\left(\frac{1 \text{ J}}{\text{kg} \cdot \text{m}^2/\text{s}^2}\right)$$

$$= -124.78 = \mathbf{-1.2 \times 10^2 \text{ J}}$$

d) $q_P = \Delta E + P\Delta V = (187.065 \text{ J}) + (124.78 \text{ J}) = 311.845 = \mathbf{3.1 \times 10^2 \text{ J}}$

e) $\Delta H = q_P = \mathbf{310 \text{ J}}$.

f) When a process occurs at constant pressure, the change in heat energy of the system can be described by a state function called enthalpy. The change in enthalpy equals the heat (q) lost at constant pressure: $\Delta H = \Delta E + P\Delta V = \Delta E - w = (q + w) - w = q_P$

6.93 a) $\left(\frac{1 \text{ cal}}{\text{g} °\text{C}}\right)\left(\frac{453.6 \text{ g}}{1 \text{ lb}}\right)\left(\frac{1.0°\text{C}}{1.8°\text{F}}\right)\left(\frac{4.184 \text{ J}}{1 \text{ cal}}\right)\left(\frac{1.00 \text{ lb} °\text{F}}{1 \text{ Btu}}\right) = 1054.368 = \mathbf{1.1 \times 10^3 \text{ J/Btu}}$

b) $E = (1.00 \text{ therm})\left(\frac{100,000 \text{ Btu}}{1 \text{ therm}}\right)\left(\frac{1054.368 \text{ J}}{\text{Btu}}\right) = 1.054368 \times 10^8 = \mathbf{1.1 \times 10^8 \text{ J}}$

c) $CH_4(g) + 2\, O_2(g) \rightarrow CO_2(g) + 2\, H_2O(g)$

$\Delta H°_{rxn} = [1 \text{ mol } (\Delta H°_f\ CO_2(g)) + 2 \text{ mol } (\Delta H°_f\ H_2O(g))] - [1 \text{ mol } (\Delta H°_f\ CH_4(g)) + 2 \text{ mol } (\Delta H°_f\ O_2(g))]$

$\Delta H°_{rxn} = [1 \text{ mol } (-393.5 \text{ kJ/mol}) + 2 \text{ mol } (-241.826 \text{ kJ/mol})] - [1 \text{ mol } (-74.87 \text{ kJ/mol}) + 2 \text{ mol } (0.0 \text{ kJ/mol})]$

$\quad = -802.282 = -802.3 \text{ kJ/mol } CH_4$

Moles $CH_4 = (1.00 \text{ therm})\left(\frac{1.054368 \times 10^8 \text{ J}}{1 \text{ therm}}\right)\left(\frac{1 \text{ kJ}}{10^3 \text{ J}}\right)\left(\frac{1 \text{ mol } CH_4}{802.282 \text{ kJ}}\right)$

$\quad = 131.4211 = \mathbf{1.3 \times 10^2 \text{ mol } CH_4}$

d) Cost $= \left(\frac{\$0.66}{\text{therm}}\right)\left(\frac{1 \text{ therm}}{131.4211 \text{ mol}}\right) = 0.005022 = \mathbf{\$\ 0.0050 / mol}$

e) Cost $= (318 \text{ gal})\left(\frac{3.78 \text{ L}}{1 \text{ gal}}\right)\left(\frac{1 \text{ mL}}{10^{-3} \text{ L}}\right)\left(\frac{1.0 \text{ g}}{\text{mL}}\right)\left(\frac{4.184 \text{ J}}{\text{g} °\text{C}}\right)((42.0 - 15.0)°\text{C})\left(\frac{1 \text{ therm}}{1.054368 \times 10^8 \text{ J}}\right)\left(\frac{\$0.66}{1 \text{ therm}}\right)$

$\quad = 0.850014 = \mathbf{\$\ 0.85}$

6.102 Combustion reactions can be written and $\Delta H°_{rxn}$ can be calculated for each hydrocarbon.

I. $CH_4(g) + 2\, O_2(g) \rightarrow CO_2(g) + 2\, H_2O(g)$

$\Delta H°_{rxn} = [1 \text{ mol } (\Delta H°_f\ CO_2(g)) + 2 \text{ mol } (\Delta H°_f\ H_2O(g))] - [1 \text{ mol } (\Delta H°_f\ CH_4(g)) + 2 \text{ mol } (\Delta H°_f\ O_2(g))]$

$\quad \Delta H°_{rxn} = [(-393.5) + 2\,(-241.826) - [(-74.87) + 2\,(0.0)] = -802.282 = -802.3 \text{ kJ per mol } CH_4(g)$

II) $C_2H_4(g) + 3\, O_2(g) \rightarrow 2\, CO_2(g) + 2\, H_2O(g)$

$\Delta H°_{rxn} = [2 \text{ mol } (\Delta H°_f\ CO_2(g)) + 2 \text{ mol } (\Delta H°_f\ H_2O(g))] - [1 \text{ mol } (\Delta H°_f\ C_2H_4(g)) + 3 \text{ mol } (\Delta H°_f\ O_2(g))]$

$\quad \Delta H°_{rxn} = [2\,(-393.5) + 2\,(-241.826)] - [(52.47) + 3\,(0.0)] = -1323.122 = -1323.1 \text{ kJ per mol } C_2H_4(g)$

III) $C_2H_6(g) + 7/2\, O_2(g) \rightarrow 2\, CO_2(g) + 3\, H_2O(g)$

$\Delta H°_{rxn} = [2 \text{ mol } (\Delta H°_f\ CO_2(g)) + 3 \text{ mol } (\Delta H°_f\ H_2O(g))] - [1 \text{ mol } (\Delta H°_f\ C_2H_6(g)) + 7/2 \text{ mol } (\Delta H°_f\ O_2(g))]$

$\quad \Delta H°_{rxn} = [2\,(-393.5) + 3\,(-241.826)] - [(-84.667) + 7/2\,(0.0)] = -1427.811 = -1427.8 \text{ kJ per mol } C_2H_6(g)$

a) A negative enthalpy change denotes an exothermic reaction, so the combustion of $C_2H_6(g)$ yields the most heat, followed by $C_2H_4(g)$ and $CH_4(g)$; **III > II > I**.

b) Use molecular weights to convert kJ/mol to kJ/g.

I) $\left(\dfrac{-802.282 \text{ kJ}}{1 \text{ mol CH}_4}\right)\left(\dfrac{1 \text{ mol CH}_4}{16.04 \text{ g CH}_4}\right) = -50.01758 = \mathbf{-50.02 \text{ kJ/g CH}_4}$

II) $\left(\dfrac{-1323.122 \text{ kJ}}{1 \text{ mol C}_2\text{H}_4}\right)\left(\dfrac{1 \text{ mol C}_2\text{H}_4}{28.05 \text{ g C}_2\text{H}_4}\right) = -47.1701 = \mathbf{-47.17 \text{ kJ/g C}_2\text{H}_4}$

III) $\left(\dfrac{-1427.811 \text{ kJ}}{1 \text{ mol C}_2\text{H}_6}\right)\left(\dfrac{1 \text{ mol C}_2\text{H}_6}{30.07 \text{ g C}_2\text{H}6_4}\right) = -47.4829 = \mathbf{-47.48 \text{ kJ/g C}_2\text{H}_4}$

On a per mass basis, CH_4 yields more heat followed by $C_2H_6(g)$ and $C_2H_4(g)$; **I > III > II**.

6.104   a) The heat of reaction is calculated from the heats of formation found in Appendix B. The ΔH°_f 's for all of the species, except $SiCl_4$, are found in Appendix B. Use reaction 3, with its given ΔH°_{rxn}, to find ΔH°_f [$SiCl_4(g)$].

(3) $SiCl_4(g) + 2H_2O(g) \rightarrow SiO_2(s) + 4HCl(g)$

$\Delta H^\circ_{rxn} = [1 \text{ mol } (\Delta H^\circ_f \ SiO_2(s)) + 4 \text{ mol } (\Delta H^\circ_f \ HCl(g))] - [1 \text{ mol}(\Delta H^\circ_f \ SiCl_4(g)) + 2 \text{ mol } (\Delta H^\circ_f \ H_2O(g))]$

$-139.5 \text{ kJ} = [-910.9 \text{ kJ}) + 4(-92.31 \text{ kJ}] - [\Delta H^\circ_f \ [SiCl_4(g)] + 2(-241.826 \text{ kJ}]$

$-139.5 \text{ kJ} = -1280.14 - [(\Delta H^\circ_f \ SiCl_4(g)) + (-483.652 \text{ kJ})]$

$1140.64 \text{ kJ} = -[(\Delta H^\circ_f \ SiCl_4(g)) + (-483.652 \text{ kJ})]$

ΔH°_f [$SiCl_4(g)$] = −656.988 kJ/mol (unrounded)

The heats of reaction for the first two steps can now be calculated.

(1) $Si(s) + 2Cl_2(g) \rightarrow SiCl_4(g)$

$\Delta H^\circ_{rxn1} = [1 \text{ mol } (\Delta H^\circ_f \ SiCl_4(g))] - [1 \text{ mol } (\Delta H^\circ_f \ Si(s)) + 2 \text{ mol } (\Delta H^\circ_f \ Cl_2(g))]$

$\Delta H^\circ_{rxn1} = [-656.988 \text{ kJ}] - [0 + 2(0)] = -656.988 = \mathbf{-657.0 \text{ kJ}}$

(2) $SiO_2(s) + 2C(graphite) + 2Cl_2(g) \rightarrow SiCl_4(g) + 2CO(g)$

$\Delta H^\circ_{rxn2} = [1 \text{ mol}(\Delta H^\circ_f \ SiCl_4(g)) + 2 \text{ mol } (\Delta H^\circ_f \ CO(g))]$

$\qquad\qquad - [1 \text{ mol } (\Delta H^\circ_f \ SiO_2(g)) + 2 \text{ mol } (\Delta H^\circ_f \ C(gr)) + 2 \text{ mol } (\Delta H^\circ_f \ Cl_2(g))]$

$\Delta H^\circ_{rxn2} = [-656.988 \text{ kJ} + 2(-110.5 \text{ kJ})] - [(-910.9 \text{ kJ}) + 2(0) + 2(0)] = 32.912 = \mathbf{32.9 \text{ kJ}}$

b) Adding reactions 2 and 3 yields:

(2) $\cancel{SiO_2(s)} + 2C(graphite) + 2Cl_2(g) \rightarrow \cancel{SiCl_4(g)} + 2CO(g)$ $\Delta H^\circ_{rxn} = 32.912 \text{ kJ}$

(3) $\cancel{SiCl_4(g)} + 2H_2O(g) \rightarrow \cancel{SiO_2(s)} + 4HCl(g)$ $\Delta H^\circ_{rxn} = -139.5 \text{ kJ}$

$\qquad 2 C(gr) + 2 Cl_2(g) + 2 H_2O(g) \rightarrow 2 CO(g) + 4 HCl(g)$ $\Delta H^\circ_{rxn2+3} = -106.588 \text{ kJ}$

$\Delta H^\circ_{rxn2+3} = -106.588 = \mathbf{-106.6 \text{ kJ}}$

Confirm this result by calculating ΔH°_{rxn} using Appendix B values.

$2 C(gr) + 2 Cl_2(g) + 2 H_2O(g) \rightarrow 2 CO(g) + 4 HCl(g)$

$\Delta H^\circ_{rxn2+3} = \{2 \Delta H^\circ_f \ [CO(g)] + 4 \Delta H^\circ_f \ [HCl(g)]\} - \{2 \Delta H^\circ_f \ [C(gr)] + 2 \Delta H^\circ_f \ [Cl_2(g)] + 2 \Delta H^\circ_f \ [H_2O(g)]\}$

$\Delta H^\circ_{rxn2+3} = 2(-110.5 \text{ kJ}) + 4(-92.31 \text{ kJ}) - [2(0) + 2(0) + 2(-241.826 \text{ kJ})] = -106.588 = \mathbf{-106.6 \text{ kJ}}$

6.107   a) $\Delta T = 819^\circ C - 0^\circ C = 819^\circ C = 819 \text{ K}$

$\qquad w = -P\Delta V = -nR\Delta T = -(1 \text{ mol}) \times (8.314 \text{ J/mol} \cdot \text{K}) \times (819 \text{ K}) = -6809.166 = \mathbf{-6.81 \times 10^3 \text{ J}}$

b) q = (mass)(c)(ΔT)

$\Delta T = \dfrac{q}{(mass)(c)} = \dfrac{6.809166 \times 10^3 \text{ J}}{\left[1 \text{ mol N}_2\left(\dfrac{28.02 \text{ g N}_2}{1 \text{ mol N}_2}\right)\right](1.00 \text{ J/g}\cdot\text{K})} = 243.01 = 243 \text{ K} = \mathbf{243^\circ C}$

6.108 Only reaction 3 contains $N_2O_4(g)$, and only reaction 1 contains $N_2O_3(g)$, so we can use those reactions as a starting point. N_2O_5 appears in both reactions 2 and 5, but note the physical states present: solid and gas. As a rough start, adding reactions 1, 3, and 5 yields the desired reactants and products, with some undesired intermediates:

Reverse (1) $N_2O_3(g) \rightarrow NO(g) + NO_2(g)$ $\Delta H°_{rxn1} = -(-39.8 \text{ kJ}) =$ 39.8 kJ

Multiply (3) by 2 $4 NO_2(g) \rightarrow 2 N_2O_4(g)$ $\Delta H°_{rxn3} = 2(-57.2 \text{ kJ}) = -114.4 \text{ kJ}$

(5) $\underline{N_2O_5(s) \rightarrow N_2O_5(g)}$ $\Delta H°_{rxn5} = (54.1 \text{ kJ}) =$ 54.1 kJ

     $N_2O_3(g) + 4NO_2(g) + N_2O_5(s) \rightarrow NO(g) + NO_2(g) + 2N_2O_4(g) + N_2O_5(g)$

To cancel out the $N_2O_5(g)$ intermediate, reverse equation 2. This also cancels out some of the undesired $NO_2(g)$ but adds $NO(g)$ and $O_2(g)$. Finally, add equation 4 to remove those intermediates:

Reverse (1) $N_2O_3(g) \rightarrow \cancel{NO(g)} + \cancel{NO_2(g)}$ $\Delta H°_{rxn1} = -(-39.8 \text{ kJ}) = 39.8 \text{ kJ}$

Multiply (3) by 2 $4\cancel{NO_2(g)} \rightarrow 2 N_2O_4(g)$ $\Delta H°_{rxn3} = 2(-57.2 \text{ kJ}) = -114.4 \text{ kJ}$

(5) $N_2O_5(s) \rightarrow \cancel{N_2O_5(g)}$ $\Delta H°_{rxn5} = 54.1 \text{ kJ}$

Reverse (2) $\cancel{N_2O_5(g)} \rightarrow \cancel{NO(g)} + \cancel{NO_2(g)} + \cancel{O_2(g)}$ $\Delta H°_{rxn2} = -(-112.5 \text{ kJ}) = 112.5$

(4) $\underline{2\cancel{NO(g)} + \cancel{O_2(g)} \rightarrow 2\cancel{NO_2(g)}}$ $\underline{\Delta H°_{rxn4} = -114.2 \text{ kJ}}$

Total: $N_2O_3(g) + N_2O_5(s) \rightarrow 2 N_2O_4(g)$ $\Delta H°_{rxn} = \textbf{-22.2 kJ}$

6.109 a) $2 NH_3(aq) + NaOCl(aq) \rightarrow N_2H_4(aq) + NaCl(aq) + H_2O(l)$

$\Delta H°_{rxn} = [1 \text{ mol } (\Delta H°_f \ N_2H_4(aq)) + 1 \text{ mol } (\Delta H°_f \ NaCl(aq)) + 1 \text{ mol } (\Delta H°_f \ H_2O(l))]$

$- [2 \text{ mol } (\Delta H°_f \ NH_3(aq)) + 1 \text{ mol } (\Delta H°_f \ NaOCl(aq))]$

Note that the Appendix B value for N_2H_4 is for the liquid state, so this term must be calculated. In addition, Appendix B does not list a value for $NaCl(aq)$, so this term must be broken down into $\Delta H°_f \ Na^+(aq)$ and $\Delta H°_f \ Cl^-(aq)$.

$-151 \text{ kJ} = [\Delta H°_f \ N_2H_4(aq) + (-239.66 \text{ kJ}) + (-167.46 \text{ kJ}) + (-285.840 \text{ kJ})] - [2(-80.83 \text{ kJ}) + (-346 \text{ kJ})]$

$\Delta H°_f \ N_2H_4(aq) = 34.3 = \textbf{34 kJ}$

b) Determine the moles of O_2 present, and then multiply by the $\Delta H°_{rxn}$ for the first reaction.

Moles of $O_2 = (5.00 \times 10^3 \text{ L})(2.50 \times 10^{-4} \text{ mol/L}) = 1.25 \text{ mol } O_2$

$N_2H_4(aq) + O_2(g) \rightarrow N_2(g) + 2 H_2O(l)$

$\Delta H°_{rxn} = [1 \text{ mol } (\Delta H°_f \ N_2(g)) + 2 \text{ mol } (\Delta H°_f \ H_2O(l))] - [1 \text{ mol } (\Delta H°_f \ N_2H_4(aq)) + 1 \text{ mol } (\Delta H°_f \ O_2(g))]$

$\Delta H°_{rxn} = [0 + 2(-285.840 \text{ kJ})] - [(34.3 \text{ kJ}) + 0] = -605.98 \text{ kJ /mol } O_2 \text{ (unrounded)}$

Heat $= (-605.98 \text{ kJ /mol } O_2)(1.25 \text{ mol } O_2) = -757.475 = \textbf{-757 kJ}$

6.111 a) The balanced chemical equation for this reaction is:

$CH_4(g) + 2 O_2(g) \rightarrow CO_2(g) + 2 H_2O(g)$

Instead of burning one mole of methane, $(25.0 \text{ g } CH_4)\left(\dfrac{1 \text{ mol } CH_4}{16.04 \text{ g } CH_4}\right) = 1.5586 \text{ mol } CH_4 \text{ (unrounded) of methane}$ are burned.

$CH_4(g) + 2 O_2(g) \rightarrow CO_2(g) + 2 H_2O(g)$

$\Delta H°_{comb} = [1 \text{ mol } (\Delta H°_f \ CO_2(g)) + 2 \text{ mol } (\Delta H°_f \ H_2O(g))] - [1 \text{ mol } (\Delta H°_f \ CH_4(g)) + 2 \text{ mol } (\Delta H°_f \ O_2(g))]$

$\Delta H°_{comb} = [1 \text{ mol } (-393.5 \text{ kJ/mol}) + 2 \text{ mol } (-241.826 \text{ kJ/mol})] - [1 \text{mol } (-74.87 \text{kJ/mol}) + 2 \text{mol } (0.0 \text{ kJ/mol})]$
$= -802.282 \text{ kJ/mol } CH_4 \text{ (unrounded)}$

$(1.5586 \text{ mol } CH_4)\left(\dfrac{-802.282 \text{ kJ}}{1 \text{ mol } CH_4}\right) = -1250.4 = \textbf{-1.25 x } 10^3 \textbf{ kJ}$

b) The heat released by the reaction is "stored" in the gaseous molecules by virtue of their specific heat capacities, c, using the equation $\Delta H = mc\Delta T$. The problem specifies heat capacities on a molar basis, so we modify the equation to use moles, instead of mass.

The gases that remain at the end of the reaction are CO_2 and H_2O. All of the methane and oxygen molecules were consumed. However, the oxygen was added as a component of air, which is 78% N_2 and 21% O_2, and there is leftover N_2.

Moles of $CO_2(g) = (1.5586 \text{ mol } CH_4)\left(\dfrac{1 \text{ mol } CO_2}{1 \text{ mol } CH_4}\right) = 1.5586$ mol

Moles of $H_2O(g) = (1.5586 \text{ mol } CH_4)\left(\dfrac{2 \text{ mol } H_2O}{1 \text{ mol } CH_4}\right) = 3.1172$ mol

Mole fraction N_2 = (79% / 100%) = 0.79
Mole fraction O_2 = (21% / 100%) = 0.21
Moles of $N_2(g)$ = (3.1172 mol O_2 used) (0.79 mol N_2 / 0.21 mol O_2) = 11.7266 mol N_2 (unrounded)
q = (mol)(c)(ΔT)
1250.4 kJ (10^3 J/ kJ) = 1.5586 mol CO_2 (57.2 J / mol°C) (T_f – 0.0)°C
$\qquad$ + 3.1172 mol H_2O (36.0 J / mol°C) (T_f – 0.0)°C
$\qquad$ + 11.7266 mol N_2 (30.5 J / mol°C) (T_f – 0.0)°C
1.2504×10^6 J = ((89.1519 + 112.2192 + 357.6613) J /°C)T_f = (559.0324 J /°C)T_f
$T_f = (1.2504 \times 10^6$ J) / (559.0324 J /°C) = 2236.72 = **2.24×10^3°C**

CHAPTER 7 QUANTUM THEORY AND ATOMIC STRUCTURE

The value for the speed of light will be 3.00×10^8 m/s except when more significant figures are necessary, in which case, 2.9979×10^8 m/s will be used.

FOLLOW–UP PROBLEMS

7.1   Plan: Given the frequency of the light, use the equation $c = \lambda\nu$ to solve for wavelength.
Solution:

$$\lambda = c / \nu = \frac{3.00 \times 10^8 \, \text{m/s}}{7.23 \times 10^{14} \, \text{s}^{-1}} \left(\frac{1 \, \text{nm}}{10^{-9} \, \text{m}} \right) = 414.938 = \textbf{415 nm}$$

$$\lambda = c / \nu = \frac{3.00 \times 10^8 \, \text{m/s}}{7.23 \times 10^{14} \, \text{s}^{-1}} \left(\frac{1 \, \text{Å}}{10^{-10} \, \text{m}} \right) = 4149.38 = \textbf{4150 Å}$$

Check: The purple region of the visible light spectrum occurs between approximately 400 to 450 nm, so 415 nm is in the correct range for wavelength of purple light.

7.2   Plan: To calculate the energy for each wavelength we use the formula $E = hc / \lambda$
Solution:

$$E = hc / \lambda = \frac{\left(6.626 \times 10^{-34} \, \text{J} \cdot \text{s} \right)(3.00 \times 10^8 \, \text{m/s})}{1 \times 10^{-8} \, \text{m}} = 1.9878 \times 10^{-17} = \textbf{2} \times \textbf{10}^{-17} \textbf{ J}$$

$$E = hc / \lambda = \frac{\left(6.626 \times 10^{-34} \, \text{J} \cdot \text{s} \right)(3.00 \times 10^8 \, \text{m/s})}{5 \times 10^{-7} \, \text{m}} = 3.9756 \times 10^{-19} = \textbf{4} \times \textbf{10}^{-19} \textbf{ J}$$

$$E = hc / \lambda = \frac{\left(6.626 \times 10^{-34} \, \text{J} \cdot \text{s} \right)(3.00 \times 10^8 \, \text{m/s})}{1 \times 10^{-4} \, \text{m}} = 1.9878 \times 10^{-21} = \textbf{2} \times \textbf{10}^{-21} \textbf{ J}$$

As the wavelength of light increases from ultraviolet (uv) to visible (vis) to infrared (ir) the energy of the light decreases.
Check: The decrease in energy with increase in wavelength follows the inverse relationship between energy and wavelength in the equation $E = hc / \lambda$.

7.3   Plan: Use the equation relating $\Delta E = -2.18 \times 10^{-18} \, \text{J} \left(\dfrac{1}{n^2_{\text{final}}} - \dfrac{1}{n^2_{\text{initial}}} \right)$ to find the energy change; a photon

in the IR (infrared) region is emitted when has n has a final value of 3. Then use $E = hc / \lambda$ to find the wavelength of the photon.
Solution:

$$\Delta E = -2.18 \times 10^{-18} \, \text{J} \left(\frac{1}{n^2_{\text{final}}} - \frac{1}{n^2_{\text{initial}}} \right)$$

$$\Delta E = -2.18 \times 10^{-18} \, \text{J} \left(\frac{1}{3^2} - \frac{1}{6^2} \right)$$

$$\Delta E = 1.8166667 \times 10^{-19} = \textbf{1.82} \times \textbf{10}^{-19} \textbf{ J}$$

$$E = \text{hc} / \lambda$$

$$\lambda = \frac{\text{hc}}{E} = \frac{\left(6.626 \times 10^{-34} \text{ J} \cdot \text{s}\right)\left(3.00 \times 10^8 \text{ m/s}\right)}{1.8166667 \times 10^{-19} \text{ J}}\left(\frac{1 \text{ Å}}{10^{-10} \text{ m}}\right) = 1.094202 \times 10^4 = \textbf{1.09} \times \textbf{10}^4 \text{ Å}$$

7.4   Plan: With the equation for the de Broglie wavelength, $\lambda = h/mu$ and the given de Broglie wavelength, calculate the electron speed.
Solution:

$$u = h / m\lambda = \frac{6.626 \times 10^{-34} \text{ J} \cdot \text{s}}{\left(9.11 \times 10^{-31} \text{ kg}\right)\left[\left(100.\text{nm}\right)\left(\frac{10^{-9} \text{ m}}{1 \text{ nm}}\right)\right]}\left(\frac{\text{kg} \cdot \text{m}^2/\text{s}^2}{\text{J}}\right) = 7273.3 = \textbf{7.27} \times \textbf{10}^3 \text{ m/s}$$

Check: Perform a rough calculation to check order of magnitude:

$$\frac{10^{-34}}{\left(10^{-31}\right)\left[\left(100.\right)\left(\frac{10^{-9}}{1}\right)\right]} = 1 \times 10^4$$

7.5   Plan: Use the equation $\Delta x \cdot m\Delta u \geq \dfrac{h}{4\pi}$ to solve for the uncertainty (Δx) in position of the baseball.

Solution:

$$\Delta x \cdot m\Delta u \geq \frac{h}{4\pi}$$

$\Delta u = 1\%$ of $u = 0.0100(44.7 \text{ m/s}) = 0.447 \text{ m/s}$

$$\Delta x \geq \frac{h}{4\pi m \Delta u} = \frac{6.626 \times 10^{-34} \text{ J} \cdot \text{s}}{4\pi \left(0.142 \text{ kg}\right)\left(0.447 \text{ m/s}\right)} = 8.3070285 \times 10^{-34} = 8.31 \times 10^{-34} \text{ m}$$

7.6   Plan: Following the rules for l (integer from 0 to $n-1$) and m_l (integer from $-l$ to $+l$) write quantum numbers for $n = 4$.
Solution:
For $n = 4$ $l = \textbf{0, 1, 2, 3}$
For $l = 0$, $m_l = \textbf{0}$
For $l = 1$, $m_l = \textbf{–1, 0, 1}$
For $l = 2$, $m_l = \textbf{–2, –1, 0, 1, 2}$
For $l = 3$, $m_l = \textbf{–3, –2, –1, 0, 1, 2, 3}$
Check: The total number of orbitals with a given n is n^2. For $n = 4$, the total number of orbitals would be 16. Adding the number of orbitals identified in the solution, gives
1 (for $l = 0$) + 3 (for $l = 1$) + 5 (for $l = 2$) + 7 (for $l = 3$) = 16 orbitals.

7.7   Plan: Identify n and l from the subshell designation and knowing the value for l, find the m_l values. The subshells are given a letter designation, in which s represents $l = 0$, p represents $l = 1$, d represents $l = 2$, f represents $l = 3$.
Solution:

Sublevel name	n value	l value	m_l values
2p	2	1	–1, 0, 1
5f	5	3	–3, –2, –1, 0, 1, 2, 3

Check: The number of orbitals for each sublevel equals $2l + 1$. Sublevel 2p should have 3 orbitals and sublevel 5f should have 7 orbitals. Both of these agree with the number of m_l values for the sublevel.

7.8   Plan: Use the rules for designating quantum numbers to fill in the blanks.
For a given n, l can be any integer from 0 to $n–1$.
For a given l, m_l can be any integer from $– l$ to $+ l$.
The subshells are given a letter designation, in which s represents $l = 0$, p represents $l = 1$, d represents $l = 2$, f represents $l = 3$.
Solution:
The completed table is:

	n	l	m_l	Name
a)	4	1	0	$4p$
b)	2	1	0	$2p$
c)	3	2	–2	$3d$
d)	2	0	0	$2s$

Check: All the given values are correct designations.

END–OF–CHAPTER PROBLEMS

7.2   a) Figure 7.3 describes the electromagnetic spectrum by wavelength and frequency. Wavelength increases from left (10^{-2} nm) to right (10^{12} nm). The trend in increasing wavelength is: **x-ray < ultraviolet < visible < infrared < microwave < radio waves**.
b) Frequency is inversely proportional to wavelength according to the equation $c = \lambda \upsilon$, so frequency has the opposite trend: **radio < microwave < infrared < visible < ultraviolet < x-ray**.
c) Energy is directly proportional to frequency according to the equation $E = h\upsilon$. Therefore, the trend in increasing energy matches the trend in increasing frequency: **radio < microwave < infrared < visible < ultraviolet < x-ray**. High-energy electromagnetic radiation disrupts cell function. It makes sense that you want to limit exposure to ultraviolet and x-ray radiation.

7.7   Plan: Wavelength is related to frequency through the equation $c = \lambda v$. Recall that a Hz is a reciprocal second, or $1/s = s^{-1}$. Assume that the number "950" has three significant figures.
Solution:
$c = \lambda v$

$$\lambda(m) = \frac{c}{v} = \frac{3.00 \times 10^8 \text{ m/s}}{(950.\text{kHz})\left(\frac{10^3 \text{ Hz}}{1 \text{ kHz}}\right)\left(\frac{s^{-1}}{\text{Hz}}\right)} = 315.8 = \textbf{316 m}$$

$$\lambda \text{ (nm)} = \frac{c}{v} = \frac{3.00 \times 10^8 \text{ m/s}}{(950.\text{ kHz})\left(\frac{10^3 \text{ Hz}}{1 \text{k Hz}}\right)\left(\frac{s^{-1}}{\text{Hz}}\right)}\left(\frac{1 \text{ nm}}{10^{-9} \text{ m}}\right) = 3.158 \times 10^{11} = \textbf{3.16 x 10}^{11} \textbf{ nm}$$

$$\lambda \text{ (Å)} = \frac{c}{v} = \frac{3.00 \times 10^8 \text{ m/s}}{(950.\text{ kHz})\left(\frac{10^3 \text{ Hz}}{1 \text{ kHz}}\right)\left(\frac{s^{-1}}{\text{Hz}}\right)}\left(\frac{1 \text{ Å}}{10^{-10} \text{ m}}\right) = 3.158 \times 10^{12} = \textbf{3.16 x 10}^{12} \textbf{ Å}$$

7.9   Plan: Frequency is related to energy through the equation $E = h v$. Note that 1 Hz = 1 s^{-1}.
Solution:
$E = h v$
$E = (6.626 \times 10^{-34} \text{ J} \cdot \text{s}) (3.8 \times 10^{10} \text{ s}^{-1}) = 2.51788 \times 10^{-23} = \textbf{2.5 x 10}^{-23} \textbf{ J}$

7.11   Since energy is directly proportional to frequency ($E = h v$) and frequency and wavelength are inversely related ($v = c/\lambda$), it follows that energy is inversely related to wavelength ($E = \frac{hc}{\lambda}$). As wavelength decreases, energy increases. In terms of increasing energy the order is **red < yellow < blue**.

7.13   <u>Plan:</u> Frequency and wavelength can be calculated using the speed of light: $c = \lambda\nu$.
         <u>Solution:</u>

$$\lambda\ (nm) = c/\nu = \frac{2.9979 \times 10^8\ m/s}{\left(22.235\ GHz\right)\left(\dfrac{10^9\ Hz}{1\ GHz}\right)\left(\dfrac{s^{-1}}{Hz}\right)}\left(\frac{1\ nm}{10^{-9}\ m}\right) = 1.3482797 \times 10^7 = \mathbf{1.3483 \times 10^7\ nm}$$

$$\lambda\ (\text{Å}) = c/\nu = \frac{2.9979 \times 10^8\ m/s}{\left(22.235\ GHz\right)\left(\dfrac{10^9\ Hz}{1\ GHz}\right)\left(\dfrac{s^{-1}}{Hz}\right)}\left(\frac{1\ \text{Å}}{10^{-10}\ m}\right) = 1.34827973 \times 10^8 = \mathbf{1.3483 \times 10^8\ \text{Å}}$$

7.16   <u>Plan:</u> a) The least energetic photon has the longest wavelength (242 nm). b) The most energetic photon has the shortest wavelength (2200 Å).
         <u>Solution:</u>

$$a)\ \nu = \frac{c}{\lambda} = \frac{3.00 \times 10^8\ m/s}{242\ nm}\left(\frac{1\ nm}{10^{-9}\ m}\right) = 1.239669 \times 10^{15} = \mathbf{1.24 \times 10^{15}\ s^{-1}}$$

$$E = \frac{hc}{\lambda} = \frac{\left(6.626 \times 10^{-34}\ J \bullet s\right)\left(3.00 \times 10^8\ m/s\right)}{242\ nm}\left(\frac{1\ nm}{10^{-9}\ m}\right) = 8.2140 \times 10^{-19} = \mathbf{8.21 \times 10^{-19}\ J}$$

$$b)\ \nu = \frac{c}{\lambda} = \frac{3.00 \times 10^8\ m/s}{2200\ \text{Å}}\left(\frac{1\ \text{Å}}{10^{-10}\ m}\right) = 1.3636 \times 10^{15} = \mathbf{1.4 \times 10^{15}\ s^{-1}}$$

$$E = \frac{hc}{\lambda} = \frac{\left(6.626 \times 10^{-34}\ J \bullet s\right)\left(3.00 \times 10^8\ m/s\right)}{2200\ \text{Å}}\left(\frac{1\ \text{Å}}{10^{-10}\ m}\right) = 9.03545 \times 10^{-19} = \mathbf{9.0 \times 10^{-19}\ J}$$

7.18   Bohr's key assumption was that the electron in an atom does not radiate energy while in a stationary state, and the electron can move to a different orbit by absorbing or emitting a photon whose energy is equal to the difference in energy between two states. These differences in energy correspond to the wavelengths in the known spectra for the hydrogen atoms. A solar system model does not allow for the movement of electrons between levels.

7.20   <u>Plan:</u> The quantum number n is related to the energy level of the electron. An electron *absorbs* energy to change from lower energy (lower n) to higher energy (higher n), giving an absorption spectrum. An electron *emits* energy as it drops from a higher energy level (higher n) to a lower one (lower n), giving an emission spectrum.
         <u>Solution:</u>
         a) **absorption**   b) **emission**   c) **emission**   d) **absorption**

7.22   The Bohr model has successfully predicted the line spectra for the H atom and Be^{3+} ion since both are one electron species. The energies could be predicted from $E_n = \dfrac{-\left(Z^2\right)\left(2.18 \times 10^{-18}\ J\right)}{n^2}$ where Z is the atomic number for the atom or ion. The line spectra for H would not match the line spectra for Be^{3+} since the H nucleus contains one proton while the Be^{3+} nucleus contains 4 protons (the Z values in the equation do not match), thus the force of attraction of the nucleus for the electron would be greater in the beryllium ion than in the hydrogen atom. This means that the pattern of lines would be similar, but at different wavelengths.

7.23   <u>Plan:</u> Calculate wavelength by substituting the given values into equation 7.3, where $n_1 = 2$ and $n_2 = 5$ because $n_2 > n_1$. Although more significant figures could be used, five significant figures are adequate for this calculation.
         <u>Solution:</u>

$$\frac{1}{\lambda} = R\left(\frac{1}{n_1^2} - \frac{1}{n_2^2}\right) \qquad R = 1.096776 \times 10^7\ m^{-1}$$

$$n_1 = 2 \qquad n_2 = 5$$

$$\frac{1}{\lambda} = R\left(\frac{1}{n_1^2} - \frac{1}{n_2^2}\right) = \left(1.096776 \times 10^7 \text{ m}^{-1}\right)\left(\frac{1}{2^2} - \frac{1}{5^2}\right) = 2303229.6 \text{ m}^{-1} \text{ (unrounded)}$$

$$\lambda \text{ (nm)} = \left(\frac{1}{2303229.6 \text{ m}^{-1}}\right)\left(\frac{1 \text{ nm}}{10^{-9} \text{ m}}\right) = 434.1729544 = \textbf{434.17 nm}$$

7.25   Plan: The Rydberg equation is needed. For the infrared series of the H atom, n_1 equals 3. The least energetic spectral line in this series would represent an electron moving from the next highest energy level, $n_2 = 4$. Although more significant figures could be used, five significant figures are adequate for this calculation.
Solution:

$$\frac{1}{\lambda} = R\left(\frac{1}{n_1^2} - \frac{1}{n_2^2}\right) = \left(1.096776 \times 10^7 \text{ m}^{-1}\right)\left(\frac{1}{3^2} - \frac{1}{4^2}\right) = 533155 \text{ m}^{-1} \text{ (unrounded)}$$

$$\lambda \text{ (nm)} = \left(\frac{1}{533155 \text{ m}^{-1}}\right)\left(\frac{1 \text{ nm}}{10^{-9} \text{ m}}\right) = 1875.627 = \textbf{1875.6 nm}$$

Check: Checking this wavelength with Figure 7.9, we find the line in the infrared series with the greatest wavelength occurs at approximately 1850 nm.

7.27   Plan: To find the transition energy, use the equation for the energy of an electron transition and multiply by Avogadro's number.
Solution:

$$\Delta E = \left(-2.18 \times 10^{-18} \text{ J}\right)\left(\frac{1}{n_{\text{final}}^2} - \frac{1}{n_{\text{initial}}^2}\right)$$

$$\Delta E = \left(-2.18 \times 10^{-18} \text{ J}\right)\left(\frac{1}{2^2} - \frac{1}{5^2}\right) = -4.578 \times 10^{-19} \text{ J/photon}$$

$$\Delta E = \left(\frac{-4.578 \times 10^{-19} \text{ J}}{\text{photon}}\right)\left(\frac{6.022 \times 10^{23} \text{ photons}}{1 \text{ mol}}\right) = -2.75687 \times 10^5 = \textbf{--2.76} \times \textbf{10}^5 \text{ J/mol}$$

The energy has a negative value since this electron transition to a lower n value is an emission of energy.

7.29   Looking at an energy chart will help answer this question.

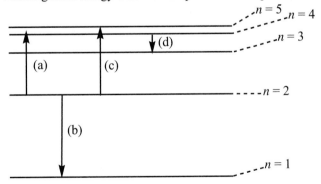

Frequency is proportional to energy so the smallest frequency will be d) $n = 4$ to $n = 3$; levels 4 and 5 have a smaller ΔE than the levels in the other transitions. The largest frequency is b) $n = 2$ to $n = 1$ since levels 1 and 2 have a larger ΔE than the levels in the other transitions. Transition a) $n = 2$ to $n = 4$ will be smaller than transition c) $n = 2$ to $n = 5$ since level 5 is a higher energy than level 4. In order of increasing frequency the transitions are **d < a < c < b.**

7.31   <u>Plan:</u> Use the Rydberg equation.
       <u>Solution:</u>

$$\lambda = \left(97.20 \text{ nm}\right)\left(\frac{10^{-9} \text{ m}}{1 \text{ nm}}\right) = 9.720 \text{ x } 10^{-8} \text{ m} \qquad \text{ground state: } n_1 = 1; \quad n_2 = ?$$

$$\frac{1}{\lambda} = \left(1.096776 \text{ x } 10^7 \text{ m}^{-1}\right)\left(\frac{1}{n_1^2} - \frac{1}{n_2^2}\right)$$

$$\frac{1}{9.72 \text{ x } 10^{-8} \text{ m}} = \left(1.096776 \text{ x } 10^7 \text{ m}^{-1}\right)\left(\frac{1}{1^2} - \frac{1}{n_2^2}\right)$$

$$0.9380 = \left(\frac{1}{1^2} - \frac{1}{n_2^2}\right)$$

$$\frac{1}{n_2^2} = 1 - 0.9380 = 0.0620$$

$$n_2^2 = 16.1$$

$$n_2 = \mathbf{4}$$

7.37   Macroscopic objects have significant mass. A large m in the denominator of $\lambda = h/mu$ will result in a very small wavelength. Macroscopic objects do exhibit a wavelike motion, but the wavelength is too small for humans to see it.

7.39   <u>Plan:</u> Use the de Broglie equation. Mass in pounds must be converted to kg and velocity in mi/h must be converted to m/s because a joule is equivalent to $kg \cdot m^2 / s^2$.
       <u>Solution:</u>

$$a) \; \lambda = \frac{h}{mu} = \frac{\left(6.626 \text{ x } 10^{-34} \text{ J} \cdot \text{s}\right)}{\left(232 \text{ lb}\right)\left(19.8 \frac{\text{mi}}{\text{h}}\right)}\left(\frac{\text{kg} \cdot \text{m}^2/\text{s}^2}{\text{J}}\right)\left(\frac{2.205 \text{ lb}}{1 \text{ kg}}\right)\left(\frac{0.62 \text{ mi}}{1 \text{ km}}\right)\left(\frac{1 \text{ km}}{10^3 \text{ m}}\right)\left(\frac{3600 \text{ s}}{1 \text{ h}}\right)$$

$$= 7.099063 \text{ x } 10^{-37} = \mathbf{7.10 \text{ x } 10^{-37} \text{ m}}$$

$$b) \; \Delta x \cdot m\Delta v \geq \frac{h}{4\pi}$$

$$\Delta x \geq h/4 \; \pi m\Delta v \geq \frac{\left(6.626 \text{ x } 10^{-34} \text{ J} \cdot \text{s}\right)}{4\pi\left(232 \text{ lb}\right)\left(\frac{0.1 \text{ mi}}{\text{h}}\right)}\left(\frac{\text{kg} \cdot \text{m}^2/\text{s}^2}{\text{J}}\right)\left(\frac{2.205 \text{ lb}}{1 \text{ kg}}\right)\left(\frac{0.62 \text{ mi}}{1 \text{ km}}\right)\left(\frac{1 \text{ km}}{10^3 \text{ m}}\right)\left(\frac{3600 \text{ s}}{1 \text{ h}}\right)$$

$$\geq 1.11855 \text{ x } 10^{-35} \geq \mathbf{1 \text{ x } 10^{-35} \text{ m}}$$

7.41 $\lambda = \dfrac{h}{mu}$

$$u = \frac{h}{m\lambda} = \frac{\left(6.626 \text{ x } 10^{-34} \text{ J} \cdot \text{s}\right)}{\left(56.5 \text{ g}\right)\left(5400 \text{ Å}\right)}\left(\frac{\text{kg} \cdot \text{m}^2/\text{s}^2}{\text{J}}\right)\left(\frac{10^3 \text{ g}}{1 \text{ kg}}\right)\left(\frac{1 \text{ Å}}{10^{-10} \text{ m}}\right) = 2.1717 \text{ x } 10^{-26} = \mathbf{2.2 \text{ x } 10^{-26} \text{ m/s}}$$

7.43   Plan: The de Broglie wavelength equation will give the mass equivalent of a photon with known wavelength and velocity. The term "mass-equivalent" is used instead of "mass of photon" because photons are quanta of electromagnetic energy that have no mass. A light photon's velocity is the speed of light, 3.00×10^8 m/s.
Solution:

$$\lambda = \frac{h}{mu}$$

$$m = \frac{h}{\lambda u} = \frac{\left(6.626 \times 10^{-34} \text{ J} \cdot \text{s}\right)}{(589 \text{ nm})\left(3.00 \times 10^8 \text{ m/s}\right)} \left(\frac{\text{kg} \cdot \text{m}^2/\text{s}^2}{\text{J}}\right)\left(\frac{1 \text{ nm}}{10^{-9} \text{ m}}\right) = 3.7498 \times 10^{-36} = \textbf{3.75 x 10}^{-36} \textbf{ kg/photon}$$

7.47   A peak in the radial probability distribution at a certain distance means that the total probability of finding the electron is greatest within a thin spherical volume having a radius very close to that distance. Since principal quantum number (n) correlates with distance from the nucleus, the peak for $n = 2$ would occur at a greater distance from the nucleus than 0.529 Å. Thus, the probability of finding an electron at 0.529 Å is much greater for the 1s orbital than for the 2s.

7.48   a) Principal quantum number, n, relates to the size of the orbital. More specifically, it relates to the distance from the nucleus at which the probability of finding an electron is greatest. This distance is determined by the energy of the electron.
b) Angular momentum quantum number, l, relates to the shape of the orbital. It is also called the azimuthal quantum number.
c) Magnetic quantum number, m_l, relates to the orientation of the orbital in space in three-dimensional space.

7.49   Plan: The following letter designations correlate with the following l quantum numbers:
$l = 0 = $ s orbital; $l = 1 = $ p orbital; $l = 2 = $ d orbital; $l = 3 = $ f orbital. Remember that allowed m_l values are $-l$ to $+l$. The number of orbitals of a particular type is given by the m_l quantum number.
Solution:
a) **one**   b) **five**   c) **three**   d) **nine**
a) There is only a single s orbital in any shell. $l = 1$ and $m_l = 0$: one value of m_l.
b) All d-orbitals consist of sets of five ($m_l = -2, -1, 0, +1, +2$).
c) All p-orbitals consist of sets of three ($m_l = -1, 0, +1$).
d) If $n = 3$, then there is a 3s (1 orbital), a 3p set (3 orbitals), and a 3d set (5 orbitals) giving $1 + 3 + 5 = 9$.

7.51   Magnetic quantum numbers (m_l) can have integer values from $-l$ to $+l$.
a) m_l: $-2, -1, 0, +1, +2$
b) m_l: 0   (if $n = 1$, then $l = 0$)
c) m_l: $-3, -2, -1, 0, +1, +2, +3$

7.53   (a) s: spherical   (b) p_x: 2 lobes along the x-axis

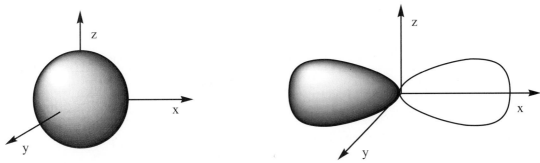

The variations in coloring of the p orbital are a consequence of the quantum mechanical derivation of atomic orbitals that are beyond the scope of this course.

7.55   Plan: The following letter designations correlate with the following l quantum numbers:
        $l = 0 =$ s orbital; $l = 1 =$ p orbital; $l = 2 =$ d orbital; $l = 3 =$ f orbital. Remember that allowed m_l values are $-l$ to $+l$.
        Solution:

sublevel	allowable m_l	# of possible orbitals
a) d ($l = 2$)	$-2, -1, 0, +1, +2$	5
b) p ($l = 1$)	$-1, 0, +1$	3
c) f ($l = 3$)	$-3, -2, -1, 0, +1, +2, +3$	7

7.57   Plan: The integer in front of the letter represents the n value. The letter designates the l value:
        $l = 0 =$ s orbital; $l = 1 =$ p orbital; $l = 2 =$ d orbital; $l = 3 =$ f orbital.
        Solution:
        a) For the 5s subshell, $n = 5$ and $l = 0$. Since $m_l = 0$, there is one orbital.
        b) For the 3p subshell, $n = 3$ and $l = 1$. Since $m_l = -1, 0, +1$, there are three orbitals.
        c) For the 4f subshell, $n = 4$ and $l = 3$. Since $m_l = -3, -2, -1, 0, +1, +2, +3$, there are seven orbitals.

7.59   Plan: Allowed values of quantum numbers: n = positive integers; $l =$ integers from 0 to n $-$ 1;
        $m_l =$ integers from $-l$ through 0 to $+l$.
        Solution:
        a) With $l = 0$, the only allowable m_l value is 0. To correct, either change l or m_l value.
        Correct $n = 2, l = 1, m_l = -1; n = 2, l = 0, m_l = 0$.
        b) Combination is allowed.
        c) Combination is allowed.
        d) With $l = 2, m_l = -2, -1, 0, +1, +2; +3$ is not an allowable m_l value. To correct, either change l or m_l value.
        Correct: $n = 5, l = 3, m_l = +3; n = 5, l = 2, m_l = 0$.

7.62   a) The mass of the electron is 9.1094×10^{-31} kg.

$$E = -\frac{h^2}{8\pi^2 m_e a_0^2 n^2} = -\frac{h^2}{8\pi^2 m_e a_0^2}\left(\frac{1}{n^2}\right)$$

$$= -\frac{\left(6.626 \times 10^{-34}\ J \cdot s\right)^2}{8\pi^2\left(9.1094 \times 10^{-31}\ kg\right)\left(52.92 \times 10^{-12}\ m\right)^2}\left(\frac{kg \cdot m^2/s^2}{J}\right)\left(\frac{1}{n^2}\right)$$

$$= -(2.17963 \times 10^{-18}\ J)\left(\frac{1}{n^2}\right) = -(2.180 \times 10^{-18}\ J)\left(\frac{1}{n^2}\right)$$

This is identical with the result from Bohr's theory. For the H atom, $Z = 1$ and Bohr's constant $= -2.18 \times 10^{-18}$ J. For the hydrogen atom, derivation using classical principles or quantum-mechanical principles yields the same constant.
b) The $n = 3$ energy level is higher in energy than the $n = 2$ level. Because the zero point of the atom's energy is defined as an electron's infinite distance from the nucleus, a larger negative number describes a lower energy level. Although this may be confusing, it makes sense that an energy *change* would be a positive number.

$$\Delta E = -(2.180 \times 10^{-18}\ J)\left(\frac{1}{2^2} - \frac{1}{3^2}\right) = -3.027778 \times 10^{-19} = \mathbf{3.028 \times 10^{-19}\ J}$$

c) $E = \dfrac{hc}{\lambda}$ $\lambda\ (m) = \dfrac{hc}{E} = \dfrac{\left(6.626 \times 10^{-34}\ J \cdot s\right)\left(2.9979 \times 10^8\ m/s\right)}{\left(3.027778 \times 10^{-19}\ J\right)} = 6.56061 \times 10^{-7} = \mathbf{6.561 \times 10^{-7}\ m}$

$$\left(6.56061 \times 10^{-7}\ m\right)\left(\frac{1\ nm}{10^{-9}\ m}\right) = 656.061 = \mathbf{656.1\ nm}$$

This is the wavelength for the observed red line in the hydrogen spectrum.

7.63   a) The lines do not begin at the origin because an electron must absorb a minimum amount of energy before it has enough energy to overcome the attraction of the nucleus and leave the atom. This minimum energy is the energy of photons of light at the threshold frequency.

b) The lines for K and Ag do not begin at the same point. The amount of energy that an electron must absorb to leave the K atom is less than the amount of energy that an electron must absorb to leave the Ag atom, where the attraction between the nucleus and outer electron is stronger than in a K atom.

c) Wavelength is inversely proportional to energy. Thus, the metal that requires a larger amount of energy to be absorbed before electrons are emitted will require a shorter wavelength of light. Electrons in Ag atoms require more energy to leave, so Ag requires a shorter wavelength of light than K to eject an electron.

d) The slopes of the line show an increase in kinetic energy as the frequency (or energy) of light is increased. Since the slopes are the same, this means that for an increase of one unit of frequency (or energy) of light, the increase in kinetic energy of an electron ejected from K is the same as the increase in the kinetic energy of an electron ejected from Ag. After an electron is ejected, the energy that it absorbs above the threshold energy becomes kinetic energy of the electron. For the same increase in energy above the threshold energy, for either K or Ag, the kinetic energy of the ejected electron will be the same.

7.66   The Bohr model has been successfully applied to predict the spectral lines for one-electron species other than H. Common one-electron species are small cations with all but one electron removed. Since the problem specifies a metal ion, assume that the possible choices are Li^{+2} or Be^{+3}, and solve Bohr's equation to verify if a whole number for Z can be calculated. Recall that the negative sign is a convention based on the zero point of the atom's energy; it is deleted in this calculation to avoid taking the square root of a negative number.

The highest-energy line corresponds to the transition from $n = 1$ to $n = \infty$.

$E = h\nu = (6.626 \times 10^{-34} \text{ Js}) (2.961 \times 10^{16} \text{ Hz}) (s^{-1}/\text{Hz}) = 1.9619586 \times 10^{-17} \text{ J (unrounded)}$

$$E = \left(2.18 \times 10^{-18} \text{ J}\right)\left(\frac{Z^2}{n^2}\right) \qquad Z = \text{charge of the nucleus}$$

$$Z^2 = \frac{En^2}{2.18 \times 10^{-18} \text{ J}} = \frac{1.9619586 \times 10^{-17}(1^2)}{2.18 \times 10^{-18} \text{ J}} = 8.99998$$

Then $Z^2 = 9$ and $Z = 3$.

Therefore, the ion is $\mathbf{Li^{2+}}$ with an atomic number of 3.

7.69   The electromagnetic spectrum shows that the visible region goes from 400 to 750 nm (4000 Å to 7500 Å). Thus, wavelengths b, c, and d are for the three transitions in the visible series with $n_{final} = 2$. Wavelength a is in the ultraviolet region of the spectrum and the ultraviolet series has $n_{final} = 1$. Wavelength e is in the infrared region of the spectrum and the infrared series has $n_{final} = 3$.

$n = ? \to n = 1$; $\lambda = 1212.7$ Å (shortest λ corresponds to the largest ΔE)

$$1/\lambda = \left(1.096776 \times 10^7 \text{ m}^{-1}\right)\left(\frac{1}{n_1^2} - \frac{1}{n_2^2}\right)$$

$$\frac{1}{1212.7 \text{ Å}}\left(\frac{1 \text{ Å}}{1 \times 10^{-10} \text{ m}}\right) = \left(1.096776 \times 10^7 \text{ m}^{-1}\right)\left(\frac{1}{1^2} - \frac{1}{n_2^2}\right)$$

$$0.7518456 = \left(\frac{1}{1^2} - \frac{1}{n_2^2}\right)$$

$$\left(\frac{1}{n_2^2}\right) = 1 - 0.7518456$$

$$\left(\frac{1}{n_2^2}\right) = 0.2481544$$

$n_2^2 = 4.0297$

$\mathbf{n_2 = 2}$ for line (a) $(n = 2 \to n = 1)$

$n = ? \rightarrow n = 3$; $\lambda = 10{,}938$ Å (longest λ corresponds to the smallest ΔE)

$$1/\lambda = \left(1.096776 \times 10^7\, \text{m}^{-1}\right)\left(\frac{1}{n_1^2} - \frac{1}{n_2^2}\right)$$

$$\frac{1}{10938\ \text{Å}}\left(\frac{1\ \text{Å}}{1 \times 10^{-10}\,\text{m}}\right) = \left(1.096776 \times 10^7\,\text{m}^{-1}\right)\left(\frac{1}{3^2} - \frac{1}{n_2^2}\right)$$

$$0.083357396 = \left(\frac{1}{3^2} - \frac{1}{n_2^2}\right)$$

$$\left(\frac{1}{n_2^2}\right) = 0.11111111 - 0.083357396$$

$$\left(\frac{1}{n_2^2}\right) = 0.027753715111$$

$$n_2^2 = 36.03121$$

$n_2 = 6$ for line (e) ($n = 6 \rightarrow n = 3$)
For the other three lines, $n_1 = 2$.
For line (d), **$n_2 = 3$** (largest $\lambda \rightarrow$ smallest ΔE).
For line (b), **$n_2 = 5$** (smallest $\lambda \rightarrow$ largest ΔE).
For line (c), **$n_2 = 4$**.

7.71   a) The energy of visible light is lower than that of UV light. Thus, metal A must be **barium**, since of the three metals listed, barium has the smallest work function indicating the attraction between barium's nucleus and outer electron is less than the attraction in tantalum or tungsten. The longest wavelength corresponds to the lowest energy (work function). $E = hc/\lambda$ thus $\lambda = hc/E$

$$\text{Ta: } \lambda = \frac{hc}{E} = \frac{\left(6.626 \times 10^{-34}\,\text{J} \cdot \text{s}\right)\left(3.00 \times 10^8\,\text{m/s}\right)}{6.81 \times 10^{-19}\,\text{J}}\left(\frac{1\ \text{nm}}{10^{-9}\,\text{m}}\right) = 291.894 = \mathbf{292\ nm}$$

$$\text{Ba: } \lambda = \frac{hc}{E} = \frac{\left(6.626 \times 10^{-34}\,\text{J} \cdot \text{s}\right)\left(3.00 \times 10^8\,\text{m/s}\right)}{4.30 \times 10^{-19}\,\text{J}}\left(\frac{1\ \text{nm}}{10^{-9}\,\text{m}}\right) = 462.279 = \mathbf{462\ nm}$$

$$\text{W: } \lambda = \frac{hc}{E} = \frac{\left(6.626 \times 10^{-34}\,\text{J} \cdot \text{s}\right)\left(3.00 \times 10^8\,\text{m/s}\right)}{7.16 \times 10^{-19}\,\text{J}}\left(\frac{1\ \text{nm}}{10^{-9}\,\text{m}}\right) = 277.6257 = \mathbf{278\ nm}$$

Metal A must be barium, because barium is the only metal that emits in the visible range (462 nm).
b) A UV range of 278 nm to 292 nm is necessary to distinguish between tantalum and tungsten.

7.74   The speed of light is necessary; however, the frequency is irrelevant.

a) Time $= \left(8.1 \times 10^7\ \text{km}\right)\left(\frac{10^3\,\text{m}}{1\ \text{km}}\right)\left(\frac{1\ \text{s}}{3.00 \times 10^8\,\text{m}}\right) = 270 = \mathbf{2.7 \times 10^2\ s}$

b) Distance $= \left(1.2\ \text{s}\right)\left(\frac{3.00 \times 10^8\,\text{m}}{\text{s}}\right) = \mathbf{3.6 \times 10^8\ m}$

7.75   a) The l value must be at least 1 for m_l to be -1, but cannot be greater than $n - 1 = 2$. Increase the l value to 1 or 2 to create an allowable combination.
b) The l value must be at least 1 for m_l to be $+1$, but cannot be greater than $n - 1 = 2$. Decrease the l value to 1 or 2 to create an allowable combination.
c) The l value must be at least 3 for m_l to be $+3$, but cannot be greater than $n - 1 = 6$. Increase the l value to 3, 4, 5, or 6 to create an allowable combination.

d) The l value must be at least 2 for m_l to be -2, but cannot be greater than $n - 1 = 3$. Increase the l value to 2 or 3 to create an allowable combination.

7.77   a) Ionization occurs when the electron is completely removed from the atom, or when $n = \infty$. We can use the equation for the energy of an electron transition to find the quantity of energy needed to remove completely the electron, called the ionization energy (IE). The charge on the nucleus must affect the IE because a larger nucleus would exert a greater pull on the escaping electron. The Bohr equation applies to H and other one-electron species. The general expression is:

$$\Delta E = \left(-2.18 \times 10^{-18}\,J\right)\left(\frac{1}{\infty^2} - \frac{1}{n_{initial}^2}\right)Z^2\left(\frac{6.022 \times 10^{23}}{1\ mol}\right)$$

$$= 1.312796 \times 10^6 = (1.31 \times 10^6\ J/mol)\,Z^2\ \text{for } n = 1$$

b) In the ground state $n = 1$, the initial energy level for the single electron in B^{4+}. Once ionized, $n = \infty$ is the final energy level. The ionization energy is calculated as follows (remember that $1/\infty = 0$). $Z = 5$ for B^{4+}.

$$\Delta E = IE = (1.312796 \times 10^6\ J/mol)(\,5^2) = 3.28199 \times 10^7 = \mathbf{3.28 \times 10^7\ J/mol}$$

c) $\Delta E = \left(-2.18 \times 10^{-18}\,J\right)\left(\frac{1}{\infty^2} - \frac{1}{n_{initial}^2}\right)Z^2 = \left(-2.18 \times 10^{-18}\,J\right)\left(\frac{1}{\infty^2} - \frac{1}{3^2}\right)Z^2$

$$= (2.42222 \times 10^{-19}\,J)Z^2\ \text{for } n = 3 \qquad Z = 2\ \text{for } He^+$$

$$\Delta E = IE = (2.42222 \times 10^{-19}\,J)(2^2) = 9.68889 \times 10^{-19}\,J$$

$$\lambda = hc/E = \frac{\left(6.626 \times 10^{-34}\,J \cdot s\right)\left(3.00 \times 10^8\,m/s\right)}{9.68889 \times 10^{-19}\,J}\left(\frac{1\ nm}{10^{-9}\ m}\right) = 205.162 = \mathbf{205\ nm}$$

d) $\Delta E = \left(-2.18 \times 10^{-18}\,J\right)\left(\frac{1}{\infty^2} - \frac{1}{n_{initial}^2}\right)Z^2 = \left(-2.18 \times 10^{-18}\,J\right)\left(\frac{1}{\infty^2} - \frac{1}{2^2}\right)Z^2$

$$= (5.45 \times 10^{-19}\,J)Z^2\ \text{for } n = 2 \qquad (Z = 4\ \text{for } Be^{3+})$$

$$\Delta E = IE = (5.45 \times 10^{-19}\,J)(4^2) = 8.72 \times 10^{-18}\,J$$

$$\lambda = hc/E = \frac{\left(6.626 \times 10^{-34}\,J \cdot s\right)\left(3.00 \times 10^8\,m/s\right)}{8.72 \times 10^{-18}\,J}\left(\frac{1\ nm}{10^{-9}\ m}\right) = 22.79587 = \mathbf{22.8\ nm}$$

7.79   <u>Plan:</u> Use the values and the equation given in the problem to calculate the appropriate values. Additional information is given in the back of the textbook.

$$r_n = \frac{n^2 h^2 \varepsilon_0}{\pi m_e e^2}$$

<u>Solution:</u>

a) $r_1 = \dfrac{1^2\left(6.626 \times 10^{-34}\,J \cdot s\right)^2\left(8.854 \times 10^{-12}\,\dfrac{C^2}{J \cdot m}\right)}{\pi\left(9.109 \times 10^{-31}\,kg\right)\left(1.602 \times 10^{-19}\,C\right)^2}\left(\dfrac{kg \cdot m^2/s^2}{J}\right) = 5.2929377 \times 10^{-11} = \mathbf{5.293 \times 10^{-11}\ m}$

b) $r_{10} = \dfrac{10^2\left(6.626 \times 10^{-34}\,J \cdot s\right)^2\left(8.854 \times 10^{-12}\,\dfrac{C^2}{J \cdot m}\right)}{\pi\left(9.109 \times 10^{-31}\,kg\right)\left(1.602 \times 10^{-19}\,C\right)^2}\left(\dfrac{kg \cdot m^2/s^2}{J}\right) = 5.2929377 \times 10^{-9} = \mathbf{5.293 \times 10^{-9}\ m}$

7.81   <u>Plan:</u> Refer to Chapter 6 for the calculation of the amount of heat energy absorbed by a substance from its heat capacity and temperature change ($q = c \times mass \times \Delta T$). Using this equation, calculate the energy absorbed by the water. This energy equals the energy from the microwave photons. The energy of each photon can be calculated from its wavelength: $E = hc/\lambda$. Dividing the total energy by the energy of each photon gives the number of photons absorbed by the water.
<u>Solution:</u>
$q = c \times mass \times \Delta T$
$q = (4.184 \text{ J/g°C}) (252 \text{ g}) (98 - 20.)°C = 8.22407 \times 10^4 \text{ J (unrounded)}$

$$E = \frac{hc}{\lambda} = \frac{\left(6.626 \times 10^{-34} \text{ J} \cdot \text{s}\right)\left(3.00 \times 10^8 \text{ m/s}\right)}{1.55 \times 10^{-2} \text{ m}} = 1.28245 \times 10^{-23} \text{ J/photon (unrounded)}$$

Number of photons $= \left(8.22407 \times 10^4 \text{ J}\right)\left(\dfrac{1 \text{ photon}}{1.28245 \times 10^{-23} \text{ J}}\right) = 6.41278 \times 10^{27} = \mathbf{6.4 \times 10^{27} \text{ photons}}$

7.83   In general, to test for overlap of the two series, compare the longest wavelength in the "n" series with the shortest wavelength in the "$n+1$" series. The longest wavelength in any series corresponds to the transition between the n_1 level and the next level above it; the shortest wavelength corresponds to the transition between the n_1 level and the $n = \infty$ level. Use:

$$\frac{1}{\lambda} = R\left(\frac{1}{n_1^2} - \frac{1}{n_2^2}\right) \qquad R = 1.096776 \times 10^7 \text{ m}^{-1}$$

$$\frac{1}{\lambda} = \left(1.096776 \times 10^7 \text{ m}^{-1}\right)\left(\frac{1}{n_1^2} - \frac{1}{n_2^2}\right)$$

a) The overlap between the $n_1 = 1$ series and the $n_1 = 2$ series would occur between the longest wavelengths for $n_1 = 1$ and the shortest wavelengths for $n_1 = 2$.
Longest wavelength in $n_1 = 1$ series has n_2 equal to 2.

$$\frac{1}{\lambda} = \left(1.096776 \times 10^7 \text{ m}^{-1}\right)\left(\frac{1}{1^2} - \frac{1}{2^2}\right)$$

$\lambda = 1.215684272 \times 10^{-7} = \mathbf{1.215684 \times 10^{-7} \text{ m}}$
Shortest wavelength in the $n_1 = 2$ series:

$$\frac{1}{\lambda} = \left(1.096776 \times 10^7 \text{ m}^{-1}\right)\left(\frac{1}{2^2} - \frac{1}{\infty^2}\right)$$

$\lambda = 3.647052817 \times 10^{-7} = \mathbf{3.647053 \times 10^{-7} \text{ m}}$
Since the longest wavelength for $n_1 = 1$ series is shorter than shortest wavelength for $n_1 = 2$ series, there is **no overlap** between the two series.
b) The overlap between the $n_1 = 3$ series and the $n_1 = 4$ series would occur between the longest wavelengths for $n_1 = 3$ and the shortest wavelengths for $n_1 = 4$.
Longest wavelength in $n_1 = 3$ series has n_2 equal to 4.

$$\frac{1}{\lambda} = \left(1.096776 \times 10^7 \text{ m}^{-1}\right)\left(\frac{1}{3^2} - \frac{1}{4^2}\right)$$

$\lambda = 1.875627163 \times 10^{-6} = \mathbf{1.875627 \times 10^{-6} \text{ m}}$
Shortest wavelength in $n_1 = 4$ series has $n_2 = \infty$.

$$\frac{1}{\lambda} = \left(1.096776 \times 10^7 \text{ m}^{-1}\right)\left(\frac{1}{4^2} - \frac{1}{\infty^2}\right)$$

$\lambda = 1.458821127 \times 10^{-6} = \mathbf{1.458821 \times 10^{-6} \text{ m}}$
Since the $n_1 = 4$ series shortest wavelength is shorter than the $n_1 = 3$ series longest wavelength, the **series do overlap**.

c) Shortest wavelength in $n_1 = 5$ series:

$$\frac{1}{\lambda} = \left(1.096776 \times 10^7 \text{ m}^{-1}\right)\left(\frac{1}{5^2} - \frac{1}{\infty^2}\right)$$

$\lambda = 2.27940801 \times 10^{-6} = \mathbf{2.279408 \times 10^{-6}}$ **m**

Similarly for the other combinations:

For $n_1 = 4$, $n_2 = 5$, $\lambda = \mathbf{4.052281 \times 10^{-6}}$ **m**
For $n_1 = 4$, $n_2 = 6$, $\lambda = \mathbf{2.625878 \times 10^{-6}}$ **m**
For $n_1 = 4$, $n_2 = 7$, $\lambda = \mathbf{2.166128 \times 10^{-6}}$ **m**

Therefore, only the first two lines of the $n_1 = 4$ series overlap with the $n_1 = 5$ series.

d) At longer wavelengths (i.e., lower energies), there is increasing overlap between the lines from different series (i.e., with different n_1 values). The hydrogen spectrum becomes more complex, since the lines begin to merge into a more-or-less continuous band, and much more care is needed to interpret the information.

7.86   <u>Plan:</u> The energy differences sought may be determined by looking at the energy changes in steps.
<u>Solution:</u>
a) The difference between levels 3 and 2 (E_{32}) may be found by taking the difference in the energies for the $3 \rightarrow 1$ transition (E_{31}) and the $2 \rightarrow 1$ transition (E_{21}).

$E_{32} = E_{31} - E_{21} = (4.854 \times 10^{-17} \text{ J}) - (4.098 \times 10^{-17} \text{ J}) = \mathbf{7.56 \times 10^{-18}}$ **J**

$$\lambda = \frac{hc}{E} = \frac{\left(6.626 \times 10^{-34} \text{ J} \cdot \text{s}\right)\left(3.00 \times 10^8 \text{ m/s}\right)}{\left(7.56 \times 10^{-18} \text{ J}\right)} = 2.629365 \times 10^{-8} = \mathbf{2.63 \times 10^{-8}} \textbf{ m}$$

b) The difference between levels 4 and 1 (E_{41}) may be found by adding the energies for the $4 \rightarrow 2$ transition (E_{42}) and the $2 \rightarrow 1$ transition (E_{21}).

$E_{41} = E_{42} + E_{21} = (1.024 \times 10^{-17} \text{ J}) + (4.098 \times 10^{-17} \text{ J}) = \mathbf{5.122 \times 10^{-17}}$ **J**

$$\lambda = \frac{hc}{E} = \frac{\left(6.626 \times 10^{-34} \text{ J} \cdot \text{s}\right)\left(3.00 \times 10^8 \text{ m/s}\right)}{\left(5.122 \times 10^{-17} \text{ J}\right)} = 3.88091 \times 10^{-9} = \mathbf{3.881 \times 10^{-9}} \textbf{ m}$$

c) The difference between levels 5 and 4 (E_{54}) may be found by taking the difference in the energies for the $5 \rightarrow 1$ transition (E_{51}) and the $4 \rightarrow 1$ transition (see part b).

$E_{54} = E_{51} - E_{41} = (5.242 \times 10^{-17} \text{ J}) - (5.122 \times 10^{-17} \text{ J}) = \mathbf{1.2 \times 10^{-18}}$ **J**

$$\lambda = \frac{hc}{E} = \frac{\left(6.626 \times 10^{-34} \text{ J} \cdot \text{s}\right)\left(3.00 \times 10^8 \text{ m/s}\right)}{\left(1.2 \times 10^{-18} \text{ J}\right)} = 1.6565 \times 10^{-7} = \mathbf{1.66 \times 10^{-7}} \textbf{ m}$$

7.88   a) The energy of the electron is a function of its speed leaving the surface of the metal. The mass of the electron is 9.109×10^{-31} kg.

$$E_K = 1/2\, mu^2 = \frac{1}{2}\left(9.109 \times 10^{-31} \text{ kg}\right)\left(6.40 \times 10^5 \text{ m/s}\right)^2\left(\frac{\text{J}}{\text{kg} \cdot \text{m}^2/\text{s}^2}\right)$$

$$= 1.86552 \times 10^{-19} = \mathbf{1.87 \times 10^{-19}} \textbf{ J}$$

b) The minimum energy required to dislodge the electron (ϕ) is a function of the incident light. In this example, the incident light is higher than the threshold frequency, so the kinetic energy of the electron, E_k, must be subtracted from the total energy of the incident light, $h\nu$, to yield the work function, ϕ. (The number of significant figures given in the wavelength requires more significant figures in the speed of light.)

$$E = hc/\lambda = \frac{\left(6.626 \times 10^{-34} \text{ J} \cdot \text{s}\right)\left(2.9979 \times 10^8 \text{ m/s}\right)}{\left(358.1 \text{ nm}\right)}\left(\frac{1 \text{ nm}}{10^{-9} \text{ m}}\right) = 5.447078 \times 10^{-19} \text{ J (unrounded)}$$

$\Phi_K = h\nu - E_K = (5.447078 \times 10^{-19} \text{ J}) - (1.86552 \times 10^{-19} \text{ J}) = 3.581558 \times 10^{-19} = \mathbf{3.58 \times 10^{-19}}$ **J**

7.90   a) Figure 7.3 indicates that the 641 nm wavelength of Sr falls in the **red** region and the 493 nm wavelength of Ba falls in the **green** region.
b) Use the formula $E = hc/\lambda$ to calculate kJ/photon. Convert the 1.00 g amounts of $BaCl_2$ ($\mathcal{M} = 208.2$ g/mol) and $SrCl_2$ ($\mathcal{M} = 158.52$ g/mol) to moles, then to atoms, and assume each atom undergoes one electron transition (which produces the colored light) to find number of photons. Multiply kJ/photon by number of photons to find total energy.
$SrCl_2$

$$\text{Number of photons} = \left(5.00 \text{ g } SrCl_2\right)\left(\frac{1 \text{ mol } SrCl_2}{158.52 \text{ g } SrCl_2}\right)\left[\frac{6.022 \times 10^{23} \text{ photons}}{1 \text{ mol } SrCl_2}\right]$$

$$= 1.8994449 \times 10^{22} \text{ photons (unrounded)}$$

$$E_{photon} = \frac{hc}{\lambda} = \frac{\left(6.626 \times 10^{-34} \text{ J} \cdot \text{s}\right)\left(3.00 \times 10^8 \text{ m/s}\right)}{641 \text{ nm}}\left(\frac{1 \text{ nm}}{10^{-9} \text{ m}}\right)\left(\frac{1 \text{ kJ}}{10^3 \text{ J}}\right)$$

$$= 3.10109 \times 10^{-22} \text{ kJ/photon (unrounded)}$$

$E_{total} = (1.8994449 \times 10^{22} \text{ photons})(3.10109 \times 10^{-22} \text{ kJ/photon}) = 5.89035 = \textbf{5.89 kJ (Sr)}$
$BaCl_2$

$$\text{Number of photons} = \left(5.00 \text{ g } BaCl_2\right)\left(\frac{1 \text{ mol } BaCl_2}{208.2 \text{ g } BaCl_2}\right)\left[\frac{6.022 \times 10^{23} \text{ photons}}{1 \text{ mol } BaCl_2}\right]$$

$$= 1.44620557 \times 10^{22} \text{ photons (unrounded)}$$

$$E_{photon} = \frac{hc}{\lambda} = \frac{\left(6.626 \times 10^{-34} \text{ J} \cdot \text{s}\right)\left(3.00 \times 10^8 \text{ m/s}\right)}{493 \text{ nm}}\left(\frac{1 \text{ nm}}{10^{-9} \text{ m}}\right)\left(\frac{1 \text{ kJ}}{10^3 \text{ J}}\right)$$

$$= 4.0320487 \times 10^{-22} \text{ kJ/photon (unrounded)}$$

$E_{total} = (1.44620557 \times 10^{22} \text{ photons})(4.0320487 \times 10^{-22} \text{ kJ/photon}) = 5.83117 = \textbf{5.83 kJ (Ba)}$

7.92   a) At this wavelength the sensitivity to absorbance of light by Vitamin A is maximized while minimizing interference due to the absorbance of light by other substances in the fish-liver oil.
b) The wavelength 329 nm lies in the **ultraviolet region** of the electromagnetic spectrum.
c) A known quantity of vitamin A (1.67×10^{-3} g) is dissolved in a known volume of solvent (250. mL) to give a standard concentration with a known response (1.018 units). This can be used to find the unknown quantity of Vitamin A that gives a response of 0.724 units. An equality can be made between the two concentration-to-absorbance ratios:

$$\frac{A_1}{C_1} = \frac{A_2}{C_2}$$

$$C_1 = \frac{A_1 C_2}{A_2} = \frac{(0.724)\left(\dfrac{1.67 \times 10^{-3} \text{ g}}{250. \text{ mL}}\right)}{(1.018)} = 4.7508 \times 10^{-6} \text{ g/mL (unrounded)}$$

$$\frac{\left(\dfrac{4.7508 \times 10^{-6} \text{ g Vitamin A}}{1 \text{ mL}}\right)(500. \text{ mL})}{(0.1232 \text{ g Oil})} = 1.92808 \times 10^{-2} = \textbf{1.93} \times \textbf{10}^{-2} \textbf{ g Vitamin A/g Oil}$$

7.97   The amount of energy is calculated from the wavelength of light:

$$E_{photon} = \frac{hc}{\lambda} = \frac{\left(6.626 \times 10^{-34} \text{ J} \cdot \text{s}\right)\left(3.00 \times 10^8 \text{ m/s}\right)}{550 \text{ nm}}\left(\frac{1 \text{ nm}}{10^{-9} \text{ m}}\right) = 3.61418 \times 10^{-19} \text{ J/photon (unrounded)}$$

$$\text{Amount of E from the bulb} = \left(75 \text{ W}\right)\left(\frac{1 \text{ J/s}}{\text{W}}\right)\left(\frac{10\%}{100\%}\right)\left(\frac{5\%}{100\%}\right) = 0.375 \text{ J/s}$$

Number of photons: $\left(\dfrac{0.375 \text{ J}}{\text{s}}\right)\left(\dfrac{1 \text{ photon}}{3.61418 \times 10^{-19} \text{ J}}\right) = 1.0376 \times 10^{18} = \mathbf{1.0 \times 10^{18} \text{ photons/s}}$

7.100   In the visible series with $n_{final} = 2$, the transitions will end in either the $2s$ or $2p$ orbitals. With the restriction that the angular momentum quantum number can change by only ± 1, the allowable transitions are from a p orbital to $2s$, from an s orbital to $2p$, and from a d orbital to $2p$. The problem specifies a change in *energy level*, so n_{init} must be 3, 4, 5, etc. (Although a change from $2p$ to $2s$ would result in a $+1$ change in l, this is not a change in energy level.) The first four transitions are as follows:

$3s \rightarrow 2p$
$3d \rightarrow 2p$
$4s \rightarrow 2p$
$3p \rightarrow 2s$

CHAPTER 8 ELECTRON CONFIGURATION AND CHEMICAL PERIODICITY

FOLLOW–UP PROBLEMS

8.1   Plan: The superscripts can be added to indicate the number of electrons in the element, and hence its identity. A horizontal orbital diagram is written for simplicity, although it does not indicate the sublevels have different energies. Based on the orbital diagram, we identify the electron of interest and determine its four quantum numbers.
Solution:
The number of electrons = 2 + 2 + 4 = 8; element = oxygen.
Orbital diagram:

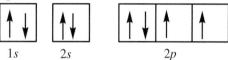

$1s$ $2s$ $2p$

The electron in the middle $2p$ orbital is the sixth electron (this electron would have entered in this position due to Hund's rule). This electron has the following quantum numbers: $n = 2$, $l = 1$ (for p orbital), $m_l = 0$, $m_s = +1/2$. By convention, $+1/2$ is assigned to the first electron in an orbital. Also, the first p orbital is assigned a a m_l value of -1, the middle p orbital a m_l value of 0 and the last p orbital a m_l value of $+1$.

8.2   Plan: The atomic number gives the number of electrons. The order of filling may be inferred by the location of the element on the periodic table. The partial orbital diagrams shows only those electrons after the preceding noble gas except those used to fill inner d and f subshells. The number of inner electrons is simply the total electrons minus those electrons in the partial orbital diagram.
Solution:
a) For Ni ($Z = 28$), the full electron configuration is $1s^2 2s^2 2p^6 3s^2 3p^6 4s^2 3d^8$.
The condensed configuration is $[\mathbf{Ar}]4s^2 3d^8$.
The partial orbital diagram for the valence electrons is

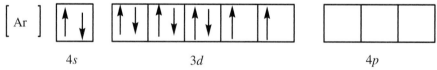

$4s$ $3d$ $4p$

There are 28 – 10(valence) = **18 inner electrons**.
b) For Sr ($Z = 38$), the full electron configuration is $1s^2 2s^2 2p^6 3s^2 3p^6 4s^2 3d^{10} 4p^6 5s^2$.
The condensed configuration is $[\mathbf{Kr}]5s^2$. The partial orbital diagram is

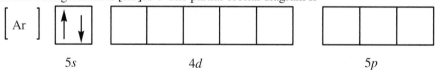

$5s$ $4d$ $5p$

There are 38 – 2(valence) = **36 inner electrons**.
c) For Po (84 electrons), the full configuration is $1s^2 2s^2 2p^6 3s^2 3p^6 4s^2 3d^{10} 4p^6 5s^2 4d^{10} 5p^6 6s^2 4f^{14} 5d^{10} 6p^4$.
The condensed configuration is $[\mathbf{Xe}]6s^2 4f^{14} 5d^{10} 6p^4$.
The partial orbital diagram represents valence electrons only; the inner electrons are those in the previous noble gas (Xe or $1s^2 2s^2 2p^6 3s^2 3p^6 4s^2 3d^{10} 4p^6 5s^2 4d^{10} 5p^6$) and *filled* transition ($5d^{10}$) and inner transition ($4f^{14}$) levels.

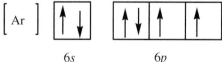

$6s$ $6p$

There are 84 – 6(valence) = **78 inner electrons**.

8.3   Plan: Locate each of the elements on the periodic table. All of these are main-group elements, so their sizes increase down and to the left in the periodic table.
Solution:
a) **Cl < Br < Se**. Cl has a smaller n than Br and Se. Br experiences a higher Z_{eff} than Se and is smaller.
b) **Xe < I < Ba**. Xe and I have the same n, but Xe experiences a higher Z_{eff} and is smaller. Ba has the highest n and is the largest.

8.4   Plan: These main-group elements must be located on the periodic table. The value of IE_1 increases toward the top (same column) and the right of the periodic table (same n).
Solution:
a) **Sn < Sb < I**. These elements have the same n, so the values increase to the right on the periodic table. Iodine has the highest IE, because its outer electron is most tightly held and hardest to remove.
b) **Ba < Sr < Ca**. These elements are in the same column, so the values decrease towards the bottom of the column. Barium's outer electron receives the most shielding; therefore, it is easiest to remove and has the lowest IE.

8.5   Plan: We must look for a large "jump" in the IE values. This jump occurs after the valence electrons have been removed. The next step is to determine the element in the designated period with the proper number of valence electrons.
Solution:
The exceptionally large jump from IE_3 to IE_4 means that the fourth electron is an inner electron. Thus, Q has three valence electrons. Since Q is in period 3, it must be **aluminum, Al: $1s^2 2s^2 2p^6 3s^2 3p^1$**.

8.6   Plan: We need to locate each element on the periodic table. Elements in the first two columns on the left or the two columns to the left of the noble gases tend to adopt ions with a noble gas configuration. Elements in the remaining columns may use either their ns and np electrons, or just their np electrons.
Solution:
a) Barium loses two electrons to be isoelectronic with Xe: Ba ([Xe] $6s^2$) $\rightarrow$ Ba^{2+} ([Xe]) + $2e^-$
b) Oxygen gains two e^- to be isoelectronic with Ne: O ([He] $2s^2 2p^4$ + $2e^-$ $\rightarrow$ O^{2-} ([Ne])
c) Lead can lose two electrons to form an "inert pair" configuration:
Pb ([Xe]$6s^2 4f^{14} 5d^{10} 6p^2$) $\rightarrow$ Pb^{2+} ([Xe]$6s^2 4f^{14} 5d^{10}$) + $2e^-$
or lead can lose four electrons form a "pseudo–noble gas" configuration:
Pb ([Xe]$6s^2 4f^{14} 5d^{10} 6p^2$) $\rightarrow$ Pb^{4+} ([Xe]$4f^{14} 5d^{10}$) + $4e^-$

8.7   Plan: Write the condensed electron configuration for each atom, being careful to note those elements, which are irregular. The charge on the cation tells how many electrons are to be removed. The electrons are removed beginning with the ns electrons. If any electrons in the final ion are unpaired, the ion is paramagnetic. If it is not obvious that there are unpaired electrons, a partial orbital diagram might help.
Solution:
a) The V atom ([Ar]$4s^2 3d^3$) loses the two $4s$ electrons and one $3d$ electron to form V^{3+} (**[Ar]$3d^2$**). There are two unpaired d electrons, so V^{3+} is **paramagnetic**.
b) The Ni atom ([Ar]$4s^2 3d^8$) loses the two $4s$ electrons to form Ni^{2+} (**[Ar]$3d^8$**). There are two unpaired d electrons, so Ni^{2+} is **paramagnetic**.

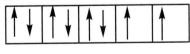

3d

c) The La atom ([Xe]$6s^2 5d^1$) loses all three valence electrons to form La^{3+} (**[Xe]**). There are no unpaired electrons, so La^{3+} is **diamagnetic**.

8.8   Plan: Locate each of the elements on the periodic table. Cations are smaller than the parent atoms, and anions are larger than the parent atoms. If the electrons are equal, anions are larger than cations. The more electrons added or removed; the greater the change in size.
        Solution:
        a) Ionic size increases down a group, so $\mathbf{F^- < Cl^- < Br^-}$.
        b) These species are isoelectronic (all have 10 electrons), so size increases with increasing negative charge:
            $\mathbf{Mg^{2+} < Na^+ < F^-}$.
        c) Ionic size increases as charge decreases for different cations of the same element, so $\mathbf{Cr^{3+} < Cr^{2+}}$.

END–OF–CHAPTER PROBLEMS

8.1   Elements are listed in the periodic table in an ordered, systematic way that correlates with a periodicity of their chemical and physical properties. The theoretical basis for the table in terms of atomic number and electron configuration does not allow for an "unknown element" between Sn and Sb.

8.3   Plan: The value should be the average of the elements above and below the one of interest.
        Solution:
        a) Predicted atomic mass (K) =
        $$\frac{Na + Rb}{2} = \frac{22.99 + 85.47}{2} = \mathbf{54.23\ amu} \qquad \text{(actual value = 39.10 amu)}$$
        b) Predicted melting point (Br_2) =
        $$\frac{Cl_2 + I_2}{2} = \frac{-101.0 + 113.6}{2} = \mathbf{6.3°C} \qquad \text{(actual value = –7.2°C)}$$

8.6   The quantum number m_s relates to just the electron; all the others describe the orbital.

8.9   Shielding occurs when inner electrons protect or shield outer electrons from the full nuclear attractive force. The effective nuclear charge is the nuclear charge an electron actually experiences. As the number of inner electrons increases, the effective nuclear charge decreases.

8.11   a) The $l = 1$ quantum number can only refer to a p orbital. These quantum numbers designate the $2p$ orbital set ($n = 2$), which hold a maximum of **6** electrons, 2 electrons in each of the three $2p$ orbitals.
        b) There are five $3d$ orbitals, therefore a maximum of **10** electrons can have the $3d$ designation.
        c) There is one $4s$ orbital which holds a maximum of **2** electrons.

8.13   a) **6** electrons can be found in the three $4p$ orbitals, 2 in each orbital.
        b) The $l = 1$ quantum number can only refer to a p orbital, and the m_l value of +1 specifies one particular p orbital, which hold a maximum of **2** electrons with the difference between the two electrons being in the m_s quantum number.
        c) **14** electrons can be found in the $5f$ orbitals ($l = 3$ designates f orbitals; there are $7f$ orbitals in a set).

8.16   Hund's rule states that electrons will fill empty orbitals in the same sublevel before filling half-filled orbitals. This lowest-energy arrangement has the maximum number of unpaired electrons with parallel spins. The correct electron configuration for nitrogen is shown in (a), which is contrasted to an incorrect configuration shown in (b). The arrows in the $2p$ orbitals of configuration (a) could alternatively all point down.
        (a) – correct   (b) – incorrect

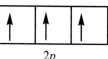

        In (a), there is one electron in each of the $2p$ orbitals; in (b), which is incorrect, 2 electrons were paired in one of the $2p$ orbitals while leaving one $2p$ orbital empty.

8.18   For elements in the same group (vertical column in periodic table), the electron configuration of the outer electrons are identical except for the n value. For elements in the same period (horizontal row in periodic table), their configurations vary because each succeeding element has one additional electron. The electron configurations are similar only in the fact that the same level (principal quantum number) is the outer level.

8.20   The total electron capacity for an energy level is $2n^2$, so the $n = 4$ energy level holds a maximum of $2(4^2) = 32$ **electrons**. A filled $n = 4$ energy level would have the following configuration: $4s^2 4p^6 4d^{10} 4f^{14}$.

8.21   Plan: Assume that the electron is in the ground state configuration and that electrons fill in a p_x–p_y–p_z order. By convention, we assign the first electron to fill an orbital with a m_s value of $+1/2$. Also by convention, $m_l = -1$ for the p_x orbital, $m_l = 0$ for the p_y orbital, and $m_l = +1$ for the p_z orbital. Also, keep in mind the following letter orbital designation for each l value: $l = 0 = s$ orbital, $l = 1 = p$ orbital, $l = 2 = d$ orbital, and $l = 3 = f$ orbital.
Solution:
a) The outermost electron in a rubidium atom would be in a $5s$ orbital (rubidium is in Row 5, Group 1). The quantum numbers for this electron are $n = 5, l = 0, m_l = 0$, and $m_s = +1/2$.
b) The S^- ion would have the configuration $[Ne]3s^2 3p^5$. The electron added would go into the $3p_z$ orbital and is the second electron in that orbital. Quantum numbers are $n = 3, l = 1, m_l = +1$, and $m_s = -1/2$.
c) Ag atoms have the configuration $[Kr]5s^1 4d^{10}$. The electron lost would be from the $5s$ orbital with quantum numbers $n = 5, l = 0, m_l = 0$, and $m_s = +1/2$.
d) The F atom has the configuration $[He]2s^2 2p^5$. The electron gained would go into the $2p_z$ orbital and is the second electron in that orbital. Quantum numbers are $n = 2, l = 1, m_l = +1$, and $m_s = -1/2$.

8.23   a) Rb: $1s^2 2s^2 2p^6 3s^2 3p^6 4s^2 3d^{10} 4p^6 5s^1$
        b) Ge: $1s^2 2s^2 2p^6 3s^2 3p^6 4s^2 3d^{10} 4p^2$
        c) Ar: $1s^2 2s^2 2p^6 3s^2 3p^6$

8.25   a) Cl: $1s^2 2s^2 2p^6 3s^2 3p^5$
        b) Si: $1s^2 2s^2 2p^6 3s^2 3p^2$
        c) Sr: $1s^2 2s^2 2p^6 3s^2 3p^6 4s^2 3d^{10} 4p^6 5s^2$

8.27   Valence electrons are those electrons beyond the previous noble gas configuration and unfilled d and f sublevels. For a condensed ground-state electron configuration, the electron configuration of the previous noble gas is shown by its element symbol in brackets, followed by the electron configuration of the energy level being filled.
a) Ti $(Z = 22)$; $[Ar]4s^2 3d^2$

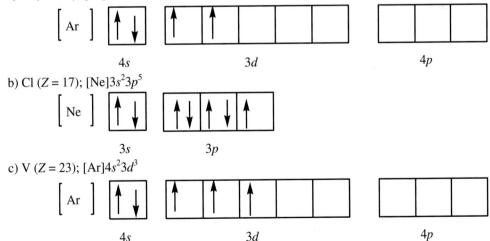

b) Cl $(Z = 17)$; $[Ne]3s^2 3p^5$

c) V $(Z = 23)$; $[Ar]4s^2 3d^3$

8.29   Valence electrons are those electrons beyond the previous noble gas configuration and unfilled d and f sublevels. For a condensed ground-state electron configuration, the electron configuration of the previous noble gas is shown by its element symbol in brackets, followed by the electron configuration of the energy level being filled.
a) Mn ($Z = 25$); $[Ar]4s^23d^5$

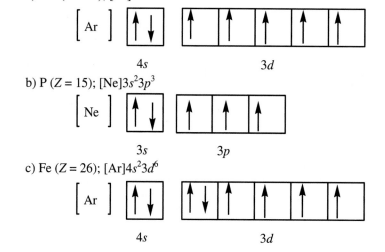

b) P ($Z = 15$); $[Ne]3s^23p^3$

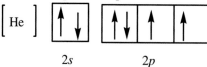

c) Fe ($Z = 26$); $[Ar]4s^23d^6$

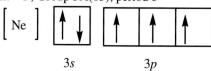

8.31   a) Element = O; Group 6A(16); period 2

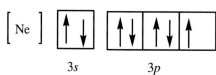

b) Element = P; Group 5A(15); period 3

8.33   a) Elemet = Cl; Group 7A(17); period 3

b) Element = As; Group 5A(15); period 4

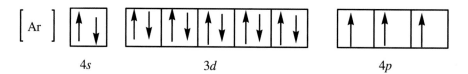

8.35   a) The orbital diagram shows the element is in period 4 ($n = 4$ as outer level). The configuration is $1s^22s^22p^63s^23p^64s^23d^{10}4p^1$ or $[Ar]4s^23d^{10}4p^1$. One electron in the p level indicates the element is in group **3A(13)**. The element is Ga.
b) The orbital diagram shows the $2s$ and $2p$ orbitals filled which would represent the last element in period 2, Ne. The configuration is $1s^22s^22p^6$ or $[He]2s^22p^6$. Filled s and p orbitals indicate the group **8A(18)**.

8.37   Inner electrons are those seen in the previous noble gas and completed transition series (*d* orbitals). Outer electrons are those in the highest energy level (highest n value). Valence electrons are the outer electrons for main-group elements; for transition metals, valence electrons also include electrons in the unfilled *d* set of orbitals. It is easiest to determine the types of electrons by writing a condensed electron configuration.
a) O (Z = 8); [He]$2s^2 2p^4$. There are **2** inner electrons (represented by [He]) and **6** outer electrons. The number of valence electrons (**6**) equals the outer electrons in this case.
b) Sn (Z = 50); [Kr]$5s^2 4d^{10} 5p^2$. There are 36 (from [Kr]) + 10 (from the filled *d* level) = **46** inner electrons. There are **4** outer electrons (highest energy level is *n* = 5) and **4** valence electrons.
c) Ca (Z = 20); [Ar]$4s^2$. There are **2** outer electrons, **2** valence electrons, and **18** inner electrons.
d) Fe (Z = 26); [Ar]$4s^2 3d^6$. There are **2** outer electrons (from *n* = 4 level), **8** valence electrons (the *d* orbital electrons count in this case because the sublevel is not full), and **18** inner electrons.
e) Se (Z = 34); [Ar]$4s^2 3d^{10} 4p^4$. There are **6** outer electrons (2 + 4 in the *n* = 4 level), **6** valence electrons (filled *d* sublevels count as inner electrons), and **28** ((18 + 10) or (34 − 6)) inner electrons.

8.39   a) The electron configuration [He]$2s^2 2p^1$ has a total of 5 electrons (3 + 2 from He configuration) which is element boron with symbol **B**. Boron is in group 3A(13). Other elements in this group are **Al, Ga, In,** and **Tl**.
b) The electrons in this element total 16, 10 from the neon configuration plus 6 from the rest of the configuration. Element 16 is sulfur, **S**, in group 6A(16). Other elements in group 6A(16) are **O, Se, Te,** and **Po**.
c) Electrons total 3 + 54 (from xenon) = 57. Element 57 is lanthanum, **La**, in group 3B(3). Other elements in this group are **Sc, Y,** and **Ac**.

8.41   a) The electron configuration [He]$2s^2 2p^2$ has a total of 6 electrons (4 + 2 from He configuration) which is element Carbon with symbol **C**; other Group 4A(14) elements include **Si, Ge, Sn,** and **Pb**.
b) Electrons total 5 + 18 (from argon) = 23 which is **Vanadium**; other Group 5B(5) elements include **Nb, Ta,** and **Db**.
c) The electrons in this element total 15, 10 from the neon configuration plus 5 from the rest of the configuration. Element 15 is **Phosphorus**; other Group 5A(15) elements include **N, As, Sb,** and **Bi**.

8.43   The ground state configuration of Na is $1s^2 2s^2 2p^6 3s^1$. Upon excitation, the $3s^1$ electron is promoted to the $3p$ level, with configuration $1s^2 2s^2 2p^6 3p^1$.

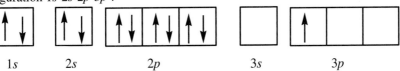

$1s$ $2s$ $2p$ $3s$ $3p$

8.50   A high, endothermic IE_1 means it is very difficult to remove the first outer electron. This value would exclude any metal, because metals lose an outer electron easily. A very negative, exothermic EA_1 suggests that this element easily gains one electron. These values indicate that the element belongs to the halogens, Group **7A(17)**, which form −1 ions.

8.53   <u>Plan:</u> Atomic size decreases up a main group and left to right across a period.
<u>Solution:</u>
a) Increasing atomic size: **K < Rb < Cs**; these three elements are all part of the same group, the alkali metals. Atomic size decreases up a main group (larger outer electron orbital), so potassium is the smallest and cesium is the largest.
b) Increasing atomic size: **O < C < Be**; these three elements are in the same period and atomic size decreases across a period (increasing effective nuclear charge), so beryllium is the largest and oxygen the smallest.
c) Increasing atomic size: **Cl < S < K**; chlorine and sulfur are in the same period so chlorine is smaller since it is further to the right in the period. Potassium is the first element in the next period so it is larger than either Cl or S.
d) Increasing atomic size: **Mg < Ca < K**; calcium is larger than magnesium because Ca is further down the alkaline earth metal group on the periodic table than Mg. Potassium is larger than calcium because K is further to the left than Ca in period 4 of the periodic table.

8.55   <u>Plan:</u> Ionization energy increases up a group and left to right across a period.
        <u>Solution:</u>
        a) **Ba < Sr < Ca** The "group" rule applies in this case. Ionization energy increases up a main group. Barium's outer electron receives the most shielding; therefore, it is easiest to remove and has the lowest IE.
        b) **B < N < Ne** These elements have the same n, so the "period" rule applies. Ionization energy increases from left to right across a period. B experiences the lowest Z_{eff} and has the lowest IE. Ne has the highest IE, because it's very difficult to remove an electron from the stable noble gas configuration.
        c) **Rb < Se < Br** IE decreases with increasing atomic size, so Rb (largest atom) has the smallest IE. Se has a lower IE than Br because IE increases across a period.
        d) **Sn < Sb < As** IE increases up a group, so Sn and Sb will have smaller IE's than As. The "period" rule applies for ranking Sn and Sb.

8.57   The successive ionization energies show a significant jump between the first and second IE's and between the third and fourth IE's. This indicates that 1) the first electron removed occupies a different orbital than the second electron removed, 2) the second and third electrons occupy the same orbital, 3) the third and fourth electrons occupy different orbitals, and 4) the fourth and fifth electrons occupy the same orbital. The electron configurations for period 2 elements range from $1s^2 2s^1$ for lithium to $1s^2 2s^2 2p^6$ for neon. The three different orbitals are $1s$, $2s$, and $2p$. The first electron is removed from the $2p$ orbital and the second from the $2s$ orbital in order for there to be a significant jump in the ionization energy between the two. The third electron is removed from the $2s$ orbital while the fourth is removed from the $1s$ orbital since IE$_4$ is much greater than IE$_3$. The configuration is **$1s^2 2s^2 2p^1$** which represents the five electrons in boron, **B**.

8.59   a) **Na** would have the highest IE$_2$ because ionization of a second electron would require breaking the stable [Ne] configuration:
        First ionization: Na ([Ne]$3s^1$) $\rightarrow$ Na$^+$ ([Ne]) + e$^-$ (low IE)
        Second ionization: Na$^+$ ([Ne]) $\rightarrow$ Na^{+2} ([He]$2s^2 2p^5$) + e$^-$ (high IE)
        b) **Na** would have the highest IE$_2$ because it has one valence electron and is smaller than K.
        c) You might think that Sc would have the highest IE$_2$, because removing a second electron would require breaking the stable, filled $4s$ shell. However, **Be** has the highest IE$_2$ because Be's small size makes it difficult to remove a second electron.

8.61   Three of the ways that metals and nonmetals differ are: 1) metals conduct electricity, nonmetals do not; 2) when they form stable ions, metal ions tend to have a positive charge, nonmetal ions tend to have a negative charge; and 3) metal oxides are ionic and act as bases, nonmetal oxides are covalent and act as acids.

8.62   Metallic character decreases up a group and decreases toward the right across a period. These trends are the same as those for atomic size and opposite those for ionization energy.

8.64   The two largest elements in group 4A, Sn and Pb, have atomic electron configurations that look like $ns^2(n-1)d^{10}np^2$. Both of these elements are metals so they will form positive ions. To reach the noble gas configuration of xenon the atoms would have to lose 14 electrons, which is not likely. Instead the atoms lose either 2 or 4 electrons to attain a stable configuration with either the ns and $(n-1)d$ filled for the 2+ ion or the $(n-1)d$ orbital filled for the 4+ ion. The Sn^{2+} and Pb^{2+} ions form by losing the two p electrons:
        Sn ([Kr]$5s^2 4d^{10} 5p^2$) $\rightarrow$ Sn^{2+} ([Kr]$5s^2 4d^{10}$) + 2 e$^-$
        Pb ([Xe]$6s^2 5d^{10} 6p^2$) $\rightarrow$ Pb^{2+} ([Xe]$6s^2 5d^{10}$) + 2 e$^-$
        The Sn^{4+} and Pb^{4+} ions form by losing the two p and two s electrons:
        Sn ([Kr]$5s^2 4d^{10} 5p^2$) $\rightarrow$ Sn^{4+} ([Kr]$4d^{10}$) + 4 e$^-$
        Pb ([Xe]$6s^2 5d^{10} 6p^2$) $\rightarrow$ Pb^{4+} ([Xe]$5d^{10}$) + 4 e$^-$
        Possible ions for tin and lead have **+2** and **+4** charges. The +2 ions form by loss of the outermost two p electrons, while the +4 ions form by loss of these and the outermost two s electrons.

8.68   Metallic behavior increases down a group and decreases across a period.
        a) **Rb** is more metallic because it is to the left and below Ca.
        b) **Ra** is more metallic because it lies below Mg in Group 2A(2).
        c) **I** is more metallic because it lies below Br in Group 7A(17).

8.70 Metallic behavior increases down a group and decreases across a period.
a) **As** should be less metallic than antimony because it lies above Sb in the same group of the periodic table.
b) **P** should be less metallic because it lies to the right of silicon in the same period of the periodic table.
c) **Be** should be less metallic since it lies above and to the right of sodium on the periodic table.

8.72 The reaction of a *nonmetal* oxide in water produces an *acidic* solution. An example of a Group 6A(16) nonmetal oxide is $SO_2(g)$: $SO_2(g) + H_2O(g) \rightarrow H_2SO_3(aq)$.

8.74 For main group elements, the most stable ions have electron configurations identical to noble gas atoms.
a) Cl: $1s^2 2s^2 2p^6 3s^2 3p^5$; Cl⁻, $\mathbf{1s^2 2s^2 2p^6 3s^2 3p^6}$, chlorine atoms are one electron short of the noble gas configuration, so a **–1** ion will form by adding an electron to have the same electron configuration as an argon atom.
b) Na: $1s^2 2s^2 2p^6 3s^1$; Na⁺, $\mathbf{1s^2 2s^2 2p^6}$, sodium atoms contain one more electron than the noble gas configuration of neon. Thus, a sodium atom loses one electron to form a **+1** ion.
c) Ca: $1s^2 2s^2 2p^6 3s^2 3p^6 4s^2$; Ca²⁺, $\mathbf{1s^2 2s^2 2p^6 3s^2 3p^6}$, calcium atoms contain two more electrons than the noble gas configuration of argon. Thus, a calcium atom loses two electrons to form a **+2** ion.

8.76 a) Elements in the main group form ions that attain a filled outer level, or noble gas configuration. In this case, Al achieves the noble gas configuration of Ne ($1s^2 2s^2 2p^6$) by losing its 3 outer shell electrons to form **Al³⁺**.
Al ($1s^2 2s^2 2p^6 3s^2 3p^1$) $\rightarrow$ Al³⁺ ($\mathbf{1s^2 2s^2 2p^6}$) + 3 e⁻
b) Sulfur achieves the noble gas configuration of Ar by gaining 2 electrons to form **S²⁻**.
S ($1s^2 2s^2 2p^6 3s^2 3p^4$) + 2e⁻ $\rightarrow$ S²⁻ ($\mathbf{1s^2 2s^2 2p^6 3s^2 3p^6}$)
c) Strontium achieves the noble gas configuration of Kr by losing 2 electrons to form **Sr²⁺**.
Sr ($1s^2 2s^2 2p^6 3s^2 3p^6 4s^2 3d^{10} 4p^6 5s^2$) $\rightarrow$ Sr²⁺ ($\mathbf{1s^2 2s^2 2p^6 3s^2 3p^6 4s^2 3d^{10} 4p^6}$) + 2 e⁻

8.78 <u>Plan:</u> To find the number of unpaired electrons look at the electron configuration expanded to include the different orientations of the orbitals, such as p_x and p_y and p_z. Remember that one electron will occupy every orbital in a set (p, d, or f) before electrons will pair in an orbital in that set. In the noble gas configurations, all electrons are paired because all orbitals are filled.
<u>Solution:</u>
a) Configuration of 2A(2) group elements: [noble gas]ns^2, **no unpaired electrons**. The electrons in the ns orbital are paired.
b) Configuration of 5A(15) group elements: [noble gas]$ns^2 np_x^1 np_y^1 np_z^1$. **Three** unpaired electrons, one each in p_x, p_y, and p_z.
c) Configuration of 8A(18) group elements: noble gas configuration $ns^2 np^6$ with no half-filled orbitals, **no unpaired electrons**.
d) Configuration of 3A(13) group elements: [noble gas]$ns^2 np^1$. There is **one** unpaired electron in one of the p orbitals.

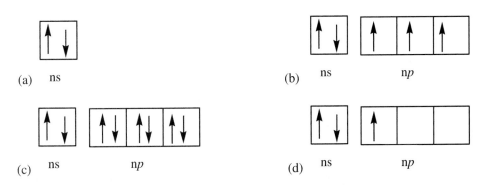

(a)   ns

(b)   ns   np

(c)   ns   np

(d)   ns   np

8.80   Substances are paramagnetic if they have unpaired electrons. This problem is more challenging if you have difficulty picturing the orbital diagram from the electron configuration. Obviously an odd number of electrons will necessitate that at least one electron is unpaired, so odd electron species are paramagnetic. However, a substance with an even number of electrons can also be paramagnetic, because even numbers of electrons do not guarantee all electrons are paired.
a) Ga ($Z = 31$) = $[Ar]4s^2 3d^{10} 4p^1$. The s and d sublevels are filled, so all electrons are paired. The lone p electron is unpaired, so this element is **paramagnetic**.
b) Si ($Z = 14$) = $[Ne]3s^2 3p^2$. This element is **paramagnetic** with two unpaired electrons because the p subshell looks like this,   not this.

| ↑ | ↑ | | | ↑↓ | | |

c) Be ($Z = 4$) = $[He]2s^2$, so it is **not paramagnetic**.
d) Te ($Z = 52$) = $[Kr]5s^2 4d^{10} 5p^4$ is **paramagnetic** with two unpaired electrons in the $5p$ set.

8.82   Substances are paramagnetic if they have unpaired electrons. This problem is more challenging if you have difficulty picturing the orbital diagram from the electron configuration. Obviously an odd number of electrons will necessitate that at least one electron is unpaired, so odd electron species are paramagnetic. However, a substance with an even number of electrons can also be paramagnetic, because even numbers of electrons do not guarantee all electrons are paired. Remember that all orbitals in a p, d, or f set will each have one electron before electrons pair in an orbital.
a) V: $[Ar]4s^2 3d^3$;   V^{3+}: $[Ar]3d^2$ Transition metals first lose the s electrons in forming ions, so to form the +3 ion a vanadium atom loses two $4s$ electrons and one $3d$ electron. **Paramagnetic**

[Ar]   [   ] | ↑ | ↑ | | | |
        $4s$ $3d$

b) Cd: $[Kr]5s^2 4d^{10}$;   **Cd^{2+}: [Kr]4d^{10}** Cadmium atoms lose two electrons from the $4s$ orbital to form the +2 ion.
Diamagnetic

[Kr]   [   ] | ↑↓ | ↑↓ | ↑↓ | ↑↓ | ↑↓ |
        $5s$ $4d$

c) Co: $[Ar]4s^2 3d^7$;   **Co^{3+}: [Ar]3d^6** Cobalt atoms lose two $4s$ electrons and one $3d$ electron to form the +3 ion.
Paramagnetic

[Ar]   [   ] | ↑↓ | ↑ | ↑ | ↑ | ↑ |
        $4s$ $3d$

d) Ag: $[Kr]5s^1 4d^{10}$;   **Ag$^+$: [Kr]4d^{10}** Silver atoms lose the one electron in the $5s$ orbital to form the +1 ion.
Diamagnetic

[Kr]   [   ] | ↑↓ | ↑↓ | ↑↓ | ↑↓ | ↑↓ |
        $5s$ $4d$

8.84   For palladium to be diamagnetic, all of its electrons must be paired. You might first write the condensed electron configuration for Pd as $[Kr]5s^2 4d^8$. However, the partial orbital diagram is not consistent with diamagnetism.

[Kr] | ↑↓ | | ↑↓ | ↑↓ | ↑↓ | ↑ | ↑ | | | | |
        $5s$ $4d$ $5p$

Promoting an s electron into the d sublevel (as in (c) $[Kr]5s^1 4d^9$) still leaves two electrons unpaired.

The only configuration that supports diamagnetism is **(b) $[Kr]4d^{10}$**.

8.86   The size of ions increases down a group. For ions that are isoelectronic (have the same electron configuration) size decreases with increasing atomic number.
a) Increasing size: $\mathbf{Li^+ < Na^+ < K^+}$, size increases down group 1A.
b) Increasing size: $\mathbf{Rb^+ < Br^- < Se^{2-}}$, these three ions are isoelectronic with the same electron configuration as krypton. Size decreases with increasing atomic number in an isoelectronic series.
c) Increasing size: $\mathbf{F^- < O^{2-} < N^{3-}}$, the three ions are isoelectronic with an electron configuration identical to neon. Size decreases with increasing atomic number in an isoelectronic series.

8.90   Ce: $[Xe]6s^2 4f^1 5d^1$ Eu: $[Xe]6s^2 4f^7$
Ce^{4+}: $[Xe]$ Eu^{2+}: $[Xe]4f^7$
Ce^{4+} has a noble-gas configuration and Eu^{2+} has a half-filled f subshell.

8.92   Plan: Write the formula of the oxoacid. Remember that in naming oxoacids (H + polyatomic ion), the suffix of the polyatomic changes: -ate becomes -ic acid and -ite becomes -ous acid. Determine the oxidation state of the nonmetal in the oxoacid; hydrogen has an O.N. of +1 and oxygen has an O.N. of –2. Based on the oxidation state of the nonmetal, and the oxidation state of the oxide ion (–2), the formula of the nonmetal oxide may be determined. The name of the nonmetal oxide comes from the formula; remember that nonmetal compounds use prefixes to indicate the number of each type of atom in the formula.
Solution:
a) hypochlorous acid = $HClO$ has Cl^+ so the oxide is $\mathbf{Cl_2O}$ = **dichlorine oxide** or **dichlorine monoxide**
b) chlorous acid = $HClO_2$ has Cl^{3+} so the oxide is $\mathbf{Cl_2O_3}$ = **dichlorine trioxide**
c) chloric acid = $HClO_3$ has Cl^{5+} so the oxide is $\mathbf{Cl_2O_5}$ = **dichlorine pentaoxide**
d) perchloric acid = $HClO_4$ has Cl^{7+} so the oxide is $\mathbf{Cl_2O_7}$ = **dichlorine heptaoxide**
e) sulfuric acid = H_2SO_4 has S^{6+} so the oxide is $\mathbf{SO_3}$ = **sulfur trioxide**
f) sulfurous acid = H_2SO_3 has S^{4+} so the oxide is $\mathbf{SO_2}$ = **sulfur dioxide**
g) nitric acid = HNO_3 has N^{5+} so the oxide is $\mathbf{N_2O_5}$ = **dinitrogen pentaoxide**
h) nitrous acid = HNO_2 has N^{3+} so the oxide is $\mathbf{N_2O_3}$ = **dinitrogen trioxide**
i) carbonic acid = H_2CO_3 has C^{4+} so the oxide is $\mathbf{CO_2}$ = **carbon dioxide**
j) phosphoric acid = H_3PO_4 has P^{5+} so the oxide is $\mathbf{P_2O_5}$ = **diphosphorus pentaoxide**

8.95   Remember that isoelectronic species have the same electron configuration.
a) A chemically unreactive Period 4 element would be Kr in Group 8A(18). Both the Sr^{2+} ion and Br^- ion are isoelectronic with Kr. Their combination results in **$SrBr_2$, strontium bromide**.
b) Ar is the Period 3 noble gas. Ca^{2+} and S^{2-} are isoelectronic with Ar. The resulting compound is **CaS, calcium sulfide**.
c) The smallest filled d subshell is the $3d$ shell, so the element must be in Period 4. Zn forms the Zn^{2+} ion by losing its two s subshell electrons to achieve a *pseudo–noble gas* configuration ($[Ar]3d^{10}$). The smallest halogen is fluorine, whose anion is F^-. The resulting compound is **ZnF_2, zinc fluoride**.
d) Ne is the smallest element in Period 2, but it is not ionizable. Li is the largest atom whereas F is the smallest atom in Period 2. The resulting compound is **LiF, lithium fluoride**.

8.96   Both the ionization energies and the electron affinities of the elements are needed.
        a) F: ionization energy = 1681 kJ/mol   electron affinity = –328 kJ/mol
                $F(g) \rightarrow F^+(g) + e^-$ $\Delta H = 1681$ kJ/mol
                $F(g) + e^- \rightarrow F^-(g)$ $\Delta H = -328$ kJ/mol
        Reverse the electron affinity reaction to give: $F^-(g) \rightarrow F(g) + e^-$ $\Delta H = +328$ kJ/mol
        Summing the ionization energy reaction with the reversed electron affinity reaction (Hess's Law):
                $F^-(g) \rightarrow F^+(g) + 2\ e^-$ $\Delta H = +328$ kJ/mol + 1681 kJ/mol = **2009 kJ/mol**
        b) Na: ionization energy = 496 kJ/mol   electron affinity = –52.9 kJ/mol
                $Na(g) \rightarrow Na^+(g) + e^-$ $\Delta H = 496$ kJ/mol
                $Na(g) + e^- \rightarrow Na^-(g)$ $\Delta H = -52.9$ kJ/mol
        Reverse the ionization reaction to give: $Na^+(g) + e^- \rightarrow Na(g)$ $\Delta H = -496$ kJ/mol
        Summing the electron affinity reaction with the reversed ionization reaction (Hess's Law):
                $Na^+(g) + 2\ e^- \rightarrow Na^-(g)$ $\Delta H = -496$ kJ/mol + –52.9 kJ/mol = –548.9 = **–549 kJ/mol**

8.98   Plan: Determine the electron configuration for iron, and then begin removing one electron at a time. Filled
        subshells give diamagnetic contributions. Any partially filled subshells give a paramagnetic contribution. The
        more unpaired electrons, the greater the attraction to a magnetic field.
        Solution:

Fe	$[Ar]4s^2 3d^6$	partially filled $3d$ = **paramagnetic**	number of unpaired electrons = 4
Fe^+	$[Ar]4s^1 3d^6$	partially filled $3d$ = **paramagnetic**	number of unpaired electrons = 5
Fe^{2+}	$[Ar]3d^6$	partially filled $3d$ = **paramagnetic**	number of unpaired electrons = 4
Fe^{3+}	$[Ar]3d^5$	partially filled $3d$ = **paramagnetic**	number of unpaired electrons = 5
Fe^{4+}	$[Ar]3d^4$	partially filled $3d$ = **paramagnetic**	number of unpaired electrons = 4
Fe^{5+}	$[Ar]3d^3$	partially filled $3d$ = **paramagnetic**	number of unpaired electrons = 3
Fe^{6+}	$[Ar]3d^2$	partially filled $3d$ = **paramagnetic**	number of unpaired electrons = 2
Fe^{7+}	$[Ar]3d^1$	partially filled $3d$ = **paramagnetic**	number of unpaired electrons = 1
Fe^{8+}	$[Ar]$	filled orbitals = **diamagnetic**	number of unpaired electrons = 0
Fe^{9+}	$[Ne]3s^2 3p^5$	partially filled $3p$ = **paramagnetic**	number of unpaired electrons = 1
Fe^{10+}	$[Ne]3s^2 3p^4$	partially filled $3p$ = **paramagnetic**	number of unpaired electrons = 2
Fe^{11+}	$[Ne]3s^2 3p^3$	partially filled $3p$ = **paramagnetic**	number of unpaired electrons = 3
Fe^{12+}	$[Ne]3s^2 3p^2$	partially filled $3p$ = **paramagnetic**	number of unpaired electrons = 2
Fe^{13+}	$[Ne]3s^2 3p^1$	partially filled $3p$ = **paramagnetic**	number of unpaired electrons = 1
Fe^{14+}	$[Ne]3s^2$	filled orbitals = **diamagnetic**	number of unpaired electrons = 0

        Fe^+ and Fe^{3+} would both be most attracted to a magnetic field. They each have 5 unpaired electrons.

CHAPTER 9 MODELS OF CHEMICAL BONDING

FOLLOW–UP PROBLEMS

9.1   Plan: First, write out the condensed electron configuration and electron-dot structure for magnesium atoms and chlorine atoms. In the formation of the ions, each magnesium atom will lose two electrons to form the +2 ion and each chlorine atom will gain one electron to form the –1 ion. Write the condensed electron configuration and electron-dot structure for each of the ions. The formula of the compound is found by combining the ions in a ratio that gives a neutral compound.
Solution:
Condensed electron configuration:
Mg ([Ne]$3s^2$) + Cl ([Ne]$3s^2 3p^5$) → Mg^{+2} ([Ne]) + Cl$^-$ ([Ne]$3s^2 3p^6$)
In order to balance the charge (or the number of electrons lost and gained) two chlorine atoms are needed:
Mg ([Ne]$3s^2$) + 2Cl ([Ne]$3s^2 3p^5$) → Mg^{+2} ([Ne]) + 2Cl$^-$ ([Ne]$3s^2 3p^6$)
Lewis electron–dot symbols:

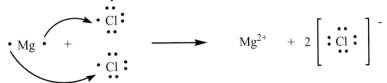

The formula of the compound would contain two chloride ions for each magnesium ion, $MgCl_2$.

9.2   Plan: a) All bonds are single bonds from silicon to a second row element. The trend in bond length can be predicted from the fact that atomic radii decrease across a row, so fluorine will be smaller than oxygen, which is smaller than carbon. Bond length will therefore decrease across the row while bond energy increases.
b) All the bonds are between two nitrogen atoms, but differ in the number of electrons shared between the atoms. The more electrons shared the shorter the bond length and the greater the bond energy.
Solution:
a) Bond length: Si–F < Si–O < Si–C   Bond strength: Si–C < Si–O < Si–F
b) Bond length: N≡N < N=N < N–N   Bond energy: N–N < N=N < N≡N
Check: (Use the tables in the chapter.)
Bond lengths from Table: Si–F, 156 pm < Si–O, 161 pm < Si–C, 186 pm
Bond energies from Table: Si–C, 301 kJ < Si–O, 368 kJ < Si–F, 565 kJ
Bond lengths from Table: N≡N, 110 pm < N=N, 122 pm < N–N, 146 pm
Bond energies from Table: N–N, 160 kJ < N=N, 418 kJ < N≡N, 945 kJ
The values from the tables agree with the order predicted.

9.3   Plan: Assume that all reactant bonds break and all product bonds form. Use the bond energy Table to find the values for the different bonds. Sum the values for the reactants, and sum the values for the products (these are all negative values). Add the sum of product to the values from the sum of the reactant values.
Solution:
Calculating $\Delta H°$ for the bonds broken:

$$1 \times N≡N = (1\ mol)\ (945\ kJ/mol) = 945\ kJ$$
$$3 \times H–H = (3\ mol)\ (432\ kJ/mol) = \underline{1296\ kJ}$$
$$\text{Sum broken} = 2241\ kJ$$

Calculating $\Delta H°$ for the bonds formed:

$$1 \times NH_3 = (3\ mol)\ (-391\ kJ/mol) = -1173\ kJ$$
$$1 \times NH_3 = (3\ mol)\ (-391\ kJ/mol) = \underline{-1173\ kJ}$$
$$\text{Sum formed} \qquad\qquad = -2346\ kJ$$
$$\Delta H°_{rxn} = 2241\ kJ + (-2346\ kJ) = -105\ kJ$$

9.4   Plan: Bond polarity can be determined from the difference in electronegativity of the two atoms. The more electronegative atom holds the δ– charge and the other atom holds the δ+ charge.
Solution:
a) From the electronegativity figure, EN of Cl = 3.0, EN of F = 4.0, EN of Br = 2.8. Cl–Cl will be the least polar since there is no difference between the electronegativity of the two chlorine atoms. Cl–F will be more polar than Cl–Br since the electronegativity difference between Cl and F (4.0 – 3.0 = 1.0) is greater than the electronegativity difference between Cl and Br (3.0 – 2.8 =0.2).

$$\overset{\delta+\ \delta-}{}\quad\overset{\delta+\ \delta-}{}$$
$$Cl\text{-}Cl < Br\text{-}Cl < Cl\text{-}F$$

b) EN of Cl = 3.0, EN of Si = 1.8, EN of P = 2.1, EN of S = 2.5. Si–Si bond is nonpolar since there is no difference between the electronegativity of the two silicon atoms and is the least polar of the four bonds. The most polar is Si–Cl with an EN difference of 1.2, next is P–Cl with an EN difference of 0.9 and next is S–Cl with an EN of 0.5.

$$\overset{\delta+\ \delta-}{}\quad\overset{\delta+\ \delta-}{}\quad\overset{\delta+\ \delta-}{}$$
$$Si\text{-}Si < S\text{-}Cl < P\text{-}Cl < Si\text{-}Cl$$

Check: For a) the order follows the increase in electronegativity going up the group of halogens. For b) the order follows the increase in electronegativity across a row. The closer to the end of row 3 (chlorine being at the end of row 3 next to the noble gas argon) that the element that is bonded to chlorine is found, the smaller the electronegativity difference of the bond. Therefore, the order of bond polarity follows the order of elements in row 3: Si, P, S.

END–OF–CHAPTER PROBLEMS

9.1   a) Larger ionization energy decreases metallic character.
b) Larger atomic radius increases metallic character.
c) Larger number of outer electrons decreases metallic character.
d) Larger effective nuclear charge decreases metallic character.

9.4   Metallic behavior increases to the left and down on the periodic table.
a) **Cs** is more metallic since it is further down the alkali metal group than Na.
b) **Rb** is more metallic since it is both to the left and down from Mg.
c) **As** is more metallic since it is further down Group 5A than N.

9.6   Ionic bonding occurs between metals and nonmetals, covalent bonding between nonmetals, and metallic bonds between metals.
a) Bond in CsF is **ionic** because Cs is a metal and F is a nonmetal.
b) Bonding in N_2 is **covalent** because N is a nonmetal.
c) Bonding in Na(s) is **metallic** because this is a monatomic, metal solid.

9.8   Ionic bonding occurs between metals and nonmetals, covalent bonding between nonmetals, and metallic bonds between metals.
a) Bonding in O_3 would be **covalent** since O is a nonmetal.
b) Bonding in $MgCl_2$ would be **ionic** since Mg is a metal and Cl is a nonmetal.
c) Bonding in BrO_2 would be **covalent** since both Br and O are nonmetals.

9.10 Lewis electron-dot symbols show valence electrons as dots. Place one dot at a time on the four sides (this method explains the structure in b) and then pair up dots until all valence electrons are used. The group number of the main group elements (Groups 1A-8A) gives the number of valence electrons. Rb is Group **1A**, Si is Group **4A**, I is Group **7A**.

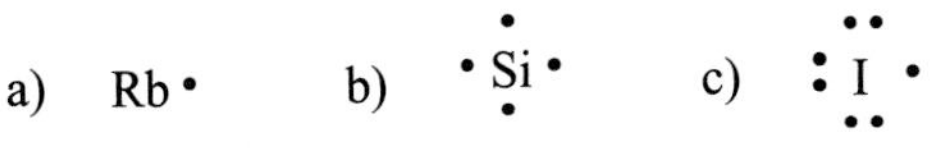

a) Rb•   b) •Si•   c) :I•

9.12

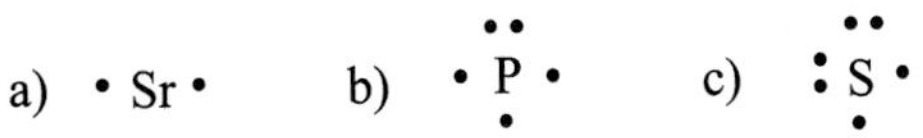

a) •Sr•   b) •P•   c) :S•

9.14 a) Assuming X is an A group element, the number of dots (valence electrons) equals the group number. Therefore, X is a 6A(16) element with 6 valence electrons. Its general electron configuration is [noble gas]ns^2np^4, where n is the energy level.
b) X has three valence electrons and is a 3A(13) element with general e⁻ configuration [noble gas]ns^2np^1.

9.20 a) Barium is a metal and loses 2 electrons to achieve a noble gas configuration:

$$Ba\ ([Xe]6s^2) \rightarrow Ba^{2+}\ ([Xe]) + 2\ e^-$$

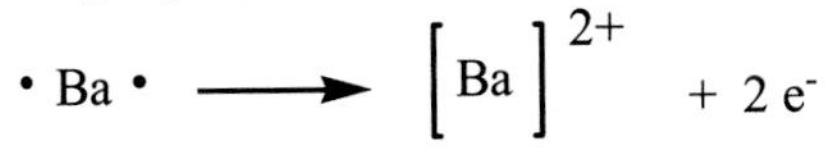

•Ba• ⟶ $[Ba]^{2+}$ + 2 e⁻

Chlorine is a nonmetal and gains 1 electron to achieve a noble gas configuration:

$$Cl\ ([Ne]3s^23p^5) + 1\ e^- \rightarrow Cl^-\ ([Ne]3s^23p^6)$$

:Cl• + 1 e⁻ ⟶ $[:Cl:]^-$

Two Cl atoms gain the two electrons lost by Ba. The ionic compound formed is $BaCl_2$.

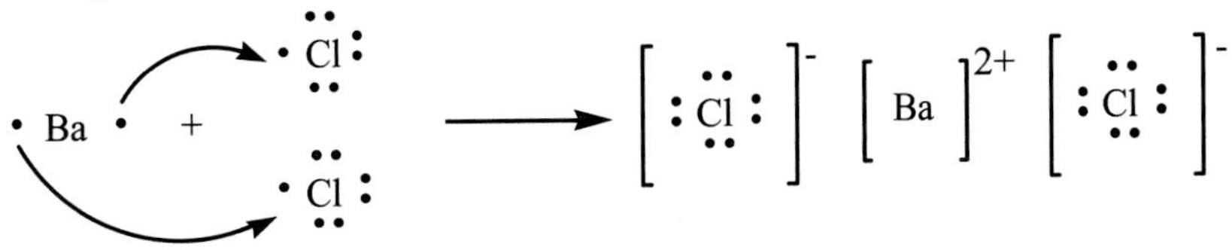

•Cl:

•Ba• +   ⟶ $[:Cl:]^-$ $[Ba]^{2+}$ $[:Cl:]^-$

•Cl:

b) $Sr\ ([Kr]5s^2) \rightarrow Sr^{2+}\ ([Kr]) + 2\ e^-$ $O\ ([He]2s^22p^4) + 2\ e^- \rightarrow O^{2-}\ ([He]2s^22p^6)$
The ionic compound formed is SrO.

•Sr• + •O: ⟶ $[Sr]^{2+}$ $[:O:]^{2-}$

c) $Al\ ([Ne]3s^23p^1) \rightarrow Al^{3+}\ ([Ne]) + 3\ e^-$ $F\ ([He]2s^22p^5) + 1\ e^- \rightarrow F^-\ ([He]2s^22p^6)$

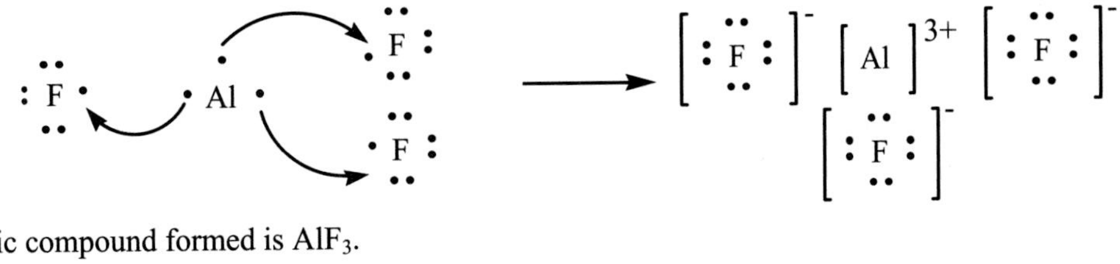

F:

:F• •Al• ⟶ $[:F:]^-$ $[Al]^{3+}$ $[:F:]^-$

•F: $[:F:]^-$

The ionic compound formed is AlF_3.

d) Rb ([Kr]5s^1) → Rb$^+$ ([Kr]) + 1 e$^-$ O ([He]2$s^2$2p^4) + 2 e$^-$ → O^{2-} ([He]2$s^2$2p^6)

The ionic compound formed is Rb$_2$O.

9.22   a) X in XF$_2$ is a cation with +2 charge since the anion is F$^-$ and there are two fluoride ions in the compound. **Group 2A(2)** metals form +2 ions.
b) X in MgX is an anion with – 2 charge since Mg^{2+} is the cation. Elements in **Group 6A(16)** form -2 ions.
c) X in X$_2$SO$_4$ must be a cation with +1 charge since the polyatomic sulfate ion has a charge of –2. X comes from **Group 1A(1)**.

9.24   a) X in X$_2$O$_3$ is a cation with +3 charge. The oxygen in this compound has a –2 charge. To produce an electrically neutral compound, 2 cations with +3 charge bond with 3 anions with –2 charge: 2(+3) + 3(–2) = 0. Elements in **Group 3A(13)** form +3 ions.
b) The carbonate ion, CO$_3^{2-}$, has a –2 charge, so X has a +2 charge. **Group 2A(2)** elements form +2 ions.
c) X in Na$_2$X has a –2 charge, balanced with the +2 overall charge from the two Na$^+$ ions. Group **6A(16)** elements gain 2 electrons to form –2 ions with a noble gas configuration.

9.26   a) **BaS** would have the higher lattice energy since the charge on each ion is twice the charge on the ions in CsCl and lattice energy is greater when ionic charges are larger.
b) **LiCl** would have the higher lattice energy since the ionic radius of Li$^+$ is smaller than that of Cs$^+$ and lattice energy is greater when the distance between ions is smaller.

9.28   a) **BaS** has the lower lattice energy because the ionic radius of Ba^{2+} is larger than Ca^{2+}. A larger ionic radius results in a greater distance between ions. The lattice energy decreases with increasing distance between ions.
b) **NaF** has the lower lattice energy since the charge on each ion (+1, –1) is half the charge on the Mg^{2+} and O^{2-} ions. Lattice energy increases with increasing ion charge.

9.30   Lattice energy: NaCl(s) → Na$^+$(aq) + Cl$^-$ (g)
Use Hess's Law: $\Delta H°$
~~Na(s) → Na(g)~~   109 kJ
~~1/2Cl2(g) → Cl(g)~~   243/2 = 121.5 kJ
~~Na(g) → Na$^+$(g) + e–~~   496 kJ
~~Cl(g) + e– → Cl$^-$ (g)~~   –349 kJ
~~NaCl(s) → Na(s) + 1/2Cl(g)~~   411 kJ (Reaction is reversed and the sign of $\Delta H°$ changed.)
NaCl(s) → Na$^+$(aq) + Cl$^-$ (g)   788.5 = **788 kJ**
The lattice energy for NaCl is less than that of LiF, which is expected since lithium and fluoride ions are smaller than sodium and chloride ions.

9.33   An analogous Born-Haber cycle has been described in Figure 9.6 for LiF. Use Hess's Law and solve for the EA of fluorine: $\Delta H°$
~~K(s) → K(g)~~   90 kJ
~~K(g) → K$^+$(g) + e$^-$~~   419 kJ
~~1/2 F$_2$(g) → F(g)~~   1/2(159) = 79.5 kJ
~~F(g) + e– → F$^-$ (g)~~   ? = EA
~~KF(s) → K(s) + 1/2 F$_2$(g)~~   569 kJ (Reverse the reaction and change the sign of $\Delta H°$.)
KF(s) → K$^+$ + F$^-$ (g)   821 kJ (Reverse the reaction and change the sign of $\Delta H°$.)

821 kJ = 90 kJ + 419 kJ + 79.5 kJ + EA + 569 kJ
EA = 821 kJ – (90 + 419 + 79.5 + 569)kJ = –336.5 = **–336 kJ**

146

9.34   When two chlorine atoms are far apart, there is no interaction between them. Once the two atoms move closer together, the nucleus of each atom attracts the electrons on the other atom. As the atoms move closer this attraction increases, but the repulsion of the two nuclei also increases. When the atoms are very close together the repulsion between nuclei is much stronger than the attraction between nuclei and electrons. The final internuclear distance for the chlorine molecule is the distance at which maximum attraction is achieved in spite of the repulsion. At this distance, the energy of the molecule is at its lowest value.

9.35   The bond energy is the energy required to overcome the attraction between H atoms and Cl atoms in one mole of HCl molecules in the gaseous state. Energy input is needed to break bonds, so bond energy is always absorbed (endothermic) and $\Delta H°_{bond\ breaking}$ is positive. The same amount of energy needed to break the bond is released upon its formation, so $\Delta H°_{bond\ forming}$ has the same magnitude as $\Delta H°_{bond\ breaking}$, but opposite in sign (always exothermic and negative).

9.39   a) **I–I < Br–Br < Cl–Cl**. Bond strength increases as the atomic radii of atoms in the bond decreases. Atomic radii decrease up a group in the periodic table, so I is the largest and Cl is the smallest of the three.
b) **S–Br < S–Cl < S–H**. Radius of H is the smallest and Br is the largest, so the bond strength for S–H is the greatest and that for S–Br is the weakest.
c) **C–N < C=N < C≡N**. Bond strength increases as the number of electrons in the bond increases. The triple bond is the strongest and the single bond is the weakest.

9.41   a) For a given pair of atoms, in this case carbon and oxygen, bond strength increases with increasing bond order. The C=O bond (bond order = 2) is stronger than the C–O bond (bond order = 1).
b) O is smaller than C so the O–H bond is shorter and stronger than the C–H bond.

9.44   Reaction between molecules requires the breaking of existing bonds and the formation of new bonds. Substances with weak bonds are more reactive than are those with strong bonds because less energy is required to break weak bonds.

9.46   For methane: $CH_4(g) + 2\ O_2(g) \rightarrow CO_2(g) + 2\ H_2O(l)$ which requires that 4 C–H bonds and 2 O=O bonds be broken and 2 C=O bonds and 4 O–H bonds be formed.
For formaldehyde: $CH_2O(g) + O_2(g) \rightarrow CO_2(g) + H_2O(l)$ which requires that 2 C–H bonds, 1 C=O bond and 1 O=O bond be broken and 2 C=O bonds and 2 O–H bonds be formed.
Methane contains more C–H bonds and fewer C=O bonds than formaldehyde. Since C–H bonds take less energy to break than C=O bonds, more energy is released in the combustion of methane than of formaldehyde.

9.48   To find the heat of reaction, add the energy required to break all the bonds in the reactants to the energy released to form all bonds in the product. Remember to use a negative sign for the energy of the bonds formed since bond formation is exothermic. The bond energy values are found in Table 9.2.
Reactant bonds broken: 1(C=C) + 4(C–H) + 1(Cl–Cl)
        = (1 mol) (614 kJ/mol) + (4 mol) (413 kJ/mol) + (1 mol) (243 kJ/mol)
        = 2509 kJ
Product bonds formed: 1(C–C) + 4(C–H) + 2(C–Cl)
        = (1 mol) (–347 kJ/mol) + (4 mol) (–413 kJ/mol) + (2 mol) (–339 kJ/mol)
        = –2677 kJ
$\Delta H°_{rxn} = \Delta H°_{bonds\ broken} + \Delta H°_{bonds\ formed}$ = 2509 kJ + (–2677 kJ) = **–168 kJ**
Note: It is correct to report the answer in kJ or kJ/mol as long as the value refers to a reactant or product with a molar coefficient of 1.

9.50 To find the heat of reaction, add the energy required to break all the bonds in the reactants to the energy released to form all bonds in the product. Remember to use a negative sign for the energy of the bonds formed since bond formation is exothermic. The bond energy values are found in Table 9.2.

The reaction:

$\Delta H^{\circ}{}_{\text{bonds broken}} = 1$ C–O= 1 mol (358 kJ/mol)
  3 C–H = 3 mol (413 kJ/mol)
  1 O–H= 1 mol (467 kJ/mol)
  1 C ≡ O = 1 mol (1070 kJ/mol)
   = 3134 kJ

$\Delta H^{\circ}{}_{\text{bonds formed}} = 3$ C–H = 3 mol (–413 kJ/mol)
  1 C–C = 1 mol (–347 kJ/mol)
  1 C=O = 1 mol (–745 kJ/mol)
  1 C–O = 1 mol (–358 kJ/mol)
  1 O–H = 1 mol (–467 kJ/mol)
   = –3156 kJ

$\Delta H^{\circ}{}_{\text{rxn}} = \Delta H^{\circ}{}_{\text{bonds broken}} + \Delta H^{\circ}{}_{\text{bonds formed}} = 3134$ kJ + (–3156 kJ) = **–22 kJ**

9.51 Add the energy required to break all the bonds in the reactants to the energy released to form all bonds in the product. Remember to use a negative sign for the energy of the bonds formed since bond formation is exothermic. The bond energy values are found in Table 9.2.

Bonds broken: 1(C=C) + 4(C–H) + 1(H–Br) = 614 + 4(413) + 363 = +2629 kJ
Bonds formed: 1(C–C) + 5(C–H) + 1(C–Br) = (–347) + 5(–413) + (–276) = –2688 kJ
$\Delta H^{\circ}{}_{\text{rxn}} = \Delta H^{\circ}{}_{\text{bonds broken}} + \Delta H^{\circ}{}_{\text{bonds formed}} = 2629 + (–2688) = $ **–59 kJ**

9.52 Electronegativity increases from left to right across a period (except for the noble gases) and increases from bottom to top within a group. Fluorine (F) and oxygen (O) are the two most electronegative elements. Cesium (Cs) and francium (Fr) are the two least electronegative elements.

9.54 The H–O bond in water is polar covalent, because the two atoms differ in electronegativity which results in an unequal sharing of the bonding electrons. A nonpolar covalent bond occurs between two atoms with identical electronegativity values where the sharing of bonding electrons is equal. Although electron sharing occurs to a very small extent in some ionic bonds, the primary force in ionic bonds is attraction of opposite charges resulting from electron transfer between the atoms.

9.57 Electronegativity increases from left to right across a period (except for the noble gases) and increases from bottom to top within a group.
a) **Si < S < O**, sulfur is more electronegative than silicon since it is located further to the right in the table. Oxygen is more electronegative than sulfur since it is located nearer the top of the table.
b) **Mg < As < P**, magnesium is the least electronegative because it lies on the left side of the periodic table and phosphorus and arsenic on the right side. Phosphorus is more electronegative than arsenic because it is higher in the table.

9.59 Electronegativity generally increases up a group and left to right across a period.
a) **N > P > Si**, nitrogen is above P in Group 5(A)15 and P is to the right of Si in period 3.
b) **As > Ga > Ca**, all three elements are in Period 4, with As the right-most element.

9.61   The arrow points toward the more electronegative atom.

a) $\overset{\longleftarrow\!+}{\text{N}\text{——}\text{B}}$ b) $\overset{+\!\longrightarrow}{\text{N}\text{——}\text{O}}$ c) none $\text{C}\text{——}\text{S}$

d) $\overset{+\!\longrightarrow}{\text{S}\text{——}\text{O}}$ e) $\overset{\longleftarrow\!+}{\text{N}\text{——}\text{H}}$ f) $\overset{+\!\longrightarrow}{\text{Cl}\text{——}\text{O}}$

9.63   The more polar bond will have a greater difference in electronegativity, ΔEN.
a) N: EN = 3.0; B: EN = 2.0; ΔEN_a = 1.0   b) N: EN = 3.0; O: EN = 3.5; ΔEN_b = 0.5
c) C: EN = 2.5; S: EN = 2.5; ΔEN_c = 0   d) S: EN = 2.5; O: EN = 3.5; ΔEN_d = 1.0
e) N: EN = 3.0; H: EN = 2.1; ΔEN_e = 0.9   f) Cl: EN = 3.0; O: EN = 3.5; ΔEN_f = 0.5
(a), (d), and (e) have greater bond polarity.

9.65   a) Bonds in S_8 are **nonpolar covalent**. All the atoms are nonmetals so the substance is covalent and bonds are nonpolar because all the atoms are of the same element and thus have the same electronegativity value. $\Delta EN = 0$.
b) Bonds in RbCl are **ionic** because Rb is a metal and Cl is a nonmetal. ΔEN is large.
c) Bonds in PF_3 are **polar covalent**. All the atoms are nonmetals so the substance is covalent. The bonds between P and F are polar because their electronegativity differs. (By 1.9 units for P–F)
d) Bonds in SCl_2 are **polar covalent**. S and Cl are nonmetals and differ in electronegativity. (By 0.5 unit for S–Cl)
e) Bonds in F_2 are **nonpolar covalent**. F is a nonmetal. Bonds between two atoms of the same element are nonpolar since $\Delta EN = 0$.
f) Bonds in SF_2 are **polar covalent**. S and F are nonmetals that differ in electronegativity. (By 1.5 units for S–F)
Increasing bond polarity: $SCl_2 < SF_2 < PF_3$

9.67   Increasing ionic character occurs with increasing ΔEN.
a) ΔEN_{HBr} = 0.7, ΔEN_{HCl} = 0.9, ΔEN_{HI} = 0.4
b) ΔEN_{HO} = 1.4, ΔEN_{CH} = 0.4, ΔEN_{HF} = 1.9
c) ΔEN_{SCl} = 0.5, ΔEN_{PCl} = 0.9, ΔEN_{SiCl} = 1.2

a) $\overset{+\!\longrightarrow}{\text{H}\text{——}\text{I}}$ < $\overset{+\!\longrightarrow}{\text{H}\text{——}\text{Br}}$ < $\overset{+\!\longrightarrow}{\text{H}\text{——}\text{Cl}}$

b) $\overset{+\!\longrightarrow}{\text{H}\text{——}\text{C}}$ < $\overset{+\!\longrightarrow}{\text{H}\text{——}\text{O}}$ < $\overset{+\!\longrightarrow}{\text{H}\text{——}\text{F}}$

c) $\overset{+\!\longrightarrow}{\text{S}\text{——}\text{Cl}}$ < $\overset{+\!\longrightarrow}{\text{P}\text{——}\text{Cl}}$ < $\overset{+\!\longrightarrow}{\text{Si}\text{——}\text{Cl}}$

9.70   a) A solid metal is a shiny solid that conducts heat, is malleable, and melts at high temperatures. (Other answers include relatively high boiling point and good conductor of electricity.)
b) Metals lose electrons to form positive ions and metals form basic oxides.

9.74   a) $C_2H_2 + 5/2 \, O_2 \rightarrow 2 \, CO_2 + H_2O$ $\Delta H_{rxn} = -1259$ kJ/mol

$\text{H–C} \equiv \text{C–H} + \dfrac{5}{2} \, \text{O=O} \rightarrow 2 \, \text{O=C=O} + \text{H–O–H}$

$\Delta H_{rxn} = [2 \, BE_{C-H} + BE_{C\equiv C} + \dfrac{5}{2} BE_{O_2}] + [4 \, (-BE_{C=O}) + 2 \, (-BE_{O-H})]$

-1259 kJ = $[2(413) + BE_{C\equiv C} + 5/2 \, (498)] + [4(-799) + 2(-467)]$
-1259 kJ = $[826 + BE_{C\equiv C} + 1245] + [-4130)]$ kJ
$BE_{C\equiv C}$ = **800. kJ/mol**   Table 9.2 lists the value as 839 kJ/mol.

149

b) heat (kJ) $= (500.0 \text{ g } C_2H_2) \left(\dfrac{1 \text{ mol } C_2H_2}{26.04 \text{ g } C_2H_2} \right) \left(\dfrac{-1259 \text{ kJ}}{1 \text{ mol } C_2H_2} \right)$

$\qquad = -2.4174347 \times 10^4 = \mathbf{-2.417 \times 10^4 \ kJ}$

c) CO_2 produced $(g) = (500.0 \text{ g } C_2H_2) \left(\dfrac{1 \text{ mol } C_2H_2}{26.04 \text{ g } C_2H_2} \right) \left(\dfrac{2 \text{ mol } CO_2}{1 \text{ mol } C_2H_2} \right) \left(\dfrac{44.01 \text{ g } CO_2}{1 \text{ mol } CO_2} \right)$

$\qquad = 1690.092 = \mathbf{1690. \ g \ CO_2}$

d) mol $O_2 = (500.0 \text{ g } C_2H_2) \left(\dfrac{1 \text{ mol } C_2H_2}{26.04 \text{ g } C_2H_2} \right) \left(\dfrac{(5/2) \text{ mol } O_2}{1 \text{ mol } C_2H_2} \right) = 48.00307 \text{ mol } O_2 \text{ (unrounded)}$

$$V = nRT / P = \dfrac{(48.00307 \text{ mol } O_2)\left(0.08206 \dfrac{L \cdot atm}{mol \cdot K} \right)(298 \text{ K})}{18.0 \text{ atm}} = 65.2145 = \mathbf{65.2 \ L \ O_2}$$

9.76   a)   1) $Mg(s) \rightarrow \cancel{Mg(g)}$ $\qquad\qquad\qquad \Delta H_1^{\circ} = 148 \text{ kJ}$

2) $1/2 \ Cl_2(g) \rightarrow \cancel{Cl(g)}$ $\qquad\qquad \Delta H_2^{\circ} = 1/2 \ (243 \text{ kJ})$

3) $\cancel{Mg(g)} \rightarrow \cancel{Mg+(g)} + \cancel{e-}$ $\qquad \Delta H_3^{\circ} = 738 \text{ kJ}$

4) $\cancel{Cl(g)} + \cancel{e^-} \rightarrow \cancel{Cl^-(g)}$ $\qquad\quad \Delta H_4^{\circ} = -349 \text{ kJ}$

$\underline{5) \ \cancel{Mg+(g)} + \cancel{Cl^-(g)} \rightarrow MgCl(s) \quad \Delta H_5^{\circ} = -783.5 \text{ kJ} \ (= -\Delta H_{lattice}^{\circ} \ (MgCl))}$

6) $Mg(s) + 1/2 \ Cl_2(g) \rightarrow MgCl(s) \quad \Delta H_f^{\circ} \ (MgCl) = ?$

$\Delta H_f^{\circ} \ (MgCl) = \Delta H_1^{\circ} + \Delta H_2^{\circ} + \Delta H_3^{\circ} + \Delta H_4^{\circ} + \Delta H_5^{\circ}$

$\qquad = 148 \text{ kJ} + 1/2 \ (243 \text{ kJ}) + 738 \text{ kJ} + (-349 \text{ kJ}) + (-783.5 \text{ kJ}) = \mathbf{-125 \ kJ}$

b) **Yes**, since ΔH°_f for MgCl is negative, $MgCl(s)$ is stable relative to its elements.

c) $2 \ MgCl(s) \rightarrow MgCl_2(s) + Mg(s)$

$\Delta H^{\circ} = [1 \text{ mol } (\Delta H_f^{\circ}, MgCl_2(s)) + 1 \text{ mol } (\Delta H_f^{\circ}, Mg(s))] - [2 \text{ mol } (\Delta H_f^{\circ}, MgCl(s))]$

$\qquad = [1 \text{ mol } (-641.6 \text{ kJ/mol}) + 1 \text{ mol } (0)] - [2 \text{ mol } (-125 \text{ kJ/mol})]$

$\qquad = -391.6 = \mathbf{-392 \ kJ}$

d) **No**, ΔH_f° for $MgCl_2$ is much more negative than that for MgCl. This makes the ΔH° value for the above reaction very negative, and the formation of $MgCl_2$ would be favored.

9.78   a) Find the bond energy for an H–I bond from Table 9.2. Calculate wavelength from this energy using the relationship from Chapter 7: $E = hc / \lambda$.

Bond energy for H–I is 295 kJ/mol (Table 9.2).

$$\lambda = hc / E = \dfrac{(6.626 \times 10^{-34} \text{ J} \cdot \text{s})(3.00 \times 10^8 \text{ m/s})}{\left(295 \dfrac{kJ}{mol} \right)\left(\dfrac{10^3 \text{ J}}{1 \text{ kJ}} \right)\left(\dfrac{mol}{6.022 \times 10^{23}} \right)} \left(\dfrac{1 \text{ nm}}{10^{-9} \text{ m}} \right) = 405.7807 = \mathbf{406 \ nm}$$

b) Calculate the energy for a wavelength of 254 nm and then subtract the energy from part a) to get the excess energy.

$$E \ (HI) = \left(295 \dfrac{kJ}{mol} \right)\left(\dfrac{10^3 \text{ J}}{1 \text{ kJ}} \right)\left(\dfrac{mol}{6.022 \times 10^{23}} \right) = 4.8987 \times 10^{-19} \text{ J (unrounded)}$$

$$E \ (254 \text{ nm}) = hc / \lambda. = \dfrac{(6.626 \times 10^{-34} \text{ J} \cdot \text{s})(3.00 \times 10^8 \text{ m/s})}{254 \text{ nm}} \left(\dfrac{1 \text{ nm}}{10^{-9} \text{ m}} \right) = 7.82598 \times 10^{-19} \text{ J}$$

Excess energy $= 7.82598 \times 10^{-19} \text{ J} - 4.8987 \times 10^{-19} \text{ J} = 2.92728 \times 10^{-19} = \mathbf{2.93 \times 10^{-19} \ J}$

c) Speed can be calculated from the excess energy since $E_k = 1/2 \, mu^2$.

$$E_k = 1/2 \, mu^2 \text{ thus, } u = \sqrt{\frac{2E}{m}} \quad m = \left(\frac{1.008 \text{ g H}}{\text{mol}}\right)\left(\frac{\text{mol}}{6.022 \times 10^{23}}\right)\left(\frac{1 \text{ kg}}{10^3 \text{ g}}\right) = 1.67386 \times 10^{-27} \text{ kg}$$

$$u = \sqrt{\frac{2(2.92728 \times 10^{-19} \text{ J})}{1.67386 \times 10^{-27} \text{ kg}}\left(\frac{\text{kg} \cdot \text{m}^2/\text{s}^2}{\text{J}}\right)} = 1.8701965 \times 10^4 = \mathbf{1.87 \times 10^4 \text{ m/s}}$$

9.82   Find the appropriate bond energies in Table 9.2. Calculate the wavelengths using $E = hc / \lambda$.
C–Cl bond energy = 339 kJ/mol.

$$\lambda = \frac{hc}{E} = \frac{\left(6.626 \times 10^{-34} \text{ J} \cdot \text{s}\right)\left(3.00 \times 10^8 \text{ m/s}\right)}{\left(339 \dfrac{\text{kJ}}{\text{mol}}\right)\left(\dfrac{10^3 \text{ J}}{1 \text{ kJ}}\right)\left(\dfrac{\text{mol}}{6.022 \times 10^{23}}\right)} = 3.53113 \times 10^{-7} = \mathbf{3.53 \times 10^{-7} \text{ m}}$$

O_2 bond energy = 498 kJ/mol

$$\lambda = \frac{hc}{E} = \frac{\left(6.626 \times 10^{-34} \text{ J} \cdot \text{s}\right)\left(3.00 \times 10^8 \text{ m/s}\right)}{\left(498 \dfrac{\text{kJ}}{\text{mol}}\right)\left(\dfrac{10^3 \text{ J}}{1 \text{ kJ}}\right)\left(\dfrac{\text{mol}}{6.022 \times 10^{23}}\right)} = 2.40372 \times 10^{-7} = \mathbf{2.40 \times 10^{-7} \text{ m}}$$

9.83   Write balanced chemical equations for the formation of each of the compounds. Obtain the bond energy of fluorine from Table 9.2 (159 kJ/mol). Determine the average bond energy from $\Delta H =$ bonds broken + bonds formed. Remember that the bonds formed (Xe–F) have negative values since bond formation is exothermic.

XeF_2 $Xe(g) + F_2(g) \rightarrow XeF_2(g)$
$\Delta H = [1 \text{ mol } F_2 \, (159 \text{ kJ/mol})] + [2 \, (-\text{Xe–F})] = -105 \text{ kJ/mol}$
Xe–F = **132 kJ/mol**

XeF_4 $Xe(g) + 2 \, F_2(g) \rightarrow XeF_4(g)$
$\Delta H = [2 \text{ mol } F_2 \, (159 \text{ kJ/mol})] + [4 \, (-\text{Xe–F})] = -284 \text{ kJ/mol}$
Xe–F = 150.5 = **150. kJ/mol**

XeF_6 $Xe(g) + 3 \, F_2(g) \rightarrow XeF_6(g)$
$\Delta H = [3 \text{ mol } F_2 \, (159 \text{ kJ/mol})] + [6 \, (-\text{Xe–F})] = -402 \text{ kJ/mol}$
Xe–F = 146.5 = **146 kJ/mol**

9.85   a) The presence of the very electronegative fluorine atoms bonded to one of the carbons makes the C–C bond polar. This polar bond will tend to undergo heterolytic rather than homolytic cleavage. More energy is required to force heterolytic cleavage.
b) Since one atom gets both of the bonding electrons in heterolytic bond breakage, this results in the formation of ions. In heterolytic cleavage a cation is formed, involving ionization energy; an anion is also formed, involving electron affinity. The bond energy of the O_2 bond is 498 kJ/mol.
ΔH = (homolytic cleavage + electron affinity + first ionization energy)
ΔH = (498/2 kJ/mol + (–141 kJ/mol) + 1314 kJ/mol) = 1422 = **1420 kJ/mol**
It would require 1420 kJ to heterolytically cleave 1 mol of O_2.

9.88   Use Hess's Law. ΔH_f^0 of SiO_2 is found in Appendix B.

1) $Si(s) \rightarrow \cancel{Si(g)}$ $\Delta H_1^o = 454$ kJ

2) $\cancel{Si(g)} \rightarrow \cancel{Si^{4+}(g)} + \cancel{4e^-}$ $\Delta H_2^o = 9949$ kJ

3) $O_2(g) \rightarrow \cancel{2O(g)}$ $\Delta H_3^o = 498$ kJ

4) $\cancel{2O(g)} + \cancel{4e^-} \rightarrow \cancel{2O^{2-}(g)}$ $\Delta H_4^o = 2(737)$ kJ

5) $\cancel{Si^{4+}(g)} + \cancel{2O^{2-}(g)} \rightarrow SiO_2(s)$ $\underline{\Delta H_5^o = \Delta H_{lattice}^o (SiO_2) = ?}$

6) $Si(s) + O_2(g) \rightarrow SiO_2(s)$ $\Delta H_f^0 (SiO_2) = -910.9$ kJ

$\Delta H_6^0 = -(\Delta H_1^o + \Delta H_2^o + \Delta H_3^o + \Delta H_4^o + \Delta H_{lattice}^o)$

-910.9 kJ $= (454$ kJ $+ 9949$ kJ $+ 498$ kJ $+ 2(737)$ kJ $+ \Delta H_{lattice}^o)$

$\Delta H_{lattice}^o = -13285.9 = \textbf{-13286 kJ}$

9.90   Use the equations $E = h\nu$, and $E = hc / \lambda$.

$$\nu = \frac{E}{h} = \frac{\left(347 \frac{kJ}{mol}\right)\left(\frac{10^3 J}{1 kJ}\right)\left(\frac{mol}{6.022 \times 10^{23}}\right)}{6.626 \times 10^{-34} J\bullet s} = 8.6963556 \times 10^{14} = \textbf{8.70} \times \textbf{10}^{\textbf{14}} \textbf{ s}^{\textbf{-1}}$$

$$\lambda = hc / E = \frac{\left(6.626 \times 10^{-34} J\bullet s\right)\left(3.00 \times 10^8 m/s\right)}{\left(347 \frac{kJ}{mol}\right)\left(\frac{10^3 J}{1 kJ}\right)\left(\frac{mol}{6.022 \times 10^{23}}\right)} = 3.44972 \times 10^{-7} = \textbf{3.45} \times \textbf{10}^{\textbf{-7}} \textbf{ m}$$

This is in the ultraviolet region of the electromagnetic spectrum.

9.92   a) $2 CH_4(g) + O_2(g) \rightarrow CH_3OCH_3(g) + H_2O(g)$

$\Delta H_{rxn} = \Sigma BE_{reactants} + \Sigma BE_{products}$

$\Delta H_{rxn} = [8 \times (BE_{C-H}) + (BE_{O=O})] + [6 (BE_{C-H}) + 2 (BE_{C-O}) + 2 (BE_{O-H})]$

$\Delta H_{rxn} = [8 (413$ kJ$) + 498$ kJ$] + [6 (-413$ kJ$) + 2 (-358$ kJ$) + 2 (-467)]$

$\Delta H_{rxn} = \textbf{-326 kJ}$

$2 CH_4(g) + O_2(g) \rightarrow CH_3CH_2OH(g) + H_2O(g)$

$\Delta H_{rxn} = \Sigma BE_{reactants} + \Sigma BE_{products}$

$\Delta H_{rxn} = [8 \times (BE_{C-H}) + (BE_{O=O})] + [5(BE_{C-H}) + (BE_{C-C}) + (BE_{C-O}) + 3(BE_{O-H})]$

$\Delta H_{rxn} = [8 (413$ kJ$) + 498$ k$] + [5 (-413$ kJ$) + (-347$ kJ$) + (-358$ kJ$) + 3 (-467)]$

$\Delta H_{rxn} = \textbf{-369 kJ}$

b) The formation of gaseous **ethanol** is more exothermic.

c) $\Delta H_{rxn} = -326$ kJ $- (-369$ kJ$) = \textbf{43 kJ}$

CHAPTER 10 THE SHAPES OF MOLECULES

FOLLOW–UP PROBLEMS

10.1   <u>Plan:</u> Follow the steps outlined in the sample problem.
<u>Solution:</u>
a) The sulfur is the central atom, as the hydrogen is never central. Each of the hydrogen atoms is placed around the sulfur. The actual positions of the hydrogen atoms are not important. The total number of valence electrons available is [2 x H(1 e$^-$)] + [1 x S(6 e$^-$)] = 8 e$^-$. Connect each hydrogen atom to the sulfur with a single bond. These bonds use 4 of the electrons leaving 4 electrons. The last 4 e$^-$ go to the sulfur because the hydrogen atoms can take no more electrons.

$$H\!-\!\overset{\displaystyle ..}{S}:$$
$$|$$
$$H$$

<u>Check:</u> Count the electrons. The sulfur has an octet (4 from two lone pairs and 4 from two bonding pairs). Each hydrogen has its 2 electrons (from the bonding pair).
<u>Solution:</u>
b) The oxygen has the lower group number so it is the central atom. Each of the fluorine atoms will be attached to the central oxygen. The total number of valence electrons available is [2 x F(7 e$^-$)] + [1 x O(6 e$^-$)] = 20 e$^-$. Connecting the two fluorine atoms to the oxygen with single bonds uses 2 x 2 = 4 e$^-$, leaving 20 – 4 = 16 e$^-$. The more electronegative fluorine atoms each need 6 electrons to complete their octets. This requires 2 x 6 = 12 e$^-$. The 4 remaining electrons go to the oxygen.

$$:\overset{\displaystyle ..}{\underset{\displaystyle ..}{F}}\!-\!\overset{\displaystyle ..}{O}:$$
$$|$$
$$:\overset{\displaystyle ..}{\underset{\displaystyle ..}{F}}:$$

<u>Check:</u> Count the electrons. Each of the three atoms has an octet.
<u>Solution:</u>
c) Both S and O have a lower group number than Cl, thus, one of these two elements must be central. Between S and O, S has the higher period number so it is the central atom. The total number of valence electrons available is [2 x Cl(7 e$^-$)] + [1 x S(6 e$^-$)] + [1 x O(6e$^-$)] = 26 e$^-$. Begin by distributing the two chlorine atoms and the oxygen atom around the central sulfur atom. Connect each of the three outlying atoms to the central sulfur with single bonds. This uses 3 x 2 = 6 e$^-$, leaving 26 – 6 = 20 e$^-$. Each of the outlying atoms still needs 6 electrons to complete their octets. Completing these octets uses 3 x 6 = 18 electrons. The remaining 2 electrons are all the sulfur needs to complete its octet.

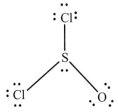

<u>Check:</u> Count the electrons. Each of the four atoms has an octet. Note: Later we will see that the presence of sulfur may lead to some complications.

10.2   <u>Plan:</u> Follow the steps outlined in the sample problem.
        <u>Solution:</u>
        a) As in CH_4O, the N and O both serve as "central" atoms. The N is placed next to the O and the H atoms are distributed around them. The N needs more electrons so it gets two of the three hydrogen atoms. You can try placing the N in the center with all the other atoms around it, but you will quickly see that you will have trouble with the oxygen. The number of valence electrons is: $[3 \times H(1\ e^-)] + [1 \times N(5\ e^-)] + [1 \times O(6\ e^-)] = 14\ e^-$. Four single bonds are needed ($4 \times 2 = 8\ e^-$). This leaves 6 electrons. The oxygen needs 4 electrons, and the nitrogen needs 2. These last 6 electrons serve as three lone pairs.

correct   incorrect

        <u>Solution:</u>
        b) The hydrogen atoms cannot be the central atoms. The problem states that there are no O–H bonds, so the oxygen must be connected to the carbon atoms. Place the O atom between the two C atoms, and then distribute the H atoms equally around each of the C atoms. The total number of valence electrons is $[6 \times H(1\ e^-)] + [2 \times C(4\ e^-)] + [1 \times O(6\ e^-)] = 20\ e^-$ Draw single bonds between each of the atoms. This creates six C–H bonds, and two C–O bonds, and uses $8 \times 2 = 16$ electrons. The four remaining electrons will become two lone pairs on the O atom to complete its octet.

        <u>Check:</u> Count the electrons. Both the C's and the O have octets. Each H has its pair.

10.3   <u>Plan:</u> Follow the steps in the example, and pay attention to the hint for CO.
        <u>Solution:</u>
        a) In CO there are a total of $[1 \times C(4\ e^-)] + [1 \times O(6\ e^-)] = 10\ e^-$. The hint states that carbon has three bonds. Since oxygen is the only other atom present, these bonds must be between the C and the O. This uses 6 of the 10 electrons. The remaining 4 electrons become two lone pairs, one pair for each of the atoms.

        <u>Check:</u> Count the electrons. Both the C and the O have octets.
        <u>Solution:</u>
        b) In HCN there are a total of $[1 \times H(1\ e^-)] + [1 \times C(4\ e^-)] + [1 \times N(5\ e^-)] = 10$ electrons. Carbon has a lower group number, so it is the central atom. Place the C between the other two atoms and connect each of the atoms to the central C with a single bond. This uses 4 of the 10 electrons, leaving 6 electrons. Distribute these 6 to nitrogen to complete its octet. However, the carbon atom is 4 electrons short of an octet. Change two lone pairs on the nitrogen atom to bonding pairs to form two more bonds between carbon and nitrogen for a total of three bonds.

        <u>Check:</u> Count the electrons. Both the C and the N have octets. The H has its pair.
        <u>Solution:</u>
        c) In CO_2 there are a total of $[1 \times C(4\ e^-)] + [2 \times O(6\ e^-)] = 16$ electrons. Carbon has a lower group number, so it is the central atom. Placing the C between the two O atoms and adding single bonds uses 4 electrons, leaving $16 - 4 = 12\ e^-$. Distributing these 12 electrons to the oxygen atoms completes those octets, but the carbon atom does not have an octet. Change one lone pair on each oxygen atom to a bonding pair to form two double bonds to the carbon atom, completing its octet.

        <u>Check:</u> Count the electrons. Both the C and the O atoms have octets.

154

10.4   Plan: Modify the structure shown in the text to produce additional resonance structures.
        Solution:
        The structure drawn in the text is shown below on the right. Just shift the double bond and electron pairs.

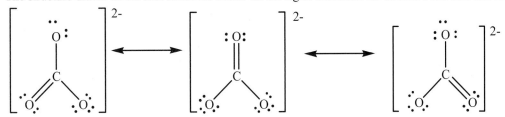

10.5   Plan: The presence of available *d* orbitals makes checking formal charges more important. Use the equation for
        formal charge: FC = # of valence electrons – [# lone electrons + ½(# bonding electrons)]
        Solution:
        a) In POCl$_3$, the P is the most likely central atom because all the other elements have higher group numbers. The
        molecule contains: [1 x P(5 e⁻)] + [1 x O(6 e⁻)] + [3 x Cl(7 e⁻)] = 32 electrons. Placing the P in the center with
        single bonds to all the surrounding atoms uses 8 electrons and gives P an octet. The remaining 24 can be split into
        12 pairs with each of the surrounding atoms receiving three pairs. At this point, structure I below, all the atoms
        have an octet. The central atom is P (smallest group number, highest period number) and can have more than an
        octet. To see how reasonable this structure is, calculate the formal (FC) for each atom. The +1 and –1 formal
        charges are not too unreasonable, however, 0 charges are better. If one of the lone pairs is moved from the O (the
        atom with the negative FC) to form a double bond to the P (the atom with the positive FC), structure II results.
        The calculated formal changes in structure II are all 0, thus, this is a better structure even though P has 10
        electrons.

```
        :Ö:                         :O:
         |                           ‖
  :Cl —— P —— Cl:          :Cl —— P —— Cl:
         |                           |
        :Cl:                        :Cl:

        I                            II
```

FC$_P$ = 5 – (0 + 1/2(8)) = +1   FC$_P$ = 5 – (0 + 1/2(10)) = 0
FC$_O$ = 6 – (6 + 1/2(2)) = –1   FC$_O$ = 6 – (4 + 1/2(4)) = 0
FC$_{Cl}$ = 7 – (6 + 1/2(2)) = 0   FC$_{Cl}$ = 7 – (6 + 1/2(2)) = 0

Check: Count the electrons. All the atoms in structure II, except P, have an octet. The P can have more than an
octet, because it has readily available *d* orbitals.
Solution:
b) In ClO$_2$, the Cl is probably the central atom because the O atoms have a lower period. The molecule contains:
[1 x Cl(7 e⁻)] + [2 x O(6 e⁻)] = 19 electrons. The presence of an odd number of electrons means that there will be
an exception to the octet rule. Placing the O atoms around the Cl and using 4 electrons to form single bonds leaves
15 electrons, 14 of which may be separated into 7 pairs. If 3 of these pairs are given to each O, and the remaining
pair plus the lone electron are given to the Cl, we have the following structure:

```
        ∙∙        ∙∙        ∙∙
      : O —— Cl —— O :
        ∙∙        ∙∙        ∙∙
```

Calculating formal charges: FC$_{Cl}$ = 7 – (3 + 1/2(4)) = +2   FC$_O$ = 6 – (6 + 1/2(2)) = –1
The +2 charge on the Cl is a little high, so other structures should be tried. Moving a lone pair from one of the O
atoms (negative FC) to form a double bond between the Cl and one of the oxygen atoms gives either structure I or
II below. If both O atoms donate a pair of electrons to form a double bond, then structure III results. The next step
is to calculate the formal charges.

155

I

$FC_{Cl} = 7 - (3 + 1/2(6)) = +1$
Left $FC_O = 6 - (4 + 1/2(4)) = 0$
Right $FC_O = 6 - (6 + 1/2(2)) = -1$

II

$FC_{Cl} = 7 - (3 + 1/2(6)) = +1$
$FC_O = 6 - (6 + 1/2(4)) = -1$
$FC_O = 6 - (4 + 1/2(2)) = 0$

III

$FC_{Cl} = 7 - (3 + 1/2(8)) = 0$
$FC_O = 6 - (4 + 1/2(8)) = 0$
$FC_O = 6 - (4 + 1/2(8)) = 0$

Check: Pick the structure with the best distribution of formal charges (structure III).

Solution:

c) XeF_4 is a noble gas compound, thus, there will be an exception to the octet rule. The molecule contains: $[1 \times Xe(8 \text{ e}^-)] + [4 \times F(7 \text{ e}^-)] = 36$ electrons. The Xe is in a higher period than F so Xe is the central atom. If it is placed in the center with a single bond to each of the four fluorine atoms, 8 electrons are used, and 28 electrons remain. The remaining electrons can be divided into 14 pairs with 3 pairs given to each F and the last 2 pairs being given to the Xe. This gives the structure:

Check: Determine the formal charges for each atom:
$FC_{Xe} = 8 - (4 + 1/2(8)) = 0$ $FC_F = 7 - (6 + 1/2(2)) = 0$

10.6   Plan: Draw a Lewis structure. Determine the electron arrangement by counting the electron pairs around the central atom.

Solution:

a) The Lewis structure for CS_2 is shown below. The central atom, C, has two pairs (double bonds only count once). The two pair arrangement is **linear** with the designation, AX_2. The absence of lone pairs on the C means there is no deviation in the bond angle (**180°**).

S══C══S

Check: Compare the results in the figures in the chapter.

Solution:

b) Even though this is a combination of a metal with a nonmetal, it may be treated as a molecular species. The Lewis structure for $PbCl_2$ is shown below. The molecule is of the AX_2E type; the central atom has three pairs of electrons (1 lone pair and two bonding pairs). This means the *electron-group arrangement* is trigonal planar (**120°**), with a lone pair giving a **bent or V-shaped** molecule. The lone pair causes the ideal bond angle to decrease to < 120°.

Check: Compare the results in the figures in the chapter.

<u>Solution:</u>
c) The Lewis structure for the CBr_4 molecule is shown above. It has the AX_4 type formula which is a perfect **tetrahedron** (with **109.5°** bond angles) because all bonds are identical, and there are no lone pairs.

<u>Check:</u> Compare the results in the figures in the chapter.
<u>Solution:</u>
d) The SF_2 molecule has the Lewis structure shown below. This is a AX_2E_2 molecule; the central atom is surrounded by four electron pairs, two of which are lone pairs and two of which are bonding pairs. The *electron group arrangement* is tetrahedral. The two lone pairs give a **V-shaped or bent** arrangement. The ideal tetrahedral bond angle is **decreased from the ideal 109.5°** value.

<u>Check:</u> Compare the results in the figures in the chapter.

10.7   <u>Plan:</u> Draw a Lewis structure. Determine the electron arrangement by counting the electron pairs around the central atom.
<u>Solution:</u>
a) The Lewis structure for the ICl_2^- is shown below. This is a AX_2E_3 type structure. The five pairs give a trigonal bipyramidal arrangement of electron groups. The presence of 3 lone pairs leads to a **linear** shape (**180°**). The usual distortions caused by lone pairs cancel in this case. In the trigonal bipyramidal geometry, lone pairs always occupy equatorial positions.

<u>Check:</u> Compare the results in the figures in the chapter.
<u>Solution:</u>
b) The Lewis structure for the ClF_3 molecule is given below. Like ICl_2^- there are 5 pairs around the central atom; however, there are only 2 lone pairs. This gives a molecule that is **T-shaped**. The presence of the lone pairs decreases the ideal bond angles to **less than 90°**.

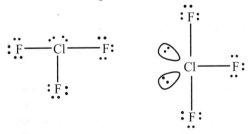

<u>Check:</u> Compare the results in the figures in the chapter.

157

Solution:
c) The SOF_4 molecule has several possible Lewis structures, two of which are shown below. In both cases, the central atom has 5 atoms attached with no lone pairs. The formal charges work out the same in both structures. The structure on the right has an equatorial double bond. Double bonds require more room than single bonds, and equatorial positions have this extra room.

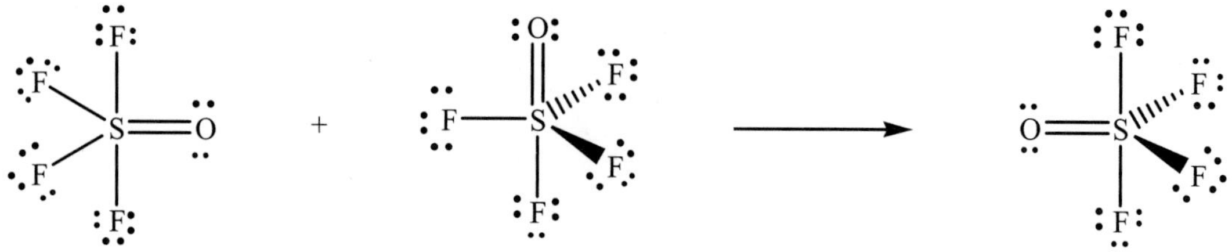

The molecule is **trigonal bipyramidal**, and the double bond causes deviation from ideal bond angles. All of the F atoms move away from the O. Thus, all angles involving the O are increased, and all other angles are decreased.
Check: Compare the results in the figures in the chapter.

10.8   Plan: Draw the Lewis structure for each of the substances, and determine the molecular geometry of each.
Solution:
a) The Lewis structure for H_2SO_4 is shown below. The double bonds ease the problem of a high formal charge on the sulfur. Sulfur is allowed to exceed an octet. The S has 4 groups around it, making it **tetrahedral**. The ideal angles around the S are **109.5°**. The double bonds move away from each other, and force the single bonds away. This opens the angle between the double bonded oxygen atoms, and results in an angle between the single bonded oxygen atoms that is less than ideal.

Solution:
b) The hydrogen atoms cannot be central so the carbons must be attached to each other. The problem states that there is a carbon-carbon triple bond. This leaves only a single bond to connect the third carbon to a triple-bonded carbon, and give that carbon an octet. The other triple-bonded carbon needs one hydrogen to complete its octet. The remaining three hydrogen atoms are attached to the single-bonded carbon, which allows it to complete its octet. This structure is shown below. The single-bonded carbon has four groups tetrahedrally around it leading to bond angles ~109.5° (little or no deviation). The triple-bonded carbons each have two groups (the triple bond counts as one electron group) so they should be **linear (180°)**.

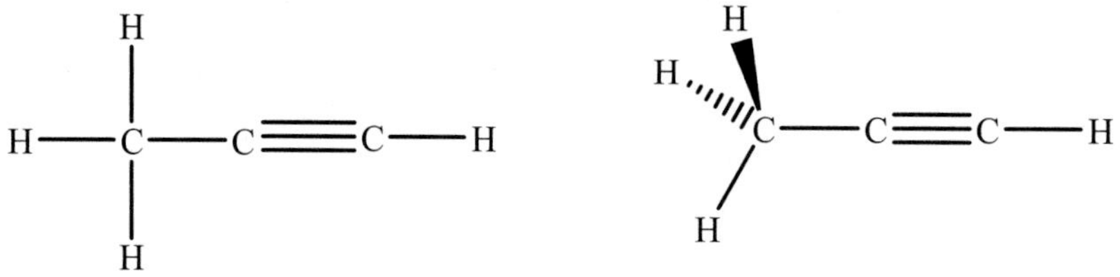

Solution:

c) Fluorine, like hydrogen, is never a central atom. Thus, the sulfurs must be bonded to each other. Each F has 3 lone pairs, and the sulfur atoms have 2 lone pairs. All 4 atoms now have an octet. This structure is shown below. Each sulfur has four groups around it, so the electron arrangements is **tetrahedral**. The presence of the lone pairs on the sulfur atoms results in a geometry that is **V-shaped or bent** and in a bond angle that is **< 109.5°**.

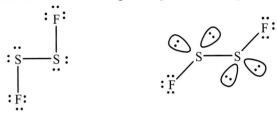

10.9   Plan: Draw the Lewis structures, predict the shapes, and then examine the positions of the bond dipoles.
Solution:
a) Dichloromethane, CH_2Cl_2, has the Lewis structure shown below. It is tetrahedral, and if the outlying atoms were identical, it would be nonpolar. However, the chlorine atoms are more electronegative than hydrogen so there is a general shift in their direction resulting in the arrows shown.

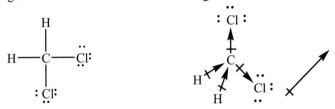

Solution:
b) Iodine oxide pentafluoride, IOF_5, has the Lewis structure shown below. The overall geometry is octahedral. All six bonds are polar, with the more electronegative O and F atoms shifting electron density away from the I. The 4 equatorial fluorines counterbalance each other. The axial F is not equivalent to the axial O. The more electronegative F results in an overall polarity in the direction of the axial F.

The lone electron pairs are left out for simplicity.
Solution:
c) Nitrogen tribromide, NBr_3, has the trigonal pyramidal Lewis structure shown below. The nitrogen is more electronegative than the Br, so the shift in electron density is towards the N, resulting in the dipole indicated below.

END–OF–CHAPTER PROBLEMS

10.1   To be the central atom in a compound, the atom must be able to simultaneously bond to at least two other atoms. **He, F, and H** cannot serve as central atoms in a Lewis structure. Helium $(1s^2)$ is a noble gas, and as such, it does not need to bond to any other atoms. Hydrogen $(1s^1)$ and fluorine $(1s^2 2s^2 2p^5)$ only need one electron to complete their valence shells. Thus, they can only bond to one other atom, and they do not have d orbitals available to expand their valence shells.

10.3   For an element to obey the octet rule it must be surrounded by 8 electrons. To determine the number of electrons present (1) count the individual electrons actually shown adjacent to a particular atom, and (2) add two times the number of bonds to that atom. Using this method the structures shown give: (a) $0 + 2(4) = 8$; (b) $2 + 2(3) = 8$; (c) $0 + 2(5) = 10$; (d) $2 + 2(3) = 8$; (e) $0 + 2(4) = 8$; (f) $2 + 2(3) = 8$; (g) $0 + 2(3) = 6$; (h) $8 + 2(0) = 8$. All the structures obey the octet rule except: c and g.

10.5   Plan: Count the valence electrons and draw Lewis structures.
Solution:
Total valence electrons: SiF_4: $[1 \times Si(4e^-)] + [4 \times F(7e^-)] = 32$; $SeCl_2$: $[1 \times Se(6e^-)] + [2 \times Cl(7\ e^-)] = 20$; COF_2: $[1 \times C(4e^-)] + [1 \times O(6e^-)] + [2 \times F(7e^-)] = 24$. The Si, Se, and the C are the central atoms, because these are the elements in their respective compounds with the lower group number (in addition, we are told C is central). Place the other atoms around the central atoms and connect each to the central atom with a single bond.
SiF_4: At this point, 8 electrons ($2e^-$ in 4 Si–F bonds) have been used with $32 - 8 = 24$ remaining; the remaining electrons are placed around the fluorines (3 pairs each). All atoms have an octet.
$SeCl_2$: The 2 bonds use $4\ e^-$ ($2e^-$ in 2 Se–Cl bonds) leaving $20 - 4 = 16\ e^-$. These $16\ e^-$ are used to complete the octets on Se and the Cl atoms.
COF_2: The 3 bonds to the C use $6\ e^-$ ($2e^-$ in 3 bonds) leaving $24 - 6 = 18\ e^-$. These 18 e– are distributed to the surrounding atoms first to complete their octets. After the 18 e– are used, the central C is 2 electrons short of an octet. Forming a double bond to the O (change a lone pair on O to a bonding pair on C) completes the C octet.

(a) SiF_4 (b) $SeCl_2$

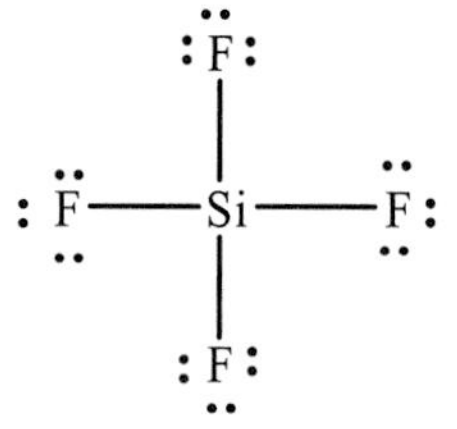

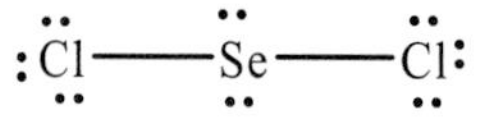

(c) COF_2

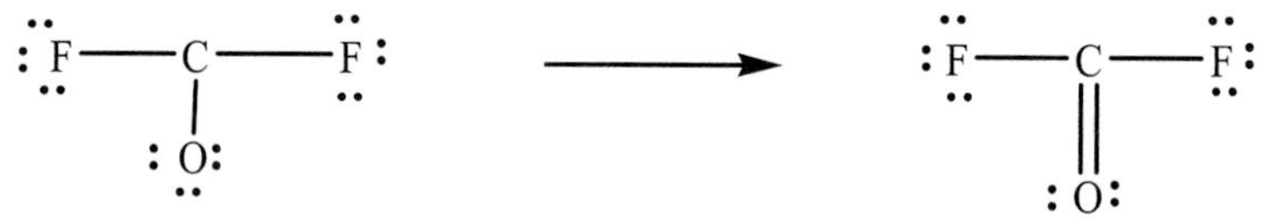

10.7   Plan: Count the valence electrons and draw Lewis structures.
Solution:
a) PF_3: $[1 \times P(5\ e^-)] + [3 \times F(7e^-)] = 26$ valence electrons. P is the central atom. Draw single bonds from P to the 3 F atoms, using $2e^- \times 3$ bonds $= 6\ e$–. Remaining e^-: $26 - 6 = 20\ e^-$. Distribute the $20\ e^-$ around the P and F atoms to complete their octets.
b) H_2CO_3: $[2 \times H(1e^-)] + [1 \times C(4e^-)] + 3 \times O(6e^-)] = 24$ valence electrons. C is the central atom with the H atoms attached to the O atoms. Place appropriate single bonds between all atoms using $2e^- \times 5$ bonds $= 10\ e^-$ so that $24 - 10 = 14\ e^-$ remain. Use these $14\ e^-$ to complete the octets of the O atoms (the H atoms already have their two electrons). After the $14\ e^-$ are used, the central C is 2 electrons short of an octet. Forming a double bond to the O that does not have an H bonded to it (change a lone pair on O to a bonding pair on C) completes the C octet.
c) CS_2: $[1 \times C(4\ e^-)] + [2 \times S(6\ e^-)] = 16$ valence electrons. C is the central atom. Draw single bonds from C to the 2 S atoms, using $2e^- \times 2$ bonds $= 4\ e^-$. Remaining e^-: $16 - 4 = 12\ e^-$. Use these $12\ e^-$ to complete the octets of the surrounding S atoms; this leaves C four electrons short of an octet. Form a double bond from each S to the C by changing a lone pair on each S to a bonding pair on C.

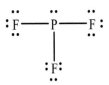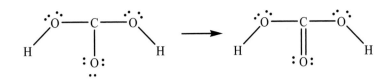

c) CS_2 (16 valence e⁻)

S══C══S

10.9   Plan: The problem asks for resonance structures, so there must be more than one answer for each part.
Solution:
a) NO_2^+ has [1 x N(5 e⁻)] + [2 x O(6 e⁻)] – 1 e⁻ (+ charge) = 16 valence electrons. Draw a single bond from N to each O, using 2e⁻ x 2 bonds = 4 e⁻; 16 – 4 = 12 e⁻ remain. Distribute these 12 e⁻ to the O atoms to complete their octets. This leaves N 4 e⁻ short of an octet. Form a double bond from each O to the N by changing a lone pair on each O to a bonding pair on N. No resonance is required as all atoms can achieve an octet with double bonds.

$$\left[\ddot{\underset{..}{O}}\!\!-\!\!N\!\!-\!\!\ddot{\underset{..}{O}}\!\! \right]^+ \longrightarrow \left[\ddot{O}\!\!=\!\!N\!\!=\!\!\ddot{O} \right]^+$$

b) NO_2F has [1 x N(5 e⁻)] + [2 x O(6 e⁻)] + [1 x F(7e⁻)] = 24 valence electrons. Draw a single bond from N to each surrounding atom, using 2e⁻ x 3 bonds = 6 e⁻; 24 – 6 = 18 e⁻ remain. Distribute these 18 e⁻ to the O and F atoms to complete their octets. This leaves N 2 e⁻ short of an octet. Form a double bond from either O to the N by changing a lone pair on O to a bonding pair on N. There are two resonance structures since a lone pair from either of the two O atoms can be moved to a bonding pair with N:

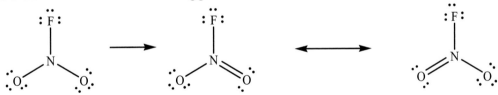

10.11   Plan: Count the valence electrons and draw Lewis structures. Additional structures are needed to show resonance.
Solution:
a) N_3^- has [3 x N(5e⁻)] + [1 e⁻(from charge)] = 16 valence electrons. Place a single bond between the nitrogen atoms. This uses 2e⁻ x 2 bonds = 4 electrons, leaving 16 – 4 = 12 electrons (6 pairs). Giving 3 pairs on each end, nitrogen gives them an octet, but leaves the central N with only 4 electrons as shown below:

$$\left[\ddot{\underset{..}{N}}\!\!-\!\!N\!\!-\!\!\ddot{\underset{..}{N}}\!\! \right]^-$$

The central N needs 4 electrons. There are three options to do this: (1) each of the end N's could form a double bond to the central N by sharing one of its pairs; (2) one of the end N's could form a triple bond by sharing two of its lone pairs; (3) the other end N could form the triple bond instead.

$$\left[\underset{..}{\overset{..}{N}}\!\!=\!\!N\!\!=\!\!\underset{..}{\overset{..}{N}} \right]^- \longleftrightarrow \left[\ddot{\underset{..}{N}}\!\!\equiv\!\!N\!\!-\!\!\ddot{\underset{..}{N}}\!\! \right]^- \longleftrightarrow \left[\ddot{\underset{..}{N}}\!\!-\!\!N\!\!\equiv\!\!\ddot{\underset{..}{N}} \right]^-$$

b) NO_2^- has $[1 \times N(5e^-)] + [2 \times O(6\ e^-)] + [1\ e^-\ (\text{from charge})] = 18$ valence electrons. The nitrogen should be the central atom with each of the oxygen atoms attached to it by a single bond ($2e^- \times 2$ bonds = 4 electrons). This leaves $18 - 4 = 14$ electrons (7 pairs). If 3 pairs are given to each O and 1 pair is given to the N, then both O's have an octet, but the N only has 6. To complete an octet the N needs to gain a pair of electrons from one O or the other (from a double bond). The resonance structures are:

$$\left[:\ddot{O}\text{—}\ddot{N}\text{—}\ddot{O}: \right]^- \longrightarrow \left[:\ddot{O}\text{—}\ddot{N}\text{=}\ddot{O} \right]^- \longleftrightarrow \left[\ddot{O}\text{=}\ddot{N}\text{—}\ddot{O}: \right]^-$$

10.13   Plan: Initially, the method used in the preceding problems may be used to establish a Lewis structure. The total of the formal charges must equal the charge on an ion or be equal to 0 for a compound. The FC only needs to be calculated once for a set of identical atoms.
Solution:
a) IF_5 has $[1 \times I(7\ e^-)] + [5 \times F(7\ e^-)] = 42$ valence electrons.
The presence of 5 F atoms around the central I means that the I will have a minimum of 10 electrons; thus, this is an exception to the octet rule. The 5 I–F bonds use $2e^- \times 5$ bonds = 10 electrons leaving $42 - 10 = 32$ electrons (16 pairs). Each F needs 3 pairs to complete an octet. The 5 F atoms use 15 of the 16 pairs, so there is 1 pair left for the central I. This gives:

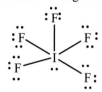

Calculating formal charges:
FC = valence electrons – (lone electrons + 1/2 (bonded electrons))
For iodine: $FC_I = 7 - (2 + 1/2(10)) = 0$ For each fluorine: $FC_F = 7 - (6 + 1/2(2)) = 0$
Total formal charge = 0 = charge on the compound.
b) AlH_4^- has $[1 \times Al(3\ e^-)] + [4 \times H(1\ e^-)] + [1e^-\ (\text{from charge})] = 8$ valence electrons.
The 4 Al–H bonds use all the electrons and Al has an octet.

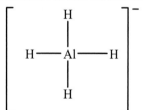

FC = valence electrons – (lone electrons + 1/2 (bonded electrons))
For aluminum: $FC_{Al} = 3 - (0 + 1/2(8)) = -1$
For each hydrogen $FC_H = 1 - (0 + 1/2(2)) = 0$

10.15   Plan: Initially, the method used in the preceding problems may be used to establish a Lewis structure. The total of the formal charges must equal the charge on an ion or be equal to 0 for a compound. The FC only needs to be calculated once for a set of identical atoms.
Solution:
a) CN^-: $[1 \times C(4\ e^-)] + [1 \times N(5\ e^-)] + [1\ e^-\ \text{from charge}] = 10$ valence electrons. Place a single bond between the carbon and nitrogen atoms. This uses $2e^- \times 1$ bond = 2 electrons, leaving $10 - 2 = 8$ electrons (4 pairs). Giving 3 pairs of electrons to the nitrogen atom completes its octet but that leaves only one pair of electrons for the carbon atom which will not have an octet. The nitrogen could form a triple bond by sharing two of its lone pairs with the carbon atom. A triple bond between the two atoms plus a lone pair on each atom satisfies the octet rule and uses all 10 electrons.

$$\left[\, :C\!\equiv\!N\!: \,\right]^{-}$$

FC = valence electrons – (lone electrons + 1/2 (bonded electrons))
$FC_C = 4 – (2 + 1/2(6)) = –1$;　　　$FC_N = 5 – (2 + 1/2(6)) = 0$
<u>Check:</u> The total formal charge equals the charge on the ion (–1).
b) ClO^-: [1 x Cl(7 e$^-$)] + [1 x O(6 e$^-$)] + [1 e$^-$ from charge] = 14 valence electrons. Place a single bond between the chlorine and oxygen atoms. This uses 2e$^-$ x 1 bond = 2 electrons, leaving 14 – 2 = 12 electrons (6 pairs). Giving 3 pairs of electrons each to the carbon and oxygen atoms completes their octets.

$$\left[\, :\ddot{C}l\!-\!\ddot{O}\!: \,\right]^{-}$$

FC = valence electrons – (lone electrons + 1/2 (bonded electrons))
$FC_{Cl} = 7 – (6 + 1/2(2)) = 0$　　　$FC_O = 6 – (6 + 1/2(2)) = –1$
<u>Check:</u> The total formal charge equals the charge on the ion (–1).

10.17　<u>Plan:</u> The general procedure is similar to the preceding problems, plus the oxidation number determination.
<u>Solution:</u>
a) BrO_3^- has [1 x Br(7 e$^-$)] + 3 x O(6 e$^-$)] + [1e– (from charge)] = 26 valence electrons.
Placing the O atoms around the central Br and forming 3 Br–O bonds uses 2e– x 3 bonds = 6 electrons and leaves 26 – 6 = 20 electrons (10 pairs). Placing 3 pairs on each O (3 x 3 = 9 total pairs) leaves 1 pair for the Br and yields structure I below. In structure I, all the atoms have a complete octet. Calculating formal charges:
$FC_{Br} = 7 – (2 + 1/2(6)) = +2$　　　$FC_O = 6 – (6 + 1/2(2)) = –1$
The FC_O is acceptable, but FC_{Br} is larger than is usually acceptable. Forming a double bond between any one of the O atoms gives Structure II. Calculating formal charges:
$FC_{Br} = 7 – (2 + 1/2(8)) = +1$　　　$FC_O = 6 – (6 + 1/2(2)) = –1$　　　$FC_O = 6 – (4 + 1/2(4)) = 0$
　　　　　　　　　　　　　　　　　　　　　　　　　　　　　　Double bonded O

The FC_{Br} can be improved further by forming a second double bond to one of the other O atoms (structure III).
$FC_{Br} = 7 – (2 + 1/2(10)) = 0$　　　$FC_O = 6 – (6 + 1/2(2)) = –1$　　　$FC_O = 6 – (4 + 1/2(4)) = 0$
　　　　　　　　　　　　　　　　　　　　　　　　　　　　　　Double bonded O atoms

Structure III has the most reasonable distribution of formal charges.

　　　　　I　　　　　　　　　　　　II　　　　　　　　　　　　III

The oxidation numbers (O.N.) are: $O.N._{Br} = +5$ and $O.N._O = –2$.　　　$\begin{array}{c}-6\\ +5 \ -2\\ BrO_3^-\end{array}$
<u>Check:</u> The total formal charge equals the charge on the ion (–1).
b) SO_3^{2-} has [1 x S(6 e$^-$)] + [3 x O(6 e$^-$)] + [2e$^-$ (from charge)] = 26 valence electrons.
Placing the O atoms around the central S and forming 3 S–O bonds uses 2e$^-$ x 3 bonds = 6 electrons and leaves 26 – 6 = 20 electrons (10 pairs). Placing 3 pairs on each O (3 x 3 = 9 total pairs) leaves 1 pair for the S and yields structure I below. In structure I all the atoms have a complete octet. Calculating formal charges:
$FC_S = 6 – (2 + 1/2(6)) = +1$; $FC_O = 6 – (6 + 1/2(2)) = –1$
The FC_O is acceptable, but FC_S is larger than is usually acceptable. Forming a double bond between any one of the O atoms (Structure II) gives:
$FC_S = 6 – (2 + 1/2(8)) = 0$　　　$FC_O = 6 – (6 + 1/2(2)) = –1$　　　$FC_O = 6 – (4 + 1/2(4)) = 0$
　　　　　　　　　　　　　　　　　　　　　　　　　　　　　　Double bonded O

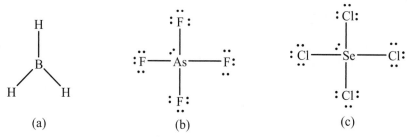

I   II   −6

Structure II has the more reasonable distribution of formal charges.   +4 −2
The oxidation numbers (O.N.) are: O.N.$_S$ = +4 and O.N.$_O$ = −2.   SO$_3^{2-}$
Check: The total formal charge equals the charge on the ion (−2).

10.19   Plan: We know that each of these compounds does not obey the octet rule.
        Solution:
        a) BH$_3$ has [1 x B(3e$^-$)] + [3 x H(1e$^-$)] = 6 valence electrons. These are used in 3 B–H bonds. The B has 6
        electrons instead of an octet; this molecule is electron deficient.
        b) AsF$_4^-$ has [1 x As(5e$^-$)] +[4 x F(7e$^-$)] + [1e$^-$ (from charge)] = 34 valence electrons. Four As–F bonds use 8
        electrons leaving 34 – 8 = 26 electrons (13 pairs). Each F needs 3 pairs to complete its octet and the remaining
        pair goes to the As. The As has an expanded octet with 10 electrons. The F cannot expand its octet.
        c) SeCl$_4$ has [1 x Se(6e$^-$)] + 4 x Cl(7e$^-$)] = 34 valence electrons. The SeCl$_4$ is isoelectronic (has the same electron
        structure) as AsF$_4^-$, and so its Lewis structure looks the same. Se has an expanded octet of 10 electrons.

        (a)   (b)   (c)

10.21   Plan: We know that each of these compounds does not obey the octet rule. Draw the Lewis structure.
        Solution:
        a) BrF$_3$ has [1 x Br(7e$^-$)] + [3 x F(7 e$^-$)] = 28 valence electrons. Placing a single bond between Br and each F
        uses 2e$^-$ x 3 bonds = 6 e$^-$, leaving 28 – 6 = 22 electrons (11 pairs). After the F atoms complete their octets with 3
        pairs each, the Br gets the last 2 lone pairs. The Br expands its octet to 10 electrons.
        b) ICl$_2^-$ has [1 x I(7e$^-$)] + [2 x Cl(7e$^-$)] + [1e$^-$ (from charge)] = 22 valence electrons. Placing a single bond
        between I and each Cl uses 2e$^-$ x 2 bond = 4e$^-$, leaving 22 – 4 = 18 electrons (9 pairs). After the Cl atoms
        complete their octets with 3 pairs each, the iodine finishes with the last 3 lone pairs. The iodine has an expanded
        octet of 10 electrons.
        c) BeF$_2$, [1 x Be(2e$^-$)] + [2 x F(7e$^-$)] = has 16 valence electrons. Placing a single bond between Be and each of
        the F atoms uses 2e$^-$ x 2 bonds = 4e$^-$, leaving 16 – 4 = 12 electrons (6 pairs).The F atoms complete their octets
        with 3 pairs each, and there are no electrons left for the Be. Formal charges work against the formation of double
        bonds. Be, with only 4 electrons, is electron deficient.

        (a)   (b)   (c)

10.23   <u>Plan:</u> Draw Lewis structures for the reactants and products.
         <u>Solution:</u>
         Beryllium chloride has the formula $BeCl_2$. $BeCl_2$ has $[1 \times Be(2e^-)] + [2 \times Cl(7e^-)] = 16$ valence electrons. 4 of these electrons are used to place a single bond between Be and each of the Cl atoms, leaving $16 - 4 = 12$ electrons (6 pairs). These 6 pairs are used to complete the octets of the Cl atoms, but Be does not have an octet – it is electron deficient.
         Chloride ion has the formula Cl^- with an octet of electrons.
         $BeCl_4^{2-}$ has $[1 \times Be(2e^-)] + [4 \times Cl(7e^-)] + [2e^-$ (from charge)$] = 32$ valence electrons. 8 of these electrons are used to place a single bond between Be and each Cl atom, leaving $32 - 8 = 24$ electrons (12 pairs). These 12 pairs complete the octet of the Cl atoms (Be already has an octet).

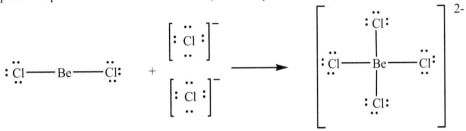

10.26   <u>Plan:</u> Use the structures in the text to determine the formal charges.
         FC = valence electrons – (lone electrons + 1/2 (bonded electrons))
         <u>Solution:</u>
         Structure **A**: $FC_C = 4 - (0 + 1/2(8)) = 0$; $FC_O = 6 - (4 + 1/2(4)) = 0$; $FC_{Cl} = 7 - (6 + 1/2(2)) = 0$
         Total FC = 0
         Structure **B**: $FC_C = 4 - (0 + 1/2(8)) = 0$; $FC_O = 6 - (6 + 1/2(2)) = -1$;
         $FC_{Cl} = 7 - (4 + 1/2(4)) = +1$ (double bonded); $FC_{Cl} = 7 - (6 + 1/2(2)) = 0$ (single bonded)
         Total FC = 0
         Structure **C**: $FC_C = 4 - (0 + 1/2(8)) = 0$; $FC_O = 6 - (6 + 1/2(2)) = -1$;
         $FC_{Cl} = 7 - (4 + 1/2(4)) = +1$ (double bonded); $FC_{Cl} = 7 - (6 + 1/2(2)) = 0$ (single bonded)
         Total FC = 0
         Structure **A** has the most reasonable set of formal charges.

10.28   The molecular shape and the electron-group arrangement are the same when there are no lone pairs on the central atom.

10.30   <u>Plan:</u> Examine a list of all possible structures, and choose the ones with four electron groups since the tetrahedral electron group arrangement has four electron groups.
         <u>Solution:</u>
         Tetrahedral AX_4
         Trigonal pyramidal AX_3E
         Bent or V-shaped AX_2E_2

10.32   <u>Plan:</u> Begin with the basic structures and redraw them.
         <u>Solution:</u>
         a) A molecule that is V-shaped has two bonds and generally has either one (AX_2E) or two (AX_2E_2) lone electron pairs.
         b) A trigonal planar molecule follows the formula AX_3 with three bonds and no lone electron pairs.
         c) A trigonal bipyramidal molecule contains five bonding pairs (single bonds) and no lone pairs (AX_5).
         d) A T-shaped molecule has three bonding groups and two lone pairs (AX_3E_2).
         e) A trigonal pyramidal molecule follows the formula AX_3E with three bonding pairs and one lone pair.
         f) A square pyramidal molecule shape follows the formula AX_5E with 5 bonding pairs and one lone pair.

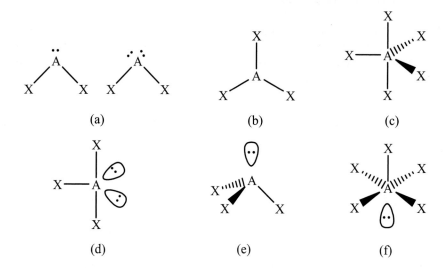

(a)   (b)   (c)

(d)   (e)   (f)

10.34   Plan: First, draw a Lewis structure, and then apply VSEPR.
Solution:
a) O_3: The molecule has [3 x O(6e$^-$)] = 18 valance electrons. 4 electrons are used to place a single bond between the oxygen atoms, leaving 18 – 4 = 14e$^-$ (7 pairs). 6 pairs are required to give the end oxygen atoms an octet; the last pair is distributed to the central oxygen, leaving this atom 2 electrons short of an octet. Form a double bond from one of the end O atoms to the central O by changing a lone pair on the end O to a bonding pair on the central O. This gives the following Lewis structure:

There are three electron groups around the central O, one of which is a lone pair. This gives a **trigonal planar** electron–group arrangement (AX$_2$E), a **bent** molecular shape, and an ideal bond angle of **120°**.
b) H_3O^+: This ion has [3 x H(1e$^-$)] + [1 x O(6e$^-$)] – [1e$^-$ (due to + charge] = 8 valence electrons. 6 electrons are used to place a single bond between O and each H, leaving 8 – 6 = 2e$^-$ (1 pair). Distribute this pair to the O atom, giving it an octet (the H atoms only get 2 electrons). This gives the following Lewis structure:

There are four electron groups around the O, one of which is a lone pair. This gives a **tetrahedral** electron-group arrangement (AX$_3$E), a **trigonal pyramidal** molecular shape, and an ideal bond angle of **109.5°**.
c) NF_3: The molecule has [1 x N(5e$^-$)] + [3 x F(7e$^-$)] = 26 valence electrons. 6 electrons are used to place a single bond between N and each F, leaving 26 – 6 = 20 e$^-$ (10 pairs). These 10 pairs are distributed to all of the F atoms and the N atoms to give each atom an octet. This gives the following Lewis structure:

There are four electron groups around the N, one of which is a lone pair. This gives a **tetrahedral** electron-group arrangement (AX$_3$E), a **trigonal pyramidal** molecular shape, and an ideal bond angle of **109.5°**.

166

10.36   <u>Plan:</u> First, draw a Lewis structure, and then apply VSEPR.
    <u>Solution:</u>
    (a) CO_3^{2-}: This ion has [1 x C(4e$^-$)] + [3 x O(6e$^-$)] + [2e$^-$ (from charge)] = 24 valence electrons. 6 electrons
    are used to place single bonds between C and each O atom, leaving 24 – 6 = 18 e$^-$ (9 pairs). These 9 pairs are
    used to complete the octets of the three O atoms, leaving C 2 electrons short of an octet.   Form a double bond
    from one of the O atoms to C by changing a lone pair on an O to a bonding pair on C. This gives the following
    Lewis structure:

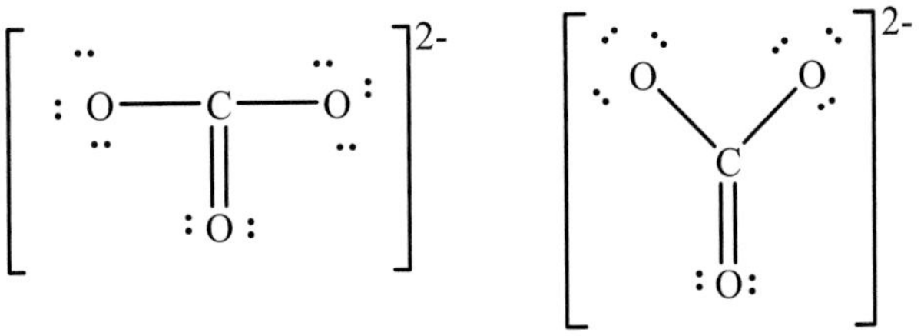

    + 2 additional resonance forms. There are three groups of electrons around the C, none of which are lone pairs.
    This gives a **trigonal planar** electron-group arrangement (AX$_3$), a **trigonal planar** molecular shape, and an ideal
    bond angle of **120°**.
    (b) SO_2: This molecule has [1 x S(6 e$^-$)] + [2 x S(6 e$^-$)] = 18 valence electrons. 4 electrons are used to place a
    single bond between S and each O atom, leaving 18 – 4 = 14 e$^-$ (7 pairs). 6 pairs are needed to complete the
    octets of the O atoms, leaving a pair of electrons for S. S needs one more pair to complete its octet. Form a
    double bond from one of the end O atoms to the S by changing a lone pair on the O to a bonding pair on the S.
    This gives the following Lewis structure:

    There are three groups of electrons around the C, one of which is a lone pair.
    This gives a **trigonal planar** electron-group arrangement (AX$_2$E), a **bent (V-shaped)** molecular shape, and an
    ideal bond angle of **120°**.
    (c) CF_4: This molecule has [1 x C(4 e$^-$)] + [4 x F(7 e$^-$)] = 32 valence electrons. 8 electrons are used to place a
    single bond between C and each F, leaving 32 – 8 = 24 e$^-$ (12 pairs). Use these 12 pairs to complete the octets of
    the F atoms (C already has an octet). This gives the following Lewis structure:

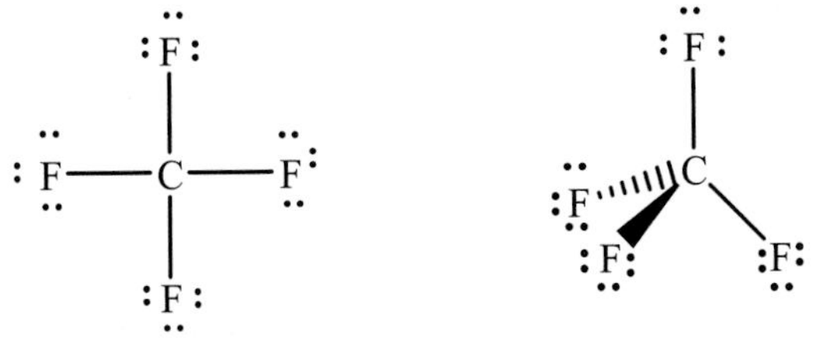

    There are four groups of electrons around the C, none of which is a lone pair.
    This gives a **tetrahedral** electron-group arrangement (AX$_4$), a **tetrahedral** molecular shape, and an ideal bond
    angle of **109.5°**.

10.38   <u>Plan:</u> Examine the structure shown, and then apply VSEPR.
    <u>Solution:</u>
    a) This structure shows three electron groups with three bonds around the central atom.
    There appears to be no distortion of the bond angles so the shape is **trigonal planar**, the classification is **AX$_3$**,
    with an ideal bond angle of **120°**.
    b) This structure shows three electron groups with three bonds around the central atom.
    The bonds are distorted down indicating the presence of a lone pair. The shape of the molecule is **trigonal
    pyramidal** and the classification is **AX$_3$E**, with an ideal bond angle of **109.5°**.

c) This structure shows five electron groups with five bonds around the central atom.
There appears to be no distortion of the bond angles so the shape is **trigonal bipyramidal** and the classification is **AX$_5$**, with ideal bond angles of **90°** and **120°**.

10.40   <u>Plan:</u> The Lewis structures must be drawn, and VSEPR applied to the structures.
<u>Solution:</u>
a) The ClO$_2^-$ ion has [1 x Cl(7 e$^-$)] + [2 x O(6 e$^-$)] + [1 e$^-$ (from charge)] = 20 valence electrons. 4 electrons are used to place a single bond between the Cl and each O, leaving 20 – 4 = 16 electrons (8 pairs). All 8 pairs are used to complete the octets of the Cl and O atoms. There are two bonds (to the O's) and two lone pairs on the Cl for a total of 4 electron groups (AX$_2$E$_2$). The structure is based on a tetrahedral electron-group arrangement with an ideal bond angle of **109.5°**. The shape is **bent** (or V-shaped). The presence of the lone pairs will cause the remaining angles to be **less than 109.5°**.
b) The PF$_5$ molecule has [1 x P(5 e$^-$)] + [5 x F(7 e$^-$)] = 40 valence electrons. 10 electrons are used to place single bonds between P and each F atom, leaving 40 – 10 = 30 e$^-$ (15 pairs). The 15 electrons are used to complete the octets of the F atoms. There are 5 bonds to the P and no lone pairs (AX$_5$). The electron-group arrangement and the shape is **trigonal bipyramidal**. The ideal bond angles are **90°** and **120°**. The absence of lone pairs means the **angles are ideal**.
c) The SeF$_4$ molecule has [1 x Se(6 e$^-$)] + [4 x F(7 e$^-$)] = 34 valence electrons. 8 electrons are used to place single bonds between Se and each F atom, leaving 34 – 8 = 26 e$^-$ (13 pairs). 12 pairs are used to complete the octets of the F atoms which leaves 1 pair of electrons. This pair is placed on the central Se atom. There are 4 bonds to the Se which also has a lone pair (AX$_4$E). The structure is based on a trigonal bipyramidal structure with ideal angles of **90°** and **120°**. The shape is **see-saw**. The presence of the lone pairs means the angles are **less than ideal**.
d) The KrF$_2$ molecule has [1 x Kr(8 e$^-$)] + [2 x F(7 e$^-$)] = 22 valence electrons. 4 electrons are used to place a single bond between the Kr atom and each F atom, leaving 22 – 4 = 18 e$^-$ (9 pairs). 6 pairs are used to complete the octets of the F atoms. The remaining 3 pairs of electrons are placed on the Kr atom. The Kr is the central atom. There are 2 bonds to the Kr and 3 lone pairs (AX$_2$E$_3$). The structure is based on a trigonal bipyramidal structure with ideal angles of 90° and 120°. The shape is **linear**. The placement of the F atoms makes their ideal bond angle to be 2 x 90° = **180°**. The placement of the lone pairs is such that they cancel each other's repulsion, thus the actual **bond angle is ideal**.

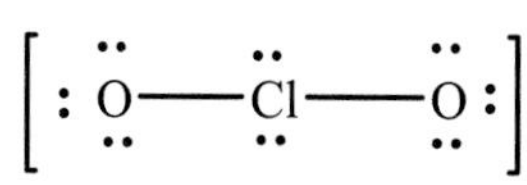

(a) 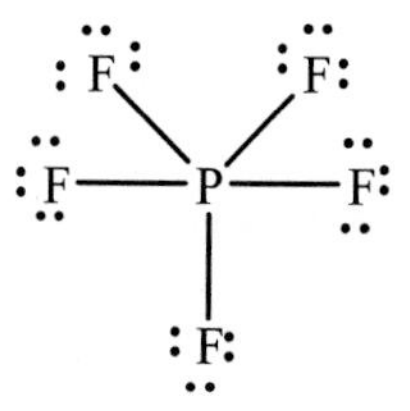   (b) 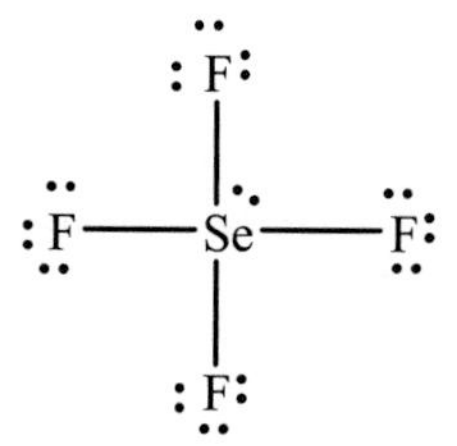   (c) 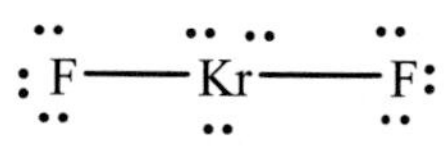   (d)

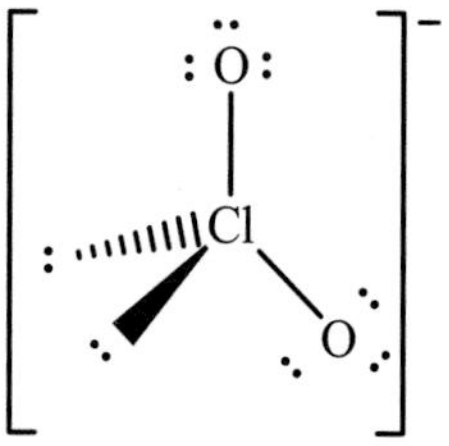

(a) 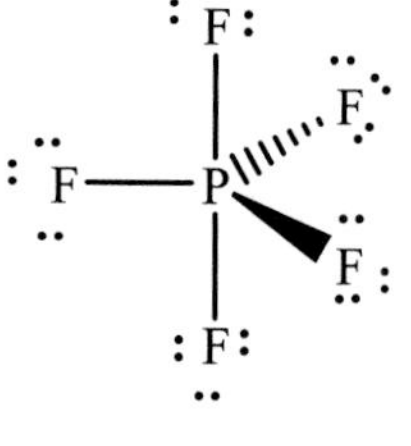   (b) 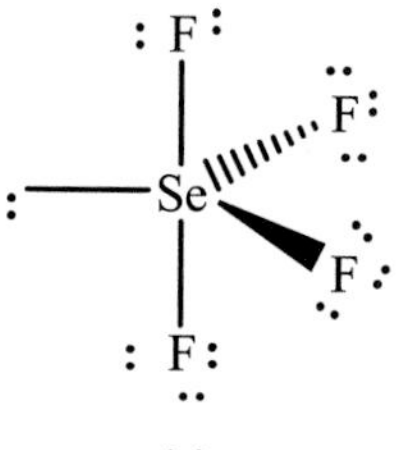   (c) 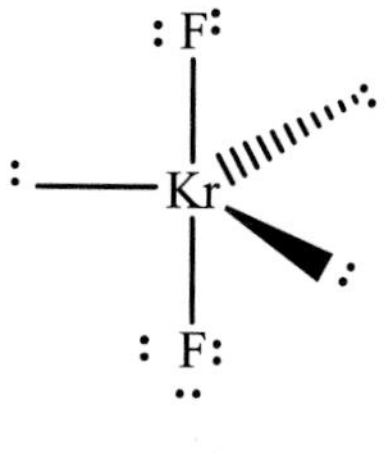   (d)

10.42   <u>Plan:</u> The Lewis structures must be drawn, and VSEPR applied to the structures.
    <u>Solution:</u>
    a) CH_3OH: This molecule has [1 x C(4 e⁻)] + [4 x H(1 e⁻)] + [1 x O(6 e⁻)] = 14 valence electrons. In the CH_3OH molecule, both carbon and oxygen serve as central atoms. (H can never be central.) Use 8 electrons to place a single bond between the C and the O atom and 3 of the H atoms and another 2 electrons to place a single bond between the O and the last H atom. This leaves 14 − 10 = 4 e⁻ (2 pairs). Use these two pairs to complete the octet of the O atom. C already has an octet and each H only gets 2 electrons. The carbon has 4 bonds and no lone pairs (AX_4), so it is **tetrahedral** with **no deviation** (no lone pairs) from the ideal angle of 109.5°. The oxygen has 2 bonds and 2 lone pairs (AX_2E_2), so it is **V-shaped** or **bent** with the angles **less than the ideal** angle of 109.5°.

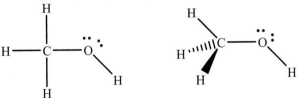

    b) N_2O_4: This molecule has [2 x N(5 e⁻)] + [4 x O(6 e⁻)] = 34 valence electrons. Use 10 electrons to place a single bond between the two N atoms and between each N and two of the O atoms. This leaves 34 − 10 = 24 e⁻ (12 pairs). Use the 12 pairs to complete the octets of the oxygen atoms. Neither N atom has an octet however. Form a double bond from one O atom to one N atom by changing a lone pair on the O to a bonding pair on the N. Do this for the other N atom as well. In the N_2O_4 molecule, both nitrogens serve as central atoms. This is the arrangement given in the problem. Both nitrogens are equivalent with 3 groups and no lone pairs (AX_3), so the arrangement is **trigonal planar** with **no deviation** (no lone pairs) from the ideal angle of 120°. The same results arise from the other resonance structures.

10.44   <u>Plan:</u> The Lewis structures must be drawn, and VSEPR applied to the structures.
    <u>Solution:</u>
    a) CH_3COOH has [2 x C(4 e⁻)] + [4 x H(1 e⁻)] + [2 x O(6 e⁻)] = 24 valence electrons. Use 14 electrons to place a single bond between all of the atoms. This leaves 24 − 14 = 10 e⁻ (5 pairs). Use these 5 pairs to complete the octets of the O atoms; the C atom bonded to the H atoms has an octet but the other C atom does not have a complete octet. Form a double bond from the O atom (not bonded to H) to the C by changing a lone pair on the O to a bonding pair on the C. In the CH_3COOH molecule, the carbons and the O with H attached serve as central atoms. The carbon bonded to the H atoms has 4 groups and no lone pairs (AX_4), so it is **tetrahedral** with **no deviation** from the ideal angle of 109.5°. The carbon bonded to the O atoms has 3 groups and no lone pairs (AX_3), so it is **trigonal planar** with **no deviation** from the ideal angle of 120°. The H bearing O has 2 bonds and 2 lone pairs (AX_2E_2), so the arrangement is **V-shaped** or **bent** with an angle **less than the ideal** values of 109.5°.

b) H_2O_2 has [2 x H(1 e$^-$)] + [2 x O(6 e$^-$)] = 14 valence electrons. Use 6 electrons to place single bonds between the O atoms and between each O atom and an H atom. This leaves 14 – 6 = 8 e$^-$ (4 pairs). Use these 4 pairs to complete the octets of the O atoms. In the H_2O_2 molecule, both oxygens serve as central atoms. Both O's have 2 bonds and 2 lone pairs (AX$_2$E$_2$), so they are **V-shaped** or **bent** with angles **less than the ideal** values of 109.5°.

10.46   Plan: First, draw a Lewis structure, and then apply VSEPR.
Solution:

| 120° | 180° | 109.5° | < 109.5° | << 109.5° |

Bond angles: **OF$_2$ < NF$_3$ < CF$_4$ < BF$_3$ < BeF$_2$**
BeF$_2$ is an AX$_2$ type molecule, so the angle is the ideal 180°. BF$_3$ is an AX$_3$ molecule, so the angle is the ideal 120°. CF$_4$, NF$_3$, and OF$_2$ all have tetrahedral electron-group arrangements of the following types: AX$_4$, AX$_3$E, and AX$_2$E$_2$, respectively. The ideal tetrahedral bond angle is 109.5°, which is present in CF$_4$. The one lone pair in NF$_3$ decreases the angle a little. The two lone pairs in OF$_2$ decrease the angle even more.

10.48   Plan: The ideal bond angles depend on the electron-group arrangement. Deviations depend on lone pairs.
Solution:
a) The C and N have 3 groups, so they are **ideally 120°**, and the O has 4 groups, so **ideally the angle is 109.5°**. The N and O have lone pairs, so the **angles are less than ideal**.
b) All central atoms have 4 pairs, so ideally all the angles are **109.5°**. The lone pairs on the O **reduce** this value.
c) The B has 3 groups (no lone pairs) leading to an **ideal bond angle of 120°**. All the O's have 4 pairs (**ideally 109.5°**), 2 of which are lone, and **reduce the angle**.

10.51   Plan: The Lewis structures are needed to predict the ideal bond angles.
Solution:
The P atoms have no lone pairs in any case so the angles are ideal.

PCl$_5$:   PCl$_4^+$:   PCl$_6^-$:

The original PCl₅ is AX₅, so the shape is trigonal bipyramidal, and the angles are 120° and 90°.
The PCl₄⁺ is AX₄, so the shape is tetrahedral, and the angles are 109.5°.
The PCl₆⁻ is AX₆, so the shape is octahedral, and the angles are 90°.
Half the PCl₅ (trigonal bipyramidal, 120° and 90°) become tetrahedral PCl₄⁺ (tetrahedral, 109.5°), and the other half become octahedral PCl₆⁻ (octahedral, 90°).

10.52   Molecules are polar if they have polar bonds that are not arranged to cancel each other. A polar bond is present any time there is a bond between elements with differing electronegativities.

10.55   <u>Plan:</u> To determine if a bond is polar, determine the electronegativity difference of the atoms participating in the bond. To determine if a molecule is polar (has a dipole moment), it must have polar bonds, and a certain shape determined by VSEPR.
<u>Solution:</u>
a) The greater the difference in electronegativity the more polar the bond:

Molecule	Bond	Electronegativities	Electronegativity difference
SCl₂	S–Cl	S = 2.5 Cl = 3.0	3.0 – 2.5 = 0.5
F₂	F–F	F = 4.0 F = 4.0	4.0 – 4.0 = 0.0
CS₂	C–S	C = 2.5 S = 2.5	2.5 – 2.5 = 0.0
CF₄	C–F	C = 2.5 F = 4.0	4.0 – 2.5 = 1.5
BrCl	Br–Cl	Br = 2.8 Cl = 3.0	3.0 – 2.8 = 0.2

The polarities of the bonds increases in the order: F–F = C–S < Br–Cl < S–Cl < C–F. Thus, **CF₄** has the most polar bonds.
b) The F₂ and CS₂ cannot be polar since they do not have polar bonds. CF₄ is a AX₄ molecule, so it is tetrahedral with the 4 polar C–F bonds arranged to cancel each other giving an overall nonpolar molecule. **BrCl has a dipole moment** since there are no other bonds to cancel the polar Br–Cl bond. **SCl₂ has a dipole moment** (is polar) because it is a bent molecule, AX₂E₂, and the S–Cl bonds both pull to one side.

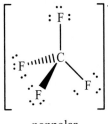

nonpolar

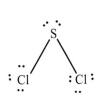

polar

10.57   <u>Plan:</u> If only 2 atoms are involved, only an electronegativity difference is needed. If there are more than 2 atoms, the structure must be determined.
<u>Solution:</u>
a) All the bonds are polar covalent. The SO₃ molecule is trigonal planar, AX₃, so the bond dipoles cancel leading to a nonpolar molecule (no dipole moment). The SO₂ molecule is bent, AX₂E, so the polar bonds both pull to one side. **SO₂ has a greater dipole moment** because it is the only one of the pair that is polar.

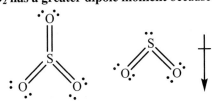

b) ICl and IF are polar, as are all diatomic molecules composed of atoms with differing electronegativities. The electronegativity difference for ICl (3.0 – 2.5 = 0.5) is less than that for IF (4.0 – 2.5 = 1.5). The greater difference means that **IF has a greater dipole moment**.

c) All the bonds are polar covalent. The SiF_4 molecule is nonpolar (has no dipole moment) because the bonds are arranged tetrahedrally, AX_4. SF_4 is AX_4E, so it has a see-saw shape, where the bond dipoles do not cancel. **SF_4 has the greater dipole moment**.

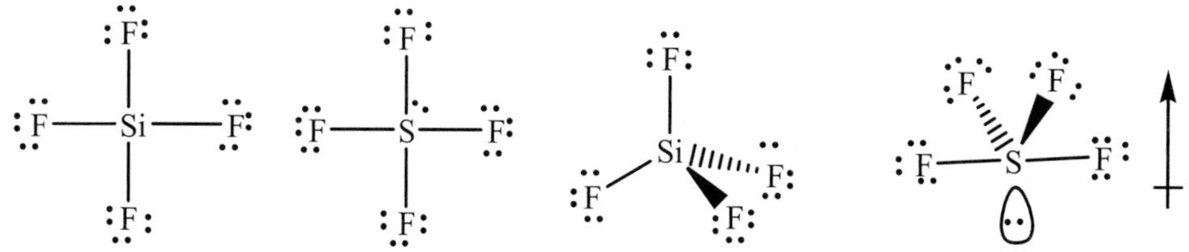

d) H_2O and H_2S have the same basic structure. They are both bent molecules, AX_2E_2, and as such, they are polar. The electronegativity difference in H_2O (3.5 – 2.1 = 1.4) is greater than the electronegativity difference in H_2S (2.5 – 2.1 = 0.4) so **H_2O has a greater dipole moment**.

10.59   Plan: Draw Lewis structures, and then apply VSEPR. A molecule has a dipole moment if polar bonds do not cancel.
        Solution:
        There are 3 possible structures for the compound $C_2H_2Cl_2$:

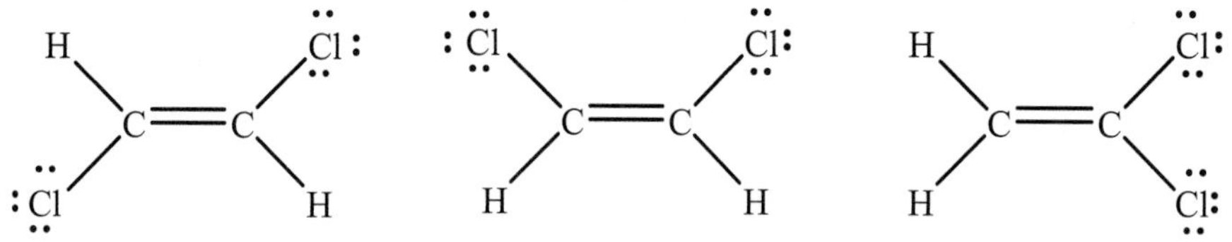

        I   **II**   **III**

        The presence of the double bond prevents rotation about the C=C bond, so the structures are "fixed." The C–Cl bonds are more polar than the C–H bonds, so the key to predicting the polarity is the positioning of the C–Cl bonds. Compound I has the C–Cl bonds arranged so that they cancel leaving I as a nonpolar molecule. Both II and III have C–Cl bonds on the same side so the bonds work together making both molecules polar. Both I and II will react with H_2 to give a compound with a Cl attached to each C (same product). Compound III will react with H_2 to give a compound with 2 Cl atoms on one C and none on the other (different product). **Compound I must be X** as it is the only one that is nonpolar (has no dipole moment). **Compound II must be Z** because it is polar and gives the same product as compound X. This means that compound III must be the remaining compound — Y. **Compound Y (III) has a dipole moment.**

10.61   Plan: The Lewis structures are needed to do this problem. For part (b), obtain bond energy values from Table 9.2.
        Solution:
        a) The H atoms cannot be central, and they are evenly distributed on the N's.
        N_2H_4 has [2 x N(5 e$^-$)] + [4 x H(1 e$^-$)] = 14 valence electrons, 10 of which are used in the bonds between the atoms. The remaining two pairs are used to complete the octets of the N atoms.
        N_2H_2 has [2 x N(5 e$^-$)] + (2 x H(1 e$^-$)] = 12 valence electrons, 6 of which are used in the bonds between the atoms. The remaining three pairs of electrons are not enough to complete the octets of both N atoms, so one lone pair is moved to a bonding pair between the N atoms.
        N_2 has [2 x N(5 e$^-$)] = 10 valence electrons, 2 of which are used to place a single bond between the two N atoms. Since only 4 pairs of electrons remain and 6 pairs are required to complete the octets, two lone pairs become bonding pairs to form a triple bond.

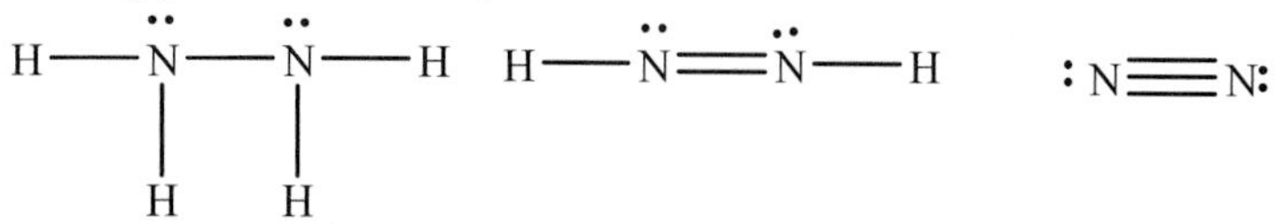

        Hydrazine   Diazene   Nitrogen

        The single (bond order = 1) **N–N bond is weaker and longer** than any of the others are. The **triple bond (bond order = 3) is stronger and shorter** than any of the others. The **double bond (bond order = 2) has an intermediate strength and length.**

b) N_4H_4 has $[4 \times N(5\ e^-)] + [4 \times H(1\ e^-)] = 24$ valence electrons, 14 of which are used for single bonds between the atoms. When the remaining 5 pairs are distributed to complete the octets, one N atom lacks two electrons. A lone pair is moved to a bonding pair for a double bond.

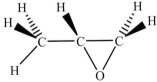

$\Delta H°_{\text{bonds broken}} = 4\ N–H = 4\ \text{mol}\ (391\ \text{kJ/mol})$
$2\ N–N\ = 2\ \text{mol}\ (160\ \text{kJ/mol})$
$1\ N=N\ = \underline{1\ \text{mol}\ (418\ \text{kJ/mol})}$
$\qquad\ = 2302\ \text{kJ}$

$\Delta H°_{\text{bonds formed}} = 4\ N–H = 4\ \text{mol}\ (–391\ \text{kJ/mol})$
$1\ N–N\ = 1\ \text{mol}\ (–160\ \text{kJ/mol})$
$1\ N≡N\ = \underline{1\ \text{mol}\ (–945\ \text{kJ/mol})}$
$\qquad\ = –2669\ \text{kJ}$

$\Delta H°_{\text{rxn}} = \Delta H°_{\text{bonds broken}} + \Delta H°_{\text{bonds formed}} = 2302\ \text{kJ} + (–2669\ \text{kJ}) = \textbf{–367 kJ}$

Note: It is correct to report the answer in kJ or kJ/mol as long as the value refers to a reactant or product with a molar coefficient of 1.

10.65 Plan: Use the Lewis structures shown in the text.
FC = valence electrons – [lone electrons + 1/2 (bonded electrons)]
Solution:
a) Formal charges for Al_2Cl_6:
$\qquad FC_{Al} = 3 – (0 + 1/2(8)) = –1$
$\qquad FC_{Cl,\ ends} = 7 – (6 + 1/2(2)) = 0$
$\qquad FC_{Cl,\ bridging} = 7 – (4 + 1/2(4)) = +1$
$\qquad$(Check: Formal charges add to zero, the charge on the compound.)
$\qquad$Formal charges for I_2Cl_6:
$\qquad FC_I = 7 – (4 + 1/2(8)) = –1$
$\qquad FC_{Cl,\ ends} = 7 – (6 + 1/2(2)) = 0$
$\qquad FC_{Cl,\ bridging} = 7 – (4 + 1/2(4)) = +1$
$\qquad$(Check: Formal charges add to zero, the charge on the compound.)
b) The aluminum atoms have no lone pairs and are AX_4, so they are tetrahedral. The 2 tetrahedral Al's cannot give a planar structure. The iodine atoms have two lone pairs each and are AX_4E_2 so they are square planar. Placing the square planar I's adjacent can give a planar molecule.

10.70 Plan: Draw the Lewis structures, and then use VSEPR to describe propylene oxide.
Solution:
(a)

a) In propylene oxide, the C atoms are all AX_4. The C atoms do not have any unshared (lone) pairs. All the ideal bond angles for the C atoms in propylene oxide are 109.5°.
b) In propylene oxide, the C that is not part of the three-membered ring should have an ideal angle. The atoms in the ring form an equilateral triangle. The angles in an equilateral triangle are 60°. The angles around the 2 carbons in the rings are reduced from the ideal 109.5° to 60°.

10.72   <u>Plan:</u> Draw an acceptable Lewis structure using both the octet rule and the minimization of formal charges.
        <u>Solution:</u>
        The molecule has [2 x Cl(7 e⁻)] + [7 x O(6 e⁻)] = 56 valence electrons. Use 16 electrons to place single bonds between the O atoms and the Cl atoms. This leaves 56 – 16 = 40 e⁻ (20 pairs). Use these 20 pairs to complete the octets of the O and Cl atoms.

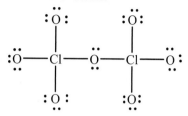

The formal charges now need to be calculated for each type of atom:
        FC = valence electrons – (lone electrons + 1/2 (bonded electrons))
        $FC_{Cl} = 7 – (0 + 1/2(8)) = +3$ $FC_O = 6 – (4 + 1/2(4)) = 0$ (bridging O)
        $FC_O = 6 – (6 + 1/2(2)) = –1$ (terminal O atoms)
The formal charges check (the total is 0), but the values for the Cl atoms are high. An attempt should be made to decrease the formal charges.
The formal charges in the Cl atoms may be reduced by the formation of double bonds. To reduce the formal charge from +3 to 0 would require three double bonds. This is allowed since Cl can exceed an octet. This gives:

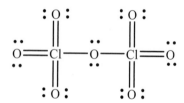

The formal charge needs to be recalculated:
        $FC_{Cl} = 7 – (0 + 1/2(14)) = 0$
        $FC_O = 6 – (4 + 1/2(4)) = 0$ (bridging O)
        $FC_O = 6 – (4 + 1/2(4)) = 0$ (terminal O atoms)
The formal charges check (the total is 0). All the values are reasonable (0).
The central bridging O is AX_2E_2, which has an ideal bond angle of 109.5°, but the lone pairs make this less than ideal.

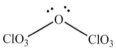

10.76   <u>Plan:</u> Assume all reactants and products are assumed to be gaseous. Ethanol burns (combusts) with O_2 to produce CO_2 and H_2O. Remember that energy is released (–) when bonds are formed.
        <u>Solution:</u>
        a) $CH_3CH_2OH(g) + 3 O_2(g) \rightarrow 2 CO_2(g) + 3 H_2O(g)$

Bonds broken:	1 C–C	1(347 kJ/mol)	Bonds formed:	2 x 2 C=O	4(– 799 kJ/mol)
	5 C–H	5(413 kJ/mol)		2 x 3 O–H	6(– 467 kJ/mol)
	1 C–O	1(358 kJ/mol)		Total:	–5998 kJ
	1 O–H	1(467 kJ/mol)			
	3 O=O	3(498 kJ/mol)			
	Total:	4731 kJ			

174

ΔH = Bonds broken + Bonds formed
 = 4731 kJ + (–5998 kJ) = **–1267 kJ** for each mole of ethanol burned.
b) If it takes 40.5 kJ/mol to vaporize the ethanol, part of the heat of combustion must be used to convert liquid ethanol to gaseous ethanol. The new value becomes:

$$\Delta H_{comb(liquid)} = -1267 \text{ kJ} + \left(1 \text{ mol}\right)\left[\frac{40.5 \text{ kJ}}{1 \text{ mol}}\right] = -1226.5 = \textbf{–1226 kJ per mol}\text{ of liquid ethanol burned}$$

c) $\Delta H_{comb(liquid)} = \sum n\Delta H_f^o$ (products) $- \sum m\Delta H_f^o$ (reactants)

$\Delta H_{comb(liquid)} = [2(\Delta H^o{}_f(CO_2(g))) + 3(\Delta H^o{}_f (H_2O(g)))] - [(\Delta H^o{}_f (C_2H_5OH(l))) + 3(\Delta H^o{}_f (O_2(g)))]$
$\Delta H_{comb(liquid)} = [2(-393.5 \text{ kJ}) + 3(-241.826 \text{ kJ})] - [(-277.63 \text{ kJ}) + 3(0 \text{ kJ})] = -1234.848 = \textbf{–1234.8 kJ}$
The two answers differ by less than 10 kJ. This is a very good agreement since average bond energies were used to calculate answers a and b.
d) $C_2H_4(g) + H_2O(g) \rightarrow CH_3CH_2OH(g)$
The Lewis structures for the reaction are:

Bonds broken:	1 x C=C	1(614 kJ/mol)	Bonds formed:	1 x C–C	1(–347 kJ/mol)
	4 x C–H	4(413 kJ/mol)		5 x C–H	5(– 413 kJ/mol)
	2 x O–H	2(467 kJ/mol)		1 x C–O	1(– 358 kJ/mol)
	Total:	3200 kJ		1 x O–H	1(– 467 kJ/mol)
				Total:	–5998 kJ

$\Delta H^o{}_{rxn} = \Delta H^o{}_{bonds\ broken} + \Delta H^o{}_{bonds\ formed} = 3200. \text{ kJ} + (-3237 \text{ kJ}) = \textbf{–37 kJ}$

10.79   Plan: Determine the empirical formula from the percent composition (assuming 100 grams of compound). Use the titration data to determine the mole ratio of acid to the NaOH. This ratio relates the number of acidic H's to the formula of the acid. Finally, combine this information to construct the Lewis structure.
Solution:

$$H \quad (2.24 \text{ g H})\left(\frac{1 \text{ mol}}{1.008 \text{ g H}}\right) = 2.222 \text{ mol H} \qquad \frac{2.222 \text{ mol}}{2.222 \text{ mol}} = 1.00$$

$$C \quad (26.7 \text{ g C})\left(\frac{1 \text{ mol}}{12.01 \text{ g C}}\right) = 2.223 \text{ mol C} \qquad \frac{2.223 \text{ mol}}{2.222 \text{ mol}} = 1.00$$

$$O \quad (71.1 \text{ g O})\left(\frac{1 \text{ mol}}{16.00 \text{ g O}}\right) = 4.444 \text{ mol O} \qquad \frac{4.444 \text{ mol}}{2.222 \text{ mol}} = 2.00$$

The empirical formula is HCO_2.
The titration required:

$$\left(\frac{0.040 \text{ mol NaOH}}{L}\right)\left(\frac{50.0 \text{ mL}}{}\right)\left(\frac{1 \text{ L}}{1000 \text{ mL}}\right)\left(\frac{1 \text{ mmol}}{0.001 \text{ mol}}\right) = 2.0 \text{ mmole NaOH}$$

Thus, the ratio is 2.0 mmole base / 1.0 mmole acid, or each acid molecule has two hydrogens to react (diprotic). The empirical formula indicates a monoprotic acid, so the formula must be doubled to: $H_2C_2O_4$.
$H_2C_2O_4$ has [2 x H(1 e⁻)] + [2 x C(4 e⁻)] + [4 x O(6 e⁻)] = 34 valence electrons to be used in the Lewis structure. 14 of these electrons are used to bond the atoms with single bonds, leaving 34 – 14 = 20 electrons or 10 pairs of electrons. When these 10 pairs of electrons are distributed to the atoms to complete octets, neither C atom has an octet; a lone pair from the oxygen without hydrogen is changed to a bonding pair on C.

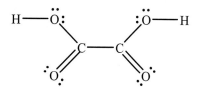

10.82   <u>Plan:</u> Write the balanced chemical equations for the reactions. Draw the Lewis structures. Calculate the heat of reaction from the bond energies, and divide the value by the number of moles of oxygen gas appearing in each reaction to get the heat of reaction per mole of oxygen.
<u>Solution:</u>

$$CH_4(g) + 2\ O_2(g) \rightarrow CO_2(g) + 2\ H_2O(g)$$

Bonds broken:	4 C-H	4(413 kJ/mol)	Bonds formed:	2 C=O	2(–799 kJ/mol)
	2 O=O	2(498 kJ/mol)		2 x 2 O-H	4(–467 kJ/mol)
	Total:	2648 kJ		Total:	–3466 kJ

$\Delta H = \Delta H°_{rxn} = \Delta H°_{bonds\ broken} + \Delta H°_{bonds\ formed}$
    $= 2648\ kJ + (–3466\ kJ) = –818\ kJ\ /\ 2\ mol\ O_2$
Per mole of O_2 = **–409 kJ/mol O_2**

$$2\ H_2S(g) + 3\ O_2(g) \rightarrow 2\ SO_2(g) + 2\ H_2O(g)$$

Bonds broken:	2 x 2 S-H	4(347 kJ/mol)	Bonds formed:	2 x 2 S=O	4(–552 kJ/mol)
	3 O=O	3(498 kJ/mol)		2 x 2 O-H	4(–467 kJ/mol)
	Total:	2882 kJ		Total:	–4076 kJ

$\Delta H°_{rxn} = \Delta H°_{bonds\ broken} + \Delta H°_{bonds\ formed}$
    $= 2882\ kJ + (–4076\ kJ) = –1194\ kJ\ /\ 3\ mol\ O_2$
Per mole of O_2 = **–398 kJ/mol O_2**

10.84   <u>Plan:</u> Draw the Lewis structure of the OH species. Find the additional bond energy information in Table 9.2.
<u>Solution:</u>
a) The OH molecule has [1 x O(6 e⁻)] + [1 x H(1 e⁻)] = 7 valence electrons to be used in the Lewis structure. 2 of these electrons are used to bond the atoms with a single bond, leaving 7 – 2 = 5 electrons. Those 5 electrons are given to oxygen. But no atom can have an octet, and one electron is left unpaired. The Lewis structure is:

· O——H

b) The formation reaction is: $1/2\ O_2(g) + 1/2\ H_2(g) \rightarrow OH(g)$. The heat of reaction is:
$\Delta H_{reaction} = \Delta H_{broken} + \Delta H_{formed}$ = 39.0 kJ
$[1/2(BE_{O=O}) + \frac{1}{2}(BE_{H–H})] + [BE_{O–H}]$ = 39.0 kJ
$[(1/2\ mol)\ (498\ kJ/mol) + (1/2\ mol)\ (432\ kJ/mol)] + [BE_{O–H}]$ = 39.0 kJ
        $BE_{O–H}$ = **–426 kJ or 426 kJ**
c) The average bond energy (from the bond energy table) is 467 kJ/mol. There are 2 O–H bonds in water for a total of 2 x 467 kJ/mol = 934 kJ. The answer to part b) accounts for 426 kJ of this, leaving:
934 kJ – 426 kJ = **508 kJ**

10.86   <u>Plan:</u> The basic Lewis structure will be the same for all species. The Cl atoms are larger than the F atoms. All of the molecules are of the type AX_5 and have trigonal bipyramid molecular shape. The equatorial positions are in the plane of the triangle and the axial positions above and below the plane of the triangle. In this molecular shape, there is more room in the equatorial positions.
   <u>Solution:</u>
   a) The F atoms will occupy the smaller axial positions first so that the larger Cl atoms can occupy the equatorial positions which are less crowded.
   b) The molecules containing only F or only Cl are nonpolar (have no dipole moment), as all the polar bonds would cancel. The molecules with one F or one Cl would be polar since there would be no corresponding bond to cancel the polarity. The presence of 2 axial F atoms means that their polarities will cancel (as would the 3 Cl atoms) giving a nonpolar molecule. The molecule with 3 F atoms is also polar.

| Polar | Non-polar | Polar | Polar | Non-polar |

10.88   <u>Plan:</u> Count the valence electrons and draw Lewis structures for the resonance forms.
   <u>Solution:</u>
   The $H_2C_2O_4$ molecule has [2 x H(1 e⁻)] + [2 x C(4 e⁻)] +[4 x O(6 e⁻)] = 34 valence electrons to be used in the Lewis structure. 14 of these electrons are used to bond the atoms with a single bond, leaving 34 –14= 20 electrons. If these 20 electrons are given to the oxygen atoms to complete their octet, the carbon atoms do not have octets. A lone pair from each of the oxygen atoms without hydrogen is changed to a bonding pair on C.
   The $HC_2O_4^-$ ion has [1 x H(1 e⁻)] + [2 x C(4 e⁻)] +[4 x O(6 e⁻)] +[1 e⁻ (from the charge)] = 34 valence electrons to be used in the Lewis structure. 12 of these electrons are used to bond the atoms with a single bond, leaving 34 – 12= 22 electrons. If these 22 electrons are given to the oxygen atoms to complete their octet, the carbon atoms do not have octets. A lone pair from two of the oxygen atoms without hydrogen is changed to a bonding pair on C. There are two resonance structures.
   The $C_2O_4^{2-}$ ion has [2 x C(4 e⁻)] +[4 x O(6 e⁻)] +[2 e⁻ (from the charge)] = 34 valence electrons to be used in the Lewis structure. 10 of these electrons are used to bond the atoms with a single bond, leaving 34 –10 = 24 electrons. If these 24 electrons are given to the oxygen atoms to complete their octets, the carbon atoms do not have octets. A lone pair from each two oxygen atoms is changed to a bonding pair on C. There are four resonance structures.

$H_2C_2O_4$:

$HC_2O_4^-$:

$C_2O_4^{2-}$:

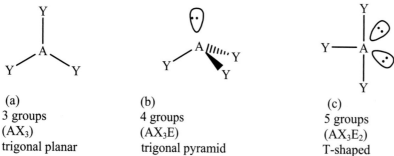

In $H_2C_2O_4$, there are 2 shorter C=O bonds and 2 longer C—O bonds.
In $HC_2O_4^-$, the C-O bonds on the side retaining the H remain as 1 long C-O and 1 shorter C=O. The C-O bonds on the other side of the molecule have resonance forms with an average bond order of 1.5, so they are intermediate in length and strength.
In $C_2O_4^{2-}$, all the carbon to oxygen bonds are resonating and have an average bond order of 1.5.

10.92   Plan: Draw the Lewis structures. Calculate the heat of reaction using the bond energies in Table 9.2.
Solution:

$$SO_3(g) + H_2SO_4(l) \rightarrow H_2S_2O_7(l)$$

Bonds broken:	5 S=O	5(552 kJ/mol)	Bonds formed:	4 S=O	4(−552 kJ/mol)
	2 S–O	2(265 kJ/mol)		4 S–O	4(−265 kJ/mol)
	2 O–H	2(467 kJ/mol)		2 O–H	2(−467 kJ/mol)
	Totals:	4224 kJ			−4202 kJ

$\Delta H^\circ_{rxn} = \Delta H^\circ_{bonds\ broken} + \Delta H^\circ_{bonds\ formed}$
$= 4224\ kJ + (-\ 4202\ kJ) = \textbf{22 kJ}$

10.93   Plan: Pick the VSEPR structures for AY_3 substances. Then determine which are polar.
Solution:
The molecular shapes that have a central atom bonded to three other atoms are trigonal planar, trigonal pyramid and T-shaped:

(a)
3 groups
(AX_3)
trigonal planar

(b)
4 groups
(AX_3E)
trigonal pyramid

(c)
5 groups
(AX_3E_2)
T-shaped

178

The trigonal planar molecules, such as (a), are nonpolar, so it cannot be AY_3. Trigonal pyramidal molecules (b) and T-shaped molecules (c) are polar, so either could represent AY_3.

10.98   <u>Plan:</u> Draw the Lewis structure of each compound. Atoms 180° apart are separated by the sum of the bond's length. Atoms not at 180° apart must have their distances determined by geometrical relationships.
<u>Solution:</u>

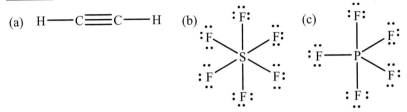

a) C_2H_2 has [2 x C(4 e^-)] + [2 x H(1 e^-)] = 10 valence electrons to be used in the Lewis structure. 6 of these electrons are used to bond the atoms with a single bond, leaving 10 – 6= 4 electrons. Giving one carbon atom the 4 electrons to complete its octet results in the other carbon atom not having an octet. The two lone pairs from the carbon with an octet are changed to two bonding pairs for a triple bond between the two carbon atoms. The molecular shape is linear. The H atoms are separated by two carbon-hydrogen bonds (109 pm) and a carbon-carbon triple bond (121 pm).
Total separation = 2 (109 pm) + 121 pm = **339 pm**
b) SF_6 has [1 x S(6 e^-)] + [6 x F(7 e^-)] = 48 valence electrons to be used in the Lewis structure. 12 of these electrons are used to bond the atoms with a single bond, leaving 48 – 12= 36 electrons. These 36 electrons are given to the fluorine atoms to complete their octets. The molecular shape is octahedral. The fluorine atoms on opposite sides of the S are separated by twice the sulfur-fluorine bond length (158 pm).
Total separation = 2 (158 pm) = **316 pm**
Adjacent fluorines are at two corners of a right triangle, with the sulfur at the 90° angle. Two sides of the triangle are equal to the sulfur-fluorine bond length (158 pm). The separation of the fluorine atoms is at a distance equal to the hypotenuse of this triangle. This length of the hypotenuse may be found using the Pythagorean Theorem (a^2 + $b^2 = c^2$). In this case a = b = 158 pm. Thus, $c^2 = (158\ pm)^2 + (158\ pm)^2$, and so c = 223.4457 = **223 pm**
c) PF_5 has [1 x P(5 e^-)] + [5 x F(7 e^-)] = 40 valence electrons to be used in the Lewis structure. 10 of these electrons are used to bond the atoms with a single bond, leaving 40 – 10= 30 electrons. These 30 electrons are given to the fluorine atoms to complete their octets. The molecular shape is trigonal bipyramid. Adjacent equatorial fluorine atoms are at two corners of a triangle with an F-P-F bond angle of 120°. The length of the P-F bond is 156 pm. If the 120° bond angle is A, then the F-F bond distance is a and the P-F bond distances are b and c. The F-F bond distance can be found using the Law of Cosines: $a^2 = b^2 + c^2 – 2bc$ (cos A).
$a^2 = (156)^2 + (156)^2 – 2(156)(156)\cos 120°$.   a = 270.1999 = **271 pm**.

CHAPTER 11 THEORIES OF COVALENT BONDING

FOLLOW–UP PROBLEMS

11.1   <u>Plan:</u> Draw a Lewis structure. Determine the number and arrangement of the electron pairs about the central atom. This process leads to the hybridization.
<u>Solution:</u>
a) In BeF_2, Be is surrounded by two electron groups (two single bonds) so hybridization around Be is *sp*.

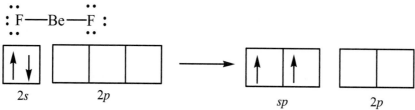

Isolated Be atom   Hybridized Be atom

b) In $SiCl_4$, Si is surrounded by four electron groups (four single bonds) so its hybridization is sp^3.

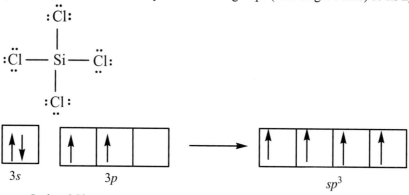

Isolated Si atom   Hybridized Si atom

c) In XeF_4, xenon is surrounded by 6 electron groups (4 bonds and 2 lone pairs) so the hybridization is sp^3d^2.

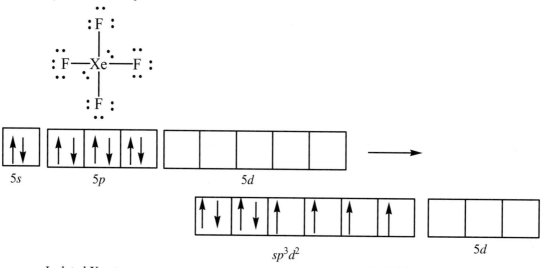

Isolated Xe atom   Hybridized Xe atom

11.2   Plan: First, determine the Lewis structure for the molecule. Then count the number of electron groups around each atom. Hybridization is *sp* if there are two groups, *sp²* if there are three groups and *sp³* if there are four groups. No hybridization occurs with only one group of electrons. The bonds are then designated as sigma or pi. A single bond is a sigma bond. A double bond consists of one sigma and one pi bond. A triple bond includes one sigma bond and two pi bonds. Hybridized orbitals overlap head on to form sigma bonds whereas pi bonds form through the sideways overlap of *p* or *d* orbitals.

Solution:

a) Hydrogen cyanide has H–C≡N: as its Lewis structure. The single bond between carbon and hydrogen is a sigma bond formed by the overlap of a hybridized *sp* orbital on carbon with the $1s$ orbital from hydrogen. Between carbon and nitrogen are three bonds. One is a sigma bond formed by the overlap of a hybridized *sp* orbital on carbon with a hybridized *sp* orbital on nitrogen. The other two bonds between carbon and nitrogen are pi bonds formed by the overlap of *p* orbitals from carbon and nitrogen. One *sp* orbital on nitrogen is filled with a lone pair of electrons.

b) The Lewis structure of carbon dioxide, CO_2, is

$$\ddot{O} = C = \ddot{O}$$

Both oxygen atoms are *sp²* hybridized (the O atoms have three electron groups – one double bond and two lone pairs) and form a sigma bond and a pi bond with carbon. The sigma bonds are formed by the overlap of a hybridized *sp* orbital on carbon with a hybridized *sp²* orbital on oxygen. The pi bonds are formed by the overlap of an oxygen *p* orbital with a carbon *p* orbital. Two *sp²* orbitals on each oxygen are filled with a lone pair of electrons.

11.3   Plan: Draw the molecular orbital diagram. Determine the bond order from calculation: BO = 1/2(# e^- in bonding orbitals – # e^- in antibonding orbitals). A bond order of zero indicates the molecule will not exist. A bond order greater than zero indicates that the molecule is at least somewhat stable and is likely to exist. H_2^{2-} has 4 electrons (1 from each hydrogen and two from the –2 charge).

Solution:

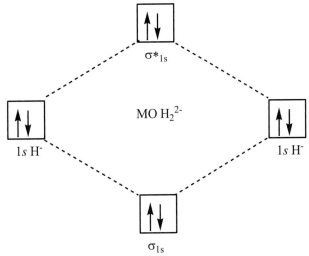

Configuration for H_2^{2-} is $(\sigma_{1s})^2(\sigma^*_{1s})^2$. Bond order of H_2^{2-} is 1/2(2 – 2) = 0. Thus, it is not likely that two H^- ions would combine to form the ion H_2^{2-}.

11.4 <u>Plan:</u> To find bond order it is necessary to determine the molecular orbital electron configuration from the total number of electrons. Bond order is calculated from the configuration as $1/2(\# e^-$ in bonding orbitals $- \#e^-$ in antibonding orbitals).
<u>Solution:</u>
F_2^{2-}: total electrons $= 9 + 9 + 2 = 20$

Configuration: $(\sigma_{1s})^2(\sigma^*_{1s})^2(\sigma_{2s})^2(\sigma^*_{2s})^2(\sigma_{2p})^2(\pi_{2p})^4(\pi^*_{2p})^4(\sigma^*_{2p})^2$

Bond order $= 1/2(10 - 10) = 0$

F_2^-: total electrons $= 9 + 9 + 1 = 19$

Configuration: $(\sigma_{1s})^2(\sigma^*_{1s})^2(\sigma_{2s})^2(\sigma^*_{2s})^2(\sigma_{2p})^2(\pi_{2p})^4(\pi^*_{2p})^4(\sigma^*_{2p})^1$

Bond order $= 1/2(10 - 9) = 0.5$

F_2: total electrons $= 9 + 9 = 18$

Configuration: $(\sigma_{1s})^2(\sigma^*_{1s})^2(\sigma_{2s})^2(\sigma^*_{2s})^2(\sigma_{2p})^2(\pi_{2p})^4(\pi^*_{2p})^4$

Bond order $= 1/2(10 - 8) = 1$

F_2^+: total electrons $= 9 + 9 - 1 = 17$

Configuration: $(\sigma_{1s})^2(\sigma^*_{1s})^2(\sigma_{2s})^2(\sigma^*_{2s})^2(\sigma_{2p})^2(\pi_{2p})^4(\pi^*_{2p})^3$

Bond order $= 1/2(10 - 7) = 1.5$

F_2^{2+}: total electrons $= 9 + 9 - 2 = 16$

Configuration: $(\sigma_{1s})^2(\sigma^*_{1s})^2(\sigma_{2s})^2(\sigma^*_{2s})^2(\sigma_{2p})^2(\pi_{2p})^4(\pi^*_{2p})^2$

Bond order $= 1/2(10 - 6) = 2$

Bond energy increases as bond order increases: $F_2^{2-} < F_2^- < F_2 < F_2^+ < F_2^{2+}$
Bond length decreases as bond energy increases, so the order of increasing bond length will be opposite that of increasing bond energy.
Increasing bond length: $F_2^{2+} < F_2^+ < F_2 < F_2^-$.
F_2^{2-} will not form a bond, so it has no bond length and is not included in the list.
<u>Check:</u> Since the highest energy orbitals are antibonding orbitals, it makes sense that the bond order increases as electrons are removed from the antibonding orbitals.

END–OF–CHAPTER PROBLEMS

11.1 Table 11.1 describes the types of shapes that form from a given set of hybrid orbitals.
a) trigonal planar: three electron groups - three hybrid orbitals: *sp²*
b) octahedral: six electron groups - six hybrid orbitals: *sp³d²*
c) linear: two electron groups - two hybrid orbitals: *sp*
d) tetrahedral: four electron groups - four hybrid orbitals: *sp³*
e) trigonal bipyramidal: five electron groups - five hybrid orbitals: *sp³d*

11.3 Carbon and silicon have the same number of valence electrons, but the outer level of electrons is $n = 2$ for carbon and $n = 3$ for silicon. Thus, silicon has $3d$ orbitals in addition to $3s$ and $3p$ orbitals available for bonding in its outer level, to form up to 6 hybrid orbitals, whereas carbon has only $2s$ and $2p$ orbitals available in its outer level to form up to 4 hybrid orbitals.

11.5 The *number* of orbitals remains the same as the number of orbitals before hybridization. The *type* depends on the orbitals mixed.
a) There are six unhybridized orbitals, and therefore **six** hybrid orbitals result. The type is *sp³d²*.
b) **Four *sp³*** hybrid orbitals form from three p and one s atomic orbitals.

11.7 To determine hybridization, draw the Lewis structure and count the number of electron groups. Hybridize that number of orbitals. Single, double and triple bonds all count as one electron group. An unshared pair of electrons or one unshared electron also counts as one electron group.
a) The three electron groups (one double bond, one lone pair and one unpaired electron) around nitrogen require three hybrid orbitals. The hybridization is *sp²*.

b)The nitrogen has three electron groups (one single bond, one double bond and one unpaired electron) , requiring three hybrid orbitals so hybridization is sp^2.

c) The nitrogen has three electron groups (one single bond, one double bond and one lone pair so hybridization is sp^2.

11.9   To determine hybridization, draw the Lewis structure and count the number of electron groups. Hybridize that number of orbitals. Single, double and triple bonds all count as one electron group. An unshared pair of electrons or one unshared electron also counts as one electron group.
a) sp^3 The Cl has four electron groups (1 lone pair, 1 lone electron, and 2 double bonds) and therefore four hybrid orbitals are required; the hybridization is sp^3. Note that in ClO_2, the π bond is formed by the overlap of d-orbitals from chlorine with p-orbitals from oxygen.

b) sp^3 The Cl has four electron groups (1 lone pair and 3 bonds) and therefore four hybrid orbitals are required; the hybridization is sp^3.

c) sp^3 The Cl has four electron groups (4 bonds) and therefore four hybrid orbitals are required; the hybridization is sp^3.

11.11   a) Silicon has four electron groups (four bonds) requiring four hybrid orbitals; four sp^3 hybrid orbitals are made from **one s and three p atomic orbitals**.

b) Carbon has two electron groups (two double bonds) requiring two hybrid orbitals; two sp hybrid orbitals are made from **one s and one p orbital**.

c) Sulfur is surrounded by 5 electron groups (4 bonding pairs, 1 lone pair) requiring five hybrid orbitals; five sp^3d hybrid orbitals are formed from **one s orbital, three p orbitals, and one d orbital**.

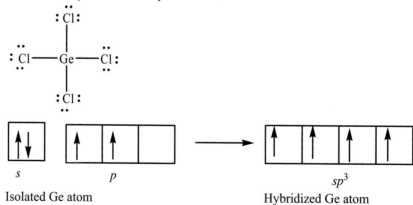

d) Nitrogen is surrounded by 4 electron groups (3 bonding pairs, 1 lone pair) requiring four hybrid orbitals; four sp^3 hybrid orbitals are formed from **one s orbital and three p orbitals**.

11.13   a) The P in PH_3 has four electron groups (1 lone pair and 3 bonds) and therefore four hybrid orbitals are required; the hybridization is sp^3. The P in the product also has four electron groups (4 bonds) and again four hybrid orbitals are required. The hybridization of P remains sp^3. There is no change in hybridization. Illustration **B** best shows the hybridization of P during the reaction as $sp^3 \rightarrow sp^3$.

b) The B in BH_3 has three electron groups (3 bonds) and therefore three hybrid orbitals are required; the hybridization is sp^2. The B in the product has four electron groups (4 bonds) and four hybrid orbitals are required. The hybridization of B is now sp^3. The hybridization of B changes from **sp^2 to sp^3**; this is best shown by illustration **A**.

11.15   a) Germanium is the central atom in $GeCl_4$. Ge has four electron groups (four bonds), requiring four hybrid orbitals. Hybridization is sp^3 around Ge.

b) Boron is the central atom in BCl_3. B has three electron groups (three bonds), requiring three hybrid orbitals. Hybridization is sp^2 around B.

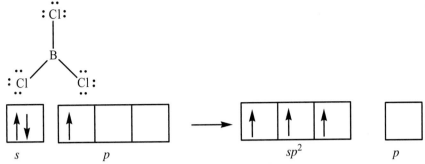

Isolated B atom   Hybridized B atom

c) Carbon is the central atom in CH_3^+. C has three electron groups (three bonds), requiring three hybrid orbitals. Hybridization is sp^2 around C.

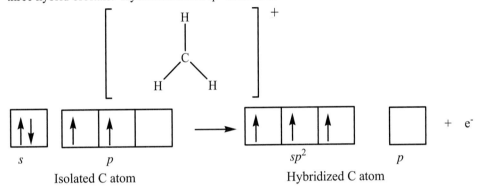

Isolated C atom   Hybridized C atom

11.17   a) In $SeCl_2$, Se is the central atom and has four electron groups (two single bonds and two lone pairs), requiring four hybrid orbitals. Se is sp^3 hybridized. Two sp^3 orbitals are filled with lone electron pairs and two sp^3 orbitals bond with the chlorine atoms.

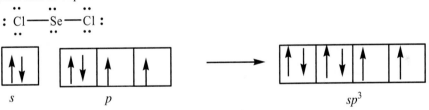

b) In H_3O^+, O is the central atom and has four electron groups (three single bonds and one lone pair, requiring four hybrid orbitals. O is sp^3 hybridized. One sp^3 orbital is filled with a lone electron pair and three sp^3 orbitals bond with the hydrogen atoms.

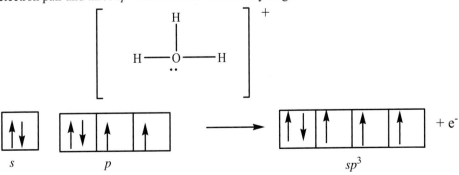

185

c) I is the central atom in IF$_4^-$ with 6 electron groups (four single bonds and two lone pairs) surrounding it. Six hybrid orbitals are required and I has sp^3d^2 hybrid orbitals. The sp^3d^2 hybrid orbitals are composed of one s orbital, three p-orbitals, and two d-orbitals. Two sp^3d^2 orbitals are filled with a lone pair and four sp^3d^2 orbitals bond with the fluorine atoms.

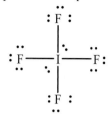

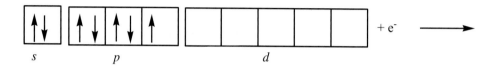

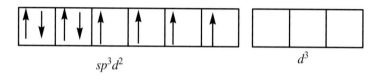

11.20   a) **False**, a double bond is one sigma (σ) and one pi (π) bond.
b) **False**, a triple bond consists of one sigma (σ) and two pi (π) bonds.
c) **True**
d) **True**
e) **False**, a π bond consists of one pair of electrons; it occurs after a σ bond has been previously formed.
f) **False**, end-to-end overlap results in a bond with electron density along the bond axis.

11.21   a) Nitrogen is the central atom in NO$_3^-$. Nitrogen has 3 surrounding electron groups (two single bonds and one double bond), so it is sp^2 hybridized. Nitrogen forms **three σ bonds** (one each for the N–O bonds) and **one π bond** (part of the N=O double bond).

b) Carbon is the central atom in CS$_2$. Carbon has 2 surrounding electron groups (two double bonds), so it is sp hybridized. Carbon forms **two σ bonds** (one each for the C–S bonds) and **two π bonds** (part of the two C=S double bonds).

$$\ddot{S}\!=\!\!=\!\!C\!=\!\!=\!\ddot{S}$$

c) Carbon is the central atom in CH$_2$O. Carbon has 3 surrounding electron groups (two single bonds and one double bond), so it is sp^2 hybridized. Carbon forms **three σ bonds** (one each for the two C–H bonds and one C–O bond) and **one π bond** (part of the C=O double bond).

11.23   a) Examine the Lewis structure. Three electron groups (one lone pair, one single bond and one double bond) surround the central N atom. Hybridization is **sp²** around nitrogen. One sigma bond exists between F and N, and one sigma and one pi bond exist between N and O. Nitrogen participates in a total of **2 σ and 1 π bonds**.

$$:\!\overset{\displaystyle\cdot\cdot}{\underset{\displaystyle\cdot\cdot}{F}}\!\!-\!\!\overset{\displaystyle\cdot\cdot}{N}\!\!=\!\!\overset{\displaystyle\cdot\cdot}{\underset{\displaystyle\cdot\cdot}{O}}$$

b) Examine the Lewis structure. Each carbon has three electron groups (two single bonds and one double bond) with **sp²** hybridization. The bonds between C and F are sigma bonds. The C–C bond consists of one sigma and one pi bond. Each carbon participates in a total of **3 σ and 1 π bonds**.

c) Examine the Lewis structure. Each carbon has two electron groups (1 single bond and one triple bond) and is **sp** hybridized with a sigma bond between the two carbons and a sigma and two pi bonds comprising each C–N triple bond. Each carbon participates in a total of **2 σ and 2 π bonds**.

$$:\!N\!\!\equiv\!\!C\!\!-\!\!C\!\!\equiv\!\!N\!:$$

11.25   The double bond in 2-butene restricts rotation of the molecule, so that *cis* and *trans* structures result. The two structures are shown below:

The carbons participating in the double bond each have three surrounding groups, so they are *sp²* hybridized. The =C-H σ bonds result from the head-on overlap of a C *sp²* orbital and a H *s* orbital. The C-CH₃ bonds are also σ bonds, resulting from the head-on overlap of an *sp²* orbital and an *sp³* orbital. The C=C bond contains 1 σ bond (head on overlap of two *sp²* orbitals) and 1 π bond (sideways overlap of unhybridized *p* orbitals). Finally, C-H bonds in the methyl (-CH₃) groups are σ bonds resulting from the overlap of C's *sp³* orbital with H's *s* orbital.

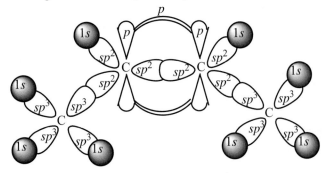

11.26   Four molecular orbitals form from the four *p* atomic orbitals. In forming molecular orbitals, the total number of molecular orbitals must equal the number of atomic orbitals. Two of the four molecular orbitals formed are bonding orbitals and two are antibonding.

11.28   a) Bonding MO's have lower energy than antibonding MO's. The bonding MO's lower energy, even lower than its constituent atomic orbitals, accounts for the stability of a molecule in relation to its individual atoms. However, the sum of energy of the MO's must equal the sum of energy of the AO's.
b) The node is the region of an orbital where the probability of finding the electron is zero, so the nodal plane is the plane that bisects the node perpendicular to the bond axis. There is no node along the bond axis (probability is positive between the two nuclei) for the bonding MO. The antibonding MO does have a nodal plane.
c) The bonding MO has higher electron density between nuclei than the antibonding MO.

11.30   a) **Two** electrons are required to fill a σ-bonding molecular orbital. Each molecular orbital requires two electrons.
b) **Two** electrons are required to fill a π-antibonding molecular orbital. There are two π-antibonding orbitals, each holding a maximum of two electrons.
c) **Four** electrons are required to fill the two σ molecular orbitals (two electrons to fill the σ-bonding and two to fill the σ-antibonding) formed from two $1s$ atomic orbitals.

11.32   a) A is the π^*_{2p} molecular orbital; B is the σ_{2p} molecular orbital; C is the π_{2p} molecular orbital; D is the σ^*_{2p} molecular orbital.
b) The π^*_{2p} molecular orbital, A, σ_{2p} molecular orbital, B, and π_{2p} molecular orbitals, C, are all occupied by at least one electron. The σ^*_{2p} molecular orbital is unoccupied.
c) A π^*_{2p} molecular orbital, A, has only one electron.

11.34   a) The molecular orbital configuration for Be_2^+ with a total of 7 electrons is $(\sigma_{1s})^2(\sigma^*_{1s})^2(\sigma_{2s})^2(\sigma^*_{2s})^1$ and bond order = $1/2(4-3) = 1/2$. With a bond order of 1/2 the Be_2^+ ion will be **stable**. (Bond order = ½(number of electrons in bonding MO – number of electrons in antibonding MO))
b) No, the ion has one unpaired electron in the σ^*_{2s} MO, so it is **paramagnetic**, not diamagnetic.
c) Valence electrons would be those in the molecular orbitals at the $n = 2$ level, so the valence electron configuration is $(\sigma_{2s})^2((\sigma^*_{2s})^1$.

11.36   The sequence of MO's for C_2 is shown in Figure 11.21 (the $(\sigma_{1s})^2$ and $(\sigma_{1s}{}^*)^2$ MO's are not shown for convenience). For each species, determine the total number of electrons, the valence molecular orbital electron configuration and bond order. Bond order = ½(number of electrons in bonding MO – number of electrons in antibonding MO)
C_2^- Total valence electrons = 4 + 4 + 1 = 9
        Valence configuration: $(\sigma_{2s})^2(\sigma^*_{2s}{}^2(\pi_{2p})^4(\sigma_{2p})^1$
        Bond order = $1/2(7-2) = 2.5$
C_2 Total valence electrons = 4 + 4 = 8
        Valence configuration: $(\sigma_{2s})^2(\sigma^*_{2s})^2(\pi_{2p})^4$
        Bond order = $1/2(6-2) = 2$
C_2^+ Total valence electrons = 4 + 4 – 1 = 7
        Valence configuration: $(\sigma_{2s})^2(\sigma^*_{2s})^2(\pi_{2p})^3$
        Bond order = $1/2(5-2) = 1.5$
a) Bond energy increases as bond order increases: $C_2^+ < C_2 < C_2^-$
b) Bond length decreases as bond energy increases, so the order of increasing bond length will be opposite that of increasing bond energy. Increasing bond length: $C_2^- < C_2 < C_2^+$

11.40   a) Each of the six C atoms in the ring has three electron groups (two single bonds and a double bond) and has sp^2 hybridization; all of the other C atoms have four electron groups (four single bonds) and have sp^3 hybridization; all of the O atoms have four electron groups (two single bonds and two lone pairs and have sp^3 hybridization; the N atom has four electron groups (three single bonds and a lone pair) and has sp^3 hybridization.
b) Each of the single bonds is a sigma bond; each of the double bonds has one sigma bond for a total of **26 sigma bonds**.

c) The ring has three double bonds each of which is composed of one sigma bond and one pi bond; so there are three pi bonds each with 2 electrons for a total of **6 pi electrons.**

11.42   a) Every single bond is a sigma bond. There is one sigma bond in each double bond as well. There are **17** σ bonds in isoniazid. Every atom-to-atom connection contains a σ bond.
b) All carbons have three surrounding electron groups (two single and one double bond), so their hybridization is sp^2. The ring N also has three surrounding electron groups (one single bond, one double bond and one lone pair), so its hybridization is also sp^2. The other two N's have four surrounding electron groups (three single bonds and one lone pair) and are sp^3 hybridized.

11.44   a) B changes from $sp^2 \rightarrow sp^3$. Boron in BF_3 has three electron groups with sp^2 hybridization. In BF_4^-, 4 electron groups surround B with sp^3 hybridization.

b) P changes from $sp^3 \rightarrow sp^3d$. Phosphorus in PCl_3 is surrounded by 4 electron groups (3 bonds to Cl and 1 lone pair) for sp^3 hybridization. In PCl_5, phosphorus is surrounded by 5 electron groups for sp^3d hybridization.

c) C changes from $sp \rightarrow sp^2$. Two electron groups surround C in C_2H_2 and 3 electron groups surround C in C_2H_4.

d) Si changes from $sp^3 \rightarrow sp^3d^2$. Four electron groups surround Si in SiF_4 and 6 electron groups surround Si in SiF_6^{2-}.

e) **No change**, S in SO_2 is surrounded by three electron groups (one single bond, one double bond and one lone pair) and in SO_3 is surrounded by 3 electron groups (two single bonds and one double bond); both have sp^2 hybridization.

11.46   Count the electron groups, both bonds and lone pairs, about each atom. (Do not forget to count double bonds only once.)

P (3 single bonds and one double bond)	AX_4	**tetrahedral**	sp^3
N (3 single bonds and one lone pair)	AX_3E	**trigonal pyramid**	sp^3
C_1 and C_2 (4 single bonds)	AX_4	**tetrahedral**	sp^3
C_3 (2 single bonds and one double bond)	AX_3	**trigonal planar**	sp^2

11.50

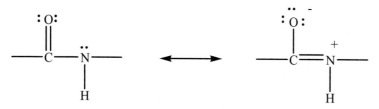

The central C is *sp* hybridized, and the other two C's are sp^2.

11.54   a) **B and D** show hybrid orbitals that are present in the molecule. B shows sp^3 hybrid orbitals, used by atoms that have four groups of electrons. In the molecule, the C atom in the CH_3 group, the S atom and the O atom all have four groups of electrons and would have sp^3 hybrid orbitals. D shows sp^2 hybrid orbitals, used by atoms that have three groups of electrons. In the molecule, the C bonded to the nitrogen atom, the C atoms involved in the C=C bond, and the nitrogen atom all have three groups of electrons and would have sp^2 hybrid orbitals.
b) The C atoms in the C≡C bond have only two electron groups and would have *sp* **hybrid orbitals**. These orbitals are not shown in the picture.
c) There are **two sets of** *sp* hybrid orbitals, **four sets of** sp^2 hybrid orbitals and **three sets of** sp^3 hybrid orbitals in the molecule.

11.56   The resonance gives the C-N bond some double bond character, which hinders rotation about the C-N bond. The C-N single bond is a σ bond; the resonance interaction exchanges a C-O π bond for a C-N π bond.

11.58   a) The 6 carbons in the ring each have three surrounding electron groups (two single bonds and one double bond) with sp^2 hybrid orbitals. The two carbons participating in the C=O bond are also sp^2 hybridized. The single carbon in the $-CH_3$ group has 4 electron groups (four single bonds) and is sp^3 hybridized. There are only 2 central oxygen atoms, one in a C–O–H configuration and the other in a C–O–C configuration. Both of these atoms have 4 surrounding electron groups (two single bonds and two lone pairs) and are sp^3 hybridized.
Summary: C in $-CH_3$: sp^3, all other C atoms (8 total): sp^2, O (2 total): sp^3
b) The **two** C=O bonds are localized; the double bonds on the ring are delocalized as in benzene.
c) Each carbon with three surrounding groups has sp^2 hybridization and trigonal planar shape; therefore, **8** carbons have this shape. Only **1** carbon in the CH_3 group has four surrounding groups with sp^3 hybridization and tetrahedral shape.

11.59   a) **Four** different isomeric fatty acids: trans-cis, cis-cis, cis-trans, trans-trans.
b) With three double bonds, there are $2^n = 2^3 = $ **8 isomers** possible.

cis-cis-cis	trans-trans-trans
cis-trans-cis	trans-cis-trans
cis-cis-trans	trans-cis-cis
cis-trans-trans	trans-trans-cis

CHAPTER 12 INTERMOLECULAR FORCES: LIQUIDS, SOLIDS, AND PHASE CHANGES

FOLLOW–UP PROBLEMS

12.1 Plan: This is a three step process: warming the ice to 0.0°C, melting the ice and warming the liquid water to 16.0°C. Use the molar heat capacities of ice and water and the relationship q = CmΔT to calculate the heat use for the warming of the ice and of the water; use the heat of fusion and the relationship q = nΔH_fus to calculate the heat required to melt the ice.
Solution:
The total heat required is the sum of three processes:
1) Warming the ice from –7.00°C to 0.00°C
$$q_1 = CmΔT = (37.6 \text{ J/mol°C}) (2.25 \text{ mol}) [0.0 – (–7.00)]°C = 592.2 \text{ J} = 0.5922 \text{ kJ}$$
2) Phase change of ice at 0.00°C to water at 0.00°C
$$q_2 = nΔH_{fus} = (2.25 \text{ mol})\left(\frac{6.02 \text{ kJ}}{\text{mol}}\right) = 13.545 \text{ kJ}$$
3) Warming the liquid from 0.00°C to 16.0°C
$$q_3 = CmΔT = (75.4 \text{ J/mol°C}) (2.25 \text{ mol}) [16.0 – (0.0)]°C = 2714.4 \text{ J} = 2.7144 \text{ kJ}$$
The three heats are positive because each process takes heat from the surroundings (endothermic). The phase change requires much more energy than the two temperature change processes. The total heat is $q_1 + q_2 + q_3$ = (0.5922 kJ + 13.545 kJ + 2.7144 kJ) = 16.8516 = **16.8 kJ**.

12.2 Plan: The variables $ΔH_{vap}$, P_1, T_1, and T_2 are given, so substitute them into the Clausius-Clapeyron equation and solve for P_2. Convert both temperatures from °C to K. Convert $ΔH_{vap}$ to J so that the units cancel with R.
Solution:
$T_1 = 34.1 + 273.15 = 307.2 \text{ K}$ $T_2 = 85.5 + 273.15 = 358.6 \text{ K}$ $P_1 = 40.1 \text{ torr}$

$$\ln\frac{P_2}{P_1} = \frac{-ΔH_{vap}}{R}\left(\frac{1}{T_2} - \frac{1}{T_1}\right)$$

$$\ln\frac{P_2}{40.1 \text{ torr}} = \frac{-40.7\frac{\text{kJ}}{\text{mol}}}{8.314 \text{ J/mol}\cdot\text{K}}\left(\frac{1}{358.6 \text{ K}} - \frac{1}{307.2 \text{ K}}\right)\left(\frac{10^3 \text{ J}}{1 \text{ kJ}}\right) = 2.28410 \text{ (unrounded)}$$

$$\frac{P_2}{40.1 \text{ torr}} = 9.81689 \text{ (unrounded)}$$

$P_2 = (9.81689) (40.1 \text{ torr}) = 393.657 = \textbf{394 torr}$
Note: Watch significant figures when subtracting numbers:
1/358.6 = 0.002789 and 1/307.2 = 0.003255
0.002789 – 0.003255 = 0.000466
In the subtraction, the number of significant figures decreased from 4 to 3.
Check: The temperature increased, so the vapor pressure should be higher.

12.3   a) The –A:····H–B– sequence is present in both of the following structures:

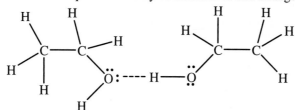

b) The –A:····H–B– sequence can only be achieved in one arrangement.

The hydrogens attached to the carbons cannot form H bonds because they do not satisfy the –A:····H–B–
sequence.
c) Hydrogen bonding is not possible because there are no O–H, N–H, or F–H bonds.

12.4   a) Both CH_3Br and CH_3F are polar molecules that experience dipole-dipole and dispersion intermolecular
interactions. Because CH_3Br (M = 94.9 g/mol) is ~3 times larger than CH_3F (M = 34.0 g/mol), dispersion forces
result in a higher boiling point (BP) for **CH_3Br**.
b) $CH_3CH_2CH_2OH$, n–propanol, forms hydrogen bonds and has dispersion forces whereas $CH_3CH_2OCH_3$, ethyl
methyl ether, has only dipole-dipole and dispersion attractions. **$CH_3CH_2CH_2OH$** has the higher BP.
c) Both C_2H_6 and C_3H_8 are nonpolar and experience dispersion forces only. **C_3H_8**, with the higher molar mass,
experiences greater dispersion forces and has the higher boiling point.

12.5   Plan: Use the relationship between radius and edge length for a body-centered cubic structure shown in Figure
12.29.
Solution:

$$a = \frac{4r}{\sqrt{3}} = \frac{4(126 \text{ pm})}{\sqrt{3}} = 290.98454 = \textbf{291 pm}$$

END–OF–CHAPTER PROBLEMS

12.1   The energy of attraction is a *potential* energy and denoted E_p. The energy of motion is *kinetic* energy and denoted
E_k. The relative strength of E_p vs. E_k determines the phase of the substance. In the gas phase, $E_p \ll E_k$ because the
gas particles experience little attraction for one another and the particles are moving very fast. In the solid phase,
$E_p \gg E_k$ because the particles are very close together and are only vibrating in place.

Two properties that differ between a gas and a solid are the volume and density. The volume of a gas expands to
fill the container it is in while the volume of a solid is constant no matter what container holds the solid. Density
of a gas is much less than the density of a solid. The density of a gas also varies significantly with temperature and
pressure changes. The density of a solid is only slightly altered by changes in temperature and pressure.
Compressibility and ability to flow are other properties that differ between gases and solids.

12.4   a) Heat of fusion refers to the change between the solid and the liquid states and heat of vaporization refers to the change between liquid and gas states. In the change from solid to liquid, the kinetic energy of the molecules must increase only enough to partially offset the intermolecular attractions between molecules. In the change from liquid to gas, the kinetic energy of the molecules must increase enough to overcome the intermolecular forces. The energy to overcome the intermolecular forces for the molecules to move freely in the gaseous state is much greater than the amount of energy needed to allow the molecules to move more easily past each other but still stay very close together.
b) The net force holding molecules together in the solid state is greater than that in the liquid state. Thus, to change solid molecules to gaseous molecules in sublimation requires more energy than to change liquid molecules to gaseous molecules in vaporization.
c) At a given temperature and pressure, the magnitude of ΔH_{vap} is the same as the magnitude of ΔH_{cond}. The only difference is in the sign: $\Delta H_{vap} = -\Delta H_{cond}$.

12.5   Intermolecular forces are the forces that exist between molecules that attract the molecules to each other; intramolecular forces exist within a molecule and are the forces holding the atoms together in the molecule.
a) **Intermolecular** — Oil evaporates when individual oil molecules can escape the attraction of other oil molecules in the liquid phase.
b) **Intermolecular** — The process of butter (fat) melting involves a breakdown in the rigid, solid structure of fat molecules to an amorphous, less ordered system. The attractions between the fat molecules are weakened, but the bonds within the fat molecules are not broken.
c) **Intramolecular** — A process called oxidation tarnishes pure silver. Oxidation is a chemical change and involves the breaking of bonds and formation of new bonds.
d) **Intramolecular** — The decomposition of O_2 molecules into O atoms requires the breaking of chemical bonds, i.e., the force that holds the two O atoms together in an O_2 molecule.
Both a) and b) are physical changes, whereas c) and d) are chemical changes. In other words, intermolecular forces are involved in physical changes while intramolecular forces are involved in chemical changes.

12.7   a) **Condensation**   The water vapor in the air condenses to liquid when the temperature drops during the night.
        b) **Fusion** (melting)   Solid ice melts to liquid water.
        c) **Evaporation**   Liquid water on clothes evaporates to water vapor.

12.9   The propane gas molecules slow down as the gas is compressed. Therefore, much of the **kinetic energy** lost by the propane molecules is released to the surroundings upon liquefaction.

12.13   In closed containers, two processes, evaporation and condensation occur simultaneously. Initially there are few molecules in the vapor phase, so more liquid molecules evaporate than gas molecules condense. Thus, the numbers of molecules in the gas phase increases causing the vapor pressure of hexane to increase. Eventually, the number of molecules in the gas phase reaches a maximum where the number of liquid molecules evaporating equals the number of gas molecules condensing. In other words, the evaporation rate equals the condensation rate. At this point, there is no further change in the vapor pressure.

12.14   a) At the critical temperature, the molecules are moving so fast that they can no longer be condensed. This temperature decreases with weaker intermolecular forces because the forces are not strong enough to overcome molecular motion. Alternatively, as intermolecular forces increase, the **critical temperature increases**.
b) As intermolecular forces increase, the **boiling point increases** because it becomes more difficult and takes more energy to separate molecules from the liquid phase.
c) As intermolecular forces increase, the **vapor pressure decreases** for the same reason given in b). At any given temperature, strong intermolecular forces prevent molecules from easily going into the vapor phase and thus vapor pressure is decreased.
d) As intermolecular forces increase, the **heat of vaporization increases** because more energy is needed to separate molecules from the liquid phase.

12.18 When water at 100°C touches skin, the heat released is from the lowering of the temperature of the water. The specific heat of water is approximately 75 J/mol•K. When steam at 100°C touches skin, the heat released is from the condensation of the gas with a heat of condensation of approximately 41 kJ/mol. Thus, the amount of heat released from gaseous water condensing will be greater than the heat from hot liquid water cooling and the burn from the steam will be worse than that from hot water.

12.19 The total heat required is the sum of three processes:
1) Warming the ice from –6.00°C to 0.00°C

$$q_1 = Cm\Delta T = (2.09 \text{ J/g°C}) (22.00 \text{ g}) [0.0 - (-6.00)]°C = 275.88 \text{ J}$$

2) Phase change of ice at 0.00°C to water at 0.00°C

$$q_2 = n\Delta H_{\text{fus}} = (22.0 \text{ g}) \left(\frac{1 \text{ mol}}{18.02 \text{ g}} \right) \left(\frac{6.02 \text{ kJ}}{\text{mol}} \right) \left(\frac{10^3 \text{ J}}{1 \text{ kJ}} \right) = 7349.6115 \text{ J}$$

3) Warming the liquid from 0.00°C to 0.500°C

$$q_3 = Cm\Delta T = (4.21 \text{ J/g°C}) (22.00 \text{ g}) [0.500 - (0.0)]°C = 46.31 \text{ J}$$

The three heats are positive because each process takes heat from the surroundings (endothermic). The phase change requires much more energy than the two temperature change processes. The total heat is $q_1 + q_2 + q_3 =$ (275.88 J + 7349.6115 J + 46.31 J) = 7671.8015 = **7.67 x 10³ J**.

12.21 The Clausius-Clapeyron equation gives the relationship between vapor pressure and temperature. Boiling point is defined as the temperature when vapor pressure of liquid equals atmospheric pressure, usually assumed to be exactly 1 atm. So, the number of significant figures in the pressure of 1 atm given is not one, but is unlimited since it is an exact number. In the calculation below, 1.00 atm is used to emphasize the additional significant figures.

$$\ln \frac{P_2}{P_1} = \frac{-\Delta H_{\text{vap}}}{R} \left(\frac{1}{T_2} - \frac{1}{T_1} \right) \qquad P_1 = 1.00 \text{ atm}; \qquad T_1 = 122°C + 273 = 395 \text{ K};$$

$$T_2 = 113°C + 273 = 386 \text{ K}; \qquad P_2 = ?$$

$$\ln \frac{P_2}{1.00 \text{ atm}} = \frac{-35.5 \frac{\text{kJ}}{\text{mol}}}{8.314 \text{ J/mol•K}} \left(\frac{1}{386 \text{ K}} - \frac{1}{395 \text{ K}} \right) \left(\frac{10^3 \text{ J}}{1 \text{kJ}} \right) = -0.2520440 \text{ (unrounded)}$$

$$\frac{P_2}{1.00 \text{ atm}} = 0.7772105 \text{ (unrounded)}$$

$$P_2 = (0.7772105) (1.00 \text{ atm}) = 0.7772105 = \textbf{0.777 atm}$$

12.23 At the boiling point, the vapor pressure equals the external pressure. Summarize the pressure and temperature variables and use the Clausius-Clapeyron equation to find the ΔH_{vap}.

$$\ln \frac{P_2}{P_1} = \frac{-\Delta H_{\text{vap}}}{R} \left(\frac{1}{T_2} - \frac{1}{T_1} \right) \qquad P_1 = (621 \text{ torr}) \left(\frac{1 \text{ atm}}{760 \text{ torr}} \right) = 0.817105 \text{ atm}$$

$$P_2 = 1.00 \text{ atm} \qquad T_1 = 85.2°C + 273.2 = 358.4 \text{ K}; \qquad T_2 = 95.6°C + 273.2 = 368.8 \text{ K};$$

$$\ln \frac{1 \text{ atm}}{0.817105 \text{ atm}} = \frac{-\Delta H_{\text{vap}}}{8.314 \text{ J/mol•K}} \left(\frac{1}{368.8 \text{ K}} - \frac{1}{358.4 \text{ K}} \right)$$

$$0.2019877 = -\Delta H_{\text{vap}}(-9.46377 \times 10^{-6})\text{J/mol}$$

$$\Delta H_{\text{vap}} = 21343.246 = \textbf{2 x 10}^4 \textbf{ J/mol}$$

(The significant figures in the answer are limited by the 1 atm in the problem.)

12.25

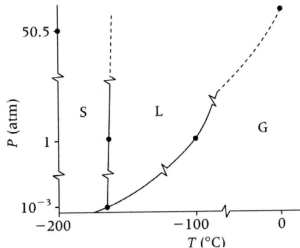

The pressure scale is distorted to represent the large range in pressures given in the problem, so the liquid-solid curve looks different from the one shown in Figure 12.9. The important features of the graph include the distinction between the gas, liquid and solid states, and the melting point T, which is located directly above the critical T. Solid ethylene is denser than liquid ethylene since the solid-liquid line slopes to the right with increasing pressure.

12.28   This problem is an application of the Clausius-Clapeyron equation. Convert the temperatures from °C to K and ΔH_{vap} from kJ/mol to J/mol to allow cancellation with the units in R.

$$\ln\frac{P_2}{P_1} = \frac{-\Delta H_{vap}}{R}\left(\frac{1}{T_2} - \frac{1}{T_1}\right) \qquad P_1 = 2.3\ \text{atm}; \qquad T_1 = 25.0°C + 273 = 298\ \text{K};$$

$$T_2 = 135°C + 273 = 408\ \text{K}; \qquad P_2 = ?$$

$$\ln\frac{P_2}{2.3\ \text{atm}} = \frac{-24.3\dfrac{\text{kJ}}{\text{mol}}}{8.314\ \text{J/mol} \cdot \text{K}}\left(\frac{1}{408\ \text{K}} - \frac{1}{298\ \text{K}}\right)\left(\frac{10^3\ \text{J}}{1\,\text{kJ}}\right) = 2.89834$$

$$\ln\frac{P_2}{2.3\ \text{atm}} = 2.644311$$

$$\frac{P_2}{2.3\ \text{atm}} = 14.07374$$

$$P_2 = (14.07374)\,(2.3\ \text{atm}) = 32.3696 = \textbf{32 atm}$$

12.32   To form hydrogen bonds, the atom bonded to hydrogen must have two characteristics: small size and high electronegativity (so that the atom has a very high electron density). With this high electron density, the attraction for a hydrogen on another molecule is very strong. Selenium is much larger than oxygen (atomic radius of 119 pm vs. 73 pm) and less electronegative than oxygen (2.4 for Se and 3.5 for O) resulting in an electron density on Se in H_2Se that is too small to form hydrogen bonds.

12.34   All particles (atoms and molecules) exhibit dispersion forces, but these are the weakest of intermolecular forces. The dipole-dipole forces in polar molecules dominate the dispersion forces.

12.37   Dispersion forces are the only forces between nonpolar substances; dipole-dipole forces exist between polar substances. Hydrogen bonds only occur in substances in which hydrogen is directly bonded to either oxygen, nitrogen or fluorine.
a) **Hydrogen bonding** will be the strongest force between methanol molecules since they contain O–H bonds. Dipole-dipole and dispersion forces also exist.

b) **Dispersion forces** are the only forces between nonpolar carbon tetrachloride molecules and, thus, are the strongest forces.

c) **Dispersion forces** are the only forces between nonpolar chlorine molecules and, thus, are the strongest forces.

12.39   Dispersion forces are the only forces between nonpolar substances; dipole-dipole forces exist between polar substances. Hydrogen bonds only occur in substances in which hydrogen is directly bonded to either oxygen, nitrogen or fluorine.

a) **Dipole-dipole** interactions will be the strongest forces between methyl chloride molecules because the C–Cl bond has a dipole moment.

b) **Dispersion** forces dominate because CH_3CH_3 (ethane) is a symmetrical nonpolar molecule.

c) **Hydrogen bonding** dominates because hydrogen is bonded to nitrogen, which is one of the 3 atoms (N, O, or F) that participate in hydrogen bonding.

12.41   Hydrogen bonds are formed when a hydrogen atom is bonded to N, O, or F.

a) The presence of an OH group leads to the formation of hydrogen bonds in $CH_3CH(OH)CH_3$. There are no hydrogen bonds in CH_3SCH_3.

b) The presence of H attached to F in **HF** leads to the formation of hydrogen bonds. There are no hydrogen bonds in HBr.

12.43   In the vaporization process, intermolecular forces between particles in the liquid phase must be broken as the particles enter the vapor phase. In other words, the question is asking for the strongest interparticle force that must be broken to vaporize the liquid.

a) **Dispersion**, because hexane, C_6H_{14}, is a nonpolar molecule.

b) **Hydrogen bonding**; hydrogen is bonded to oxygen in water. A single water molecule can engage in as many as four hydrogen bonds.

c) **Dispersion**, although the individual Si-Cl bonds are polar, the molecule has a symmetrical, tetrahedral shape and is therefore nonpolar.

12.45   Polarizability increases down a group and decreases from left to right because as atomic size increases, polarizability increases.

a) **Iodide ion** has greater polarizability than the bromide ion because the iodide ion is larger. The electrons can be polarized over a larger volume in a larger atom or ion.

b) **Ethene ($CH_2=CH_2$)** has greater polarizability than ethane (CH_3CH_3) because π orbitals are more easily polarized than σ orbitals.

c) **H_2Se** has greater polarizability than water because the selenium atom is larger than the oxygen atom.

12.47   Weaker attractive forces result in a higher vapor pressure because the molecules have a smaller energy barrier in order to escape the liquid and go into the gas phase.

a) **C_2H_6** C_2H_6 is a smaller molecule exhibiting weaker dispersion forces than C_4H_{10}.

b) **CH_3CH_2F** CH_3CH_2F has no H–F bonds (F is bonded to C, not to H), so it only exhibits dipole-dipole forces, which are weaker than the hydrogen bonding in CH_3CH_2OH.

c) **PH_3** PH_3 has weaker intermolecular forces (dipole-dipole) than NH_3 (hydrogen bonding).

12.49   The weaker the intermolecular forces, the lower the boiling point.
a) **HCl** would have a lower boiling point than LiCl because the dipole-dipole intermolecular forces between hydrogen chloride molecules in the liquid phase are weaker than the significantly stronger ionic forces holding the ions in lithium chloride together.
b) **PH_3**, would have a lower boiling point than NH_3 because the intermolecular forces in PH_3 are weaker than those in NH_3. Hydrogen bonding exists between NH_3 molecules but weaker dipole-dipole forces hold PH_3 molecules together.
c) **Xe** would have a lower boiling point than iodine. Both are nonpolar with dispersion forces, but the forces between xenon atoms would be weaker than those between iodine molecules are since the iodine molecules are more polarizable because of their larger size.

12.51   The molecule in the pair with the weaker intermolecular attractive forces will have the lower boiling point.
a) **C_4H_8**, the cyclic molecule, cyclobutane, has less surface area exposed, so its dispersion forces are less than the straight chain molecule, C_4H_{10}.
b) **PBr_3**, the dipole–dipole forces of phosphorous tribromide are weaker than the ionic forces of sodium bromide.
c) **HBr**, the dipole-dipole forces of hydrogen bromide are weaker than the hydrogen bonding forces of water.

12.53   The trend in both atomic size and electronegativity predicts that the trend in increasing strength of hydrogen bonds   is N-H < O-H < F-H. As the atomic size decreases and electronegativity increases, the electron density of the atom increases. High electron density strengthens the attraction to a hydrogen atom on another molecule. Fluorine is the smallest of the three and the most electronegative, so its hydrogen bonds would be the strongest. Oxygen is smaller and more electronegative than nitrogen, so hydrogen bonds for water would be stronger than hydrogen bonds for ammonia.

12.55   The ethylene glycol molecules have two sites (two –OH groups) which can hydrogen bond; the propanol only one.

12.56   The molecules at the surface are attracted to one another and to those molecules in the bulk of the liquid. Since this force is directed downwards and sideways, it tends to "tighten the skin."

12.57   The shape of the drop depends upon the competing cohesive forces (attraction of molecules within the drop itself) and adhesive forces (attraction between molecules in the drop and the molecules of the waxed floor). If the cohesive forces are strong and outweigh the adhesive forces, the drop will be as spherical as gravity will allow. If, on the other hand, the adhesive forces are significant, the drop will spread out. Both water (hydrogen bonding) and mercury (metallic bonds) have strong cohesive forces, whereas cohesive forces in oil (dispersion) are relatively weak. Figure 12.20 explains the difference between these forces when *glass*, not *wax*, is used. Neither water nor mercury will have significant adhesive forces to the nonpolar wax molecules, so these drops will remain nearly spherical. The adhesive forces between the oil and wax can compete with the weak, cohesive forces of the oil (dispersion) and so the oil drop spreads out.

12.59   Surface tension is defined as the energy needed to increase the surface area by a given amount, so units of energy per surface area describe this property.

12.61   The stronger the intermolecular force, the greater the surface tension. All three molecules exhibit hydrogen bonding, but the extent of hydrogen bonding increases with the number of O–H bonds present in each molecule. Molecule b) with 3 O–H groups can form more hydrogen bonds than molecule c) with 2 O–H groups, which in turn can form more hydrogen bonds than molecule a) with only 1 O–H group.
$CH_3CH_2CH_2OH$ < $HOCH_2CH_2OH$ < $HOCH_2CH(OH)CH_2OH$
The greater the number of hydrogen bonds, the greater the energy requirements.

12.63   Viscosity is a measure of the resistance of a liquid to flow, and is greater for molecules with stronger intermolecular forces. The stronger the force attracting the molecules to each other, the harder it is for one molecule to move past another. Thus, the substance will not flow easily if the intermolecular force is strong. The ranking of decreasing viscosity is the opposite of that for increasing surface tension (Problem 12.61).
$HOCH_2CH(OH)CH_2OH$ > $HOCH_2CH_2OH$ > $CH_3CH_2CH_2OH$

12.68   Water is a good solvent for polar and ionic substances and a poor solvent for nonpolar substances. Water is a polar molecule and dissolves polar substances because their intermolecular forces are of similar strength. Water is also able to dissolve ionic compounds and keep ions separated in solution through ion-dipole interactions. Nonpolar substances will not be very soluble in water since their dispersion forces are much weaker than the hydrogen bonds in water. A solute whose intermolecular attraction to a solvent molecule is less than the attraction between two solvent molecules will not dissolve because its attraction cannot replace the attraction between solvent molecules.

12.69   A single water molecule can form 4 hydrogen bonds. The two hydrogen atoms form a hydrogen bond each to oxygen atoms on neighboring water molecules. The two lone pairs on the oxygen atom form hydrogen bonds with hydrogen atoms on neighboring molecules.

12.72   Water exhibits strong capillary action, which allows it to be easily absorbed by the plant's roots and transported to the leaves.

12.78   The unit cell is a simple cubic cell. According to the bottom row in Figure 12.27, two atomic radii (or one atomic diameter) equal the width of the cell.

12.81   The energy gap is the energy difference between the highest filled energy level (valence band) and the lowest unfilled energy level (conduction band). In conductors and superconductors, the energy gap is zero because the valence band overlaps the conduction band. In semiconductors, the energy gap is small but greater than zero. In insulators, the energy gap is large and thus insulators do not conduct electricity.

12.83   The density of a solid depends on the atomic mass of the element (greater mass = greater density), the atomic radius (how many atoms can fit in a given volume) and the type of unit cell, which determines the packing efficiency (how much of the volume is occupied by empty space).

12.84   The simple cubic structure unit cell contains 1 atom since the atoms at the 8 corners are shared by eight cells for a total of 8 atoms x 1/8 atom per cell = 1 atom.; the body-centered cell also has an atom in the center, for a total of 2 atoms; the face-centered cell has 6 atoms in the faces which are shared by two cells: 6 atoms x ½ atom per cell = 3 atoms plus another atom from the 8 corners for a total of 4 atoms.
         a) Ni is **face-centered cubic** since there are 4 atoms/unit cell.
         b) Cr is **body-centered cubic** since there are 2 atoms/unit cell.
         c) Ca is **face-centered cubic** since there are 4 atoms/unit cell.

12.86   a) **Yes**, there is a change in unit cell from CdO in a sodium chloride structure to CdSe in a zinc blende structure.
         b) **Yes**, the coordination number of Cd does change from 6 in the CdO unit cell to 4 in the CdSe unit cell.

12.88   a) Nickel forms a **metallic solid** since nickel is a metal whose atoms are held together by metallic bonds.
         b) Fluorine forms a **molecular solid** since the F_2 molecules are held together by dispersion forces.
         c) Methanol forms a **molecular solid** since the CH_3OH molecules are held together by hydrogen bonds.
         d) Tin forms a **metallic solid** since nickel is a metal whose atoms are held together by metallic bonds.
         e) Silicon is in the same group as carbon, so it exhibits similar bonding properties. Since diamond and graphite are both **network covalent** solids, it makes sense that Si forms the same type of bonds.
         f) Xe is an **atomic** solid since individual atoms are held together by dispersion forces.

12.90   Figure P12.90 shows the face-centered cubic array of zinc blende, ZnS. Both ZnS and ZnO have a 1:1 ion ratio, so the ZnO unit cell will also contain **four** Zn^{2+} ions.

12.92   a) To determine the number of Zn^{2+} ions and Se^{2-} ions in each unit cell count the number of ions at the corners, faces, and center of unit cell. Looking at selenide ions, there is one ion at each corner and one ion on each face. The total number of selenide ions is 1/8 (8 corner ions) + 1/2 (6 face ions) = **4 Se^{2-} ions**. There are also **4 Zn^{2+} ions** due to the 1:1 ratio of Se ions to Zn ions.

b) Mass of unit cell = (4 x mass of Zn atom) + (4 x mass of Se atom)

$\quad\quad\quad\quad\quad\quad\quad\quad$ = (4 x 65.41 amu) + (4 x 78.96 amu) = **577.48 amu**

c) Given the mass of one unit cell and the ratio of mass to volume (density) divide the mass, converted to grams (conversion factor is 1 amu = 1.66054 x 10^{-24} g), by the density to find the volume of the unit cell.

$$\text{Volume} = \left(\frac{cm^3}{5.42 \text{ g}}\right)\left(\frac{1.66054 \times 10^{-24} \text{ g}}{1 \text{ amu}}\right)(577.48 \text{ amu}) = 1.76924 \times 10^{-22} = \textbf{1.77 x } 10^{-22} \textbf{ cm}^3$$

d) The volume of a cube equals (length of side)^3.

Side = $\sqrt[3]{1.76924 \times 10^{-22} \text{ cm}^3}$ = 5.6139 x 10^{-8} = **5.61 x 10^{-8} cm**

12.94 To classify a substance according to its electrical conductivity, first locate it on the periodic table as a metal, metalloid, or nonmetal. In general, metals are conductors, metalloids are semiconductors, and nonmetals are insulators.
a) Phosphorous is a nonmetal and an **insulator**.
b) Mercury is a metal and a **conductor**.
c) Germanium is a metalloid in Group 4A(14) and is beneath carbon and silicon in the periodic table. Pure germanium crystals are **semiconductors** and are used to detect gamma rays emitted by radioactive materials. Germanium can also be doped with phosphorous (similar to the doping of silicon) to form an n-type semiconductor or be doped with lithium to form a p-type semiconductor.

12.96 First, classify the substance as an insulator, conductor, or semiconductor. The electrical conductivity of conductors decreases with increasing temperature, whereas that of semiconductors increases with temperature. Temperature increases have little impact on the electrical conductivity of insulators.
a) Antimony, Sb, is a metalloid, so it is a semiconductor. Its electrical conductivity **increases** as the temperature increases.
b) Tellurium, Te, is a metalloid, so it is a semiconductor. Its electrical conductivity **increases** as temperature increases.
c) Bismuth, Bi, is a metal, so it is a conductor. Its electrical conductivity **decreases** as temperature increases.

12.98 To calculate an approximate radius for polonium, first multiply the molar mass of Po by the inverse of the density of Po to find the volume per mole Po *metal*.

$$\text{Volume} = \left(\frac{cm^3}{9.142 \text{ g}}\right)\left(\frac{209 \text{ g}}{1 \text{ mol Po}}\right) = 22.861518 \text{ cm}^3/\text{mol Po (unrounded)}$$

Volume/mol of Po atoms = volume/mol Po x packing efficiency
The packing efficiency in the simple cubic unit cell is 52%.
Volume/mol of Po atoms = 22.861518 cm^3/mol Po x 0.52 = 11.887999 cm^3/mol Po atoms

$$\text{Volume of one Po atom} = \left(\frac{11.88799 \text{ cm}^3}{1 \text{ mol Po atoms}}\right)\left(\frac{1 \text{ mol Po atoms}}{6.022 \times 10^{23} \text{ Po atoms}}\right) = 1.97409 \times 10^{-23} \text{ cm}^3/\text{atom}$$

Use the volume of a sphere to find the radius of the Po atom:

$$V = \frac{4}{3}\pi r^3$$

$$r = \sqrt[3]{\frac{3V}{4\pi}} = \sqrt[3]{\frac{3(1.97409 \times 10^{-23} \text{ cm}^3)}{4\pi}} = 1.676588 \times 10^{-8} = \textbf{1.68 x } 10^{-8} \textbf{ cm}$$

12.105 A substance whose physical properties are the same in all directions is called isotropic; otherwise, the substance is anisotropic. Liquid crystals flow like liquids but have a degree of order that gives them the anisotropic properties of a crystal.

12. 111 Germanium and silicon are elements in group 4A with 4 valence electrons. If germanium or silicon is doped with an atom with more than 4 valence electrons, an n-type semiconductor is produced. If it is doped with an atom with fewer than 4 valence electrons, a p-type semiconductor is produced.
a) Phosphorus has 5 valence electrons so an **n-type semiconductor** will form by doping Ge with P.
b) Indium has 3 valence electrons so a **p-type semiconductor** will form by doping Si with In.

12.113 The degree of polymerization is the number of monomer, or repeat units that are bonded together in the polymer chain. Since the molar mass of one monomer unit, $C_6H_5CHCH_2$, is 104.14 g/mol, there are (3.5 x 10^5 g/mol / 104.14 g/mol) = 3361 monomer units. Reporting in correct significant figures, **n = 3.4 x 10^3**.

12.115 The length of the repeat unit, l_0, is given (0.252 pm). Calculate the degree of polymerization, n, by dividing the molar mass of the polymer chain ($\mathcal{M}$ = 2.8 x 10^5 g/mol) by the molar mass of one monomer unit (CH_3CHCH_2, $\mathcal{M}$ = 42.08 g/mol). Substitute these values into the equation below:
The degree of polymerization = (2.8 x 10^5 g/mol) / (42.08 g/mol) = 6653.99 (unrounded)

$$\text{Radius of gyration} = R_g = \sqrt{\frac{nl_0^2}{6}}$$

$$R_g = \sqrt{\frac{(6653.99)(0.252 \text{ pm})^2}{6}} = 8.39201 = \textbf{8.4 pm}$$

Note that the length of the repeat unit may actually be 0.252 nm, which leads to a radius of gyration = 7.9 nm.

12.118 At 22°C the vapor pressure of water is 19.8 torr (from Table 5.3).
a) Once compressed, the N_2 gas would still be saturated with water. The vapor pressure depends on the temperature, which has not changed. Therefore, the partial pressure of water in the compressed gas remains the same at **19.8 torr**.
b) In order to maintain the vapor pressure of water at 19.8 torr, some of the water must condense to liquid. The volume is now one-half the original volume (5.00 L /2 = 2.50 L). The mass of water contained in the "lost" volume may be calculated.

$$\text{Moles } H_2O \text{ at } 22°C = n = PV / RT = \frac{(19.8 \text{ torr})(2.50 \text{ L})}{\left(0.0821\dfrac{L \cdot atm}{mol \cdot K}\right)\left((273+22)K\right)}\left(\frac{1 \text{ atm}}{760 \text{ torr}}\right) = 0.002689 \text{ mol } H_2O$$

$$\text{Mass } H_2O \text{ at } 22°C = (0.002689 \text{ mol } H_2O)\left(\frac{18.02 \text{ g } H_2O}{1 \text{ mol } H_2O}\right) = 0.0484558 = \textbf{0.0485 g } H_2O$$

12.122 The lowest energy will be for a photon striking an **edge** (4.48 eV).

$$\lambda = \frac{hc}{E} = \frac{(6.626 \text{ x } 10^{-34} \text{ J} \cdot s)(3.00 \text{ x } 10^8 \text{ m/s})}{(4.48 \text{ eV})}\left(\frac{1 \text{ eV}}{1.602 \text{ x } 10^{-19} \text{ J}}\right)\left(\frac{1 \text{ nm}}{10^{-9} \text{ m}}\right) = 276.9696 = \textbf{277 nm}$$

12.125 The Clausius-Clapeyron equation can be solved for the temperature that will give a vapor pressure of 5.0x10^{-5} torr.

$$\ln\frac{P_2}{P_1} = \frac{-\Delta H_{vap}}{R}\left(\frac{1}{T_2} - \frac{1}{T_1}\right) \qquad P_1 = 1.20 \text{ x } 10^{-3} \text{ torr;} \qquad T_1 = 20.0°C + 273 = 293 \text{ K;}$$

$$P_2 = 5.0 \text{ x } 10^{-5} \text{ torr;} \qquad T_2 = ?$$

$$\ln\frac{5.0 \text{ x } 10^{-5} \text{ torr}}{1.20 \text{ x } 10^{-3} \text{ torr}} = \frac{-59.1\dfrac{kJ}{mol}}{8.314 \text{ J/mol} \cdot K}\left(\frac{1}{T_2} - \frac{1}{293 \text{ K}}\right)\left(\frac{10^3 \text{ J}}{1 kJ}\right)$$

$$-3.17805 = -7108.49\left(\frac{1}{(T)K} - \frac{1}{(293)K}\right) \text{ (unrounded)}$$

$(-3.17805) / (-7108.49) = 4.47078 \times 10^{-4} = \left(\dfrac{1}{(T)K} - \dfrac{1}{(293)K} \right)$ (unrounded)

$4.47078 \times 10^{-4} + 1 / 293 = 1 / T$

$T = 259.064 = \mathbf{259\ K}$

12.127 a) Use the volume of the liquid water and the density of water to find the moles of water present in the volume of the greenhouse. Find the pressure of this number of moles of water vapor using the Ideal Gas Equation.

$$P = \dfrac{nRT}{V} = \dfrac{\left[4.20\,L \left(\dfrac{1\ mL}{10^{-3}\ L} \right)\left(\dfrac{1.00\ g}{mL} \right)\left(\dfrac{1\ mol\ H_2O}{18.02\ g\ H_2O} \right) \right]\left(0.0821 \dfrac{L \cdot atm}{mol \cdot K} \right)\left((273 + 26)K \right)}{(256\ m^3)}\left(\dfrac{10^{-3}\ m^3}{1\ L} \right)$$

$= 2.223496 \times 10^{-2} = \mathbf{2.23 \times 10^{-2}\ atm}$

b) 25.2 torr is needed to saturate the air, so the volume of liquid water needed is:

$$V = (4.20\ L)\left(\dfrac{25.2\ torr}{2.223496 \times 10^{-2}\ atm} \right)\left(\dfrac{1\ atm}{760\ torr} \right) = 6.26325 = \mathbf{6.26\ L\ H_2O}$$

12.130 a) Both furfuryl alcohol and 2–furoic acid can form hydrogen bonds since these two molecules have hydrogen directly bonded to oxygen.

2-furoic acid

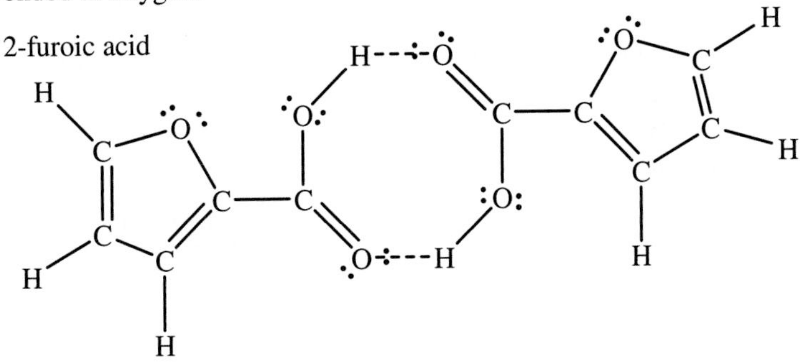

furfuryl alcohol

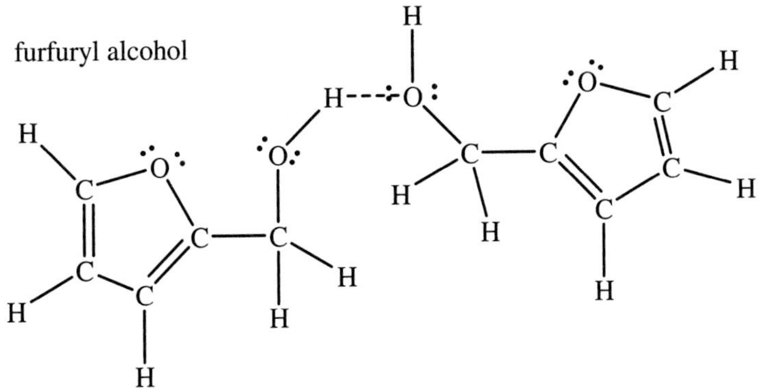

b) Both furfuryl alcohol and 2–furoic acid can form internal hydrogen bonds by forming a hydrogen bond between the O–H and the O in the ring.

2-furoic acid

furfuryl alcohol

12.131   Determine the total mass of water, with a partial pressure of 31.0 torr, contained in the air. Then calculate the total mass of water, with a partial pressure of 10.0 torr, contained in the air. The difference between these two values is the amount of water removed. The mass of water and the heat of condensation are necessary to find the amount of energy removed from the water.

a) $n = PV / RT$

Mass H_2O (31.0 torr) = mol H_2O x (18.02 g/mol)

$$= \frac{(31.0 \text{ torr})(2.4 \times 10^6 \text{ m}^3)}{\left(0.0821 \dfrac{\text{L•atm}}{\text{mol•K}}\right)((273.2 + 22.0)\text{K})} \left(\frac{1 \text{ atm}}{760 \text{ torr}}\right)\left(\frac{1 \text{ L}}{10^{-3} \text{ m}^3}\right)\left(\frac{18.02 \text{ g } H_2O}{1 \text{ mol } H_2O}\right)\left(\frac{1 \text{ kg}}{10^3 \text{ g}}\right)\left(\frac{1 \text{ ton}}{10^3 \text{ kg}}\right)$$

= 72.7871 tons H_2O (unrounded)

Mass H_2O (10.0 torr) = mol H_2O x (18.02 g/mol)

$$= \frac{(10.0 \text{ torr})(2.4 \times 10^6 \text{ m}^3)}{\left(0.0821 \dfrac{\text{L•atm}}{\text{mol•K}}\right)((273.2 + 22.0)\text{K})} \left(\frac{1 \text{ atm}}{760 \text{ torr}}\right)\left(\frac{1 \text{ L}}{10^{-3} \text{ m}^3}\right)\left(\frac{18.02 \text{ g } H_2O}{1 \text{ mol } H_2O}\right)\left(\frac{1 \text{ kg}}{10^3 \text{ g}}\right)\left(\frac{1 \text{ ton}}{10^3 \text{ kg}}\right)$$

= 23.4797 tons H_2O (unrounded)

Mass of water removed = (72.7871 tonnes H_2O) – (23.4797 tonnes H_2O) = 49.3074 = **49.3 tons H_2O**

b) The heat of condensation for water is –40.7 kJ/mol.

$$\text{Heat} = (49.3074 \text{ tons } H_2O)\left(\frac{10^3 \text{ kg}}{1 \text{ ton}}\right)\left(\frac{10^3 \text{ g}}{1 \text{ kg}}\right)\left(\frac{1 \text{ mol } H_2O}{18.02 \text{ g } H_2O}\right)\left(\frac{-40.7 \text{ kJ}}{1 \text{ mol } H_2O}\right)$$

$$= -1.113658 \times 10^8 = \mathbf{-1.11 \times 10^8 \ kJ}$$

12.132   At the boiling point, the vapor pressure equals the atmospheric pressure, so the two boiling points can be used to find the heat of vaporization for amphetamine using the Clausius-Clapeyron equation.

$$\ln \frac{P_2}{P_1} = \frac{-\Delta H_{vap}}{R}\left(\frac{1}{T_2} - \frac{1}{T_1}\right)$$

$P_1 = 760$ torr $T_1 = 201°C = (273 + 201) = 474$ K
$P_2 = 13$ torr $T_2 = 83°C = (273 + 83) = 356$ K
$\Delta H_{vap} = ?$

$$\ln \frac{13 \text{ torr}}{760 \text{ torr}} = \frac{-\Delta H_{vap}}{8.314 \text{ J/mol•K}}\left(\frac{1}{356 \text{ K}} - \frac{1}{474 \text{ K}}\right)$$

-4.068369076 = –0.000084109 ΔH_{vap} (unrounded)

ΔH_{vap} = (–4.068369076) / (–0.000084109) = 48370.199 J/mol (unrounded)

Use the Clausius-Clapeyron equation to find the vapor pressure of amphetamine at 20.°C.

$$\ln \frac{P_2}{P_1} = \frac{-\Delta H_{vap}}{R}\left(\frac{1}{T_2} - \frac{1}{T_1}\right)$$

$P_1 = 13$ torr $\qquad\qquad T_1 = 83°C = (273 + 83) = 356$ K
$P_2 = ?$ $\qquad\qquad\qquad T_2 = 20.°C = (273 + 20.) = 293$ K
$\Delta H_{vap} = 48370.199$ J/mol

$$\ln \frac{P_2}{13 \text{ torr}} = \frac{-48370.199\dfrac{\text{J}}{\text{mol}}}{8.314 \text{ J/mol} \cdot \text{K}}\left(\frac{1}{293 \text{ K}} - \frac{1}{356 \text{ K}}\right) = -3.5139112 \text{ (unrounded)}$$

$$\frac{P_2}{13 \text{ torr}} = 0.0297802 \text{ (unrounded)}$$

$P_2 = (0.0297802)(13 \text{ torr})(1 \text{ atm} / 760 \text{ torr}) = 5.09398 \times 10^{-4}$ atm (unrounded)
At this pressure and a temperature of 20.°C, use the ideal gas equation to calculate the concentration of amphetamine in the air. Use a volume of 1 m^3. Moles from the ideal gas equation times the molar mass gives the mass in a cubic meter.

$$\text{Moles} = n = \frac{PV}{RT} = \frac{\left(5.09398 \times 10^{-4}\text{ atm}\right)\left(1 \text{ m}^3\right)}{\left(0.0821\dfrac{\text{L}\cdot\text{atm}}{\text{mol}\cdot\text{K}}\right)(293 \text{ K})}\left(\frac{1 \text{ L}}{10^{-3} \text{ m}^3}\right) = 0.02117612 \text{ mol amphetamine}$$

$$\left(0.02117612 \text{ mol}\right)\left(\frac{135.20 \text{ g amphetamine}}{1 \text{ mol}}\right) = 2.8630114 = \textbf{2.9 g/m}^3$$

12.137 a) To heat the substance to the melting point:
Warming the substance from –40.°C to –20.°C
$\qquad\qquad q_1 = Cm\Delta T = (1.0 \text{ J/g°C}) (25 \text{ g}) [-20. - (-40.)]°C = 500$ J

Time required: $\left(500 \text{ J}\right)\left(\dfrac{1 \text{ min}}{450 \text{ J}}\right) = \textbf{1.1 min}$

b) To melt the sample:
Phase change of the substance from solid at –20.°C to liquid at –20.°C

$$q_2 = n\Delta H_{fus} = \left(25 \text{ g}\right)\left(\frac{180. \text{ J}}{1 \text{ g}}\right) = 4500 \text{ J}$$

Time required: $\left(4500 \text{ J}\right)\left(\dfrac{1 \text{ min}}{450 \text{ J}}\right) = \textbf{10. min}$

c) Warming the liquid from –20.°C to 85°C
$\qquad\qquad q_3 = Cm\Delta T = (2.5 \text{ J/g°C}) (25 \text{ g}) [85 - (-20.)]°C = 6562.5$ J

Time required: $\left(6562.5 \text{ J}\right)\left(\dfrac{1 \text{ min}}{450 \text{ J}}\right) = 14.583 = \textbf{15 min}$

Phase change of the substance from liquid at 85°C to gas at 85°C

$$q_4 = n\Delta H_{vap} = \left(25 \text{ g}\right)\left(\frac{500. \text{ J}}{1 \text{ g}}\right) = 12500 \text{ J}$$

Time required: $\left(12500 \text{ J}\right)\left(\dfrac{1 \text{ min}}{450 \text{ J}}\right) = 27.78 = \textbf{28 min}$

Warming the liquid from 85°C to 100.°C

$$q_5 = Cm\Delta T = (0.5 \text{ J/g°C})(25 \text{ g})[100 - 85]°C = 187.5 \text{ J}$$

Time required: $(187.5 \text{ J})\left(\dfrac{1 \text{ min}}{450 \text{ J}}\right) = 0.417 = \mathbf{0.4 \text{ min}}$

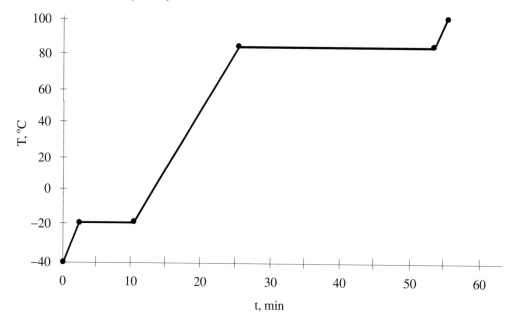

12.138 Balanced chemical equations are necessary. See the section on ceramic materials for the reactions. These equations may be combined to produce an overall equation with an overall yield. From this point on it becomes a straight stoichiometry problem with the aid of the ideal gas equation.

Step 1: $B(OH)_3(s) + 3 \text{ NH}_3(g) \rightarrow B(NH_2)_3(s) + 3 \text{ H}_2O(g)$

Step 2: $B(NH_2)_3(s) \xrightarrow{\Delta} BN(s) + 2 \text{ NH}_3(g)$

Yields are 85.5% for step 1 and 86.8% for step 2.

Overall reaction: $B(OH)_3(s) + NH_3(g) \rightarrow BN(s) + 3 \text{ H}_2O(g)$

Overall fractional yield = (85.5% / 100%) (86.8% / 100%) = 0.74214 (unrounded)

Find the limiting reactant. Since the overall reaction has a 1:1 mole ratio, the reactant with the fewer moles will be limiting.

$$\text{Moles } B(OH)_3 = \left(1.00 \text{ t } B(OH)_3\right)\left(\frac{10^3 \text{ kg}}{1 \text{ t}}\right)\left(\frac{10^3 \text{ g}}{1 \text{ kg}}\right)\left(\frac{1 \text{ mol } B(OH)_3}{61.83 \text{ g } B(OH)_3}\right)$$

$$= 1.6173378 \times 10^4 \text{ mol } B(OH)_3 \text{ (unrounded)}$$

$$\text{Moles NH}_3 = PV / RT = \frac{\left(3.07 \times 10^3 \text{ kPa}\right)\left(12.5 \text{ m}^3\right)}{\left(0.0821 \dfrac{\text{L·atm}}{\text{mol·K}}\right)(275 \text{ K})}\left(\frac{1 \text{ atm}}{101.325 \text{ kPa}}\right)\left(\frac{1 \text{ L}}{10^{-3} \text{ m}^3}\right)$$

$$= 1.6774744 \times 10^4 \text{ mol NH}_3 \text{ (unrounded)}$$

Thus, $B(OH)_3$ is limiting.

$$\text{Mass BN} = \left(1.6173378 \times 10^4 \text{ mol } B(OH)_3\right)\left(\frac{1 \text{ mol BN}}{1 \text{ mol } B(OH)_3}\right)\left(\frac{24.82 \text{ g BN}}{1 \text{ mol BN}}\right)\left(\frac{74.214\%}{100\%}\right) = 2.97912 \times 10^5$$

$$= \mathbf{2.98 \times 10^5 \text{ g BN}}$$

12.141 A body-centered unit cell contains two atoms, which would weigh 2 (22.99 amu) = **45.98 amu**.

CHAPTER 13 THE PROPERTIES OF MIXTURES: SOLUTIONS AND COLLOIDS

FOLLOW–UP PROBLEMS

13.1   Plan: Compare the intermolecular forces in the solutes with the intermolecular forces in the solvent. The intermolecular forces for the more soluble solute will be more similar to the intermolecular forces in the solvent than the forces in the less soluble solute.
Solution:
a) 1,4–Butanediol is more soluble in water than butanol. Intermolecular forces in water are primarily hydrogen bonding. The intermolecular forces in both solutes also involve hydrogen bonding with the hydroxyl groups. Compared to butanol, each 1,4–butanediol molecule will form more hydrogen bonds with water because 1,4–butanediol contains two hydroxyl groups in each molecule, whereas butanol contains only one –OH group. Since 1,4–butanediol has more hydrogen bonds to water than butanol, it will be more soluble than butanol.
b) Chloroform is more soluble in water than carbon tetrachloride because chloroform is a polar molecule and carbon tetrachloride is nonpolar. Polar molecules are more soluble in water, a polar solvent.

13.2   Plan: Solubility of a gas can be found from Henry's law: $S_{gas} = k_H \times P_{gas}$. The problem gives k_H for N_2 but not its partial pressure. To calculate the partial pressure, use the relationship from Chapter 5: $P_{gas} = X_{gas} \times P_{total}$ where X represents the mole fraction of the gas.
Solution:
To find partial pressure use the 78% N_2 given for the mole fraction:
$P_{N_2} = 0.78 \times 1 \text{ atm} = 0.78 \text{ atm}$

Use Henry's law to find solubility at this partial pressure:
$S_{N_2} = 7 \times 10^{-4} \text{ mol/L•atm} \times 0.78 \text{ atm} = 5.46 \times 10^{-4} \text{ mol/L} = \textbf{5 x 10}^{-4} \textbf{ mol/L}$

Check: The solubility should be close to the value of k_H since the partial pressure is close to 1 atm.

13.3   Plan: Molality (m) is defined as amount (mol) of solute per kg of solvent. Use the molality and the mass of solvent given to calculate the amount of glucose. Then convert amount (mol) of glucose to mass (g) of glucose.
Solution:
Mass of solvent must be converted to kg: 563 g x (1 kg/1000 g) = 0.563 kg

$$(0.563 \text{ kg}) \left(\frac{2.40 \times 10^{-2} \text{ mol C}_6\text{H}_{12}\text{O}_6}{1 \text{ kg}} \right) \left(\frac{180.16 \text{ C}_6\text{H}_{12}\text{O}_6}{1 \text{ mol C}_6\text{H}_{12}\text{O}_6} \right) = 2.4343 = \textbf{2.43 g C}_6\textbf{H}_{12}\textbf{O}_6$$

Check: To check these results estimate the molality from the calculated mass of glucose and given mass of solvent. The number of moles of glucose is approximately (2.4 g) (1 mol / 180 g) = 0.013 mol glucose. Using the calculated value for amount of glucose and the given mass of solvent, calculate the molality: (0.013) / (0.56 kg) = 0.023 m, which is close to the given value of 0.0240 m.

13.4   Plan: Mass percent is the mass (g) of solute per 100 g of solution. For each alcohol, divide the mass of the alcohol by the total mass. Multiply this number by 100 to obtain mass percent. To find mole fraction, first find the amount (mol) of each alcohol, then divide by the total moles.
Solution:
Mass percent:

$$\text{Mass\% propanol} = \frac{35.0 \text{ g}}{(35.0 + 150.) \text{g}} \times 100\% = 18.9189 = \textbf{18.9\% propanol}$$

$$\text{Mass\% ethanol} = \frac{150. \text{ g}}{(35.0 + 150.) \text{g}} \times 100\% = 81.08108 = \textbf{81.1\% ethanol}$$

Mole fraction:

$$\text{Moles propanol} = \left(35.0 \text{ g } C_3H_7OH\right)\left(\frac{1 \text{ mol } C_3H_7OH}{60.09 \text{ g } C_3H_7OH}\right) = 0.5824596 \text{ mol propanol (unrounded)}$$

$$\text{Moles ethanol} = \left(150. \text{ g } C_2H_5OH\right)\left(\frac{1 \text{ mol } C_2H_5OH}{46.07 \text{ g } C_2H_5OH}\right) = 3.2559149 \text{ mol ethanol (unrounded)}$$

$$X_{propanol} = \frac{0.5824596 \text{ mol propanol}}{0.5824596 \text{ mol propanol} + 3.2559149 \text{ mol ethanol}} = 0.151746 = \mathbf{0.152}$$

$$X_{ethanol} = \frac{3.2559149 \text{ mol ethanol}}{0.5824596 \text{ mol propanol} + 3.2559149 \text{ mol ethanol}} = 0.84825 = \mathbf{0.848}$$

Check: Since the mass of ethanol is 4 times that of propanol, the mass percent of ethanol should be about 4 times greater. 81% is about 4 times 19%. The mole fraction should be similar, but the fraction that is ethanol should be slightly greater than the mass fraction of 0.81 – and it is at 0.85.

13.5   Plan: To find the mass percent, molality and mole fraction of HCl, the following is needed.
   1) Moles of HCl in 1 L solution (from molarity)
   2) Mass of HCl in 1 L solution (from molarity times molar mass of HCl)
   3) Mass of 1 L solution (from volume times density)
   4) Mass of solvent in 1 L solution (by subtracting mass of solute from mass of solution)
   5) Moles of solvent (by multiplying mass of solvent by molar mass of water)
Mass percent is calculated by dividing mass of HCl by mass of solution.
Molality is calculated by dividing moles of HCl by mass of solvent in kg.
Mole fraction is calculated by dividing mol of HCl by the sum of mol of HCl plus mol of solvent.
Solution:
Assume the volume is exactly 1 L.

$$1) \text{ Mole HCl in 1 L solution} = \left((1.0 \text{ L})\left(\frac{11.8 \text{ mol HCl}}{1.0 \text{ L}}\right)\right) = 11.8 \text{ mol HCl}$$

$$2) \text{ Mass HCl in 1 L solution} = \left(11.8 \text{ mol HCl}\right)\left(\frac{36.46 \text{ g HCl}}{1 \text{ mol HCl}}\right) = 430.228 \text{ g HCl}$$

$$3) \text{ Mass of 1 L solution} = \left(1.0 \text{ L}\right)\left(\frac{1 \text{ mL}}{10^{-3} \text{ L}}\right)\left(\frac{1.190 \text{ g}}{1 \text{ mL}}\right) = 1190. \text{ g solution}$$

4) Mass of solvent in 1 L solution = 1190. g – 430.228 g = 759.772 g solvent (H_2O)

$$\left(759.772 \text{ g}\right)\left(\frac{1 \text{ kg}}{10^3 \text{ g}}\right) = 0.759772 \text{ kg solvent}$$

$$5) \text{ Mole of solvent in 1 L solution} = \left(759.772 \text{ g } H_2O\right)\left(\frac{1 \text{ mol } H_2O}{18.02 \text{ g } H_2O}\right) = 42.1627 \text{ mol solvent}$$

$$\text{Mass percent of HCl} = \frac{\text{Mass HCl}}{\text{Mass solution}}(100) = \frac{430.228 \text{ g HCl}}{1190 \text{ g solution}}(100) = 36.1536 = \mathbf{36.2\%}$$

$$\text{Molality of HCl} = \frac{\text{Mole HCl}}{\text{Kg solvent}} = \frac{11.8 \text{ mol HCl}}{0.759772 \text{ kg}} = 15.530975 = \mathbf{15.5 \textit{ m}}$$

$$\text{Mole fraction of HCl} = \frac{\text{Mol HCl}}{\text{Mol HCl} + \text{Mol } H_2O} = \frac{11.8 \text{ mol}}{11.8 \text{ mol} + 42.1627 \text{ mol}} = 0.21866956 = \mathbf{0.219}$$

Check: The mass of HCl is about 1/3 of the mass of the solution, so 36% is in the ballpark for the mass% HCl. Since the moles of HCl is about 1/4 the moles of water, the mole fraction of HCl should come out around 0.2.

13.6   <u>Plan:</u> Raoult's law states that the vapor pressure of a solvent is proportional to the mole fraction of the solvent: $P_{solvent} = X_{solvent} \times P°_{solvent}$. To calculate the drop in vapor pressure, a similar relationship is used with the mole fraction of the solute substituted for that of the solvent.
<u>Solution:</u>
Mole fraction of aspirin in methanol:

$$\frac{(2.00 \text{ g})\left(\dfrac{1 \text{ mol aspirin}}{180.15 \text{ g aspirin}}\right)}{(2.00 \text{ g})\left(\dfrac{1 \text{ mol aspirin}}{180.15 \text{ g aspirin}}\right) + (50.0 \text{ g})\left(\dfrac{1 \text{ mol methanol}}{32.04 \text{ g methanol}}\right)}$$

$X_{aspirin} = 7.06381 \times 10^{-3}$ (unrounded)
$\Delta P = X_{aspirin} \, P°_{methanol} = (7.06381 \times 10^{-3})(101 \text{ torr}) = 0.71344 = \textbf{0.713 torr}$
<u>Check:</u> The drop in vapor pressure is expected to be insignificant since the amount of aspirin in solution is so small – only 0.01 mol aspirin in 1.5 mol solvent.

13.7   <u>Plan:</u> The question asks for the concentration of ethylene glycol that would prevent freezing at 0.00°F. First, convert 0.00°F to °C. The change in freezing point of water will then be this temperature subtracted from the freezing point of pure water, 0.00°C. Use this value for ΔT in $\Delta T_f = 1.86°C/m \times$ molality of solution and solve for molality.
<u>Solution:</u>
Temperature conversion

$$T\ (°C) = \left(T(°F) - 32°F\right)\left(\frac{5°C}{9°F}\right) = \left(0.00°F - 32°F\right)\left(\frac{5°C}{9°F}\right) = -17.7778°C \text{ (unrounded)}$$

$$\Delta T_f = m\, K_f$$
$$m = \frac{\Delta T_f}{K_f} = \frac{-17.7778°C}{1.86°C/m} = 9.557956 = \textbf{9.56 } \boldsymbol{m}$$

The minimum concentration of ethylene glycol would have to be 9.56 m in order to prevent the water from freezing at 0.00°F.
<u>Check:</u> In Sample Problem 13.7, a 3.6 m solution lowered the freezing point by about 7 degrees. Thus, to lower the temperature by 18 degrees, the concentration must be 18/7 times 3.6 or 9.3 m. Calculated value of 9.56 m is close to the estimated value.

13.8   <u>Plan:</u> The osmotic pressure is calculated as the product of molarity, temperature, and the gas constant.
<u>Solution:</u>

$$\Pi = MRT = \left(\frac{0.30 \text{ mol}}{L}\right)\left(0.0821\frac{L \cdot atm}{mol \cdot K}\right)\left((273 + 37)K\right) = 7.6353 = \textbf{7.6 atm}$$

<u>Check:</u> The answer, 7.6 atm, is a significant osmotic pressure, which is expected from a relatively concentrated solution of 0.3 M.

13.9   <u>Plan:</u> Calculate the molarity of the solution, then calculate the osmotic pressure of the magnesium chloride solution. Do not forget that $MgCl_2$ is a strong electrolyte, and ionizes to yield three ions per formula unit. This will result in a pressure three times as great as a nonelectrolyte (glucose) solution of equal concentration.
<u>Solution:</u>

a)   Liters of solution = $(100. + 0.952 \text{ g solution})\left(\dfrac{1 \text{ mL}}{1.006 \text{ g}}\right)\left(\dfrac{10^{-3} \text{ L}}{1 \text{ mL}}\right) = 0.1003499$ L (unrounded)

Moles $MgCl_2$ = $(0.952 \text{ g MgCl}_2)\left(\dfrac{1 \text{ mol MgCl}_2}{95.21 \text{ g MgCl}_2}\right) = 9.9989 \times 10^{-3}$ mol $MgCl_2$ (unrounded)

Molarity $MgCl_2$ = $\dfrac{\text{Mol MgCl}_2}{\text{Volume solution}} = \dfrac{0.0099989 \text{ mol}}{0.1003499 \text{ L}} = 0.099640\ M$ $MgCl_2$ (unrounded)

$$\Pi = iMRT = 3\left(\frac{0.099640 \text{ mol}}{L}\right)\left(0.0821\frac{L \cdot atm}{mol \cdot K}\right)\left((273.2 + 20.0)K\right) = 7.1955 = \textbf{7.20 atm}$$

b) The magnesium chloride solution has a higher osmotic pressure, thus, liquid will diffuse from the glucose solution into the magnesium chloride solution. The glucose side will lower and the magnesium chloride side will rise. **Scene C** represents this situation.
Check: Repeat the calculations using approximate values ($M = 0.1$, R = 0.08, and T = 300).

END–OF–CHAPTER PROBLEMS

13.2   When a salt such as NaCl dissolves, ion-dipole forces cause the ions to separate, and many water molecules cluster around each of them in hydration shells. Ion-dipole forces hold the first shell. Additional shells are held by hydrogen bonding to inner shells.

13.4   **Sodium stearate** would be a more effective soap because the hydrocarbon chain in the stearate ion is longer than the chain in the acetate ion. A soap forms suspended particles called micelles with the polar end of the soap interacting with the water solvent molecules and the nonpolar ends forming a nonpolar environment inside the micelle. Oils dissolve in the nonpolar portion of the micelle. Thus, a better solvent for the oils in dirt is a more nonpolar substance. The long hydrocarbon chain in the stearate ion is better at dissolving oils in the micelle than the shorter hydrocarbon chain in the acetate ion.

13.7   **a**) A more concentrated solution will have more solute dissolved in the solvent. Potassium nitrate, KNO_3, is an ionic compound and therefore soluble in a polar solvent like water. Potassium nitrate is not soluble in the nonpolar solvent CCl_4. Because potassium nitrate dissolves to a greater extent in water, **KNO_3 in H_2O** will result in the more concentrated solution.

13.9   To identify the strongest type of intermolecular force, check the formula of the solute and identify the forces that could occur. Then look at the formula for the solvent and determine if the forces identified for the solute would occur with the solvent. The strongest force is ion-dipole followed by dipole-dipole (including H bonds). Next in strength is ion-induced dipole force and then dipole-induced dipole force. The weakest intermolecular interactions are dispersion forces.
a) **Ion-dipole forces** are the strongest intermolecular forces in the solution of the ionic substance cesium chloride in polar water.
b) **Hydrogen bonding** (type of dipole-dipole force) is the strongest intermolecular force in the solution of polar propanone (or acetone) in polar water.
c) **Dipole-induced dipole forces** are the strongest forces between the polar methanol and nonpolar carbon tetrachloride.

13.11   a) **Hydrogen bonding** occurs between the H atom on water and the lone electron pair on the O atom in dimethyl ether (CH_3OCH_3). However, none of the hydrogen atoms on dimethyl ether participates in hydrogen bonding because the C–H bond does not have sufficient polarity.
b) The dipole in water induces a dipole on the Ne(g) atom, so **dipole-induced dipole** interactions are the strongest intermolecular forces in this solution.
c) Nitrogen gas and butane are both nonpolar substances, so **dispersion forces** are the principal attractive forces.

13.13 $CH_3CH_2OCH_2CH_3$ is polar with dipole-dipole interactions as the dominant intermolecular forces. Examine the solutes to determine which has intermolecular forces more similar to those for the diethyl ether. This solute is the one that would be more soluble.
a) **HCl** would be more soluble since it is a covalent compound with dipole-dipole forces, whereas NaCl is an ionic solid. Dipole-dipole forces between HCl and diethyl ether are more similar to the dipole forces in diethyl ether than the ion-dipole forces between NaCl and diethyl ether.
b) **CH_3CHO** (acetaldehyde) would be more soluble. The dominant interactions in H_2O are hydrogen bonding, a stronger type of dipole-dipole force. The dominant interactions in CH_3CHO are dipole-dipole. The solute-solvent interactions between CH_3CHO and diethyl ether are more similar to the solvent intermolecular forces than the forces between H_2O and diethyl ether.

c) **CH₃CH₂MgBr** would be more soluble. CH₃CH₂MgBr has a polar end (–MgBr) and a nonpolar end (CH₃CH₂–), whereas MgBr₂ is an ionic compound. The nonpolar end of CH₃CH₂MgBr and diethyl ether would interact with dispersion forces, while the polar end of CH₃CH₂MgBr and the dipole in diethyl ether would interact with dipole-dipole forces. Recall, that if the polarity continues to increase, the bond will eventually become ionic. There is a continuous sequence from nonpolar covalent to ionic.

13.16   Gluconic acid is a very polar molecule because it has –OH groups attached to every carbon. The abundance of –OH bonds allows gluconic acid to participate in extensive H–bonding with water, hence its great solubility in water. On the other hand, caproic acid has a 5–carbon, nonpolar, hydrophobic ("water hating") tail that does not easily dissolve in water. The dispersion forces in the nonpolar tail are more similar to the dispersion forces in hexane, hence its greater solubility in hexane.

13.18   The nitrogen bases hydrogen bond to their complimentary bases. The flat, N-containing bases stack above each other, which allow extensive interaction through dispersion forces. The exterior negatively charged sugar-phosphate chains form ion-dipole and hydrogen bonds to the aqueous surroundings, but this is of minor importance to the structure.

13.20   Dispersion forces are present between the nonpolar portions of the molecules within the bilayer. Polar groups are present to hydrogen bond or to form dipole-dipole interactions with the aqueous surroundings.

13.25   For a general solvent, the energy changes needed to separate solvent into particles ($\Delta H_{solvent}$), and that needed to mix the solvent and solute particles (ΔH_{mix}) would be combined to obtain $\Delta H_{solution}$.

13.29   This compound would be very soluble in water. A large exothermic value in $\Delta H_{solution}$ (enthalpy of solution) means that the solution has a much lower energy state than the isolated solute and solvent particles, so the system tends to the formation of the solution. Entropy that accompanies dissolution always favors solution formation. Entropy becomes important when explaining why solids with an endothermic $\Delta H_{solution}$ (and higher energy state) are still soluble in water.

13.30

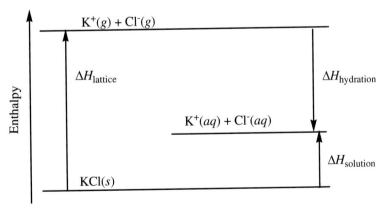

$\Delta H_{solution} > 0$ (endothermic)

Lattice energy values are always positive as energy is required to separate the ions from each other. Hydration energy values are always negative as energy is released when intermolecular forces between ions and water form. Since the heat of solution for KCl is endothermic, the lattice energy must be greater than the hydration energy for an overall input of energy.

13.32   Charge density is the ratio of an ion's charge (regardless of sign) to its volume. An ion's volume is related to its radius. For ions whose charges have the same sign (+ or –), ion size decreases as a group in the periodic table is ascended and as you proceed from left to right in the periodic table. Charge density increases with increasing charge and increases with decreasing size.
a) Both ions have a +1 charge, but the volume of **Na⁺** is smaller, so it has the greater charge density.

b) Sr^{2+} has a greater ionic charge and a smaller size (because it has a greater Z_{eff}), so it has the greater charge density.
c) Na^+ has a smaller ion volume than Cl^-, so it has the greater charge density.
d) O^{2-} has a greater ionic charge and similar ion volume, so it has the greater charge density.
e) OH^- has a smaller ion volume than SH^- (O is smaller than S), so it has the greater charge density.
f) Mg^{2+} has the higher charge density because it has a smaller ion volume.
g) Mg^{2+} has the higher charge density because it has both a smaller ion volume and greater charge.
h) CO_3^{2-} has the higher charge density because it has both a smaller ion volume and greater charge.

13.34   The ion with the greater charge density will have the larger $\Delta H_{hydration}$.
a) Na^+ would have a larger $\Delta H_{hydration}$ than Cs^+ since its charge density is greater than that of Cs^+.
b) Sr^{2+} would have a larger $\Delta H_{hydration}$ than Rb^+.
c) Na^+ would have a larger $\Delta H_{hydration}$ than Cl^-.
d) O^{2-} would have a larger $\Delta H_{hydration}$ than F^-.
e) OH^- would have a larger $\Delta H_{hydration}$ than SH^-.
f) Mg^{2+} would have a larger $\Delta H_{hydration}$ than Ba^{2+}.
g) Mg^{2+} would have a larger $\Delta H_{hydration}$ than Na^+.
h) CO_3^{2-} would have a larger $\Delta H_{hydration}$ than NO_3^-.

13.36   a) The two ions in potassium bromate are K^+ and BrO_3^-. The heat of solution for ionic compounds is
$\Delta H_{soln} = \Delta H_{lattice} + \Delta H_{hydr of the ions}$. Therefore, the combined heats of hydration for the ions is ($\Delta H_{soln} - \Delta H_{lattice}$) or
41.1 kJ/mol – 745 kJ/mol = –703.9 = **–704 kJ/mol**.
b) K^+ ion contributes more to the heat of hydration because it has a smaller size and, therefore, a greater charge density.

13.38   Entropy increases as the possible states for a system increases.
a) Entropy **increases** as the gasoline is burned. Gaseous products at a higher temperature form.
b) Entropy **decreases** as the gold is separated from the ore. Pure gold has only the arrangement of gold atoms next to gold atoms, while the ore mixture has a greater number of possible arrangements among the components of the mixture.
c) Entropy **increases** as a solute dissolves in the solvent.

13.41   Add a pinch of the solid solute to each solution. A saturated solution contains the maximum amount of dissolved solute at a particular temperature. When additional solute is added to this solution, it will remain undissolved. An unsaturated solution contains less than the maximum amount of dissolved solute and so will dissolve added solute. A supersaturated solution is unstable and addition of a "seed" crystal of solute causes the excess solute to crystallize immediately, leaving behind a saturated solution.

13.44   a) Increasing pressure for a gas **increases** the solubility of the gas according to Henry's law.
b) Increasing the volume of a gas causes a decrease in its pressure (Boyle's Law), which **decreases** the solubility of the gas.

13.46   a) Solubility for a gas is calculated from Henry's law: $S_{gas} = k_H \times P_{gas}$. S_{gas} is expressed as mol/L, so convert moles of O_2 to mass of O_2 using the molar mass.
$$S_{gas} = k_H \times P_{gas}$$
$$S_{gas} = \left(1.28 \times 10^{-3}\, \frac{mol}{L \cdot atm}\right)(1.00\ atm) = 1.28 \times 10^{-3}\ mol/L$$
$$\left(\frac{1.28 \times 10^{-3}\ mol\ O_2}{L}\right)\left(\frac{32.0\ g\ O_2}{1\ mol\ O_2}\right)(2.50\ L) = 0.1024 = \textbf{0.102 g } O_2$$

b) The amount of gas that will dissolve in a given volume decreases proportionately with the partial pressure of the gas, so

$$S_{gas} = \left(1.28 \times 10^{-3} \frac{mol}{L \cdot atm}\right)(0.209 \text{ atm}) = 2.6752 \times 10^{-4} \text{ mol/L}$$

$$\left(\frac{2.6752 \times 10^{-4} \text{ mol } O_2}{L}\right)\left(\frac{32.0 \text{ g } O_2}{1 \text{ mol } O_2}\right)(2.50 \text{ L}) = 0.0214016 = \textbf{0.0214 g } O_2$$

13.49   Solubility for a gas is calculated from Henry's law: $S_{gas} = k_H \times P_{gas}$.
         $S_{gas} = (3.7 \times 10^{-2} \text{ mol/L} \cdot atm)(5.5 \text{ atm}) = 0.2035 = \textbf{0.20 mol/L}$

13.52   Refer to the Table 13.5 for the different methods of expressing concentration.
         a) **Molarity** and **parts-by-volume** (% w/v or % v/v) include the volume of the solution.
         b) **Parts-by-mass** (% w/w) include the mass of solution directly. (Others may involve the mass indirectly.)
         c) **Molality** includes the mass of the solvent.

13.54   Converting between molarity and molality involves conversion between volume of solution and mass of solution. Both of these quantities are given so interconversion is possible. To convert to mole fraction requires that the mass of solvent be converted to moles of solvent. Since the identity of the solvent is not given, conversion to mole fraction is not possible if the molar mass is not known.

13.56   Convert the masses to moles and the volumes to liters and use the definition of molarity: $M = \dfrac{\text{mol of solute}}{V(L) \text{ of solution}}$

         a) Molarity = $\left(\dfrac{32.3 \text{ g } C_{12}H_{22}O_{11}}{100. \text{ mL}}\right)\left(\dfrac{1 \text{ mL}}{10^{-3} \text{ L}}\right)\left(\dfrac{1 \text{ mol } C_{12}H_{22}O_{11}}{342.30 \text{ g } C_{12}H_{22}O_{11}}\right) = 0.943617 = \textbf{0.944 } \textit{M } \textbf{C}_{12}\textbf{H}_{22}\textbf{O}_{11}$

         b) Molarity = $\left(\dfrac{5.80 \text{ g } LiNO_3}{505 \text{ mL}}\right)\left(\dfrac{1 \text{ mL}}{10^{-3} \text{ L}}\right)\left(\dfrac{1 \text{ mol } LiNO_3}{68.95 \text{ g } LiNO_3}\right) = 0.166572 = \textbf{0.167 } \textit{M } \textbf{LiNO}_3$

13.58   Dilution calculations can be done using $M_{conc}V_{conc} = M_{dil}V_{dil}$
         a) $M_{conc} = 0.240 \, M \text{ NaOH} \quad V_{conc} = 78.0 \text{ mL} \quad M_{dil} = ? \quad V_{dil} = 0.250 \text{ L}$

         $$M_{dil} = M_{conc} V_{conc} / V_{dil} = \frac{(0.240 \text{ M})(78.0 \text{ mL})}{(0.250 \text{ L})}\left(\frac{10^{-3} \text{ L}}{1 \text{ mL}}\right) = 0.07488 = \textbf{0.0749 } \textit{M}$$

         b) $M_{conc} = 1.2 \, M \text{ HNO}_3 \quad V_{conc} = 38.5 \text{ mL} \quad M_{dil} = ? \quad V_{dil} = 0.130 \text{ L}$

         $$M_{dil} = M_{conc} V_{conc} / V_{dil} = \frac{(1.2 \text{ M})(38.5 \text{ mL})}{(0.130 \text{ L})}\left(\frac{10^{-3} \text{ L}}{1 \text{ mL}}\right) = 0.355385 = \textbf{0.36 } \textit{M}$$

13.60   a) Find the number of moles KH_2PO_4 needed to make 365 mL of this solution. Convert moles to mass using the molar mass of KH_2PO_4 (Molar mass = 136.09 g/mol)

         $$\text{Mass } KH_2PO_4 = (365 \text{ mL})\left(\frac{10^{-3} \text{ L}}{1 \text{ mL}}\right)\left(\frac{8.55 \times 10^{-2} \text{ mol } KH_2PO_4}{L}\right)\left(\frac{136.09 \text{ g } KH_2PO_4}{1 \text{ mol } KH_2PO_4}\right)$$

         $$= 4.24703 = 4.25 \text{ g } KH_2PO_4$$
         Add **4.25 g KH₂PO₄** to enough water to make 365 mL of aqueous solution.
         b) Use the relationship $M_{conc}V_{conc} = M_{dil}V_{dil}$ to find the volume of 1.25 M NaOH needed.
         $\quad M_{conc} = 1.25 \, M \text{ NaOH} \quad V_{conc} = ? \quad M_{dil} = 0.335 \, M \text{ NaOH} \quad V_{dil} = 465 \text{ mL}$

         $$V_{conc} = M_{dil} V_{dil} / M_{conc} = \frac{(0.335 \text{ M})(465 \text{ mL})}{(1.25 \text{ M})} = 124.62 = \textbf{125 mL}$$

         Add **125 mL** of 1.25 M NaOH to enough water to make 465 mL of solution.

13.62   a) To find the mass of KBr needed, find the moles of KBr in 1.40 L of a 0.288 M solution and convert to grams using molar mass of KBr.

$$\text{Mass KBr} = (1.40\ \text{L})\left(\frac{0.288\ \text{mol KBr}}{\text{L}}\right)\left(\frac{119.00\ \text{g KBr}}{1\ \text{mol KBr}}\right) = 47.9808 = 48.0\ \text{g KBr}$$

To make the solution, weigh **48.0 g KBr** and then dilute to 1.40 L with distilled water.

b) To find the volume of the concentrated solution that will be diluted to 255 mL, use $M_{conc}V_{conc} = M_{dil}V_{dil}$ and solve for V_{conc}.

$$M_{conc} = 0.264\ M\ \text{LiNO}_3 \quad V_{conc} = ? \quad M_{dil} = 0.0856\ M\ \text{LiNO}_3 \quad V_{dil} = 255\ \text{mL}$$

$$V_{conc} = M_{dil}\ V_{dil}\ /\ M_{conc} = (0.0856\ M)\ (255\ \text{mL})\ /\ (0.264\ M) = 82.68182 = 82.7\ \text{mL}$$

To make the 0.0856 M solution, measure **82.7 mL** of the 0.264 M solution and add distilled water to make a total of 255 mL.

13.64   Molality, $m = \dfrac{\text{moles of solute}}{\text{kg of solvent}}$

a) m glycine $= \dfrac{85.4\ \text{g Glycine}\left(\dfrac{1\ \text{mol Glycine}}{75.07\ \text{g Glycine}}\right)}{(1.270\ \text{kg})} = 0.895752 = \mathbf{0.896\ \textit{m}\ glycine}$

b) m glycerol $= \dfrac{8.59\ \text{g Glycerol}\left(\dfrac{1\ \text{mol Glycerol}}{92.09\ \text{g Glycerol}}\right)}{(77.0\ \text{g})}\left(\dfrac{10^3\ \text{g}}{1\ \text{kg}}\right) = 1.2114 = \mathbf{1.21\ \textit{m}\ glycerol}$

13.66   Molality, $m = \dfrac{\text{moles of solute}}{\text{kg of solvent}}$ Use density to convert volume to mass.

$$m\ \text{benzene} = \dfrac{(44.0\ \text{mL C}_6\text{H}_6)\left(\dfrac{0.877\ \text{g}}{\text{mL}}\right)\left(\dfrac{1\ \text{mol C}_6\text{H}_6}{78.11\ \text{g C}_6\text{H}_6}\right)}{(167\ \text{mL C}_6\text{H}_{14})\left(\dfrac{0.660\ \text{g}}{\text{mL}}\right)}\left(\dfrac{10^3\ \text{g}}{1\ \text{kg}}\right) = 4.48214 = \mathbf{4.48\ \textit{m}\ C_6H_6}$$

13.68   a) The total mass of the <u>solution</u> is 3.10×10^2 g, so mass$_{solute}$ + mass$_{solvent}$ = 3.10×10^2 g.

$$\text{Mass of C}_2\text{H}_6\text{O}_2 \text{ in } 1000\ \text{g (1 kg) of H}_2\text{O} = \left(\dfrac{0.125\ \text{mol C}_2\text{H}_6\text{O}_2}{1\ \text{kg H}_2\text{O}}\right)\left(\dfrac{62.07\ \text{g C}_2\text{H}_6\text{O}_2}{1\ \text{mol C}_2\text{H}_6\text{O}_2}\right)$$

$$= 7.75875\ \text{g C}_2\text{H}_6\text{O}_2 \text{ in } 1000\ \text{g H}_2\text{O}$$

Grams of this solution = 1000 g H_2O + 7.75875 g $C_2H_6O_2$ = 1007.75875 g

$$\text{Mass C}_2\text{H}_6\text{O}_2 = \left(\dfrac{7.75875\ \text{g C}_2\text{H}_6\text{O}_2}{1007.75875\ \text{g solution}}\right)(3.10 \times 10^2\ \text{g solution}) = = 2.386695\ \text{g C}_2\text{H}_6\text{O}_2$$

Mass$_{solvent}$ = 3.10×10^2 g – mass$_{solute}$ = 3.10×10^2 g – 2.386695 g $C_2H_6O_2$ = 307.613305 = 308 g H_2O

Therefore, **add 2.39 g C$_2$H$_6$O$_2$ to 308 g of H$_2$O** to make a 0.125 m solution.

b) This is a disguised dilution problem. First, determine the amount of solute in your target solution:

$$\left(\dfrac{2.20\%}{100\%}\right)(1.20\ \text{kg}) = 0.0264\ \text{kg HNO}_3 \text{ (solute)}$$

Then determine the amount of the concentrated acid solution needed to get 0.0264 kg solute:

$$\left(\dfrac{52.0\%}{100\%}\right)(\text{mass needed}) = 0.0264\ \text{kg}$$

Mass solute needed = 0.050769 = 0.0508 kg

Mass solvent = Mass solution – Mass solute = 1.20 kg – 0.050769 kg = 1.149231 = 1.15 kg

Add 0.0508 kg of the 52.0% (w/w) HNO$_3$ to 1.15 kg H$_2$O to make 1.20 kg of 2.20% (w/w) HNO$_3$.

13.70   a) Mole fraction is moles of isopropanol per total moles.

$$X_{isopropanol} = \frac{0.35 \text{ mol Isopropanol}}{(0.35 + 0.85) \text{ mol}} = 0.2916667 = \mathbf{0.29} \text{ (Notice that mole fractions have no units.)}$$

b) Mass percent $= \dfrac{\text{mass of solute}}{\text{mass of solution}}(100)$. From the mole amounts, find the masses

of isopropanol and water:

$$\text{Mass isopropanol} = (0.35 \text{ mol C}_3\text{H}_7\text{OH})\left(\frac{60.09 \text{ g C}_3\text{H}_7\text{OH}}{1 \text{ mol C}_3\text{H}_7\text{OH}}\right) = 21.0315 \text{ g isopropanol}$$

$$\text{Mass water} = (0.85 \text{ mol H}_2\text{O})\left(\frac{18.02 \text{ g H}_2\text{O}}{1 \text{ mol H}_2\text{O}}\right) = 15.317 \text{ g water (unrounded)}$$

$$\text{Percent isopropanol} = \frac{(21.0315 \text{ g Isopropanol})}{(21.0315 + 15.317) \text{ g}} \times 100\% = 57.860710 = \mathbf{58\%} \text{ isopropanol}$$

c) Molality of isopropanol is moles of isopropanol per kg of solvent.

$$\text{Molality isopropanol} = \frac{0.35 \text{ mol Isopropanol}}{15.317 \text{ g Water}}\left(\frac{10^3 \text{ g}}{1 \text{ kg}}\right) = 22.85043 = \mathbf{23\ } \textit{m} \text{ isopropanol}$$

13.72   The density of water is 1.00 g/mL. The mass of water is:

$$\text{Mass of water} = (0.500 \text{ L})\left(\frac{1 \text{ mL}}{10^{-3} \text{ L}}\right)\left(\frac{1.00 \text{ g}}{1 \text{ mL}}\right)\left(\frac{1 \text{ kg}}{10^3 \text{ g}}\right) = 0.500 \text{ kg}$$

$$0.400 \ \textit{m} \text{ CsBr} = \frac{\text{moles CsBr}}{0.500 \text{ kg H}_2\text{O}}$$

Moles CsBr = 0.2000 mol (unrounded)

$$\text{Mass of CsBr} = (0.200 \text{ mol CsBr})\left(\frac{212.8 \text{ g CsBr}}{1 \text{ mol CsBr}}\right) = 42.56 = \mathbf{42.6 \text{ g CsBr}}$$

Mass H_2O = Mass of solution – Mass of CsBr =

$$(0.500 \text{ kg})\left(\frac{10^3 \text{ g}}{1 \text{ kg}}\right) - 42.56 \text{ g CsBr} = 457.44 \text{ g H}_2\text{O}$$

$$\text{Moles H}_2\text{O} = (457.44 \text{ g H}_2\text{O})\left(\frac{1 \text{ mol H}_2\text{O}}{18.02 \text{ g H}_2\text{O}}\right) = 25.38513 \text{ mol H}_2\text{O}$$

$$X_{CsBr} = \frac{\text{mol CsBr}}{\text{mol CsBr} + \text{mol H}_2\text{O}} = \frac{0.2000 \text{ mol CsBr}}{(0.2000 + 25.38513) \text{ mol}} = 7.817 \times 10^{-3} = \mathbf{7.82 \times 10^{-3}}$$

$$\text{Percent CsBr} = \frac{\text{mass of CsBr}}{\text{mass of solution}}(100) = \frac{(42.56 \text{ g CsBr})}{(42.56 + 457.44) \text{ g}} \times 100\% = 8.512 = \mathbf{8.51\%} \text{ CsBr}$$

13.74   The information given is 8.00 mass % NH_3 solution with a density of 0.9651 g/mL.
For convenience, choose exactly 100.00 grams of solution.
Determine some fundamental quantities:

$$\text{Mass of NH}_3 = (100 \text{ g solution})\left(\frac{8.00\% \text{ NH}_3}{100\% \text{ solution}}\right) = 8.00 \text{ g NH}_3$$

$$\text{Mass H}_2\text{O} = \text{mass of solution} - \text{mass NH}_3 = (100.00 - 8.00) \text{ g} = 92.00 \text{ g H}_2\text{O}$$

$$\text{Moles NH}_3 = (8.00 \text{ g NH}_3)\left(\frac{1 \text{ mol NH}_3}{17.03 \text{ g NH}_3}\right) = 0.469759 \text{ mol NH}_3 \text{ (unrounded)}$$

$$\text{Moles H}_2\text{O} = \left(92.00 \text{ g H}_2\text{O}\right)\left(\frac{1 \text{ mol H}_2\text{O}}{18.02 \text{ g H}_2\text{O}}\right) = 5.1054 \text{ mol H}_2\text{O (unrounded)}$$

$$\text{Volume solution} = \left(100.00 \text{ g solution}\right)\left(\frac{1 \text{ mL solution}}{0.9651 \text{ g solution}}\right)\left(\frac{10^{-3} \text{ L}}{1 \text{ mL}}\right) = 0.103616 \text{ L (unrounded)}$$

Using the above fundamental quantities and the definitions of the various units:

$$\text{Molality} = \frac{\text{Moles solute}}{\text{kg solvent}} = \left(\frac{0.469759 \text{ mol NH}_3}{92.00 \text{ g H}_2\text{O}}\right)\left(\frac{10^3 \text{ g}}{1 \text{ kg}}\right) = 5.106076 = \textbf{5.11 } \textit{m} \textbf{ NH}_3$$

$$\text{Molarity} = \frac{\text{Moles solute}}{\text{L solution}} = \left(\frac{0.469759 \text{ mol NH}_3}{0.103616 \text{ L}}\right) = 4.53365 = \textbf{4.53 } \textit{M} \textbf{ NH}_3$$

$$\text{Mole fraction} = \text{X} = \frac{\text{Moles substance}}{\text{total moles}} = \frac{0.469759 \text{ mol NH}_3}{\left(0.469759 + 5.1054\right)\text{mol}} = 0.084259 = \textbf{0.0843}$$

13.76 $$\text{ppm} = \left(\frac{\text{mass solute}}{\text{mass solution}}\right) \text{ x } 10^6$$

The mass of 100.0 L of waste water solution is $\left(100.0 \text{ L solution}\right)\left(\frac{1 \text{ mL}}{10^{-3} \text{ L}}\right)\left(\frac{1.001 \text{ g}}{1 \text{ mL}}\right) = 1.001 \text{ x } 10^5 \text{ g.}$

$$\text{ppm Ca}^{2+} = \left(\frac{0.25 \text{ g Ca}^{2+}}{1.001 \text{ x } 10^5 \text{ g solution}}\right) \text{ x } 10^6 = 2.49750 = \textbf{2.5 ppm Ca}^{2+}$$

$$\text{ppm Mg}^{2+} = \left(\frac{0.056 \text{ g Mg}^{2+}}{1.001 \text{ x } 10^5 \text{ g solution}}\right) \text{ x } 10^6 = 0.5594406 = \textbf{0.56 ppm Mg}^{2+}$$

13.80   The "strong" in "strong electrolyte" refers to the ability of an electrolyte solution to conduct a large current. This conductivity occurs because solutes that are strong electrolytes dissociate completely into ions when dissolved in water.

13.82   The boiling point temperature is higher and the freezing point temperature is lower for the solution compared to the solvent because the addition of a solute lowers the freezing point and raises the boiling point of a liquid.

13.85   A dilute solution of an electrolyte behaves more ideally than a concentrated one. With increasing concentration, the effective concentration deviates from the molar concentration because of ionic attractions.. Thus, the more dilute **0.050 *m* NaF** solution has a boiling point closer to its predicted value.

13.88   Strong electrolytes are substances that produce a large number of ions when dissolved in water; strong acids and bases and soluble salts are strong electrolytes. Weak electrolytes produce few ions when dissolved in water; weak acids and bases are weak electrolytes. Nonelectrolytes produce no ions when dissolved in water. Molecular compounds other than acids and bases are nonelectrolytes.
a) **Strong electrolyte** When hydrogen chloride is bubbled through water, it dissolves and dissociates completely into H^+ (or H_3O^+) ions and Cl^- ions. HCl is a strong acid.
b) **Strong electrolyte** Potassium nitrate is a soluble salt.
c) **Nonelectrolyte** Glucose solid dissolves in water to form individual $C_6H_{12}O_6$ molecules, but these units are not ionic and therefore do not conduct electricity. Glucose is a molecular compound.
d) **Weak electrolyte** Ammonia gas dissolves in water, but is a weak base that forms few NH_4^+ and OH^- ions.

13.90 To count solute particles in a solution of an ionic compound, count the number of ions per mole and multiply by the number of moles in solution. For a covalent compound, the number of particles equals the number of molecules.

a) $\left(\dfrac{0.3 \text{ mol KBr}}{L}\right)\left(\dfrac{2 \text{ mol particles}}{1 \text{ mol KBr}}\right)(1 \text{ L}) = \textbf{0.6 mol of particles}$

Each KBr forms one K^+ ion and and one Br^- ion, 2 particles for each KBr.

b) $\left(\dfrac{0.065 \text{ mol HNO}_3}{L}\right)\left(\dfrac{2 \text{ mol particles}}{1 \text{ mol HNO}_3}\right)(1 \text{ L}) = \textbf{0.13 mol of particles}$

HNO_3 is a strong acid that forms $H^+(H_3O^+)$ ions and NO_3^- ions in aqueous solution.

c) $\left(\dfrac{10^{-4} \text{ mol KHSO}_4}{L}\right)\left(\dfrac{2 \text{ mol particles}}{1 \text{ mol KHSO}_4}\right)(1 \text{ L}) = \textbf{2 x } 10^{-4} \textbf{ mol of particles}$

Each $KHSO_4$ forms 1 K^+ ion and 1 HSO_4^- ion in aqueous solution, 2 particles for each $KHSO_4$.

d) $\left(\dfrac{0.06 \text{ mol C}_2\text{H}_5\text{OH}}{L}\right)\left(\dfrac{1 \text{ mol particles}}{1 \text{ mol C}_2\text{H}_5\text{OH}}\right)(1 \text{ L}) = \textbf{0.06 mol of particles}$

Ethanol is not an ionic compound so each molecule dissolves as one particle. The number of moles of particles is the same as the number of moles of molecules, **0.06 mol** in 1 L.

13.92 The magnitude of freezing point depression is directly proportional to molality.

a) Molality of $CH_3OH = \dfrac{(11.0 \text{ g CH}_3\text{OH})}{(100. \text{ g H}_2\text{O})}\left(\dfrac{1 \text{ mol CH}_3\text{OH}}{32.04 \text{ g CH}_3\text{OH}}\right)\left(\dfrac{10^3 \text{ g}}{1 \text{ kg}}\right) = 3.4332085 = 3.43 \ m \text{ CH}_3\text{OH}$

Molality of $CH_3CH_2OH = \dfrac{(22.0 \text{ g CH}_3\text{CH}_2\text{OH})}{(200. \text{ g H}_2\text{O})}\left(\dfrac{1 \text{ mol CH}_3\text{CH}_2\text{OH}}{46.07 \text{ g CH}_3\text{CH}_2\text{OH}}\right)\left(\dfrac{10^3 \text{ g}}{1 \text{ kg}}\right)$

$= 2.387671 = 2.39 \ m \text{ CH}_3\text{CH}_2\text{OH}$

The molality of methanol, CH_3OH, in water is 3.43 m whereas the molality of ethanol, CH_3CH_2OH, in water is 2.39 m. Thus, **CH_3OH/H_2O solution** has the lower freezing point.

b) Molality of $H_2O = \dfrac{(20.0 \text{ g H}_2\text{O})}{(1.00 \text{ kg CH}_3\text{OH})}\left(\dfrac{1 \text{ mol H}_2\text{O}}{18.02 \text{ g H}_2\text{O}}\right) = 1.10988 = 1.11 \ m \text{ H}_2\text{O}$

Molality of $CH_3CH_2OH = \dfrac{(20.0 \text{ g CH}_3\text{CH}_2\text{OH})}{(1.00 \text{ kg CH}_3\text{OH})}\left(\dfrac{1 \text{ mol CH}_3\text{CH}_2\text{OH}}{46.07 \text{ g CH}_3\text{CH}_2\text{OH}}\right) = 0.434122 = 0.434 \ m \text{ CH}_3\text{CH}_2\text{OH}$

The molality of H_2O in CH_3OH is 1.11 m, whereas CH_3CH_2OH in CH_3OH is 0.434 m. Therefore, **H_2O/CH_3OH solution** has the lower freezing point.

13.94 To rank the solutions in order of increasing osmotic pressure, boiling point, freezing point, and vapor pressure, convert the molality of each solute to molality of particles in the solution. The higher the molality of particles, the higher the osmotic pressure, the higher the boiling point, the lower the freezing point, and the lower the vapor pressure at a given temperature.

(I) $(0.100 \ m \text{ NaNO}_3)\left(\dfrac{2 \text{ mol particles}}{1 \text{ mol NaNO}_3}\right) = 0.200 \ m \text{ ions}$

$NaNO_3$ consists of Na^+ ions and NO_3^- ions, 2 particles for each $NaNO_3$.

(II) $(0.100 \ m \text{ glucose})\left(\dfrac{1 \text{ mol particles}}{1 \text{ mol glucose}}\right) = 0.100 \ m \text{ molecules}$

Ethanol is not an ionic compound so each molecule dissolves as one particle. The number of moles of particles is the same as the number of moles of molecules.

(III) $(0.100 \ m \text{ CaCl}_2)\left(\dfrac{3 \text{ mol particles}}{1 \text{ mol CaCl}_2}\right) = 0.300 \ m \text{ ions}$

$CaCl_2$ consists of Ca^{+2} ions and Cl^- ions, 3 particles for each $CaCl_2$.

a) Osmotic pressure: $\Pi_{II} < \Pi_I < \Pi_{III}$
b) Boiling point: $\mathbf{bp_{II} < bp_I < bp_{III}}$
c) Freezing point: $\mathbf{fp_{III} < fp_I < fp_{II}}$
d) Vapor pressure at 50°C: $\mathbf{vp_{III} < vp_I < vp_{II}}$

13.96   The mol fraction of solvent affects the vapor pressure according to the equation: $P_{solvent} = X_{solvent}P^\circ_{solvent}$

$$\text{Moles } C_3H_8O_3 = \left(34.0 \text{ g } C_3H_8O_3\right)\left(\frac{1 \text{ mol } C_3H_8O_3}{92.09 \text{ g } C_3H_8O_3}\right) = 0.369204 \text{ mol } C_3H_8O_3 \text{ (unrounded)}$$

$$\text{Moles } H_2O = \left(500.0 \text{ g } H_2O\right)\left(\frac{1 \text{ mol } H_2O}{18.02 \text{ g } H_2O}\right) = 27.7469 \text{ mol } H_2O \text{ (unrounded)}$$

$$X_{solvent} = \frac{\text{mol } H_2O}{\text{mol } H_2O + \text{mol glycerol}} = \frac{27.7469 \text{ mol } H_2O}{27.7469 \text{ mol } H_2O + 0.369204 \text{ mol glycerol}} = 0.9868686$$

$P_{solvent} = X_{solvent}P^\circ_{solvent} = (0.9868686)(23.76 \text{ torr}) = 23.447998 = \mathbf{23.4 \text{ torr}}$

13.98   The change in freezing point is calculated from $\Delta T_f = iK_f m$, where K_f is 1.86°C/m for aqueous solutions, i is the van't Hoff factor, and m is the molality of particles in solution. Since urea is a covalent compound and does not ionize in water, i = 1. Once ΔT_f is calculated, the freezing point is determined by subtracting it from the freezing point of pure water (0.00°C).
$\Delta T_f = iK_f m = (1)(1.86°C/m)(0.251 m) = 0.46686°C$ (unrounded)
The freezing point is $0.00°C - 0.46686°C = -0.46686 = \mathbf{-0.467°C}$.

13.100   The boiling point of a solution is increased relative to the pure solvent by the relationship $\Delta T_b = iK_b m$. Vanillin is a nonelectrolyte so i = 1. The molality must be calculated, and K_b is given (1.22°C/m).

$$\text{Molality of vanillin} = \frac{\text{moles of vanillin}}{\text{kg of solvent (ethanol)}} = \frac{\left(6.4 \text{ g Vanillin}\right)\left(\frac{1 \text{ mol Vanillin}}{152.14 \text{ g Vanillin}}\right)}{\left(50.0 \text{ g Ethanol}\right)}\left(\frac{10^3 \text{ g}}{1 \text{ kg}}\right)$$

$= 0.841330354 \, m$ vanillin (unrounded)
$\Delta T_b = iK_b m = (1)(1.22°C/m)(0.841330354 m) = 1.02642°C$ (unrounded)
The boiling point is $78.5°C + 1.02642°C = 79.52642 = \mathbf{79.5°C}$.

13.102   The molality of the solution can be determined from the relationship $\Delta T_f = iK_f m$ with the value 1.86°C/m inserted for K_f and i = 1 for the nonelectrolyte ethylene glycol (ethylene glycol is a covalent compound that will form one particle per molecule when dissolved). Convert the freezing point of the solution to °C and find ΔT_f:
$°C = (5/9)(°F - 32.0) = (5/9)((-12.0)°F - 32.0) = -24.44444°C$ (unrounded)
$\Delta T_f = (0.00 - (-24.44444))°C = 24.44444°C$

$$m = \frac{\Delta T_f}{K_f} = \frac{24.44444°C}{1.86 \, °C/m} = 13.14217 \, m \text{ (unrounded)}$$

Ethylene glycol will be abbreviated as EG.
Multiply the molality by the given mass of solvent to find the mass of ethylene glycol that must be in solution.

$$\text{Mass ethylene glycol} = \left(\frac{13.14217 \text{ mol EG}}{1 \text{ kg } H_2O}\right)\left(14.5 \text{ kg } H_2O\right)\left(\frac{62.07 \text{ g EG}}{1 \text{ mol EG}}\right)$$

$= 1.18282 \times 10^4 = \mathbf{1.18 \times 10^4 \text{ g ethylene glycol}}$
To prevent the solution from freezing, dissolve a minimum of 1.13×10^4 g ethylene glycol in 14.5 kg water.

13.104   Convert the mass percent to molality and use $\Delta T = iK_b m$ to find the van't Hoff factor.
a) Assume exactly 100 grams of solution.

$$\text{Mass NaCl} = \left(100 \text{ g solution}\right)\left(\frac{1.00\% \text{ NaCl}}{100\% \text{ solution}}\right) = 1.00 \text{g NaCl}$$

$$\text{Moles NaCl} = \left(1.00 \text{ g NaCl}\right)\left(\frac{1 \text{ mol NaCl}}{58.44 \text{ g NaCl}}\right) = 0.0171116 \text{ mol NaCl}$$

Mass H_2O = 100.00 g solution – 1.00g NaCl = 99.00 g H_2O

$$\text{Molality NaCl} = \frac{\text{mole NaCl}}{\text{kg } H_2O} = \frac{0.0171116 \text{ mol NaCl}}{99.00 \text{ g } H_2O}\left(\frac{10^3 \text{ g}}{1 \text{ kg}}\right) = 0.172844 = \mathbf{0.173 \textit{ m} \text{ NaCl}}$$

Calculate ΔT = (0.000 – (–0.593))°C = 0.593°C

$$\Delta T_f = iK_f m$$

$$i = \frac{\Delta T_f}{K_f m} = \frac{0.593\,^{\circ}C}{\left(1.86°C/m\right)\left(0.172844 \ m\right)} = 1.844537 = \mathbf{1.84}$$

The value of i should be close to 2 because NaCl dissociates into 2 particles when dissolving in water.

b) For acetic acid, CH_3COOH:

Assume exactly 100 grams of solution.

$$\text{Mass } CH_3COOH = \left(100 \text{ g solution}\right)\left(\frac{0.500\% \ CH_3COOH}{100\% \text{ solution}}\right) = 0.500 \text{ g } CH_3COOH$$

$$\text{Moles } CH_3COOH = \left(0.500 \text{ g } CH_3COOH\right)\left(\frac{1 \text{ mol } CH_3COOH}{60.05 \text{ g } CH_3COOH}\right) = 0.0083264 \text{ mol } CH_3COOH$$

Mass H_2O = 100.00 g solution – 0.500 g CH_3COOH = 99.500 g H_2O

$$\text{Molality } CH_3COOH = \frac{\text{mole } CH_3COOH}{\text{kg } H_2O} = \frac{0.0083264 \text{ mol } CH_3COOH}{99.500 \text{ g } H_2O}\left(\frac{10^3 \text{ g}}{1 \text{ kg}}\right)$$

$$= 0.083682 = \mathbf{0.0837 \textit{ m} \ CH_3COOH}$$

Calculate ΔT = (0.000 – (–0.159))°C = 0.159°C

$$\Delta T_f = iK_f m$$

$$i = \frac{\Delta T_f}{K_f m} = \frac{0.159\,^{\circ}C}{\left(1.86°C/m\right)\left(0.083682 \ m\right)} = 1.02153 = \mathbf{1.02}$$

Acetic acid is a weak acid and dissociates to a small extent in solution, hence a van't Hoff factor that is close to 1.

13.106 Osmotic pressure is calculated from the molarity of particles, gas constant and temperature. Convert the mass of sucrose to moles using the molar mass, and then to molarity. Sucrose is a nonelectrolyte so i = 1.

$$T = (273 + 20.)K = 293 \text{ K}$$

$$\text{Molarity} = \frac{\text{moles of sucrose}}{\text{volume of solution}} = \frac{\left(3.55 \text{ g sucrose}\right)\left(\frac{1 \text{ mol sucrose}}{342.30 \text{ g sucrose}}\right)}{1.0 \text{ L}} = 1.0371 \times 10^{-2} \text{ M sucrose}$$

$$\Pi = iMRT = (1)\,(1.0371 \times 10^{-2} \text{ mol/L})\,(0.0821 \text{ L·atm/mol·K})\,(293 \text{ K}) = 0.2494778 = \mathbf{0.249 \text{ atm}}$$

A pressure greater than 0.249 atm must be applied to obtain pure water from a 3.55 g/L solution.

13.108 The pressure of each compound is proportional to its mole fraction according to Raoult's law: $P_A = X_A P°_A$

$$X_{CH_2Cl_2} = \frac{\text{moles } CH_2Cl_2}{\text{moles } CH_2Cl_2 + \text{mol } CCl_4} = \frac{1.60 \text{ mol}}{1.60 + 1.10 \text{ mol}} = 0.592593$$

$$X_{CCl_4} = \frac{\text{moles } CCl_4}{\text{moles } CH_2Cl_2 + \text{mol } CCl_4} = \frac{1.10 \text{ mol}}{1.60 + 1.10 \text{ mol}} = 0.407407$$

$$P_A = X_A P°_A$$
$$= (0.592593)\,(352 \text{ torr}) = 208.593 = \mathbf{209 \text{ torr } CH_2Cl_2}$$
$$= (0.407407)\,(118 \text{ torr}) = 48.0740 = \mathbf{48.1 \text{ torr } CCl_4}$$

13.109 The fluid inside a bacterial cell is **both a solution and a colloid**. It is a solution of ions and small molecules and a colloid of large molecules, proteins, and nucleic acids.

13.113 Soap micelles have nonpolar "tails" pointed inward and anionic "heads" pointed outward. The charges on the "heads" on one micelle repel the "heads" on a neighboring micelle because the charges are the same. This repulsion between soap micelles keeps them from coagulating.
Soap is more effective in **freshwater** than in seawater because the divalent cations in seawater combine with the anionic "head" to form an insoluble precipitate.

13.117 To find the volume of seawater needed, substitute the given information into the equation that describes the ppb concentration, account for extraction efficiency, and convert mass to volume using the density of seawater.

$$1.1 \times 10^{-2} \text{ ppb} = \frac{\text{mass Gold}}{\text{mass seawater}} \times 10^9 = \frac{31.1 \text{ g Au}}{\text{mass seawater}} \times 10^9$$

$$\text{Mass seawater} = \left[\frac{31.1 \text{ g}}{1.1 \times 10^{-2}} \times 10^9 \right] = 2.827273 \times 10^{12} \text{ g} \quad \text{(with 100\% efficiency)}$$

$$\text{Mass of seawater} = \left(2.827273 \times 10^{12} \text{ g}\right)\left(\frac{100\%}{81.5\%}\right) = 3.46905 \times 10^{12} \text{ g seawater (unrounded) (81.5\% efficiency)}$$

$$\text{Volume seawater} = \left(3.46905 \times 10^{12} \text{ g}\right)\left(\frac{1 \text{ mL}}{1.025 \text{ g}}\right)\left(\frac{10^{-3} \text{ L}}{1 \text{ mL}}\right) = 3.384439 \times 10^9 = \mathbf{3.4 \times 10^9 \text{ L}}$$

13.121 Convert mass of O_2 dissolved to moles of O_2. Use the density to convert the 1 kg mass of solution to volume in L.
0.0°C

$$\left(\frac{14.5 \text{ mg O}_2}{1 \text{ kg H}_2\text{O}}\right)\left(\frac{10^{-3} \text{ g}}{1 \text{ mg}}\right)\left(\frac{1 \text{ mol O}_2}{32.00 \text{ g O}_2}\right)\left(\frac{1 \text{ kg}}{10^3 \text{ g}}\right)\left(\frac{0.99987 \text{ g}}{\text{mL}}\right)\left(\frac{1 \text{ mL}}{10^{-3} \text{ L}}\right)$$

$$= 4.53066 \times 10^{-4} = \mathbf{4.53 \times 10^{-4} \text{ M O}_2}$$

20.0°C

$$\left(\frac{9.07 \text{ mg O}_2}{1 \text{ kg H}_2\text{O}}\right)\left(\frac{10^{-3} \text{ g}}{1 \text{ mg}}\right)\left(\frac{1 \text{ mol O}_2}{32.00 \text{ g O}_2}\right)\left(\frac{1 \text{ kg}}{10^3 \text{ g}}\right)\left(\frac{0.99823 \text{ g}}{\text{mL}}\right)\left(\frac{1 \text{ mL}}{10^{-3} \text{ L}}\right)$$

$$= 2.829358 \times 10^{-4} = \mathbf{2.83 \times 10^{-4} \text{ M O}_2}$$

40.0°C

$$\left(\frac{6.44 \text{ mg O}_2}{1 \text{ kg H}_2\text{O}}\right)\left(\frac{10^{-3} \text{ g}}{1 \text{ mg}}\right)\left(\frac{1 \text{ mol O}_2}{32.00 \text{ g O}_2}\right)\left(\frac{1 \text{ kg}}{10^3 \text{ g}}\right)\left(\frac{0.99224 \text{ g}}{\text{mL}}\right)\left(\frac{1 \text{ mL}}{10^{-3} \text{ L}}\right)$$

$$= 1.996883 \times 10^{-4} = \mathbf{2.00 \times 10^{-4} \text{ M O}_2}$$

13.123 a) First, find the molality from the freezing point depression and then use the molality, given mass of solute and volume of water to calculate the molar mass of the solute compound. Assume the solute is a nonelectrolyte (i = 1).
$\Delta T_f = iK_f m = (0.000 - (-0.201)) = 0.201°C$

$$m = \frac{\Delta T_f}{K_f i} = \frac{0.201°C}{(1.86°C/m)(1)} = 0.1080645 \ m \text{ (unrounded)}$$

$$m = \frac{\text{moles solute}}{\text{kg solvent}} \qquad (25.0 \text{ mL})\left(\frac{1.00 \text{ g}}{1 \text{ mL}}\right)\left(\frac{1 \text{ kg}}{10^3 \text{ g}}\right) = 0.0250 \text{ kg water}$$

moles solute = (m)(kg solvent) = (0.1080656 m)(0.0250 kg) = 0.0027016 mol

$$\text{molar mass} = \frac{0.243 \text{ g}}{0.0027016 \text{ mol}} = 89.946698 = \mathbf{89.9 \text{ g/mol}}$$

218

b) Assume that 100.00 g of the compound gives 53.31 g carbon, 11.18 g hydrogen and 100.00 – 53.31 – 11.18 = 35.51 g oxygen.

$$\text{Moles C} = (53.31 \text{ g C})\left(\frac{1 \text{ mol C}}{12.01 \text{ g C}}\right) = 4.43880 \text{ mol C (unrounded)}; \qquad \frac{4.43880}{2.219375} = 2$$

$$\text{Moles H} = (11.18 \text{ g H})\left(\frac{1 \text{ mol H}}{1.008 \text{ g H}}\right) = 11.09127 \text{ mol H (unrounded)}; \qquad \frac{11.09127}{2.219375} = 5$$

$$\text{Moles O} = (35.51 \text{ g O})\left(\frac{1 \text{ mol O}}{16.00 \text{ g O}}\right) = 2.219375 \text{ mol O (unrounded)}; \qquad \frac{2.219375}{2.219375} = 1$$

Dividing the values by the lowest amount of moles (2.219375) gives an **empirical formula of C_2H_5O** with molar mass 45.06 g/mol.
Since the molar mass of the compound , 89.9 g/mol from part (a), is twice the molar mass of the empirical formula, the **molecular formula is $2(C_2H_5O)$ or $C_4H_{10}O_2$**.
c) There is more than one example in each case. Possible Lewis structures:

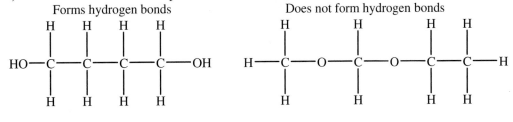

Forms hydrogen bonds   Does not form hydrogen bonds

13.127 a) $$\left(\frac{4.50 \times 10^{-5} \text{ mol F}^-}{L}\right)(5000.\text{ L})\left(\frac{1 \text{ mol NaF}}{1 \text{ mol F}^-}\right)\left(\frac{41.99 \text{ g NaF}}{1 \text{ mol NaF}}\right) = 9.44775 = \textbf{9.45 g NaF}$$

b) $$\left(\frac{4.50 \times 10^{-5} \text{ mol F}^-}{L}\right)(2.0\text{ L})\left(\frac{19.00 \text{ g F}^-}{1 \text{ mol F}}\right) = \textbf{0.0017 g F}^-$$

13.131 a) Use the boiling point elevation of 0.45°C to calculate the molality of the solution. Then, with molality, the mass of solute, and volume of water calculate the molar mass.

$\Delta T = iK_b m \qquad i = 1$ (nonelectrolyte)
$\Delta T = (100.45 - 100.00)°C = 0.45°C$

$$m = \frac{\Delta T_b}{K_b i} = \frac{0.45°\text{ C}}{(0.512°\text{ C/}m)(1)} = 0.878906 \; m = 0.878906 \text{ mol/kg (unrounded)}$$

$$m = \frac{\text{moles solute}}{\text{kg solvent}} \qquad (25.0 \text{ mL})\left(\frac{0.997 \text{ g}}{1 \text{ mL}}\right)\left(\frac{1 \text{ kg}}{10^3 \text{ g}}\right) = 0.0249250 \text{ kg water}$$

moles solute = (m)(kg solvent) = $(0.878906 \; m)(0.0249250 \text{ kg}) = 0.0219067$ mol

$$\text{molar mass} = \frac{1.50 \text{ g}}{0.0219067 \text{ mol}} = 68.4722 = \textbf{68 g/mol}$$

b) The molality calculated would be the moles of ions per kg of solvent. If the compound consists of three ions the molality of the compound would be 1/3 of 0.878906 m and the calculated molar mass would be three times greater: 3 x 68.4722 = 205.417 = **2.1 x 10^2 g/mol**.
c) The molar mass of CaN_2O_6 is 164.10 g/mol. This molar mass is less than the 2.1 x 10^2 g/mol calculated when the compound is assumed to be a strong electrolyte and is greater than the 68 g/mol calculated when the compound is assumed to be a nonelectrolyte. Thus, the compound is an electrolyte, since it dissociates into ions in solution. However, the ions do not dissociate completely in solution.

d) Use the molar mass of CaN_2O_6 to calculate the molality of the compound. Then calculate i in the boiling point elevation formula.

$$m = \left(\frac{1.50 \text{ g } CaN_2O_6}{25.0 \text{ mL}}\right)\left(\frac{mL}{0.997 \text{ g}}\right)\left(\frac{10^3 \text{ g}}{1 \text{ kg}}\right)\left(\frac{1 \text{ mol}}{164.10 \text{ g } CaN_2O_6}\right) = 0.3667309 \text{ } m \text{ (unrounded)}$$

$$\Delta T = iK_b m$$

$$i = \frac{\Delta T_b}{K_b m} = \frac{(0.45°C)}{(0.512°/m)(0.3667309 \text{ m})} = 2.396597 = \mathbf{2.4}$$

13.135   a) From the osmotic pressure, the molarity of the solution can be found. The ratio of mass per volume to moles per volume gives the molar mass of the compound.

$$\Pi = MRT$$

$$M = \frac{\Pi}{RT} = \frac{(0.340 \text{ torr})}{\left(0.0821\dfrac{L \cdot atm}{mol \cdot K}\right)((273 + 25) K)}\left(\frac{1 \text{ atm}}{760 \text{ torr}}\right) = 1.828546 \text{ x } 10^{-5} \text{ M (unrounded)}$$

$$(M)(V) = \text{moles}$$

$$\text{Moles} = \left(1.828546 \text{ x } 10^{-5} \text{ M}\right)(30.0 \text{ mL})\left(\frac{10^{-3} \text{ L}}{1 \text{ mL}}\right) = 5.48564 \text{ x } 10^{-7} \text{ mol}$$

$$\text{Molar mass} = \frac{(10.0 \text{ mg})\left(\dfrac{10^{-3} \text{ g}}{1 \text{ mg}}\right)}{5.48564 \text{ x } 10^{-7} \text{ mol}} = 1.82294 \text{ x } 10^4 = \mathbf{1.82 \text{ x } 10^4 \text{ g/mol}}$$

b) To find the freezing point depression, the molarity of the solution must be converted to molality. Then use $\Delta T_f = iK_f m$. (i = 1)

Mass solvent = mass of solution – mass of solute

$$\text{Mass solvent} = \left[(30.0 \text{ mL})\left(\frac{0.997 \text{ g}}{1 \text{ mL}}\right) - (10.0 \text{ mg})\left(\frac{10^{-3} \text{ g}}{1 \text{ mg}}\right)\right]\left(\frac{1 \text{ kg}}{10^3 \text{ g}}\right) = 0.0299 \text{ kg}$$

$$\text{Moles solute} = \left(\frac{1.828546 \text{ x } 10^{-5} \text{ mol}}{L}\right)\left(\frac{10^{-3} \text{ L}}{1 \text{ mL}}\right)(30.0 \text{ mL}) = 5.485638 \text{ x } 10^{-7} \text{ mol (unrounded)}$$

$$\text{Molality} = \frac{5.485638 \text{ x } 10^{-7} \text{ mol}}{0.0299 \text{ kg}} = 1.83466 \text{ x } 10^{-5} \text{ } m \text{ (unrounded)}$$

$$\Delta T_f = iK_f m = (1)(1.86°C/m)(1.83466 \text{ x } 10^{-5} \text{ } m) = 3.412 \text{ x } 10^{-5} = \mathbf{3.41 \text{ x } 10^{-5 \circ}C}$$

(So the solution would freeze at $0 - (3.41 \text{ x } 10^{-5 \circ}C) = -3.41 \text{ x } 10^{-5 \circ}C$.)

13.136   Henry's law expresses the relationship between gas pressure and the gas solubility (S_{gas}) in a given solvent. Use Henry's law to solve for pressure (assume that the constant (k_H) is given at 21°C), use the ideal gas law to find moles per unit volume, and convert moles/L to ng/L.

$$S_{gas} = k_H P_{gas}$$

$$P_{gas} = S_{gas} / k_H = \left(\frac{0.65 \text{ mg/L}}{0.033 \text{ mol/L} \cdot atm}\right)\left(\frac{10^{-3} \text{ g}}{1 \text{ mg}}\right)\left(\frac{1 \text{ mol } C_2H_2Cl_2}{96.94 \text{ g } C_2H_2Cl_2}\right)$$

$$= 2.3446799 \text{ x } 10^{-4} \text{ atm (unrounded)}$$

$$PV = nRT$$

$$n / V = P / RT = \frac{\left(2.3446799 \text{ x } 10^{-4} \text{ atm}\right)}{\left(0.0821\dfrac{L \cdot atm}{mol \cdot K}\right)((273 + 21) K)} = 9.7130 \text{ x } 10^{-6} \text{ mol/L (unrounded)}$$

$$\left(\frac{9.7130 \text{ x } 10^{-6} \text{ mol } C_2H_2Cl_2}{L}\right)\left(\frac{96.94 \text{ g } C_2H_2Cl_2}{1 \text{ mol } C_2H_2Cl_2}\right)\left(\frac{1 \text{ ng}}{10^{-9} \text{ g}}\right) = 9.41578 \text{ x} 10^5 = \mathbf{9.4 \text{ x } 10^5 \text{ ng/L}}$$

13.140 a) Assume a concentration of 1 mol/m3 for both ethanol and 2-butoxyethanol in the detergent solution. Then, from Henry's Law, the partial pressures of the two substances can be calculated:

$$S_{gas} = k_H \times P_{gas}$$
$$P_{gas} = S_{gas} / k_H$$

$$P_{ethanol} = \left(\frac{1 \text{ mol}}{m^3}\right)\left(\frac{5 \times 10^{-6} \text{ atm} \cdot m^3}{mol}\right) = 5 \times 10^{-6} \text{ atm}$$

$$P_{2\text{-butoxyethanol}} = \left(\frac{1 \text{ mol}}{m^3}\right)\left(\frac{1.6 \times 10^{-6} \text{ atm} \cdot m^3}{mol}\right) = 1.6 \times 10^{-6} \text{ atm}$$

$$\%_{2\text{-butoxyethanol}} = (5\%)\left(\frac{1.6 \times 10^{-6} \text{ atm 2-butoxyethanol}}{5 \times 10^{-6} \text{ atm ethanol}}\right) = 1.6\%$$

"Down-the-drain" factor is 0.016 = **0.02**

b) $k_{Hethanol} = \left(\frac{5 \times 10^{-6} \text{ atm} \cdot m^3}{mol}\right)\left(\frac{1 \text{ L}}{10^{-3} m^3}\right) = \textbf{5} \times \textbf{10}^{-3} \textbf{ atm} \cdot \textbf{L/mol}$

c) $k_{Hethanol} = \left(\frac{0.64 \text{ Pa} \cdot m^3}{mol}\right)\left(\frac{1 \text{ atm}}{1.01325 \times 10^5 \text{ Pa}}\right) = \textbf{6.3} \times \textbf{10}^{-6} \textbf{ atm} \cdot \textbf{m}^3 \textbf{/mol}$

Considering the single significant figure in the measured value of 5×10^{-6}, the agreement is good.

13.142 Molality is defined as moles of solute per kg of solvent, so 0.150 m means 0.150 mol $NaHCO_3$ per kg of water. The total mass of the solution would be 1 kg + (0.150 mol x molar mass of $NaHCO_3$).

$$0.150 \, m = (0.150 \text{ mol } NaHCO_3) / (1 \text{ kg solvent}) = \frac{(0.150 \text{ mol } NaHCO_3)\left(\dfrac{84.01 \text{ g } NaHCO_3}{1 \text{ mol } NaHCO_3}\right)}{1 \text{ kg}}\left(\frac{1 \text{ kg}}{10^3 \text{ g}}\right)$$

$$= 12.6015 \text{ g } NaHCO_3 / 1000 \text{ g solvent}$$

$$\left(\frac{12.6015 \text{ g } NaHCO_3}{(1000 + 12.6015) \text{ g Solution}}\right)(250. \text{ g Solution}) = 3.111 \text{ g } NaHCO_3 \text{ (unrounded)}$$

Grams H_2O = 250. g – 3.111 g = 246.889 g H_2O

To make 250. g of a 0.150 m solution of $NaHCO_3$, **weigh 3.11 g $NaHCO_3$ and dissolve in 247 g water.**

13.146 Molarity is moles solute/L solution and molality is moles solute/kg solvent.
Multiplying molality by concentration of solvent in kg solvent per liter of solution gives molarity:

$$\frac{\text{mol of solute}}{\text{L of solution}} = \left(\frac{\text{mol of solute}}{\text{kg solvent}}\right)\left(\frac{\text{kg solvent}}{\text{L of solution}}\right)$$

$$M = m\left(\frac{\text{kg solvent}}{\text{L of solution}}\right)$$

For a very dilute solution, the assumption that mass of solvent $\cong$ mass of solution is valid. This equation then becomes

$$M = m \text{ (kg solvent/L solution)} = m \times d_{solution}$$

Thus, for very dilute solutions **molality x density = molarity**.
In an aqueous solution, the liters of solution have approximately the same value as the kg of solvent because the density of water is close to 1 kg/L, so $m = M$.

13.148 a) $M \, SO_2 = P \, k_H = (2.0 \times 10^{-3} \text{ atm}) (1.23 \text{ mol/L} \cdot \text{atm}) = 2.46 \times 10^{-3} = \textbf{2.5} \times \textbf{10}^{-3} \, \textbf{\textit{M} SO}_2$
b) The base reacts with the sulfur dioxide to produce calcium sulfite. The reaction of sulfur dioxide makes "room" for more sulfur dioxide to dissolve.

13.153   Mass percents:

Iodine in chloroform

$[2.7 \text{ g } I_2 / (2.7 + 100.0)\text{g}] \times 100\% = 2.6290 = \textbf{2.6\% } \textbf{I}_2$

Iodine in carbon tetrachloride

$[2.5 \text{ g } I_2 / (2.5 + 100.0)\text{g}] \times 100\% = 2.4390 = \textbf{2.4\% } \textbf{I}_2$

Iodine in carbon disulfide

$[16 \text{ g } I_2 / (16 + 100.0)\text{g}] \times 100\% = 13.793 = \textbf{14\% } \textbf{I}_2$

Mole fraction:

Iodine in chloroform

Moles $I_2 = (2.7 \text{ g } I_2) (1 \text{ mol } I_2 / 253.8 \text{ g } I_2) = 0.010638 \text{ mol } I_2$ (unrounded)

Moles solvent $= (100.0 \text{ g } CHCl_3) (1 \text{ mol } CHCl_3 / 119.37 \text{ g } CHCl_3)$

$= 0.8377314 \text{ mol } CHCl_3$(unrounded)

Mole fraction $= (0.010638 \text{ mol } I_2) / [(0.010638) + (0.8377314)] \text{ mol} = 0.01245966 = \textbf{0.012}$

Iodine in carbon tetrachloride

Moles $I_2 = (2.5 \text{ g } I_2) (1 \text{ mol } I_2 / 253.8 \text{ g } I_2) = 0.009850 \text{ mol } I_2$ (unrounded)

Moles solvent $= (100.0 \text{ g } CCl_4) (1 \text{ mol } CCl_4 / 153.81 \text{ g } CCl_4)$

$= 0.650152785 \text{ mol } CCl_4$ (unrounded)

Mole fraction $= (0.009850 \text{ mol } I_2) / [(0.009850) + (0.650152785)] \text{ mol} = 0.014924 = \textbf{0.015}$

Iodine in carbon disulfide

Moles $I_2 = (16 \text{ g } I_2) (1 \text{ mol } I_2 / 253.8 \text{ g } I_2) = 0.0630418 \text{ mol } I_2$ (unrounded)

Moles solvent $= (100.0 \text{ g } CS_2) (1 \text{ mol } CS_2 / 76.15 \text{ g } CS_2) = 1.3131976 \text{ mol } CS_2$ (unrounded)

Mole fraction $= (0.0630418 \text{ mol } I_2) / [(0.0630418) + (1.3131976)] \text{ mol} = 0.045807 = \textbf{0.046}$

Molality:

Moles of iodine were calculated in part (b). Kilograms of solvent = 100.0 g (1 kg / 10^3 g) = 0.1000 kg in all cases.

Iodine in chloroform

Molality $= (0.010638 \text{ mol } I_2) / (0.1000 \text{ kg}) = 0.10638 = \textbf{0.11 } \textit{\textbf{m}} \textbf{ I}_2$

Iodine in carbon tetrachloride

Molality $= (0.009850 \text{ mol } I_2) / (0.1000 \text{ kg}) = 0.09850 = \textbf{0.098 } \textit{\textbf{m}} \textbf{ I}_2$

Iodine in carbon disulfide

Molality $= (0.0630418 \text{ mol } I_2) / (0.1000 \text{ kg}) = 0.630418 = \textbf{0.63 } \textit{\textbf{m}} \textbf{ I}_2$

13.157   a) Find the molarity of C_6H_{14} so that the molarity of O_2 can be found from its mole fraction. Then Henry's Law can be used to find k_H.

$$\text{Molarity of } C_6H_{14} = (1 \text{ L}) \left(\frac{1 \text{ mL}}{10^{-3} \text{ L}}\right) \left(\frac{1.674 \text{ g } C_6F_{14}}{1 \text{ mL}}\right) \left(\frac{1 \text{ mol } C_6F_{14}}{338 \text{ g } C_6F_{14}}\right) = 4.9527 \text{ mol/L}$$

$$X_{O_2} = \frac{\text{mol of } O_2}{\text{mol of } O_2 + \text{mol of } C_6F_{14}}$$

$$4.28 \times 10^{-3} = \frac{x \text{ mol } O_2}{4.9527 \text{ mol}}$$

Moles of $O_2 = 0.021198$ (The moles of O_2 is small enough to ignore in the denominator.)

$S_{gas} = k_H \times P_{gas}$

$k_H = S_{gas} / P_{gas}$

$k_H = 0.021198 \text{ mol/L}(101,325 \text{ Pa})(101,325 \text{ Pa/1 atm}) = 0.021198 = \textbf{0.0212 mol/L•atm}$

b) $S_{gas} = k_H \times P_{gas}$ $k_H = \dfrac{1}{756.7 \text{ L•atm/mol}} = 1.322 \times 10^{-3} \text{ mol/L•atm}$

$P_{O_2} = 0.2095 \times 1 \text{ atm} = 0.2095 \text{ atm}$ since O_2 is 20.95% of air

$S_{gas} = (1.322 \times 10^{-3} \text{ mol/L•atm})(0.2095 \text{ atm}) = 2.76959 \times 10^{-4} \text{ mol/L}$

$$\text{ppm} = \left(\frac{2.76959 \times 10^{-4} \text{ mol}}{\text{L}}\right) \left(\frac{32.0 \text{ g } O_2}{1 \text{ mol } O_2}\right) = 8.88627 \times 10^{-3} \text{ g/L} = (8.88627 \times 10^{-3} \text{ g/1000 g})(1 \times 10^{6})$$

$= 8.895688 = \textbf{8.90 ppm}$

c) k_H: **$C_6F_{14} > C_6H_{14} >$ ethanol > water**
To dissolve oxygen in a solvent, the solvent molecules must be moved apart to make room for the gas. The strong intermolecular forces in the solvent, the more difficult it is to separate solvent particles and the lower the solubility of the gas. Both C_6F_{14} and C_6H_{14} have weak dispersion forces with C_6H_{14} having the weaker forces due to the electronegative fluorine atoms repelling each other. Both ethanol and water are held together by strong hydrogen bonds with those bonds being stronger in water as the boiling point indicates.

13.159 a) **Yes**, the phases of water can still coexist at some temperature and can therefore establish equilibrium.
b) The triple point would occur at a lower pressure and **lower temperature** because the dissolved air solute lowers the vapor pressure of the solvent.
c) **Yes**, this is possible because the gas-solid phase boundary exists below the new triple point.
d) **No**, the presence of the solute lowers the vapor pressure of the liquid.

13.161 a) Concentration = $\left(\dfrac{28 \text{ mL}}{7.0 \text{ L}}\right)\left(\dfrac{40\%}{100\%}\right)\left(\dfrac{0.789 \text{ g}}{\text{mL}}\right)\left(\dfrac{22\%}{100\%}\right)\left(\dfrac{10^{-3} \text{ L}}{1 \text{ mL}}\right) = 2.77728 \times 10^{-4} = \mathbf{2.8 \times 10^{-4} \text{ g/mL}}$

b) $\left(\dfrac{28 \text{ mL}}{2.77728 \times 10^{-4} \text{ g/mL}}\right)\left(8.0 \times 10^{-4} \text{ g/mL}\right) = 80.6545 = \mathbf{81 \text{ mL}}$

CHAPTER 14 PERIODIC PATTERNS IN THE MAIN-GROUP ELEMENTS

FOLLOW–UP PROBLEMS

There are no follow–up problems for this chapter.

END–OF–CHAPTER PROBLEMS

14.2   a) Z_{eff}, the effective nuclear charge, increases across a period and decreases down a group.
b) As you move to the right across a period, the atomic size decreases, the IE_1 increases, and the electronegativity increases, all because of the increased Z_{eff}.

14.3   a) Polarity is the molecular property that is responsible for the difference in boiling points between iodine monochloride and bromine. The stronger the intermolecular forces holding molecules together, the higher the boiling point. More polar molecules have stronger intermolecular forces and higher boiling points. Molecular polarity is a result of the difference in electronegativity, an atomic property, of the atoms in the molecule.
b) ICl is a polar molecule while Br_2 is nonpolar. The dipole-dipole forces between polar molecules are stronger than the dispersion forces between nonpolar molecules. The boiling point of the polar ICl is higher than the boiling point of Br_2.

14.7   The leftmost element in a period is an alkali metal [except for hydrogen, Group 1A(1)]. As the other bonding atom moved to the right, the bonding would change from metallic to polar covalent to ionic. The atomic properties (IE, EA, and electronegativity) of the leftmost element and the "other" element become increasingly different, causing changes in the character of the bond between the two elements.

14.9   The most stable ion of elements in groups 1A, 2A, and 3A is a cation with an electron configuration of the noble gas from the previous period. The most stable ion of the elements in groups 5A, 6A, and 7A is an anion with electron configuration of the noble gas of the same period. For ions in the nth period, the outer level electrons for the cations are in the $n-1$ level, whereas the outer level electrons for the anions are in the n level. Ions with an $n-1$ outer level are smaller than ions with n as the outer level since the size of the electron cloud increases as the principle quantum number, n, increases. Thus, the cations will be significantly smaller than the anions in a period.

As an example, look at period 3. The cations Na^+, Mg^{2+}, and Al^{3+} have electron configurations $1s^22s^22p^6$, isoelectronic with neon. The anions P^{3-}, S^{2-}, and Cl^- have electron configurations $1s^22s^22p^63s^23p^6$, isoelectronic with argon. Figure 8.29 shows the cations decrease in size from 102 pm for Na^+ to 54 pm for Al^{3+}, then a jump to 212 pm for the first anion, P^{3-}. Moving across the anions, the ionic radius decreases from the 212 pm for P^{3-} to 181 pm for Cl^-. The irregularity in the trend occurs in the change from positively charged ions to negatively charged ions.

14.11   a) Element oxides are either ionic or covalent molecules. Elements with low electronegativity (metals) form ionic oxides, whereas elements with relatively high electronegativity (nonmetals) form covalent oxides. The greater the electronegativity of the element, the more covalent the bonding is in its oxide.
b) Oxide acidity increases to the right across a period and decreases down a group, so increasing acidity directly correlates with increasing electronegativity.

14.14 Network covalent solids differ from molecular solids in that they are harder and have higher melting points than molecular solids. A network solid is held together by covalent bonds while intermolecular forces hold molecules in molecular solids together. Covalent bonds are much stronger than intermolecular forces and require more energy to break. To melt a network solid, more energy (higher temperature) is required to break the covalent bonds than is required to break the intermolecular forces in the molecular solid, so the network solid has a higher melting point. The covalent bonds also hold the atoms in a much more rigid structure than the intermolecular forces hold the molecules. Thus, a network solid is much harder than a molecular solid.

14.15 a) **Mg < Sr < Ba**   Atomic size increases down a group.
       b) **Na < Al < P** IE_1 increases across a period.
       c) **Se < Br < Cl**   Electronegativity increases across a period and up a group.
       d) **Ga < Sn < Bi**   Ga has three valence electrons, whereas Sn has four valence electrons and Bi has five valence p electrons.

14.17 a) Chlorine and bromine will not form ionic compounds since they are both nonmetals that form anions.
       b) Sodium and bromine will form an ionic compound because sodium is a metal that forms a cation and bromine is a nonmetal that forms an anion.
       c) Phosphorus and selenium will not form ionic compounds since they are both nonmetals that form anions.
       d) Hydrogen and barium will form barium hydride, an ionic compound. Hydrogen reacts with reactive metals, such as Ba, to form ionic hydrides and reacts with nonmetals to form covalent hydrides.
       Compounds formed in cases **b and d** are ionic.

14.19 Recall in the Lewis structure model, bond order is the number of electron pairs shared between 2 bonded atoms (bond order is defined differently in molecular orbital theory). A bond order of 2 refers to two shared pairs of electrons in a double bond. There are several possibilities, with carbon dioxide being one example.

$$\overset{..}{\underset{..}{O}}=C=\overset{..}{\underset{..}{O}}$$

14.21 Bond length increases as the size of the atoms in the bond increases. Silicon, carbon, and germanium are all in group 4A and chlorine, fluorine, and bromine are all in group 7A. Size increases down a group, so carbon is the smallest and germanium the largest in group 4A and fluorine is the smallest and bromine the largest in group 7A. The order of increasing bond length follows increasing size of atoms: $CF_4 < SiCl_4 < GeBr_4$.

14.23 a) $Na^+ < F^- < O^{2-}$ In an isoelectronic series, ionic size decreases due to an increase in Z_{eff}.
       b) $Cl^- < 2^{2-} < P^{3-}$ This series is also isoelectronic. Recall that anions are larger than their respective atoms due to electron repulsion; the more electrons added, the greater the repulsion and the larger space they will occupy.

14.25 The difference in the three species is in the number of electrons in the outer level. As the number of electrons increase, electron repulsion increases which increases the size of the electron cloud. Thus, O^{2-} will be the largest of the three since it contains the most electrons. In order of increasing size, the rank is $O < O^- < O^{2-}$.

14.27 a) Metals form basic oxides, whereas nonmetals form acidic oxides. **Calcium oxide** (CaO) will form the more basic oxide.
$$CaO(s) + H_2O(l) \rightarrow Ca^{2+}(aq) + 2\ OH^-\ (aq)$$
$$SO_3(g) + H_2O(l) \rightarrow H_2SO_4(aq) \rightarrow 2\ H^+(aq) + SO_4^{2-}\ (aq)$$
       b) Oxide basicity increases down a group as the element becomes less electronegative. **Barium oxide** (BaO) produces the more basic solution.
       c) Oxide basicity decreases to the right across a period because the element becomes less electronegative. **Carbon dioxide** (CO_2) produces the more basic solution.
       d) **Potassium oxide** (K_2O) produces the more basic solution because K is more metallic (lower electronegativity) than P.

14.29 Covalent character increases as the electronegativity difference between the atoms decreases.
       a) **Lithium chloride** will have more covalent character than potassium chloride because Li is more electronegative than K (electronegativity increases up a group) so the electronegativity difference with chlorine will be smaller for LiCl than for KCl.

b) **Phosphorus trichloride** will have more covalent character than aluminum chloride because P is more electronegative than Al (electronegativity increases across a period) so the electronegativity difference with chlorine will be smaller for PCl_3 than for $AlCl_3$.

c) **Nitrogen trichloride** will have more covalent character than arsenic trichloride since the electronegativity of N is closer to the electronegativity of Cl. To answer this question, the trends in electronegativity are not sufficient since the electronegativity needs to be compared with that of chlorine. From Figure 9.20, the electronegativity values are 3.0 for Cl, 3.0 for N, and 2.0 for As. Δelectronegativity for NCl_3 is 0 and Δelectronegativity for $AsCl_3$ is 1.0.

14.31   a) **Ar < Na < Si**. Ar forms an atomic solid held together only by dispersion forces. Na is a metallic solid and Si is a network covalent solid, requiring the most energy to separate.

       b) **Cs < Rb < Li**. The heat of fusion, ΔH_{fus}, is the amount of energy needed to melt the solid. As melting point increases, so does ΔH_{fus}. All three are metallic solids, so the intermolecular forces are metallic bonds. Li has the highest melting point because smaller atoms attract delocalized electrons more strongly and produce stronger metallic bonds.

14.33   Ionization energy is defined as the energy required to remove the outermost electron from an atom. The further the outermost electron is from the nucleus, the less energy is required to remove it from the attractive force of the nucleus. In hydrogen, the outermost electron is in the $n = 1$ level and in lithium the outermost electron is in the $n = 2$ level. Therefore, the outermost electron in lithium requires less energy to remove, resulting in a lower ionization energy.

14.35   a) **NH_3** will hydrogen bond because H is attached to one of three atoms (N, O, and F) necessary for hydrogen bonding.

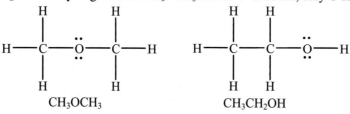

       b) **CH_3CH_2OH** will hydrogen bond. CH_3OCH_3 has no O–H bonds, only C-H bonds.

14.37   a) 2 Al(s) + 6 HCl(aq) $\rightarrow$ 2AlCl$_3$(aq) + 3 H$_2$(g)
       Active metals displace hydrogen from HCl by reducing the H$^+$ to H$_2$.
       b) LiH(s) + H$_2$O(l) $\rightarrow$ LiOH(aq) + H$_2$(g)
       In water, H$^-$ (here in LiH) reacts as a strong base to form H$_2$ and OH$^-$.

14.39   a)

Na = +1	B = +3	H = −1 in NaBH$_4$
Al = +3	B = +3	H = −1 in Al(BH$_4$)$_3$
Li = +1	Al = +3	H = −1 in LiAlH$_4$

b) The polyatomic ion in $NaBH_4$ is $[BH_4]^-$. There are $[1 \times B(3e^-)] + [4 \times H(1e^-)] + [1e^-$ from charge$] = 8$ valence electrons. All eight electrons are required to form the four bonds from the 4 hydrogen atoms to the boron atom. Boron is the central atom and has four surrounding electron groups; therefore, its shape is **tetrahedral**.

14.42   For period 2 elements in the first four groups, the number of covalent bonds equals the number of electrons in the outer level, so it increases from 1 covalent bond for lithium in group 1A(1) to 4 covalent bonds for carbon in group 4A(14). For the rest of period 2 elements, the number of covalent bonds equals the difference between 8 and the number of electrons in the outer level. So for nitrogen, $8 - 5 = 3$ covalent bonds; for oxygen, $8 - 6 = 2$ covalent bonds; for fluorine, $8 - 7 = 1$ covalent bond; and for neon, $8 - 8 = 0$, no bonds.
For elements in higher periods, the same pattern exists but with exceptions for groups 3A(13) to 7A(17) when an expanded octet allows for more covalent bonds.

14.45   a) E must have an oxidation of +3 to form an oxide E_2O_3 or fluoride EF_3. E is in group **3A(13) or 3B(3)**.
b) If E were in Group 3B(3), the oxide and fluoride would have more ionic character because 3B elements have lower electronegativity than 3A element. The Group 3B(3) oxides would be more basic.

14.48   a) Alkali metals generally lose electrons (act as **reducing agents**) in their reactions.
b) Alkali metals have relatively low ionization energies, meaning they easily lose the outermost electron. The electron configurations of alkali metals have one more electron than a noble gas configuration, so losing an electron gives a stable electron configuration.
c) $2\ Na(s) + 2\ H_2O(l) \rightarrow 2\ Na^+(aq) + 2\ OH^-(aq) + H_2(g)$
    $2\ Na(s) + Cl_2(g) \rightarrow 2\ NaCl(s)$

14.50   a) Density increases down a group. The increasing atomic size (volume) is not offset by the increasing size of the nucleus (mass), so m/V increases.
b) Ionic size increases down a group. Electron shells are added down a group, so both atomic and ionic size increase.
c) E–E bond energy decreases down a group. Shielding of the outer electron increases as the atom gets larger, so the attraction responsible for the E–E bond decreases.
d) IE_1 decreases down a group. Increased shielding of the outer electron is the cause of the decreasing IE_1.
e) ΔH_{hydr} decreases down a group. ΔH_{hydr} is the heat released when the metal salt dissolves in, or is hydrated by, water. Hydration energy decreases as ionic size increases.
Increasing down: a and b;   **Decreasing down: c, d, and e**

14.52   Peroxides are oxides in which oxygen has a –1 oxidation state. Sodium peroxide has the formula Na_2O_2 and is formed from the elements Na and O_2.
                    $2\ Na(s) + O_2(g) \rightarrow Na_2O_2(s)$

14.54   The problem specifies that an alkali halide is the desired product. The alkali metal is K (comes from potassium carbonate, $K_2CO_3(s)$) and the halide is I (comes from hydroiodic acid, $HI(aq)$). Treat the reaction as a double displacement reaction.
        $K_2CO_3(s) + 2\ HI(aq) \rightarrow 2\ KI(aq) + H_2CO_3(aq)$
However, $H_2CO_3(aq)$ is unstable and decomposes to $H_2O(l)$ and $CO_2(g)$, so the final reaction is:
        $K_2CO_3(s) + 2\ HI(aq) \rightarrow 2\ KI(aq) + H_2O(l) + CO_2(g)$

14.58   Metal atoms are held together by metallic bonding, a sharing of valence electrons. Alkaline earth metal atoms have one more valence electron than alkali metal atoms, so the number of electrons shared is greater. Thus, metallic bonds in alkaline earth metals are stronger than in alkali metals. Melting requires overcoming the metallic bonds. To overcome the stronger alkaline earth metal bonds requires more energy (higher temperature) than to overcome the alkali earth metal bonds.
First ionization energy, density, and boiling points will be larger for alkaline earth metals than for alkali metals.

14.59   a) A base forms when a basic oxide, such as CaO, is added to water.
$$CaO(s) + H_2O(l) \rightarrow Ca(OH)_2(s)$$
b) Alkaline earth metals reduce O_2 to form the oxide.
$$2\ Ca(s) + O_2(g) \rightarrow 2\ CaO(s)$$

14.62   Beryllium is generally the exception to properties exhibited by other alkaline earth elements.
a) Here, Be does not behave like other alkaline earth metals: $BeO(s) + H_2O(l) \rightarrow NR$.
The oxides of alkaline earth metals are strongly basic, but BeO is amphoteric. BeO will react with both acids and bases to form salts, but an amphoteric substance does not react with water.
b) Here, Be does behave like other alkaline earth metals:
$$BeCl_2(l) + 2\ Cl^-(solvated) \rightarrow BeCl_4{}^{2-}(solvated)$$
Each chloride ion donates a lone pair of electrons to form a covalent bond with the Be in $BeCl_2$. Metal ions form similar covalent bonds with ions or molecules containing a lone pair of electrons. The difference in beryllium is that the orbital involved in the bonding is a p-orbital, whereas in metal ions it is usually the d-orbitals that are involved.

14.65   The electron removed in Group 2A(2) atoms is from the outer level s orbital, whereas in Group 3A(13) atoms the electron is from the outer level p orbital. For example, the electron configuration for Be is $1s^2 2s^2$ and for B is $1s^2 2s^2 2p^1$. It is easier to remove the p electron of B than the s electron of Be, because the energy of a p-orbital is slightly higher than that of the s orbital from the same level. Even though the atomic size decreases from increasing Z_{eff}, the IE decreases from 2A(2) to 3A(13).

14.66   a) Compounds of Group 3A(13) elements, like boron, have only six electrons in their valence shell when combined with halogens to form three bonds. Having six electrons, rather than an octet, results in an "electron deficiency."
b) As an electron deficient central atom, B is trigonal planar. Upon accepting an electron pair to form a bond, the shape changes to tetrahedral.
$$BF_3(g) + NH_3(g) \rightarrow F_3B\text{-}NH_3(g)$$
$$B(OH)_3(aq) + OH^-(aq) \rightarrow B(OH)_4{}^-(aq)$$

14.68   Acidity of oxides increases up a group: **$In_2O_3 < Ga_2O_3 < Al_2O_3$**.

14.70   Halogens typically have a -1 oxidation state in metal-halide combinations, so the apparent oxidation state of $Tl = +3$. However, the anion $I_3{}^-$ combines with Tl in the $+1$ oxidation state. The anion $I_3{}^-$ has [3 x (I)7e$^-$] + [1e$^-$ from the charge] = 22 valence electrons; 4 of these electrons are used to form the two single bonds between iodine atoms and 16 electrons are used to give every atom an octet. The remaining two electrons belong to the central I atom; therefore the central iodine has five electron groups (2 single bonds and 3 lone pairs) and has a general formula of AX_2E_3. The electrons are arranged in a trigonal bipyramid with the three lone pairs in the trigonal plane. It is a linear molecule with bond angles = $180°$. $(Tl^{3+})\ (I^-)_3$ does not exist because of the low strength of the Tl–I bond.
O.N. = +3 (apparent); = +1 (actual)

$$\left[:\ddot{\underset{\cdot\cdot}{I}}\text{---}\ddot{I}\text{---}\ddot{\underset{\cdot\cdot}{I}}: \right]^-$$

14.75   a) $B_2O_3(s) + 2\ NH_3(g) \rightarrow 2\ BN(s) + 3\ H_2O(g)$
   b) $\Delta H_{rxn} = \Sigma \Delta H_f(\text{products}) - \Sigma \Delta H_f(\text{reactants})$
   $= [2\ \text{mol}\ \Delta H_f\ BN(s) + 3\ \text{mol}\ \Delta H_f\ H_2O(g)] - [1\ \text{mol}\ \Delta H_f\ B_2O(s) + 2\ \text{mol}\ \Delta H_f\ NH_3(g)]$
   $= [2\ \text{mol}\ (-254\ \text{kJ/mol}) + 3\ (-241.826\ \text{mol kJ/mol})] - [1\ \text{mol}\ (-1272\ \text{kJ/mol}) + 2\ \text{mol}\ (-45.9\ \text{kJ/mol})]$
   $= 130.322 = \mathbf{1.30 \times 10^2\ kJ}$

   c) Mass borax

$$= (1.0\ \text{kg BN})\left(\frac{10^3\ \text{g}}{1\ \text{kg}}\right)\left(\frac{1\ \text{mol BN}}{24.82\ \text{g BN}}\right)\left(\frac{1\ \text{mol B}}{1\ \text{mol BN}}\right)\left(\frac{1\ \text{mol Na}_2\text{B}_4\text{O}_7 \cdot 10\ \text{H}_2\text{O}}{4\ \text{mol B}}\right)\left(\frac{381.38\ \text{g}}{1\ \text{mol}}\right)\left(\frac{100\%}{72\%}\right)$$

$$= 5.335359 \times 10^3 = \mathbf{5.3 \times 10^3\ g\ borax}$$

14.76   Oxide basicity is greater for the oxide of a metal atom. Tin(IV) oxide is more basic than carbon dioxide since tin has more metallic character than carbon.

14.78   a) IE_1 values generally decrease down a group.
   b) The increase in Z_{eff} from Si to Ge is larger than the increase from C to Si because more protons have been added. Between C and Si an additional 8 protons have been added, whereas between Si and Ge an additional 18 (includes the protons for the d-block) protons have been added. The same type of change takes place when going from Sn to Pb, when the 14 f-block protons are added.
   c) Group 3A(13) would show greater deviations because the single p electron receives no shielding effect offered by other p electrons.

14.81   Atomic size increases moving down a group. As atomic size increases, ionization energy decreases so that it is easier to form a positive ion. An atom that is easier to ionize exhibits greater metallic character.

14.83

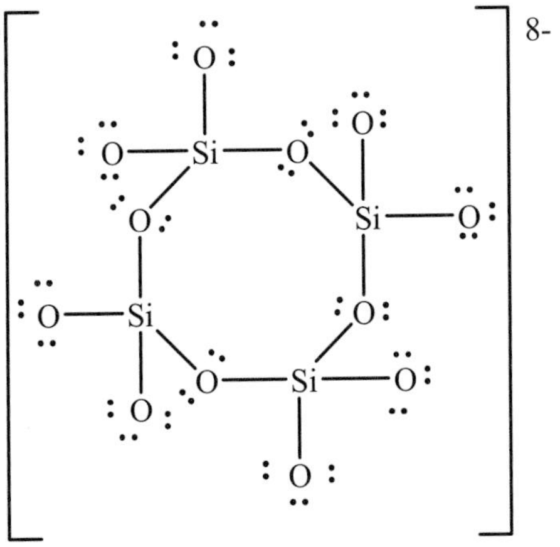

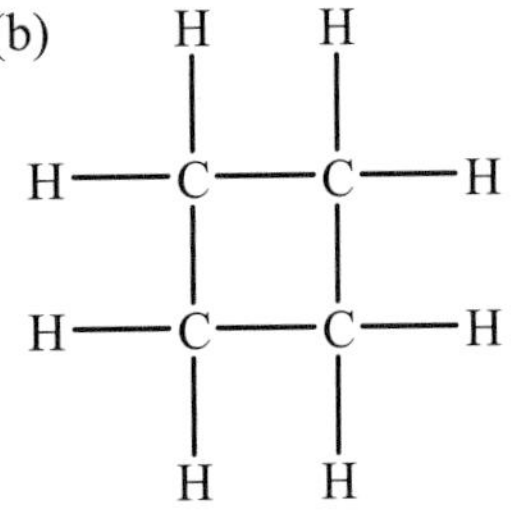

There is another answer possible for C_4H_8.

14.85

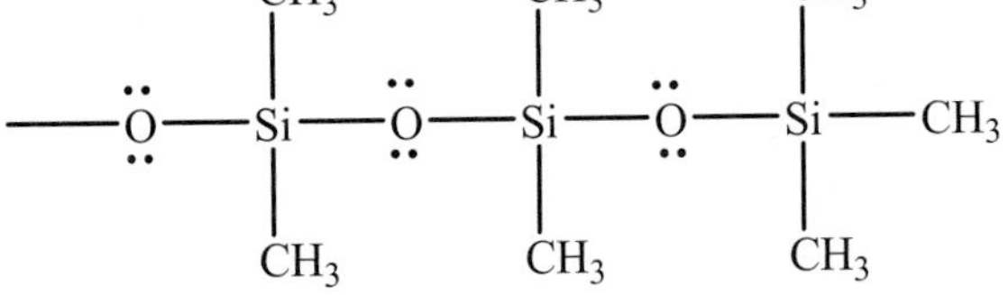

14.88   a) **Diamond, C**, a network covalent solid of carbon
   b) **Calcium carbonate, $CaCO_3$** (Brands that use this compound as an antacid also advertise them as an important source of calcium).
   c) **Carbon dioxide, CO_2**, is the most widely known greenhouse gas; CH_4 is also implicated.
   d) **Silicone $[(CH_3)_2SiO]_n$**

e) **Carborundum**, SiC
f) **Carbon monoxide, CO**, is formed in combustion when the amount of O_2 (air) is limited.
g) **Lead, Pb** (in old plumbing as lead solder, and in paint as a pigment)

14.92   a) In Group 5A(15), all elements except bismuth have a range of oxidation states from –3 to +5.
b) For nonmetals, the range of oxidation states is from the lowest at group number (A) – 8, which is 5 – 8 = –3 for Group 5A, to the highest equal to the group number (A), which is +5 for Group 5A.

14.95   Arsenic is less electronegative than phosphorus, which is less electronegative than nitrogen. Therefore, arsenic acid is the weakest acid and nitric acid is the strongest. Order of increasing strength: $H_3AsO_4 < H_3PO_4 < HNO_3$

14.97   a) With excess oxygen arsenic will form the oxide with arsenic in its highest possible oxidation state, +5.
$$4 \, As(s) + 5 \, O_2(g) \rightarrow 2 \, As_2O_5(s)$$
b) Trihalides are formed by direct combination of the elements (except N).
$$2 \, Bi(s) + 3 \, F_2(g) \rightarrow 2 \, BiF_3(s)$$
c) Metal phosphides, arsenides, and antimonides react with water to form Group 5A hydrides.
$$Ca_3As_2(s) + 6 \, H_2O(l) \rightarrow 3 \, Ca(OH)_2(s) + 2 \, AsH_3(g)$$

14.99   a) Aluminum is not as active a metal as Li or Mg, so heat is needed to drive this reaction.
$$N_2(g) + 2 \, Al(s) \xrightarrow{\Delta} 2 \, AlN(s)$$
b) The 5A halides react with water to form the oxoacid with the same oxidation state as the original halide.
$$PF_5(g) + 4 \, H_2O(l) \rightarrow H_3PO_4(aq) + 5 \, HF(g)$$

14.101 From the Lewis structure, the phosphorus has 5 electron groups for a trigonal bipyramidal molecular shape. In this shape, the three groups in the equatorial plane have greater bond angles (120°) than the two groups above and below this plane (90°). The chlorine atoms would occupy the planar sites where there is more space for the larger atoms.

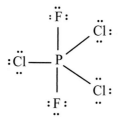

14.105 Set the atoms into the positions described, and then complete the Lewis structures.
a) and b) N_2O_2 has [2 x 5e^- (N) + 2 x 6e^- (O)] = 22 valence electrons. Six of these electrons are used to make the single bonds between the atoms, leaving 22 – 6 = 16 electrons. Since 20 electrons are needed to complete the octets of all of the atoms, two double bonds are needed.
c) N_2O_3 has [2 x 5e^- (N) + 3 x 6e^- (O)] = 28 valence electrons. Eight of these electrons are used to make the single bonds between the atoms, leaving 28 – 8 = 20 electrons. Since 24 electrons are needed to complete the octets of all of the atoms, two double bonds are needed.
d) NO^+ has [1x 5e^- (N) + 1 x 6e^- (O) – 1 e^- (due to the + charge)] = 10 valence electrons. Two of these electrons are used to make the single bond between the atoms, leaving 10 – 2 = 8 electrons. Since 12 electrons are needed to complete the octets of both atoms, a triple bond is needed.
NO_3^- has [1x 5e^- (N) + 3 x 6e^- (O) + 1 e^- (due to the – charge)] = 24 valence electrons. Six of these electrons are used to make the single bond between the atoms, leaving 24 – 6 = 18 electrons. Since 20 electrons are needed to complete the octets of all of the atoms, a double bond is needed.

(a)

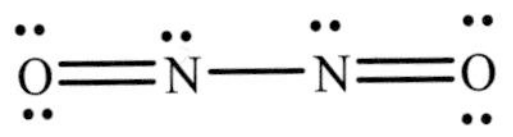

$$O = \overset{..}{N} - \overset{..}{N} = O$$

(b)

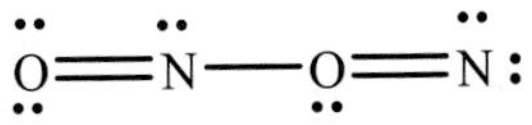

$$O = N - \overset{..}{O} = \overset{..}{N} :$$

(c)

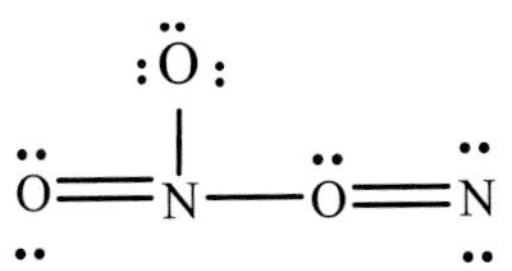

$$O = \overset{\overset{\displaystyle :O:}{|}}{N} - \overset{..}{O} = N$$

(d)

$$\left[: N \equiv O : \right]^{+} \qquad \left[O = N \overset{\displaystyle \overset{..}{O} :}{\underset{\displaystyle :O:}{\diagdown}} \right]^{-}$$

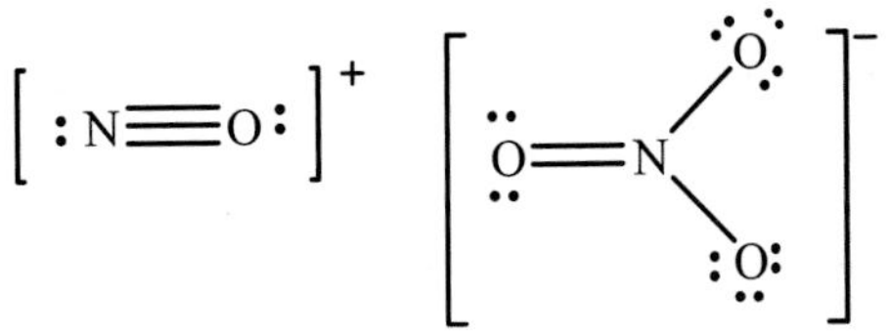

14.107 a) Thermal decomposition of KNO_3 at low temperatures:
$$2\ KNO_3(s) \xrightarrow{\Delta} 2\ KNO_2(s) + O_2(g)$$
b) Thermal decomposition of KNO_3 at high temperatures:
$$4\ KNO_3(s) \xrightarrow{\Delta} 2\ K_2O(s) + 2\ N_2(g) + 5\ O_2(g)$$

14.109 a) Both groups have elements that range from gas to metalloid to metal. Thus, their boiling points and conductivity vary in similar ways down a group.
b) The degree of metallic character and methods of bonding vary in similar ways down a group.
c) Both P and S have allotropes and both bond covalently with almost every other nonmetal.
d) Both N and O are diatomic gases at normal temperatures and pressures. Both N and O have very low melting and boiling points.
e) Oxygen, O_2, is a reactive gas whereas nitrogen, N_2, is not. Nitrogen can exist in multiple oxidation states, whereas oxygen has 2 oxidation states.

14.111 a) To decide what type of reaction will occur, examine the reactants. Notice that sodium hydroxide is a strong base. Is the other reactant an acid? If we separate the salt, sodium hydrogen sulfate, into the two ions, Na^+ and HSO_4^-, then it is easier to see the hydrogen sulfate ion as the acid. The sodium ions could be left out for the net ionic reaction.
$$NaHSO_4(aq) + NaOH(aq) \rightarrow Na_2SO_4(aq) + H_2O(l)$$
b) As mentioned in the book, hexafluorides are known to exist for sulfur. These will form when excess fluorine is present.
$$S_8(s) + 24\ F_2(g) \rightarrow 8\ SF_6(g)$$
c) Group 6A(16) elements, except oxygen, form hydrides in the following reaction.
$$FeS(s) + 2\ HCl(aq) \rightarrow H_2S(g) + FeCl_2(aq)$$
d) Tetraiodides, but not hexaiodides, of tellurium are known.
$$Te(s) + 2\ I_2(s) \rightarrow TeI_4(s)$$

14.113 a) Se is a nonmetal; its oxide is **acidic**.
b) N is a nonmetal; its oxide is **acidic**.
c) K is a metal; its oxide is **basic**.
d) Be is an alkaline earth metal, but all of its bonds are covalent; its oxide is **amphoteric**.
e) Ba is a metal; its oxide is **basic**.

14.115 Acid strength of binary acids increases down a group: **$H_2O < H_2S < H_2Te$**.

14.118 a) **SF_6, sulfur hexafluoride**
b) **O_3, ozone**
c) **SO_3, sulfur trioxide** (+6 oxidation state)
d) **SO_2, sulfur dioxide**
e) **H_2SO_4, sulfuric acid**
f) **$Na_2S_2O_3 \cdot 5H_2O$, sodium thiosulfate pentahydrate**
g) **H_2S, hydrogen sulfide**

14.120 $S_2F_{10}(g) \rightarrow SF_4(g) + SF_6(g)$
O.N. of S in S_2F_{10}: $- (10 \times -1 \text{ for F}) / 2 = +5$
O.N. of S in SF_4: $- (4 \times -1 \text{ for F}) = +4$
O.N. of S in SF_6: $- (6 \times -1 \text{ for F}) = +6$

14.122 a) Bonding with very electronegative elements: **+1, +3, +5, +7**. Bonding with other elements: **–1**
b) The electron configuration for Cl is $[Ne]3s^23p^5$. By adding one electron to form Cl^-, Cl achieves an octet similar to the noble gas Ar. By forming covalent bonds, Cl completes or expands its octet by maintaining its electrons paired in bonds or lone pairs.
c) Fluorine only forms the –1 oxidation state because its small size and no access to *d*-orbitals prevent it from forming multiple covalent bonds. Fluorine's high electronegativity also prevents it from sharing its electrons.

14.124 a) The Cl–Cl bond is stronger than the Br–Br bond since the chlorine atoms are smaller than the bromine atoms, so the shared electrons are held more tightly by the two nuclei.
b) The Br–Br bond is stronger than the I–I bond since the bromine atoms are smaller than the iodine atoms.
c) The Cl–Cl bond is stronger than the F–F bond. The fluorine atoms are smaller than the chlorine but they are so small that electron-electron repulsion of the lone pairs decreases the strength of the bond.

14.126 a) A substance that disproportionates serves as both an oxidizing and reducing agent. Assume that OH^- serves as the base. Write the reactants and products of the reaction, and balance like a redox reaction.
$$3\, Br_2(l) + 6\, OH^-\, (aq) \rightarrow 5\, Br^-\, (aq) + BrO_3^-\, (aq) + 3\, H_2O(l)$$
b) In the presence of base, instead of water, only the oxy*anion* (not oxo*acid*) and fluoride (not *hydro*fluoride) form. No oxidation or reduction takes place, because Cl maintains its +5 oxidation state and F maintains its –1 oxidation state.
$$ClF_5(l) + 6\, OH^-\, (aq) \rightarrow 5\, F^-\, (aq) + ClO_3^-\, (aq) + 3\, H_2O(l)$$

14.127 a) $2\, Rb(s) + Br_2(l) \rightarrow 2\, RbBr(s)$
b) $I_2(s) + H_2O(l) \rightarrow HI(aq) + HIO(aq)$
c) $Br_2(l) + 2\, I^-(aq) \rightarrow I_2(s) + 2\, Br^-(aq)$
d) $CaF_2(s) + H_2SO_4(l) \rightarrow CaSO_4(s) + 2\, HF(g)$

14.129 Acid strength increases with increasing electronegativity of the central atom and increasing number of oxygen atoms. Therefore, **HIO < HBrO < HClO < HClO$_2$**

14.133 $I_2 < Br_2 < Cl_2$, since Cl_2 is able to oxidize Re to the +6 oxidation state, Br_2 only to +5, and I_2 only to +4.

14.134 **Helium** is the second most abundant element in the universe. **Argon** is the most abundant noble gas in Earth's atmosphere, the third most abundant constituent after N_2 and O_2.

14.136 Whether a boiling point is high or low is a result of the strength of the forces between particles. Dispersion forces, the weakest of all the intermolecular forces, hold atoms of noble gases together. Only a relatively low temperature is required for the atoms to have enough kinetic energy to break away from the attractive force of other atoms and go into the gas phase. The boiling points are so low that all the noble gases are gases at room temperature.

14.140

$H_3O^+(g) \rightarrow \cancel{H^+(g)} + \cancel{H_2O(g)}$	$\Delta H = +720$ kJ
$\cancel{H_2O(l)} + \cancel{H^+(g)} \rightarrow H_3O^+(aq)$	$\Delta H = -1090$ kJ
$\underline{\cancel{H_2O(g)} \rightarrow \cancel{H_2O(l)}}$	$\underline{\Delta H = -40.7 \text{ kJ } (= -\Delta H_{vap}\ (H_2O))}$
Overall: $H_3O^+(g) \rightarrow H_3O^+(aq)$	$\Delta H = -410.7 = \mathbf{-411}$ **kJ**

14.145 a) No oxidation state changes occur in this reaction; S maintains a +4 oxidation state. The prefix "hexa-" denotes 6, in this case 6 waters of hydration. Start by balancing H, then balance O.
$$MgCl_2\cdot 6H_2O(s) + 6\, SOCl_2(l) \rightarrow MgCl_2(s) + 6\, SO_2(g) + 12\, HCl(g)$$

b) Sulfur is the central atom because it is the least electronegative element. The molecule has [1 x 6e⁻ (S) + 1 x 6e⁻ (O) + 2 x 7e⁻ (Cl)] = 26 valence e⁻. Six electrons are used for the three single bonds between the atoms, leaving 26 – 6 = 20e⁻. The remaining 20e⁻ electrons are distributed to the atoms to give each an octet. This results in structure **I**. An analysis of formal charges shows that structure **II** is more stable.

Formal charge = # of valence electrons – [# of unshared electrons + ½(#bonding electrons)]

I   II

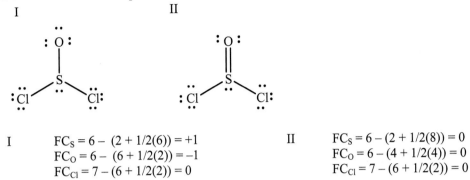

I $FC_S = 6 - (2 + 1/2(6)) = +1$ II $FC_S = 6 - (2 + 1/2(8)) = 0$
         $FC_O = 6 - (6 + 1/2(2)) = -1$ $FC_O = 6 - (4 + 1/2(4)) = 0$
         $FC_{Cl} = 7 - (6 + 1/2(2)) = 0$ $FC_{Cl} = 7 - (6 + 1/2(2)) = 0$

14.146   a) Alkali metals have an outer electron configuration of ns^1. The first electron lost by the metal is the ns electron, giving the metal a noble gas configuration. Second ionization energies for alkali metals are high because the electron being removed is from the next lower energy level and electrons in a lower level are more tightly held by the nucleus. The metal would also lose its noble gas configuration.

b) The reaction is $2 CsF_2(s) \rightarrow 2 CsF(s) + F_2(g)$.
You know the ΔH_f for the formation of CsF:
$$Cs(s) + 1/2F_2(g) \rightarrow CsF(s) \qquad \Delta H_f = -530 \text{ kJ/mol}$$
You also know the ΔH_f for the formation of CsF_2:
$$Cs(s) + F_2(g) \rightarrow CsF_2(s) \qquad \Delta H_f = -125 \text{ kJ/mol}$$
To obtain the heat of reaction for the breakdown of CsF_2 to CsF, combine the formation reaction of CsF with the reverse of the formation reaction of CsF_2, both multiplied by 2:

$2 Cs(s) + F_2(g) \rightarrow 2 CsF(s)$ $\Delta H = 2 \times (-530 \text{ kJ})$
$2 CsF_2(s) \rightarrow 2Cs(s) + 2 F_2(g)$ $\Delta H = 2 \times (+125 \text{ kJ})$ (Note sign change)
$2 CsF_2(s) \rightarrow 2 CsF(s) + F_2(g)$ $\Delta H = -810 \text{ kJ}$

810 kJ of energy are released when 2 moles of CsF_2 convert to 2 moles of CsF, so heat of reaction for one mole of CsF is –810/2 or **–405 kJ/mol**.

14.148   a) Empirical formula HNO has a molar mass of 31.02, so hyponitrous acid with a molar mass of 62.04 g/mol which is twice 31.02 would have a molecular formula twice the empirical formula, $H_2N_2O_2$. In addition, the molecular formula of nitroxyl would be the same as the empirical formula, HNO, since the molar mass of nitroxyl is the same as the empirical formula.

b) $H_2N_2O_2$ has [2 x 1e⁻ (H) + 2 x 5e⁻ (N) + 2 x 6e⁻ (O)] = 24 valence e⁻. Ten electrons are used for single bonds between the atoms, leaving 24 – 10 = 14e⁻. 16 electrons are needed to give every atom an octet; since only 16 electrons are available, one double bond (between the N atoms) is needed.

HNO has [1 x 1e⁻ (H) + 1 x 5e⁻ (N) + 1 x 6e⁻ (O)] = 12 valence e⁻. Four electrons are used for single bonds between the atoms, leaving 12 – 4 = 8e⁻. 10 electrons are needed to give every atom an octet; since only 8 electrons are available, one double bond is needed between the N and O atoms.

H——Ö——N══N——Ö——H   H——N══Ö

c) In both hyponitrous acid and nitroxyl, the nitrogens are surrounded by 3 electron groups (one single bond, one double bond and one unshared pair), so the electron arrangement is trigonal planar and the molecular shape is **bent**.

d)   cis   trans

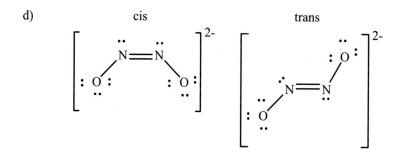

14.151   Determine the volume of a unit cell:

$$\text{Volume} = \left(\frac{cm^3}{1.693\ g}\right)\left(\frac{12.01\ g\ C}{1\ mol\ C}\right)\left(\frac{1\ mol\ C}{6.022\ x\ 10^{23}\ C\ atoms}\right)\left(\frac{240\ C\ atoms}{1\ unit\ cell}\right)$$

$$= 2.8272\ x\ 10^{-21} = \mathbf{2.827\ x\ 10^{-21}\ cm^3}$$

Since the cell is cubic, the length of an edge is the cube root of the volume:
Cell edge = $(2.8272\ x\ 10^{-21}\ cm^3)^{1/3} = 1.414009\ x\ 10^{-7} = \mathbf{1.414\ x\ 10^{-7}\ cm}$
The low density arises because even though the spheres are closest packed, the spheres are hollow.

14.154   Carbon monoxide and carbon dioxide would be formed from the reaction of coke (carbon) with the oxygen in the air. The nitrogen in the producer gas would come from the nitrogen already in the air. So, the calculation of mass of product is based on the mass of CO and CO_2 that can be produced from 1.75 metric tons of coke.
Using 100 grams of sample, the percentages simply become grams.
Since 5.0 grams of CO_2 is produced for each 25 grams of CO, we can calculate a mass ratio of carbon that produces each:

$$\frac{(25\ g\ CO)\left(\dfrac{12.01\ g\ C}{28.01\ g\ CO}\right)}{(5.0\ g\ CO_2)\left(\dfrac{12.01\ g\ C}{44.01\ g\ CO_2}\right)} = 7.85612\ /\ 1\ \text{(unrounded)}$$

Using the ratio of carbon that reacts as 7.85612:1, the total C reacting is 7.85612 + 1 = 8.85612. The mass fraction of the total carbon that produces CO is 7.85612/8.85612 and the mass fraction of the total carbon reacting that produces CO_2 is 1.00/8.85612. To find the mass of CO produced from 1.75 metric tons of carbon with an 87% yield:

     Mass CO = (7.85612 / 8.85612) (1.75 ton) (28.01 ton CO / 12.01 ton C) (87% / 100%)
              = 3.1498656 ton CO (unrounded)
     Mass CO_2 = (1 / 8.85612) (1.75 ton) (44.01 ton CO_2 / 12.01 ton C) (87% / 100%)
              = 0.6299733 ton CO_2 (unrounded)
The mass of CO and CO_2 represent a total of 30% (= 100% – 70.% N_2) of the mass of the producer gas, so the total mass would be (3.1498656 + 0.6299733) (100% / 30%) = 12.59946 = **13 metric tons.**

14.156   In a disproportionation reaction, a substance acts as both a reducing agent and oxidizing agent because an atom within the substance reacts to form atoms with higher and lower oxidation states.
        0   –1   –1/3
a) $I_2(s) + KI(aq) \rightarrow KI_3(aq)$
I in I_2 reduces to I in KI_3. I in KI oxidizes to I in KI_3. This is not a disproportionation reaction since different substances have atoms that reduce or oxidize. The *reverse* direction would be a disproportionation reaction because a single substance (I in KI) both oxidizes and reduces.
       +4   +5   +3
b) 2 $ClO_2(g) + H_2O(l) \rightarrow HClO_3(aq) + HClO_2(aq)$
Yes, ClO_2 disproportionates, as the chlorine reduces from +4 to +3 and oxidizes from +4 to +5.
       0   –1   +1
c) $Cl_2(g) + 2\ NaOH(aq) \rightarrow NaCl(aq) + NaClO(aq) + H_2O(l)$
Yes, Cl_2 disproportionates, as the chlorine reduces from 0 to –1 and oxidizes from 0 to +1.

$$\overset{-3}{} \quad \overset{+3}{} \qquad \overset{0}{}$$

d) $NH_4NO_2(s) \rightarrow N_2(g) + 2\ H_2O(g)$

Yes, NH_4NO_2 disproportionates; the ammonium (NH_4^+) nitrogen oxidizes from –3 to 0, and the nitrite (NO_2^-) nitrogen reduces from +3 to 0.

$$\overset{+6}{} \qquad\qquad\qquad\quad \overset{+7}{} \qquad \overset{+4}{}$$

e) $3\ MnO_4^{2-}(aq) + 2\ H_2O(l) \rightarrow 2\ MnO_4^-(aq) + MnO_2(s) + 4\ OH^-(aq)$

Yes, MnO_4^{2-} disproportionates, the manganese oxidizes from +6 to +7 and reduces from +6 to +4.

$$\overset{+1}{} \qquad \overset{+3}{} \qquad\qquad \overset{0}{}$$

f) $3\ AuCl(s) \rightarrow AuCl_3(s) + 2\ Au(s)$

Yes, AuCl disproportionates, the gold oxidizes from +1 to +3 and reduces from +1 to 0.

14.158   a) Group **5A(15)** elements have 5 valence electrons and typically form three bonds with a lone pair to complete the octet. An example is NH_3.

b) Group **7A(17)** elements readily gain an electron causing the other reactant to be oxidized. They form monatomic ions of formula X^- and oxoanions. Examples would be Cl^- and ClO^-.

c) Group **6A(16)** elements have six valence electrons and gain a complete octet by forming two covalent bonds. An example is H_2O.

d) Group **1A(1)** elements are the strongest reducing agents because they most easily lose an electron. As the least electronegative and most metallic of the elements, they are not likely to form covalent bonds. Group **2A(2)** have similar characteristics. Thus, either Na or Ca could be an example.

e) Group **3A(13)** elements have only three valence electrons to share in covalent bonds, but with an empty orbital they can accept an electron pair from another atom. Boron would be an example of an element of this type.

f) Group **8A(18)**, the noble gases, are the least reactive of all the elements. Xenon is an example that forms compounds, while helium does not form compounds.

14.160   a) The Lewis structure has 72 valence e^-.

b) Three lone pairs appear on N ring atoms.

c) The bond order is 1.5. The molecule has a resonance structure, similar to the "alternating" bonds of benzene. The single-double bonds do not really alternate, but rather the valence electrons are delocalized on the ring, resulting in a bond order of 1.5.

14.162   Find ΔH_{rxn} for the reaction $2\ BrF(g) \rightarrow Br_2(g) + F_2(g)$. Apply Hess's Law to the equations given:

1)	$3\ BrF(g) \rightarrow Br_2(g) + BrF_3(l)$	$\Delta H_{rxn} = -125.3$ kJ
2)	$5\ BrF(g) \rightarrow 2\ Br_2(g) + BrF_5(l)$	$\Delta H_{rxn} = -166.1$ kJ
3)	$BrF_3(l) + F_2(g) \rightarrow BrF_5(l)$	$\Delta H_{rxn} = -158.0$ kJ

Reverse equations 1 and 3, and add to equation 2:

1)	$Br_2(g) + \cancel{BrF_3(l)} \rightarrow 3\ BrF(g)$	$\Delta H_{rxn} = +125.3$ kJ (note sign change)
2)	$5\ BrF(g) \rightarrow 2\ Br_2(g) + \cancel{BrF_5(l)}$	$\Delta H_{rxn} = -166.1$ kJ
3)	$\cancel{BrF_5(l)} \rightarrow \cancel{BrF_3(l)} + F_2(g)$	$\Delta H_{rxn} = +158.0$ kJ (note sign change)

Total:	$2\ BrF(g) \rightarrow Br_2(g) + F_2(g)$	$\boldsymbol{\Delta H_{rxn} = +117.2}$ **kJ**

14.165   Write a balanced chemical equation and use mole relationships to finish the problem.

$$6\ Cl_2(g) + 2\ CS_2(l) \xrightarrow{\text{cat.}} CCl_4(l) + CCl_3SCl(s) + S_2Cl_2(s) + SCl_2(s)$$

In this reaction, the only source of carbon is CS_2. The number of grams of carbon in the products must equal the number of grams of carbon in the reactant.

Mass of carbon from CCl_4 = $(50.0\ g\ CCl_4)\left(\dfrac{1\ mol\ CCl_4}{153.81\ g\ CCl_4}\right)\left(\dfrac{1\ mol\ C}{1\ mol\ CCl_4}\right)\left(\dfrac{12.01\ g\ C}{1\ mol\ C}\right)$

= 3.904167 g C (unrounded)

Mass of carbon from CCl_3SCl = $(0.500\ g\ CCl_3SCl)\left(\dfrac{1\ mol\ CCl_3SCl}{185.88\ g\ CCl_3SCl}\right)\left(\dfrac{1\ mol\ C}{1\ mol\ CCl_3SCl}\right)\left(\dfrac{12.01\ g\ C}{1\ mol\ C}\right)$

= 0.032305788 g C (unrounded)

Total mass of carbon = (3.904167 + 0.032305788) g C = 3.93647 g C (unrounded)

Mass of CS_2 = $(3.93647\ g\ C)\left(\dfrac{1\ mol\ C}{12.01\ g\ C}\right)\left(\dfrac{1\ mol\ CS_2}{1\ mol\ C}\right)\left(\dfrac{76.15\ g\ CS_2}{1\ mol\ CS_2}\right)$

= 24.95938 = **25.0 g CS_2**

It is not possible to determine how much S_2Cl_2 and SCl_2 are formed. The total number of grams of sulfur in products can be calculated, but without knowing the ratio of S_2Cl_2 to SCl_2, no further determination can be made. Knowing the amount of $Cl_2(g)$ would also allow for the determination of the amount of SCl_2 and S_2Cl_2 made.

14.170   To answer these questions, draw an initial structure for $Al_2Cl_7^-$:

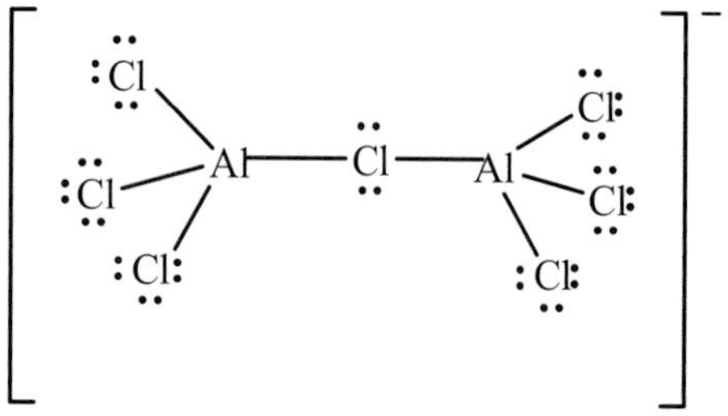

a) Aluminum uses its $3s$ and $3p$ valence orbitals to form sp^3 hybrid orbitals for bonding.
b) With formula AX_4, the shape is **tetrahedral** around each aluminum atom.
c) Since the Al–Cl–Al bond is linear (180° Al–Cl–Al bond angle), the central atom must be sp hybridized.
d) The sp hybridization suggests that there are no lone pairs on the central chlorine. Instead, the extra four electrons participate in bonding with the empty d-orbitals on the aluminum to form double bonds between the chlorine and each aluminum.

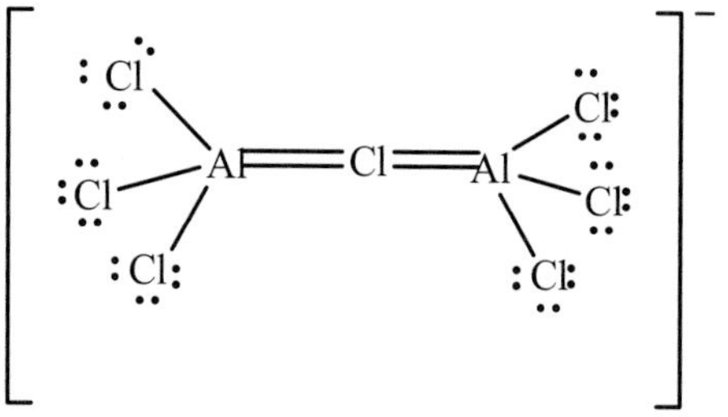

14.172 The Lewis structures are

The nitronium ion (NO_2^+) has a linear shape because the central N atom has 2 surrounding electron groups, which achieve maximum repulsion at 180°. Both the nitrite ion (NO_2^-) and nitrogen dioxide (NO_2) have a central N surrounded by three electron groups. The electron group arrangement would be trigonal planar with an ideal bond angle of 120°. The bond angle in NO_2^- is more compressed than that in NO_2 since the lone pair of electrons in NO_2^- takes up more space than the lone electron in NO_2. Therefore the bond angle in NO_2^- is smaller (115°) than that of NO_2 (134°).

14.174 $U(s) + 3\ ClF_3(l) \rightarrow UF_6(l) + 3\ ClF(g)$
To find the limiting reactant, calculate the amount of UF_6 that would form from each reactant.
(1 metric ton = 1 t = 1000 kg)
From U:

$$\text{Mole } UF_6 = \left(1.00\ \text{t Ore}\right)\left(\frac{10^3\ \text{kg}}{1\ \text{t}}\right)\left(\frac{10^3\ \text{g}}{1\ \text{kg}}\right)\left(\frac{1.55\%}{100\%}\right)\left(\frac{1\ \text{mol U}}{238.0\ \text{g U}}\right)\left(\frac{1\ \text{mol } UF_6}{1\ \text{mol U}}\right)$$

$$= 65.12605\ \text{mol } UF_6 \text{ (unrounded)}$$

From ClF_3:

$$\text{Mole } UF_6 = \left(12.75\ \text{L}\right)\left(\frac{1\ \text{mL}}{10^{-3}\ \text{L}}\right)\left(\frac{1.88\ \text{g } ClF_3}{1\ \text{mL}}\right)\left(\frac{1\ \text{mol } ClF_3}{92.45\ \text{g } ClF_3}\right)\left(\frac{1\ \text{mol } UF_6}{3\ \text{mol } ClF_3}\right)$$

$$= 86.42509\ \text{mol } UF_6 \text{ (unrounded)}$$

Since the amount of uranium will produce less uranium hexafluoride, it is the limiting reactant.
Mass UF_6 = (65.12605 mol UF_6) (352.0 g UF_6 / 1 mol UF_6) = 2.2924 x 10^4 = **2.29 x 10^4 g UF_6**

14.179 Determine the electron configuration of each species. Partially filled orbitals lead to paramagnetism. If there are an odd number of electrons present, the species will be paramagnetic.

O^+	$1s^2 2s^2 2p^3$ **paramagnetic**		odd number of electrons
O^-	$1s^2 2s^2 2p^5$ **paramagnetic**		odd number of electrons
O^{2-}	$1s^2 2s^2 2p^6$ **diamagnetic**		all orbitals filled (all electrons paired)
O^{2+}	$1s^2 2s^2 2p^2$ **paramagnetic**		Two of the $2p$ orbitals have one electron each. These electrons are spin parallel (Hund's rule).

14.180 a) Determine the mass of As in a formula mass of the compound.

$$\% \text{ As in } CuHAsO_3 = \frac{74.92\ \text{g As}}{187.48\ \text{g } CuHAsO_3} \times 100\% = 39.96160 = \textbf{39.96\% As}$$

$$\% \text{ As in } (CH_3)_3As = \frac{74.92\ \text{g As}}{120.02\ \text{g } (CH_3)_3As} \times 100\% = 62.4229 = \textbf{62.42\% As}$$

b) $$\text{Mass As} = \left(\frac{0.50\ \text{mg As}}{\text{m}^3}\right)\left(\frac{10^{-3}\ \text{g}}{1\ \text{mg}}\right)\big((12.35\ \text{m})(7.52\ \text{m})(2.98\ \text{m})\big) = 0.13838\ \text{g As}$$

$$\text{Mass } CuHAsO_3 = \left(0.13838\ \text{g As}\right)\left(\frac{100\ \text{g } CuHAsO_3}{39.96160\ \text{g As}}\right) = 0.346282 = \textbf{0.35 g } CuHAsO_3$$

CHAPTER 15 ORGANIC COMPOUNDS AND THE ATOMIC PROPERTIES OF CARBON

FOLLOW–UP PROBLEMS

15.1   a) Plan: For compounds with seven carbons, first start with 7 C atoms in a straight chain. Then make all arrangements with 6 carbons in a straight chain and 1 carbon branched off the chain. Examine the structures to make sure they are all different. Continue in the same manner with 5 C atoms in chain and 2 branched off the chain, then 4 C atoms in chain and 3 branched off, and 3 C atoms in chain and 4 branched off. Examine all structures to guarantee there are no duplicates.
Solution:
Seven carbons in chain:

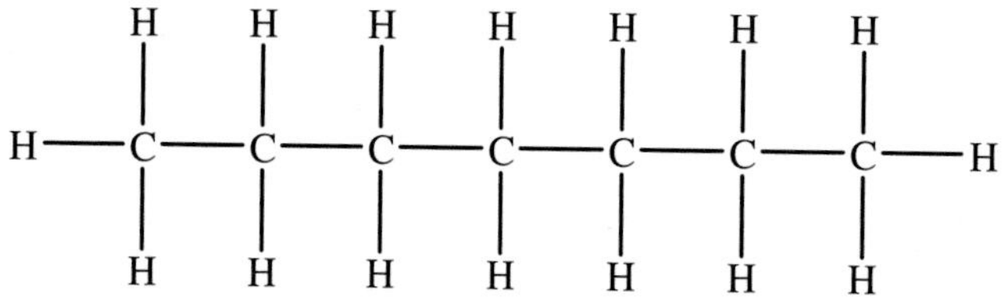

(1)

Six carbons in chain:

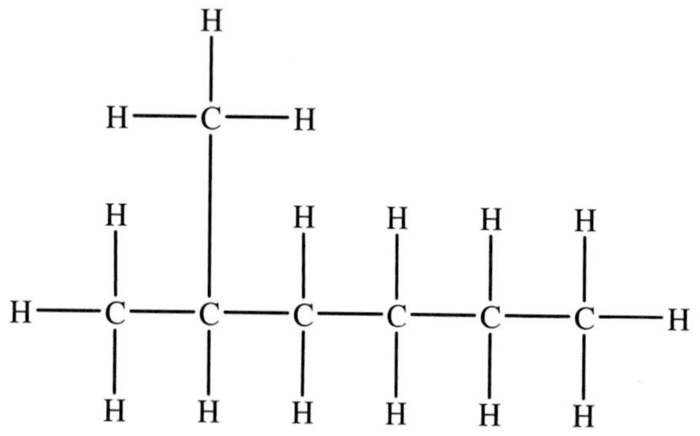

(2) 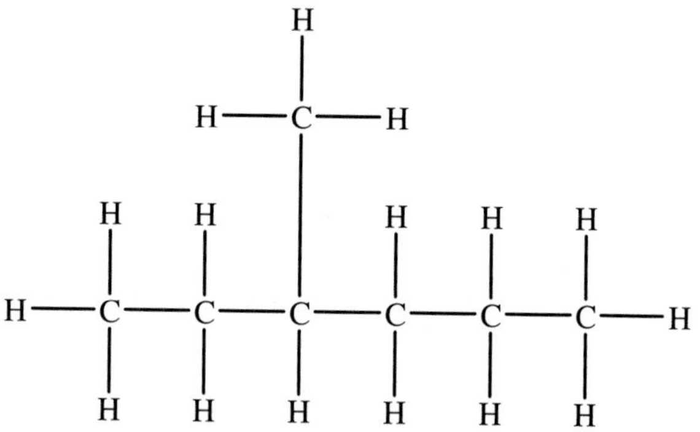   (3)

If the methyl group (-CH$_3$) is moved to the fourth carbon from the left, the structure is the same as (3). If the methyl group is moved to the fifth carbon from the left, the structure is the same as (2). In addition, if the methyl group is moved to the sixth carbon from the left, the resulting structure has 7 carbons in a chain and is the same as (1).

Five carbons in chain:

```
        H
        |
   H —  C — H
        |
   H    H    H    H
   |    |    |    |
H— C  — C  — C  — C  — C —H
   |    |    |    |
   H    H    H    H
        |
   H —  C — H
        |
        H
```
(4)

```
        H
        |
   H —  C — H
        |
   H    H    H    H
   |    |    |    |
H— C  — C  — C  — C  — C —H
   |    |    |    |
   H    H    H    H
        |
   H —  C — H
        |
        H
```
(5)

```
        H
        |
   H —  C — H
        |
   H    H    H    H
   |    |    |    |
H— C  — C  — C  — C  — C —H
   |    |    |    |
   H    H    H    H
             |
        H  — C — H
             |
             H
```
(6)

```
             H
             |
        H  — C — H
             |
   H    H    H    H    H
   |    |    |    |    |
H— C  — C  — C  — C  — C —H
   |    |    |    |    |
   H    H    H    H    H
             |
        H  — C — H
             |
             H
```
(7)

```
        H
        |
   H —  C — H
        |
   H —  C — H
        |
   H    H    H    H    H
   |    |    |    |    |
H— C  — C  — C  — C  — C —H
   |    |    |    |    |
   H    H    H    H    H
```
(8)

Starting with a methyl group on the second carbon from the left, add another methyl group in a systematic pattern. Structure (4) has the second methyl on the second carbon, (5) on the third carbon, (6) on the 4th carbon. These are all the unique possibilities with the first methyl group on the second carbon. Any other variation will produce a structure identical to a structure already drawn. Next, move the first methyl group to the third carbon where the only unique placement for the second methyl group is on the third carbon as shown in structure (7). Which structure above would be the same as placing the first methyl group on the third carbon and the second methyl group on the fourth carbon?

Are there any more arrangements with a 5-carbon chain? Think of attaching the two extra carbons as an ethyl group ($-CH_2CH_3$). If the ethyl group is attached to the second carbon from the left, the longest chain becomes 6 carbons instead of 5 and the structure is the same as (3). If the ethyl group is instead attached to the third carbon, the longest chain is still 5 carbons and the ethyl group is a side chain to give structure (8).

4 carbons in chain:

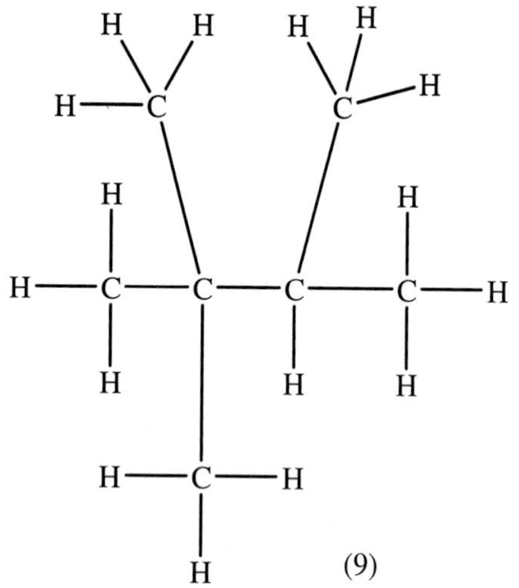

(9)

Only this arrangement of 3 methyl groups branching off the two inner carbons in a 4C chain produces a unique structure.

Attaching 4 methyl groups off a 3C chain is impossible without making the chain longer. So, the above nine structures are all the arrangements.

Check: Build these structures with a model set (make the structures without the hydrogen atoms to look only at the carbon chains) and try to change one into another by rotating the molecule, but not by breaking any bonds. If you have to break a bond to convert from one structure to another, then the structures are different.

b) Plan: For a five-carbon compound start with 5 C atoms in a chain and place the triple bond in as many unique places as possible.

Solution:

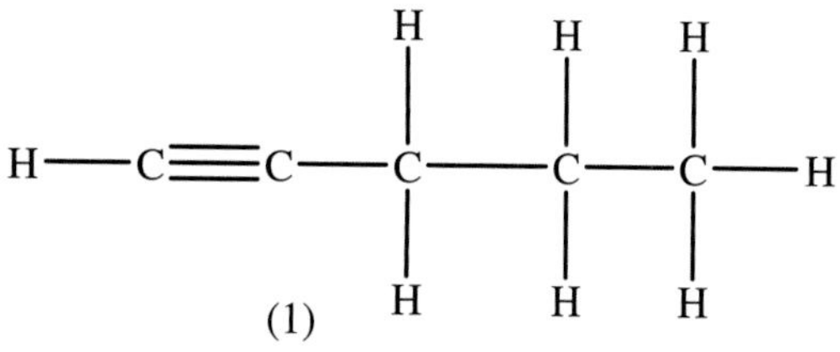

(1) 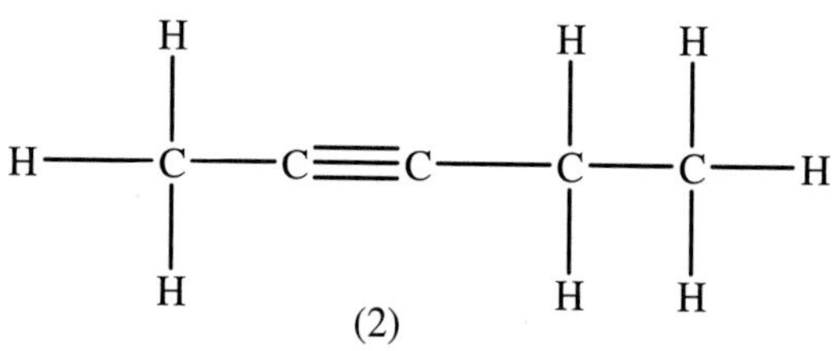   (2)

4 C atoms in chain: There are two unique placements for the triple bond: one between the first and second carbons and one between the second and third carbons in the chain.

The fifth carbon is added as a methyl group branched off the chain. With the triple bond between the first and second carbons, the methyl group is attached to the third carbon to give structure (3).

240

$$H-C\equiv C-\underset{\underset{H}{|}}{\overset{\overset{H}{|}}{C}}-\underset{\underset{H}{|}}{\overset{\overset{H}{|}}{C}}-H$$

with a CH_2 (as $H-\overset{|}{\underset{|}{C}}-H$, i.e. CH_3) group attached to the third carbon

(3)

With the triple bond between the second and third carbons, the methyl group cannot be attached to either the second or the third carbons because that would create 5 bonds to a carbon and carbon has only 4 bonds. Thus, a unique structure cannot be formed with a 4 C chain where the triple bond is between the second and third carbons. There are only 3 structures with 5 carbon atoms, one triple bond, and no rings.

15.2   Plan: Analyze the name for chain length and side groups, then draw the structure.
Solution:
a)   3–ethyl–3–methyloctane. The end of the name, octane, indicates an 8 C chain (oct- represents 8 C) with only single bonds between the carbons (-ane represents alkanes, only single bonds). 3–ethyl means an ethyl group ($-CH_2CH_3$) attached to carbon #3 and 3–methyl means a methyl group ($-CH_3$) attached to carbon #3.

$$H_3C-CH_2-\underset{\underset{\underset{CH_3}{|}}{\underset{CH_2}{|}}}{\overset{\overset{CH_3}{|}}{C}}-CH_2-CH_2-CH_2-CH_2-CH_3$$

b) 1–ethyl–3–propylcyclohexane. The hexane indicates 6 C chain (hex-) and only single bonds (-ane). The cyclo- indicates that the 6 carbons are in a ring. The 1–ethyl indicates that an ethyl group ($-CH_2CH_3$) is attached to carbon #1. Select any carbon atom in the ring as carbon #1 since all the carbon atoms in the ring are equivalent. The 3-propyl indicates that a propyl group ($-CH_2CH_2CH_3$) is attached to carbon #3.

Cyclohexane ring with CH_2CH_3 attached to one carbon (top) and $CH_2CH_2CH_3$ attached to another carbon (bottom right).

c) 3,3–diethyl–1–hexyne. The 1–hexyne indicates a 6 C chain with a triple bond (-yne) between C #1 and C #2. The 3,3–diethyl means two (di-) ethyl groups (–CH_2CH_3) both attached to C #3.

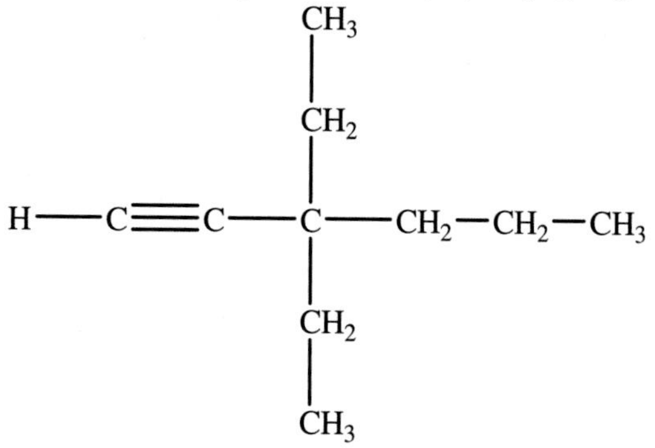

$$H-C\equiv C-\underset{\underset{\displaystyle CH_3}{\overset{\displaystyle |}{\underset{\displaystyle |}{CH_2}}}}{\overset{\overset{\displaystyle CH_3}{\overset{\displaystyle |}{\overset{\displaystyle CH_2}{|}}}}{C}}-CH_2-CH_2-CH_3$$

d) *trans*–3–methyl–3–heptene. The 3–heptene indicates a 7 C chain (hept-) with one double bond (-ene) between the 3rd and 4th carbons. The 3–methyl indicates a methyl group (–CH_3) attached to carbon #3. The *trans* indicates that the arrangement around the two carbons in the double bond gives the two smaller groups on opposite sides of the double bond. Therefore, the smaller group on the third carbon (which would be the methyl group) is above the double bond while the smaller group, H, on the fourth carbon is below the double bond.

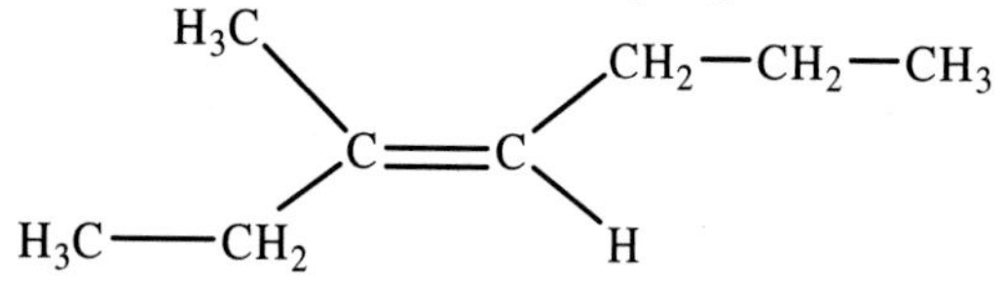

$$\underset{\displaystyle H_3C-CH_2}{\overset{\displaystyle H_3C}{\diagdown}}C=C\underset{\displaystyle H}{\overset{\displaystyle CH_2-CH_2-CH_3}{\diagup}}$$

15.3   Plan:

a) An addition reaction involves breaking a multiple bond, in this case the double bond in 2–butene, and adding the other reactant to the carbons in the double bond. The reactant Cl_2 will add –Cl to one of the carbons and –Cl to the other carbon.

b) A substitution reaction involves removing one atom or group from a carbon chain and replacing it with another atom or group. For 1–bromopropane the bromine will be replaced by hydroxide.

c) An elimination reaction involves removing two atoms or groups, one from each of two adjacent carbon atoms, and forming a double bond between the two carbon atoms. For 2–methyl–2–propanol, the –OH group from the center carbon and a hydrogen from one of the terminal carbons will be removed and a double bond formed between the two carbon atoms.

Solution:

(a)

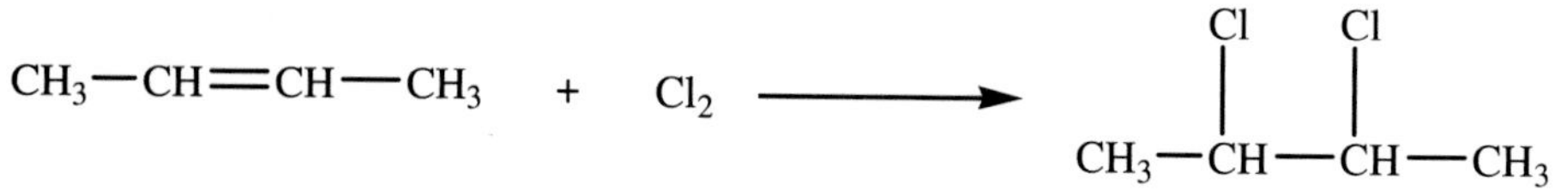

$$CH_3-CH=CH-CH_3 \; + \; Cl_2 \longrightarrow CH_3-\underset{\overset{\displaystyle |}{\displaystyle Cl}}{CH}-\underset{\overset{\displaystyle |}{\displaystyle Cl}}{CH}-CH_3$$

(b)

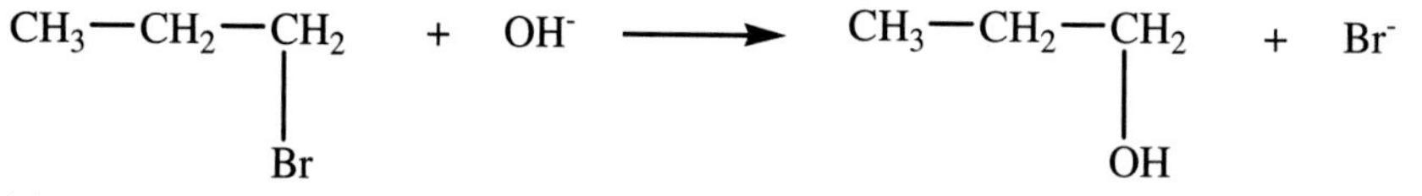

$$CH_3-CH_2-\underset{\underset{\displaystyle Br}{\displaystyle |}}{CH_2} \; + \; OH^- \longrightarrow CH_3-CH_2-\underset{\underset{\displaystyle OH}{\displaystyle |}}{CH_2} \; + \; Br^-$$

(c)

$$CH_3-\underset{\underset{\displaystyle O-H}{\underset{\displaystyle |}{\displaystyle |}}}{\overset{\overset{\displaystyle CH_3}{\displaystyle |}}{C}}-CH_3 \longrightarrow CH_3-\underset{\overset{\displaystyle |}{\displaystyle CH_3}}{\overset{\displaystyle CH_3}{C}}=CH_2 \; + \; H_2O$$

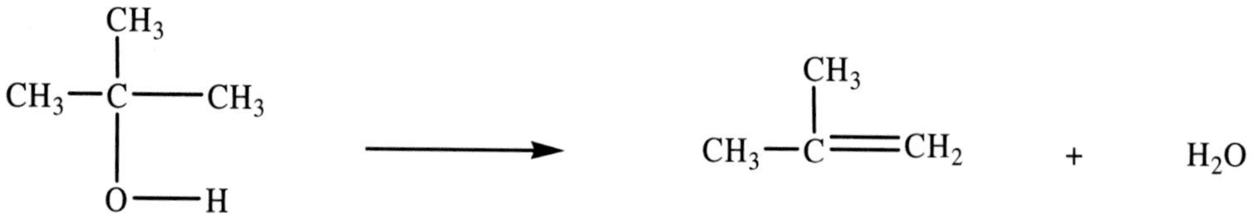

15.4   Plan: Reaction a) is an elimination reaction where H and Cl are eliminated from the organic reactant and a double bond is formed. Determine which carbons were involved in formation of the double bond and those two carbons would have been bonded to the H and Cl. Reaction b) is an oxidation reaction because $Cr_2O_7^{2-}$ and H_2SO_4 are oxidizing agents. An alcohol group (–OH) is oxidized to an acid group (–COOH), so determine which carbon in the product is in the acid group and attach an alcohol group to it for the reactant.
Solution:
a) The organic reactant must contain a single bond between the second and third carbons and the second and third carbons each have an additional group: an H atom on one carbon and a Cl atom on the other carbon.

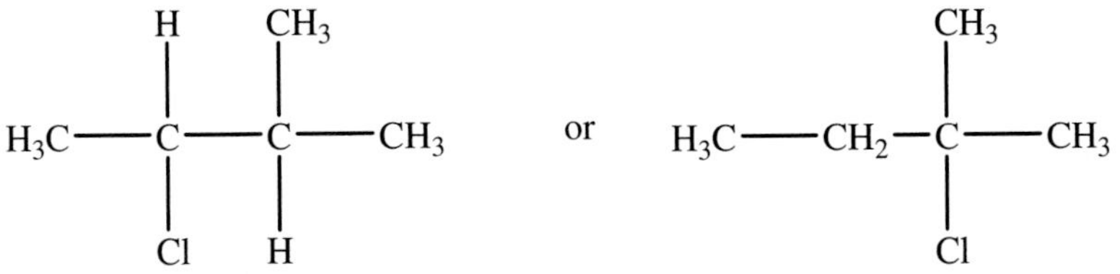

H₃C—C—C—CH₃   with H, CH₃ above and Cl, H below   or   H₃C—CH₂—C—CH₃   with CH₃ above and Cl below

b) The third carbon from the left in the product contains the acid group. Therefore, in the reactant this carbon should have an alcohol group.

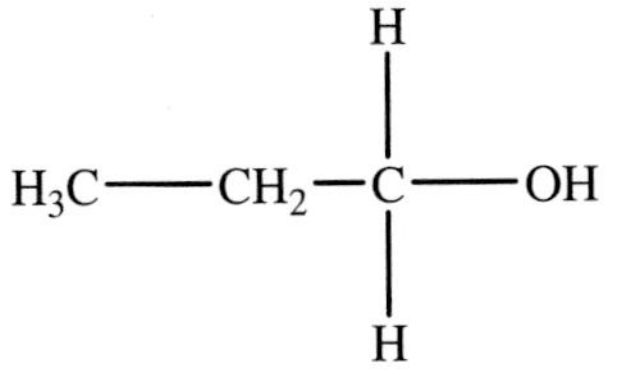

H₃C—CH₂—C—OH   with H above and H below

Check: When the reactants are combined, the products are identical to those given.

15.5   Plan: The oxidizing agents in reaction a) indicate that the ketone group (=O) has been oxidized from an alcohol. To form the reactant, replace the C=O with an alcohol group, (C-OH). In reaction b) the reactants CH_3CH_2—Li and H_2O indicate that a ketone or aldehyde reacts to form the alcohol. To form the reactant find the carbon with the alcohol group, remove the ethyl group that came from CH_3CH_2—Li and change the -OH bond to a carbonyl group, =O.
Solution:
a) In the reactant, a hydrogen is lost from the oxygen and another hydrogen is lost from the adjacent carbon. A double bond forms between the carbon and the oxygen.

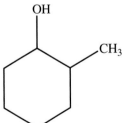

b) In the reactant, the first carbon off the ring will have a carbonyl group (-C=O) in place of the alcohol group (-OH) and the ethyl group (-CH₂CH₃).

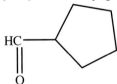

Check: When the reactants are combined, the products are identical to those given.

15.6   Plan: a) The product is an ester because it contains the unit:

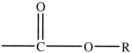

Reacting an alcohol with an acid forms an ester. Since the reactant shown is an alcohol the other reactant must be the acid. Take the –OR group off the carbon in the ester product and replace with a hydroxyl group to get the structure of the acid.

b) The product is an amide because it contains the unit:

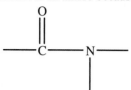

An amide forms from the reaction between an ester and an amine. This is the indicated reaction because the other product is an alcohol that results when the –OR group on the ester breaks off. To draw the reactants first take the –NH—R group off the amide and add another hydrogen to the nitrogen to form the amine. Attach the –OR group to the remaining carbonyl (–C=O) group of the amide molecule so that an ester linkage is formed.

Solution:

a) The –OR group attached to the oxygen in the product is –OCH₃. Removing this and adding the –OH gives the acid reactant:

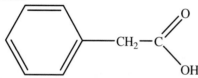

b) Take the –NHCH₂CH₃ group off the amide product, attach a hydrogen to the nitrogen to make the amine reactant, and attach the –OCH₃ from the alcohol product in its place to make the ester reactant.

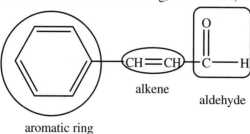

ester   amine

Check: When the reactants are combined, the products are identical to those given.

15.7   Plan: Examine the structures for known functional groups: alkenes (C=C), alkynes (C≡C), haloalkanes C–X, where X is a halogen), alcohols (C–OH), esters (–COOR), ethers (C–O–C), amines (NR₃), carboxylic acids (–COOH), amides (–CONR₂), aldehydes (O=CHR), and ketones (O=CR₂ where R cannot be a H).

Solution:

a) The structure contains an aromatic ring, alkene bond, and an aldehyde group.

alkene

aldehyde

aromatic ring

b) The structure contains an amide and a haloalkane.

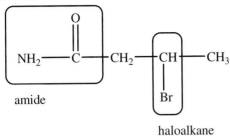

amide

haloalkane

END–OF–CHAPTER PROBLEMS

15.3   a) Carbon's electronegativity is midway between the most metallic and nonmetallic elements of period 2. To attain a filled outer shell, carbon forms covalent bonds to other atoms in molecules (e.g., methane, CH_4), network covalent solids (e.g., diamond) and polyatomic ions (e.g., carbonate, CO_3^{2-}).
b) Since carbon has 4 valence shell electrons, it forms four covalent bonds to attain an octet.
c) Two noble gas configurations, He and Ne, are equally near carbon's configuration. To reach the He configuration, the carbon atom must lose 4 electrons, requiring too much energy to form the C^{4+} cation. This is confirmed by the fact that the value of the ionization energy for carbon is very high. To reach the Ne configuration, the carbon atom must gain 4 electrons, also requiring too much energy to form the C^{4-} anion. The fact that a carbon anion is unlikely to form is supported by carbon's electron affinity. The other possible ions would not have a stable noble gas configuration.
d) Carbon is able to bond to itself extensively because carbon's small size allows for closer approach and greater orbital overlap. The greater orbital overlap results in a strong, stable bond.
e) The C–C bond is short enough to allow the sideways overlap of unhybridized p-orbitals on neighboring C atoms. The sideways overlap of p-orbitals results in double and triple bonds.

15.4   a) The elements that most frequently bond to carbon are other carbon atoms, hydrogen, oxygen, nitrogen, phosphorus, sulfur, and the halogens, F, Cl, Br, and I.
b) In organic compounds, heteroatoms are defined as atoms of any element other than carbon and hydrogen. The elements **O, N, P, S, F, Cl, Br, and I** listed in part a) are heteroatoms.
c) Elements more electronegative than carbon are N, O, F, Cl, and Br. Elements less electronegative than carbon are H and P. Sulfur and iodine have the same electronegativity as carbon.
d) The more types of atoms that can bond to carbon, the greater the variety of organic compounds that are possible.

15.7   Chemical reactivity occurs when unequal sharing of electrons in a covalent bond results in regions of high and low electron density. The C–H, C–C, and C–I bonds are unreactive because electron density is shared equally between the two atoms. The **C=O** bond is reactive because oxygen is more electronegative than carbon and the electron rich pi bond is above and below the C–O bond axis, making it very attractive to electron-poor atoms. The **C–Li** bond is also reactive because the bond polarity results in an electron-rich region around carbon and an electron-poor region around Li.

15.8   a) An alkane is an organic compound consisting of carbon and hydrogen in which there are no multiple bonds between carbons, only single bonds. A cycloalkane is an alkane in which the carbon chain is arranged in a ring. An alkene is a hydrocarbon with at least one double bond between two carbons. An alkyne is a hydrocarbon with at least one triple bond between two carbons.
b) The general formula for an alkane is C_nH_{2n+2}.
The general formula for a cycloalkane is C_nH_{2n}. Elimination of two hydrogen atoms is required to form the additional bond between carbons in the ring.
For an alkene, assuming only one double bond, the general formula is C_nH_{2n}. When a double bond is formed in an alkane, two hydrogen atoms are removed.
For an alkyne, assuming only one triple bond, the general formula is C_nH_{2n-2}. Forming a triple bond from a double bond causes the loss of two hydrogen atoms.

c) For hydrocarbons, "saturated" is defined as a compound that cannot add more hydrogen. An unsaturated hydrocarbon contains multiple bonds that react with H_2 to form single bonds. The **alkanes and cycloalkanes** are saturated hydrocarbons since they contain only single C–C bonds.

15.11   An asymmetric molecule has no plane of symmetry.
a) A circular clock face numbered 1 to 12 o'clock is **asymmetric**. Imagine that the clock is cut in half, from 12 to 6 or from 9 to 3. The one-half of the clock could never be superimposed on the other half, so the halves are not identical. Another way to visualize symmetry is to imagine cutting an object in half and holding the half up to a mirror. If the original object is "re-created" in the mirror, then the object has a plane of symmetry.
b) A football is symmetric and has two planes of symmetry — one axis along the length and one axis along the fattest part of the football.
c) A dime is **asymmetric**. Either cutting it in half or slicing it into two thin diameters results in two pieces that cannot be superimposed on one another.
d) A brick, assuming that it is perfectly shaped, is symmetric and has three planes of symmetry at right angles to each other.
e) A hammer is symmetric and has one plane of symmetry, slicing through the metal head and down through the handle.
f) A spring is **asymmetric**. Every coil of the spring is identical to the one before it, so a spring can be cut in half and the two pieces can be superimposed on one another by sliding (not flipping) the second half over the first. However, if the cut spring is held up to a mirror, the resulting image is not the same as the uncut spring. Disassemble a ballpoint pen and cut the spring inside to verify this explanation.

15.14   To draw the possible skeletons, it is useful to have a systematic approach to make sure no structures are missed.
a) Since there are 7 C atoms but only a 6 C chain, there is one C branch off of the chain. First, draw the skeleton with the double bond between the first and second carbons and place the branched carbon in all possible positions starting with C #2. Then move the double bond to between the second and third carbon and place the branched carbon in all possible positions. Then move the double bond to between the third and fourth carbons and place the branched carbon in all possible positions. The double bond does not need to be moved further in the chain since the placement between the second and third carbon is equivalent to placement between the fourth and fifth carbons and placement between the first and second carbons is equivalent to placement between the fifth and sixth carbons. The other position to consider for the double bond is between the branched carbon and the 6 C chain.
Double bond between first and second carbons:

Double bond between second and third carbons:

Double bond between third and fourth carbons:

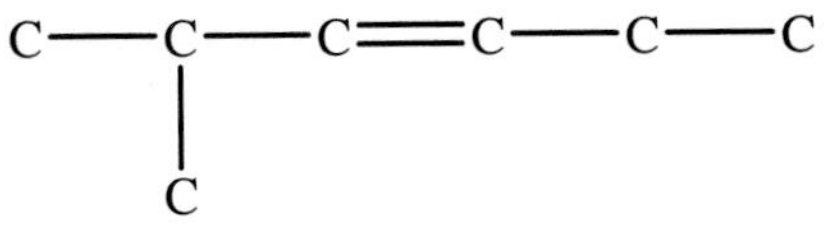

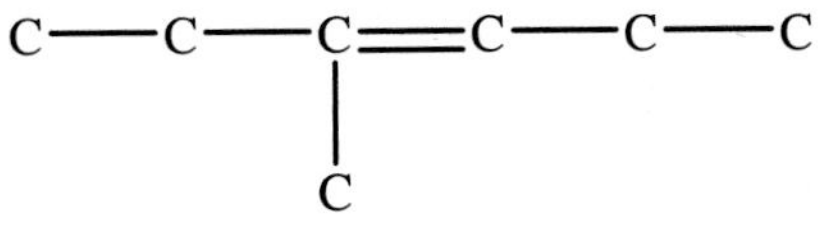

C—C—C=C—C—C   C—C—C=C—C—C
    | |
    C   C

Double bond between branched carbon and chain:

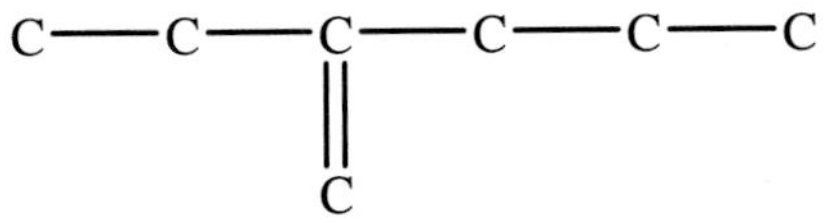

C—C—C—C—C—C
        ‖
        C

The total number of unique skeletons is 11. To determine if structures are the same, build a model of one skeleton and see if you can match the structure of the other skeleton by rotating the model and without breaking any bonds. If bonds must be broken to make the other skeleton, the structures are not the same.

b) The same approach can be used here with placement of the double bond first between C #1 and C #2, then between C #2 and C #3. Since there are 7 C atoms but only 5 C atoms in the chain, there are two C branches.

Double bond between first and second carbons:

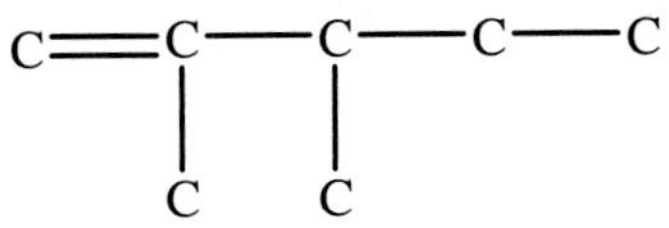

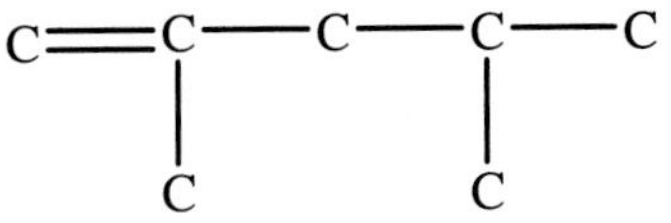

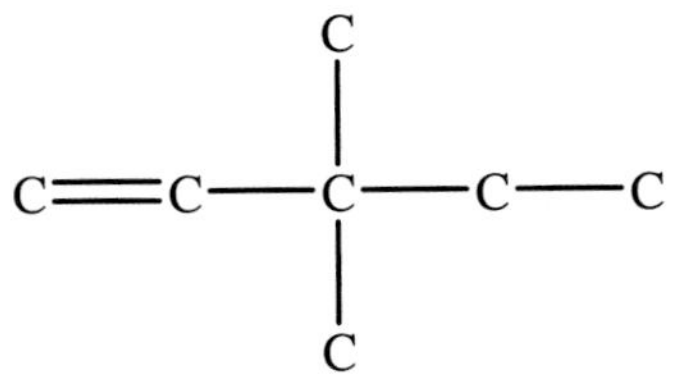

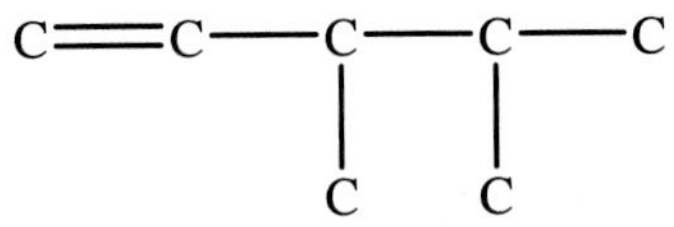

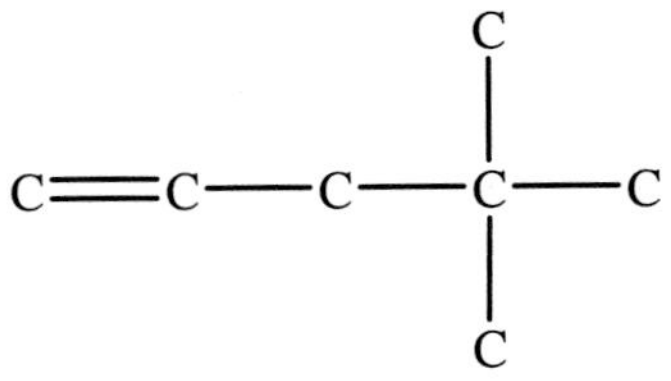

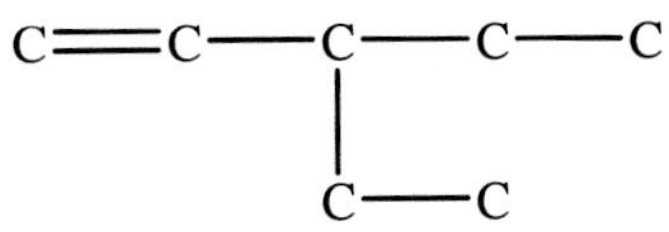

C=C—C—C—C   C=C—C—C—C
     | | | |
     C C   C   C

          C
          |
C=C—C—C—C   C=C—C—C—C
        | | |
        C   C C

          C
          |
C=C—C—C—C   C=C—C—C—C
          | |
          C   C—C

Double bond between second and third carbons:

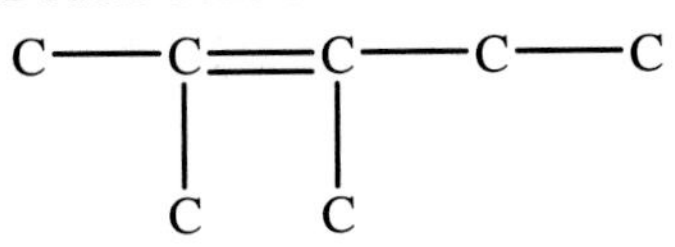

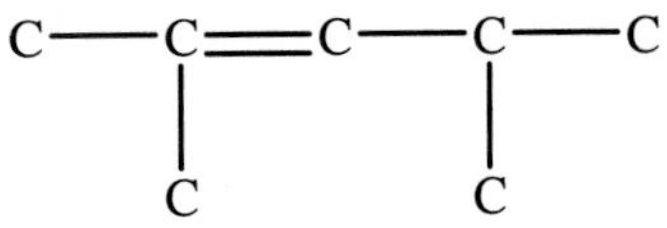

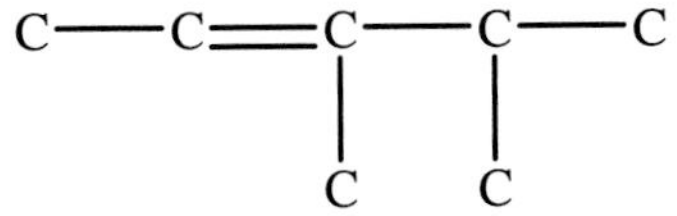

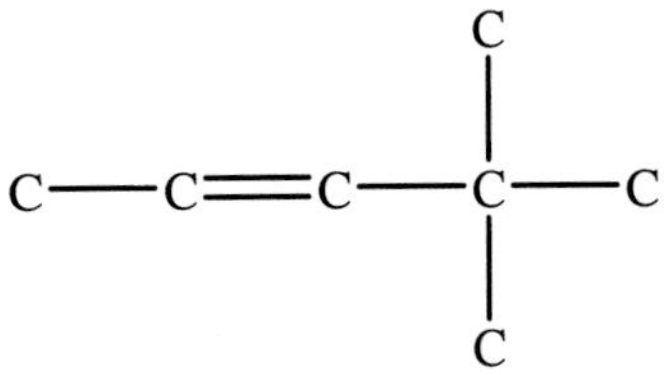

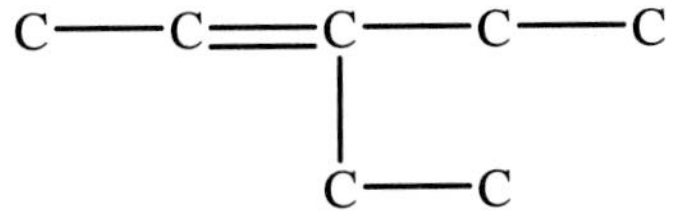

C—C=C—C—C   C—C=C—C—C
        | | | |
        C C   C   C

                                C
                                |
C—C=C—C—C   C—C=C—C—C
         | | |
         C C   C

C—C=C—C—C
        |
        C—C

c) Five of the carbons are in the ring and two are branched off the ring. Remember that all the carbons in the ring are equivalent and there are two groups bonded to each carbon in the ring.

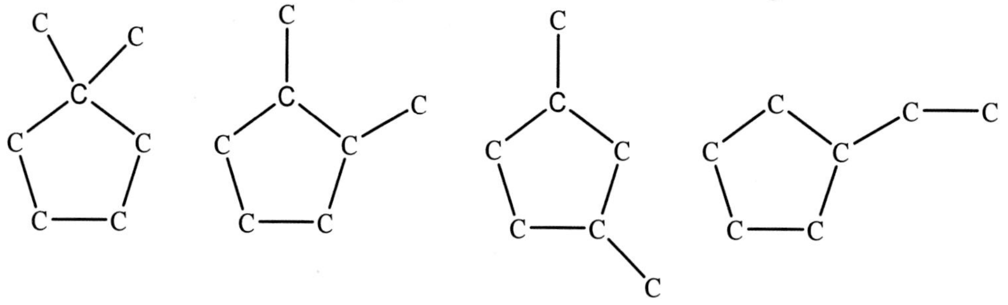

15.16 Add hydrogen atoms to make a total of four bonds to each carbon.

(a)

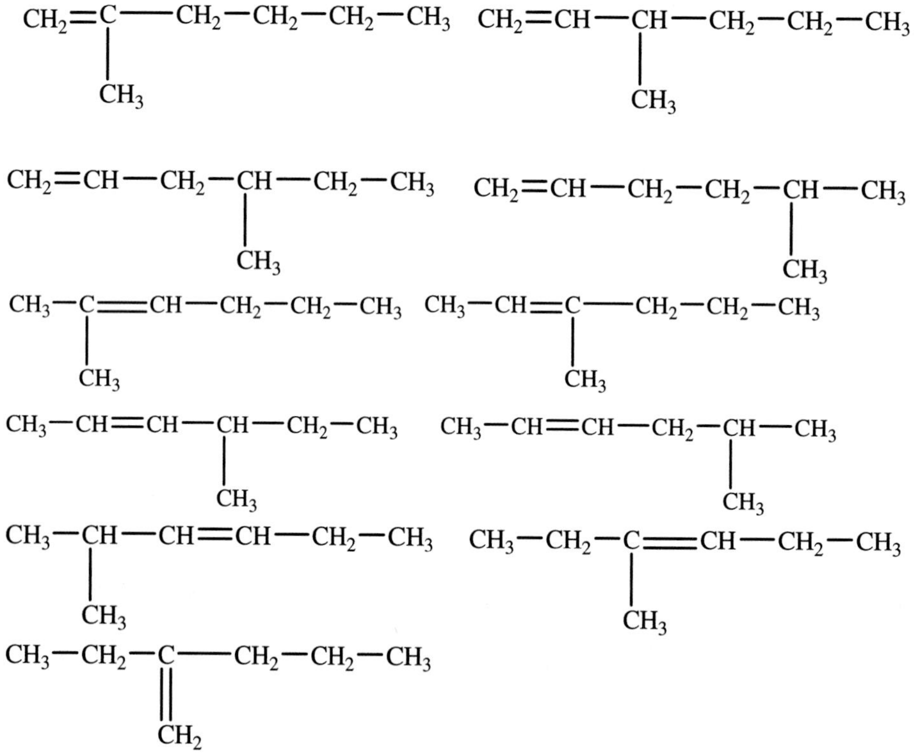

CH_2=C—CH_2—CH_2—CH_2—CH_3 CH_2=CH—CH—CH_2—CH_2—CH_3
$\qquad$ | $\qquad\qquad\qquad\qquad\qquad\qquad\qquad\qquad$ |
$\qquad$ CH_3 $\qquad\qquad\qquad\qquad\qquad\qquad\qquad\qquad$ CH_3

CH_2=CH—CH_2—CH—CH_2—CH_3 CH_2=CH—CH_2—CH_2—CH—CH_3
$\qquad\qquad\qquad\qquad$ | $\qquad\qquad\qquad\qquad\qquad\qquad\qquad\qquad$ |
$\qquad\qquad\qquad\qquad$ CH_3 $\qquad\qquad\qquad\qquad\qquad\qquad\qquad\qquad$ CH_3

CH_3—C=CH—CH_2—CH_2—CH_3 CH_3—CH=C—CH_2—CH_2—CH_3
$\qquad$ | $\qquad\qquad\qquad\qquad\qquad\qquad\qquad\qquad\qquad$ |
$\qquad$ CH_3 $\qquad\qquad\qquad\qquad\qquad\qquad\qquad\qquad\quad$ CH_3

CH_3—CH=CH—CH—CH_2—CH_3 CH_3—CH=CH—CH_2—CH—CH_3
$\qquad\qquad\qquad\qquad$ | $\qquad\qquad\qquad\qquad\qquad\qquad\qquad\qquad\qquad$ |
$\qquad\qquad\qquad\qquad$ CH_3 $\qquad\qquad\qquad\qquad\qquad\qquad\qquad\qquad\qquad$ CH_3

CH_3—CH—CH=CH—CH_2—CH_3 CH_3—CH_2—C=CH—CH_2—CH_3
$\qquad$ | $\qquad\qquad\qquad\qquad\qquad\qquad\qquad\qquad\qquad\qquad$ |
$\qquad$ CH_3 $\qquad\qquad\qquad\qquad\qquad\qquad\qquad\qquad\qquad\qquad$ CH_3

CH_3—CH_2—C—CH_2—CH_2—CH_3
$\qquad\qquad\quad$ ‖
$\qquad\qquad\quad$ CH_2

(b)

CH_2=C—CH—CH_2—CH_3 $\qquad\qquad$ CH_2=C—CH_2—CH—CH_3
$\qquad$ | $\quad$ | $\qquad\qquad\qquad\qquad\qquad\qquad$ | $\qquad\qquad$ |
$\qquad$ CH_3 CH_3 $\qquad\qquad\qquad\qquad\qquad\qquad$ CH_3 $\qquad$ CH_3

$\qquad\qquad\qquad$ CH_3
$\qquad\qquad\qquad$ |
CH_2=CH—C—CH_2—CH_3 $\qquad$ CH_2=CH—CH—CH—CH_3
$\qquad\qquad\quad$ | $\qquad\qquad\qquad\qquad\qquad\qquad\qquad\qquad$ | $\quad$ |
$\qquad\qquad\quad$ CH_3 $\qquad\qquad\qquad\qquad\qquad\qquad\qquad\quad$ CH_3 CH_3

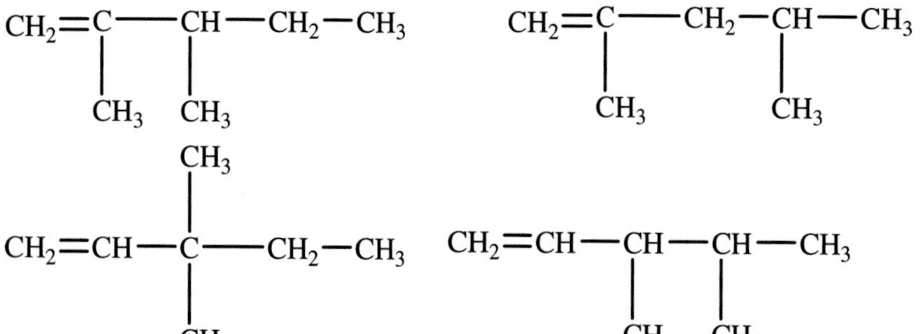

$$CH_2{=}CH{-}CH_2{-}\underset{\underset{\displaystyle CH_3}{|}}{\overset{\overset{\displaystyle CH_3}{|}}{C}}{-}CH_3 \qquad\qquad CH_2{=}CH{-}\underset{\underset{\displaystyle CH_2{-}CH_3}{|}}{CH}{-}CH_2{-}CH_3$$

$$CH_3{-}\underset{\underset{\displaystyle CH_3}{|}}{C}{=}\underset{\underset{\displaystyle CH_3}{|}}{C}{-}CH_2{-}CH_3 \qquad\qquad CH_3{-}\underset{\underset{\displaystyle CH_3}{|}}{C}{=}CH{-}\underset{\underset{\displaystyle CH_3}{|}}{CH}{-}CH_3$$

$$CH_3{-}CH{=}\underset{\underset{\displaystyle CH_3}{|}}{C}{-}\underset{\underset{\displaystyle CH_3}{|}}{CH}{-}CH_3 \qquad\qquad CH_3{-}CH{=}CH{-}\underset{\underset{\displaystyle CH_3}{|}}{\overset{\overset{\displaystyle CH_3}{|}}{C}}{-}CH_3$$

$$CH_3{-}CH{=}\underset{\underset{\displaystyle CH_2{-}CH_3}{|}}{C}{-}CH_2{-}CH_3$$

(c)

15.18 a) The second carbon from the left in the chain is bonded to five groups. Removing one of the groups gives a correct structure.

$$CH_3{-}\underset{\underset{\displaystyle CH_3}{|}}{\overset{\overset{\displaystyle CH_3}{|}}{C}}{-}CH_2{-}CH_3$$

b) The first carbon in the chain has five bonds, so remove one of the hydrogens on this carbon.

$$H_2C{=}CH{-}CH_2{-}CH_3$$

c) The second carbon in the chain has five bonds, so move the ethyl group from the second carbon to the third. To do this, a hydrogen atom must be removed from the third carbon atom.

$$HC{\equiv}C{-}\underset{\underset{\underset{\displaystyle CH_3}{|}}{\overset{\displaystyle CH_2}{|}}}{CH}{-}CH_3$$

d) Structure is correct.

15.20   a) *Octane* denotes an eight carbon alkane chain. A methyl group (–CH₃) is located at the second and third carbon position from the left.

$$CH_3-\underset{\displaystyle CH_3}{\overset{\displaystyle CH_3}{CH}}-\underset{\displaystyle CH_3}{CH}-CH_2-CH_2-CH_2-CH_2-CH_3$$

b) *Cyclohexane* denotes a six-carbon ring containing only single bonds. Numbering of the carbons on the ring could start at any point, but typically, numbering starts at the top carbon atom of the ring for convenience. The ethyl group (–CH₂CH₃) is located at position 1 and the methyl group is located at position 3.

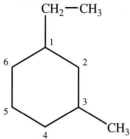

c) The longest continuous chain contains seven carbon atoms, so the root name is "hept." The molecule contains only single bonds, so the suffix is "ane." Numbering the carbon chain from the left results in side groups (methyl groups) at positions 3 and 4. Numbering the carbon chain from the other will result in side groups at positions 4 and 5. Since the goal is to obtain the lowest numbering position for a side group, the correct name is **3,4–dimethylheptane**. Note that the prefix "di" is used to denote that two methyl side groups are present in this molecule.

d) At first glance, this molecule looks like a 4 carbon *ring*, but the two –CH₃ groups mean that they cannot be bonded to each other. Instead, this molecule is a 4–carbon *chain*, with two methyl groups (dimethyl) located at the position 2 carbon. The correct name is **2,2–dimethylbutane**.

15.22   a) 4–methylhexane means a 6 C chain with a methyl group on the 4th carbon:

$$\underset{6}{CH_3}-\underset{5}{CH_2}-\underset{4}{CH_2}-\underset{3}{\overset{\displaystyle CH_3}{CH}}-\underset{2}{CH_2}-\underset{1}{CH_3}$$

Numbering from the end carbon to give the lowest value for the methyl group gives correct name of 3–methylhexane.

b) 2–ethylpentane means a 5 C chain with an ethyl group on the 2nd carbon:

$$\underset{1}{CH_3}-\underset{2}{\overset{\displaystyle CH_3}{\underset{\displaystyle \vert}{\overset{\displaystyle \vert}{\underset{\displaystyle CH_2}{CH}}}}}-\underset{3\ \ 4\ \ 5}{CH_2CH_2CH_3}$$

Numbering the longest chain gives the correct name, **3–methylhexane**.

$$\overset{1\ \ CH_3}{\underset{\displaystyle \vert}{}}$$
$$\overset{2\ \ CH_2}{\underset{\displaystyle \vert}{}}$$
$$CH_3-\underset{3}{CH}-\underset{4}{CH_2}-\underset{5}{CH_2}-\underset{6}{CH_3}$$

c) 2–methylcyclohexane means a 6 C ring with a methyl group on carbon #2:

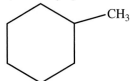

In a ring structure, whichever carbon is bonded to the methyl group is automatically assigned as carbon #1. Since this is automatic, it is not necessary to specify 1–methyl in the name. Correct name is **methylcyclohexane**.

d) 3,3–methyl–4–ethyloctane means an 8 C chain with 2 methyl groups attached to the 3rd carbon and one ethyl group to the 4th carbon.

$$CH_3-CH_2-\underset{\underset{CH_3}{|}}{\overset{\overset{CH_3}{|}}{C}}-\underset{\underset{\underset{CH_2-CH_3}{}}{\overset{|}{}}}{CH}-CH_2-CH_2-CH_2-CH_3$$

Numbering is good for this structure, but the fact that there are two methyl groups must be indicated by the prefix di- in addition to listing 3,3. The branch names appear in alphabetical order. Correct name is **4–ethyl–3,3–dimethyloctane**.

15.24   A carbon atom is chiral if it is attached to four different groups. The circled atoms below are chiral.

(a)   (b)

Both can exhibit optical activity.

15.26   An optically active compound will contain at least one chiral center, a carbon with four distinct groups bonded to it.
a) This compound is a 6 carbon chain with a Br on the third carbon. 3–bromohexane is optically active because carbon #3 has four distinct groups bonded to it: 1) –Br, 2) –H, 3) –CH$_2$CH$_3$, 4) –CH$_2$CH$_2$CH$_3$.

b) This compound is a 5 carbon chain with a Cl and a methyl (CH$_3$) group on the third carbon. 3–chloro–3–methylpentane is not optically active because no carbon has four distinct groups. The third carbon has three distinct groups: 1) –Cl, 2) –CH$_3$ 3) two –CH$_2$CH$_3$ groups.

c) This compound is a 4 carbon chain with Br atoms on the first and second carbon atoms and a methyl group on the second carbon. 1,2–dibromo–2–methylbutane is optically active because the 2nd carbon is chiral, bonded to the four groups: 1) –CH$_2$Br, 2) –CH$_3$, 3) –Br, 4) –CH$_2$CH$_3$.

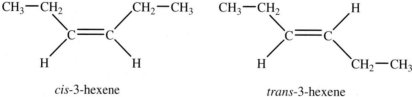

15.28   Geometric isomers are defined as compounds with the same atom sequence but different arrangements of the atoms in space. The *cis-trans* geometric isomers occur when rotation is restricted around a bond, as in a double bond or a ring structure, and when two different groups are bonded to each atom in the restricted bond.
a) Both carbons in the double bond are bonded to two distinct groups, so geometric isomers will occur. The double bond occurs at position 2 in a five-carbon chain.

           cis-2-pentene   *trans*-2-pentene

b) *Cis-trans* geometric isomerism occurs about the double bond. The ring is named as a side group (cyclohexyl) occurring at position 1 on the propene main chain.

           cis-1-cyclohexylpropene   *trans*-1-cyclohexylpropene

c) No geometric isomers occur because the left carbon participating in the double bond is attached to two identical methyl (–CH$_3$) groups.

15.30   a) The structure of propene is CH$_2$=CH–CH$_3$. The first carbon that is involved in the double bond is bonded to two of the same type of group, hydrogen. Geometric isomers will not occur in this case.
b) The structure of 3–hexene is CH$_3$CH$_2$CH=CHCH$_2$CH$_3$. Both carbons in the double bond are bonded to two distinct groups, so geometric isomers will occur.

          cis-3-hexene   *trans*-3-hexene

c) The structure of 1,1–dichloroethene is CCl$_2$=CH$_2$. Both carbons in the double bond are bonded to two identical groups, so no geometric isomers occur.

d) The structure of 1,2–dichloroethene is CHCl=CHCl. Each carbon in the double bond is bonded to two distinct groups, so geometric isomers do exist.

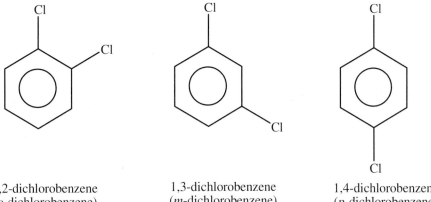

cis-1,2-dichloroethene   *trans*-1,2-dichlorethene

15.32   Benzene is a planar, aromatic hydrocarbon. It is commonly depicted as a hexagon with a circle in the middle to indicate that the π bonds are delocalized around the ring and that all ring bonds are identical. The *ortho, meta, para* naming system is used to denote the location of attached groups in benzene compounds only, not other ring structures like the cycloalkanes.

1,2-dichlorobenzene   1,3-dichlorobenzene   1,4-dichlorobenzene
(*o*-dichlorobenzene)   (*m*-dichlorobenzene)   (*p*-dichlorobenzene)

15.34   Analyzing the name gives benzene as the base structure with the following groups bonded to it: 1) on carbon #1 a hydroxy group, –OH; 2) on carbons #2 and #6 a *tert*-butyl group, –C(CH₃)₃; and 3) on carbon #4 a methyl group, –CH₃.

15.35   The compound 2–methyl–3–hexene has *cis-trans* isomers.

cis-2-methyl-3-hexene   *trans*-2-methyl-3-hexene

253

The compound 2–methyl–2–hexene does not have *cis-trans* isomers because the #2 carbon atom is attached to two identical methyl (–CH$_3$) groups:

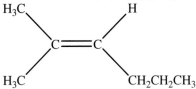

2-methyl-2-hexene

15.39 In an addition reaction, a double bond is broken to leave a single bond, and in an elimination reaction, a double bond is formed from a single bond. A double bond consists of a σ bond and a π **bond**. It is the π bond that breaks in the addition reaction and that forms in the elimination reaction.

15.41 a) HBr is removed in this **elimination reaction**, and an unsaturated product is formed.
b) This is an **addition reaction** in which hydrogen is added to the double bond, resulting in a saturated product.

15.43 a) Water (H$_2$O or H and OH) is added to the double bond:

$$CH_3CH_2CH=CHCH_2CH_3 + H_2O \xrightarrow{H^+} CH_3CH_2CH_2CHCH_2CH_3$$
$$\underset{OH}{|}$$

b) H and Br are eliminated from the molecule, resulting in a double bond:
$$CH_3CHBrCH_3 + CH_3CH_2OK \rightarrow CH_3CH=CH_2 + CH_3CH_2OH + KBr$$
c) Two chlorine atoms are substituted for two hydrogen atoms in ethane:
$$CH_3CH_3 + 2Cl_2 \xrightarrow{h\upsilon} CHCl_2CH_3 + 2 HCl$$

15.45 To decide whether an organic compound is oxidized or reduced in a reaction relie on the rules in the chapter:
A C atom is oxidized when it forms more bonds to O or fewer bonds to H because of the reaction.
A C atom is reduced when it forms fewer bonds to O or more bonds to H because of the reaction.
a) The C atom is **oxidized** because it forms more bonds to O.
b) The C atom is **reduced** because it forms more bonds to H.
c) The C atom is **reduced** because it forms more bonds to H.

15.47 a) The reaction CH$_3$CH=CHCH$_2$CH$_2$CH$_3$ → CH$_2$CH(OH)—CH(OH)CHCH$_2$CH$_2$CH$_3$ shows the second and third carbons in the chain gaining a bond to oxygen: C-O-H. Therefore, the 2-hexene compound has been **oxidized**.
b) The reaction shows that each carbon atom in the cyclohexane loses a bond to hydrogen to form benzene. Fewer bonds to hydrogen in the product indicates **oxidation**.

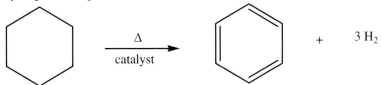

15.50 a) The structures for chloroethane and methylethylamine are given below. The compound **methylethylamine** is more soluble due to its ability to form H-bonds with water. Recall that N–H, O–H, or F–H bonds are required for H-bonding.

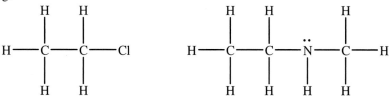

b) The compound 1–butanol is able to H-bond with itself (shown below) because it contains covalent oxygen-hydrogen bonds in the molecule. Diethylether molecules contain no O–H covalent bonds and experience dipole-dipole interactions instead of H-bonding as intermolecular forces. Therefore, **1–butanol** has a higher melting point because H-bonds are stronger intermolecular forces than dipole-dipole attractions.

$$CH_3-CH_2-CH_2-CH_2-\overset{..}{\underset{\underset{H}{|}}{O}}: \text{||||||||||} \quad \overset{H}{\underset{\underset{CH_2-CH_2-CH_2-CH_3}{\diagup}}{\searrow}}\overset{..}{\underset{..}{O}}$$

Hydrogen Bond

$$H_3C-\overset{\overset{H}{|}}{\underset{\underset{H}{|}}{C}}-O-\overset{\overset{H}{|}}{\underset{\underset{H}{|}}{C}}-CH_3$$

diethyl ether

c) **Propylamine** has a higher boiling point because it contains N–H bonds necessary for hydrogen bonding. Trimethylamine is a tertiary amine with no N–H bonds, and so its intermolecular forces are weaker.

$$CH_3-CH_2-CH_2-\overset{..}{\underset{\underset{H}{|}}{N}}-H \qquad CH_3-\overset{..}{\underset{\underset{CH_3}{|}}{N}}-CH_3$$

Propylamine   Trimethylamine

15.53 The C=C bond is nonpolar while the C=O bond is polar, since oxygen is more electronegative than carbon. Both bonds react by addition. In the case of addition to a C=O bond, an electron-rich group will bond to the carbon and an electron-poor group will bond to the oxygen, resulting in one product. In the case of addition to an alkene, the carbons are identical, or nearly so, so there will be no preference for which carbon bonds to the electron-poor group and which bonds to the electron-rich group. This may lead to two isomeric products, depending on the structure of the alkene.

When water is added to a double bond, the hydrogen is the electron-poor group and hydroxyl is the electron-rich group. For a compound with a carbonyl group, only one product results as H bonds to the O atom in the double bond and –OH bonds to the carbon atom in the double bond:

$$CH_3-CH_2-\overset{}{\underset{\underset{O}{\|}}{C}}-CH_3 \quad + \quad \overset{O}{\underset{\underset{H \qquad H}{}}{\diagup\diagdown}} \quad \longrightarrow \quad CH_3-CH_2-\overset{\overset{OH}{|}}{\underset{\underset{OH}{|}}{C}}-CH_3$$

However, when water adds to a C=C, two products result since the OH can bond to either carbon in the double bond:

$$CH_3-CH_2-CH=CH_2 \quad + \quad \underset{H \qquad H}{\overset{O}{\wedge}} \quad \longrightarrow$$

$$CH_3-CH_2-\underset{\underset{OH}{|}}{CH}-CH_3$$

+

$$CH_3-CH_2-CH_2-\underset{\underset{OH}{|}}{CH_2}$$

In this reaction, very little of the second product forms.

15.56   Esters and acid anhydrides form through **dehydration-condensation** reactions. Dehydration indicates that the other product is **water**. In the case of ester formation, condensation refers to the combination of the carboxylic acid and the alcohol. The ester forms and water is the other product.

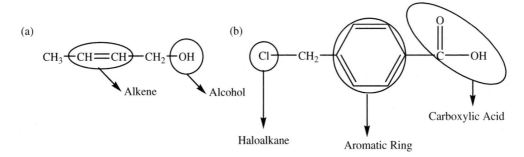

Water Removed   + H_2O

15.58   a) Halogens, except iodine, differ from carbon in electronegativity and form a single bond with carbon. The organic compound is an **alkyl halide**.
b) Carbon forms triple bonds with itself and nitrogen. For the bond to be polar, it must be between carbon and nitrogen. The compound is a **nitrile**.
c) **Carboxylic acids** contain a double bond to oxygen and a single bond to oxygen. Carboxylic acids dissolve in water to give acidic solutions.
d) Oxygen is commonly double–bonded to carbon. A carbonyl group (C=O) that is at the end of a chain is found in an **aldehyde**.

15.60

(a)   (b)

$$CH_3-(CH=CH)-CH_2-(OH)$$

Alkene   Alcohol

$$(Cl)-CH_2-\text{(aromatic ring)}-\underset{O}{\overset{O}{\|}}C-OH$$

Haloalkane   Aromatic Ring   Carboxylic Acid

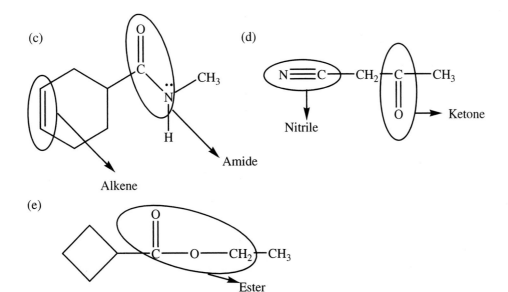

(c)

CH₃

Alkene

Amide

(d)

Nitrile

Ketone

(e)

Ester

15.62 To draw all structural isomers, begin with the longest chain possible. For C₅H₁₂O, the longest chain is 5 carbons:

CH₂—CH₂—CH₂—CH₂—CH₃ CH₃—CH—CH₂—CH₂—CH₃ CH₃—CH₂—CH—CH₂—CH₃
| | |
OH   OH   OH

The three structures represent all the unique positions for the alcohol group on a 5 C chain.
Next, use a 4 C chain and attach to side groups, –OH and –CH₃.

CH₂—CH—CH₂—CH₃   CH₂—CH₂—CH—CH₃
| | | |
OH   CH₃   OH   CH₃
                                                CH₃

CH₃—CH—CH—CH₃   CH₃—C—CH₂—CH₃
| | |
OH   CH₃   OH

And use a 3 C chain with three side groups:

CH₃
|
CH₃—C—CH₂—OH
|
CH₃

The total number of different structures is eight.

15.64 First, draw all primary amines (formula R–NH₂). Next, draw all secondary amines (formula R–NH–R'). There is only one possible tertiary amine structure (formula R₃–N). Eight amines with the formula C₄H₁₁N exist.

CH₃—CH₂—CH₂—CH₂—NH₂   CH₃—CH₂—CH—NH₂
                                              |
                                             CH₃

CH₃—CH—CH₂—NH₂   CH₃—CH₂—CH₂—NH—CH₃
|
CH₃

$$CH_3-\underset{\underset{\displaystyle CH_3}{|}}{CH}-NH-CH_3$$

$$CH_3-CH_2-\underset{\underset{\displaystyle CH_3}{|}}{N}-CH_3$$

$$CH_3-\underset{\underset{\displaystyle CH_3}{|}}{\overset{\overset{\displaystyle CH_3}{|}}{C}}-NH_2$$

$$CH_3-CH_2-NH-CH_2-CH_3$$

15.66   a) With mild oxidation, an alcohol group is oxidized to a carbonyl group. The product is 2-butanone.

$$CH_3-\underset{\underset{\displaystyle O}{\parallel}}{C}-CH_2-CH_3$$

b) With mild oxidation, an aldehyde is oxidized to a carboxylic acid. The product is 2-methylpropanoic acid.

$$CH_3-\underset{\underset{\displaystyle CH_3}{|}}{CH}-\underset{\underset{\displaystyle O}{\parallel}}{C}-OH$$

c) Mild oxidation of an alcohol produces a carbonyl group. The product is cyclopentanone.

15.68   a) This reaction is a dehydration-condensation reaction.

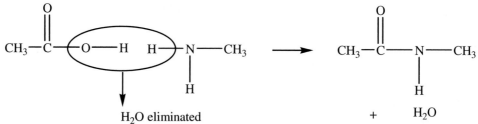

H_2O eliminated

$+$ H_2O

b) An alcohol and a carboxylic acid undergo dehydration-condensation to form an ester.

$$CH_3-CH_2-CH_2-\underset{\underset{\displaystyle O}{\parallel}}{C}-\boxed{O-H \quad H-O}-\underset{\underset{\displaystyle CH_3}{|}}{\overset{\overset{\displaystyle CH_3}{|}}{CH}}$$

H_2O Eliminated

$$CH_3-CH_2-CH_2-\underset{\underset{\displaystyle O}{\parallel}}{C}-O-\underset{\underset{\displaystyle CH_3}{|}}{\overset{\overset{\displaystyle CH_3}{|}}{CH}}$$

c) This reaction is also an ester formation through dehydration-condensation.

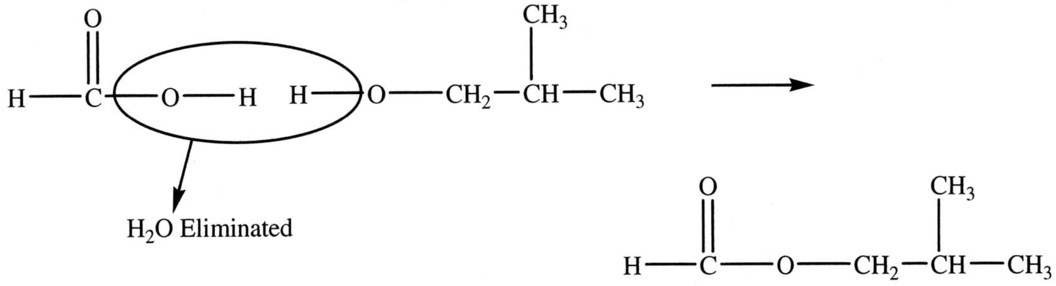

H₂O Eliminated

15.70 To break an ester apart, break the –C–O– single bond and add water (–H and –OH) as shown.

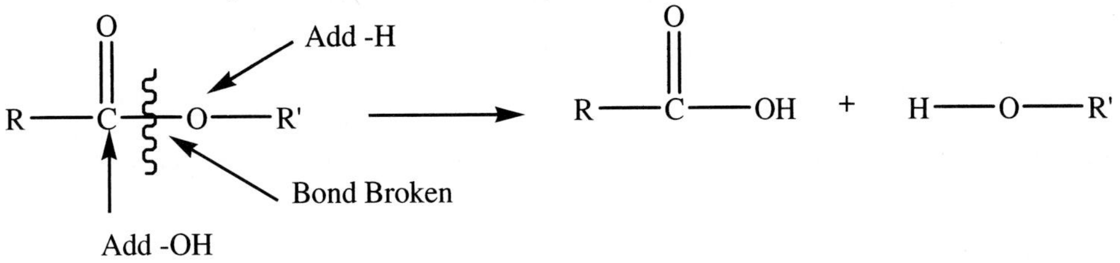

Add -H

Bond Broken

Add -OH

(a)

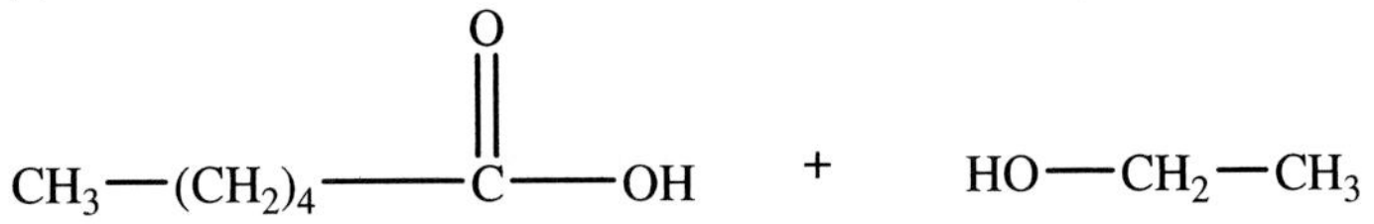

$CH_3-(CH_2)_4$—C—OH   +   HO—CH_2—CH_3

(b)

C—OH   +   HO—CH_2—CH_2—CH_3

(c)

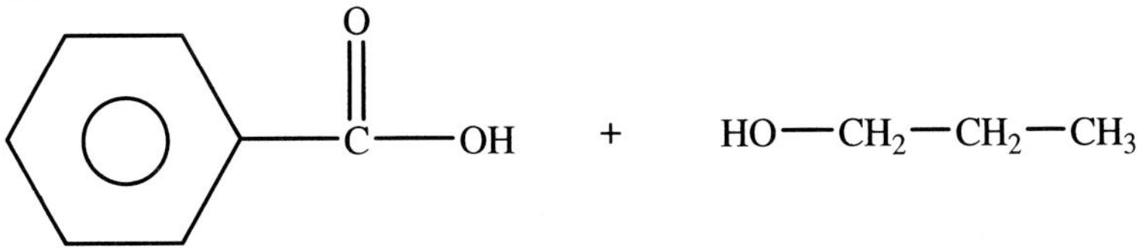

CH_3—CH_2—OH   +   HO—C—CH_2—CH_2—

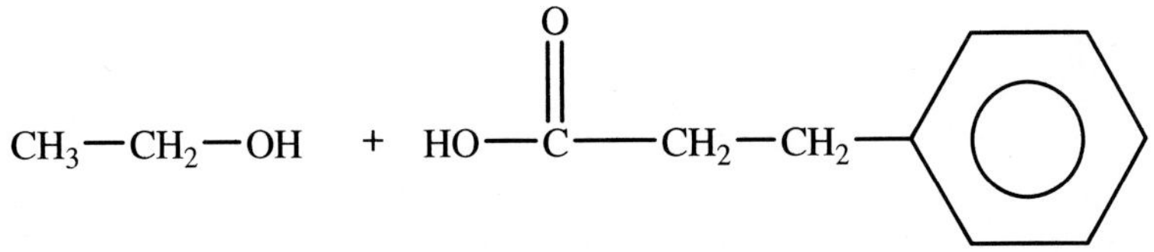

15.72 a) Substitution of Br⁻ occurs by the stronger base, OH⁻. Then a substitution reaction between the alcohol and carboxylic acid produces an ester:

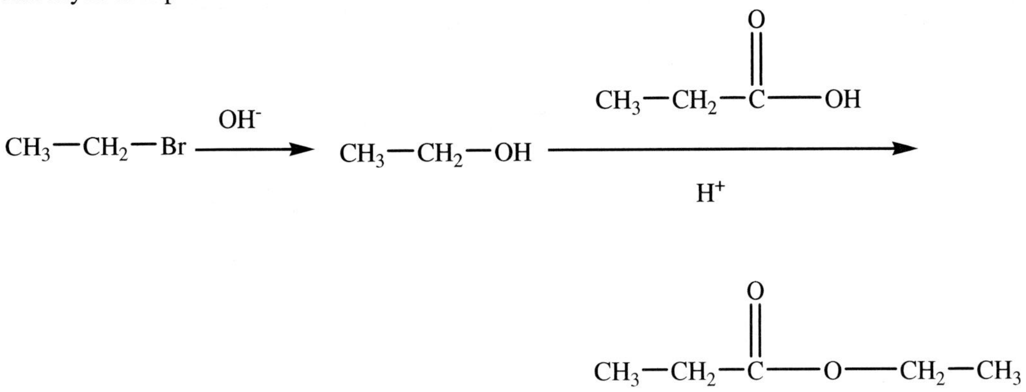

CH_3—CH_2—Br $\xrightarrow{OH^-}$ CH_3—CH_2—OH $\xrightarrow[H^+]{CH_3-CH_2-C-OH}$

CH_3—CH_2—C—O—CH_2—CH_3

b) The strong base, CN⁻, substitutes for Br. The nitrile is then hydrolyzed to a carboxylic acid.

$$CH_3-CH_2-\underset{\underset{\displaystyle Br}{|}}{CH}-CH_3 \quad \xrightarrow{CN^-} \quad CH_3-CH_2-\underset{\underset{\displaystyle C\equiv N}{|}}{CH}-CH_3$$

$$\xrightarrow{H_3O^+, H_2O} \quad CH_3-CH_2-\underset{\underset{\displaystyle \overset{\displaystyle OH}{|}\,C=O}{|}}{CH}-CH_3$$

15.74   a) The product is an ester and the given reactant is an alcohol. An alcohol reacts with a carboxylic acid to make an ester. To identify the acid, break the single bond between the carbon and oxygen in the ester group.

$$CH_3-CH_2-O-\overset{\overset{\displaystyle O}{\|}}{C}-CH_2-CH_3$$

Portion from Alcohol   Portion from Carboxylic Acid

The missing reactant is **propanoic acid**.

b) To form an amide, an amine must react with an ester to replace the -O-R group. To identify the amine that must be added, break the C-N bond in the amide. The amine is **ethylamine**.

$$CH_3-CH_2-NH-\overset{\overset{\displaystyle O}{\|}}{C}-CH_3$$

Portion from Amine   Carboxylic Acid Portion from Ester

15.78   **Addition reactions** and **condensation reactions** are the two reactions that lead to the two types of synthetic polymers that are named for the reactions that form them.

15.81   Polyethylene comes in a range of strengths and flexibilities. The intermolecular dispersion forces (also called London forces) that attract the long, unbranched chains of high-density polyethylene (HDPE) are strong due to the large size of the polyethylene chains. Low-density polyethylene (LDPE) has increased branching that prevents packing and weakens intermolecular dispersion forces.

15.83   Nylon is formed by the condensation reaction between an **amine and a carboxylic acid** resulting in an amide bond. Polyester is formed by the condensation reaction between a **carboxylic acid and an alcohol** to form an ester bond.

15.84   Both PVC and polypropylene are addition polymers. The general formula for creating repeating units from a
monomer is given in the chapter.
(a)

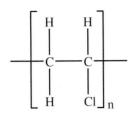

(b)

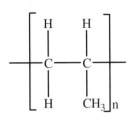

15.88   a) Amino acids form **condensation** polymers, called proteins.
b) Alkenes form **addition** polymers, the simplest of which is polyethylene.
c) Simple sugars form **condensation** polymers, called polysaccharides.
d) Mononucleotides form **condensation** polymers, called nucleic acids.

15.90   The amino acid sequence in a protein determines its shape and structure, which determine its function.

15.93   The DNA base sequence contains an information template that is carried by the RNA base sequence (messenger
and transfer) to create the protein amino acid sequence. In other words, the DNA sequence determines the RNA
sequence, which determines the protein amino acid sequence.

15.94   Locate the specific amino acids in Figure 15.28.
a) alanine   b) histidine   c) methionine

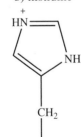

CH_3

S

CH_2

CH_2

15.96   a) A tripeptide contains three amino acids and two peptide (amide) bonds.

$$H_2N-CH(CH_2-C(=O)-OH)-C(=O)-OH \qquad H_2N-CH(CH_2-\text{imidazole})-C(=O)-OH \qquad H_2N-CH(CH_2-\text{indole})-C(=O)-OH$$

aspartic acid   histidine   tryptophan

Join the three acids to give the tripeptide; water is produced:

$$H_2N-CH(CH_2-C(=O)-OH)-C(=O)-NH-CH(CH_2-\text{imidazole})-C(=O)-NH-CH(CH_2-\text{indole})-C(=O)-OH$$

b) Repeat the preceding procedure, with charges on the terminal groups.

glycine   cysteine   tyrosine

$$H_3\overset{+}{N}-CH(H)-C(=O)-NH-CH(CH_2-SH)-C(=O)-NH-CH(CH_2-\text{phenol})-C(=O)-O^-$$

15.98   Base A always pairs with Base T; Base C always pairs with Base G.
a) Complementary DNA strand is **AATCGG**.
b) Complementary DNA strand is **TCTGTA**.

15.100 Uracil (U) substitutes for thymine (T) in RNA. Therefore, A pairs with U. The G-C pair and A-T pair remain unchanged. The RNA sequence is derived from the DNA template sequence **ACAATGCCT**. A three-base sequence constitutes a word, and each word translates into an amino acid. There are three words in the sequence, so **three** amino acids are coded in the sequence.

15.102 a) Both side chains are part of the amino acid **cysteine**. Two cysteine R groups can form a **disulfide bond** (covalent bond).
b) The R group ($-(CH_2)_4-NH_3^+$) is found in the amino acid **lysine** and the R group ($-CH_2COO^-$) is found in the amino acid **aspartic acid**. The positive charge on the amine group in lysine is attracted to the negative charge on the acid group in aspartic acid to form a **salt link**.
c) The R group in the amino acid **asparagine** and the R group in the amino acid **serine**. The –NH– and –OH groups will **hydrogen bond**.
d) Both the R group $-CH(CH_3)-CH_3$ from **valine** and the R group $C_6H_5-CH_2-$ from **phenylalanine** are nonpolar, so their interaction is **is through dispersion forces.**

15.104 The desired reaction is the displacement of Br^- by OH^-.

$$CH_3-\underset{\underset{Br}{|}}{CH}-CH_2-CH_3 \xrightarrow{OH^-} CH_3-\underset{\underset{OH}{|}}{CH}-CH_2-CH_3$$

$$\xrightarrow{OH^-} CH_3-CH=CH-CH_3$$

However, the elimination of HBr also occurs in the presence of a strong base to give the second product. A simple test to detect the presence of a double bond is the addition of liquid Br_2. Bromine is reddish-brown in color and adds easily to the double bond to form a brominated alkane, which has no color. The reddish-brown color of liquid Br_2 disappears if added to an alkene.

15.108 a) Perform an acid-catalyzed dehydration of the alcohol (elimination), followed by bromination of the double bond (addition of Br_2):

$$CH_3-CH_2-CH_2-OH \xrightarrow{H^+} CH_3-CH=CH_2$$

$$CH_3-CH=CH_2 \ + \ Br_2 \longrightarrow CH_3-\underset{\underset{Br}{|}}{CH}-\underset{\underset{Br}{|}}{CH_2}$$

b) The product is an ester, so a carboxylic acid is needed to prepare the ester. First, oxidize one mole of ethanol to acetic acid:

$$CH_3-CH_2-OH \xrightarrow[H^+]{Cr_2O_7^{2-}} CH_3-\overset{\overset{O}{\|}}{C}-OH$$

Then, react one mole of acetic acid with a second mole of ethanol to form the ester:

$$CH_3-CH_2-OH \ \ HO-\overset{\overset{O}{\|}}{C}-CH_3 \xrightarrow{H^+} CH_3-CH_2-O-\overset{\overset{O}{\|}}{C}-CH_3$$

H_2O eliminated

15.110 The two structures are:

$$H_2N\!-\!CH_2\!-\!CH_2\!-\!CH_2\!-\!CH_2\!-\!CH_2\!-\!NH_2 \qquad H_2N\!-\!CH_2\!-\!CH_2\!-\!CH_2\!-\!CH_2\!-\!NH_2$$

cadaverine putrescine

The addition of cyanide, $[C\equiv N]^-$, to form a nitrile is a convenient way to increase the length of a carbon chain:

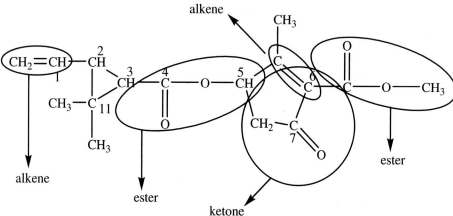

The nitrile can then be reduced to an amine through the addition of hydrogen ($NaBH_4$ is a hydrogen-rich reducing agent):

$$\text{reduction} \longrightarrow H_2N\!-\!CH_2\!-\!CH_2\!-\!CH_2\!-\!CH_2\!-\!NH_2$$

15.111 a) Functional groups in jasmolin II

alkene

ester

alkene

ketone

ester

b) Carbon atoms surrounded by four electron regions (4 single bonds) are sp^3 hybridized. Carbon atoms surrounded by three electron regions (2 single bonds and 1 double bond) are sp^2 hybridized.

Carbon 1 is sp^2 hybridized. Carbon 2 is sp^3 hybridized.
Carbon 3 is sp^3 hybridized. Carbon 4 is sp^2 hybridized.
Carbon 5 is sp^3 hybridized. Carbon 6 and 7 are sp^2 hybridized.

c) Carbons 2, 3, and 5 are chiral centers as they are each bonded to four different groups.

15.116 The resonance structures show that the bond between carbon and nitrogen will have some double bond character that restricts rotation around the bond.

15.120 a) The hydrolysis process requires the addition of water to break the peptide bonds.
b) Convert each mass to moles, and divide the moles by the smallest number to determine molar ratio, and thus relative numbers of the amino acids.

(3.00 g gly) / (75.07 g/mol) = 0.0399627 mol glycine (unrounded)
(0.90 g ala) / (89.10 g/mol) = 0.010101010 mol alanine (unrounded)
(3.70 g val) / (117.15 g/mol) = 0.0315834 mol valine (unrounded)
(6.90 g pro) / (115.13 g/mol) = 0.0599322 mol proline (unrounded)
(7.30 g ser) / (105.10 g/mol) = 0.0694576 mol serine (unrounded)
(86.00 g arg) / (174.21 g/mol) = 0.493657 mol arginine (unrounded)

Divide by the smallest value (0.01010214 mol alanine), and round to a whole number.

(0.0399627 mol glycine) / (0.010101010 mol) = **4**
(0.01010214 mol alanine) / (0.010101010 mol) = **1**
(0.0315834 mol valine) / (0.010101010 mol) = **3**
(0.0599322 mol proline) / (0.010101010 mol) = **6**
(0.0694576 mol serine) / (0.010101010 mol) = **7**
(0.493657 mol arginine) / (0.010101010 mol) = **49**

c) Minimum $\mathcal{M}$ = (4 x 75.07 g/mol) + (1 x 89.09 g/mol) + (3 x 117.15 g/mol) + (6 x 115.13 g/mol)
+ (7 x 105.09 g/mol) + (49 x 174.20 g/mol) = **10,700 g/mol**

15.122 When 2-butanone is reduced, equal amounts of both isomers are produced because the reaction does not favor the production of one over the other. In a 1:1 mixture of the two stereoisomers, the light rotated to the right by the other isomer cancels the light rotated to the left by one isomer. The mixture is not optically active since there is no net rotation of light. The two stereoisomers of 2-butanol are shown below.

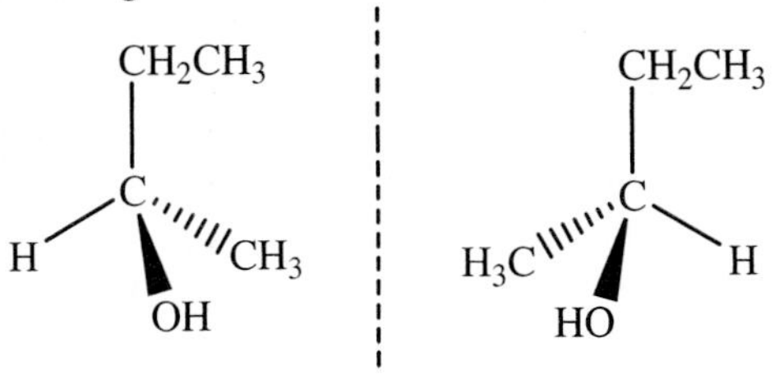

CHAPTER 16 KINETICS: RATES AND MECHANISMS OF CHEMICAL REACTIONS

FOLLOW-UP PROBLEMS

16.1   a) <u>Plan:</u> The balanced equation is $4 NO(g) + O_2(g) \rightarrow 2 N_2O_3$. Choose O_2 as the reference because its coefficient is 1. Four molecules of NO (nitrogen monoxide) are consumed for every one O_2 molecule, so the rate of O_2 disappearance is 1/4 the rate of NO decrease. By similar reasoning, the rate of O_2 disappearance is 1/2 the rate of N_2O_3 (dinitrogen trioxide) increase.
<u>Solution:</u>

$$\text{Rate} = -\frac{1}{4}\frac{\Delta[NO]}{\Delta t} = -\frac{\Delta[O_2]}{\Delta t} = +\frac{1}{2}\frac{\Delta[N_2O_3]}{\Delta t}$$

b) <u>Plan:</u> Because NO is decreasing; its rate of concentration change is negative. Substitute the negative value into the expression and solve for $\Delta[O_2]/\Delta t$.
<u>Solution:</u>

$$\text{Rate} = -\frac{1}{4}\left(1.60 \times 10^{-4}\,\text{mol/L} \cdot \text{s}\right) = -\frac{\Delta[O_2]}{\Delta t}$$

$$\frac{\Delta[O_2]}{\Delta t} = \textbf{-4.00} \times \textbf{10}^{\textbf{-5}}\textbf{ mol/L·s}$$

<u>Check:</u> a) Verify that the general formula shown in Equation 16.2 produces the same result when a = 4, b = 1, and c = 2. b) The rate should be negative because O_2 is disappearing. The rate can also be expressed as $-\Delta[O_2]/\Delta t = 4.00 \times 10^{-5}$ mol/L·sec, where the negative sign on the left side of the equation denote reactant disappearance as well. It makes sense that the rate of NO disappearance is 4 times greater than the rate of O_2 disappearance, because 4 molecules of NO disappear for every one O_2 molecule.

16.2   The exponent of $[Br^-]$ is 1, so the reaction is **first order with respect to Br^-**. Similarly, the reaction is **first order with respect to BrO_3^-**, and **second order with respect to H^+**. The overall reaction order is $(1 + 1 + 2) = 4$, or **fourth order overall**.

16.3   <u>Solution:</u> Assume that the rate law takes the general form rate $= k[H_2]^m[I_2]^n$. To find how the rate varies with respect to $[H_2]$, find two experiments in which $[H_2]$ changes but $[I_2]$ remains constant. Take the ratio of rate laws for Experiments 1 and 3 to find m.

$$\frac{\text{rate 3}}{\text{rate 1}} = \frac{[H_2]_3^m}{[H_2]_1^m}$$

$$\frac{9.3 \times 10^{-23}}{1.9 \times 10^{-23}} = \frac{[0.0550]_3^m}{[0.0113]_1^m}$$

$$4.8947 = (4.8672566)^m$$

Therefore, m = 1
If the reaction order was more complex, an alternate method of solving for *m* is:
log (4.8947) = *m* log (4.8672566); *m* = log (4.8947)/log (4.8672566) = 1
Take the ratio of rate laws for Experiments 2 and 4 to find n.

$$\frac{\text{rate } 4}{\text{rate } 2} = \frac{[I_2]_4^n}{[I_2]_2^n}$$

$$\frac{1.9 \times 10^{-22}}{1.1 \times 10^{-22}} = \frac{[0.0056]_4^n}{[0.0033]_2^n}$$

$$1.72727 = (1.69697)^n$$

Therefore, n = 1

The rate law is rate = $k[H_2][I_2]$ and is second order overall.

Check: The rate is first order with respect to each reactant. If one reactant doubles, the rate doubles; if one reactant triples, the rate triples. If *both* reactants double (each reactant x 2), the rate quadruples (2 x 2). Comparing reactions 1 and 4, the $[H_2]$ doubles (x 2) and the $[I_2]$ increases by 5 times (x 5). Accordingly, the rate increases by 10 (2 x 5). This exercise verifies that the determined rate law exponents are correct.

16.4   Plan: The reaction is second order in X and zero order in Y. For part a), compare the two amounts of reactant X. For part b), compare the two rate values.
Solution:
a) Since the rate law is rate = $k[X]^2$, the reaction is zero order in Y. In Experiment 2, the amount of reactant X has not changed from Experiment 1 and the amount of Y has doubled. The rate is not affected by the doubling of Y since Y is zero order. Since the amount of X is the same, the rate has not changed. The initial rate of Experiment 2 is also **0.25 x 10^{-5} mol/L·s**.
b) The rate of Experiment 3 is four times the rate in Experiment 1. Since the reaction is second order in X, the concentration of X must have doubled to cause a four-fold increase in rate. There should be 6 black spheres in Experiment 3.

$$\frac{1.0 \times 10^{-5}}{0.25 \times 10^{-5}} = \frac{[x]^2}{[3]^2}$$

$$4 = \frac{[x]^2}{9}$$

x = 6 black spheres

16.5   Plan: The rate expression indicates that the reaction order is two (exponent of [HI] = 2), so use the integrated second-order law. Substitute the given concentrations and the rate constant into the expression and solve for time.
Solution:

$$\frac{1}{[A]_t} - \frac{1}{[A]_0} = kt$$

$$\frac{1}{[0.00900 \text{ mol/L}]_t} - \frac{1}{[0.0100 \text{ mol/L}]_0} = (2.4 \times 10^{-21} \text{ L/mol·s}) \, t$$

$$t = \frac{\dfrac{1}{0.00900 \text{ mol/L}} - \dfrac{1}{0.0100 \text{ mol/L}}}{2.4 \times 10^{-21} \text{ L/mol} \cdot \text{s}}$$

$$t = 4.6296 \times 10^{21} \text{ s} = \mathbf{4.6 \times 10^{21} \text{ s}}$$

Check: The problem indicates this decomposition reaction is very slow, so it should take a long time to decrease the HI concentration. The incredibly small rate constant also indicates that the reaction rate is very slow. A hundred thousand billion years is indeed a long time.

16.6   <u>Plan:</u> Find the half-life by examining the time needed to reduce the X to one-fourth the original amount. Use the relationship between k and $t_{1/2}$ to find the rate constant, k.
<u>Solution:</u>
The amount of X at time = 0 is reduced to one-fourth that amount in 6.0 minutes. Since one-fourth X remains after 6.0 minutes, one-half X would remain after **3.0 minutes**.

$$8\,X \quad \longrightarrow \quad 4\,X \quad \longrightarrow \quad 2\,X$$
$$\qquad\quad 3.0\ \text{min} \qquad\qquad 3.0\ \text{min}$$

$$k = \frac{\ln 2}{t_{1/2}}$$

$$k = \frac{\ln 2}{3.0\ \text{min}} = 0.2310491 = \mathbf{0.23\ min^{-1}}$$

16.7   <u>Plan:</u> Rearrange the first-order half-life equation to solve for k.
<u>Solution:</u> $k = \ln 2 / t_{1/2} = \ln 2 / 13.1\ h = 0.052911998 = \mathbf{0.0529\ h^{-1}}$
<u>Check:</u> Verify that the units on k reflect the data given.

16.8   <u>Plan:</u> Activation energy, rate constant at T_1, and a second temperature, T_2, are given. Substitute these values into the Arrhenius equation and solve for k_2, the rate constant at T_2.
<u>Solution:</u> Rearranging Equation 16.8 to solve for k_2:

$$\ln \frac{k_2}{k_1} = -\frac{E_a}{R}\left(\frac{1}{T_2} - \frac{1}{T_1}\right)$$

$$k_1 = 0.286\ \text{L/mol}\bullet\text{s} \qquad T_1 = 500.\ \text{K} \qquad E_a = 1.00 \times 10^2\ \text{kJ/mol}$$
$$k_2 = ?\ \text{L/mol}\bullet\text{s} \qquad T_2 = 490.\ \text{K}$$

$$\ln \frac{k_2}{0.286\ \text{L/mol}\bullet\text{s}} = -\frac{1.00 \times 10^2\ \text{kJ/mol}}{8.314\ \text{J/mol}\bullet\text{K}}\left(\frac{1}{490.\text{K}} - \frac{1}{500.\text{K}}\right)\left(\frac{10^3\ \text{J}}{1\ \text{kJ}}\right)$$

$$\ln \frac{k_2}{0.286\ \text{L/mol}\bullet\text{s}} = -0.49093$$

$$k_2 = (0.612054)(0.286\ \text{L/mol}\bullet\text{s}) = 0.175047423 = \mathbf{0.175\ L/mol\bullet s}$$

<u>Check:</u> The temperature only changes by 10 K degrees, so it is likely that the rate constant at T_2 would not change by orders of magnitude; 0.175 s^{-1} is in the same "ballpark" as 0.286 s^{-1}. Furthermore, the rate should decrease at the lower temperature, so the lower rate of 0.175 s^{-1} at 490 K is expected.

16.9   <u>Plan:</u> Begin by using Sample Problem 16.9 as a guide for labeling the diagram.
<u>Solution:</u> The reaction energy diagram indicates that $O(g) + H_2O(g) \rightarrow 2\ OH(g)$ is an endothermic process, because the energy of the product is higher than the energy of the reactants. The highest point on the curve indicates the transition state. In the transition state, an oxygen atom forms a bond with one of the hydrogen atoms on the H_2O molecule (hashed line) and the O-H bond (dashed line) in H_2O weakens. $E_{a(fwd)}$ is the sum of ΔH_{rxn} and $E_{a(rev)}$.
Transition state:

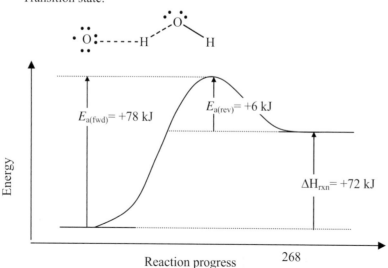

16.10   Plan: Sum the elementary steps to obtain the overall equation. Think of the individual steps as reactions along the "reaction progress" axis, i.e., sum each equation as written (this is different from the application of Hess's law in which reactions are reversed as necessary to reach a final reaction).
Solution:

	b) Molecularity	c) Rate law
(1) $2 \, NO(g) \rightarrow N_2O_2(g)$	2	$rate_1 = k_1[NO]^2$
(2) $2[H_2(g) \rightarrow 2 \, H(g)]$	1	$rate_2 = k_2[H_2]$
(3) $N_2O_2(g) + H(g) \rightarrow N_2O(g) + HO(g)$	2	$rate_3 = k_3[N_2O_2][H]$
(4) $2[HO(g) + H(g) \rightarrow H_2O(g)]$	2	$rate_4 = k_4[HO][H]$
(5) $H(g) + N_2O(g) \rightarrow HO(g) + N_2(g)$	2	$rate_5 = k_5[H][N_2O]$

a) $2 \, NO(g) + 2 \, H_2(g) \rightarrow 2 \, H_2O(g) + N_2(g)$
Check: The overall equation is balanced.
b) Sep 2 is unimolecular; all of the other steps are bimolecular.
c) $rate_1 = k_1[NO]^2$; $rate_2 = k_2[H_2]$; $rate_3 = k_3[N_2O_2][H]$; $rate_4 = k_4[HO][H]$; $rate_5 = k_5[H][N_2O]$

END-OF-CHAPTER PROBLEMS

16.2   Rate is proportional to concentration. An increase in pressure will increase the number of gas molecules per unit volume. In other words, the gas concentration increases due to increased pressure, so the **reaction rate increases**. Increased pressure also causes more collisions between gas molecules.

16.3   The addition of more water will dilute the concentrations of all solutes dissolved in the reaction vessel. If any of these solutes are reactants, the **rate of the reaction will decrease**.

16.5   An increase in temperature affects the rate of a reaction by increasing the number of collisions, but more importantly the energy of collisions increases. As the energy of collisions increase, more collisions result in reaction (i.e., reactants → products), so the **rate of reaction increases.**

16.8   a) For most reactions, the rate of the reaction changes as a reaction progresses. The instantaneous rate is the rate at one point, or instant, during the reaction. The average rate is the average of the instantaneous rates over a period of time. On a graph of reactant concentration versus time of reaction, the instantaneous rate is the slope of the tangent to the curve at any one point. The average rate is the slope of the line connecting two points on the curve. The closer together the two points (shorter the time interval), the more closely the average rate agrees with the instantaneous rate.
b) The initial rate is the instantaneous rate at the point on the graph where time = 0, that is when reactants are mixed.

16.10   At time t = 0, no product has formed, so the B(g) curve must start at the origin. Reactant concentration (A(g)) decreases with time; product concentration (B(g)) increases with time. Many correct graphs can be drawn. Two examples are shown below. The graph on the left shows a reaction that proceeds nearly to completion, i.e., [products] >> [reactants] at the end of the reaction. The graph on the right shows a reaction that does not proceed to completion, i.e., [reactants] > [products] at reaction end.

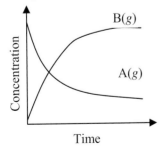 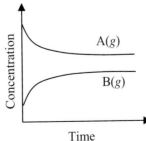

16.12   a) The average rate from t = 0 to t = 20.0 s is proportional to the slope of the line connecting these two points:

$$\text{Rate} = -\frac{1}{2}\frac{\Delta[AX_2]}{\Delta t} = -\frac{1}{2}\frac{(0.0088\text{ M} - 0.0500\text{ M})}{(20.0\text{ s} - 0\text{ s})} = 0.00103 = \mathbf{0.0010\ \textit{M}/s}$$

The negative of the slope is used because rate is defined as the change in product concentration with time. If a reactant is used, the rate is the negative of the change in reactant concentration. The 1/2 factor is included to account for the stoichiometric coefficient of 2 for AX_2 in the reaction.

b)

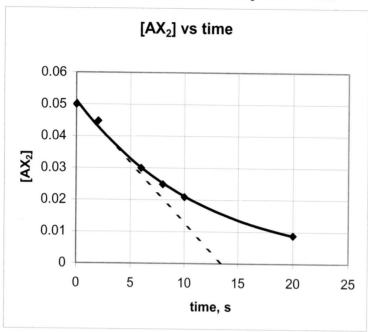

The slope of the tangent to the curve (dashed line) at t = 0 is approximately –0.004 M/s. This initial rate is greater than the average rate as calculated in part (a). The **initial rate is greater than the average rate** because rate decreases as reactant concentration decreases.

16.14   Use Equation 16.2 to describe the rate of this reaction in terms of reactant disappearance and product appearance.

$$\text{Rate} = -\frac{1}{2}\frac{\Delta[A]}{\Delta t} = \frac{\Delta[B]}{\Delta t} = \frac{\Delta[C]}{\Delta t}$$

A negative sign is used for the rate in terms of reactant A since A is reacting and [A] is decreasing over time. Positive signs are used for the rate in terms of products B and C since B and C are being formed and [B] and [C] increase over time.

Reactant A decreases twice as fast as product C increases because 2 molecules of A disappear for every molecule of C that appears.

$$-\frac{1}{2}\frac{\Delta[A]}{\Delta t} = \frac{\Delta[C]}{\Delta t}$$

$$(2\text{ mol C/L}\cdot\text{s})\left(\frac{2\text{ mol A/L}\cdot\text{s}}{1\text{ mol C/L}\cdot\text{s}}\right) = -4\text{ mol/L}\cdot\text{s}$$

The negative value indicates that [A] is decreasing as the reaction progresses. The rate of reaction is always expressed as a positive number, so [A] is decreasing at a rate of **4 mol/L•s**.

16.16 $\text{Rate} = -\dfrac{\Delta[A]}{\Delta t} = -\dfrac{1}{2}\dfrac{\Delta[B]}{\Delta t} = \dfrac{\Delta[C]}{\Delta t}$

Rate is defined as the change in product concentration with time. If a reactant is used, the rate is the negative of the change in reactant concentration. The 1/2 factor is included for reactant B to account for the stoichiometric coefficient of 2 for B in the reaction.

$-\dfrac{\Delta[A]}{\Delta t} = -\dfrac{1}{2}\dfrac{\Delta[B]}{\Delta t} = -\dfrac{1}{2}\dfrac{(0.50\ \text{mol/L})}{s} = -0.25\ \text{mol/L} \cdot \text{s (unrounded)}$

[B] decreases twice as fast as [A], so [A] is decreasing at a rate of 1/2 (0.5 mol/L·s) or **0.2 mol/L·s**.

16.18 A term with a negative sign is a reactant; a term with a positive sign is a product. The inverse of the fraction becomes the coefficient of the molecule:

$2\ N_2O_5(g) \rightarrow 4\ NO_2(g) + O_2(g)$

16.21 $\text{Rate} = -\dfrac{\Delta[N_2]}{\Delta t} = -\dfrac{1}{3}\dfrac{\Delta[H_2]}{\Delta t} = \dfrac{1}{2}\dfrac{\Delta[NH_3]}{\Delta t}$

16.22 a) $\text{Rate} = -\dfrac{1}{3}\dfrac{\Delta[O_2]}{\Delta t} = \dfrac{1}{2}\dfrac{\Delta[O_3]}{\Delta t}$

b) Use the mole ratio in the balanced equation:

$\left(\dfrac{2.17 \times 10^{-5}\ \text{mol } O_2\ /\text{L} \cdot \text{s}}{}\right)\left(\dfrac{2\ \text{mol } O_3\ /\text{L} \cdot \text{s}}{3\ \text{mol } O_2 /\text{L} \cdot \text{s}}\right) = \textbf{1.45} \times \textbf{10}^{-5}\ \textbf{mol/L} \cdot \textbf{s}$

16.23 a) k is the rate constant, the proportionality constant in the rate law. k represents the fraction of successful collisions which includes the fraction of collisions with sufficient energy and the fraction of collisions with correct orientation. k is a constant that varies with temperature.

b) m represents the order of the reaction with respect to [A] and n represents the order of the reaction with respect to [B]. The order is the exponent in the relationship between rate and reactant concentration and defines how reactant concentration influences rate.

The order of a reactant does not necessarily equal its stoichiometric coefficient in the balanced equation. If a reaction is an elementary reaction, meaning the reaction occurs in only one step, then the orders and stoichiometric coefficients are equal. However, if a reaction occurs in a series of elementary reactions, called a mechanism, then the rate law is based on the slowest elementary reaction in the mechanism. The orders of the reactants will equal the stoichiometric coefficients of the reactants in the slowest elementary reaction but may not equal the stoichiometric coefficients in the overall reaction.

c) For the rate law rate $= k[A][B]^2$ substitute in the units:

$\text{Rate (mol/L} \cdot \text{min)} = k[A]^1[B]^2$

$k = \dfrac{\text{Rate}}{[A]^1[B]^2} = \dfrac{\text{mol/L} \cdot \text{min}}{\left[\dfrac{\text{mol}}{L}\right]^1\left[\dfrac{\text{mol}}{L}\right]^2} = \dfrac{\text{mol/L} \cdot \text{min}}{\dfrac{\text{mol}^3}{L^3}}$

$k = \dfrac{\text{mol}}{L \cdot \text{min}}\left(\dfrac{L^3}{\text{mol}^3}\right)$

$k = \textbf{L}^2\textbf{/mol}^2 \cdot \textbf{min}$

16.25 a) The **rate doubles**. If rate $= k[A]^1$ and [A] is doubled, then the rate law becomes rate $= k[2 \times A]^1$. The rate increases by 2^1 or 2.

b) The **rate decreases by a factor of four**. If rate $= k[1/2 \times B]^2$, then rate decreases to $(1/2)^2$ or 1/4 of its original value.

c) The **rate increases by a factor of nine**. If rate $= k[3 \times C]^2$, then rate increases to 3^2 or 9 times its original value.

16.26   The order for each reactant is the exponent on the reactant concentration in the rate law. The orders with respect to $[BrO_3^-]$ and to $[Br^-]$ are both 1. The order with respect to $[H^+]$ is 2. The overall reaction order is the sum of each reactant order: $1 + 1 + 2 = 4$.
first order with respect to BrO_3^-, first order with respect to Br^-, second order with respect to H^+, fourth order overall

16.28   a) The rate is first order with respect to $[BrO_3^-]$. If $[BrO_3^-]$ is doubled, rate $= k[2 \times BrO_3^-]$, then rate increases to 2^1 or 2 times its original value. The rate **doubles.**
b) The rate is first order with respect to $[Br^-]$. If $[Br^-]$ is halved, rate $= k[1/2 \times Br^-]$, then rate decreases by a factor of $1/2^1$ or 1/2 times its original value. The rate is **halved.**
c) The rate is second order with respect to $[H^+]$. If $[H^+]$ is quadrupled, rate $= k[4 \times H^+]^2$, then rate increases to 4^2 or **16 times** its original value.

16.30   The order for each reactant is the exponent on the reactant concentration in the rate law. The order with respect to **$[NO_2]$ is 2**, and the order with respect to **$[Cl_2]$ is 1**. The overall order is the sum of the orders of individual reactants: $2 + 1 = $ **3 for the overall order**.

16.32   a) The rate is second order with respect to $[NO_2]$. If $[NO_2]$ is tripled, rate $= k[3 \times NO_2]^2$, then rate increases to 3^2 or 9 times its original value. The rate **increases by a factor of 9.**
b) The rate is second order with respect to $[NO_2]$ and first order with respect to $[Cl_2]$. If $[NO_2]$ and $[Cl_2]$ are doubled, rate $= k[2 \times NO_2]^2[2 \times Cl_2]^1$, then the rate increases by a factor of $2^2 \times 2^1 = $ **8.**
c) The rate is first order with respect to $[Cl_2]$. If Cl_2 is halved, rate $= k[1/2 \times Cl_2]^1$, then rate decreases to ½ times its original value. The rate is **halved.**

16.34   a) To find the order for reactant A, first identify the reaction experiments in which [A] changes but [B] is constant. Set up a proportionality: $\dfrac{rate_{exp\,1}}{rate_{exp\,2}} = \left(\dfrac{[A]_{exp\,1}}{[A]_{exp\,2}}\right)^m$. Fill in the values given for rates and concentrations and solve for m, the order with respect to [A]. Repeat the process to find the order for reactant B. Use experiments 1 and 2 (or 3 and 4 would work) to find the order with respect to [A].

$$\frac{rate_{exp\,1}}{rate_{exp\,2}} = \left(\frac{[A]_{exp\,1}}{[A]_{exp\,2}}\right)^m$$

$$\frac{45.0 \text{ mol/L} \cdot \text{min}}{5.00 \text{ mol/L} \cdot \text{min}} = \left(\frac{0.300 \text{ mol/L}}{0.100 \text{ mol/L}}\right)^m$$

$9.00 = (3.00)^m$
$\log (9.00) = m \log (3.00)$
$m = 2$

Using experiments 3 and 4 also gives **2nd order with respect to [A].**
Use experiments 1 and 3 with [A] = 0.100 M or 2 and 4 with [A] = 0.300 M to find order with respect to [B].

$$\frac{rate_{exp\,1}}{rate_{exp\,3}} = \left(\frac{[B]_{exp\,1}}{[B]_{exp\,3}}\right)^n$$

$$\frac{10.0 \text{ mol/L} \cdot \text{min}}{5.00 \text{ mol/L} \cdot \text{min}} = \left(\frac{0.200 \text{ mol/L}}{0.100 \text{ mol/L}}\right)^n$$

$2.00 = (2.00)^n$
$\log (2.00) = n \log (2.00)$
$n = 1$
The reaction is **first order with respect to [B].**
b) The rate law, without a value for k, is **rate $= k[A]^2[B]$**.
c) Using experiment 1 to calculate k (the date from any of the experiments can be used):

$$k = \frac{rate}{[A]^2[B]} = \frac{5.00 \text{ mol/L} \cdot \text{min}}{[0.100 \text{ mol/L}]^2[0.100 \text{ mol/L}]} = \textbf{5.00 x } 10^3 \text{ L}^2/\text{mol}^2 \cdot \text{min}$$

16.36 a) A first order rate law follows the general expression, rate $= k[A]$. The reaction rate is expressed as a change in concentration per unit time with units of M/time or mol/L·time. Since [A] has units of M (the brackets stand for concentration), then k **has units of time^{-1}**:

$$\text{Rate} = k[A]$$

$$\frac{\text{mol}}{\text{L} \cdot \text{time}} = \frac{1}{\text{time}} \frac{\text{mol}}{\text{L}} \qquad \text{or} \qquad \frac{\text{mol}}{\text{L} \cdot \text{time}} = \text{time}^{-1} \frac{\text{mol}}{\text{L}}$$

b) Second order: Rate $= k[A]^2$

$$\frac{\text{mol}}{\text{L} \cdot \text{time}} = k \left(\frac{\text{mol}}{\text{L}} \right)^2$$

$$k = \frac{\dfrac{\text{mol}}{\text{L} \cdot \text{time}}}{\dfrac{\text{mol}^2}{\text{L}^2}} = \textbf{L/mol·time} \text{ or } M^{-1}\,\text{time}^{-1}$$

c) Third order: rate $= k[A]^3$

$$\frac{\text{mol}}{\text{L} \cdot \text{time}} = k \left(\frac{\text{mol}}{\text{L}} \right)^3$$

$$k = \frac{\dfrac{\text{mol}}{\text{L} \cdot \text{time}}}{\dfrac{\text{mol}^3}{\text{L}^3}} = \textbf{L}^2\textbf{/mol}^2\textbf{·time} \text{ or } M^{-2}\,\text{time}^{-1}$$

d) 5/2 order: rate $= k[A]^{5/2}$

$$\frac{\text{mol}}{\text{L} \cdot \text{time}} = k \left(\frac{\text{mol}}{\text{L}} \right)^{5/2}$$

$$k = \frac{\dfrac{\text{mol}}{\text{L} \cdot \text{time}}}{\dfrac{\text{mol}^{5/2}}{\text{L}^{5/2}}} = \textbf{L}^{3/2}\textbf{/mol}^{3/2}\textbf{·time} \text{ or } M^{-3/2}\,\text{time}^{-1}$$

16.39 The integrated rate law can be used to plot a graph. If the plot of [reactant] versus time is linear, the order is zero. If the plot of ln [reactant] versus time is linear, the order is first. If the plot of inverse concentration (1/[reactant]) versus time is linear, the order is second.
a) The reaction is **first order** since ln[reactant] versus time is linear.
b) The reaction is **second order** since 1 / [reactant] versus time is linear.
c) The reaction is **zero order** since [reactant] versus time is linear.

16.41 The rate expression indicates that this reaction is second order overall (the order of [AB] is 2), so use the second order integrated rate law to find time. ([AB] = 1/3 [AB]$_0$ = 1/3 (1.50 M) = 0.500 M)

$$\frac{1}{[AB]_t} - \frac{1}{[AB]_0} = kt$$

$$t = \frac{\left(\dfrac{1}{[AB]_t} - \dfrac{1}{[AB]_t} \right)}{k}$$

$$t = \frac{\left(\dfrac{1}{0.500\ M} - \dfrac{1}{1.50\ M} \right)}{0.2\ \text{L/mol} \cdot \text{s}}$$

$$t = 6.6667 = \textbf{7 s}$$

16.43   a) The given information is the amount that has reacted in a specified amount of time. With this information, the integrated rate law must be used to find a value for the rate constant. For a first order reaction, the rate law is $\ln [A]_t = \ln [A]_0 - kt$. Using the fact that 50% has decomposed, let $[A]_0 = 1$ M and then $[A]_t = 50\%$ of 1 M = 0.5 M:

$\ln [A]_t = \ln [A]_0 - kt$

$\ln [0.5] = \ln [1] - k(10.5 \text{ min})$

$-0.693147 = 0 - k(10.5 \text{ min})$

$0.693147 = k(10.5 \text{ min})$

$k = 0.0660 \text{ min}^{-1}$

Alternatively, 50.0% decomposition means that one half-life has passed. Thus, the first order half-life equation may be used:

$$t_{1/2} = \frac{\ln 2}{k} \qquad\qquad k = \frac{\ln 2}{t_{1/2}} = \frac{\ln 2}{10.5 \text{ min}} = 0.066014 = \mathbf{0.0660 \text{ min}^{-1}}$$

b) Use the value for k calculated in part a. Let $[A]_0 = 1$ M; since 75% of A has decomposed, 25% of A remains and $[A]_t = 25\%$ of $[A]_0$ or $0.25\{A\}_0 = 0.25[1] = 0.25$ M.

$\ln [A]_t = \ln [A]_0 - kt$

$$\frac{\ln[A]_t - \ln[A]_0}{-k} = t$$

$$\frac{\ln[0.25] - \ln[1]}{-0.0660 \text{ min}^{-1}} = t$$

$t = 21.0045 = \mathbf{21.0 \text{ min}}$

If you recognize that 75.0% decomposition means that two half-lives have passed, then t = 2 (10.5 min) = **21.0 min**

16.45   In a first order reaction, $\ln[NH_3]$ versus time is a straight line with slope equal to k. A new data table is constructed below.

Note that additional significant figures are retained in the calculations.

x-axis (time, s)	$[NH_3]$	y-axis ($\ln[NH_3]$)
0	4.000 M	1.38629
1.000	3.986 M	1.38279
2.000	3.974 M	1.37977

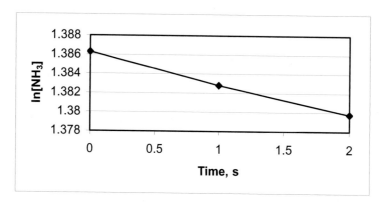

k = slope = rise/run = $(y_2 - y_1) / (x_2 - x_1)$

$k = (1.37977 - 1.38629) / (2.000 - 0) = (0.00652) / (2.) = 3.260 \times 10^{-3} \text{ s}^{-1} = \mathbf{3 \times 10^{-3} \text{ s}^{-1}}$

(Note that the starting time is not exact, and hence, limits the significant figures.)

b) $t_{1/2} = \ln 2 / k = \ln 2 / (3.260 \times 10^{-3} \text{ s}^{-1}) = 212.62 = \mathbf{2 \times 10^2 \text{ s}}$

16.47 The Arrhenius equation, $k = Ae^{-E_a/RT}$, can be used directly to solve for activation energy at a specified temperature if the rate constant, k, and the frequency factor, A, are known. However, the frequency factor is usually not known. To find E_a without knowing A, rearrange the Arrhenius equation to put it in the form of a linear plot: $\ln k = \ln A - E_a / RT$ where the y value is $\ln k$ and the x value is $1 / T$. Measure the rate constant at a series of temperatures and plot $\ln k$ versus $1 / T$. The slope equals $-E_a / R$.

16.49 Substitute the given values into Equation 16.9 and solve for k_2.

$k_1 = 4.7 \times 10^{-3} \text{ s}^{-1}$ $T_1 = 25°C = (273 + 25) = 298 \text{ K}$
$k_2 = ?$ $T_2 = 75°C = (273 + 75) = 348 \text{ K}$
$E_a = 33.6 \text{ kJ/mol} = 33600 \text{ J/mol}$

$$\ln \frac{k_2}{k_1} = -\frac{E_a}{R}\left(\frac{1}{T_2} - \frac{1}{T_1}\right)$$

$$\ln \frac{k_2}{4.7 \times 10^{-3} \text{ s}^{-1}} = -\frac{33600 \text{ J/mol}}{8.314 \text{ J/mol} \cdot \text{K}}\left(\frac{1}{348 \text{ K}} - \frac{1}{298 \text{ K}}\right)$$

$$\ln \frac{k_2}{4.7 \times 10^{-3} \text{ s}^{-1}} = 1.948515 \text{ (unrounded)} \qquad \text{Raise each side to } e^x$$

$$\frac{k_2}{4.7 \times 10^{-3} \text{ s}^{-1}} = 7.0182577$$

$k_2 = (4.7 \times 10^{-3} \text{ s}^{-1})(7.0182577) = 0.0329858 = \mathbf{0.033 \text{ s}^{-1}}$

16.53 **No,** collision frequency is not the only factor affecting reaction rate. The collision frequency is a count of the total number of collisions between reactant molecules. Only a small number of these collisions lead to a reaction. Other factors that influence the fraction of collisions that lead to reaction are the energy and orientation of the collision. A collision must occur with a minimum energy (activation energy) to be successful. In a collision, the orientation — which ends of the reactant molecules collide — must bring the reacting atoms in the molecules together in order for the collision to lead to a reaction.

16.56 **No,** for 4×10^{-5} moles of EF to form, every collision must result in a reaction and no EF molecule can decompose back to AB and CD. Neither condition is likely. In principle, all reactions are reversible, so some EF molecules decompose. Even if all AB and CD molecules did combine, the reverse decomposition rate would result in an amount of EF that is less than 4×10^{-5} moles.

16.57 Collision frequency is proportional to the velocity of the reactant molecules. At the same temperature, both reaction mixtures have the same average kinetic energy, but not the same velocity. Kinetic energy equals $1/2 \, mv^2$, where m is mass and v velocity. The methylamine ($N(CH_3)_3$) molecule has a greater mass than the ammonia molecule, so methylamine molecules will collide less often than ammonia molecules, because of their slower velocities. Collision energy thus is less for the $N(CH_3)_3(g) + HCl(g)$ reaction than for the $NH_3(g) + HCl(g)$ reaction. Therefore, the **rate of the reaction between ammonia and hydrogen chloride is greater** than the rate of the reaction between methylamine and hydrogen chloride.
The fraction of successful collisions also differs between the two reactions. In both reactions the hydrogen from HCl is bonding to the nitrogen in NH_3 or $N(CH_3)_3$. The difference between the reactions is in how easily the H can collide with the N, the correct orientation for a successful reaction. The groups (H) bonded to nitrogen in ammonia are less bulky than the groups bonded to nitrogen in trimethylamine (CH_3). So, collisions with correct orientation between HCl and NH_3 occur more frequently than between HCl and $N(CH_3)_3$ and $NH_3(g) + HCl(g) \rightarrow NH_4Cl(s)$ occurs at a higher rate than $N(CH_3)_3(g) + HCl(g) \rightarrow (CH_3)_3NHCl(s)$. Therefore, the **rate of the reaction between ammonia and hydrogen chloride is greater** than the rate of the reaction between methylamine and hydrogen chloride.

16.58   Each A particle can collide with three B particles, so (4 x 3) = **12 unique collisions** are possible.

16.60   The fraction of collisions with a specified energy is equal to the $e^{-Ea/RT}$ term in the Arrhenius equation.

$$f = e^{-Ea/RT} \qquad (25°C = 273 + 25 = 298 \text{ K}) \qquad -E_a/RT = -\frac{100 \text{ kJ/mol}}{(8.314 \text{ J/mol} \cdot \text{K})(298 \text{ K})}\left(\frac{10^3 \text{ J}}{1 \text{ kJ}}\right)$$

$$-E_a/RT = -40.362096$$

Fraction $= e^{-Ea/RT} = e^{-40.362096} = 2.9577689 \times 10^{-18} = \mathbf{2.96 \times 10^{-18}}$

16.62   a) The reaction is exothermic, so the energy of the reactants is higher than the energy of the products.

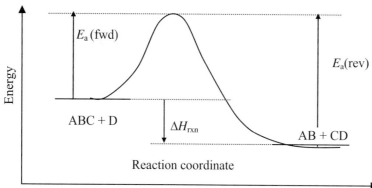

b) $E_{a(rev)} = E_{a(fwd)} - \Delta H_{rxn} = 215 \text{ kJ/mol} - (-55 \text{ kJ/mol}) = \mathbf{2.70 \times 10^2 \text{ kJ/mol}}$
c)

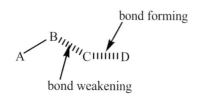

16.64   a) The reaction is endothermic since the enthalpy change is positive.

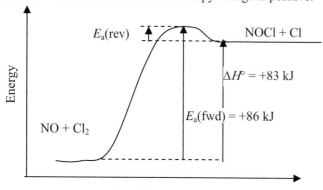

b) Activation energy for the reverse reaction: $E_a(rev) = E_a(fwd) - \Delta H° = 86 \text{ kJ} - 83 \text{ kJ} = \mathbf{3 \text{ kJ}}$.
c) To draw the transition state, look at structures of reactants and products:

$$:\ddot{C}l-\ddot{C}l: \quad + \quad .\ddot{N}=\ddot{O}: \quad \longrightarrow \quad :\ddot{C}l. \quad + \quad$$

276

The collision must occur between one of the chlorines and the nitrogen. The transition state would have weak bonds between the nitrogen and chlorine and between the two chlorines.

16.65   The rate of an overall reaction depends on the slowest step. Each individual step's reaction rate can vary widely, so the rate of the slowest step, and hence the overall reaction, will be **slower than the average of the individual rates** because the average contains faster rates as well as the rate-determining step.

16.69   A bimolecular step (a collision between two particles) is more reasonable physically than a termolecular step (a collision involving three particles) because the likelihood that two reactant molecules will collide with the proper energy and orientation is much greater than the likelihood that three reactant molecules will collide simultaneously with the proper energy and orientation.

16.70   **No**, the overall rate law must contain reactants only (no intermediates) and is determined by the slow step. If the first step in a reaction mechanism is slow, the rate law for that step is the overall rate law.

16.72   a) The overall reaction can be obtained by adding the three steps together:

$$(1) \; A(g) + B(g) \rightarrow X(g) \quad \text{fast}$$
$$(2) \; X(g) + C(g) \rightarrow Y(g) \quad \text{slow}$$
$$\underline{(3) \; Y(g) \rightarrow D(g) \quad \text{fast}}$$

Total: $A(g) + B(g) + \cancel{X(g)} + C(g) + \cancel{Y(g)} \rightarrow \cancel{X(g)} + \cancel{Y(g)} + D(g)$
$$A(g) + B(g) + C(g) \rightarrow D(g)$$

b) Intermediates appear in the mechanism first as products, then as reactants. Both X and Y are intermediates in the given mechanism. Intermediate X is produced in the first step and consumed in the second step; intermediate Y is produced in the second step and consumed in the third step. Notice that neither X nor Y were included in the overall reaction.

c)

Step:	Molecularity	Rate law
$A(g) + B(g) \rightarrow \; X(g)$	bimolecular	$\text{rate}_1 = k_1[A][B]$
$X(g) + C(g) \rightarrow Y(g)$	bimolecular	$\text{rate}_2 = k_2[X][C]$
$Y(g) \rightarrow D(g)$	unimolecular	$\text{rate}_3 = k_3[Y]$

d) **Yes**, the mechanism is consistent with the actual rate law. The slow step in the mechanism is the second step with rate law: rate = $k_2[X][C]$. Since X is an intermediate, it must be replaced by using the first step. For an equilibrium, $\text{rate}_{\text{forward rxn}} = \text{rate}_{\text{reverse rxn}}$. For step 1 then, $k_1[A][B] = k_{-1}[X]$. Rearranging to solve for [X] gives [X] = $(k_1 / k_{-1})[A][B]$. Substituting this value for [X] into the rate law for the second step gives the overall rate law as rate = $(k_2 k_1 / k_{-1})[A][B][C]$ which is identical to the actual rate law with $k = k_2 k_1 / k_{-1}$.
e) **Yes**, The one step mechanism $A(g) + B(g) + C(g) \rightarrow D(g)$ would have a rate law of rate = $k[A][B][C]$, which is the actual rate law.

16.74   Nitrosyl bromide is NOBr(g). The reactions sum to the equation 2 NO(g) + Br_2(g) → 2 NOBr(g), so criterion 1 (elementary steps must add to overall equation) is satisfied. Both elementary steps are bimolecular and chemically reasonable, so criterion 2 (steps are physically reasonable) is met. The reaction rate is determined by the slow step; however, rate expressions do not include reaction intermediates ($NOBr_2$). Derive the rate law. The slow step in the mechanism is the second step with rate law: rate = $k_2[NOBr_2][NO]$. Since $NOBr_2$ is an intermediate, it must be replaced by using the first step. For an equilibrium like Step 1, $\text{rate}_{\text{forward rxn}} = \text{rate}_{\text{reverse rxn}}$.
        Solve for [$NOBr_2$] in Step 1:
        Rate_1 (forward) = Rate_1 (reverse)
        $k_1[NO][Br_2] = k_{-1}[NOBr_2]$
        $[NOBr_2] = (k_1 / k_{-1})[NO][Br_2]$
    Rate of the slow step: $\text{Rate}_2 = k_2[NOBr_2][NO]$

Substitute the expression for [NOBr$_2$] into this equation, the slow step:
$$\text{Rate}_2 = k_2(k_1 / k_{-1})[\text{NO}][\text{Br}_2][\text{NO}]$$
Combine the separate constants into one constant: $k = k_2(k_1 / k_{-1})$
$$\text{Rate}_2 = k[\text{NO}]^2[\text{Br}_2]$$
The derived rate law equals the known rate law, so criterion 3 is satisfied. The proposed mechanism is valid.

16.77   **No**, a catalyst changes the mechanism of a reaction to one with lower activation energy. Lower activation energy means a faster reaction. An increase in temperature does not influence the activation energy, but instead increases the fraction of collisions with sufficient energy to react.

16.78   a) **No**, by definition, a catalyst is a substance that increases reaction rate without being consumed. The spark provides energy that is absorbed by the H$_2$ and O$_2$ molecules to achieve the threshold energy needed for reaction.
b) **Yes**, the powdered metal acts like a heterogeneous catalyst, providing a surface upon which the O$_2$-H$_2$ reaction becomes more favorable because the activation energy is lowered.

16.82   a) Water does not appear as a reactant in the rate-determining step.
Note that as a solvent in the reaction, the concentration of the water is assumed not to change even though some water is used up as a reactant. This assumption is valid as long as the solute concentrations are low ($\sim$1 *M* or less). So, even if water did appear as a reactant in the rate-determining step, it would not appear in the rate law. See rate law for step 2 below.
b)   Rate law for step (1):   rate$_1$ = k_1[(CH$_3$)$_3$CBr]
         Rate law for step (2):   rate$_2$ = k_2[(CH$_3$)$_3$C$^+$]
         Rate law for step (3):   rate$_3$ = k_3[(CH$_3$)$_3$COH$_2$$^+$]
c) The intermediates are (CH$_3$)$_3$C$^+$ and (CH$_3$)$_3$COH$_2$$^+$.
d) The rate-determining step is the slow step, (1). The rate law for this step agrees with the actual rate law with $k = k_1$.

16.85   The activation energy can be calculated using Equation 16.9. Although the rate constants, k_1 and k_2, are not expressly stated, the relative times give an idea of the rate. The reaction rate is proportional to the rate constant. At T$_1$ = 20°C = (273 + 20) = 293 K, the rate of reaction is 1 apple/4 days while at T$_2$ = 0°C = (273 + 0) = 273 K, the rate is 1 apple/16 days. Therefore, rate$_1$ = 1 apple/4 days and rate$_2$ = 1 apple/16 days are substituted for k_1 and k_2, respectively.

$$k_1 = 1/4 \qquad\qquad T_1 = 293 \text{ K}$$
$$k_2 = 1/16 \qquad\qquad T_2 = 273 \text{ K}$$
$$E_a = ?$$

$$\ln\frac{k_2}{k_1} = -\frac{E_a}{R}\left(\frac{1}{T_2} - \frac{1}{T_1}\right)$$

$$E_a = -\frac{R\left(\ln\dfrac{k_2}{k_1}\right)}{\left(\dfrac{1}{T_2} - \dfrac{1}{T_1}\right)} = -\frac{\left(8.314\dfrac{\text{J}}{\text{mol} \cdot \text{K}}\right)\left(\ln\dfrac{(1/16)}{(1/4)}\right)}{\left(\dfrac{1}{273 \text{ K}} - \dfrac{1}{293 \text{ K}}\right)}$$

E_a = 4.6096266 x 10^4 J/mol = **4.61 x 10^4 J/mol**
         The significant figures are based on the Kelvin temperatures.

16.89   Use the given rate law, and enter the given values:
         rate = k[H$^+$] [sucrose]
         [H$^+$]$_i$ = 0.01 *M*   [sucrose]$_i$ = 1.0 *M*
         The glucose and fructose are not in the rate law, so they may be ignored.
a) The rate is first-order with respect to [sucrose]. The [sucrose] is changed from 1.0 M to 2.5 M, or is increased by a factor of 2.5/1.0 or 2.5. Then the rate = k[H$^+$][2.5 x sucrose]; the rate **increases by a factor of 2.5.**
b) The [sucrose] is changed from 1.0 M to 0.5 M, or is decreased by a factor of 0.5/1.0 or 0.5. Then the rate = k[H$^+$][0.5 x sucrose]; the rate decreases by a factor of ½ or **half the original rate.**

c) The rate is first-order with respect to [H$^+$]. The [H$^+$] is changed from 0.01 M to 0.0001 M, or is decreased by a factor of 0.0001/0.01 or 0.01. Then the rate = k[0.01 x H$^+$][sucrose]; the rate **decreases by a factor of 0.01.** Thus, the reaction will decrease to **1/100 the original.**

d) [sucrose] decreases from 1.0 M to 0.1 M, or by a factor of (0.1 M / 1.0 M) = 0.1 [H$^+$] increases from 0.01 M to 0.1 M, or by a factor of (0.1 M / 0.01 M) = 10. Then the rate will increase by
k[10 x H$^+$][0.1 x sucrose]= 1.0 times as fast. Thus, there will be **no change.**

16.92   The overall order is equal to the sum of the individual orders, i.e., the sum of the exponents equals 11:
11 = 1 + 4 + 2 + m + 2, so m = 2.
The reaction is **second order** with respect to NAD.

16.94   Rate is proportional to the rate constant, so if the rate constant increases by a certain factor, the rate increases by the same factor. Thus, to calculate the change in rate the Arrhenius equation can be used and substitute rate$_{cat}$/rate$_{uncat}$ = k_{cat}/k_{uncat}.
$$k = Ae^{-E_a/RT}$$

$$\frac{k_2}{k_1} = \frac{Ae^{-E_{a2}/RT}}{Ae^{-E_{a1}/RT}} = e^{(E_{a1} - E_{a2})/RT}$$

$$\ln\frac{k_2}{k_1} = \frac{E_{a1} - E_{a2}}{RT} = \frac{5 \text{ kJ/mol}}{\left(8.314\frac{J}{mol \cdot K}\right)((273 + 37)K)}\left(\frac{10^3 J}{1 \text{ kJ}}\right) = 1.93998 \text{ (unrounded)}$$

$$\frac{k_2}{k_1} = 6.95861 = 7$$

The rate of the enzyme-catalyzed reaction occurs at a rate **7 times faster** than the rate of the uncatalyzed reaction at 37°C.

16.96   First, find the rate constant, k, for the reaction by solving the first order half-life equation for k. Then use the first-order integrated rate law expression to find t, the time for decay.
Rearrange $t_{1/2} = \dfrac{\ln 2}{k}$ to $k = \dfrac{\ln 2}{t_{1/2}}$

$$k = \frac{\ln 2}{12 \text{ yr}} = 5.7762 \times 10^{-2} \text{ yr}^{-1} \text{ (unrounded)}$$

Use the first-order integrated rate law:
$$\ln [A]_t = \ln [A]_0 - kt$$
$$\frac{\ln[A]_t - \ln[A]_0}{-k} = t$$
$$\frac{\ln[10. \text{ ppbm}] - \ln[275 \text{ ppbm}]}{-5.7762 \times 10^{-2} \text{ yr}^{-1}} = t$$
$$t = 57.3765798 = \textbf{57 yr}$$

16.99   a) The rate constant can be determined from the slope of the integrated rate law plot. To find the correct order, the data should be plotted as 1) [sucrose] versus time – linear for zero order, 2) ln[sucrose] versus time – linear for first order, and 3) 1/[sucrose] versus time – linear for second order.
All three graphs are linear, so picking the correct order is difficult. One way to select the order is to compare correlation coefficients (R^2) — you may or may not have experience with this. The best correlation coefficient is the one closest to a value of 1.00. Based on this selection criterion, the plot of ln[sucrose] vs. time for the first order reaction is the best.
Another method when linearity is not obvious from the graphs is to examine the reaction and decide which order fits the reaction. For the reaction of one molecule of sucrose with one molecule of liquid water, the rate law would most likely include sucrose with an order of one and would not include water.

The plot for a first order reaction is described by the equation $\ln[A]_t = -kt + \ln[A]_0$. The slope of the plot of $\ln[\text{sucrose}]$ versus t equals $-k$. The equation for the straight line in the first order plot is $y = -0.21x - 0.6936$. So, $k = -(-0.21\ \text{h}^{-1}) = \mathbf{0.21\ h^{-1}}$.

Half-life for a first-order reaction equals $\ln 2/k$, so $t_{1/2} = \ln 2\ /\ 0.21\ \text{h}^{-1} = 3.3007 = \mathbf{3.3\ h}$.

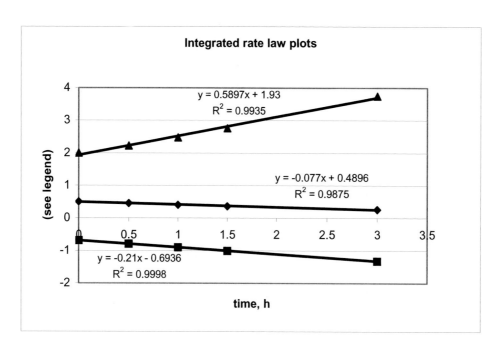

Legend: ♦ y–axis is [sucrose]
■ – axis is $\ln[\text{sucrose}]$
▲ – y–axis is $1/[\text{sucrose}]$

b) If 75% of the sucrose has been reacted, 25% of the sucrose remains. The time to hydrolyze 75% of the sucrose can be calculated by substituting 1.0 M for $[A]_0$ and 0.25 M for $[A]_t$ in the integrated rate law equation:
$\ln[A]_t = (-0.21\ \text{h}^{-1})t + \ln[A]_0$
$\ln[A]_t - \ln[A]_0 = (-0.21\text{h}^{-1})t$
$\ln[0.25] - \ln[1.0] = (-0.21\ \text{h}^{-1})t$
$t = 6.6014 = \mathbf{6.6\ h}$

c) The reaction might be second-order overall with first-order in sucrose and first-order in water. If the concentration of sucrose is relatively low, the concentration of water remains constant even with small changes in the amount of water. This gives an apparent zero-order reaction with respect to water. Thus, the reaction appears to be first order overall because the rate does not change with changes in the amount of water.

16.101 a) To find concentration at a later time, given starting concentration and the rate constant, use an integrated rate law expression. Since the units on k are s^{-1}, this is a first order reaction. Use the first-order integrated rate law:
$\ln[N_2O_5]_t = \ln[N_2O_5]_0 - kt$

$\ln[N_2O_5]_t = \ln[1.58\ \text{mol/L}] - (2.8 \times 10^{-3}\text{s}^{-1})(5.00\ \text{min})\left(\dfrac{60\ \text{s}}{1\ \text{min}}\right)$

$\ln[N_2O_5]_t = -0.382578$
$[N_2O_5]_t = 0.68210 = \mathbf{0.68\ mol/L}$

b) Fraction decomposed $= [(1.58 - 0.68210)\ M] / (1.58\ M) = 0.56829 = \mathbf{0.57}$

16.104 To solve this problem, a clear picture of what is happening is useful. Initially only N_2O_5 is present at a pressure of 125 kPa. Then a reaction takes place that consumes the gas N_2O_5 and produces the gases NO_2 and O_2. The balanced equation gives the change in the number of moles of gas as N_2O_5 decomposes. Since the number of moles of gas is proportional to the pressure, this change mirrors the change in pressure. The total pressure at the end, 178 kPa, equals the sum of the partial pressures of the three gases.

Balanced equation: $N_2O_5(g) \rightarrow 2\ NO_2(g) + 1/2\ O_2(g)$

Therefore, for each mole of dinitrogen pentaoxide 2.5 moles of gas are produced.

$$N_2O_5(g) \rightarrow 2\ NO_2(g) + 1/2\ O_2(g)$$

	$N_2O_5(g)$	$2\ NO_2(g)$	$1/2\ O_2(g)$	
Initial P (kPa)	125	0	0	total $P_{initial}$ = 125 kPa
Final P (kPa)	125 – x	2 x	1/2 x	total P_{final} = 178 kPa

Solve for x:

$$P_{N_2O_5} + P_{NO_2} + P_{O_2} = (125 - x) + 2\ x + 1/2\ x = 178$$

x = 35.3333 kPa (unrounded)

Partial pressure of NO_2 equals 2 x = 2 (35.3333) = 70.667 = **71 kPa**.

Check: Substitute values for all partial pressures to find total final pressure:

(125 – 35.3333) + (2 x 35.3333) + ((1/2) x 35.3333) = 178 kPa

The result agrees with the given total final pressure.

16.108 $k_1 = 1\ egg\ /\ 4.8\ min$ $T_1 = (273.2 + 90.0)\ K = 363.2\ K$

$k_2 = 1\ egg\ /\ 4.5\ min$ $T_2 = (273.2 + 100.0)\ K = 373.2\ K$ $E_a = ?$

The number of eggs (1) is exact, and has no bearing on the significant figures.

$$\ln \frac{k_2}{k_1} = -\frac{E_a}{R}\left(\frac{1}{T_2} - \frac{1}{T_1}\right)$$

$$E_a = -\frac{R\left(\ln \dfrac{k_2}{k_1}\right)}{\left(\dfrac{1}{T_2} - \dfrac{1}{T_1}\right)} = -\frac{\left(8.314\dfrac{J}{mol \cdot K}\right)\left(\ln \dfrac{(1\ egg/4.5\ min)}{(1\ egg/4.8\ min)}\right)}{\left(\dfrac{1}{373.2\ K} - \dfrac{1}{363.2\ K}\right)}$$

$E_a = 7.2730$ x 10^3 J/mol = **7.3 x 10^3 J/mol**

16.110 a) Starting with the fact that rate of formation of O (rate of step 1) equals the rate of consumption of O (rate of step 2), set up an equation to solve for [O] using the given values of k_1, k_2, [NO_2], and [O_2].

$rate_1 = rate_2$

$k_1[NO_2] = k_2[O][O_2]$

$$[O] = \frac{k_1[NO_2]}{k_2[O_2]} = \frac{\left(6.0\ x\ 10^{-3}\ s^{-1}\right)\left[4.0\ x\ 10^{-9}\ M\right]}{\left(1.0\ x\ 10^6\ L/mol \cdot s\right)\left[1.0\ x\ 10^{-2}\ M\right]} = \textbf{2.4 x } 10^{-15}\ \textbf{M}$$

b) Since the rate of the two steps is equal, either can be used to determine rate of formation of ozone.

$rate_2 = k_2[O][O_2] = (1.0$ x 10^6 L/mol•s) (2.4 x 10^{-15} M) (1.0 x 10^{-2} M) = **2.4 x 10^{-11} mol/L•s**

16.113 This problem involves the first-order integrated rate law (ln [A]$_t$ / [A]$_0$ = –kt). The temperature must be part of the calculation of the rate constant. The concentration of the ammonium ion is directly related to the ammonia concentration.

a) [NH_3]$_0$ = 3.0 mol/m^3 [NH_3]$_t$ = 0.35 mol/m^3 T = 20°C $k_1 = 0.47\ e^{0.095(T-15°C)}$

$\ln[NH_3]_t = -kt + \ln[NH_3]_0$

$$t = \frac{\ln[NH_3]_0 - \ln[NH_3]_t}{k}$$

$$t = \frac{\ln[3.0\ mol/m^3] - \ln[0.35\ mol/m^3]}{0.47e^{0.095(20 - 15°C)}}$$

t = 2.84272 = **2.8 d**

b) Repeating the calculation at the different temperature:

$[NH_3]_0 = 3.0$ mol/m^3 $[NH_3]_t = 0.35$ mol/m^3 $T = 10°C$ $k_1 = 0.47\ e^{0.095(T-15°C)}$

$$t = \frac{\ln[NH_3]_0 - \ln[NH_3]_t}{k}$$

$$t = \frac{\ln[3.0\ mol/m^3] - \ln[0.35\ mol/m^3]}{0.47e^{0.095(10-15°C)}}$$

$t = 7.35045 = $ **7.4 d**

c) For NH_4^+ the rate $= k_1[NH_4^+]$

From the balanced chemical equation:

$$-\frac{\Delta\left[NH_4^+\right]}{\Delta t} = -\frac{1}{2}\frac{\Delta\left[O_2\right]}{\Delta t}$$

Thus, for O_2: rate $= 2\ k_1\ [NH_4^+]$

$$Rate = (2)\left(0.47e^{0.095(20-15)}\right)\left[3.0\ mol/m^3\right] = 4.5346 = \textbf{4.5 mol/m}^3$$

-16.116   a) There are a number of dilution calculations using $M_iV_i = M_fV_f$ in the form: $M_f = M_iV_i / V_f$

$M_f\ S_2O_3^{2-} = [(10.0\ mL)\ (0.0050\ M)] / 50.0\ mL = 0.0010\ M\ S_2O_3^{2-}$

$$-\frac{\Delta[I_2]}{\Delta t} = -\frac{1}{2}\frac{\Delta\left[S_2O_3^{2-}\right]}{\Delta t} = [1/2(0.0010\ M)] / time = [0.00050\ M] / time = Rate$$

Average rates:

$Rate_1 = (0.00050\ M) / 29.0\ s = 1.724\ x\ 10^{-5} = \textbf{1.7 x 10}^{-5}\ \textbf{\textit{M}s}^{-1}$

$Rate_2 = (0.00050\ M) / 14.5\ s = 3.448\ x\ 10^{-5} = \textbf{3.4 x 10}^{-5}\ \textbf{\textit{M}s}^{-1}$

$Rate_3 = (0.00050\ M) / 14.5\ s = 3.448\ x\ 10^{-5} = \textbf{3.4 x 10}^{-5}\ \textbf{\textit{M}s}^{-1}$

b) $M_f\ KI = M_f\ I^- = [(10.0\ mL)\ (0.200\ M)] / 50.0\ mL = 0.0400\ M\ I^-$ (Experiment 1)

$M_f\ I^- = [(20.0\ mL)\ (0.200\ M)] / 50.0\ mL = 0.0800\ M\ I^-$ (Experiments 2 and 3)

$M_f\ Na_2S_2O_8 = M_f\ S_2O_8^{2-} = [(20.0\ mL)\ (0.100\ M)] / 50.0\ mL = 0.0400\ M\ S_2O_8^{2-}$ (Experiments 1 and 2)

$M_f\ S_2O_8^{2-} = [(10.0\ mL)\ (0.100\ M)] / 50.0\ mL = 0.0200\ M\ S_2O_8^{2-}$ (Experiment 3)

Generic rate law equation: $Rate = k\ [I^-]^m[S_2O_8^{2-}]^n$

$$\frac{Rate_3}{Rate_2} = \frac{k_3\left[I^-\right]^m\left[S_2O_8^{2-}\right]^n}{k_2\left[I^-\right]^m\left[S_2O_8^{2-}\right]^n}$$ Molarity of I^- is constant

$$\frac{3.4\ x\ 10^{-5}\ Ms^{-1}}{3.4\ x\ 10^{-5}\ Ms^{-1}} = \frac{\left[0.0200\right]^n}{\left[0.0400\right]^n}$$

$1.0 = (0.500)^n$

$n = 0$

The reaction is **zero order with respect to $S_2O_8^{2-}$**.

$$\frac{Rate_2}{Rate_1} = \frac{k_2\left[I^-\right]^m\left[S_2O_8^{2-}\right]^n}{k_1\left[I^-\right]^m\left[S_2O_8^{2-}\right]^n}$$ Molarity of $S_2O_8^{2-}$ is constant

$$\frac{3.4\ x\ 10^{-5}\ Ms^{-1}}{1.7\ x\ 10^{-5}\ Ms^{-1}} = \frac{\left[0.0800\right]^m}{\left[0.0400\right]^m}$$

$2.0 = (2.00)^m$

$m = 1$

The reaction is **first order with respect to I^-**.

c) $Rate = k\ [I^-]$

$k = Rate / [I^-]$ Using Experiment 2 (unrounded rate value)

$k = (3.448\ x\ 10^{-5}\ Ms^{-1}) / (0.0800\ M) = 4.31\ x\ 10^{-4} = \textbf{4.3 x 10}^{-4}\ \textbf{s}^{-1}$

d) $Rate = \textbf{(4.3 x 10}^{-4}\ \textbf{s}^{-1}\textbf{)}\ \textbf{[I}^-\textbf{]}$

16.119   a) Treat the concentration of the cells as you would molarity. The increasing cell density changes the integrated rate law from $-kt$ to $+kt$.

$\ln[A]_t = kt + \ln[A]_0$

$\ln[A]_t = (0.035 \text{ min}^{-1})(2 \text{ h})(60 \text{ min/h}) + \ln[1.0 \times 10^3]$

$\ln[A]_t = 11.107755$

$[A]_t = 6.6686 \times 10^4 = \mathbf{7 \times 10^4 \text{ cells/L}}$

b) $\ln[A]_t = kt + \ln[A]_0$

$\ln[2.0 \times 10^3] = (0.035 \text{ min}^{-1})(t) + \ln[1.0 \times 10^3]$

$t = 19.804 = \mathbf{2.0 \times 10^1 \text{ min}}$

16.122   a) Use the Monod equation:

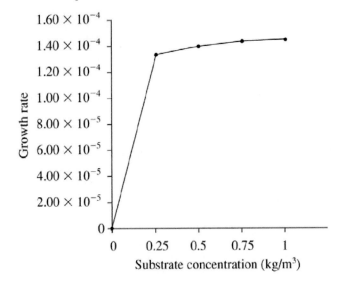

$$\mu = \frac{\mu_{max} S}{K_s + S} = \frac{(1.5 \times 10^{-4} \text{ s}^{-1})(0.25 \text{ kg/m}^3)}{(0.03 \text{ kg/m}^3) + (0.25 \text{ kg/m}^3)} = 1.34 \times 10^{-4} \text{ s}^{-1}$$

$$\mu = \frac{\mu_{max} S}{K_s + S} = \frac{(1.5 \times 10^{-4} \text{ s}^{-1})(0.50 \text{ kg/m}^3)}{(0.03 \text{ kg/m}^3) + (0.50 \text{ kg/m}^3)} = 1.42 \times 10^{-4} \text{ s}^{-1}$$

$$\mu = \frac{\mu_{max} S}{K_s + S} = \frac{(1.5 \times 10^{-4} \text{ s}^{-1})(0.75 \text{ kg/m}^3)}{(0.03 \text{ kg/m}^3) + (0.75 \text{ kg/m}^3)} = 1.44 \times 10^{-4} \text{ s}^{-1}$$

$$\mu = \frac{\mu_{max} S}{K_s + S} = \frac{(1.5 \times 10^{-4} \text{ s}^{-1})(1.0 \text{ kg/m}^3)}{(0.03 \text{ kg/m}^3) + (1.0 \text{ kg/m}^3)} = 1.46 \times 10^{-4} \text{ s}^{-1}$$

b) Again, use the Monod equation:

$$\mu = \frac{\mu_{max} S}{K_s + S} = \frac{(1.5 \times 10^{-4} \text{ s}^{-1})(0.30 \text{ kg/m}^3)}{(0.03 \text{ kg/m}^3) + (0.30 \text{ kg/m}^3)} = 1.3636 \times 10^{-4} \text{ s}^{-1} \text{ (unrounded)}$$

$\mu = k$, and change $-kt$ to $+kt$ to get the growth rate.

$\ln[A]_t = kt + \ln[A]_0$

$\ln[A]_t = (1.3636 \times 10^{-4} \text{ s}^{-1})(1.0 \text{ h})(3600 \text{ s/h}) + \ln[5.0 \times 10^3]$

$\ln[A]_t = 9.008089$

$[A]_t = 8.16890 \times 10^3 = \mathbf{8.2 \times 10^3 \text{ cells/m}^3}$

c) Again, use the Monod equation:

$$\mu = \frac{\mu_{max}S}{K_s + S} = \frac{\left(1.5 \times 10^{-4}\ s^{-1}\right)\left(0.70\ kg/m^3\right)}{\left(0.03\ kg/m^3\right)+\left(0.70\ kg/m^3\right)} = 1.438356 \times 10^{-4}\ s^{-1}\ \text{(unrounded)}$$

$\mu = k$, and change $-kt$ to $+kt$ to get the growth rate.

$\ln[A]_t = kt + \ln[A]_0$

$\ln[A]_t = (1.438356 \times 10^{-4}\ s^{-1})\ (1.0\ h)\ (3600\ s/h) + \ln[5.0 \times 10^3]$

$\ln[A]_t = 9.03500$

$[A]_t = 8.39172 \times 10^3 = \textbf{8.4} \times \textbf{10}^{\textbf{3}}\ \textbf{cells/m}^{\textbf{3}}$

16.126   a) $t_{1/2} = \ln 2\ /\ k = \ln 2\ /\ (8.7 \times 10^{-3}\ s^{-1}) = 79.672 = \textbf{8.0} \times \textbf{10}^{\textbf{1}}\ \textbf{s}$

b) If 40.% has decomposed, 60.% remain.

$$\ln\frac{[\text{Acetone}]_t}{[\text{Acetone}]_0} = -\,kt$$

$$\ln\frac{[60.\%]_t}{[100\%]_0} = -\,(8.7 \times 10^{-3}\ s^{-1})\,t$$

$t = 58.71558894 = \textbf{59 s}$

c) If 90.% has decomposed, 10.% remain.

$$\ln\frac{[10.\%]_t}{[100\%]_0} = -\,(8.7 \times 10^{-3}\ s^{-1})\,t$$

$t = 264.66 = \textbf{2.6} \times \textbf{10}^{\textbf{2}}\ \textbf{s}$

CHAPTER 17 EQUILIBRIUM: THE EXTENT OF CHEMICAL REACTIONS

FOLLOW-UP PROBLEMS

17.1 Plan: First, balance the equations and then use the coefficients to write the reaction quotient. Products appear in numerator of the reaction quotient and reactants appear in the denominator; coefficients in the balanced reaction become exponents.
Solution:
a) Balanced equation: $4 NH_3(g) + 5 O_2(g) \leftrightarrows 4 NO(g) + 6 H_2O(g)$

Reaction quotient: $Q_c = \dfrac{[NO]^4 [H_2O]^6}{[NH_3]^4 [O_2]^5}$

b) Balanced equation: $3 NO(g) \leftrightarrows N_2O(g) + NO_2(g)$

Reaction quotient: $Q_c = \dfrac{[N_2O][NO_2]}{[NO]^3}$

Check: Are products in numerator and reactants in denominator? Are all reactants and products expressed as concentrations by using brackets? Are all exponents equal to the coefficient of the reactant or product in the balanced equation?

17.2 Plan: Add the individual steps to find the overall equation. Write the reaction quotient for each step and the overall equation. Multiply the reaction quotients for each step and cancel terms to obtain the overall reaction quotient.
Solution:
(1) $Br_2(g) \leftrightarrows 2 Br(g)$
(2) $Br(g) + H_2(g) \leftrightarrows HBr(g) + H(g)$
(3) $\underline{H(g) + Br(g) \leftrightarrows HBr(g)}$
$Br_2(g) + \cancel{2\ Br(g)} + H_2(g) + \cancel{H(g)} \leftrightarrows \cancel{2\ Br(g)} + 2 HBr(g) + \cancel{H(g)}$
Canceling the reactants leaves the overall equation as $Br_2(g) + H_2(g) \leftrightarrows 2 HBr(g)$.
Write the reaction quotients for each step:

$$Q_{c1} = \dfrac{[Br]^2}{[Br_2]} \qquad Q_{c2} = \dfrac{[HBr][H]}{[Br][H_2]} \qquad Q_{c3} = \dfrac{[HBr]}{[H][Br]}$$

and for the overall equation:

$$Q_c = \dfrac{[HBr]^2}{[H_2][Br_2]}$$

Multiplying the individual Q_c's and canceling terms gives

$$Q_{c1}\, Q_{c2}\, Q_{c3} = \dfrac{[Br]^2}{[Br_2]} \times \dfrac{[HBr][H]}{[Br][H_2]} \times \dfrac{[HBr]}{[H][Br]} = \dfrac{[HBr]^2}{[H_2][Br_2]}$$

The product of the multiplication of the three individual reaction quotients equals the overall reaction quotient.

17.3 Plan: When a reaction is multiplied by a factor, the equilibrium constant is raised to a power equal to the factor.
Solution:
a) All coefficients have been multiplied by the factor 1/2 so the equilibrium constant should be raised to the 1/2 power (which is the square root).
For reaction a) $K_c = (7.6 \times 10^8)^{1/2} = 2.7568 \times 10^4 = \mathbf{2.8 \times 10^4}$

b) The reaction has been reversed so the new K_c is the inverse of the original equilibrium constant. The reaction has also been multiplied by a factor of 2/3, so the inverse of the original K_c must be raised to the 2/3 power.

For reaction b) $K_c = \left(\dfrac{1}{7.6 \times 10^8} \right)^{\frac{2}{3}} = 1.20076 \times 10^{-6} = \mathbf{1.2 \times 10^{-6}}$

Check: The order of magnitude of each answer can be estimated by examining the exponents: for part a) $(10^8)^{1/2} = 10^4$ and for part b) rounding 7.6×10^8 to 10^9 gives $1/10^9 = 10^{-9}$ and $(10^{-9})^{2/3} = 10^{-6}$. These estimates are within the same order of magnitude as the answers under solution.

17.4   Plan: K_p and K_c for a reaction are related through the ideal gas equation as shown in $K_P = K_c(RT)^{\Delta n}$. Find Δn_{gas}, the change in the number of moles of gas between reactants and products (calculated as products minus reactants). Then, use the given K_c to solve for K_p.
Solution: The total number of product moles of gas is 1 and the total number of reactant moles of gas is 2.
   $\Delta n = 1 - 2 = -1$.
   $K_P = K_c(RT)^{\Delta n}$
   $K_P = 1.67[(0.08206)\,(500.)]^{-1}$
   $K_P = 0.040701925 = \mathbf{4.07 \times 10^{-2}}$
Check: Although equilibrium constants do not have units, it is useful to check this type of calculation by doing unit analysis. For this purpose add units for the equilibrium constants and substitute M for mol/L in gas constant.
   $K_c = (M/M{\bullet}M) = M$ and $K_p = (atm/atm{\bullet}atm) = atm$
   $K_p = (M)\,(atm/M{\bullet}K)\,(K)^{-1} = atm$
The units check out to convert from M to atm.

17.5   Plan: Write the reaction quotient for the reaction and calculate Q_c for each circle. Compare Q_c to K_c determine the direction needed to reach equilibrium. If $Q_c > K_c$, reactants are forming. If $Q_c < K_c$, products are forming.
Solution:

The reaction quotient is $\dfrac{[Y]}{[X]}$.

Circle 1: $Q_c = \dfrac{[Y]}{[X]} = \dfrac{[3]}{[9]} = 0.33$

Since $Q_c < K_c$ $(0.33 < 1.4)$, the reaction will shift to the **right** to reach equilibrium

Circle 2: $Q_c = \dfrac{[Y]}{[X]} = \dfrac{[7]}{[5]} = 1.4$

Since $Q_c = K_c$ $(1.4 = 1.4)$, there is **no change** in the reaction direction. The reaction is at equilibrium now.

Circle 3: $Q_c = \dfrac{[Y]}{[X]} = \dfrac{[8]}{[4]} = 2.0$

Since $Q_c > K_c$ $(2.0 > 1.4)$, the reaction will shift to the **left** to reach equilibrium

17.6   Plan: To decide whether CH_3Cl or CH_4 are forming while the reaction system moves toward equilibrium, calculate Q_p and compare it to K_p. If $Q_p > K_p$, reactants are forming. If $Q_p < K_p$, products are forming.
Solution:

$$Q_P = \frac{P_{CH_3Cl}P_{HCl}}{P_{CH_4}P_{Cl_2}} = \frac{(0.24\,atm)(0.47\,atm)}{(0.13\,atm)(0.035\,atm)} = 24.7912 = 25$$

K_p for this reaction is given as 1.6×10^4. Q_p is smaller than K_p $(Q_p < K_p)$ so more products will form. **CH_3Cl is one** of the products forming.

Check: Since number of moles of gas is proportional to pressure, a new calculation of Q_p can be done with increased pressures for products and decreased pressures for reactants. If the conclusion that formation of products moves the reaction system closer to equilibrium then the new Q_p should be closer to K_p than the 25 found for the original Q_p.
Increasing and decreasing by 0.01 atm:

$$Q_P = \frac{(0.25\,\text{atm})(0.48\,\text{atm})}{(0.12\,\text{atm})(0.025\,\text{atm})} = 40$$

40 is closer to 16000 than 25, so the answer that CH_3Cl is forming is correct.

17.7   Plan: The information given includes the equilibrium constant, initial pressures of both reactants and equilibrium pressure for one reactant. First, set up a reaction table showing initial partial pressures for reactants and 0 for product. The change to get to equilibrium is to react some of reactants to form some product. Use the equilibrium quantity for O_2 and the expression for O_2 at equilibrium to solve for the change. From the change find the equilibrium partial pressure for NO and NO_2. Calculate K_p using the equilibrium values.
Solution:

Pressures (atm)	2 NO(g) +	O_2(g)	$\leftrightarrows$	2 NO_2(g)
Initial	1.000	1.000		0
Change	−2 x	−x		+2 x
Equilibrium	1.000 − 2 x	1.000 − x		2 x

At equilibrium $P_{O_2} = 0.506$ atm $= 1.000 - x$; so $x = 1.000 - 0.506 = 0.494$ atm

$$P_{NO} = 1.000 - 2\,x = 1.000 - 2(0.494) = 0.012 \text{ atm}$$

$$P_{NO_2} = 2\,x = 2\,(0.494) = 0.988 \text{ atm}$$

Use the equilibrium pressures to calculate K_p.

$$K_P = \frac{P_{NO_2}^2}{P_{NO}^2 P_{O_2}} = \frac{(0.988)^2}{(0.012)^2(0.506)} = 1.339679 \times 10^4 = 1.3 \times 10^4$$

Check: One way to check is to plug the equilibrium expressions for NO and NO_2 into the equilibrium expression along with 0.506 atm for the equilibrium pressure of O_2. Then solve for x and make sure the value of x by this calculation is the same as the 0.494 atm calculated above.

$$1.3 \times 10^4 = \frac{(2x)^2(0.506)}{(1.000 - 2x)^2}$$

$$(1.3 \times 10^4) / 0.506 = 2.569 \times 10^4 = \frac{(2x)^2(0.506)}{(1.000 - 2x)^2}$$

$$\sqrt{2.569 \times 10^4} = \frac{(2x)}{(1.000 - 2x)}$$

$$x = 0.497$$

The two values for x agree to give $K_p = 1.3 \times 10^4$.

17.8   Plan: Convert K_c to K_p for the reaction. Write the equilibrium expression for K_p and insert the atmospheric pressures for P_{N_2} and P_{O_2} as their equilibrium values. Solve for P_{NO}.

Solution: The conversion of K_c to K_p: $K_p = K_c(RT)^{\Delta n}$. For this reaction $\Delta n = 0$, so $K_p = K_c$.

$$K_P = \frac{P_{N_2} P_{O_2}}{P_{NO}^2} = K_c = 2.3 \times 10^{30} = \frac{(0.781)(0.209)}{x^2}$$

$$x = 2.6640 \times 10^{-16} = 2.7 \times 10^{-16} \text{ atm}$$

The equilibrium partial pressure of NO in the atmosphere is **2.7×10^{-16} atm**.
Check: Insert the pressure of NO and calculate the equilibrium constant to check the result.

$$\frac{(0.781)(0.209)}{(2.6640 \times 10^{-16})^2} = 2.300 \times 10^{30} = 2.3 \times 10^{30}$$

17.9   Plan: Find the initial molarity of HI. Set up a reaction table and use the variables to find equilibrium concentrations in the equilibrium expression.
Solution:

$$M_{HI} = \frac{\text{moles HI}}{\text{volume}} = \frac{2.50 \text{ mol}}{10.32 \text{ L}} = 0.242248$$

Concentration (M)	2 HI(g)	$\leftrightarrows$	H$_2$(g)	+	I$_2$(g)	
Initial	0.242248		0		0	
Change	−2x		+x		+x	
Equilibrium	0.242248 − 2x		x		x	(unrounded values)

Set up equilibrium expression:

$$K_c = 1.26 \times 10^{-3} = \frac{[H_2][I_2]}{[HI]^2} = \frac{[x][x]}{[0.242248 - 2x]^2} \qquad \text{Take the square root of each side}$$

$$3.54965 \times 10^{-2} = \frac{[x]}{[0.242248 - 2x]}$$

x = 8.59895 × 10^{-3} − 7.0993 × 10^{-2} x
x = 8.02895 × 10^{-3} = 8.03 × 10^{-3}
[H$_2$] = [I$_2$] = **8.02 × 10^{-3} M**
Check: Plug equilibrium concentrations into the equilibrium expression and calculate K_c.
[HI] = 0.242 − 2(8.02 × 10^{-3}) = 0.226 M
(8.02 × 10^{-3})^2 / (0.226)^2 = 1.2593 × 10^{-3} = 1.26 × 10^{-3}

17.10   Plan: First set up the reaction table, then set up the equilibrium expression. To solve for variable, x, first assume that x is negligible with respect to initial concentration of I$_2$. Check the assumption by calculating the% error. If the error is greater than 5%, calculate x using the quadratic equation. The next step is to use x to determine the equilibrium concentrations of I$_2$ and I.
Solution: [I$_2$]$_{init}$ = 0.50 mol I$_2$/2.5 L = 0.20 M
a) Equilibrium at 600 K

Concentration (M)	I$_2$(g)	$\leftrightarrows$	2 I(g)
Initial	0.20		0
Change	− x		+2 x
Equilibrium	0.20 − x		2 x

Equilibrium expression: $\quad K_c = \dfrac{[I]^2}{[I_2]} = 2.94 \times 10^{-10}$

$$\frac{[2x]^2}{[0.20 - x]} = 2.94 \times 10^{-10} \qquad \text{Assume x is negligible so } 0.20 - x \approx 0.20$$

$$\frac{[2x]^2}{[0.20]} = 2.94 \times 10^{-10}$$

4 x^2 = (2.94 × 10^{-10}) (0.20); x = 3.834 × 10^{-6} = 3.8 × 10^{-6}
Check the assumption by calculating the % error: 3.8 × 10^{-6} ÷ 0.20 × 100% = 1.9x10^{-3}% which is smaller than 5%, so the assumption is valid.
At equilibrium [I]$_{eq}$ = 2x = 2(3.8 × 10^{-6}) = 7.668 × 10^{-6} = **7.7 × 10^{-6} M** and
[I$_2$]$_{eq}$ = 0.20 − x = 0.20 − 3.834 × 10^{-6} = 0.199996 = **0.20 M**

b) Equilibrium at 2000 K

Equilibrium expression: $K_c = \dfrac{[I]^2}{[I_2]} = 0.209$

$\dfrac{[2x]^2}{[0.20 - x]} = 0.209$ Assume x is negligible so 0.20 – x is approximately 0.20

$\dfrac{[2x]^2}{[0.20]} = 0.209$

$4x^2 = (0.209)(0.20)$; $x = 0.102225 = 0.102$

Check the assumption by calculating the % error: $0.102 \div 0.20 = 51\%$ which is larger than 5% so the assumption is not valid. Solve using quadratic equation.

$\dfrac{[2x]^2}{[0.20 - x]} = 0.209$

$4x^2 + 0.209\,x - 0.0418 = 0$

$$x = \dfrac{-0.209 \pm \sqrt{(0.209)^2 - 4(4)(-0.0418)}}{2(4)} = 0.0793857 \text{ or } -0.1316$$

Choose the positive value, x = 0.079
At equilibrium $[I]_{eq} = 2x = 2(0.079) = 0.15877 = \mathbf{0.16\ M}$ and
$[I_2]_{eq} = 0.20 - x = 0.20 - 0.079 = 0.12061 = \mathbf{0.12\ M}$
Check: Calculate equilibrium constants using calculated equilibrium concentrations:
a) $K = (7.7\times10^{-6})^2/(0.20) = 2.9645 \times 10^{-10} = 3.0 \times 10^{-10}$
b) $K = (0.16)^2/(0.12) = 0.21333 = 0.213$

17.11   Plan: Calculate the initial concentrations of each substance. For part (a), calculate Q_c and compare to given K_c. If $Q_c > K_c$ then the reaction proceeds to the left to make reactants from products. If $Q_c < K_c$ then the reaction proceeds to right to make products from reactants. For part (b), use the result of part (a) and the given equilibrium concentration of PCl_5 to find the equilibrium concentrations of PCl_3 and Cl_2.
Solution:
Initial concentrations: $[PCl_5] = 0.1050\ mol/0.5000\ L = 0.2100\ M$
$[PCl_3] = [Cl_2] = 0.0450\ mol/0.5000\ L = 0.0900\ M$

a) $Q_c = \dfrac{[PCl_3][Cl_2]}{[PCl_5]} = \dfrac{[0.0900][0.0900]}{[0.2100]} = 0.038571 = 0.0386$

Q_c, 0.0386, is less than K_c, 0.042, so the reaction will proceed to the **right** to make more products.
b) To reach equilibrium, concentrations will increase for the products, PCl_3 and Cl_2, and decrease for the reactant, PCl_5.
$[PCl_5] = 0.2065 = 0.2100 - x$; $x = 0.0035$
$[PCl_3] = [Cl_2] = 0.0900 + x = 0.0900 + 0.0035 = \mathbf{0.0935\ M}$
Check: Calculate equilibrium constants using the calculated equilibrium concentrations:

$K_c = \dfrac{[PCl_3][Cl_2]}{[PCl_5]} = \dfrac{[0.0935][0.0935]}{[0.2065]} = 0.042335 = 0.0423$

17.12   Plan: Examine each change for its impact on Q_c. Then decide how the system would respond to re-establish equilibrium.

Solution: $$Q_c = \frac{[SiF_4][H_2O]^2}{[HF]^4}$$

a) Decreasing $[H_2O]$ leads to $Q_c < K_c$, so the reaction would shift to make more products from reactants. Therefore, the SiF_4 concentration, as a product, would **increase**.
b) Adding liquid water to this system at a temperature above the boiling point of water would result in an increase in the concentration of water vapor. The increase in $[H_2O]$ increases Q_c to make it greater than K_c To re-establish equilibrium products will be converted to reactants and the $[SiF_4]$ will **decrease**.
c) Removing the reactant HF increases Q_c, which causes the products to react to form more reactants. Thus, $[SiF_4]$ **decreases**.
d) Removal of a solid product has no impact on the equilibrium; $[SiF_4]$ **does not change**.
Check: Look at each change and decide which direction the equilibrium would shift using Le Châtelier's principle to check the changes predicted above.
         a) Remove product, equilibrium shifts to right.
         b) Add product, equilibrium shifts to left.
         c) Remove reactant, equilibrium shifts to left.
         d) Remove solid reactant, equilibrium does not shift.

17.13   Plan: Changes in pressure (and volume) affect the concentration of gaseous reactants and products. A decrease in pressure, i.e. increase in volume, favors the production of more gas molecules whereas an increase in pressure favors the production of fewer gas molecules. Examine each reaction to decide whether more or fewer gas molecules will result from producing more products. If more gas molecules result, then the pressure should be increased (volume decreased) to reduce product formation. If fewer gas molecules result, then pressure should be decreased to produce more reactants.
Solution:
a) In $2\,SO_2(g) + O_2(g) \rightleftharpoons 2\,SO_3(g)$ three molecules of gas form two molecules of gas, so there are fewer gas molecules in the product. **Decreasing pressure** (increasing volume) will decrease the product yield.
b) In $4\,NH_3(g) + 5\,O_2(g) \rightleftharpoons 4\,NO(g) + 6\,H_2O(g)$ 9 molecules of reactant gas convert to 10 molecules of product gas. **Increasing pressure** (decreasing volume) will favor the reaction direction that produces fewer moles of gas: towards the reactants and away from products.
c) In $CaC_2O_4(s) \rightleftharpoons CaCO_3(s) + CO(g)$ there are no reactant gas molecules and one product gas molecule. The yield of the products will decrease when volume decreases, which corresponds to a **pressure increase**.

17.14   Plan: A decrease in temperature favors the exothermic direction of an equilibrium reaction. First, identify whether the forward or reverse reaction is exothermic from the given enthalpy change. $\Delta H < 0$ means the forward reaction is endothermic, and $\Delta H > 0$ means the reverse reaction is exothermic. If the forward reaction is exothermic then a decrease in temperature will shift the equilibrium to make more products from reactants and increase K_p. If the reverse reaction is exothermic then a decrease in temperature will shift the equilibrium to make more reactants from products and decrease K_p.
Solution:
a) $\Delta H < 0$ so the forward reaction is exothermic. A decrease in temperature increases the partial pressure of products and decreases the partial pressures of reactants, so P_{H_2} **decreases**. With increases in product pressures and decreases in reactant pressures, K_p **increases**.
b) $\Delta H > 0$ so the reverse reaction is exothermic. A decrease in temperature decreases the partial pressure of products and increases the partial pressures of reactants, so P_{N_2} **increases**. K_p **decreases** with decrease in product pressures and increase in reactant pressures.
c) $\Delta H < 0$ so the forward reaction is exothermic. Decreasing temperature **increases** P_{PCl_3} and **increases** K_p.

Solve the problem for increasing instead of decreasing temperature and make sure the answers are opposite.

a) Increased temperature favors the reverse, endothermic reaction leading to an increase in reactant pressures, so P_{H_2} increases and K_p decreases.

b) Increased temperature favors the forward, endothermic reaction leading to a decrease in reactant pressures. P_{N_2} decreases and K_p increases.

c) Both P_{PCl_5} and K_p decrease with increased temperature.

17.15   Plan: Given the balanced equilibrium equation, it is possible to set up the appropriate equilibrium expression (Q_c). For the equation given $\Delta n = 0$ meaning that $K_p = K_c$. The value of K may be found for scene 1, and values for Q may be determined for the other two scenes. The reaction will shift towards the reactant side if $Q > K$, and the reaction will shift towards the product side if $Q < K$. The reaction is exothermic, thus, heat may be considered a product. Increasing the temperature adds a product and decreasing the temperature removes a product.
Solution:
a) K_p requires the equilibrium value of P for each gas. The pressure may be found from $P = nRT / V$.

$$K_p = \frac{P_{CD}^2}{P_{C_2}P_{D_2}} = \frac{\left(\dfrac{n_{CD}RT}{V}\right)^2}{\left(\dfrac{n_{C_2}RT}{V}\right)\left(\dfrac{n_{D_2}RT}{V}\right)}$$

This equation may be simplified because for the sample R, T, and V are constant. Using scene 1:

$$K_p = \frac{n_{CD}^2}{n_{C_2}n_{D_2}} = \frac{(4)^2}{(2)(2)} = 4$$

b) Scene 2: $Q_p = \dfrac{n_{CD}^2}{n_{C_2}n_{D_2}} = \dfrac{(6)^2}{(1)(1)} = 36$

$Q > K$ so the reaction will shift to the left (towards the reactants).

Scene 3: $Q_p = \dfrac{n_{CD}^2}{n_{C_2}n_{D_2}} = \dfrac{(2)^2}{(3)(3)} = 0.44$

$Q < K$ so the reaction will shift to the right (towards the products).
c) Increasing the temperature is equivalent to adding a product (heat) to the equilibrium. The reaction will shift to consume the added heat. The reaction will shift to the left (towards the reactants). However, since there are 2 moles of gas on each side of the equation, the shift has **no effect** on total moles of gas.
Check: a) Remember to use the coefficients as exponents, and to multiply, not add the values for the different reactants. b) The relationship between Q and K determines the direction the reaction will precede.

END-OF-CHAPTER PROBLEMS

17.1   If the rate of the forward reaction exceeds the rate of reverse reaction, products are formed faster than they are consumed. The change in reaction conditions results in more products and less reactants. A change in reaction conditions can result from a change in concentration or a change in temperature. If concentration changes, product concentration increases while reactant concentration decreases, but the K_c remains unchanged because the *ratio* of products and reactants remains the same. If the increase in the forward rate is due to a change in temperature, the rate of the reverse reaction also increases. The equilibrium ratio of product concentration to reactant concentration is no longer the same. Since the rate of the forward reaction increases more than the rate of the reverse reaction, K_c increases (numerator, [products], is larger and denominator, [reactants], is smaller).

$$K_c = \frac{[products]}{[reactants]}$$

17.7   The equilibrium constant expression is $K = [O_2]$ (we do not include solid substances in the equilibrium expression). If the temperature remains constant, K remains constant. If the initial amount of Li_2O_2 present was sufficient to reach equilibrium, the amount of O_2 obtained will be constant, regardless of how much $Li_2O_2(s)$ is present.

17.8   On the graph, the concentration of HI increases at twice the rate that H_2 decreases because the stoichiometric ratio in the balanced equation is 1 H_2: 2 HI.
        Q for a reaction is the ratio of concentrations of products to concentrations of reactants. As the reaction progresses the concentration of reactants H_2 and I_2 decrease and the concentration of product HI increases, which means that Q increases as a function of time.

$$H_2(g) + I_2(g) \leftrightarrows 2\ HI(g) \qquad Q = \frac{[HI]^2}{[H_2][I_2]}$$

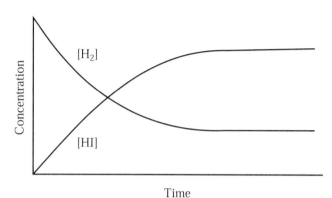

The value of Q increases as a function of time until it reaches the value of K.
b) No, Q would still increase with time because the $[I_2]$ would decrease in exactly the same way as $[H_2]$ decreases.

17.11   Yes, the Q's for the two reactions do differ. The balanced equation for the first reaction is
        $3/2\ H_2(g) + 1/2\ N_2(g) \leftrightarrows NH_3(g)$ (1)
        The coefficient in front of NH_3 is fixed at 1 mole according to the description. In the second reaction, the coefficient in front of N_2 is fixed at 1 mole.
        $3\ H_2(g) + N_2(g) \leftrightarrows 2\ NH_3(g)$ (2)
        The reaction quotients for the two equations and their relationship are:

$$Q_1 = \frac{[NH_3]}{[H_2]^{3/2}[N_2]^{1/2}} \qquad\qquad Q_2 = \frac{[NH_3]^2}{[H_2]^3[N_2]}$$

$$Q_2 = Q_1^2$$

17.12   Check that correct coefficients from balanced equation are included as exponents in the mass action expression.
        a) $4\ NO(g) + O_2(g) \leftrightarrows 2\ N_2O_3(g)$

$$Q_c = \frac{[N_2O_3]^2}{[NO]^4[O_2]}$$

        b) $SF_6(g) + 2\ SO_3(g) \leftrightarrows 3\ SO_2F_2(g)$

$$Q_c = \frac{[SO_2F_2]^3}{[SF_6][SO_3]^2}$$

c) $2\ SClF_5(g) + H_2(g) \leftrightarrows S_2F_{10}(g) + 2\ HCl(g)$

$$Q_c = \frac{[S_2F_{10}][HCl]^2}{[SClF_5]^2[H_2]}$$

17.14 Check that correct coefficients from the balanced equation are included as exponents in the mass action expression.

a) $2\ NO_2Cl(g) \leftrightarrows 2\ NO_2(g) + Cl_2(g)$

$$Q_c = \frac{[NO_2]^2[Cl_2]}{[NO_2Cl]^2}$$

b) $2\ POCl_3(g) \leftrightarrows 2\ PCl_3(g) + O_2(g)$

$$Q_c = \frac{[PCl_3]^2[O_2]}{[POCl_3]^2}$$

c) $4\ NH_3(g) + 3\ O_2(g) \leftrightarrows 2\ N_2(g) + 6\ H_2O(g)$

$$Q_c = \frac{[N_2]^2[H_2O]^6}{[NH_3]^4[O_2]^3}$$

17.16 The K for the original reaction is $K_c = \dfrac{[H_2]^2[S_2]}{[H_2S]^2}$

a) The given reaction $1/2\ S_2(g) + H_2(g) \leftrightarrows H_2S(g)$ is the reverse reaction of the original reaction multiplied by a factor of 1/2. The equilibrium constant for the reverse reaction is the inverse of the original constant. When a reaction is multiplied by a factor, K_c for the new equation is equal to the K_c of the original equilibrium raised to a power equal to the factor. For the reaction given in part a), take $(1/K)^{1/2}$.

$$K_{c\,(a)} = (1 / K_c)^{1/2} = \frac{[H_2S]}{[S_2]^{1/2}[H_2]}$$

$$K_{c\,(a)} = (1 / 1.6 \times 10^{-2})^{1/2} = 7.90569 = \mathbf{7.9}$$

b) The given reaction $5\ H_2S(g) \leftrightarrows 5\ H_2(g) + 5/2\ S_2(g)$ is the original reaction multiplied by 5/2. Take the original K to the 5/2 power to find K of given reaction.

$$K_{c\,(b)} = (1 / K_c)^{5/2} = \frac{[H_2]^5[S_2]^{5/2}}{[H_2S]^5}$$

$$K_{c\,(b)} = (1.6 \times 10^{-2})^{5/2} = 3.23817 \times 10^{-5} = \mathbf{3.2 \times 10^{-5}}$$

17.18 The concentration of solids and pure liquids do not change, so their concentration terms are not written in the reaction quotient expression.

a) $2\ Na_2O_2(s) + 2\ CO_2(g) \leftrightarrows 2\ Na_2CO_3(s) + O_2(g)$

$$Q_c = \frac{[O_2]}{[CO_2]^2}$$

b) $H_2O(l) \leftrightarrows H_2O(g)$

$Q_c = [H_2O(g)]$ Only the gaseous water is used. The "(g)" is for emphasis.

c) $NH_4Cl(s) \leftrightarrows NH_3(g) + HCl(g)$

$Q_c = [NH_3][HCl]$

17.20   Make sure not to include solids and liquids in reaction quotient.
a) $2 NaHCO_3(s) \leftrightarrows Na_2CO_3(s) + CO_2(g) + H_2O(g)$
$$Q_c = [CO_2][H_2O]$$
b) $SnO_2(s) + 2 H_2(g) \leftrightarrows Sn(s) + 2 H_2O(g)$

$$Q_c = \frac{[H_2O]^2}{[H_2]^2}$$

c) $H_2SO_4(l) + SO_3(g) \leftrightarrows H_2S_2O_7(l)$

$$Q_c = \frac{1}{[SO_3]}$$

17.23   a) The balanced equations and corresponding reaction quotients are given below. Note the second equation must be multiplied by 2 to get the appropriate overall equation.

(1) $Cl_2(g) + F_2(g) \leftrightarrows$ ~~2 ClF(g)~~ $\qquad Q_1 = \dfrac{[ClF]^2}{[Cl_2][F_2]}$

(2) ~~2 ClF(g)~~ $+ 2 F_2(g) \leftrightarrows 2 ClF_3(g)$ $\qquad Q_2 = \dfrac{[ClF_3]^2}{[ClF]^2 [F_2]^2}$

Overall: $Cl_2(g) + 3 F_2(g) \leftrightarrows 2 ClF_3(g)$ $\qquad Q_{overall} = \dfrac{[ClF_3]^2}{[Cl_2][F_2]^3}$

b) The reaction quotient for the overall reaction, $Q_{overall}$, determined from the reaction is:

$$Q_{overall} = \frac{[ClF_3]^2}{[Cl_2][F_2]^3}$$

$$Q_{overall} = Q_1 Q_2^2 = \frac{[ClF]^2}{[Cl_2][F_2]} \times \frac{[ClF_3]^2}{[ClF]^2 [F_2]^2} = \frac{[ClF_3]^2}{[Cl_2][F_2]^3}$$

17.25 K_c and K_p are related by the equation $K_p = K_c(RT)^{\Delta n}$, where Δn represents the change in the number of moles of gas in the reaction (moles gaseous products – moles gaseous reactants). When Δn is zero (no change in number of moles of gas), the term $(RT)^{\Delta n}$ equals 1 and $K_c = K_p$. When Δn is not zero, meaning that there is a change in the number of moles of gas in the reaction, then $K_c \neq K_p$.

17.26   a) $K_p = K_c(RT)^{\Delta n}$. Since Δn = number of moles gaseous products – number of moles gaseous reactants, Δn is a positive integer for this reaction. If Δn is a positive integer, then $(RT)^{\Delta n}$ is greater than 1. Thus, K_c is multiplied by a number that is greater than 1 to give K_p. **K_c is smaller than K_p.**
b) Assuming that $RT > 1$ (which occurs when $T > 12.2$ K, because 0.0821 (R) x 12.2 = 1), **$K_p > K_c$** if the number of moles of gaseous products exceeds the number of moles of gaseous reactants. **$K_p < K_c$** when the number of moles of gaseous reactants exceeds the number of moles of gaseous product.

17.27   a) Number of moles of <u>gaseous</u> reactants = 0; Number of moles of <u>gaseous</u> products = 3; $\Delta n = 3 - 0 = $ **3**
b) Number of moles of <u>gaseous</u> reactants = 1; Number of moles of <u>gaseous</u> products = 0; $\Delta n = 0 - 1 = $ **–1**
c) Number of moles of <u>gaseous</u> reactants = 0; Number of moles of <u>gaseous</u> products = 3; $\Delta n = 3 - 0 = $ **3**

17.29   First, determine Δn for the reaction and then calculate K_c using $K_p = K_c(RT)^{\Delta n}$.
a) Δn = Number of product gas moles – Number of reactant gas moles = $1 - 2 = -1$

$$K_c = \frac{K_p}{(RT)^{\Delta n}} = \frac{3.9 \times 10^{-2}}{[(0.0821)(1000.)]^{-1}} = 3.2019 = \mathbf{3.2}$$

b) Δn = Number of product gas moles – Number of reactant gas moles = 1 – 1 = 0

$$K_c = \frac{K_p}{(RT)^{\Delta n}} = \frac{28.5}{[(0.0821)(500.)]^0} = \mathbf{28.5}$$

17.31   First, determine Δn for the reaction and then calculate K_p using $K_p = K_c(RT)^{\Delta n}$.
a) Δn = Number of product gas moles – Number of reactant gas moles = 2 – 1 = 1
$$K_p = K_c(RT)^{\Delta n} = (6.1 \times 10^{-3})\,[(0.0821)\,(298)]^1 = 0.14924 = \mathbf{0.15}$$
b) Δn = Number of product gas moles – Number of reactant gas moles = 2 – 4 = – 2
$$K_p = K_c(RT)^{\Delta n} = (2.4 \times 10^{-3})\,[(0.0821)\,(1000.)]^{-2} = 3.5606 \times 10^{-7} = \mathbf{3.6 \times 10^{-7}}$$

17.33   When $Q < K$, the reaction proceeds to the **right** to form more products. The reaction quotient and equilibrium constant are determined by [products] / [reactants]. For Q to increase and reach the value of K, the concentration of products (numerator) must increase in relation to the concentration of reactants (denominator).

17.35   To decide if the reaction is at equilibrium, calculate Q_p and compare it to K_p. If $Q_p = K_p$, then the reaction is at equilibrium. If $Q_p > K_p$, then the reaction proceeds to the left to produce more reactants. If $Q_p < K_p$, then the reaction proceeds to the right to produce more products.

$$Q_p = \frac{P_{H_2} P_{Br_2}}{P_{HBr}^2} = \frac{(0.010)(0.010)}{(0.20)^2} = 2.5 \times 10^{-3} > K_p = 4.18 \times 10^{-9}$$

$Q_p > K_p$, thus, the reaction is **not** at equilibrium and will proceed to the **left** (towards the reactants). Thus, the numerator will decrease in size as products are consumed and the denominator will increase in size as more reactant is produced. Q_p will decrease until $Q_p = K_p$.

17.38   At equilibrium, equal concentrations of $CFCl_3$ and HCl exist, regardless of starting reactant concentrations. The equilibrium concentrations of $CFCl_3$ and HCl would still be equal if unequal concentrations of CCl_4 and HF were used. This occurs only when the two products have the same coefficients in the balanced equation. Otherwise, more of the product with the larger coefficient will be produced.

17.40   a) The approximation applies when the change in concentration from initial to equilibrium is so small that it is insignificant. This occurs when K is small and initial concentration is large.
b) This approximation will not work when the change in concentration is greater than 5%. This can occur when [reactant]_initial is very small, or when [reactant]_change is relatively large due to a large K.

17.41   Since all <u>equilibrium</u> concentrations are given in molarities and the reaction is balanced, construct an equilibrium expression and substitute the equilibrium concentrations to find K_c.

$$K_c = \frac{[HI]^2}{[H_2][I_2]} = \frac{\left[1.87 \times 10^{-3}\right]^2}{\left[6.50 \times 10^{-5}\right]\left[1.06 \times 10^{-3}\right]} = 50.753 = \mathbf{50.8}$$

17.43   The reaction table requires that the initial $[PCl_5]$ be calculated: $[PCl_5]$ = 0.15 mol/2.0 L = 0.075 M
Since there is a 1:1:1 mole ratio in this reaction:
x = $[PCl_5]$ reacting (–x), and the amount of PCl_3 and of Cl_2 forming (+x).

Concentration (M)	$PCl_5(g)$	⇄	$PCl_3(g)$	+	$Cl_2(g)$
Initial	0.075		0		0
Change	–x		+x		+x
Equilibrium	0.075 – x		x		x

17.45   Two of the three <u>equilibrium</u> pressures are known. Construct an equilibrium expression and solve for P_{NOCl}.

$$K_p = 6.5 \times 10^4 = \frac{P_{NOCl}^2}{P_{NO}^2 P_{Cl_2}}$$

$$6.5 \times 10^4 = \frac{P_{NOCl}^2}{(0.35)^2 (0.10)}$$

$$P_{NOCl} = \sqrt{\left(6.5 \times 10^4\right)(0.35)^2 (0.10)} = 28.2179 = \textbf{28 atm}$$

A high pressure for NOCl is expected because the large value of K_p indicates that the reaction proceeds largely to the right, i.e., to the formation of products.

17.47   The ammonium hydrogen sulfide will decompose to produce hydrogen sulfide and ammonia gas until $K_p = 0.11$:
        $NH_4HS(s) \leftrightarrows H_2S(g) + NH_3(g)$
        $x = [NH_4HS]$ reacting (−x), and the amount of H_2S and of NH_3 forming (+x) since there is a 1:1:1 mole ratio between the reactant and products.
(It is not necessary to calculate the molarity of NH_4HS since, as a solid, it is not included in the equilibrium expression.)

Concentration (M)	$NH_4HS(s)$	$\leftrightarrows H_2S(g)$ +	$NH_3(g)$
Initial	—	0	0
Change	−x	+x	+x
Equilibrium	—	x	x

$$K_p = 0.11 = (P_{H_2S})(P_{NH_3}) \qquad \text{(The solid } NH_4HS \text{ is not included.)}$$
$$0.11 = (x)(x)$$
$$x = P_{NH_3} = = 0.33166 = \textbf{0.33 atm}$$

17.49   The initial concentrations of N_2 and O_2 are (0.20 mol/1.0 L) = 0.20 M and (0.15 mol/1.0 L) = 0.15 M, respectively.
        $N_2(g) + O_2(g) \leftrightarrows 2 NO(g)$ There is a 1:1:2 mole ratio between reactants and products.

Concentration (M)	$N_2(g)$	$O_2(g)$ $\leftrightarrows$	$2 NO(g)$	
Initial	0.20	0.15	0	
Change	−x	−x	+ 2 x	(1:1:2 mole ratio)
Equilibrium	0.20 − x	0.15 − x	2 x	

$$K_c = 4.10 \times 10^{-4} = \frac{[NO]^2}{[N_2][O_2]} = \frac{[2x]^2}{[0.20 - x][0.15 - x]}$$

$$\text{Assume } 0.20\, M - x \approx 0.20\, M \qquad \text{and} \qquad 0.15\, M - x \approx 0.15\, M$$

$$4.10 \times 10^{-4} = \frac{4x^2}{[0.20][0.15]}$$

$x = 1.753568 \times 10^{-3}\, M$ (unrounded)
$[NO] = 2x = 2 (1.753568 \times 10^{-3}\, M) = 3.507136 \times 10^{-3} = \textbf{3.5} \times \textbf{10}^{-3}\, \textbf{M}$
(Since $(1.8 \times 10^{-3}) / (0.15) < 0.05$, the assumption is OK.)

17.51   Initial concentrations:
        $[HI] = (0.0244 \text{ mol}) / (1.50 \text{ L}) = 0.0162667$
        $[H_2] = (0.00623 \text{ mol}) / (1.50 \text{ L}) = 0.0041533$
        $[I_2] = (0.00414 \text{ mol}) / (1.50 \text{ L}) = 0.00276\, M$
$2 HI(g) \leftrightarrows \quad H_2(g) + I_2(g)$ There is a 2:1:1 mole ratio between reactants and products.

Concentration (M)	$2 HI(g)$ $\leftrightarrows$	$H_2(g)$ +	$I_2(g)$	
Initial	0.0162667	0.0041533	0.00276	
Change	−2 x	+x	+x	2:1:1 mole ratio
Equilibrium	0.0162667 − 2 x	0.0041533 + x	0.00276 + x	

$$[H_2]_{eq} = 0.00467 = 0.0041533 + x$$
$$x = 0.0005167 \ M \text{ (unrounded)}$$
$$[I_2]_{eq} = 0.00276 + x = 0.0032767 = \mathbf{0.00328 \ M \ I_2}$$
$$[HI]_{eq} = 0.0162667 - 2x = 0.0152333 = \mathbf{0.0152 \ M \ HI}$$

17.53 Construct a reaction table, using $[ICl]_{init} = (0.500 \text{ mol} / 5.00 \text{ L}) = 0.100 \ M$, and substitute the equilibrium concentrations into the equilibrium expression. There is a 2:1:1 mole ratio between reactant and products:

Concentration (M)	2 ICl(g)	$\rightleftarrows$	I$_2$(g)	+	Cl$_2$(g)	
Initial	0.100		0		0	
Change	$-2x$		$+x$		$+x$	(2:1:1 mole ratio)
Equilibrium	$0.100 - 2x$		x		x	

$$K_c = 0.110 = \frac{[I_2][Cl_2]}{[ICl]^2} = \frac{[x][x]}{[0.100 - 2x]^2}$$

$$0.110 = \frac{[x]^2}{[0.100 - 2x]^2}$$

Take the square root of each side:

$$0.331662 = \frac{[x]}{[0.100 - 2x]}$$

$$x = 0.0331662 - 0.663324 \, x$$
$$1.663324 \, x = 0.0331662$$
$$x = 0.0199397$$
$$[I_2]_{eq} = [Cl_2]_{eq} = x = 0.0199397 = \mathbf{0.0200 \ M}$$
$$[ICl]_{eq} = 0.100 - 2x = 0.100 - 2 \, (0.0199397) = 0.0601206 = \mathbf{0.060 \ M \ ICl}$$

17.55 $4 \text{ NH}_3(g) + 3 \text{ O}_2(g) \rightleftarrows 2 \text{ N}_2(g) + 6 \text{ H}_2\text{O}(g)$
To find the equilibrium constant, determine the equilibrium concentrations of each reactant and product and insert into the equilibrium expression. Since the volume is 1.00 L, the concentrations are equal to the number of moles present.

Concentration (M)	4 NH$_3$(g)	+	3 O$_2$(g)	$\rightleftarrows$	2 N$_2$(g)	+	6 H$_2$O(g)
Initial	0.0150		0.0150		0		0
Change	$-4x$		$-3x$		$+2x$		$+6x$
Equilibrium	$0.0150 - 4x$		$0.0150 - 3x$		$+2x$		$+6x$

All intermediate concentration values are unrounded.
$$[N_2]_{eq} = 2x = 1.96 \times 10^{-3} \ M$$
$$x = (1.96 \times 10^{-3} \ M)/2 = 9.80 \times 10^{-4} \ M$$
$$[H_2O]_{eq} = 6x = 6 \, (9.80 \times 10^{-4}) = 5.8800 \times 10^{-3} \ M$$
$$[NH_3]_{eq} = 0.0150 - 4x = 0.0150 - 4(9.80 \times 10^{-4}) = 1.1080 \times 10^{-2} \ M$$
$$[O_2]_{eq} = 0.0150 - 3x = 0.0150 - 3(9.80 \times 10^{-4}) = 1.2060 \times 10^{-2} \ M$$

$$K_c = \frac{[N_2]^2 [H_2O]^6}{[NH_3]^4 [O_2]^3} = \frac{[1.96 \times 10^{-3}]^2 [5.8800 \times 10^{-3}]^6}{[1.1080 \times 10^{-2}]^4 [1.2060 \times 10^{-2}]^3} = 6.005859 \times 10^{-6} = \mathbf{6.01 \times 10^{-6}}$$

If values for concentrations were rounded to calculate K_c, the answer is 5.90×10^{-6}.

17.58 Equilibrium position refers to the specific concentrations or pressures of reactants and products that exist at equilibrium, whereas equilibrium constant refers to the overall ratio of equilibrium concentrations and not to specific concentrations. Changes in reactant concentration cause changes in the specific equilibrium concentrations of reactants and products (equilibrium position), but not in the equilibrium constant.

17.59 A positive ΔH_{rxn} indicates that the reaction is endothermic, and that heat is consumed in the reaction:
$$NH_4Cl(s) + \textbf{heat} \leftrightarrows NH_3(g) + HCl(g)$$
a) The addition of heat (high temperature) causes the reaction to proceed to the right to counterbalance the effect of the added heat. Therefore, more products form at a higher temperature and container (**B**) with the largest number of product molecules best represents the mixture.
b) When heat is removed (low temperature), the reaction shifts to the left to produce heat to offset that disturbance. Therefore, NH_3 and HCl molecules combine to form more reactant and container (**A**) with the smallest number of product gas molecules best represents the mixture.

17.63 a) Equilibrium position shifts **toward products**. Adding a reactant (CO) causes production of more products as the system will act to reduce the increase in reactant by proceeding toward the product side, thereby consuming additional CO.
b) Equilibrium position shifts **toward products**. Removing a product (CO_2) causes production of more products as the system acts to replace the removed product.
c) Equilibrium position **does not shift**. The amount of a solid reactant or product does not impact the equilibrium as long as there is some solid present.
d) Equilibrium position shifts **toward reactants**. When product is added, the system will act to reduce the increase in product by proceeding toward the reactant side, thereby consuming additional CO_2; dry ice is solid carbon dioxide that sublimes to carbon dioxide gas. At very low temperatures, CO_2 solid will not sublime, but since the reaction lists carbon dioxide as a gas, the assumption that sublimation takes place is reasonable.

17.65 An increase in container volume results in a decrease in pressure (Boyle's Law). Le Châtelier's principle states that the equilibrium will shift in the direction that forms more moles of gas to offset the decrease in pressure.
a) **More F** forms (2 moles of gas) and **less F_2** (1 mole of gas) is present as the reaction shifts towards the right.
b) **More C_2H_2 and H_2** form (4 moles of gas) and **less CH_4** (2 moles of gas) is present as the reaction shifts towards the right.

17.67 Decreasing container volume increases the pressure (Boyle's Law). Le Châtelier's principle states that the equilibrium will shift in the direction that forms fewer moles of gas to offset the increase in pressure.
a) There are 2 moles of reactant gas (H_2 and Cl_2) and 2 moles of product gas (HCl). Since there is the same number of reactant and product gas moles, there is **no effect** on the amounts of reactants or products in this reaction.
b) There are 3 moles of reactant gases (H_2 and O_2) and 0 moles of product gas. The reaction will shift to the right to produce fewer moles of gas to offset the increase in pressure. **H_2 and O_2 will decrease** from their initial values before the volume was changed. **More H_2O** will form because of the shift in equilibrium position.

17.69 The purpose of adjusting the volume is to cause a shift in equilibrium to the right for increased product yield.
a) Because the number of reactant gaseous moles ($4H_2$) equals the product gaseous moles ($4H_2O$), a change in volume will have **no effect** on the yield.
b) The moles of gaseous product (2 CO) exceed the moles of gaseous reactant (1 O_2). A decrease in pressure favors the reaction direction that forms more moles of gas, so **increase** the reaction vessel volume.

17.71 An increase in temperature (addition of heat) causes a shift in the equilibrium <u>away</u> from the side of the reaction with heat.
a) $CO(g) + 2H_2(g) \leftrightarrows CH_3OH(g) + heat \qquad \Delta H_{rxn} = -90.7$ kJ
A negative ΔH_{rxn} indicates an exothermic reaction. The equilibrium shifts to the left, away from heat, towards the reactants, so amount of product **decreases**.
b) $C(s) + H_2O(g) + heat \leftrightarrows CO(g) + H_2(g) \qquad \Delta H_{rxn} = 131$ kJ
A positive ΔH_{rxn} indicates an endothermic reaction. The equilibrium shifts to the right, away from heat, towards the products, so amounts of products **increase**.
c) $2NO_2(g) + heat \leftrightarrows 2NO(g) + O_2(g)$
The reaction is endothermic. The equilibrium shifts to the right, away from heat, towards the product, so amounts of products **increase**.

d) $2C(s) + O_2(g) \rightleftharpoons 2CO(g) + $ heat

The reaction is exothermic. The equilibrium shifts to the left, away from heat, towards the reactants; amount of product **decreases**.

17.73   The van't Hoff equation shows how the equilibrium constant is affected by a change in temperature. Substitute the given variables into the equation and solve for K_2.

$$K_{298} = K_1 = 1.80 \qquad T_1 = 298 \text{ K}$$
$$K_{500} = K_2 = ? \qquad T_2 = 500. \text{ K} \qquad R = 8.314 \text{ J/mol} \cdot \text{K}$$

$$\Delta H^\circ_{rxn} = \left(\frac{0.32 \text{ kJ}}{1 \text{ mol DH}}\right)(2 \text{ mol DH})\left(\frac{10^3 \text{ J}}{1 \text{ kJ}}\right) = 6.4 \times 10^2 \text{ J}$$

$$\ln\frac{K_2}{K_1} = -\frac{\Delta H^\circ_{rxn}}{R}\left(\frac{1}{T_2} - \frac{1}{T_1}\right)$$

$$\ln\frac{K_2}{1.80} = -\frac{6.4 \times 10^2 \text{ J}}{8.314 \text{ J/mol} \cdot \text{K}}\left(\frac{1}{500. \text{ K}} - \frac{1}{298 \text{ K}}\right)$$

$$\ln\frac{K_2}{1.80} = 0.104360$$

$$K_2 / 1.80 = 1.11000 \qquad\qquad K_2 = (1.80)(1.11000) = 1.99800 = \mathbf{2.0}$$

17.76   a) $SO_2(g) + \frac{1}{2} O_2(g) \rightleftharpoons SO_3(g) + $ heat

The forward reaction is exothermic (ΔH_{rxn} is negative), so it is favored by **lower temperatures**. Lower temperatures will cause a shift to the right, the heat side of the reaction. There are fewer moles of gas as products (1 SO_3) than as reactants (1 $SO_2(g) + \frac{1}{2} O_2$), so products are favored by **higher pressure**. High pressure will cause a shift in equilibrium to the side with the fewer moles of gas.

b) Addition of O_2 would **decrease** Q ($Q = \dfrac{[SO_3]}{[SO_2][O_2]^{1/2}}$) and have **no impact on K**.

c) To enhance yield of SO_3, a low temperature is used. Reaction rates are slower at lower temperatures, so a catalyst is used **to speed up the reaction**.

17.79   a) Initially, a total of 6.00 volumes of gas (5.00 N_2 + 1.00 O_2) exist at 5.00 atm and 1850. K. The volumes are proportional to moles of gas, so volume fraction equals mole fraction. The initial partial pressures of the two gases are:

$$P_{N_2} + P_{H_2} = 5.00 \text{ atm}$$

$$P_{N_2} = X_{N_2} P_{tot} = (5.00 / 6.00)(5.00 \text{ atm}) = 4.16667 \text{ atm (unrounded)}$$

$$P_{O_2} = X_{O_2} P_{tot} = (1.00 / 6.00)(5.00 \text{ atm}) = 0.83333 \text{ atm (unrounded)}$$

The reaction table describing equilibrium is written as follows:

P (atm)	$N_2(g)$	+	$O_2(g)$	$\rightleftharpoons$	$2 NO(g)$	
Initial	4.16667		0.83333		0	
Change	$-x$		$-x$		$+2x$	1:1:2 mole ratio
Equilibrium	$(4.16667 - x)$		$(0.83333 - x)$		$2x$	

Since K_p is small, assume that $[N_2]_{eq} = 4.16667 - x = 4.16667$ and $[H_2]_{eq} = 0.83333 - x = 0.83333$.

$$K_p = \frac{P^2_{NO}}{P_{N_2} P_{O_2}} = \frac{(2x)^2}{(4.16667)(0.83333)} = 1.47 \times 10^{-4}$$

$x = 1.12962 \times 10^{-2}$ (unrounded)

Check assumption: percent error $= (1.12962 \times 10^{-2} / 4.16667)\,100\% = 0.27\%$. The assumption is good.

Therefore, $[NO]_{eq} = 2x = 2(1.12962 \times 10^{-2}) = 2.259 \times 10^{-2} = \mathbf{2.26 \times 10^{-2} \text{ atm}}$

b) Convert atm (P) to concentration (mol/L) using the ideal gas law, assuming that it is applicable at 1850. K.

$n / V = P / RT$

$$\frac{\left(2.259 \times 10^{-2} \text{ atm}\right)}{\left(0.0821 \dfrac{\text{L} \cdot \text{atm}}{\text{mol} \cdot \text{K}}\right)(1850 \text{ K})} \left(\frac{30.01 \text{ g NO}}{1 \text{ mol NO}}\right)\left(\frac{1 \text{ μg}}{10^{-6} \text{ g}}\right) = 4.46342 \times 10^3 = \mathbf{4.46 \times 10^3 \text{ μg/L}}$$

17.81   Set up reaction table:

Concentration (M)	$NH_2COONH_4(s)$ ⇆	$2NH_3$	+	CO_2
Initial	7.80 g	0		0
Change	—	+ 2 x		+ x
Equilibrium	—	2 x		x

The solid is irrelevant (as long as some is present) and is not included in the K_c expression.

$K_c = [NH_3]^2[CO_2]$

$K_c = 1.58 \times 10^{-8} = (2x)^2(x)$

$x = 1.580759 \times 10^{-3} M$ (unrounded)

Total concentration of gases $= 2(1.580759 \times 10^{-3} M) + 1.580759 \times 10^{-3} M = 4.742277 \times 10^{-3} M$ (unrounded)

To find total pressure use the ideal gas equation:

$P = nRT / V = (n/V) RT = MRT$

$P = (4.742277 \times 10^{-3} M) (0.0821 \text{ L} \cdot \text{atm/mol} \cdot \text{K})[(273 + 250.)\text{K}] = 0.203625 = \mathbf{0.204 \text{ atm}}$

17.84   a) You are given a value of K_c but the amounts of reactant and product is given in units of pressure. Convert K_c to K_p.

$K_p = K_c(RT)^{\Delta n}$ $\Delta n = 1 - 2 = -1$ (1 mol of product, C_2H_5OH and 2 mol of reactants, $C_2H_4 + H_2O$)

$K_p = K_c(RT)^{-1} = (9 \times 10^3)[(0.0821 \text{ L} \cdot \text{atm/mol} \cdot \text{K}) (600. \text{ K})]^{-1} = 1.8270 \times 10^2$ (unrounded)

Substitute the given values into the equilibrium expression and solve for $P_{C_2H_4}$.

$$K_p = \frac{P_{C_2H_5OH}}{P_{C_2H_4}P_{H_2O}} = \frac{200.}{P_{C_2H_4}(400.)} = 1.8270 \times 10^2$$

$P_{C_2H_4} = 2.7367 \times 10^{-3} = \mathbf{3 \times 10^{-3} \text{ atm}}$

b) The forward direction, towards the production of ethanol, produces the smaller number of moles of gas and is favored by **high pressure**. A **low temperature** favors an exothermic reaction.

c) Use van't Hoff's equation to find K_c at a different temperature.

$K_1 = 9 \times 10^3$ $T_1 = 600. \text{ K}$ $\Delta H^{\circ}_{rxn} = (-47.8 \text{ kJ/mol}) (10^3 \text{ J/1 kJ}) = -4.78 \times 10^4 \text{ J/mol}$

$K_2 = ?$ $T_2 = 450. \text{ K}$ $R = 8.314 \text{ J/mol} \cdot \text{K}$

$$\ln \frac{K_2}{K_1} = -\frac{\Delta H^{\circ}_{rxn}}{R}\left(\frac{1}{T_2} - \frac{1}{T_1}\right)$$

$$\ln \frac{K_2}{9 \times 10^3} = -\frac{-4.78 \times 10^4 \text{ J}}{8.314 \text{ J/mol} \cdot \text{K}}\left(\frac{1}{450. \text{ K}} - \frac{1}{600. \text{ K}}\right)$$

$$\ln \frac{K_2}{9 \times 10^3} = 3.1940769$$

$K_2 / 9 \times 10^3 = 24.38765$

$K_2 = (9 \times 10^3) (24.38765) = 2.1949 \times 10^5 = \mathbf{2 \times 10^5}$

d) **No**, condensing the C_2H_5OH would not increase the yield. Ethanol has a lower boiling point (78.5°C) than water (100°C). Decreasing the temperature to condense the ethanol would also condense the water, so moles of gas from each side of the reaction are removed. The direction of equilibrium (yield) is unaffected when there is no net change in the number of moles of gas.

17.89 a) You are given a value of K_c but the amounts of reactant and product is given in units of pressure. Convert K_c to K_p.

$$K_p = K_c(RT)^{\Delta n} \qquad \Delta n = 2 - 3 = -1 \quad (2 \text{ mol of product, } SO_3 \text{ and 3 mol of reactants, } 2\ SO_2 + O_2)$$

$$K_p = K_c(RT)^{\Delta n} = K_c(RT)^{-1} = (1.7 \times 10^8)[(0.0821\ L\cdot atm\ /\ mol\cdot K)\ (600.\ K)]^{-1} = 3.451 \times 10^6 \text{ unrounded)}$$

$$K_p = \frac{P_{SO_3}^2}{P_{SO_2}^2\ P_{O_2}} = \frac{(300.)^2}{P_{SO_2}^2\ (100.)} = 3.451 \times 10^6$$

$$P_{SO_2} = 0.016149 = \textbf{0.016 atm}$$

b) Create a reaction table that describes the reaction conditions. Since the volume is 1.0 L, the moles equals the molarity. Note the 2:1:2 mole ratio between SO_2:O_2:SO_3

Concentration (M)	2 $SO_2(g)$	+	$O_2(g)$	$\leftrightarrows$	2 $SO_3(g)$	
Initial	0.0040		0.0028		0	
Change	$-2\ x$		$-x$		$+2\ x$	(2:1:2 mole ratio)
Equilibrium	$0.0040 - 2\ x$		$0.0028 - x$		$2\ x = 0.0020$	(given)

x = 0.0010, therefore:

$[SO_2] = 0.0040 - 2\ x = 0.0040 - 2(0.0010) = 0.0020\ M$
$[O_2] = 0.0028 - x = 0.0028 - 0.0010 = 0.0018\ M$
$[SO_3] = 2(0.0010) = 0.0020\ M$

Substitute equilibrium concentrations into the equilibrium expression and solve for K_c.

$$K_c = \frac{[SO_3]^2}{[SO_2]^2[O_2]} = \frac{[0.0020]^2}{[0.0020]^2[0.0018]} = 555.5556 = \textbf{5.6} \times \textbf{10}^2$$

The pressure of SO_2 is estimated using the concentration of SO_2 and the ideal gas law (although the ideal gas law is not well behaved at high pressures and temperatures).

$$P_{SO_2} = \frac{nRT}{V} = \frac{(0.0020\ mol)\left(0.0821\dfrac{L\cdot atm}{mol\cdot K}\right)(1000.\ K)}{(1.0\ L)} = 0.1642 = \textbf{0.16 atm}$$

17.91 The equilibrium constant for the reaction is $K_p = P_{CO_2} = 0.220$ atm

The amount of calcium carbonate solid in the container at the first equilibrium equals the original amount, 0.100 mol, minus the amount reacted to form 0.220 atm of carbon dioxide. The moles of $CaCO_3$ reacted is equal to the number of moles of carbon dioxide produced. Use the pressure of CO_2 and the Ideal Gas Equation to calculate the moles of $CaCO_3$ reacted:

Moles $CaCO_3$ = moles CO_2 = n = PV / RT

$$n = \frac{(0.220\ atm)(10.0\ L)}{\left(0.0821\dfrac{L\cdot atm}{mol\cdot K}\right)(385\ K)} = 0.06960\ mol\ CaCO_3 \text{ lost (unrounded)}$$

0.100 mol $CaCO_3$ − 0.06960 mol $CaCO_3$ = 0.0304 mol $CaCO_3$ at first equilibrium

As more carbon dioxide gas is added, the system returns to equilibrium by shifting to the left to convert the added carbon dioxide to calcium carbonate to maintain the partial pressure of carbon dioxide at 0.220 atm (K_p). Convert the added 0.300 atm of CO_2 to moles using the Ideal Gas Equation. The moles of CO_2 reacted equals the moles of $CaCO_3$ formed.

Moles $CaCO_3$ = moles CO_2 = n = PV / RT

$$n = \frac{(0.300\ atm)(10.0\ L)}{\left(0.0821\dfrac{L\cdot atm}{mol\cdot K}\right)(385\ K)} = 0.09491\ mol\ CaCO_3 \text{ formed (unrounded)}$$

Add the moles of $CaCO_3$ formed in the second equilibrium to the moles of $CaCO_3$ at the first equilibrium position and convert to grams.

$$\text{Mass } CaCO_3 = \left(0.0304\ mol + 0.09491\ mol\ CaCO_3\right)\left(\frac{100.09\ g\ CaCO_3}{1\ mol\ CaCO_3}\right) = 12.542 = \textbf{12.5 g } CaCO_3$$

17.95 $S_2F_{10}(g) \rightleftharpoons SF_4(g) + SF_6(g)$

The reaction is described by the following equilibrium expression:

$$K_c = \frac{[SF_4][SF_6]}{[S_2F_{10}]}$$

At the first equilibrium, $[S_2F_{10}] = 0.50\ M$ and $[SF_4] = [SF_6] = x$. Therefore, the concentrations of SF_4 and SF_6 can be expressed mathematically as

$$[SF_4] = [SF_6] = \sqrt{0.50K_c}$$

At the second equilibrium, $[S_2F_{10}] = 2.5\ M$ and $[SF_4] = [SF_6] = x$. Therefore, the concentrations of SF_4 and SF_6 can be expressed mathematically as

$$[SF_4] = [SF_6] = \sqrt{2.5K_c}$$

Thus, the concentrations of SF_4 and SF_6 increase by a factor of

$$\frac{\sqrt{2.5K_c}}{\sqrt{0.50K_c}} = \frac{\sqrt{2.5}}{\sqrt{0.50}} = 2.236 = \textbf{2.2}$$

17.97 a) Calculate the partial pressures of oxygen and carbon dioxide because volumes are proportional to moles of gas, so volume fraction equals mole fraction. Assume that the amount of carbon monoxide gas is small relative to the other gases, so the total volume of gases equals $V_{CO_2} + V_{O_2} + V_{N_2} = 10.0 + 1.00 + 50.0 = 61.0$.

$$P_{CO_2} = \left(\frac{10.0\ \text{mol}\ CO_2}{61.0\ \text{mol gas}}\right)(4.0\ \text{atm}) = 0.6557377\ \text{atm (unrounded)}$$

$$P_{O_2} = \left(\frac{1.00\ \text{mol}\ O_2}{61.0\ \text{mol gas}}\right)(4.0\ \text{atm}) = 0.06557377\ \text{atm (unrounded)}$$

a) Use the partial pressures and given K_p to find P_{CO}.

$$2\ CO_2(g) \rightleftharpoons 2\ CO(g) + O_2(g)$$

$$K_p = \frac{P_{CO}^2 P_{O_2}}{P_{CO_2}^2} = \frac{P_{CO}^2\,(0.06557377)}{(0.6557377)^2} = 1.4 \times 10^{-28}$$

$$P_{CO} = 3.0299 \times 10^{-14} = \textbf{3.0} \times \textbf{10}^{-14}\ \textbf{atm}$$

b) Convert partial pressure to moles per liter using the ideal gas equation, and then convert moles of CO to grams.

$$n_{CO}/V = P/RT = \frac{\left(3.0299 \times 10^{-14}\ \text{atm}\right)}{\left(0.0821\dfrac{\text{L} \cdot \text{atm}}{\text{mol} \cdot \text{K}}\right)(800\ \text{K})} = 4.6131 \times 10^{-16}\ \text{mol/L (unrounded)}$$

$$\left(\frac{4.6131 \times 10^{-16}\ \text{mol CO}}{\text{L}}\right)\left(\frac{28.01\ \text{g CO}}{1\ \text{mol CO}}\right)\left(\frac{1\ \text{pg}}{10^{-12}\ \text{g}}\right) = 0.01292 = \textbf{0.013 pg CO/L}$$

17.100 a) Write a reaction table given that P_{CH_4} (init) $= P_{CO_2}$ (init) $= 10.0$ atm, substitute equilibrium values into the equilibrium expression, and solve for P_{H_2}.

Pressure (atm)	$CH_4(g)\ +$	$CO_2(g) \rightleftharpoons$	$2\ CO(g)\ +$	$2\ H_2(g)$
Initial	10.0	10.0	0	0
Change	$-x$	$-x$	$+2x$	$+2x$
Equilibrium	$10.0 - x$	$10.0 - x$	$2x$	$2x$

$$K_p = \frac{P_{CO}^2 P_{H_2}^2}{P_{CH_4} P_{CO_2}} = \frac{(2x)^2 (2x)^2}{(10.0 - x)(10.0 - x)} = \frac{(2x)^4}{(10.0 - x)^2} = 3.548 \times 10^6 \quad \text{(take square root of each side)}$$

$$\frac{(2x)^2}{(10.0 - x)} = 1.8836135 \times 10^3$$

A quadratic is necessary:

$4 x^2 + (1.8836135 \times 10^3 \, x) - 1.8836135 \times 10^4 = 0$ (unrounded)

$a = 4 \quad b = 1.8836135 \times 10^3 \quad c = -1.8836135 \times 10^4$

$$x = \frac{-b \pm \sqrt{b^2 - 4ac}}{2a}$$

$$x = \frac{-1.8836135 \times 10^3 \pm \sqrt{\left(1.8836135 \times 10^3\right)^2 - 4(4)\left(-1.8836135 \times 10^4\right)}}{2(4)}$$

$x = 9.796209$ (unrounded)

P (hydrogen) $= 2x = 2\,(9.796209) = 19.592418$ atm (unrounded)

If the reaction proceeded entirely to completion, the partial pressure of H_2 would be 20.0 atm (pressure is proportional to moles, and twice as many moles of H_2 form for each mole of CH_4 or CO_2 that reacts).

The percent yield is $\dfrac{19.592418 \text{ atm}}{20.0 \text{ atm}}(100\%) = 97.96209 = \textbf{98.0\%}$.

b) Repeat the calculations for part (a) with the new K_p value. The reaction table is the same.

$$K_p = \frac{P_{CO}^2 P_{H_2}^2}{P_{CH_4} P_{CO_2}} = \frac{(2x)^2 (2x)^2}{(10.0 - x)(10.0 - x)} = \frac{(2x)^4}{(10.0 - x)^2} = 2.626 \times 10^7$$

$$\frac{(2x)^2}{(10.0 - x)} = 5.124451 \times 10^3$$

A quadratic is needed:

$4 x^2 + (5.124451 \times 10^3 - 5.124451 \times 10^4 = 0$ (unrounded)

$a = 4 \qquad b = 5.124451 \times 10^3 \quad c = -5.124451 \times 10^4$

$$x = \frac{-5.124451 \times 10^3 \pm \sqrt{\left(5.124451 \times 10^3\right)^2 - 4(4)\left(-5.124451 \times 10^4\right)}}{2(4)}$$

$x = 9.923138$ (unrounded)

P (hydrogen) $= 2x = 2\,(9.923138) = 19.846276$ atm (unrounded)

If the reaction proceeded entirely to completion, the partial pressure of H_2 would be 20.0 atm (pressure is proportional to moles, and twice as many moles of H_2 form for each mole of CH_4 or CO_2 that reacts).

The percent yield is $\dfrac{19.846276 \text{ atm}}{20.0 \text{ atm}}(100\%) = 99.23138 = \textbf{99.0\%}$.

c) van't Hoff equation:

$$\ln \frac{K_{1200}}{K_{1300}} = -\frac{\Delta H}{R}\left(\frac{1}{T_2} - \frac{1}{T_1}\right)$$

$$\ln \frac{3.548 \times 10^6}{2.626 \times 10^7} = -\frac{\Delta H}{\left(8.314 \dfrac{J}{mol \cdot K}\right)}\left(\frac{1}{1200.\ K} - \frac{1}{1300.\ K}\right)$$

$-2.0016628 = -\Delta H(7.710195 \times 10^{-6})$

$\Delta H = -2.0016628 \, / \, (-7.710195 \times 10^{-6}) = 2.5961247 \times 10^5 = \textbf{2.60} \times \textbf{10}^5 \textbf{ J/mol}$

(The subtraction of the 1/T terms limits the answer to three significant figures.)

17.102 a) Multiply the second equation by 2 to cancel the moles of CO produced in the first reaction. K_p for the second reaction is then $(K_p)^2$.

$$\begin{array}{ll} 2\,CH_4(g) + O_2(g) \leftrightarrows 2\,\cancel{CO}(g) + 4\,H_2(g) & K_p = 9.34 \times 10^{28} \\ \underline{2\,\cancel{CO}(g) + 2\,H_2O(g) \leftrightarrows 2\,CO_2(g) + 2\,H_2(g)} & K_p = (1.374)^2 = 1.888 \\ 2\,CH_4(g) + O_2(g) + 2\,H_2O(g) \leftrightarrows 2\,CO_2(g) + 6\,H_2(g) & \end{array}$$

b) $K_p = (9.34 \times 10^{28})\,(1.888) = 1.76339 \times 10^{29} = \textbf{1.76} \times \textbf{10}^{29}$

c) $\Delta n = 8 - 5 = 3$ (8 moles of product gas – 5 moles of reactant gas)

$$K_c = \frac{K_p}{[RT]^{\Delta n}}$$

$$K_c = \frac{1.76339 \times 10^{29}}{[(0.0821\ atm\square L/mol\square K)(1000)]^3} = 3.18654 \times 10^{23} = \mathbf{3.19 \times 10^{23}}$$

d) The initial total pressure is given as 30. atm. To find the final pressure use relationship between pressure and number of moles of gas: $n_{initial}/P_{initial} = n_{final}/P_{final}$

Total mol of gas initial = 2.0 mol CH_4 + 1.0 mol O_2 + 2.0 mol H_2O = 5.0 mol

Total mol of gas final = 2.0 mol CO_2 + 6.0 mol H_2 = 8.0 mol   (from mole ratios)

$$P_{final} = \frac{30.\ atm\ reactants}{} \left(\frac{8\ mol\ products}{5\ mol\ reactants} \right) = \mathbf{48\ atm}$$

17.104   a) Careful reading of the problem indicates that the given K_p occurs at 1000. K and that the initial pressure of N_2 is 200. atm.   Log $K_p = -43.10$; $K_p = 10^{-43.10} = 7.94328 \times 10^{-44}$

Pressure (atm)	$N_2(g)$	$\leftrightarrows$	$2N(g)$
Initial	200.		0
Change	$-x$		$+2x$
Equilibrium	$200-x$		$2x$

$$K_p = \frac{(P_N)^2}{(P_{N_2})} = 7.94328 \times 10^{-44}$$

$$\frac{(2x)^2}{(200. - x)} = 7.94328 \times 10^{-44} \qquad \text{Assume } 200. - x \cong 200.$$

$$P_N = 2x = \sqrt{(200.)(7.94328 \times 10^{-44})} = 3.985795 \times 10^{-21} = \mathbf{4.0 \times 10^{-21}}$$

b) Log $K_p = -17.30$; $K_p = 10^{-17.30} = 5.01187 \times 10^{-18}$

Pressure (atm)	$H_2(g)$	$\leftrightarrows$	$2H(g)$
Initial	600.		0
Change	$-x$		$+2x$
Equilibrium	$600-x$		$2x$

$$K_p = \frac{(P_H)^2}{(P_{H_2})} = 5.01187 \times 10^{-18}$$

$$\frac{(2x)^2}{(600. - x)} = 5.01187 \times 10^{-18} \qquad \text{Assume } 600. - x \cong 600.$$

$$P_H = 2x = \sqrt{(600.)(5.01187 \times 10^{-18})} = 5.48372 \times 10^{-8} = \mathbf{5.5 \times 10^{-8}}$$

c) Convert pressures to moles using the ideal gas law. PV = nRT. Convert moles to atoms using Avogadro's number.

$$\text{Moles N / V} = P / RT = \frac{\left(3.985795 \times 10^{-21}\ atm\right)}{\left(0.0821 \dfrac{L \cdot atm}{mol \cdot K}\right)(1000.\,K)} \left(\frac{6.022 \times 10^{23}\ atoms}{mol} \right)$$

$$= 29.2356 = \mathbf{29\ N\ atoms/L}$$

$$\text{Moles H / V} = P / RT = \frac{\left(5.48372 \times 10^{-8}\ atm\right)}{\left(0.0821 \dfrac{L \cdot atm}{mol \cdot K}\right)(1000.\,K)} \left(\frac{6.022 \times 10^{23}\ atoms}{mol} \right)$$

$$= 4.022 \times 10^{14} = \mathbf{4.0 \times 10^{14}\ H\ atoms/L}$$

d) The more reasonable step is **$N_2(g) + H(g) \rightarrow NH(g) + N(g)$. With only 29 N atoms in 1.0 L, the first reaction would produce virtually no NH(g) molecules. There are orders of magnitude more N_2 molecules than N atoms, so the second reaction is the more reasonable step.

17.105   a) $3 H_2(g) + N_2(g) \rightleftarrows 2 NH_3(g)$ The mole ratio $H_2:N_2 = 3:1$; at equilibrium, if $N_2 = x$, $H_2 = 3x$; $P_{NH_3} = 50.$ atm

$$K_p = \frac{\left(P_{NH_3}\right)^2}{\left(P_{N_2}\right)\left(P_{H_2}\right)^3} = 1.00 \times 10^{-4}$$

$$K_p = \frac{(50.)^2}{(x)(3x)^3} = 1.00 \times 10^{-4}$$

$x = 31.02016 = $ **31 atm N_2**
$3x = 3 (31.02016) = 93.06048 = $ **93 atm H_2**
$P_{total} = P_{nitrogen} + P_{hydrogen} + P_{ammonia} = (31.02016 \text{ atm}) + (93.06048 \text{ atm}) + (50. \text{ atm})$
$= 174.08064 = $ **174 atm total**

b) The mole ratio $H_2:N_2 = 6:1$; at equilibrium, if $N_2 = x$, $H_2 = 6x$; $P_{NH_3} = 50.$ atm

$$K_p = \frac{(50.)^2}{(x)(6x)^3} = 1.00 \times 10^{-4}$$

$x = 18.44 = $ **18 atm N_2**
$6x = 6 (18.44) = 110.64 = $ **111 atm H_2**
$P_{total} = P_{nitrogen} + P_{hydrogen} + P_{ammonia} = (18.44 \text{ atm}) + (110.64 \text{ atm}) + (50. \text{ atm})$
$= 179.09 = $ **179 atm total**

This is not a valid argument. The total pressure in (b) is greater than in (a) to produce the same amount of NH_3.

17.108   a) Equilibrium partial pressures for the reactants, nitrogen and oxygen, can be assumed to equal their initial partial pressures because the equilibrium constant is so small that very little nitrogen and oxygen will react to form nitrogen monoxide. After calculating the equilibrium partial pressure of nitrogen monoxide, test this assumption by comparing the partial pressure of nitrogen monoxide with that of nitrogen and oxygen.

Pressure (atm)	$N_2(g)$	+	$O_2(g)$	$\rightleftarrows$	$2NO(g)$
Initial	0.780		0.210		0
Change	$-x$		$-x$		$+2x$
Equilibrium	$0.780 - x$		$0.210 - x$		$2x$

$$K_p = \frac{\left(P_{NO}\right)^2}{\left(P_{N_2}\right)\left(P_{O_2}\right)} = 4.35 \times 10^{-31}$$

$$\frac{(2x)^2}{(0.780 - x)(0.210 - x)} = 4.35 \times 10^{-31} \quad \text{Assume x is small because } K \text{ is small.}$$

$$\frac{(2x)^2}{(0.780)(0.210)} = 4.35 \times 10^{-31}$$

$x = 1.33466 \times 10^{-16}$ (unrounded)

Based on the small amount of nitrogen monoxide formed, the assumption that the partial pressures of nitrogen and oxygen change to an insignificant degree holds.

$P_{nitrogen}$ (equilibrium) $= (0.780 - 1.33466 \times 10^{-16})$ atm $= $ **0.780 atm N_2**
P_{oxygen} (equilibrium) $= (0.210 - 1.33466 \times 10^{-16})$ atm $= $ **0.210 atm O_2**
P_{NO} (equilibrium) $= 2 (1.33466 \times 10^{-16})$ atm $= 2.66932 \times 10^{-16} = $ **2.67 $\times 10^{-16}$ atm NO**

b) The total pressure is the sum of the three partial pressures:
$$0.780 \text{ atm} + 0.210 \text{ atm} + 2.67 \times 10^{-16} \text{ atm} = \textbf{0.990 atm}$$

c) $K_p = K_c(RT)^{\Delta n}$ $\Delta n = 2$ mol NO product – 2 mol reactant (1 N_2 + 1 O_2) = 0

$K_p = K_c(RT)^0$

$K_c = K_p =$ **4.35 x 10^{-31}** because there is no net increase or decrease in the number of moles of gas in the course of the reaction.

17.110   a) Use the equation $K_p = K_c(RT)^{\Delta n}$ $\Delta n = 2$ mol product (1 H_2 + 1 I_2) – 2 mol reactant (HI) = 0

$K_p = K_c(RT)^0 = 1.26$ x $10^{-3}(RT)^0$ = **1.26 x 10^{-3}**

b) The equilibrium constant for the reverse reaction is simply the inverse of the equilibrium constant for the forward reaction:

$K_c' = 1 / K_c = 1 / (1.26$ x $10^{-3}) = 793.65 =$ **794**

c) $\Delta H^\circ_{rxn} = \Sigma[\Delta H^\circ_f(\text{products})] - \Sigma[\Delta H^\circ_f(\text{reactants})]$

$\Delta H^\circ_{rxn} = [\Delta H^\circ_f(H_2(g)) + \Delta H^\circ_f(I_2(g))] - [2\ \Delta H^\circ_f(HI(g))]$

$\Delta H^\circ_{rxn} = [1$ mol (0 kJ/mol) + 1 mol (0 kJ/mol)] – [2 mol (25.9 kJ/mol)]

$\Delta H^\circ_{rxn} =$ **– 51.8 kJ**

d) $\ln\dfrac{K_2}{K_1} = -\dfrac{\Delta H^\circ_{rxn}}{R}\left(\dfrac{1}{T_2} - \dfrac{1}{T_1}\right)$ $K_1 = 1.26$ x 10^{-3}; $K_2 = 2.0$ x 10^{-2}, $T_1 = 298$ K; $T_2 = 729$ K

$\ln\dfrac{2.0 \text{ x } 10^{-2}}{1.26 \text{ x } 10^{-3}} = -\dfrac{\Delta H^\circ_{rxn}}{8.314 \text{ J/mol} \cdot \text{K}}\left(\dfrac{1}{729 \text{ K}} - \dfrac{1}{298 \text{ K}}\right)$

$\Delta H^\circ_{rxn} = 1.1585$ x $10^4 =$ **1.2 x 10^4 J/mol**

17.113   a) The enzyme that inhibits F is the one that catalyzes the reaction that produces F. The enzyme is number **5** that catalyzed the reaction from E to F.

b) Enzyme **8** is inhibited by I.

c) If F inhibited enzyme 1, then neither branch of the reaction would take place once enough F was produced.

d) If F inhibited enzyme 6, then the second branch would not take place when enough F was made.

17.115   The reaction is: $CO(g) + H_2O(g) \rightleftharpoons CO_2(g) + H_2(g)$

a) Set up a table with the initial CO and initial H_2 = 0.100 mol / 20.00 L = 0.00500 M.

	CO	H_2O	CO_2	H_2
Initial	0.00500 M	0.00500 M	0	0
Change	–x	–x	+x	+x
Equilibrium	0.00500 – x	0.00500 – x	x	x

$[CO]_{equilibrium} = 0.00500 - x = 2.24$ x $10^{-3}\ M = [H_2O]$ (given in problem)

x = 0.00276 $M = [CO_2] = [H_2]$

$K_c = \dfrac{[CO_2][H_2]}{[CO][H_2O]} = \dfrac{[0.00276\ ][0.00276\ ]}{[0.00224][0.00224]} = 1.518176 =$ **1.52**

b) $M_{total} = [CO] + [H_2O] + [CO_2] + [H_2] = (0.00224\ M) + (0.00224\ M) + (0.00276\ M) + (0.00276\ M)$

= 0.01000 M

$n_{total} = M_{total}\ V = (0.01000$ mol/L) (20.00 L) = 0.2000 mol total

$P_{total} = n_{total}RT / V = \dfrac{(0.2000\,\text{mol})\left(0.08206\ \dfrac{\text{L} \cdot \text{atm}}{\text{mol} \cdot \text{K}}\right)\big((273 + 900.)\text{K}\big)}{(20.00\ \text{L})} = 0.9625638 =$ **0.9626 atm**

c) Initially, an equal number of moles must be added = **0.2000 mol CO**

d) Set up a table with the initial concentrations equal to the final concentrations from part (a), and then add 0.2000 mol CO / 20.00 L = 0.01000 M to compensate for the added CO.

	CO	H_2O	CO_2	H_2
Initial	0.00224 M	0.00224 M	0.00276 M	0.00276 M
Added CO	0.01000 M			
Change	–x	–x	+x	+x
Equilibrium	0.01224 – x	0.00224 – x	0.00276 + x	0.00276 + x

$$K_c = \frac{[CO_2][H_2]}{[CO][H_2O]} = \frac{[0.00276 + x][0.00276 + x]}{[0.01224 - x][0.00224 - x]} = 0.9625638$$

$$\frac{\left[7.6176 \times 10^{-6} + 5.52 \times 10^{-3}\, x + x^2\right]}{2.74176 \times 10^{-5} - 1.448 \times 10^{-2}\, x + x^2} = 0.9625638$$

$7.6176 \times 10^{-6} + 5.52 \times 10^{-3}\, x + x^2 = (0.9625638)(2.74176 \times 10^{-5} - 1.1448 \times 10^{-2}\, x + x^2)$

$7.6176 \times 10^{-6} + 5.52 \times 10^{-3}\, x + x^2 = 2.6391189 \times 10^{-5} - 1.3937923 \times 10^{-2}\, x + 0.9625638\, x^2$

$0.0374362\, x^2 + 1.9457923 \times 10^{-2}\, x - 1.8773589 \times 10^{-5} = 0$

$a = 0.0374362 \quad b = 1.9457923 \times 10^{-2} \quad c = -1.8773589 \times 10^{-5}$

$$x = \frac{-b \pm \sqrt{b^2 - 4ac}}{2a}$$

$$x = \frac{-(1.9457923 \times 10^{-2}) \pm \sqrt{\left(1.9457923 \times 10^{-2}\right)^2 - 4(0.0374362)\left(-1.8773589 \times 10^{-5}\right)}}{2(0.0374362)}$$

$x = 9.6304567 \times 10^{-4}$ (unrounded)

$[CO] = 0.01224 - x = 0.01224 - (9.6304567 \times 10^{-4}) = 0.011276954 = \mathbf{0.01128\ \mathit{M}}$

CHAPTER 18 ACID-BASE EQUILIBRIA

FOLLOW-UP PROBLEMS

18.1   a) Chloric acid, **HClO₃**, is the stronger acid because acid strength increases as the number of O atoms in the acid increases.
b) Hydrochloric acid, **HCl**, is one of the strong hydrohalic acids whereas acetic acid, CH_3COOH, is a weak carboxylic acid.
c) Sodium hydroxide, **NaOH**, is a strong base because Na is a Group 1A(1) metal. Methylamine, CH_3NH_2, is an organic amine and, therefore, a weak base.

18.2   <u>Plan:</u> The product of $[H_3O^+]$ and $[OH^-]$ remains constant at 25°C because the value of K_w is constant at a given temperature. Use $K_w = [H_3O^+][OH^-] = 1.0 \times 10^{-14}$ to solve for $[H_3O^+]$.
<u>Solution:</u> Calculating $[H_3O^+]$:

$$[H_3O^+] = \frac{K_w}{[OH^-]} = \frac{1.0 \times 10^{-14}}{6.7 \times 10^{-2}} = 1.4925 \times 10^{-13} = \mathbf{1.5 \times 10^{-13}\,M}$$

Since $[OH^-] > [H_3O^+]$, the solution is **basic**.
<u>Check:</u> The solution is neutral when either species concentration is $1 \times 10^{-7}\,M$. Since the problem specifies an $[OH^-]$ much greater than this, the solution is basic.

18.3   <u>Plan:</u> NaOH is a strong base that dissociates completely in water. Subtract pH from 14.00 to find the pOH, and calculate inverse logs of pH and pOH to find $[H_3O^+]$ and $[OH^-]$, respectively.
<u>Solution:</u> pH + pOH = 14.00
    pOH = 14.00 − 9.52 = **4.48**
    pH = −log $[H_3O^+]$
$[H_3O^+] = 10^{-pH} = 10^{-9.52} = 3.01995 \times 10^{-10} = \mathbf{3.0 \times 10^{-10}\,M}$
    pOH = −log $[OH^-]$
    $[OH^-] = 10^{-pOH} = 10^{-4.48} = 3.3113 \times 10^{-5} = \mathbf{3.3 \times 10^{-5}\,M}$
<u>Check:</u> At 25°C, $[H_3O^+][OH^-]$ should equal 1.0×10^{-14}. $(3.0 \times 10^{-10})(3.3 \times 10^{-5}) = 9.9 \times 10^{-15} \approx 1.0 \times 10^{-14}$. The significant figures in the concentrations reflect the number of significant figures after the decimal point in the pH and pOH values.

18.4   <u>Plan:</u> Identify the conjugate pairs by first identifying the species that donates H^+ (the acid) in either reaction direction. The other reactant accepts the H^+ and is the base. The acid has one more H and +1 greater charge than its conjugate base.
<u>Solution:</u>
a) CH_3COOH has one more H^+ than CH_3COO^-. H_3O^+ has one more H^+ than H_2O. Therefore, CH_3COOH and H_3O^+ are the acids, and CH_3COO^- and H_2O are the bases. The conjugate acid/base pairs are **CH_3COOH/CH_3COO^- and H_3O^+/H_2O.**
b) H_2O donates a H^+ and acts as the acid. F^- accepts the H^+ and acts as the base. In the reverse direction, HF acts as the acid and OH^- acts as the base. The conjugate acid/base pairs are **H_2O/OH^- and HF/F⁻.**

18.5   a) The following equation describes the dissolution of ammonia in water:

$NH_3(g)$	+	$H_2O(l)$	$\leftrightarrows$	$NH_4^+(aq)$	+	$OH^-(aq)$
weak base		weak acid	$\leftarrow$	stronger acid		strong base

Ammonia is a known weak base, so it makes sense that it accepts a H^+ from H_2O. The reaction arrow indicates that the equilibrium lies to the left because the question states, "you smell ammonia" (NH_4^+ and OH^- are odorless). NH_4^+ and OH^- are the stronger acid and base, so the reaction proceeds to the formation of the weaker acid and base.

b) The addition of excess HCl results in the following equation:

$$NH_3(g) \quad + \quad H_3O^+(aq; \text{ from HCl}) \quad \leftrightarrows \quad NH_4^+(aq) \quad + \quad H_2O(l)$$

stronger base   strong acid $\rightarrow$ weak acid   weak base

HCl is a strong acid and is much stronger than NH_4^+. Similarly, NH_3 is a stronger base than H_2O. The reaction proceeds to produce the weak acid and base, and thus the odor from NH_3 disappears.

c) The solution in (b) is mostly NH_4^+ and H_2O. The addition of excess NaOH results in the following equation:

$$NH_4^+(aq) \quad + \quad OH^-(aq; \text{ from NaOH}) \quad \leftrightarrows \quad NH_3(g) \quad + \quad H_2O(l)$$

stronger acid   strong base $\rightarrow$ weak base   weak acid

NH_4^+ and OH^- are the stronger acid and base, respectively, and drive the reaction towards the formation of the weaker base and acid, $NH_3(g)$ and H_2O, respectively. The reaction direction explains the return of the ammonia odor.

18.6   There are more HB molecules than there are HA molecules, so the equilibrium lies to the right and $K_c > 1$. The products dominate.

18.7   Plan: Write a balanced equation for the dissociation of NH_4^+ in water. Using the given information, construct a table that describes the initial and equilibrium concentrations. Construct an equilibrium expression and make assumptions where possible to simplify the calculations.
Solution:

	$NH_4^+(aq) +$	$H_2O(l)$	$\leftrightarrows$	$H_3O^+(aq)$	$+$	$NH_3(g)$
Initial	0.2 M	———		0		0
Change	$-x$	———		$+x$		$+x$
Equilibrium	$0.2 - x$	———		x		x

The initial concentration of $NH_4^+ = 0.2\ M$ because each mole of NH_4Cl completely dissociates to form one mole of NH_4^+.

$$x = [H_3O^+] = [NH_3] = 10^{-pH} = 10^{-5.0} = 1.0 \times 10^{-5}\ M$$

$$K_a = \frac{[NH_3][H_3O^+]}{[NH_4^+]} = \frac{x^2}{(0.2 - x)} = \frac{(1.0 \times 10^{-5})(1.0 \times 10^{-5})}{(0.2 - 1.0 \times 10^{-5})}$$

$$= 5 \times 10^{-10}$$

Check: The small value of K_a is consistent with the value expected for a weak acid.

18.8   Plan: Write a balanced equation for the dissociation of HOCN in water. Using the given information, construct a table that describes the initial and equilibrium concentrations. Construct an equilibrium expression and solve the quadratic expression for x, the concentration of H_3O^+.
Solution:

	HOCN(aq) +	$H_2O(l)$	$\leftrightarrows$	$H_3O^+(aq)$	$+$	$OCN^-(aq)$
Initial	0.10 M	———		0		0
Change	$-x$	———		$+x$		$+x$
Equilibrium	$0.10 - x$	———		x		x

$$K_a = \frac{[OCN^-][H_3O^+]}{[HOCN]} = \frac{x^2}{(0.10 - x)} = 3.5 \times 10^{-4}$$

In this example, the dissociation of HOCN is not negligible in comparison to the initial concentration: Therefore, $[HOCN]_{eq} = 0.10 - x$ and the equilibrium expression is solved using the quadratic formula.

$$x^2 = 3.5 \times 10^{-4}(0.10 - x)$$
$$x^2 = 3.5 \times 10^{-5} - 3.5 \times 10^{-4}\ x$$
$$x^2 + 3.5 \times 10^{-4}\ x - 3.5 \times 10^{-5} = 0 \qquad (a\ x^2 + b\ x + c = 0)$$
$$a = 1 \quad b = 3.5 \times 10^{-4} \quad c = -3.5 \times 10^{-5}$$

$$x = \frac{-b \pm \sqrt{b^2 - 4ac}}{2a}$$

$$x = \frac{-3.5 \times 10^{-4} \pm \sqrt{\left(3.5 \times 10^{-4}\right)^2 - 4(1)\left(-3.5 \times 10^{-5}\right)}}{2(1)}$$

$x = 5.7436675 \times 10^{-3} = \mathbf{5.7 \times 10^{-3}\ \textit{M}\ H_3O^+}$

pH $= -\log [H_3O^+] = -\log [5.7436675 \times 10^{-3}] = 2.2408 = \mathbf{2.24}$

18.9   There is no single correct scene. When the solution is diluted, the percent dissociation of the acid HB increases. Any scene in which the total number of HB + B$^-$ is less than in the original solution yet the relative number of HB dissociated is greater would be correct. In the original scene, 2 of 12 HB molecules, or 16.7%, have dissociated. One possible scene after dilution could show 8 HB molecules, 2 B$^-$ and 2 H$_3$O$^+$ ions. Then the dissociation is 2 B$^-$/10 HB = 20% dissociation.

18.10   Plan: Write the balanced equation and corresponding equilibrium expression for each dissociation reaction. Calculate the equilibrium concentrations of all species and convert [H$_3$O$^+$] to pH. Find the equilibrium constant values from Appendix C, $K_{a1} = 5.6 \times 10^{-2}$ and $K_{a2} = 5.4 \times 10^{-5}$.
Solution:

HOOC–COOH(aq) + H$_2$O(l) $\leftrightarrows$ HOOC–COO$^-$ (aq) + H$_3$O$^+$(aq)

$$K_{a1} = \frac{\left[HC_2O_4^-\right]\left[H_3O^+\right]}{\left[H_2C_2O_4\right]} = 5.6 \times 10^{-2}$$

HOOC–COO$^-$ (aq) + H$_2$O(l) $\leftrightarrows$ $^-$OOC–COO$^-$ (aq) + H$_3$O$^+$(aq)

$$K_{a2} = \frac{\left[C_2O_4^{2-}\right]\left[H_3O^+\right]}{\left[HC_2O_4^-\right]} = 5.4 \times 10^{-5}$$

Assumptions:
1) Since $K_{a1} \gg K_{a2}$, the first dissociation produces almost all of the H$_3$O$^+$, so [H$_3$O$^+$]$_{eq}$ = [H$_3$O$^+$] from C$_2$H$_2$O$_4$.
2) Since [HA]$_{init}$/K_{a1} = 0.150 / 5.6x10^{-2} < 400, solve the first equilibrium expression using the quadratic equation.

	HOOC–COOH(aq)	+ H$_2$O(l)	$\leftrightarrows$	H$_3$O$^+$(aq)	+	HOOC–COO$^-$ (aq)
Initial	0.150 M	———		0		0
Change	–x	———		+x		+x
Equilibrium	0.150 – x	———		x		x

$$K_{a1} = \frac{\left[HC_2O_4^-\right]\left[H_3O^+\right]}{\left[H_2C_2O_4\right]} = \frac{x^2}{(0.150 - x)} = 5.6 \times 10^{-2}$$

$$x^2 + 5.6 \times 10^{-2}\,x - 8.4 \times 10^{-3} = 0 \qquad (a\,x^2 + b\,x + c = 0)$$

$$x = \frac{-5.6 \times 10^{-2} \pm \sqrt{\left(5.6 \times 10^{-2}\right)^2 - 4(1)\left(-8.4 \times 10^{-3}\right)}}{2(1)}$$

x = 0.067833 M H$_3$O$^+$ (unrounded)
Therefore, **[H$_3$O$^+$] = [HC$_2$O$_4^-$] = 0.068 M** and **pH** $= -\log (0.067833) = 1.16856 = \mathbf{1.17}$. The oxalic acid concentration at equilibrium is [H$_2$C$_2$O$_4$]$_{init}$ – [H$_2$C$_2$O$_4$]$_{dissoc}$ = 0.150 – 0.067833 = 0.82167 = **0.082 M**.
Solve for [C$_2$O$_4^{2-}$] by rearranging the K_{a2} expression:
[C$_2$O$_4^{2-}$] = K_{a2} [HC$_2$O$_4^-$] / [H$_3$O$^+$] = [(5.4 x 10^{-5}) (0.067833)] / (0.067833) = **5.4 x 10^{-5} M**
Check: The K_{a1} indicates that this weak acid undergoes significant dissociation, so it makes sense that the pH is strongly acidic. The C$_2$O$_4^{2-}$ concentration should be very small in comparison to the HC$_2$O$_4^-$ concentration, because the second dissociation occurs to a much smaller extent.

18.11   Plan: Pyridine contains a nitrogen atom that accepts H^+ from water to form OH^- ions in aqueous solution. Write a balanced equation and equilibrium expression for the reaction, convert pK_b to K_b, make simplifying assumptions (if valid) and solve for $[OH^-]$. Calculate $[H_3O^+]$ using $[H_3O^+][OH^-] = 1.0 \times 10^{-14}$ and convert to pH.
Solution: $C_5H_5N(aq) + H_2O(l) \rightleftharpoons C_5H_5NH^+(aq) + OH^-(aq)$
$K_b = = 10^{-pK_b} = 10^{-8.77} = 1.69824 \times 10^{-9}$

$$K_b = \frac{\left[C_5N_5NH^+\right]\left[OH^-\right]}{\left[C_5N_5N\right]} = 1.69824 \times 10^{-9}$$

Assumptions:
1) Since $K_b \gg K_w$, disregard the $[OH^-]$ that results from the autoionization of water.
2) Since K_b is small, $[C_5H_5N] = 0.10 - x \approx 0.10$.

$$K_b = \frac{\left[C_5N_5NH^+\right]\left[OH^-\right]}{\left[C_5N_5N\right]} = \frac{x^2}{(0.10)} = 1.69824 \times 10^{-9} \text{ (unrounded)}$$

$x = 1.303165 \times 10^{-5} = 1.3 \times 10^{-5} M = [OH^-] = [C_5H_5NH^+]$

Since $\dfrac{[OH^-]}{[C_5H_5N_5]}(100) = \dfrac{1.303265 \times 10^{-5}}{0.10}(100) = 0.01313$ which $< 5\%$, the assumption that the dissociation of $C_5H_5N_5$ is small is valid.

$$[H_3O^+] = \frac{K_w}{[OH^-]} = \frac{1.0 \times 10^{-14}}{1.303165 \times 10^{-5}} = 7.67362 \times 10^{-10} M \text{ (unrounded)}$$

pH $= -\log (7.67362 \times 10^{-10}) = 9.1149995 = \mathbf{9.11}$
Check: Since pyridine is a weak base, a pH > 7 is expected.

18.12   Plan: The hypochlorite ion, ClO^-, acts as a weak base in water. Write a balanced equation and equilibrium expression for this reaction. The K_b of ClO^- is calculated from the K_a of its conjugate acid, hypochlorous acid, HClO (from Appendix C, $K_a = 2.9 \times 10^{-8}$). Make simplifying assumptions (if valid), solve for $[OH^-]$, convert to $[H_3O^+]$ and calculate pH.
Solution:

	$ClO^-(aq)$ +	$H_2O(l)$	$\rightleftharpoons$	$HClO(aq)$ +	$OH^-(aq)$
Initial	0.20 M	———		0	0
Change	$-x$	———		$+x$	$+x$
Equilibrium	$0.20 - x$	———		x	x

$$K_b = \frac{[HClO]\left[OH^-\right]}{\left[ClO^-\right]}$$

$$K_b = \frac{K_w}{K_a} = \frac{1.0 \times 10^{-14}}{2.9 \times 10^{-8}} = 3.448276 \times 10^{-7} \text{ (unrounded)}$$

Assumptions:
1) Since $K_b \gg K_w$, the dissociation of water provides a negligible amount of $[OH^-]$.
2) Since K_b is very small, $[ClO^-]_{eq} = 0.20 - x \approx 0.2$.

$$K_b = \frac{[HClO]\left[OH^-\right]}{\left[ClO^-\right]} = \frac{x^2}{(0.20)} = 3.448276 \times 10^{-7}$$

$x = 2.6261 \times 10^{-4}$
Therefore, $[HClO] = [OH^-] = 2.6 \times 10^{-4} M$.

Since $\dfrac{[OH^-]}{[ClO^-]}(100) = \dfrac{2.6261 \times 10^{-4}}{0.20}(100) = 0.1313$ which $< 5\%$, the assumption that the dissociation of ClO^- is small is valid.

$$[H_3O^+] = \frac{1.0 \times 10^{-14}}{2.6261 \times 10^{-4}} = 3.8079 \times 10^{-11} M \text{ and pH} = -\log (3.8079 \times 10^{-11}) = 10.4193 = \mathbf{10.42}$$

Check: Since hypochlorite ion is a weak base, a pH > 7 is expected.

18.13 a) The ions are K^+ and ClO_2^-; the K^+ is from the strong base KOH, and does not react with water. The ClO_2^- is from the weak acid $HClO_2$, so it reacts with water to produce OH^- ions. Since the base is strong and the acid is weak, the salt derived from this combination will produce a **basic** solution.
K^+ does not react with water

$$ClO_2^-(aq) + H_2O(l) \rightleftharpoons HClO_2(aq) + OH^-(aq)$$

b) The ions are $CH_3NH_3^+$ and NO_3^-; $CH_3NH_3^+$ is derived from the weak base methylamine, CH_3NH_2. Nitrate ion, NO_3^-, is derived from the strong acid HNO_3 (nitric acid). A salt derived from a weak base and strong acid produces an **acidic** solution.
NO_3^- does not react with water.

$$CH_3NH_3^+(aq) + H_2O(l) \rightleftharpoons CH_3NH_2(aq) + H_3O^+(aq)$$

c) The ions are Cs^+ and I^-. Cesium ion is derived from cesium hydroxide, CsOH, which is a strong base because Cs is a Group 1A(1) metal. Iodide ion is derived from hydroiodic acid, HI, a strong hydrohalic acid. Since both the base and acid are strong, the salt derived from this combination will produce a **neutral** solution.
Neither Cs^+ nor I^- react with water.

18.14 a) The two ions that comprise this salt are cupric ion, Cu^{2+}, and acetate ion, CH_3COO^-. Metal ions are acidic in water. Assume that the hydrated cation is $Cu(H_2O)_6^{2+}$. The K_a is found in Appendix C.

$$Cu(H_2O)_6^{2+}(aq) + H_2O(l) \rightleftharpoons Cu(H_2O)_5OH^+(aq) + H_3O^+(aq) \qquad K_a = 3 \times 10^{-8}$$

Acetate ion acts likes a base in water. The K_b is calculated from the K_a of acetic acid (1.8×10^{-5}):

$$K_b = \frac{K_w}{K_a} = \frac{1.0 \times 10^{-14}}{1.8 \times 10^{-5}} = 5.6 \times 10^{-10}.$$

$$CH_3COO^-(aq) + H_2O(l) \rightleftharpoons CH_3COOH(aq) + OH^-(aq) \qquad\qquad K_b = 5.6 \times 10^{-10}$$

$Cu(H_2O)_6^{2+}$ is a better proton donor than CH_3COO^- is a proton acceptor (i.e., $K_a > K_b$), so a solution of $Cu(CH_3COO)_2$ is **acidic**.
b) The two ions that comprise this salt are ammonium ion, NH_4^+, and fluoride ion, F^-. Ammonium ion is the acid of NH_3 with $K_a = \dfrac{K_w}{K_b} = \dfrac{1.0 \times 10^{-14}}{1.76 \times 10^{-5}} = 5.7 \times 10^{-10}.$

$$NH_4^+(aq) + H_2O(l) \rightleftharpoons NH_3(aq) + H_3O^+(aq) \qquad\qquad K_a = 5.7 \times 10^{-10}$$

Fluoride ion is the base with $K_b = \dfrac{K_w}{K_a} = \dfrac{1.0 \times 10^{-14}}{6.8 \times 10^{-4}} = 1.5 \times 10^{-11}.$

$$F^-(aq) + H_2O(l) \rightleftharpoons HF(aq) + OH^-(aq) \qquad\qquad K_b = 1.5 \times 10^{-11}$$

Since $K_a > K_b$, a solution of NH_4F is **acidic**.
c) The ions are K^+ and HSO_3^-; the K^+ is from the strong base KOH, and does not react with water. The HSO_3^- can react as an acid:

$$HSO_3^-(aq) + H_2O(l) \rightleftharpoons SO_3^{2-}(aq) + H_3O^+(aq) \qquad\qquad K_a = 6.5 \times 10^{-8}$$

HSO_3^- can also react as a base. Its K_b value can be found by using the Ka of its conjugate acid, H_2SO_3.

$$HSO_3^-(aq) + H_2O(l) \rightleftharpoons H_2SO_3(aq) + OH^-(aq) \qquad\qquad K_a = \frac{K_w}{K_b} = \frac{1.0 \times 10^{-14}}{1.4 \times 10^{-2}} = 7.1 \times 10^{-13}$$

Since $K_a > K_b$, a solution of HSO_3^- is **acidic**.

18.15 a)

trigonal planar   tetrahedral

Hydroxide ion, OH^-, donates an electron pair and is the Lewis base; $Al(OH)_3$ accepts the electron pair and is the Lewis acid. Note the change in geometry caused by the formation of the adduct.

b)

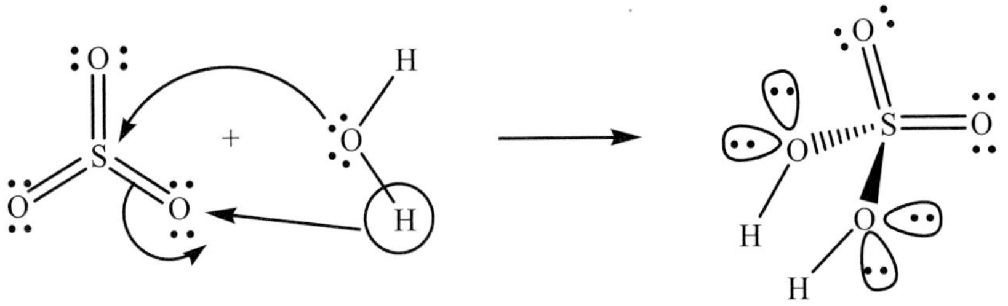

Sulfur trioxide accepts the electron pair and is the Lewis acid. Water donates an electron pair and is the Lewis base.

c)

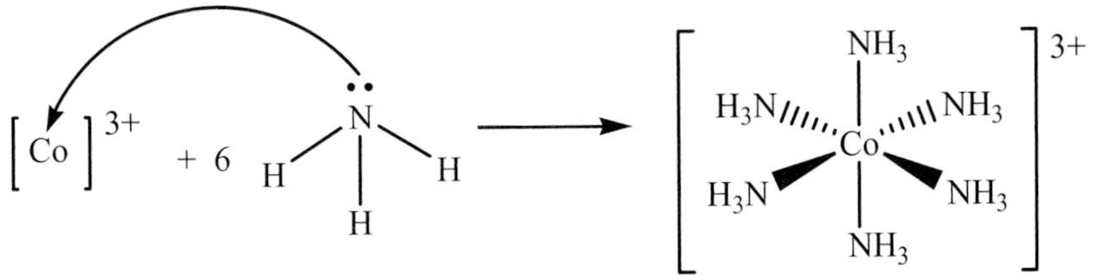

Co^{3+} accepts six electron pairs and is the Lewis acid. Ammonia donates an electron pair and is the Lewis base.

END-OF-CHAPTER PROBLEMS

18.2   All Arrhenius acids contain hydrogen and produce hydronium ion (H_3O^+) in aqueous solution. All Arrhenius bases contain an OH group and produce hydroxide ion (OH^-) in aqueous solution. Neutralization occurs when each H_3O^+ molecule combines with an OH^- molecule to form 2 molecules of H_2O. Chemists found that the ΔH_{rxn} was independent of the combination of strong acid with strong base. In other words, the reaction of any strong base with any strong acid always produced 56 kJ/mol ($\Delta H = -56$ kJ/mol). This was consistent with Arrhenius's hypothesis describing neutralization, because all other counter ions (those present from the dissociation of the strong acid and base) were spectators and did not participate in the overall reaction.

18.4   Strong acids and bases dissociate completely into their ions when dissolved in water. Weak acids only partially dissociate. The characteristic property of all weak acids is that a significant number of the acid molecules are not dissociated. For a strong acid, the concentration of hydronium ions produced by dissolving the acid is equal to the initial concentration of the undissociated acid. For a weak acid, the concentration of hydronium ions produced when the acid dissolves is less than the initial concentration of the acid.

18.5   a) Water, H_2O, is an **Arrhenius acid** because it produces H_3O^+ ion in aqueous solution. Water is also an Arrhenius base because it produces the OH^- ion as well.
b) Calcium hydroxide, $Ca(OH)_2$ is a base, not an acid.
c) Phosphorous acid, H_3PO_3, is a weak **Arrhenius acid**. It is weak because the number of O atoms equals the number of ionizable Hatoms.
d) Hydroiodic acid, HI, is a strong **Arrhenius acid**.

18.7   Barium hydroxide, $Ba(OH)_2$, and potassium hydroxide, KOH, (**b and d**) are Arrhenius bases because they contain hydroxide ions and form OH^- when dissolved in water. H_3AsO_4 and HOC,l a) and c), are Arrhenius acids, not bases.

18.9   a) $HCN(aq) + H_2O(l) \leftrightarrows H_3O^+(aq) + CN^- (aq)$

$$K_a = \frac{\left[CN^-\right]\left[H_3O^+\right]}{\left[HCN\right]}$$

b) $HCO_3^-(aq) + H_2O(l) \leftrightarrows H_3O^+(aq) + CO_3^{2-}(aq)$

$$K_a = \frac{\left[CO_3^{2-}\right]\left[H_3O^+\right]}{\left[HCO_3^-\right]}$$

c) $HCOOH(aq) + H_2O(l) \leftrightarrows H_3O^+(aq) + HCOO^-(aq)$

$$K_a = \frac{\left[HCOO^-\right]\left[H_3O^+\right]}{\left[HCOOH\right]}$$

18.11   a) $HNO_2(aq) + H_2O(l) \leftrightarrows H_3O^+(aq) + NO_2^-(aq)$

$$K_a = \frac{\left[NO_2^-\right]\left[H_3O^+\right]}{\left[HNO_2\right]}$$

b) $CH_3COOH(aq) + H_2O(l) \leftrightarrows H_3O^+(aq) + CH_3COO^-(aq)$

$$K_a = \frac{\left[CH_3COO^-\right]\left[H_3O^+\right]}{\left[CH_3COOH\right]}$$

c) $HBrO_2(aq) + H_2O(l) \leftrightarrows H_3O^+(aq) + BrO_2^-(aq)$

$$K_a = \frac{\left[BrO_2^-\right]\left[H_3O^+\right]}{\left[HBrO_2\right]}$$

18.13   Appendix C lists the K_a values. The larger the K_a value, the stronger the acid. Hydroiodic acid, HI, is not shown in Appendix C because K_a approaches infinity for strong acids and is not meaningful. Therefore, HI is the strongest acid and acetic acid, CH_3COOH, is the weakest: **$CH_3COOH < HF < HIO_3 < HI$**.

18.15   a) Arsenic acid, H_3AsO_4, is a **weak acid**. The number of O atoms is 4, which exceeds the number of ionizable H atoms, 3, by one. This identifies H_3AsO_4 as a weak acid.
b) Strontium hydroxide, $Sr(OH)_2$, is a **strong base**. Soluble compounds containing OH^- ions are strong bases. Sr is a Group 2 metal.
c) HIO is a **weak acid**. The number of O atoms is 1, which is equal to the number of ionizable H atoms identifying HIO as a weak acid.
d) Perchloric acid, $HClO_4$, is a **strong acid**. $HClO_4$ is one example of the type of strong acid in which the number of O atoms exceeds the number of ionizable H atoms by more than 2.

18.17   a) Rubidium hydroxide, RbOH, is a **strong base**, because Rb is a Group 1A(1) metal.
b) Hydrobromic acid, HBr, is a **strong acid**, because it is one of the listed hydrohalic acids.
c) Hydrogen telluride, H_2Te, is a **weak acid**, because H is not bonded to an oxygen or halide.
d) Hypochlorous acid, HClO, is a **weak acid**. The number of O atoms is 1, which is equal to the number of ionizable H atoms identifying HIO as a weak acid.

18.22   The lower the concentration of hydronium (H_3O^+) ions, the higher the pH:
a) At equal concentrations, the acid with the larger K_a will ionize to produce more hydronium ions than the acid with the smaller K_a. The solution of an **acid with the smaller $K_a = 4\times10^{-5}$** has a lower $[H_3O^+]$ and higher pH.
b) pK_a is equal to $-\log K_a$. The smaller the K_a, the larger the pK_a is. So the **acid with the larger pK_a, 3.5,** has a lower $[H_3O^+]$ and higher pH.
c) **Lower concentration** of the same acid means lower concentration of hydronium ions produced. The 0.01 *M* solution has a lower $[H_3O^+]$ and higher pH.
d) At the same concentration, strong acids dissociate to produce more hydronium ions than weak acids. The 0.1 *M* solution of a **weak acid** has a lower $[H_3O^+]$ and higher pH.
e) Bases produce OH^- ions in solution, so the concentration of hydronium ion for a solution of a base solution is lower than that for a solution of an acid. The 0.01 *M* **base solution** has the higher pH.
f) pOH equals $-\log [OH^-]$. At 25°C, the equilibrium constant for water ionization, K_w, equals 1×10^{-14} so $14 = pH + pOH$. As pOH decreases, pH increases. The solution of **pOH = 6.0** has the higher pH.

18.23   a) This problem can be approached two ways. Because NaOH is a strong base, the $[OH^-]_{eq} = [NaOH]_{init}$.
One method involves calculating $[H_3O^+]$ using from $K_w = [H_3O^+][OH^-]$, then calculating pH from the relationship
$pH = -\log [H_3O^+]$. The other method involves calculating pOH and then using $pH + pOH = 14.00$ to calculate
pH.
  First method:

$$[H_3O^+] = \frac{K_w}{[OH^-]} = \frac{1.0 \times 10^{-14}}{0.0111} = 9.0090 \times 10^{-13} \, M \text{ (unrounded)}$$

$$pH = -\log [H_3O^+] = -\log (9.0090 \times 10^{-13}) = 12.04532 = \textbf{12.05}$$

Second method:

$$pOH = -\log [OH^-] = -\log (0.0111) = 1.954677 \text{ (unrounded)}$$
$$pH = 14.00 - pOH = 14.00 - 1.954677 = 12.04532 = \textbf{12.05}$$

With a pH > 7, the solution is **basic**.
   b) There are again two acceptable methods analogous to those in part a; only one will be used here.
     For a strong acid:

$$[H_3O^+] = [HCl] = 1.35 \times 10^{-3} \, M$$
$$pH = -\log (1.35 \times 10^{-3}) = 2.869666 \text{ (unrounded)}$$
$$pOH = 14.00 - 2.869666 = 11.1303334 = \textbf{11.13}.$$

With a pH < 7, the solution is **acidic**.

18.25   a) HI is a strong acid, so $[H_3O^+] = [HI] = 6.14 \times 10^{-3} \, M$.

$$pH = -\log (6.14 \times 10^{-3}) = 2.211832 = \textbf{2.212}. \text{ Solution is \textbf{acidic}}.$$

   b) $Ba(OH)_2$ is a strong base, so $[OH^-] = 2 \times [Ba(OH)_2] = 2 \, (2.55 \, M) = 5.10 \, M$

$$pOH = -\log (5.10) = -0.70757 = \textbf{-0.708}. \text{ Solution is \textbf{basic}}.$$

18.27   a) $[H_3O^+] = 10^{-pH} = 10^{-9.85} = 1.4125375 \times 10^{-10} = \textbf{1.4} \times \textbf{10}^{\textbf{-10}} \, \textbf{M H}_3\textbf{O}^+$

$$pOH = 14.00 - pH = 14.00 - 9.85 = \textbf{4.15}$$
$$[OH^-] = 10^{-pOH} = 10^{-4.15} = 7.0794578 \times 10^{-5} = \textbf{7.1} \times \textbf{10}^{\textbf{-5}} \, \textbf{M OH}^-$$

   b) $pH = 14.00 - pOH = 14.00 - 9.43 = \textbf{4.57}$

$$[H_3O^+] = 10^{-pH} = 10^{-4.57} = 2.691535 \times 10^{-5} = \textbf{2.7} \times \textbf{10}^{\textbf{-5}} \, \textbf{M H}_3\textbf{O}^+$$
$$[OH^-] = 10^{-pOH} = 10^{-9.43} = 3.7153523 \times 10^{-10} = \textbf{3.7} \times \textbf{10}^{\textbf{-10}} \, \textbf{M OH}^-$$

18.29   a) $[H_3O^+] = 10^{-pH} = 10^{-4.77} = 1.69824 \times 10^{-5} = \textbf{1.7} \times \textbf{10}^{\textbf{-5}} \, \textbf{M H}_3\textbf{O}^+$

$$pOH = 14.00 - pH = 14.00 - 4.77 = \textbf{9.23}$$
$$[OH^-] = 10^{-pOH} = 10^{-9..23} = 5.8884 \times 10^{-10} = \textbf{5.9} \times \textbf{10}^{\textbf{-10}} \, \textbf{M OH}^-$$

   b) $pH = 14.00 - pOH = 14.00 - 5.65 = \textbf{8.35}$

$$[H_3O^+] = 10^{-pH} = 10^{-8.35} = 4.46684 \times 10^{-9} = \textbf{4.5} \times \textbf{10}^{\textbf{-9}} \, \textbf{M H}_3\textbf{O}^+$$
$$[OH^-] = 10^{-pOH} = 10^{-5.65} = 2.23872 \times 10^{-6} = \textbf{2.2} \times \textbf{10}^{\textbf{-6}} \, \textbf{M OH}^-$$

18.31   The pH is increasing, so the solution is becoming more basic. Therefore, OH^- ion is added to increase the pH.
Since 1 mole of H_3O^+ will react with 1 mole of OH^-, the difference in $[H_3O^+]$ would be equal to the $[OH^-]$ added.

$$[H_3O^+] = 10^{-pH} = 10^{-3.15} = 7.07945 \times 10^{-4} \, M \, H_3O^+ \text{ (unrounded)}$$
$$[H_3O^+] = 10^{-pH} = 10^{-3.65} = 2.23872 \times 10^{-4} \, M \, H_3O^+ \text{ (unrounded)}$$

Add $(7.07945 \times 10^{-4} \, M - 2.23872 \times 10^{-4} \, M) = 4.840737 \times 10^{-4} = \textbf{4.8} \times \textbf{10}^{\textbf{-4}} \, \textbf{mol of OH}^- \, \textbf{per liter}$.

18.33   The pH is increasing so the solution is becoming more basic. Therefore, OH^- ion is added to increase the pH.
Since 1 mole of H_3O^+ reacts with 1 mole of OH^-, the difference in $[H_3O^+]$ would be equal to the $[OH^-]$ added.

$$[H_3O^+] = 10^{-pH} = 10^{-4.52} = 3.01995 \times 10^{-5} \, M \, H_3O^+ \text{ (unrounded)}$$
$$[H_3O^+] = 10^{-pH} = 10^{-5.25} = 5.623413 \times 10^{-6} \, M \, H_3O^+ \text{ (unrounded)}$$
$$3.01995 \times 10^{-5} \, M - 5.623413 \times 10^{-6} \, M = 2.4576 \times 10^{-5} \, M \, OH^- \text{ must be added}.$$

$$\frac{2.4576 \times 10^{-5} \, mol}{L} (5.6 \, L) = 1.3763 \times 10^{-4} = \textbf{1.4} \times \textbf{10}^{\textbf{-4}} \, \textbf{mol of OH}^-$$

18.36   a) Heat is absorbed in an endothermic process: $2 \, H_2O(l) + heat \rightarrow H_3O^+(aq) + OH^-(aq)$. As the temperature
increases, the reaction shifts to the formation of products. Since the products are in the numerator of the K_w
expression, rising temperature **increases** the value of K_w.

b) Given that the pH is 6.80, the $[H_3O^+]$ can be calculated. The problem specifies that the solution is neutral, meaning $[H_3O^+] = [OH^-]$. A new K_w can then be calculated.
$[H_3O^+] = 10^{-pH} = 10^{-6.80} = 1.58489 \times 10^{-7}\ M\ H_3O^+ = \mathbf{1.6 \times 10^{-7}}\ \mathbf{M}\ \mathbf{[H_3O^+] = [OH^-]}$
$K_w = [H_3O^+][OH^-] = (1.58489 \times 10^{-7})(1.58489 \times 10^{-7}) = 2.511876 \times 10^{-14} = \mathbf{2.5 \times 10^{-14}}$
For a neutral solution: pH = pOH = 6.80

18.37   The Brønsted-Lowry theory defines acids as proton donors and bases as proton acceptors, while the Arrhenius definition looks at acids as containing ionizable H atoms and at bases as containing hydroxide ions. In both definitions, an acid produces hydronium ions and a base produces hydroxide ions when added to water. Ammonia, NH_3, and carbonate ion, CO_3^{2-}, are two Brønsted-Lowry bases that are not Arrhenius bases because they do not contain hydroxide ions. Brønsted-Lowry acids must contain an ionizable H atom in order to be proton donors, so a Brønsted-Lowry acid that is not an Arrhenius acid cannot be identified. (Other examples are also acceptable.)

18.40   An amphiprotic substance can lose a proton to act as an acid or gain a proton to act as a base. In the presence of a strong base (OH^-), the dihydrogen phosphate ion acts like an acid by donating hydrogen:
$H_2PO_4^-(aq) + OH^-(aq) \rightarrow H_2O(aq) + HPO_4^{2-}(aq)$
In the presence of a strong acid (HCl), the dihydrogen phosphate ion acts like a base by accepting hydrogen:
$H_2PO_4^-(aq) + HCl(aq) \rightarrow H_3PO_4(aq) + Cl^-(aq)$

18.41   a) When phosphoric acid is dissolved in water, a proton is donated to the water and dihydrogen phosphate ions are generated.
$$H_3PO_4(aq) + H_2O(l) \rightleftharpoons H_2PO_4^-(aq) + H_3O^+(aq)$$
$$K_a = \frac{[H_3O^+][H_2PO_4^-]}{[H_3PO_4]}$$
b) Benzoic acid is an organic acid and has only one proton to donate from the carboxylic acid group. The H atoms bonded to the benzene ring are not acidic hydrogens.
$$C_6H_5COOH(aq) + H_2O(l) \rightleftharpoons C_6H_5COO^-(aq) + H_3O^+(aq)$$
$$K_a = \frac{[H_3O^+][C_6H_5COO^-]}{[C_6H_5COOH]}$$
c) Hydrogen sulfate ion donates a proton to water and forms the sulfate ion.
$$HSO_4^-(aq) + H_2O(l) \rightleftharpoons SO_4^{2-}(aq) + H_3O^+(aq)$$
$$K_a = \frac{[H_3O^+][SO_4^{2-}]}{[HSO_4^-]}$$

18.43   To derive the conjugate base, remove one H and decrease the charge by 1. Since each formula is neutral, the conjugate base will have a charge of -1.
a) Cl^- b) HCO_3^- c) OH^-

18.45   To derive the conjugate acid, add an H and increase the charge by 1.
a) NH_4^+ b) NH_3 c) $C_{10}H_{14}N_2H^+$

18.47   The acid donates the proton to form its conjugate base; the base accepts a proton to form its conjugate acid:
a)   HCl   + H_2O $\rightleftharpoons$ Cl^- + H_3O^+
               acid   base   conjugate base   conjugate acid
        Conjugate acid/base pairs: HCl/Cl^- and H_3O^+/H_2O

b)　　$HClO_4$ + H_2SO_4 ⇌ ClO_4^- + $H_3SO_4^+$
　　acid　　　　base　　　　conjugate base　conjugate acid
　　Conjugate acid/base pairs: $HClO_4/ClO_4^-$ and $H_3SO_4^+/H_2SO_4$
　　Note: Perchloric acid is able to protonate another strong acid, H_2SO_4, because perchloric acid is a
　　stronger acid. ($HClO_4$'s oxygen atoms exceed its hydrogen atoms by one more than H_2SO_4.)

c)　　HPO_4^{2-} + H_2SO_4 ⇌ $H_2PO_4^-$ + HSO_4^-
　　base　　　　acid　　　　conjugate acid　conjugate base
　　Conjugate acid/base pairs: H_2SO_4/HSO_4^- and $H_2PO_4^-/HPO_4^{2-}$

18.49　The acid donates the proton to form its conjugate base; the base accepts a proton to
　　　form its conjugate acid:

a)　　NH_3 + H_3PO_4 ⇌ NH_4^+ + $H_2PO_4^-$
　　base　　　　acid　　　　conjugate acid　conjugate base
　　Conjugate acid-base pairs: $H_3PO_4/H_2PO_4^-$; NH_4^+/NH_3

b)　　CH_3O^- + NH_3 ⇌ CH_3OH + NH_2^-
　　base　　　　acid　　　　conjugate acid　conjugate base
　　Conjugate acid-base pairs: NH_3/NH_2^-; CH_3OH/CH_3O^-

c)　　HPO_4^{2-} + HSO_4^- ⇌ $H_2PO_4^-$ + SO_4^{2-}
　　base　　　　acid　　　　conjugate acid　conjugate base
　　Conjugate acid-base pairs: HSO_4^-/SO_4^{2-}; $H_2PO_4^-/HPO_4^{2-}$

18.51　Write total ionic equations and then remove the spectator ions to write the net ionic equations. The (aq) subscript
　　　denotes that each species is soluble and dissociates in water.
　　a) Na⁺(aq) + OH^-(aq) + Na⁺(aq) + $H_2PO_4^-$(aq) ⇌ H_2O(l) + 2 Na⁺(aq) + HPO_4^{2-}(aq)
　　　Net: OH^-(aq) + $H_2PO_4^-$(aq) ⇌ H_2O(l) + HPO_4^{2-}(aq)
　　　　　base　　　　acid　　　　　　conjugate acid　conjugate base
　　　Conjugate acid/base pairs: $H_2PO_4^-$/ HPO_4^{2-} and H_2O/OH^-

　　b) K⁺(aq) + HSO_4^-(aq) + 2 K⁺(aq) + CO_3^{2-}(aq) ⇌ 2 K⁺(aq) + SO_4^{2-}(aq) + K⁺(aq) + HCO_3^-(aq)
　　　Net: HSO_4^-(aq) + CO_3^{2-}(aq) ⇌ SO_4^{2-}(aq) + HCO_3^-(aq)
　　　　　acid　　　　base　　　　conjugate base　　　conjugate acid
　　　Conjugate acid/base pairs: HSO_4^-/SO_4^{2-} and HCO_3^-/CO_3^{2-}

18.53　The conjugate pairs are H_2S (acid)/HS^- (base) and HCl (acid)/Cl^- (base). The reactions involve reacting one acid
　　　from one conjugate pair with the base from the other conjugate pair. Two reactions are possible:
　　　　(1) $HS^- + HCl$ ⇌ $H_2S + Cl^-$ 　and 　(2) $H_2S + Cl^-$ ⇌ $HS^- + HCl$
　　　The first reaction is the reverse of the second. To decide which will have an equilibrium constant greater than 1,
　　　look for the stronger acid producing a weaker acid. HCl is a strong acid and H_2S a weak acid. The reaction that
　　　favors the products ($K_c > 1$) is the first one where the strong acid produces the weak acid. Reaction (2) with a
　　　weaker acid forming a stronger acid favors the reactants and $K_c < 1$.

18.55　a)　　HCl + NH_3 ⇌ NH_4^+ + Cl^-
　　　　strong acid　　　stronger base　　　weak acid　　　weaker base
　　　　HCl is ranked above NH_4^+ in Figure 18.9 and is the stronger acid. NH_3 is ranked above Cl^- and is the
　　　　stronger base. NH_3 is shown as a "stronger" base because it is stronger than Cl^-, but is not considered a
　　　　"strong" base. The reaction proceeds towards the production of the weaker acid and base, i.e., the
　　　　reaction as written proceeds to the right and $K_c > 1$. The stronger acid is more likely to donate a proton
　　　　than the weaker acid.

　　　b)　　H_2SO_3 + NH_3 ⇌ HSO_3^- + NH_4^+
　　　　stronger acid　　　stronger base　　　weaker base　　　weaker acid
　　　　H_2SO_3 is ranked above NH_4^+ and is the stronger acid. NH_3 is a stronger base than HSO_3^-. The reaction
　　　　proceeds towards the production of the weaker acid and base, i.e., the reaction as written proceeds to the
　　　　right and $K_c > 1$.

18.57   a) NH_4^+ + HPO_4^{2-} $\leftrightarrows$ NH_3 + $H_2PO_4^-$
weaker acid   weaker base   stronger base   stronger acid
$K_c < 1$ The reaction proceeds towards the production of the weaker acid and base, i.e., the reaction as written proceeds to the left.

b) HSO_3^- + HS^- $\leftrightarrows$ H_2SO_3 + S^{2-}
weaker base   weaker acid   stronger acid   stronger base
$K_c < 1$ The reaction proceeds towards the production of the weaker acid and base, i.e., the reaction as written proceeds to the left.

18.59   a) The concentration of a strong acid is **very different** before and after dissociation since a strong acid exhibits 100% dissociation. After dissociation, the concentration of the strong acid approaches 0, or $[HA] \approx 0$.
b) A weak acid dissociates to a very small extent (<<100%), so the acid concentration after dissociation is **nearly the same** as before dissociation.
c) Same as (b), but the percent, or extent, of dissociation is greater than in (b).
d) Same as (a)

18.60   **No**, HCl and CH_3COOH are never of equal strength because HCl is a strong acid with $K_a > 1$ and CH_3COOH is a weak acid with $K_a < 1$. The K_a of the acid, not the concentration of H_3O^+ in a solution of the acid, determines the strength of the acid.

18.63   Butanoic acid dissociates according to the following equation:
$$CH_3CH_2CH_2COOH(aq) + H_2O(l) \leftrightarrows H_3O^+(aq) + CH_3CH_2CH_2COO^-(aq)$$

Initial:	0.15 M	0	0
Change:	$-x$	$+x$	$+x$
Equilibrium:	$0.15 - x$	$+x$	$+x$

According to the information given in the problem, $[H_3O^+]eq = 1.51 \times 10^{-3}\ M = x$
Thus, $[H_3O^+] = [CH_3CH_2CH_2COO^-] = 1.51 \times 10^{-3}\ M$
$[CH_3CH_2CH_2COOH] = (0.15 - x) = (0.15 - 1.51 \times 10^{-3})\ M = 0.14849\ M$

$$K_a = \frac{\left[H_3O^+\right]\left[CH_3CH_2CH_2COO^-\right]}{\left[CH_3CH_2CH_2COOH\right]}$$

$$K_a = \frac{\left(1.51 \times 10^{-3}\right)\left(1.51 \times 10^{-3}\right)}{\left(0.14849\right)} = 1.53552 \times 10^{-5} = \mathbf{1.5 \times 10^{-5}}$$

18.65   For a solution of a weak acid, the acid dissociation equilibrium determines the concentrations of the weak acid, its conjugate base and H_3O^+. The acid dissociation reaction for HNO_2 is:

Concentration	$HNO_2(aq)$	+	$H_2O(l)$	$\leftrightarrows$	$H_3O^+(aq)$	+	$NO_2^-(aq)$
Initial	0.60		—		0		0
Change	$-x$				$+x$		$+x$
Equilibrium	$0.60 - x$				x		x

(The H_3O^+ contribution from water has been neglected.)

$$K_a = 7.1 \times 10^{-4} = \frac{\left[H_3O^+\right]\left[NO_2^-\right]}{\left[HNO_2\right]}$$

$$K_a = 7.1 \times 10^{-4} = \frac{(x)(x)}{(0.60 - x)} \qquad \text{Assume x is small compared to 0.60: } 0.60 - x = 0.60$$

$$K_a = 7.1 \times 10^{-4} = \frac{(x)(x)}{(0.60)}$$

x = 0.020639767 (unrounded)
Check assumption: (0.020639767 / 0.60) x 100% = 3.4% error, so the assumption is valid.
$[H_3O^+] = [NO_2^-] = \mathbf{2.1 \times 10^{-2}\ M}$

The concentration of hydroxide ion is related to concentration of hydronium ion through the equilibrium for water: $2 H_2O(l) \rightleftharpoons H_3O^+(aq) + OH^-(aq)$ with $K_w = 1.0 \times 10^{-14}$

$[OH^-] = 1.0 \times 10^{-14} / 0.020639767 = 4.84502 \times 10^{-13} = \textbf{4.8} \times \textbf{10}^{-13} \textbf{\textit{M} OH}^-$

$[OH^-] = 1.0 \times 10^{-14} / 0.02258 = 4.42869796 \times 10^{-13} = \textbf{4.4} \times \textbf{10}^{-13} \textbf{\textit{M} OH}^-$

18.67   Write a balanced chemical equation and equilibrium expression for the dissociation of chloroacetic acid and convert pK_a to K_a.
$K_a = 10^{-pKa} = 10^{-2.87} = 1.34896 \times 10^{-3}$ (unrounded)

Concentration	$ClCH_2COOH(aq) + H_2O(l) \rightleftharpoons$	$H_3O^+(aq) +$	$ClCH_2COO^-(aq)$
Initial	1.25	0	0
Change	−x	+x	+x
Equilibrium	1.25 − x	x	x

$$K_a = 1.34896 \times 10^{-3} = \frac{\left(H_3O^+\right)\left(ClCH_2COO^-\right)}{\left(ClCH_2COOH\right)}$$

$$K_a = 1.34896 \times 10^{-3} = \frac{(x)(x)}{(1.25 - x)} \quad \text{Assume x is small compared to 1.25.}$$

$$K_a = 1.34896 \times 10^{-3} = \frac{(x)(x)}{(1.25)}$$

x = 0.04106336 (unrounded)
Check assumption: (0.04106336 / 1.25) x 100% = 3.3%. The assumption is good.
$[H_3O^+] = [ClCH_2COO^-] = \textbf{0.041 \textit{M}}$
$[ClCH_2COOH] = 1.25 - 0.04106336 = 1.20894 = \textbf{1.21 \textit{M}}$
$pH = -\log [H_3O^+] = -\log (0.04106336) = 1.3865 = \textbf{1.39}$

18.69   Percent dissociation refers to the amount of the initial concentration of the acid that dissociates into ions. Use the percent dissociation to find the concentration of acid dissociated. HA will be used as the formula of the acid.
a) The concentration of acid dissociated is equal to the equilibrium concentrations of A^- and H_3O^+. Then pH and $[OH^-]$ are determined from $[H_3O^+]$.

$$\text{Percent HA} = \frac{\text{Dissociated Acid}}{\text{Initial Acid}} \times 100\%$$

$$3.0\% = \frac{x}{0.20}(100)$$

[Dissociated Acid] = x = 6.0×10^{-3} M

Concentration	$HA(aq) + H_2O(l) \rightleftharpoons$	$H_3O^+(aq) +$	$A^-(aq)$
Initial:	0.20	0	0
Change:	−x	+x	+x
Equilibrium:	0.20 − x	x	x

[Dissociated Acid] = x = $[H_3O^+]$ = $\textbf{6.0} \times \textbf{10}^{-3}$ \textbf{\textit{M}}
$pH = -\log [H_3O^+] = -\log (6.0 \times 10^{-3}) = 2.22185 = \textbf{2.22}$

$$[OH^-] = \frac{K_w}{\left[H_3O^+\right]} = \frac{1.0 \times 10^{-14}}{6.0 \times 10^{-3}} = 1.6666667 \times 10^{-12} = \textbf{1.7} \times \textbf{10}^{-12} \textbf{\textit{M}}$$

$pOH = -\log [OH^-] = -\log (1.6666667 \times 10^{-12}) = 11.7782 = \textbf{11.78}$
b) In the equilibrium expression, substitute the concentrations above and calculate K_a.

$$K_a = \frac{\left[H_3O^+\right]\left[A^-\right]}{[HA]} = \frac{\left(6.0 \times 10^{-3}\right)\left(6.0 \times 10^{-3}\right)}{\left(0.20 - 6.0 \times 10^{-3}\right)} = 1.85567 \times 10^{-4} = \textbf{1.9} \times \textbf{10}^{-4}$$

18.71   Calculate the molarity of HX by dividing moles by volume. Convert pH to $[H_3O^+]$ and substitute into the equilibrium expression.

$$\text{Concentration of HX} = \left(\frac{0.250 \text{ mol}}{655 \text{ mL}}\right)\left(\frac{1 \text{ mL}}{10^{-3} \text{ L}}\right) = 0.381679 \ M$$

Concentration	$HX(aq)$ + $H_2O(l)$	$\rightleftharpoons$	$H_3O^+(aq)$ +	X^- (aq)
Initial:	0.381679		0	0
Change:	–x		+x	+x
Equilibrium:	0.381679 – x		x	x

$[H_3O^+] = 10^{-pH} = 10^{-3.54} = 2.88403 \times 10^{-4} \ M = x$

Thus, $[H_3O^+] = [X^-] = 2.88403 \times 10^{-4} \ M$, and $[HX] = (0.381679 - 2.88403 \times 10^{-4}) \ M$

$$K_a = \frac{\left[H_3O^+\right]\left[X^-\right]}{\left[HX\right]} = \frac{\left(2.88403 \times 10^{-4}\right)\left(2.88403 \times 10^{-4}\right)}{\left(0.381679 - 2.88403 \times 10^{-4}\right)} = 2.18087 \times 10^{-7} = \mathbf{2.2 \times 10^{-7}}$$

18.73   a) Begin with a reaction table then, use the K_a expression as in earlier problems.

Concentration	$HZ(aq)$ +	$H_2O(l)$	$\rightleftharpoons$	$H_3O^+(aq)$ +	$Z^-(aq)$
Initial	0.075	—		0	0
Change	–x			+x	+x
Equilibrium	0.075 – x			x	x

(The H_3O^+ contribution from water has been neglected.)

$$K_a = 2.55 \times 10^{-4} = \frac{\left[H_3O^+\right]\left[Z^-\right]}{\left[HZ\right]}$$

$$K_a = 2.55 \times 10^{-4} = \frac{(x)(x)}{(0.075 - x)} \qquad \text{Assume x is small compared to 0.075.}$$

$$K_a = 2.55 \times 10^{-4} = \frac{(x)(x)}{(0.075)}$$

$[H_3O^+] = x = 4.3732 \times 10^{-3}$ (unrounded)

Check assumption: $(4.3732 \times 10^{-3} / 0.075) \times 100\% = 6\%$ error, so the assumption is not valid.

Since the error is greater than 5%, it is not acceptable to assume x is small compared to 0.045, and it is necessary to use the quadratic equation.

$$K_a = 2.55 \times 10^{-4} = \frac{(x)(x)}{(0.075 - x)}$$

$$x^2 + 2.55 \times 10^{-4} x - 1.9125 \times 10^{-5} = 0$$

$$a = 1 \qquad b = 2.55 \times 10^{-4} \qquad c = -1.9125 \times 10^{-5}$$

$$x = \frac{-b \pm \sqrt{b^2 - 4ac}}{2a}$$

$$x = \frac{-(2.55 \times 10^{-4}) \pm \sqrt{\left(2.55 \times 10^{-4}\right)^2 - 4(1)\left(-1.9125 \times 10^{-5}\right)}}{2(1)}$$

$x = 0.00425$ or -0.004503 (unrounded)

(The –0.004503 value is not possible.)

$pH = -\log [H_3O^+] = -\log (0.00425) = 2.3716 = \mathbf{2.37}$

b) Begin this part like part a.

Concentration	$HZ(aq)$ +	$H_2O(l)$	$\rightleftharpoons$	$H_3O^+(aq)$ +	$Z^-(aq)$
Initial	0.045	—		0	0
Change	–x			+x	+x
Equilibrium	0.045 – x			x	x

(The H_3O^+ contribution from water has been neglected.)

$$K_a = 2.55 \times 10^{-4} = \frac{[H_3O^+][Z^-]}{[HZ]}$$

$$K_a = 2.55 \times 10^{-4} = \frac{(x)(x)}{(0.045 - x)} \qquad \text{Assume x is small compared to } 0.045.$$

$$K_a = 2.55 \times 10^{-4} = \frac{(x)(x)}{(0.045)}$$

$[H_3O^+] = x = 3.3875 \times 10^{-3}$ (unrounded)

Check assumption: $(3.3875 \times 10^{-3} / 0.045) \times 100\% = 7.5\%$ error, so the assumption is not valid.
Since the error is greater than 5%, it is not acceptable to assume x is small compared to 0.045, and it is necessary to use the quadratic equation.

$$K_a = 2.55 \times 10^{-4} = \frac{(x)(x)}{(0.045 - x)}$$

$x^2 = (2.55 \times 10^{-4})(0.045 - x) = 1.1475 \times 10^{-5} - 2.55 \times 10^{-4} x$
$x^2 + 2.55 \times 10^{-4} x - 1.1475 \times 10^{-5} = 0$

$$a = 1 \qquad b = 2.55 \times 10^{-4} \qquad c = -1.1475 \times 10^{-5}$$

$$x = \frac{-b \pm \sqrt{b^2 - 4ac}}{2a}$$

$$x = \frac{-2.55 \times 10^{-4} \pm \sqrt{(2.55 \times 10^{-4})^2 - 4(1)(-1.1475 \times 10^{-5})}}{2(1)}$$

$x = 3.3899 \times 10^{-3} \, M \, H_3O^+$

$$[OH^-] = [OH^-] = \frac{K_w}{[H_3O^+]} = \frac{1.0 \times 10^{-14}}{3.3899 \times 10^{-3}} = 2.94994 \times 10^{-12} \, M$$

$pOH = -\log [OH^-] = -\log (2.94994 \times 10^{-12}) = 11.5302 = \mathbf{11.53}$

18.75 a) Begin with a reaction table, then use the K_a expression as in earlier problems.

Concentration	HY(aq)	+	H₂O(l)	⇌	H₃O⁺(aq)	+	Y⁻ (aq)
Initial	0.175		—		0		0
Change	−x				+x		+x
Equilibrium	0.175 − x				x		x

(The H_3O^+ contribution from water has been neglected.)

$$K_a = 1.50 \times 10^{-4} = \frac{(H_3O^+)(Y^-)}{(HY)}$$

$$K_a = 1.50 \times 10^{-4} = \frac{(x)(x)}{(0.175 - x)} \qquad \text{Assume x is small compared to } 0.175.$$

$$K_a = 1.50 \times 10^{-4} = \frac{(x)(x)}{(0.175)}$$

$[H_3O^+] = x = 5.1235 \times 10^{-3}$ (unrounded)

Check assumption: $(5.1235 \times 10^{-3} / 0.175) \times 100\% = 3\%$ error, so the assumption is valid.

$pH = -\log [H_3O^+] = -\log (5.1235 \times 10^{-3}) = 2.29043 = \mathbf{2.290}$

b) Begin this part like part a.

Concentration	HX(aq)	+	H₂O(l)	⇌	H₃O⁺(aq)	+	X⁻ (aq)
Initial	0.175		—		0		0
Change	−x				+x		+x
Equilibrium	0.175 − x				x		x

(The H_3O^+ contribution from water has been neglected.)

$$K_a = 2.00 \times 10^{-2} = \frac{\left(H_3O^+\right)\left(X^-\right)}{\left(HX\right)}$$

$$K_a = 2.00 \times 10^{-2} = \frac{(x)(x)}{(0.175 - x)} \qquad \text{Assume x is small compared to 0.175.}$$

$$K_a = 2.00 \times 10^{-2} = \frac{(x)(x)}{(0.175)}$$

$[H_3O^+] = x = 5.9161 \times 10^{-2}$ (unrounded)
Check assumption: $(5.9161 \times 10^{-2} / 0.175) \times 100\% = 34\%$ error, so the assumption is not valid.
Since the error is greater than 5%, it is not acceptable to assume x is small compared to 0.175, and it is necessary to use the quadratic equation.

$$K_a = 2.00 \times 10^{-2} = \frac{(x)(x)}{(0.175 - x)}$$

$x^2 = = (2.00 \times 10^{-2})\,(0.175 - x) = 0.0035 - 2.00 \times 10^{-2}\,x$
$x^2 + 2.00 \times 10^{-2}\,x - 0.0035 = 0$
$\qquad a = 1 \qquad b = 2.00 \times 10^{-2} \qquad c = -0.0035$

$$x = \frac{-b \pm \sqrt{b^2 - 4ac}}{2a}$$

$$x = \frac{-2.00 \times 10^{-2} \pm \sqrt{\left(2.00 \times 10^{-2}\right)^2 - 4(1)(-0.0035)}}{2(1)}$$

$x = 5.00 \times 10^{-2}\ M\ H_3O^+$
$[OH^-] = K_w / [H_3O^+] = (1.0 \times 10^{-14}) / (5.00 \times 10^{-2}) = 2.00 \times 10^{-13}\ M$
$pOH = -\log [OH^-] = -\log (2.00 \times 10^{-13}) = 12.69897 = \textbf{12.699}$

18.77   First, find the concentration of benzoate ion at equilibrium. Then use the initial concentration of benzoic acid and equilibrium concentration of benzoate to find % dissociation.

Concentration	$C_6H_5COOH(aq)$	+ $H_2O(l)$	$\leftrightarrows$	$H_3O^+(aq)$	+ $C_6H_5COO^-(aq)$
Initial	0.55	—		0	0
Change	−x			+x	+x
Equilibrium	0.55 − x			x	x

$$K_a = 6.3 \times 10^{-5} = \frac{\left[H_3O^+\right]\left[C_6H_5COO^-\right]}{\left[C_6H_5COOH\right]}$$

$$K_a = 6.3 \times 10^{-5} = \frac{[x][x]}{[0.55 - x]} \qquad \text{Assume x is small compared to 0.55.}$$

$$K_a = 6.3 \times 10^{-5} = \frac{[x][x]}{[0.55]}$$

$x = 5.8864 \times 10^{-3}$ (unrounded)
Check assumption: $(5.8864 \times 10^{-3} / 0.55) \times 100\% = 1\%$ error, so the assumption is valid.

$$\text{Percent } C_6H_5COOH \text{ Dissociated} = \frac{\text{Dissociated Acid}}{\text{Initial Acid}} \times 100\%$$

$$\text{Percent } C_6H_5COOH \text{ Dissociated} = \frac{x}{0.55} \times 100\% = \frac{5.8864 \times 10^{-3}}{0.55} \times 100\% = 1.07025 = \textbf{1.1\%}$$

18.79   Write balanced chemical equations and corresponding equilibrium expressions for dissociation of hydrosulfuric acid (H_2S).

$$H_2S(aq) + H_2O(l) \leftrightarrows H_3O^+(aq) + HS^-(aq)$$

$$K_{a1} = 9 \times 10^{-8} = \frac{\left[H_3O^+\right]\left[HS^-\right]}{\left[H_2S\right]}$$

$$HS^-(aq) + H_2O(l) \leftrightarrows H_3O^+(aq) + S^{2-}(aq)$$

$$K_{a2} = 1 \times 10^{-17} = \frac{\left[H_3O^+\right]\left[S^{2-}\right]}{\left[HS^-\right]}$$

Assumptions:

1) Since $K_{a1} \gg K_{a2}$, assume that almost all of the H_3O^+ comes from the first dissociation.

2) Since K_{a1} is so small, assume that the dissociation of H_2S is negligible and $[H_2S]_{eq} = 0.10 - x \approx \textbf{0.10 M}$.

Concentration	$H_2S(aq)$ + $H_2O(l)$	$\leftrightarrows$	$H_3O^+(aq)$ +	$HS^-(aq)$
Initial	0.10	—	0	0
Change	−x		+x	+x
Equilibrium	0.10 − x		x	x

$$K_{a1} = 9 \times 10^{-8} = \frac{\left[H_3O^+\right]\left[HS^-\right]}{\left[H_2S\right]}$$

$$K_{a1} = 9 \times 10^{-8} = \frac{[x][x]}{[0.10 - x]} \qquad \text{Assume x is small compared to 0.10.}$$

$$K_{a1} = 9 \times 10^{-8} = \frac{[x][x]}{[0.10]}$$

$x = 9.48683 \times 10^{-5}$ (unrounded)

$[H_3O^+] = [HS^-] = x = \textbf{9} \times \textbf{10}^{-5}\ \textbf{\textit{M}}$

$pH = -\log [H_3O^+] = -\log (9.48683 \times 10^{-5}) = 4.0228787 = \textbf{4.0}$

$[OH^-] = K_w / [H_3O^+] = (1.0 \times 10^{-14}) / (9.48683 \times 10^{-5}) = 1.05409 \times 10^{-10} = \textbf{1} \times \textbf{10}^{-10}\ \textbf{\textit{M}}$

$pOH = -\log [OH^-] = -\log (1.05409 \times 10^{-10}) = 9.9771 = \textbf{10.0}$

$[H_2S] = (0.10 - 9.48683 \times 10^{-5})\ M = 0.099905 = \textbf{0.10}\ \textbf{\textit{M}}$

Concentration is limited to one significant figure because K_a is given to only one significant figure. The pH is given to what appears to be 2 significant figures because the number before the decimal point (4) represents the exponent and the number after the decimal point represents the significant figures in the concentration. Calculate $[S^{2-}]$ by using the K_{a2} expression and assuming that $[HS^-]$ and $[H_3O^+]$ come mostly from the first dissociation. This new calculation will have a new x value.

Concentration	$HS^-(aq)$ +	$H_2O(l)$	$\leftrightarrows$	$H_3O^+(aq)$	+	$S^{2-}(aq)$
Initial	9.48683×10^{-5}			9.48683×10^{-5}		0
Change	−x			+x		+x
Equilibrium	$9.48683 \times 10^{-5} - x$			$9.48683 \times 10^{-5} + x$		x

$$K_{a2} = 1 \times 10^{-17} = \frac{\left[H_3O^+\right]\left[S^{2-}\right]}{\left[HS^-\right]}$$

$$K_{a2} = 1 \times 10^{-17} = \frac{\left(9.48683 \times 10^{-5} + x\right)(x)}{\left(9.48683 \times 10^{-5} - x\right)}$$

$$K_{a2} = 1 \times 10^{-17} = \frac{\left(9.48683 \times 10^{-5}\right)(x)}{\left(9.48683 \times 10^{-5}\right)} \qquad \text{Assume x is small compared to } 9.48683 \times 10^{-5}.$$

$x = [S^{2-}] = \textbf{1} \times \textbf{10}^{-17}\ \textbf{\textit{M}}$

The small value of x means that it is not necessary to recalculate the $[H_3O^+]$ and $[HS^-]$ values.

18.82　First, find the concentration of formate ion at equilibrium. Then use the initial concentration of formic acid and equilibrium concentration of formate to find % dissociation.

Concentration	$HCOOH(aq)$	+	$H_2O(l)$	$\rightleftharpoons$	$H_3O^+(aq)$	+	$HCOO^-(aq)$
Initial	0.75				0		0
Change	$-x$				$+x$		$+x$
Equilibrium	$0.75 - x$				x		x

$$K_a = 1.8 \times 10^{-4} = \frac{\left[H_3O^+\right]\left[HCOO^-\right]}{\left[HCOOH\right]}$$

$$K_a = 1.8 \times 10^{-4} = \frac{(x)(x)}{(0.75 - x)} \qquad \text{Assume x is small compared to 0.75.}$$

$$K_a = 1.8 \times 10^{-4} = \frac{(x)(x)}{(0.75)}$$

$x = 1.161895 \times 10^{-2}$ (unrounded)

Check assumption: $(1.161895 \times 10^{-2} / 0.75) \times 100\% = 2\%$ error, so the assumption is valid.

$$\text{Percent HCOOH Dissociated} = \frac{\text{Dissociated Acid}}{\text{Initial Acid}} \times 100\%$$

$$\text{Percent HCOOH Dissociated} = \frac{x}{0.75} \times 100\% = \frac{1.161895 \times 10^{-2}}{0.75} \times 100\% = 1.54919 = \textbf{1.5\%}$$

18.83　All Brønsted-Lowry bases contain at least one lone pair of electrons. This lone pair binds with an H^+ and allows the base to act as a proton-acceptor.

18.86　a) A base accepts a proton from water in the base dissociation reaction:

$$C_5H_5N(aq) + H_2O(l) \rightleftharpoons OH^-(aq) + C_5H_5NH^+(aq)$$

$$K_b = \frac{\left[C_5H_5NH^+\right]\left[OH^-\right]}{\left[C_5H_5N\right]}$$

b) The primary reaction is involved in base dissociation of carbonate ion is:

$$CO_3^{2-}(aq) + H_2O(l) \rightleftharpoons OH^-(aq) + HCO_3^-(aq)$$

$$K_b = \frac{\left[HCO_3^-\right]\left[OH^-\right]}{\left[CO_3^{2-}\right]}$$

The bicarbonate can then also dissociate as a base, but this occurs to an insignificant amount in a solution of carbonate ions.

18.88　a) Hydroxylamine has a lone pair of electrons on the nitrogen atom that acts like the Lewis base:

$$HONH_2(aq) + H_2O(l) \rightleftharpoons OH^-(aq) + HONH_3^+(aq)$$

$$K_b = \frac{\left[HONH_3^+\right]\left[OH^-\right]}{[HONH_2]}$$

b) The hydrogen phosphate ion contains oxygen atoms with lone pairs of electron that act as proton acceptors.

$$HPO_4^{2-}(aq) + H_2O(l) \rightleftharpoons H_2PO_4^-(aq) + OH^-(aq)$$

$$K_b = \frac{\left[H_2PO_4^-\right]\left[OH^-\right]}{\left[HPO_4^{2-}\right]}$$

18.90   The formula of dimethylamine has two methyl (CH_3-) groups attached to a nitrogen:

$$CH_3 - \overset{\cdot\cdot}{\underset{|}{N}} - H$$
$$CH_3$$

The nitrogen has a lone pair of electrons that will accept the proton from water in the base dissociation reaction:
The value for the dissociation constant is from Appendix C.

Concentration	$(CH_3)_2NH(aq) + H_2O(l) \rightleftharpoons OH^-(aq) + (CH_3)_2NH_2^+(aq)$		
Initial	0.070	0	0
Change	$-x$	$+x$	$+x$
Equilibrium	$0.070 - x$	x	x

$$K_b = \frac{\left[(CH_3)_2 NH_2^+\right]\left[OH^-\right]}{\left[(CH_3)_2 NH\right]} = 5.9 \times 10^{-4}$$

$$K_b = \frac{[x][x]}{[0.070 - x]} = 5.9 \times 10^{-4} \qquad \text{Assume } 0.070 - x = 0.070.$$

$$\frac{[x][x]}{[0.070]} = 5.9 \times 10^{-4}$$

$x = 6.4265 \times 10^{-3}\ M$
Check assumption: $(6.4265 \times 10^{-3} / 0.070) \times 100\% = 9\%$ error, so the assumption is invalid.
The problem will need to be solved as a quadratic.

$$\frac{[x][x]}{[0.070 - x]} = 5.9 \times 10^{-4}$$

$x^2 = (5.9 \times 10^{-4})(0.070 - x) = 4.13 \times 10^{-5} - 5.9 \times 10^{-4}\ x$
$x^2 + 5.9 \times 10^{-4}\ x - 4.13 \times 10^{-5} = 0$
$\qquad\qquad a = 1 \qquad b = 5.9 \times 10^{-4} \qquad c = -4.13 \times 10^{-5}$

$$x = \frac{-b \pm \sqrt{b^2 - 4ac}}{2a}$$

$$x = \frac{-5.9 \times 10^{-4} \pm \sqrt{\left(5.9 \times 10^{-4}\right)^2 - 4(1)\left(-4.13 \times 10^{-5}\right)}}{2(1)} = 6.13827 \times 10^{-3}\ M\ OH^-\ \text{(unrounded)}$$

$$[H_3O]^+ = \frac{K_w}{\left[OH^-\right]} = \frac{1.0 \times 10^{-14}}{6.13827 \times 10^{-3}} = 1.629124 \times 10^{-12}\ M\ H_3O^+\ \text{(unrounded)}$$

$$pH = -\log[H_3O^+] = -\log(1.629124 \times 10^{-12}) = 11.7880 = \textbf{11.79}$$

18.92   Write a balanced equation and equilibrium expression for the reaction. Make simplifying assumptions (if valid),
solve for $[OH^-]$, convert to $[H_3O^+]$ and calculate pH.

Concentration	$HOCH_2CH_2NH_2(aq) + H_2O(l) \rightleftharpoons OH^-(aq) + HOCH_2CH_2NH_3^+(aq)$		
Initial	0.25	0	0
Change	$-x$	$+x$	$+x$
Equilibrium	$0.25 - x$	x	x

$$K_b = \frac{\left[HOCH_2CH_2NH_3^+\right]\left[OH^-\right]}{\left[HOCH_2CH_2NH_2\right]} = 3.2 \times 10^{-5}$$

$$K_b = \frac{[x][x]}{[0.25 - x]} = 3.2 \times 10^{-5} \qquad \text{Assume } x \text{ is small compared to 0.25.}$$

$$K_b = 3.2 \times 10^{-5} = \frac{(x)(x)}{(0.25)}$$

$x = 2.8284 \times 10^{-3}\ M\ OH^-$ (unrounded)
Check assumption: $(2.8284 \times 10^{-3} / 0.25) \times 100\% = 1\%$ error, so the assumption is valid.
$[H_3O]^+ = K_w / [OH^-] = (1.0 \times 10^{-14}) / (2.8284 \times 10^{-3}) = 3.535568 \times 10^{-12}\ M\ H_3O^+$ (unrounded)
$pH = -\log [H_3O^+] = -\log (3.535568 \times 10^{-12}) = 11.4515 = \textbf{11.45}$

18.94   a) Acetate ion, CH_3COO^-, is the conjugate base of acetic acid, CH_3COOH. The K_b for acetate ion is related to the K_a for acetic acid through the equation $K_w = K_a \times K_b$.
K_b of $CH_3COO^- = K_w / K_a = (1.0 \times 10^{-14}) / (1.8 \times 10^{-5}) = 5.55556 \times 10^{-10} = \textbf{5.6} \times \textbf{10}^{\textbf{-10}}$
b) Anilinium ion is the conjugate acid of aniline so the K_a for anilinium ion is related to the K_b of aniline by the relationship $K_w = K_a \times K_b$.
K_a of $C_6H_5NH_3^+ = K_w / K_b = (1.0 \times 10^{-14}) / (4.0 \times 10^{-10}) = \textbf{2.5} \times \textbf{10}^{\textbf{-5}}$

18.96   a) The K_a of chlorous acid, $HClO_2$, is reported in Appendix C. $HClO_2$ is the conjugate acid of chlorite ion, ClO_2^-. The K_b for chlorite ion is related to the K_a for chlorous acid through the equation $K_w = K_a \times K_b$, and $pK_b = -\log K_b$.

K_b of $ClO_2^- = \dfrac{K_w}{K_a} = \dfrac{1.0 \times 10^{-14}}{1.1 \times 10^{-2}} = 9.0909 \times 10^{-13}$ (unrounded)

$pK_b = -\log (9.0909 \times 10^{-13}) = 12.04139 = \textbf{12.04}$
b) The K_b of dimethylamine, $(CH_3)_2NH$, is reported in Appendix C. $(CH_3)_2NH$ is the conjugate base of $(CH_3)_2NH_2^+$. The K_a for $(CH_3)_2NH_2^+$ is related to the K_b for $(CH_3)_2NH$ through the equation $K_w = K_a \times K_b$, and $pK_a = -\log K_a$.

K_a of $(CH_3)_2NH_2^+ = \dfrac{K_w}{K_b} = \dfrac{1.0 \times 10^{-14}}{5.9 \times 10^{-4}} = 1.694915 \times 10^{-11}$ (unrounded)

$pK_a = -\log (1.694915 \times 10^{-11}) = 10.77085 = \textbf{10.77}$

18.98   a) Potassium cyanide, when placed in water, dissociates into potassium ions, K^+, and cyanide ions, CN^-. Potassium ion is the conjugate acid of a strong base, KOH, so K^+ does not react with water. Cyanide ion is the conjugate base of a weak acid, HCN, so it does react with the base dissociation reaction:
$$CN^-(aq) + H_2O(l) \leftrightarrows HCN(aq) + OH^-(aq)$$
To find the pH first set up a reaction table and use K_b for CN^- to calculate $[OH^-]$.

Concentration (M)	$CN^-(aq)$	+ $H_2O(l)$	$\leftrightarrows$	$HCN(aq)$	+ $OH^-(aq)$
Initial	0.150	—		0	0
Change	$-x$	—		$+x$	$+x$
Equilibrium	$0.150 - x$	—		x	x

K_b of $CN^- = \dfrac{K_w}{K_a} = \dfrac{1.0 \times 10^{-14}}{6.2 \times 10^{-10}} = 1.612903 \times 10^{-5}$ (unrounded)

$K_b = \dfrac{[HCN][OH^-]}{[CN^-]} = 1.612903 \times 10^{-5}$

$K_b = \dfrac{[x][x]}{[0.150 - x]} = 1.612903 \times 10^{-5}$ Assume x is small compared to 0.150: $0.150 - x = 0.150$.

$K_b = 1.612903 \times 10^{-5} = \dfrac{(x)(x)}{(0.150)}$

$x = 1.555 \times 10^{-3}\ M\ OH^-$ (unrounded)
Check assumption: $(1.555 \times 10^{-3} / 0.150) \times 100\% = 1\%$ error, so the assumption is valid.

$[H_3O]^+ = \dfrac{K_w}{OH^-} = \dfrac{1.0 \times 10^{-14}}{1.555 \times 10^{-3}} = 6.430868 \times 10^{-12}\ M\ H_3O^+$ (unrounded)

$pH = -\log [H_3O^+] = -\log (6.430868 \times 10^{-12}) = 11.19173 = \textbf{11.19}$

b) The salt triethylammonium chloride in water dissociates into two ions: $(CH_3CH_2)_3NH^+$ and Cl^-. Chloride ion is the conjugate base of a strong acid so it will not influence the pH of the solution. Triethylammonium ion is the conjugate acid of a weak base, so the acid dissociation reaction below determines the pH of the solution.

Concentration (M)	$(CH_3CH_2)_3NH^+(aq)$	$+ H_2O(l)$	$\leftrightarrows$	$(CH_3CH_2)_3N(aq)$	$+ H_3O^+(aq)$
Initial	0.40	—		0	0
Change	$-x$	—		$+x$	$+x$
Equilibrium	$0.40 - x$	—		x	x

$$K_a \text{ of } (CH_3CH_2)_3NH^+ = \frac{K_w}{K_b} = \frac{1.0 \times 10^{-14}}{5.2 \times 10^{-4}} = 1.9230769 \times 10^{-11} \text{ (unrounded)}$$

$$K_a = 1.9230769 \times 10^{-11} = \frac{\left[H_3O^+\right]\left[(CH_3CH_2)_3N\right]}{\left[(CH_3CH_2)_3NH^+\right]}$$

$$K_a = 1.9230769 \times 10^{-11} = \frac{(x)(x)}{(0.40 - x)} \qquad \text{Assume x is small compared to 0.40: } 0.40 - x = 0.40.$$

$$K_a = 1.9230769 \times 10^{-11} = \frac{(x)(x)}{(0.40)}$$

$[H_3O^+] = x = 2.7735 \times 10^{-6}$ (unrounded)
Check assumption: $(2.7735 \times 10^{-6} / 0.40) \times 100\% = 0.0007\%$ error, so the assumption is valid.
pH $= -\log [H_3O^+] = -\log (2.7735 \times 10^{-6}) = 5.55697 = \mathbf{5.56}$

18.100 a) The formate ion, $HCOO^-$, acts as the base shown by the following equation:
$$HCOO^-(aq) + H_2O(l) \leftrightarrows HCOOH(aq) + OH^-(aq)$$
Because HCOOK is a soluble salt, $[HCOO^-] = [HCOOK]$. The potassium ion is from a strong base; therefore, it will not affect the pH, and can be ignored.

Concentration (M)	$HCOO^-(aq)$	$+ H_2O(l)$	$\leftrightarrows$	$HCOOH(aq)$	$+ OH^-(aq)$
Initial	0.65	—		0	0
Change	$-x$	—		$+x$	$+x$
Equilibrium	$0.65 - x$	—		x	x

K_b of $HCOO^- = K_w / K_a = (1.0 \times 10^{-14}) / (1.8 \times 10^{-4}) = 5.55556 \times 10^{-11}$

$$K_b = \frac{\left[HCOOH\right]\left[OH^-\right]}{\left[HCOO^-\right]} = 5.55556 \times 10^{-11}$$

$$K_b = \frac{[x][x]}{[0.65 - x]} = 5.55556 \times 10^{-11} \qquad \text{Assume x is small compared to 0.65.}$$

$$K_b = 5.55556 \times 10^{-11} = \frac{(x)(x)}{(0.65)} = 6.00925 \times 10^{-6} \, M \, OH^- \text{ (unrounded)}$$

Check assumption: $(6.00925 \times 10^{-6} / 0.65) \times 100\% = 0.0009\%$ error, so the assumption is valid.
$[H_3O^+] = K_w / [OH^-] = (1.0 \times 10^{-14}) / (6.00925 \times 10^{-6}) = 1.66410 \times 10^{-9} \, M \, H_3O^+$ (unrounded)
pH $= -\log [H_3O^+] = -\log (1.66410 \times 10^{-9}) = 8.7788 = \mathbf{8.78}$

b) The ammonium ion, NH_4^+, acts as an acid shown by the following equation:
$$NH_4^+(aq) + H_2O(l) \leftrightarrows H_3O^+(aq) + NH_3(aq)$$
Because NH_4Br is a soluble salt, $[NH_4^+] = [NH_4Br]$. The bromide ion is from a strong acid; therefore, it will not affect the pH, and can be ignored.

Concentration (M)	$NH_4^+(aq)$	$+ H_2O(l)$	$\leftrightarrows$	$NH_3(aq)$	$+ H_3O^+(aq)$
Initial	0.85	—		0	0
Change	$-x$	—		$+x$	$+x$
Equilibrium	$0.85 - x$	—		x	x

K_a of $NH_4^+ = K_w / K_b = (1.0 \text{ x } 10^{-14}) / (1.76 \text{ x } 10^{-5}) = 5.681818 \text{ x } 10^{-10}$ (unrounded)

$$K_a = 5.681818 \text{ x } 10^{-10} = \frac{\left[H_3O^+\right]\left[NH_3\right]}{\left[NH_4^+\right]}$$

$$K_a = 5.681818 \text{ x } 10^{-10} = \frac{[x][x]}{[0.85 - x]} \qquad \text{Assume x is small compared to 0.85.}$$

$$K_a = 5.681818 \text{ x } 10^{-10} = \frac{[x][x]}{[0.85]}$$

$[H_3O^+] = x = 2.1976 \text{ x } 10^{-5}$ (unrounded)

Check assumption: $(2.1976 \text{ x } 10^{-5} / 0.85) \text{ x } 100\% = 0.003\%$ error, so the assumption is valid.

$pH = -\log [H_3O^+] = -\log (2.1976 \text{ x } 10^{-5}) = 4.65805 = \mathbf{4.66}$

18.102　First, calculate the initial molarity of ClO^-. Then, set up reaction table with base dissociation of OCl^-:

$$[ClO^-] = \left(\frac{1 \text{ mL Solution}}{10^{-3} \text{ L Solution}}\right)\left(\frac{1.0 \text{ g Solution}}{1 \text{ mL Solution}}\right)\left(\frac{6.5\% \text{ NaOCl}}{100\% \text{ Solution}}\right)\left(\frac{1 \text{ mol NaOCl}}{74.44 \text{ g NaOCl}}\right)\left(\frac{1 \text{ mol OCl}^-}{1 \text{ mol NaOCl}}\right)$$

$= 0.873186 \; M \; OCl^-$ (unrounded)

The sodium ion is from a strong base; therefore, it will not affect the pH, and can be ignored.

Concentration (M)	$OCl^-(aq)$	$+$	$H_2O(l)$	$\leftrightharpoons$	$HOCl(aq)$	$+$	$OH^-(aq)$
Initial	0.873186		—		0		0
Change	$-x$		—		$+x$		$+x$
Equilibrium	$0.873186 - x$		—		x		x

$$K_b \text{ of } OCl^- = \frac{K_w}{K_a} = \frac{1.0 \text{ x } 10^{-14}}{2.9 \text{ x } 10^{-8}} = 3.448275862 \text{ x } 10^{-7}$$

$$K_b = \frac{\left[HOCl\right]\left[OH^-\right]}{\left[ClO^-\right]} = 3.448275862 \text{ x } 10^{-7}$$

$$K_b = \frac{[x][x]}{[0.873186 - x]} = 3.448275862 \text{ x } 10^{-7} \qquad \text{Assume x is small compared to 0.873186.}$$

$$K_b = 3.448275862 \text{ x } 10^{-7} = \frac{(x)(x)}{(0.873186)}$$

$x = 5.4872 \text{ x } 10^{-4} = \mathbf{5.5 \text{ x } 10^{-4} \; M \; OH^-}$

Check assumption: $(5.4872 \text{ x } 10^{-4} / 0.873186) \text{ x } 100\% = 0.06\%$ error, so the assumption is valid.

$$[H_3O^+] = \frac{K_w}{[OH^-]} = \frac{1.0 \text{ x } 10^{-14}}{5.4872 \text{ x } 10^{-4}} = 1.82242 \text{ x } 10^{-11} \; M \; H_3O^+ \text{ (unrounded)}$$

$pH = -\log [H_3O^+] = -\log (1.82242 \text{ x } 10^{-11}) = 10.73935 = \mathbf{10.74}$

18.104　Electronegativity increases left to right across a period. As the nonmetal becomes more electronegative, the acidity of the binary hydride increases. The electronegative nonmetal attracts the electrons more strongly in the polar bond, shifting the electron density away from H^+ and making the H^+ more easily transferred to a surrounding water molecule to make H_3O^+.

18.107　The two factors that explain the greater acid strength of $HClO_4$ are:
1) Chlorine is more electronegative than iodine, so chlorine more strongly attracts the electrons in the bond with oxygen. This makes the H in $HClO_4$ less tightly held by the oxygen than the H in HIO.
2) Perchloric acid has more oxygen atoms than HIO, which leads to a greater shift in electron density from the hydrogen atom to the oxygens making the H in $HClO_4$ more susceptible to transfer to a base.

18.108   For oxyacids, acid strength increases with increasing number of oxygen atoms and increasing electronegativity of the nonmetal in the acid. For binary acids, acid strength increases with increasing electronegativity across a row and increases with increasing size of the nonmetal in a column.
a) Selenic acid, **H$_2$SeO$_4$**, is the stronger acid because it contains more oxygen atoms.
b) Phosphoric acid, **H$_3$PO$_4$**, is the stronger acid because P is more electronegative than As.
c) Hydrotelluric acid, **H$_2$Te**, is the stronger acid because Te is larger than S and so the Te-H bond is weaker.

18.110   For oxyacids, acid strength increases with increasing number of oxygen atoms and increasing electronegativity of the nonmetal in the acid. For binary acids, acid strength increases with increasing electronegativity across a row and increases with increasing size of the nonmetal in a column.
a) **H$_2$Se**, hydrogen selenide, is a stronger acid than H$_3$As, arsenic hydride, because Se is more electronegative than As.
b) **B(OH)$_3$**, boric acid also written as H$_3$BO$_3$, is a stronger acid than Al(OH)$_3$, aluminum hydroxide, because boron is more electronegative than aluminum.
c) **HBrO$_2$**, bromous acid, is a stronger acid than HBrO, hypobromous acid, because there are more oxygen atoms in HBrO$_2$ than in HBrO.

18.112   Acidity increases as the value of K_a increases. Determine the ion formed from each salt and compare the corresponding K_a values from Appendix C.
a) Copper(II) sulfate, CuSO$_4$, contains Cu^{2+} ion with $K_a = 3 \times 10^{-8}$. Aluminum sulfate, Al$_2$(SO$_4$)$_3$, contains Al^{3+} ion with $K_a = 1 \times 10^{-5}$. The concentrations of Cu^{2+} and Al^{3+} are equal, but the K_a of Al$_2$(SO$_4$)$_3$ is almost three orders of magnitude greater. Therefore, **0.05 M Al$_2$(SO$_4$)$_3$** is the stronger acid and would have the lower pH.
b) Zinc chloride, ZnCl$_2$, contains the Zn^{2+} ion with $K_a = 1 \times 10^{-9}$. Lead chloride, PbCl$_2$, contains the Pb^{2+} ion with $K_a = 3 \times 10^{-8}$. Since both solutions have the same concentration, and K_a (Pb^{2+}) > K_a (Zn^{2+}), **0.1 M PbCl$_2$** is the stronger acid and would have the lower pH.

18.114   A higher pH (more basic solution) results when an acid has a lower K_a (from Appendix C).
a) The **Ni(NO$_3$)$_2$** solution has a higher pH than the Co(NO$_3$)$_2$ solution because K_a of Ni^{2+} (1 x 10^{-10}) is smaller than the K_a of Co^{2+} (2 x 10^{-10}). Note that nitrate ions are the conjugate bases of a strong acid and therefore do not influence the pH of the solution.
b) The **Al(NO$_3$)$_3$** solution has a higher pH than the Cr(NO$_3$)$_2$ solution because K_a of Al^{3+} (1 x 10^{-5}) is smaller than the K_a of Cr^{3+} (1 x 10^{-4}).

18.117   Sodium fluoride, NaF, contains the cation of a strong base, NaOH, and anion of a weak acid, HF. This combination yields a salt that is basic in aqueous solution as the F$^-$ ion acts as a base:
$$F^-(aq) + H_2O(l) \leftrightarrows HF(aq) + OH^-(aq)$$
Sodium chloride, NaCl, is the salt of a strong base, NaOH, and strong acid, HCl. This combination yields a salt that is neutral in aqueous solution as neither Na$^+$ or Cl$^-$ react in water to change the [H$^+$].

18.119   For each salt, first break into the ions present in solution and then determine if either ion acts as a weak acid or weak base to change the pH of the solution. Cations are neutral if they are from a strong base; other cations will be weakly acidic. Anions are neutral if they are from a strong acid; other anions are weakly basic.
a) KBr(s) $\xrightarrow{H_2O}$ K$^+$(aq) + Br$^-$(aq)
   K$^+$ is the conjugate acid of a strong base, so it does not influence pH.
   Br$^-$ is the conjugate base of a strong acid, so it does not influence pH.
   Since neither ion influences the pH of the solution, it will remain at the **neutral** pH of pure water.
b) NH$_4$I(s) $\xrightarrow{H_2O}$ NH$_4^+$(aq) + I$^-$(aq)
   NH$_4^+$ is the conjugate acid of a weak base, so it will act as a weak acid in solution and produce H$_3$O$^+$ as represented by the acid dissociation reaction:
$$NH_4^+(aq) + H_2O(l) \leftrightarrows NH_3(aq) + H_3O^+(aq)$$
   I$^-$ is the conjugate base of a strong acid, so it will not influence the pH.
   The production of H$_3$O$^+$ from the ammonium ion makes the solution of NH$_4$I **acidic**.

c) $KCN(s) \xrightarrow{\text{H}_2\text{O}} K^+(aq) + CN^-(aq)$

K^+ is the conjugate acid of a strong base, so it does not influence pH.

CN^- is the conjugate base of a weak acid, so it will act as a weak base in solution and impact pH by the base dissociation reaction:

$$CN^-(aq) + H_2O(l) \rightleftharpoons HCN(aq) + OH^-(aq)$$

Hydroxide ions are produced in this equilibrium so solution will be **basic**.

18.121   For each salt, first break into the ions present in solution and then determine if either ion acts as a weak acid or weak base to change the pH of the solution. Cations are neutral if they are from a strong base; other cations will be weakly acidic. Anions are neutral if they are from a strong acid; other anions are weakly basic.

a) The two ions that comprise sodium carbonate, Na_2CO_3, are sodium ion, Na^+, and carbonate ion, CO_3^{2-}.

$$Na_2CO_3(s) \xrightarrow{\text{H}_2\text{O}} 2\,Na^+(aq) + CO_3^{2-}(aq)$$

Sodium ion is derived from the strong base NaOH. Carbonate ion is derived from the weak acid HCO_3^-. A salt derived from a strong base and a weak acid produces a **basic** solution.

Na^+ does not react with water.

$$CO_3^{2-}(aq) + H_2O(l) \rightleftharpoons HCO_3^-(aq) + OH^-(aq)$$

b) The two ions that comprise calcium chloride, $CaCl_2$, are calcium ion, Ca^{2+}, and chloride ion, Cl^-

$$CaCl_2(s) \xrightarrow{\text{H}_2\text{O}} Ca^{2+}(aq) + 2\,Cl^-(aq)$$

Calcium ion is derived from the strong base $Ca(OH)_2$. Chloride ion is derived from the strong acid HCl. A salt derived from a strong base and strong acid produces a **neutral** solution.

Neither Ca^{2+} nor Cl^- reacts with water.

c) The two ions that comprise cupric nitrate, $Cu(NO_3)_2$, are the cupric ion, Cu^{2+}, and the nitrate ion, NO_3^-.

$$Cu(NO_3)_2(s) \xrightarrow{\text{H}_2\text{O}} Cu^{2+}(aq) + 2\,NO_3^-(aq)$$

Small metal ions are acidic in water (assume the hydration of Cu^{2+} is 6):

$$Cu(H_2O)_6^{2+}(aq) + H_2O(l) \rightleftharpoons Cu(H_2O)_5OH^+(aq) + H_3O^+(aq)$$

Nitrate ion is derived from the strong acid HNO_3. Therefore, NO_3^- does not react with water. A solution of cupric nitrate is **acidic**.

18.123   For each salt, first break into the ions present in solution and then determine if either ion acts as a weak acid or weak base to change the pH of the solution. Cations are neutral if they are from a strong base; other cations will be weakly acidic. Anions are neutral if they are from a strong acid; other anions are weakly basic.

a) A solution of strontium bromide is **neutral** because Sr^{2+} is the conjugate acid of a strong base, $Sr(OH)_2$ and Br^- is the conjugate base of a strong acid, HBr, so neither change the pH of the solution.

b) A solution of barium acetate is **basic** because CH_3COO^- is the conjugate base of a weak acid and therefore forms OH^- in solution whereas Ba^{2+} is the conjugate acid of a strong base, $Ba(OH)_2$, and does not influence solution pH. The base dissociation reaction of acetate ion is

$$CH_3COO^-(aq) + H_2O(l) \rightleftharpoons CH_3COOH(aq) + OH^-(aq)$$

c) A solution of dimethylammonium bromide is **acidic** because $(CH_3)_2NH_2^+$ is the conjugate acid of a weak base and therefore forms H_3O^+ in solution whereas Br^- is the conjugate base of a strong acid and does not influence the pH of the solution. The acid dissociation reaction for methylammonium ion is

$$(CH_3)_2NH_2^+(aq) + H_2O(l) \rightleftharpoons (CH_3)_2NH(aq) + H_3O^+(aq)$$

18.125   For each salt, first break into the ions present in solution and then determine if either ion acts as a weak acid or weak base to change the pH of the solution. Cations are neutral if they are from a strong base; other cations will be weakly acidic. Anions are neutral if they are from a strong acid; other anions are weakly basic.

a) The two ions that comprise ammonium phosphate, $(NH_4)_3PO_4$, are the ammonium ion, NH_4^+, and the phosphate ion, PO_4^{3-}.

$NH_4^+(aq) + H_2O(l) \rightleftharpoons NH_3(aq) + H_3O^+(aq)$ $K_a = K_w / K_b\,(NH_3) = 5.7 \times 10^{-10}$

$PO_4^{3-}(aq) + H_2O(l) \rightleftharpoons HPO_4^{2-}(aq) + OH^-(aq)$ $K_b = K_w / K_{a3}\,(H_3PO_4) = 2.4 \times 10^{-2}$

A comparison of K_a and K_b is necessary since both ions are derived from a weak base and weak acid. The K_a of NH_4^+ is determined by using the K_b of its conjugate base, NH_3 (Appendix C). The K_b of PO_4^{3-} is determined by using the K_a of its conjugate acid, HPO_4^{2-}. The K_a of HPO_4^{2-} comes from K_{a3} of H_3PO_4 (Appendix C). Since $K_b > K_a$, a solution of $(NH_4)_3PO_4$ is **basic**.

b) The two ions that comprise sodium sulfate, Na_2SO_4, are sodium ion, Na^+, and sulfate ion, SO_4^{2-}. The sodium ion is derived from the strong base NaOH. The sulfate ion is derived from the weak acid, HSO_4^-.

$$SO_4^{2-}(aq) + H_2O(l) \leftrightarrows HSO_4^-(aq) + OH^-(aq)$$

A solution of sodium sulfate is **basic**.

c) The two ions that comprise lithium hypochlorite, LiClO, are lithium ion, Li^+, and hypochlorite ion, ClO^-. Lithium ion is derived from the strong base LiOH. Hypochlorite ion is derived from the weak acid, HClO (hypochlorous acid).

$$ClO^-(aq) + H_2O(l) \leftrightarrows HClO(aq) + OH^-(aq)$$

A solution of lithium hypochlorite is **basic**.

18.127 a) Order of increasing pH: $\mathbf{Fe(NO_3)_2 < KNO_3 < K_2SO_3 < K_2S}$ (assuming concentrations equivalent)

Iron(II) nitrate, $Fe(NO_3)_2$, is an acidic solution because the iron ion is a small, highly charged metal ion that acts as a weak acid and nitrate ion is the conjugate base of a strong acid, so it does not influence pH.

Potassium nitrate, KNO_3, is a neutral solution because potassium ion is the conjugate acid of a strong base and nitrate ion is the conjugate base of a strong acid, so neither influences solution pH.

Potassium sulfite, K_2SO_3, and potassium sulfide, K_2S, are similar in that the potassium ion does not influence solution pH but the anions do because they are conjugate bases of weak acids. K_a for HSO_3^- is 6.5×10^{-8}, so K_b for SO_3^- is 1.5×10^{-7}, which indicates that sulfite ion is a weak base. K_a for HS^- is 1×10^{-17} from Table 18.5, so sulfide ion has a K_b equal to 1×10^3. Sulfide ion is thus a strong base. The solution of a strong base will have a greater concentration of hydroxide ions (and higher pH) than a solution of a weak base of equivalent concentrations.

b) In order of increasing pH: $\mathbf{NaHSO_4 < NH_4NO_3 < NaHCO_3 < Na_2CO_3}$

In solutions of ammonium nitrate, only the ammonium will influence pH by dissociating as a weak acid:

$$NH_4^+(aq) + H_2O(l) \leftrightarrows NH_3(aq) + H_3O^+(aq)$$
$$\text{with } K_a = 1.0 \times 10^{-14} / 1.8 \times 10^{-5} = 5.6 \times 10^{-10}$$

Therefore, the solution of ammonium nitrate is acidic.

In solutions of sodium hydrogen sulfate, only HSO_4^- will influence pH. The hydrogen sulfate ion is amphoteric so both the acid and base dissociations must be evaluated for influence on pH. As a base, HSO_4^- is the conjugate base of a strong acid, so it will not influence pH. As an acid, HSO_4^- is the conjugate acid of a weak base, so the acid dissociation applies

$$HSO_4^-(aq) + H_2O(l) \leftrightarrows SO_4^{2-}(aq) + H_3O^+(aq) \quad K_{a2} = 1.2 \times 10^{-2}$$

In solutions of sodium hydrogen carbonate, only the HCO_3^- will influence pH and it, like HSO_4^-, is amphoteric:

As an acid: $HCO_3^-(aq) + H_2O(l) \leftrightarrows CO_3^{2-}(aq) + H_3O^+(aq)$
        $K_a = 4.7 \times 10^{-11}$, the second K_a for carbonic acid from Table 18.5
As a base: $HCO_3^-(aq) + H_2O(l) \leftrightarrows H_2CO_3(aq) + OH^-(aq)$
        $K_b = 1.0 \times 10^{-14} / 4.5 \times 10^{-7} = 2.2 \times 10^{-8}$

Since $K_b > K_a$, a solution of sodium hydrogen carbonate is basic.

In a solution of sodium carbonate, only CO_3^{2-} will influence pH by acting as a weak base:

$$CO_3^{2-}(aq) + H_2O(l) \leftrightarrows HCO_3^-(aq) + OH^-(aq)$$
$$K_b = 1.0 \times 10^{-14} / 4.7 \times 10^{-11} = 2.1 \times 10^{-4}$$

Therefore, the solution of sodium carbonate is basic.

Two of the solutions are acidic. Since the K_a of HSO_4^- is greater than that of NH_4^+, the solution of sodium hydrogen sulfate has a lower pH than the solution of ammonium nitrate, assuming the concentrations are relatively close.

Two of the solutions are basic. Since the K_b of CO_3^{2-} is greater than that of HCO_3^-, the solution of sodium carbonate has a higher pH than the solution of sodium hydrogen carbonate, assuming concentrations are not extremely different.

18.129 Both methoxide ion and amide ion produce OH^- in aqueous solution. In water, the strongest base possible is OH^-. Since both bases produce OH^- in water, both bases appear equally strong.

$$CH_3O^-(aq) + H_2O(l) \rightarrow OH^-(aq) + CH_3OH(aq)$$
$$NH_2^-(aq) + H_2O(l) \rightarrow OH^-(aq) + NH_3(aq)$$

18.131 Ammonia, NH_3, is a more basic solvent than H_2O. In a more basic solvent, weak acids like HF act like strong acids and are 100% dissociated.

18.133 A Lewis acid is defined as an electron pair acceptor, while a Brønsted-Lowry acid is a proton donor. If only the proton in a Brønsted-Lowry acid is considered, then every Brønsted-Lowry acid fits the definition of a Lewis acid since the proton is accepting an electron pair when it bonds with a base. There are Lewis acids that do not include a proton, so all Lewis acids are not Brønsted-Lowry acids.
A Lewis base is defined as an electron pair donor and a Brønsted-Lowry base is a proton acceptor. In this case, the two definitions are essentially the same.

18.134 a) **No**, a weak Brønsted-Lowry base is not necessarily a weak Lewis base. For example, water molecules solvate metal ions very well:
$$Zn^{2+}(aq) + 4\ H_2O(l) \leftrightarrows Zn(H_2O)_6^{2+}(aq)$$
Water is a very weak Brønsted-Lowry base, but forms the Zn complex fairly well and is a reasonably strong Lewis base.
b) The **cyanide ion** has a lone pair to donate from either the C or the N, and donates an electron pair to the $Cu(H_2O)_6^{2+}$ complex. It is the Lewis base for the forward direction of this reaction. In the reverse direction, **water** donates one of the electron pairs on the oxygen to the $Cu(CN)_4^{2-}$ and is the Lewis base.
c) Because $K_c > 1$, the reaction proceeds in the direction written (left to right) and is driven by the stronger Lewis base, the **cyanide ion**.

18.137 a) Cu^{2+} is a **Lewis acid** because it accepts electron pairs from molecules such as water.
b) Cl^- is a **Lewis base** because it has lone pairs of electrons it can donate to a Lewis acid.
c) Tin(II) chloride, $SnCl_2$, is a compound with a structure similar to carbon dioxide, so it will act as a **Lewis acid** to form an additional bond to the tin.
d) Oxygen difluoride, OF_2, is a **Lewis base** with a structure similar to water, where the oxygen has lone pairs of electrons that it can donate to a Lewis acid.

18.139 a) The boron atom in boron trifluoride, BF_3, is electron deficient (has 6 electrons instead of 8) and can accept an electron pair; it is a **Lewis acid**.
b) The sulfide ion, S^{2-}, can donate any of four electron pairs and is a **Lewis base**.
c) The Lewis dot structure for the sulfite ion, SO_3^{2-} shows lone pairs on the sulfur and on the oxygens. The sulfur atom has a lone electron pair that it can donate more easily than the electronegative oxygen in the formation of an adduct. The sulfite ion is a **Lewis base**.
d) Sulfur trioxide, SO_3, acts as a **Lewis acid**.

18.141 a) Sodium ion is the Lewis acid because it is accepting electron pairs from water, the Lewis base.

Na^+	+	$6\ H_2O$	$\leftrightarrows$	$Na(H_2O)_6^+$
Lewis acid		Lewis base		adduct

b) The oxygen from water donates a lone pair to the carbon in carbon dioxide. Water is the Lewis base and carbon dioxide the Lewis acid.

CO_2	+	H_2O	$\leftrightarrows$	H_2CO_3
Lewis acid		Lewis base		adduct

c) Fluoride ion donates an electron pair to form a bond with boron in BF_4^-. The fluoride ion is the Lewis base and the boron trifluoride is the Lewis acid.

F^-	+	BF_3	$\leftrightarrows$	BF_4^-
Lewis base		Lewis acid		adduct

18.143 a) Since neither H^+ nor OH^- is involved, this is not an Arrhenius acid-base reaction. Since there is no exchange of protons, this is not a Brønsted-Lowry reaction. This reaction is only classified as **Lewis acid-base reaction**, where Ag^+ is the acid and NH_3 is the base.
b) Again, no OH^- is involved, so this is not an Arrhenius acid-base reaction. This is an exchange of a proton, from H_2SO_4 to NH_3, so it is a **Brønsted-Lowry acid-base reaction**. Since the Lewis definition is most inclusive, anything that is classified as a Brønsted-Lowry (or Arrhenius) reaction is automatically classified as a **Lewis acid-base reaction**.
c) This is not an acid-base reaction.
d) For the same reasons listed in (a), this reaction is only classified as **Lewis acid-base reaction**, where $AlCl_3$ is the acid and Cl^- is the base.

18.147 Calculate the $[H_3O^+]$ using the pH values given. Determine the value of K_w from the pK_w given. The $[H_3O^+]$ is combined with the K_w value at 37°C to find $[OH^-]$ using $K_w = [H_3O^+][OH^-]$.

$$K_w = 10^{-pK_w} = 10^{-13.63} = 2.3442 \times 10^{-14} \text{ (unrounded)}$$
$$K_w = [H_3O^+][OH^-] = 2.3442 \times 10^{-14} \text{ at } 37°C$$

$[H_3O^+]$ range

High value (low pH) $= 10^{-pH} = 10^{-7.35} = 4.4668 \times 10^{-8} = 4.5 \times 10^{-8} \ M \ H_3O^+$

Low value (high pH) $= 10^{-pH} = 10^{-7.45} = 3.5481 \times 10^{-8} = 3.5 \times 10^{-8} \ M \ H_3O^+$

Range: **3.5×10^{-8} to $4.5 \times 10^{-8} \ M \ H_3O^+$**

$[OH^-]$ range

$$K_w = [H_3O^+][OH^-] = 2.3442 \times 10^{-14} \text{ at } 37°C$$

$$[OH^-] = \frac{K_w}{[H_3O]^+}$$

High value (high pH) $= \dfrac{2.3442 \times 10^{-14}}{3.5481 \times 10^{-8}} = 6.6069 \times 10^{-7} = 6.6 \times 10^{-7} \ M \ OH^-$

Low value (low pH) $= \dfrac{2.3442 \times 10^{-14}}{4.4668 \times 10^{-8}} = 5.24805 \times 10^{-7} = 5.2 \times 10^{-7} \ M \ OH^-$

Range: **5.2×10^{-7} to $6.6 \times 10^{-7} \ M \ OH^-$**

18.148 a) Acids will vary in the amount they dissociate (acid strength) depending on the acid-base character of the solvent. Water and methanol have different acid-base characters.

b) The K_a is the measure of an acid's strength. A stronger acid has a smaller pK_a. Therefore, phenol is a stronger acid in water than it is in methanol. In other words, water more readily accepts a proton from phenol than does methanol, i.e., methanol is a weaker base than water.

c) $C_6H_5OH(solvated) + CH_3OH(l) \rightleftharpoons CH_3OH_2^+(solvated) + C_6H_5O^- (solvated)$

The term "solvated" is analogous to "aqueous." "Aqueous" would be incorrect in this case because the reaction does not take place in water.

d) In the autoionization process, one methanol molecule is the proton donor while another methanol molecule is the proton acceptor.

$$CH_3OH(l) + CH_3OH(l) \rightleftharpoons CH_3O^- (solvated) + CH_3OH_2^+(solvated)$$

In this equation "(solvated)" indicates that the molecules are solvated by methanol.
The equilibrium constant for this reaction is the autoionization constant of methanol:

$$K = [CH_3O^-][CH_3OH_2^+]$$

18.151 a) $SnCl_4$ is the Lewis acid accepting an electron pair from $(CH_3)_3N$, the Lewis base.

b) Tin is the element in the Lewis acid accepting the electron pair. The electron configuration of tin is $[Kr]5s^24d^{10}5p^2$. The four bonds to tin are formed by sp^3 hybrid orbitals, which completely fill the $5s$ and $5p$ orbitals. The **5d** orbitals are empty and available for the bond with trimethylamine.

18.152 Hydrochloric acid is a strong acid that completely dissociates in water. Therefore, the concentration of H_3O^+ is the same as the starting acid concentration: $[H_3O^+] = [HCl]$. The original solution pH:

$$pH = -\log (1.0 \times 10^{-5}) = \textbf{5.00} = \textbf{pH}.$$

A 1:10 dilution means that the chemist takes 1 mL of the $1.0 \times 10^{-5} \ M$ solution and dilutes it to 10 mL (or dilute 10 mL to 100 mL). The chemist then dilutes the diluted solution in a 1:10 ratio, and repeats this process for the next two successive dilutions. $M_1V_1 = M_2V_2$ can be used to find the molarity after each dilution:

Dilution 1: $M_1V_1 = M_2V_2$

$(1.0 \times 10^{-5} \ M)(1.0 \ mL) = (x)(10. \ mL)$

$[H_3O^+]_{HCl} = 1.0 \times 10^{-6} \ M \ H_3O^+$

$pH = -\log (1.0 \times 10^{-6}) = \textbf{6.00}.$

Dilution 2:

$(1.0 \times 10^{-6} \ M)(1.0 \ mL) = (x)(10. \ mL)$

$[H_3O^+]_{HCl} = 1.0 \times 10^{-7} \ M \ H_3O^+$

Once the concentration of strong acid is close to the concentration of H_3O^+ from water autoionization, the $[H_3O^+]$ in the solution does not equal the initial concentration of the strong acid. The calculation of $[H_3O^+]$ must be based on the water ionization equilibrium:

$H_2O(l) + H_2O(l) \rightleftharpoons H_3O^+(aq) + OH^-(aq)$ with $K_w = 1.0 \times 10^{-14}$ at 25°C.

The dilution gives an initial $[H_3O^+]$ of 1.0×10^{-7} M. Assuming that the initial concentration of hydroxide ions is zero, a reaction table is set up.

Concentration (M)	$2\,H_2O(l)$	$\rightleftharpoons$	$H_3O^+(aq)$	$+$	$OH^-(aq)$
Initial	—		1×10^{-7}		0
Change	—		$+x$		$+x$
Equilibrium	—		$1 \times 10^{-7} + x$		x

$K_w = [H_3O^+]\,[OH^-] = (1 \times 10^{-7} + x)\,(x) = 1.0 \times 10^{-14}$

Set up as a quadratic equation: $x^2 + 1.0 \times 10^{-7}\,x - 1.0 \times 10^{-14} = 0$

$$a = 1 \quad b = 1.0 \times 10^{-7} \quad c = -1.0 \times 10^{-14}$$

$$x = \frac{-1.0 \times 10^{-7} \pm \sqrt{\left(1.0 \times 10^{-7}\right)^2 - 4(1)\left(-1.0 \times 10^{-14}\right)}}{2(1)}$$

$x = 6.1803 \times 10^{-8}$ (unrounded)

$[H_3O^+] = (1.0 \times 10^{-7} + x)\,M = (1.0 \times 10^{-7} + 6.1803 \times 10^{-8})\,M = 1.61803 \times 10^{-7}\,M\ H_3O^+$ (unrounded)

pH $= -\log [H_3O^+] = -\log (1.61803 \times 10^{-7}) = 6.79101 = \mathbf{6.79}$

Dilution 3:

$(1.0 \times 10^{-7}\,M)\,(1.0\ mL) = (x)(10.\ mL)$

$[H_3O^+]_{HCl} = 1.0 \times 10^{-8}\,M\ H_3O^+$

The dilution gives an initial $[H_3O^+]$ of 1.0×10^{-8} M. Assuming that the initial concentration of hydroxide ions is zero, a reaction table is set up.

Concentration (M)	$2\,H_2O(l)$	$\rightleftharpoons$	$H_3O^+(aq)$	$+$	$OH^-(aq)$
Initial	—		1×10^{-8}		0
Change	—		$+x$		$+x$
Equilibrium	—		$1 \times 10^{-8} + x$		x

$K_w = [H_3O^+]\,[OH^-] = (1 \times 10^{-8} + x)\,(x) = 1.0 \times 10^{-14}$

Set up as a quadratic equation: $x^2 + 1.0 \times 10^{-8}\,x - 1.0 \times 10^{-14} = 0$

$$a = 1 \quad b = 1.0 \times 10^{-8} \quad c = -1.0 \times 10^{-14}$$

$$x = \frac{-1.0 \times 10^{-8} \pm \sqrt{\left(1.0 \times 10^{-8}\right)^2 - 4(1)\left(-1.0 \times 10^{-14}\right)}}{2(1)}$$

$x = 9.51249 \times 10^{-8}$ (unrounded)

$[H_3O^+] = (1.0 \times 10^{-8} + x)\,M = (1.0 \times 10^{-8} + 9.51249 \times 10^{-8})\,M = 1.051249 \times 10^{-7}\,M\ H_3O^+$ (unrounded)

pH $= -\log [H_3O^+] = -\log (1.051249 \times 10^{-7}) = 6.97829 = \mathbf{6.98}$

Dilution 4:

$(1.0 \times 10^{-8}\,M)\,(1.0\ mL) = (x)(10.\ mL)$

$[H_3O^+]_{HCl} = 1.0 \times 10^{-9}\,M\ H_3O^+$

The dilution gives an initial $[H_3O^+]$ of 1.0×10^{-9} M. Assuming that the initial concentration of hydroxide ions is zero, a reaction table is set up.

Concentration (M)	$2\,H_2O(l)$	$\rightleftharpoons$	$H_3O^+(aq)$	$+$	$OH^-(aq)$
Initial	—		1×10^{-9}		0
Change	—		$+x$		$+x$
Equilibrium	—		$1 \times 10^{-9} + x$		x

$K_w = [H_3O^+]\,[OH^-] = (1 \times 10^{-9} + x)\,(x) = 1.0 \times 10^{-14}$

Set up as a quadratic equation: $x^2 + 1.0 \times 10^{-9}\,x - 1.0 \times 10^{-14} = 0$

$$a = 1 \quad b = 1.0 \times 10^{-9} \quad c = -1.0 \times 10^{-14}$$

$$x = \frac{-1.0 \times 10^{-9} \pm \sqrt{\left(1.0 \times 10^{-9}\right)^2 - 4(1)\left(-1.0 \times 10^{-14}\right)}}{2(1)}$$

$x = 9.95012 \times 10^{-8}$ (unrounded)

$[H_3O^+] = (1.0 \times 10^{-9} + x)\,M = (1.0 \times 10^{-9} + 9.95012 \times 10^{-8})\,M = 1.00512 \times 10^{-7}\,M\ H_3O^+$ (unrounded)

pH $= -\log [H_3O^+] = -\log (1.00512 \times 10^{-7}) = 6.99778 = \mathbf{7.00}$

As the HCl solution is diluted, the pH of the solution becomes closer to 7.0. Continued dilutions will not significantly change the pH from 7.0. Thus, a solution with a basic pH cannot be made by adding acid to water.

18.155 Compare the contribution of each acid by calculating the concentration of H_3O^+ produced by each. For 3% hydrogen peroxide, first find initial molarity of H_2O_2, assuming the density is 1.00 g/mL (the density of water).

$$M\ H_2O_2 = \left(\frac{1.00\ g}{mL}\right)\left(\frac{3\%\ H_2O_2}{100\%}\right)\left(\frac{1\ mol\ H_2O_2}{34.02\ g\ H_2O_2}\right)\left(\frac{1\ mL}{10^{-3}\ L}\right) = 0.881834\ M\ H_2O_2\ (unrounded)$$

Find K_a from pK_a: $K_a = 10^{-pKa} = 10^{-11.75} = 1.778279 \times 10^{-12}$ (unrounded)

	H_2O_2	+	H_2O	⇌	H_3O^+	+	HO_2^-
Initial	0.881834		—		0		0
Change	−x		—		+x		+x
Equilibrium	0.881834 − x		—		x		x

$$K_a = 1.778279 \times 10^{-12} = \frac{\left[H_3O^+\right]\left[HO_2^-\right]}{\left[H_2O_2\right]}$$

$$K_a = 1.778279 \times 10^{-12} = \frac{(x)(x)}{(0.881834 - x)} \qquad \text{Assume x is small compared to 0.881834.}$$

$$K_a = 1.778279 \times 10^{-12} = \frac{(x)(x)}{(0.881834)}$$

$[H_3O^+] = x = 1.2522567 \times 10^{-6}$ (unrounded)

Check assumption: $(1.2522567 \times 10^{-6} / 0.881834) \times 100\% = 0.0001\%$ error, so the assumption is valid.

$$M\ H_3PO_4 = \left(\frac{1.00\ g}{mL}\right)\left(\frac{0.001\%\ H_3PO_4}{100\%}\right)\left(\frac{1\ mol\ H_3PO_4}{97.99\ g\ H_3PO_4}\right)\left(\frac{1\ mL}{10^{-3}\ L}\right)$$

$$= 1.0205 \times 10^{-4}\ M\ H_3PO_4\ (unrounded)$$

From Appendix C, K_a for phosphoric acid is 7.2×10^{-3}. The subsequent K_a values may be ignored. In this calculation x is not negligible since the initial concentration of acid is less than the K_a.

	H_3PO_4	+	H_2O	⇌	H_3O^+	+	$H_2PO_4^-$
Initial	1.0205×10^{-4}		—		0		0
Change	−x		—		+x		+x
Equilibrium	1.0205×10^{-4}		—		x		x

$$K_a = 7.2 \times 10^{-3} = \frac{\left[H_3O^+\right]\left[H_2PO_4^-\right]}{\left[H_3PO_4\right]}$$

$$K_a = 7.2 \times 10^{-3} = \frac{[x][x]}{\left[1.0205 \times 10^{-4} - x\right]}$$

The problem will need to be solved as a quadratic.
$x^2 = (7.2 \times 10^{-3})(1.0205 \times 10^{-4} - x) = 7.3476 \times 10^{-7} - 7.2 \times 10^{-3}\ x$
$x^2 + 7.2 \times 10^{-3}\ x - 7.3476 \times 10^{-7} = 0$
$\qquad a = 1 \qquad b = 7.2 \times 10^{-3} \qquad c = -7.3476 \times 10^{-7}$

$$x = \frac{-7.2 \times 10^{-3} \pm \sqrt{\left(7.2 \times 10^{-3}\right)^2 - 4(1)\left(-7.3476 \times 10^{-7}\right)}}{2(1)}$$

$x = 1.00643 \times 10^{-4}\ M\ H_3O^+$ (unrounded)

The concentration of hydronium ion produced by the phosphoric acid, $1 \times 10^{-4}\ M$, is greater than the concentration produced by the hydrogen peroxide, $1 \times 10^{-6}\ M$. Therefore, the **phosphoric acid** contributes more H_3O^+ to the solution.

18.159 Determine the hydrogen ion concentration from the pH. The molarity and the volume will give the number of moles, and with the aid of Avogadro's number, the number of ions may be found.

$$M\ H_3O^+ = 10^{-pH} = 10^{-6.2} = 6.30957 \times 10^{-7}\ M \text{ (unrounded)}$$

$$\left(\frac{6.30957 \times 10^{-7}\ mol\ H_3O^+}{L}\right)\left(\frac{10^{-3}\ L}{1\ mL}\right)\left(\frac{1250.mL}{d}\right)\left(\frac{7\ d}{1\ wk}\right)\left(\frac{6.022 \times 10^{23}\ H_3O^+}{1\ mol\ H_3O^+}\right) = 3.32467 \times 10^{18} = \mathbf{3 \times 10^{18}\ H_3O^+}$$

The pH has only one significant figure, and limits the significant figures in the final answer.

18.161 Autoionization involves the transfer of an ion, usually H^+, from one molecule to another. This produces a solvated cation and anion characteristic of the liquid.

a) $CH_3OH(l) + CH_3OH(l) \leftrightarrows CH_3OH_2^+(solvated) + CH_3O^-(solvated)$

$$K_{met} = [CH_3OH_2^+]\,[CH_3O^-] = (x)\,(x) = 2 \times 10^{-17}$$

$$x = [CH_3O^-] = \sqrt{2 \times 10^{-17}} = 4.4721 \times 10^{-9} = \mathbf{4 \times 10^{-9}\ M\ CH_3O^-}$$

b) $2\ NH_2CH_2CH_2NH_2(l) \leftrightarrows NH_2CH_2CH_2NH_3^+(solvated) + NH_2CH_2CH_2NH^-(solvated)$

$$[NH_2CH_2CH_2NH_3^+] = x = 2 \times 10^{-8}\ M$$

$$K_{ed} = [NH_2CH_2CH_2NH_3^+]\,[NH_2CH_2CH_2NH^-] = (x)\,(x)$$

$$K_{ed} = (2 \times 10^{-8})\,(2 \times 10^{-8}) = \mathbf{4 \times 10^{-16}}$$

18.163 Determine the K_b using the relationship: $K_b = 10^{pK} = 10^{-5.91} = 1.23027 \times 10^{-6}$

	TRIS(aq) + H₂O(l)	⇄	OH⁻(aq)	+ HTRIS⁺(aq)
Initial	0.075	—	0	0
Change	−x	—	+x	+x
Equilibrium	0.075 − x	—	x	x

$$K_b = 1.23027 \times 10^{-6} = \frac{\left[HTRIS^+\right]\left[OH^-\right]}{\left[TRIS\right]}$$

$$K_b = 1.23027 \times 10^{-6} = \frac{[x][x]}{[0.075 - x]} \qquad \text{Assume x is small compared to 0.075.}$$

$$K_b = 1.23027 \times 10^{-6} = \frac{[x][x]}{[0.075]}$$

$$x = [OH^-] = 3.03760 \times 10^{-4}\ M\ OH^- \text{ (unrounded)}$$

Check assumption: $[3.03760 \times 10^{-4} / 0.075] \times 100\% = 0.40\%$, therefore the assumption is good.

$$[H_3O^+] = K_w / [OH^-] = (1.0 \times 10^{-14}) / (3.03760 \times 10^{-4}) = 3.292073 \times 10^{-11}\ M \text{ (unrounded)}$$

$$pH = -\log[H_3O^+] = -\log(3.292073 \times 10^{-11}) = 10.4825 = \mathbf{10.48}$$

18.165 The pH is dependent on the *molar* concentration of H_3O^+. Convert % w/v to molarity, and use the K_a of acetic acid to determine $[H_3O^+]$ from the equilibrium expression.

Convert % w/v to molarity using the molecular weight of acetic acid (CH_3COOH):

$$\text{Molarity} = \left(\frac{5.0\ g\ CH_3COOH}{100\ mL\ Solution}\right)\left(\frac{1\ mol\ CH_3COOH}{60.05\ g\ CH_3COOH}\right)\left(\frac{1\ mL}{10^{-3}\ L}\right) = 0.832639\ M\ CH_3COOH \text{ (unrounded)}$$

Acetic acid dissociates in water according to the following equation and equilibrium expression:

	CH₃COOH(aq) + H₂O(l)	⇄	CH₃COO⁻(aq)	+ H₃O⁺(aq)
Initial	0.832639	—	0	0
Change	−x	—	+x	+x
Equilibrium	0.832639 − x	—	x	x

$$K_a = 1.8 \times 10^{-5} = \frac{\left[H_3O^+\right]\left[CH_3COO^-\right]}{\left[CH_3COOH\right]}$$

$$K_a = 1.8 \times 10^{-5} = \frac{[x][x]}{[0.832639 - x]} \qquad \text{Assume x is small compared to 0.832639.}$$

$$K_a = 1.8 \times 10^{-5} = \frac{[x][x]}{[0.832639]}$$

$x = 3.871 \times 10^{-3} = [H_3O^+]$ (unrounded)

Check assumption: $[3.871 \times 10^{-3} / 0.832639] \times 100\% = 0.46\%$, therefore the assumption is good.

$pH = -\log [H_3O^+] = -\log (3.871 \times 10^{-3}) = 2.4121768 = \mathbf{2.41}$

18.168 Assuming that the pH in the specific cellular environment is equal to the optimum pH for the enzyme, the hydronium ion concentrations are $[H_3O^+] = 10^{-pH}$

Salivary amylase, mouth: $[H_3O^+] = 10^{-6.8} = 1.58489 \times 10^{-7} = \mathbf{2 \times 10^{-7}\,M}$

Pepsin, stomach: $[H_3O^+] = 10^{-2.0} = \mathbf{1 \times 10^{-2}\,M}$

Trypsin, pancreas: $[H_3O^+] = 10^{-9.5} = 3.162 \times 10^{-10} = \mathbf{3 \times 10^{-10}\,M}$

18.172   The freezing point depression equation is required to determine the molality of the solution.
$$\Delta T = [0.00 - (-1.93°C)] = 1.93°C = = iK_f m$$
Temporarily assume $i = 1$.
$$m = \frac{\Delta T}{iK_f} = \frac{1.93°C}{(1)(1.86°C/m)} = 1.037634\ m = 1.037634\ M \text{ (unrounded)}$$

This molality is the total molality of all species in the solution, and is equal to their molarity.
From the equilibrium:

$$ClCH_2COOH(aq) + H_2O(l) \leftrightarrows H_3O^+(aq) + ClCH_2COO^-(aq)$$

	ClCH$_2$COOH	H$_3$O$^+$	ClCH$_2$COO$^-$
Initial	1.000 M	x	x
Change	−x	+x	+x
Equilibrium	1.000 − x	x	x

The total concentration of all species is:

$[ClCH_2COOH] + [H_3O^+] + [ClCH_2COO^-] = 1.037634\ M$

$[1.000 - x] + [x] + [x] = 1.000 + x = 1.037634\ M$

$x = 0.037634\ M$ (unrounded)

$$K_a = \frac{[H_3O^+][CH_3COO^-]}{[CH_3COOH]}$$

$$K_a = \frac{(0.037634)(0.037634)}{(1.000 - 0.037634)} = 0.0014717 = \mathbf{0.00147}$$

18.174   a) The two ions that comprise this salt are Ca^{2+} (derived from the strong base $Ca(OH)_2$) and $CH_3CH_2COO^-$ (derived from the weak acid, propionic acid, CH_3CH_2COOH). A salt derived from a strong base and weak acid produces a **basic** solution.

Ca^{2+} does not react with water.

$$CH_3CH_2COO^-(aq) + H_2O(l) \leftrightarrows CH_3CH_2COOH(aq) + OH^-(aq)$$

b) Calcium propionate is a soluble salt and dissolves in water to yield two propionate ions:

$$Ca(CH_3CH_2COO)_2(s) + H_2O(l) \rightarrow Ca^{2+}(aq) + 2\ CH_3CH_2COO^-(aq)$$

The molarity of the solution is

$$\text{Molarity} = \left(\frac{8.75\ g\ Ca(CH_3CH_2COO)_2}{0.500\ L} \right)\left(\frac{1\ mol\ Ca(CH_3CH_2COO)_2}{186.22\ g\ Ca(CH_3CH_2COO)_2} \right)\left(\frac{2\ mol\ CH_3CH_2COO^-}{1\ mol\ Ca(CH_3CH_2COO)_2} \right)$$

$= 0.1879497\ M\ CH_3CH_2COO^-$ (unrounded)

	CH$_3$CH$_2$COO$^-$	+ H$_2$O	$\leftrightarrows$ CH$_3$CH$_2$COOH	+ OH$^-$
Initial	0.1879497 M		0	0
Change	−x		+x	+x
Equilibrium	0.1879497 − x		x	x

$K_b = K_w / K_a = (1.0 \times 10^{-14}) / (1.3 \times 10^{-5}) = 7.6923 \times 10^{-10}$ (unrounded)

$$K_b = 7.6923 \times 10^{-10} = \frac{[CH_3CH_2COOH][OH^-]}{[CH_3CH_2COO^-]}$$

$$K_b = 7.6923 \times 10^{-10} = \frac{(x)(x)}{(0.1879497 - x)} \qquad \text{Assume x is small compared to 0.1879497.}$$

$$K_b = 7.6923 \times 10^{-10} = \frac{(x)(x)}{(0.1879497)}$$

$x = 1.20239988 \times 10^{-5} = [OH^-]$ (unrounded)

Check assumption: $[1.20239988 \times 10^{-5} / 0.1879497] \times 100\% = 0.006\%$, therefore the assumption is good.

$[H_3O^+] = K_w / [OH^-] = (1.0 \times 10^{-14}) / (1.20239988 \times 10^{-5}) = 8.31671 \times 10^{-10}\ M\ H_3O^+$ (unrounded)

$pH = -\log [H_3O^+] = -\log (8.31671 \times 10^{-10}) = 9.0800 = \mathbf{9.08}$

18.180   Putrescine can be abbreviated to $NH_2(CH_2)_4NH_2$. Its reaction in water is written as follows:

$NH_2(CH_2)_4NH_2(aq) + H_2O(l)\ \leftrightarrows\ NH_2(CH_2)_4NH_3^+(aq)\ +\ OH^-(aq)$
$0.10 - x$ x x

$x = [OH^-] = 2.1 \times 10^{-3}$

$$K_b = \frac{[NH_2(CH_2)_4\ NH_3^+][OH^-]}{[NH_2(CH_2)_4\ NH_2]} = \frac{[2.1 \times 10^{-3}][2.1 \times 10^{-3}]}{[0.10 - 2.1 \times 10^{-3}]} = 4.5045965 \times 10^{-5} = \mathbf{4.5 \times 10^{-5}}$$

18.183   a) The concentration of oxygen is higher in the lungs so the equilibrium shifts to the right.
b) In an oxygen deficient environment, the equilibrium would shift to the left to release oxygen.
c) A decrease in the $[H_3O^+]$ concentration would shift the equilibrium to the right. More oxygen is absorbed, but it will be more difficult to remove the O_2.
d) An increase in the $[H_3O^+]$ concentration would shift the equilibrium to the left. Less oxygen is absorbed, but it will be easier to remove the O_2.

18.186   a) Calculate the molarity of the solution

$$M = \left(\frac{7.500\ g\ CH_3CH_2COOH}{100.0\ mL\ solution}\right)\left(\frac{1\ mL}{10^{-3}\ L}\right)\left(\frac{1\ mol\ CH_3CH_2COOH}{74.08\ g\ CH_3CH_2COOH}\right)$$

$= 1.012419 = \mathbf{1.012\ M\ CH_3CH_2COOH}$

b) The freezing point depression equation is required to determine the molality of the solution.

$\Delta T = iK_f m = [0.000 - (-1.890°C)] = 1.890°C$

Temporarily assume i = 1.

$m = \Delta T / iK_f = (1.890°C) / [(1)(1.86°C/m)] = 1.016129032\ m = 1.016129032\ M$ (unrounded)

This molality is the total molality of all species in the solution, and is equal to their molarity.

From the equilibrium:

$CH_3CH_2COOH(aq) + H_2O(l)\ \leftrightarrows\ H_3O^+(aq)\ +\ CH_3CH_2COO^-(aq)$
$1.012419 - x$ x x

The total concentration of all species is:

$[ClCH_2COOH] + [H_3O^+] + [ClCH_2COO^-] = 1.016129032\ M$
$[1.012419 - x] + [x] + [x] = 1.012419 + x = 1.016129032\ M$
$x = 0.00371003 = \mathbf{0.004\ M\ CH_3CH_2COO^-}$

c) The percent dissociation is the amount dissociated (x from part b) divided by the original concentration from part a.

$$\text{Percent dissociation} = \frac{(0.00371003\ M)}{(1.012419\ M)} \times 100\% = 0.366452 = \mathbf{0.4\%}$$

18.187 Note that both pK_b values only have one significant figure. This will limit the final answers.

$K_{b \text{ (tertiary amine N)}} = 10^{-pK_b} = 10^{-5.1} = 7.94328 \times 10^{-6}$ (unrounded)

$K_{b \text{ (aromatic ring N)}} = 10^{-pK_b} = 10^{-9.7} = 1.995262 \times 10^{-10}$ (unrounded)

a) Ignoring the smaller K_b:

$$C_{20}H_{24}N_2O_2(aq) + H_2O(l) \rightleftharpoons OH^-(aq) + HC_{20}H_{24}N_2O_2^+(aq)$$

Initial	1.6×10^{-3} M	0	0
Change	$-x$	$+x$	$+x$
Equilibrium	$1.6 \times 10^{-3} - x$	x	x

$$K_b = \frac{\left[HC_{20}H_{24}N_2O_2^+\right]\left[OH^-\right]}{\left[H_2C_{20}H_{24}N_2O_2\right]} = 7.94328 \times 10^{-6}$$

$$K_b = \frac{[x][x]}{\left[1.6 \times 10^{-3} - x\right]} = 7.94328 \times 10^{-6}$$

The problem will need to be solved as a quadratic.

$x^2 = (7.94328 \times 10^{-6})(1.6 \times 10^{-3} - x) = 1.27092 \times 10^{-8} - 7.94328 \times 10^{-6} x$

$x^2 + 7.94328 \times 10^{-6} x - 1.27092 \times 10^{-8} = 0$

$\qquad a = 1 \quad b = 7.94328 \times 10^{-6} \quad c = -1.27092 \times 10^{-8}$

$$x = \frac{-b \pm \sqrt{b^2 - 4ac}}{2a}$$

$$x = \frac{-7.94328 \times 10^{-6} \pm \sqrt{\left(7.94328 \times 10^{-6}\right)^2 - 4(1)\left(-1.27092 \times 10^{-8}\right)}}{2(1)} = 1.08833 \times 10^{-4} \ M \ OH^-$$

$$[H_3O]^+ = \frac{K_w}{[OH^-]} = \frac{1.0 \times 10^{-14}}{1.08833 \times 10^{-4}} = 9.18838955 \times 10^{-11} \ M \ H_3O^+ \text{ (unrounded)}$$

$$pH = -\log [H_3O^+] = -\log (9.18838955 \times 10^{-11}) = 10.03676 = \textbf{10.0}$$

b) (Assume the aromatic N is unaffected by the tertiary amine N.) Use the K_b value for the aromatic nitrogen.

$$C_{20}H_{24}N_2O_2(aq) + H_2O(l) \rightleftharpoons OH^-(aq) + HC_{20}H_{24}N_2O_2^+(aq)$$

$\quad 1.6 \times 10^{-3} - x \qquad\qquad\qquad x \qquad\qquad x$

$$K_b = \frac{\left[HC_{20}H_{24}N_2O_2^+\right]\left[OH^-\right]}{\left[C_{20}H_{24}N_2O_2\right]} = 1.995262 \times 10^{-10}$$

$$K_b = \frac{[x][x]}{\left[1.6 \times 10^{-3} - x\right]} = 1.995262 \times 10^{-10} \quad \text{Assume x is small compared to } 1.6 \times 10^{-3}.$$

$$K_b = \frac{[x][x]}{\left[1.6 \times 10^{-3}\right]} = 1.995262 \times 10^{-10}$$

$x = 5.6501 \times 10^{-7} \ M \ OH^-$ (unrounded)

The hydroxide ion from the smaller K_b is much smaller than the hydroxide ion from the larger K_b (compare the powers of ten in the concentration).

c) $HC_{20}H_{24}N_2O_2^+(aq) + H_2O(l) \rightleftharpoons H_3O^+(aq) + C_{20}H_{24}N_2O_2(aq)$

$$K_a = \frac{K_w}{K_b} = \frac{1.0 \times 10^{-14}}{7.94328 \times 10^{-6}} = 1.2589 \times 10^{-9} \text{ (unrounded)}$$

$$K_a = 1.2589 \times 10^{-9} = \frac{\left[H_3O^+\right]\left[C_{20}H_{24}N_2O_2\right]}{\left[HC_{20}H_{24}N_2O_2^+\right]}$$

$$K_a = 1.2589 \times 10^{-9} = \frac{(x)(x)}{(0.33 - x)} \quad \text{Assume x is small compared to } 0.33.$$

$$K_a = 1.2589 \times 10^{-9} = \frac{(x)(x)}{(0.33)}$$

$[H_3O^+] = x = 2.038248 \times 10^{-5}$ (unrounded)

Check assumption: $(2.038248 \times 10^{-5} / 0.33) \times 100\% = 0.006\%$. The assumption is good.

pH $= -\log [H_3O^+] = -\log (2.038248 \times 10^{-5}) = 4.69074 = $ **4.7**

d) Quinine hydrochloride will be indicated as QHCl.

$$M = \left(\frac{1.5\%}{100\%}\right)\left(\frac{1.0 \text{ g}}{\text{mL}}\right)\left(\frac{1 \text{ mL}}{10^{-3} \text{ L}}\right)\left(\frac{1 \text{ mol QHCl}}{360.87 \text{ g QHCl}}\right) = 0.041566 \ M \text{ (unrounded)}$$

$$K_a = 1.2589 \times 10^{-9} = \frac{\left[H_3O^+\right]\left[C_{20}H_{24}N_2O_2\right]}{\left[HC_{20}H_{24}N_2O_2^+\right]}$$

$$K_a = 1.2589 \times 10^{-9} = \frac{(x)(x)}{(0.041566 - x)} \qquad \text{Assume x is small compared to 0.041566.}$$

$$K_a = 1.2589 \times 10^{-9} = \frac{(x)(x)}{(0.041566)}$$

$[H_3O^+] = x = 7.23377 \times 10^{-6}$ (unrounded)

Check assumption: $(7.23377 \times 10^{-6} / 0.041566) \times 100\% = 0.02\%$. The assumption is good.

pH $= -\log [H_3O^+] = -\log (7.23377 \times 10^{-6}) = 5.1406 = $ **5.1**

CHAPTER 19 IONIC EQUILIBRIA IN AQUEOUS SYSTEMS

19.1 Plan: The problems are both equilibria with the initial concentration of reactant and product given. For part (a), set up a reaction table for the dissociation of HF. Set up an equilibrium expression and solve for $[H_3O^+]$, assuming that the change in [HF] and $[F^-]$ is negligible. Check this assumption after finding $[H_3O^+]$. Convert $[H_3O^+]$ to pH. For part (b), first find the concentration of OH^- added. Then, use the neutralization reaction to find the change in initial [HF] and $[F^-]$. Repeat the solution in part (a) to find pH.

a) Solution:

Concentration (M)	HF(aq)	+	$H_2O(l)$	$\leftrightarrows$	$F^-(aq)$	+	$H_3O^+(aq)$
Initial	0.50		—		0.45		0
Change	– x		—		+ x		+ x
Equilibrium	0.50 – x		—		0.45 + x		x

Assumptions: 1) initial $[H_3O^+]$, from water, at $1.0 \times 10^{-7}\ M$, can be assumed to be zero and 2) x is negligible with respect to 0.50 M and 0.45 M.

$$K_a = \frac{\left[H_3O^+\right]\left[F^-\right]}{[HF]}$$

$$[H_3O^+] = K_a \frac{[HF]}{\left[F^-\right]} = \left(6.8 \times 10^{-4}\right)\frac{[0.50]}{[0.45]} = 7.5556 \times 10^{-4} = 7.6 \times 10^{-4}$$

Check assumptions:
1) $1.0 \times 10^{-7} \ll 7.6 \times 10^{-4}$ Thus, the assumption to set the initial concentration of H_3O^+ to zero is valid.
2) Percent error in assuming x is negligible: $(7.5556 \times 10^{-4}/0.45) \times 100 = 0.17\%$. The error is less than 5%, so the assumption is valid.

Solve for pH:
pH = –log (7.5556×10^{-4}) = 3.12173 = **3.12**
The other method to calculate the pH of a buffer is to use the Henderson-Hasselbalch equation:

$$pH = pK_a + \log\frac{[base]}{[acid]}$$

$$pH = pK_a + \log\frac{[F^-]}{[HF]} \qquad pK_a = -\log K_a = -\log(6.8 \times 10^{-4}) = 3.16749$$

$$pH = 3.16749 + \log\frac{[0.45]}{[0.50]}$$

pH = 3.12173 = **3.12**

Check: Since [HF] and $[F^-]$ are similar, the pH should be close to pK_a, which equals log (6.8×10^{-4}) = 3.17. The pH should be slightly less (more acidic) than pK_a because [HF] > $[F^-]$. The calculated pH of 3.12 is slightly less than pK_a of 3.17.

b) Solution: What is the initial molarity of the OH^- ion?

$$\left(\frac{0.40\,g\ NaOH}{L}\right)\left(\frac{1\,mol\ NaOH}{40.00\,g\ NaOH}\right)\left(\frac{1\,mol\ OH^-}{1\,mol\ NaOH}\right) = 0.010\ M\ OH^-$$

Set up reaction table for neutralization of 0.010 M OH^- (note the quantity of water is irrelevant).

Concentration (M)	HF(aq)	+	$OH^-(aq)$	$\rightarrow$	$F^-(aq)$	+	$H_2O(l)$
Before addition	0.50		—		0.45		—
Addition	—		0.010		—		—
Change	– 0.010		– 0.010		+ 0.010		—
After addition	0.49		0		0.46		—

Following the same solution path with the same assumptions as part (a):

$$[H_3O^+] = K_a \frac{[HF]}{[F^-]} = \left(6.8 \times 10^{-4}\right)\frac{[0.49]}{[0.46]} = 7.2434782 \times 10^{-4} = 7.2 \times 10^{-4}\ M$$

Check assumptions:
1) $7.2 \times 10^{-4} \gg 1.0 \times 10^{-7}$, the assumption is valid.
2) $(7.2 \times 10^{-4}/0.46)100 = 0.16\%$, which is less than the 5% maximum → assumption acceptable.
Solve for pH:
pH = $-\log (7.2434782 \times 10^{-4}) = 3.14005 = $ **3.14**
or using the Henderson-Hasselbalch equation:

$$pH = pK_a + \log\frac{[F^-]}{[HF]} = 3.16749 + \log\frac{[0.46]}{[0.49]}$$

$$pH = 3.14005 = \textbf{3.14}$$

Check: With addition of base, the pH should increase and it does, from 3.12 to 3.14. However, the pH should still be slightly less than pK_a: 3.14 is still less than 3.17.

19.2   Plan: For high buffer capacity, the components of a buffer should be concentrated and the concentrations of the base and acid components should be similar.
Solution:
a) The buffer has a much larger amount of weak acid than of the conjugate weak base. Addition of strong base would convert some HB into B^- to make the ration $[B^-]/\{HB]$ closer to 1 according to the reaction
    $HB(aq)$ + $OH^-(aq)$ → $B^-(aq)$ + $H_2O(l)$
b) The buffer with the highest possible buffer capacity would have 4HB particles and 4B$^-$ particles. Addition of strong base would convert 3 HB particles to 3 B$^-$ particles so that there are 4 of each of the weak acid and weak base.

19.3   Plan: Sodium benzoate is a salt so it dissolves in water to form Na^+ ions and $C_6H_5COO^-$ ions. Only the benzoate ion is involved in the buffer system represented by the equilibrium:
    $C_6H_5COOH(aq) + H_2O(l)$ ⇌ $C_6H_5COO^-(aq) + H_3O^+(aq)$
Given in the problem are the volume and pH of the buffer and the concentration of the base, benzoate ion. The question asks for the mass of benzoic acid to add to the sodium benzoate solution. First, find the concentration of C_6H_5COOH needed to make a buffer with a pH of 4.25. Multiply the volume by the concentration to find moles of C_6H_5COOH and use the molar mass to find grams of benzoic acid.
Solution: From the given pH, calculate $[H_3O^+]$:
    $[H_3O^+] = 10^{-4.25} = 5.62341 \times 10^{-5}\ M$ (unrounded)
The concentration of benzoic acid is calculated from the Henderson-Hasselbalch equation:

$$pH = pK_a + \log\frac{[C_6H_5COO^-]}{[C_6H_5COOH]}$$

$pKa = -\log Ka = -\log 6.3 \times 10^{-5} = 4.20066$

$$4.25 = 4.20066 + \log\frac{[0.050]}{[x]}$$

$$0.04934 = \log\frac{[0.050]}{[x]} \qquad \text{Raise each side to } 10^x.$$

$$1.1203146 = \frac{[0.050]}{[x]}$$

$$x = 4.46303 \times 10^{-2}\ M$$

The number of moles of benzoic acid is determined from the concentration and volume:
$(4.46303 \times 10^{-2}\ M\ C_6H_5COOH) \times (5.0\ L) = 0.22315\ mol\ C_6H_5COOH$ (unrounded)
Mass of C_6H_5COOH:

$$\left(0.22315\ mol\ C_6H_5COOH\right)\left(\frac{1\ mol\ C_6H_5COOH}{122.12\ g\ C_6H_5COOH}\right) = 27.2511 = \textbf{27 g}\ C_6H_5COOH$$

Prepare a benzoic acid/benzoate buffer by dissolving 27 g of C_6H_5COOH into 5.0 L of 0.050 M C_6H_5COONa. Using a pH meter, adjust the pH to 4.25 with strong acid or base.

Check: pK_a for benzoic acid is 4.20. The buffer pH is slightly greater than the pK_a, so the concentration of the acid should be slightly less than the concentration of the base. The calculation confirms this with $[C_6H_5COOH]$ at 0.045 M, which is slightly less than $[C_6H_5COO^-]$ at 0.050 M.

19.4   Plan: The titration is of a weak acid, HBrO, with a strong base, NaOH. The reactions involved are:
1) Neutralization of weak acid with strong base:

$$HBrO(aq) + OH^-(aq) \rightarrow H_2O(l) + BrO^-(aq)$$

Note that the reaction goes to completion and produces the conjugate base, BrO^-.
2) The weak acid and its conjugate base are in equilibrium based on the acid dissociation reaction:
a) $HBrO(aq) + H_2O(l) \rightleftharpoons BrO^-(aq) + H_3O^+(aq)$ or
b) the base dissociation reaction:

$$BrO^-(aq) + H_2O(l) \rightleftharpoons HBrO(aq) + OH^-(aq)$$

The pH of the solution is controlled by the relative concentrations of HBrO and BrO^-.
For each step in the titration, first think about what is present initially in the solution. Then use the two reactions to determine solution pH.
It is useful in a titration problem to first determine at what volume of titrant the equivalence point occurs.
e) Use the pH values from (a) – (d) to plot the titration curve.

Solution:
a) Before any base is added, the solution contains only HBrO and water. Equilibrium reaction 2a applies and pH can be found in the same way as for a weak acid solution:

Concentration (M)	$HBrO(aq)$	+	$H_2O(aq)$	$\rightleftharpoons$	$BrO^-(aq)$	+	$H_3O^+(aq)$
Initial	0.2000		—		0		0
Change	– x		—		+ x		+ x
Equilibrium	0.2000 – x		—		x		x

$$K_a = \frac{\left[H_3O^+\right]\left[BrO^-\right]}{\left[HBrO\right]} = 2.3 \times 10^{-9} = \frac{[x][x]}{\left[0.2000 - x\right]}$$

Assume 0.2000 – x = 0.2000 M

$$2.3 \times 10^{-9} = \frac{[x][x]}{\left[0.2000\right]}$$

x = $[H_3O^+]$ = 2.144761 x 10^{-5} M (unrounded)
pH = –log $[H_3O^+]$ = –log (2.144761 x 10^{-5}) = 4.66862 = **4.67**
Check the assumption by calculating the percent error in $[HBrO]_{eq}$.
Percent error = (2.144761 x 10^{-5} /0.2000)100 = 0.01%, this is well below the 5% maximum.
b) When [HBrO] = $[BrO^-]$, the solution contains significant concentrations of both the weak acid and its conjugate base. For the concentrations to be equal, half of the equivalence point volume has to have been added. Calculate the concentration of the acid and base and use the equilibrium expression for reaction 2a to find pH.
Since [HBrO] = $[BrO^-]$ their ratio equals 1.

$$K_a = \frac{\left[H_3O^+\right]\left[BrO^-\right]}{\left[HBrO\right]}$$

$$[H_3O^+] = K_a \frac{\left[HBrO\right]}{\left[BrO^-\right]} = \left(2.3 \times 10^{-9}\right)\frac{[1]}{[1]} = 2.3 \times 10^{-9} \, M$$

pH = –log $[H_3O^+]$ = –log (2.3 x 10^{-9}) = 8.63827 = **8.64**
Note that when [HBrO] = $[BrO^-]$, the titration is at the midpoint (half the volume to the equivalence point) and pH = pK_a.
c) At the equivalence point, the total number of moles of HBrO present initially in solution equals the number of moles of base added. Therefore, reaction 1 goes to completion to produce that number of moles of BrO^-. The solution consists of BrO^- and water. Calculate the concentration of BrO^-, and then find the pH using the base dissociation equilibrium, reaction 2b.

First, find equivalence point volume of NaOH.

$$\left(\frac{0.2000 \text{ mol HBrO}}{\text{L}}\right)\left(\frac{10^{-3} \text{ L}}{1 \text{ mL}}\right)(20.00 \text{ mL})\left(\frac{1 \text{ mol NaOH}}{1 \text{ mol HBrO}}\right)\left(\frac{1 \text{ L}}{0.1000 \text{ mol NaOH}}\right)\left(\frac{1 \text{ mL}}{10^{-3} \text{ L}}\right) = 40.00 \text{ mL NaOH added}$$

All of the HBrO present at the beginning of the titration is neutralized and converted to BrO^- at the equivalence point. Calculate the concentration of BrO^-.

Initial moles of HBrO: $(0.2000 \, M)(0.02000 \text{ L}) = 0.004000$ mol

Moles of added NaOH: $(0.1000 \, M)(0.04000 \text{ L}) = 0.004000$ mol

Amount (mol)	HBrO(aq) +	OH$^-$(aq)	→ H$_2$O(l) +	BrO$^-$(aq)
Before addition	0.004000 mol	—	—	0
Addition	—	0.004000 mol	—	—
Change	– 0.004000 mol	– 0.004000 mol	—	+0.004000 mol
After addition	0	0	—	0.004000 mol

At the equivalence point, 40.00 mL of NaOH solution has been added (see calculation above) to make the total volume of the solution (20.00 + 40.00) mL = 60.00 mL.

$$[BrO^-] = \left(\frac{0.004000 \text{ mol BrO}^-}{60.00 \text{ mL}}\right)\left(\frac{1 \text{ mL}}{10^{-3} \text{ L}}\right) = 0.06666667 \, M \text{ (unrounded)}$$

Set up reaction table with reaction 2b, since only BrO^- and water are present initially:

Concentration (M)	BrO$^-$(aq) +	H$_2$O(l) ⇌	HBrO(aq) +	OH$^-$(aq)
Initial	0.06666667	—	0	0
Change	– x	—	+ x	+ x
Equilibrium	0.06666667 – x	—	x	x

$K_b = K_w/K_a = (1.0 \times 10^{-14} / 2.3 \times 10^{-9}) = 4.347826 \times 10^{-6}$ (unrounded)

$$K_b = \frac{\left[OH^-\right]\left[HBrO\right]}{\left[BrO^-\right]} = \frac{[x][x]}{[0.06666667 - x]} = 4.347826 \times 10^{-6}$$

Assume that x is negligible, since $[BrO^-] \gg K_b$.

$$4.347826 \times 10^{-6} = \frac{[x][x]}{[0.06666667]}$$

$x = [OH^-] = 5.3838191 \times 10^{-4} = 5.4 \times 10^{-4} \, M$

Check the assumption by calculating the percent error in $[BrO^-]_{eq}$.

percent error = $(5.3838191 \times 10^{-4}/0.06666667)100 = 0.8\%$, which is well below the 5% maximum.

pOH = –log $(5.3838191 \times 10^{-4}) = 3.26891$

pH = 14 – pOH = 14 – 3.26891 = 10.73109 = **10.73**

d) After the equivalence point, the concentration of excess strong base determines the pH. Find the concentration of excess base and use it to calculate the pH.

Initial moles of HBrO: $(0.2000 \, M)(0.02000 \text{ L}) = 0.004000$ mol

Moles of added NaOH: 2×0.004000 mol = 0.008000 mol NaOH

Amount (mol)	HBrO(aq) +	OH$^-$(aq)	→ H$_2$O(l) +	BrO$^-$(aq)
Before addition	0.004000 mol	—	—	0
Addition	—	0.008000 mol	—	—
Change	– 0.004000 mol	– 0.004000 mol	—	+0.004000 mol
After addition	0	0.004000 mol	—	0.004000 mol

Excess NaOH: 0.004000 mol

Volume of added NaOH: $(0.0080000 \text{ mol NaOH})\left(\frac{1 \text{ L}}{0.1000 \text{ mol NaOH}}\right)\left(\frac{1 \text{ mL}}{10^{-3} \text{ L}}\right) = 80.00 \text{ mL}$

Total volume: 20.00 mL + 80.00 mL = 100.0 mL

[NaOH] = 0.004000 mol NaOH/0.1000 L = 0.0400 M

pOH = –log $(0.0400) = 1.3979$

pH = 14 – pOH = 14 – 1.3979 = **12.60**

e) Plot the pH values calculated in the preceding parts of this problem as a function of the volume of titrant.

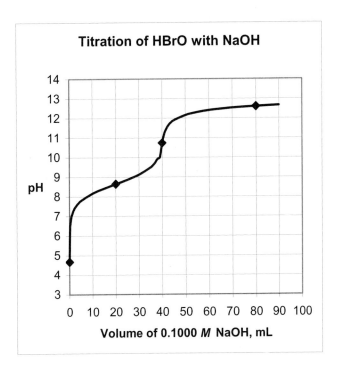

Titration of HBrO with NaOH

Check: The plot and pH values follow the pattern for a weak acid vs. strong base titration. The pH at the midpoint of the titration does equal pK_a. The equivalence point should be, and is, greater than 7.

19.5   Plan: Write the formula of the salt and the reaction showing the equilibrium of a saturated solution. The ion-product expression can be written from the stoichiometry of the solution reaction.
Solution:
a) The formula of calcium sulfate is $CaSO_4$. The equilibrium reaction is

$CaSO_4(s) \leftrightarrows Ca^{2+}(aq) + SO_4{}^{2-}(aq)$
Ion–product expression: $K_{sp} = [Ca^{2+}][SO_4{}^{2-}]$
b) Chromium(III) carbonate is $Cr_2(CO_3)_3$.
$Cr_2(CO_3)_3(s) \leftrightarrows 2\,Cr^{3+}(aq) + 3\,CO_3{}^{2-}(aq)$
Ion–product expression: $K_{sp} = [Cr^{3+}]^2[CO_3{}^{2-}]^3$
c) Magnesium hydroxide is $Mg(OH)_2$.
$Mg(OH)_2(s) \leftrightarrows Mg^{2+}(aq) + 2\,OH^-(aq)$
Ion–product expression: $K_{sp} = [Mg^{2+}][OH^-]^2$
d) Arsenic(III) sulfide is As_2S_3.
$As_2S_3(s) \leftrightarrows 2\,As^{3+}(aq) + 3\,S^{2-}(aq)$
$3\{S^{2-}(aq) + H_2O(l) \leftrightarrows HS^-(aq) + OH^-(aq)\}$
$As_2S_3(s) + 3\,H_2O(l) \leftrightarrows 2\,As^{3+}(aq) + 3\,HS^-(aq) + 3\,OH^-(aq)$
The second equilibrium must be considered in this case because its equilibrium constant is large, so essentially all the sulfide ion is converted to HS^- and OH^-.
Ion-product expression: $K_{sp} = [As^{3+}]^2[HS^-]^3[OH^-]^3$
Check: All equilibria agree with the formulas of salts.

19.6   Plan: Calculate the solubility of CaF_2 as molarity and use molar ratios to find the molarity of Ca^{2+} and F^- dissolved in solution. Calculate K_{sp} from $[Ca^{2+}]$ and $[F^-]$ using the ion-product expression.
Solution: Convert the solubility to molar solubility:

$$\left(\frac{1.5 \times 10^{-4} \text{ g } CaF_2}{10.0 \text{ mL}}\right)\left(\frac{1 \text{ mL}}{10^{-3} \text{ L}}\right)\left(\frac{1 \text{ mol } CaF_2}{78.08 \text{ g } CaF_2}\right) = 1.9211 \times 10^{-4} \text{ } M \text{ } CaF_2 \text{ (unrounded)}$$

$[Ca^{2+}] = [CaF_2] = 1.9211 \times 10^{-4} \text{ } M$ because there is 1 mol of calcium ions in each mol of CaF_2.
$[F^-] = 2[CaF_2] = 3.8422 \times 10^{-4} \text{ } M$ because there are 2 mol of fluoride ions in each mol of CaF_2.
The solubility equilibrium is:
$CaF_2(s) \rightleftharpoons Ca^{2+}(aq) + 2 F^-(aq)$ $K_{sp} = [Ca^{2+}][F^-]^2$
Calculate K_{sp} using the solubility product expression from above and the saturated concentrations of calcium and fluoride ions.
$K_{sp} = [Ca^{2+}][F^-]^2 = (1.9211 \times 10^{-4})(3.8422 \times 10^{-4})^2 = 2.836024 \times 10^{-11} = \mathbf{2.8 \times 10^{-11}}$
The K_{sp} for CaF_2 is 2.8×10^{-11} at 18°C.
Check: Appendix C reports a K_{sp} value at 25°C for CaF_2 of 3.2×10^{-11}. The calculated value at 18°C is very close to this literature value.

19.7   Plan: Write the solubility reaction for $Mg(OH)_2$ and set up a reaction table, where S is the unknown molar solubility of the Mg^{2+} ion. Use the ion-product expression to solve for the concentration of $Mg(OH)_2$ in a saturated solution (also called the solubility of $Mg(OH)_2$).
Solution:

Concentration (M)	$Mg(OH)_2(s) \rightleftharpoons$	$Mg^{2+}(aq)$ +	$2 OH^-(aq)$
Initial	—	0	0
Change	—	+ S	+ 2 S
Equilibrium	—	S	2 S

$K_{sp} = [Mg^{2+}][OH^-]^2 = (S)(2 S)^2 = 4 S^3 = 6.3 \times 10^{-10}$
$S = 5.4004114 \times 10^{-4} = 5.4 \times 10^{-4} \text{ } M \text{ } Mg(OH)_2$
The solubility of $Mg(OH)_2$ is equal to S, the concentration of magnesium ions at equilibrium, so the molar solubility of magnesium hydroxide in pure water is $\mathbf{5.4 \times 10^{-4} \text{ } M}$.
Check: To check the calculation, work it backwards (also a good technique when using problems as study tools). From the calculated solubility, find K_{sp}.
$[Mg^{2+}] = 5.4 \times 10^{-4} \text{ } M$ and $[OH^-] = 2(5.4 \times 10^{-4} \text{ } M) = 1.08 \times 10^{-4} \text{ } M$
$K_{sp} = (5.4 \times 10^{-4})(1.08 \times 10^{-4})^2 = 6.5 \times 10^{-10}$
The backwards calculation gives back given K_{sp}.

19.8   Plan: Write the solubility reaction of $BaSO_4$. For part (a) set up a reaction table in which $[Ba^{2+}] = [SO_4^{2-}] = S$, which also equals the solubility of $BaSO_4$. Then, solve for S using the ion-product expression. For part (b), there is an initial concentration of sulfate, so set up the reaction table including this initial $[SO_4^{2-}]$. Solve for the solubility, S, which equals $[Ba^{2+}]$ at equilibrium.
Solution:
a) Set up reaction table.

Concentration (M)	$BaSO_4(s) \rightleftharpoons$	$Ba^{2+}(aq)$ +	$SO_4^{2-}(aq)$
Initial	—	0	0
Change	—	+ S	+ S
Equilibrium	—	S	S

$K_{sp} = [Ba^{2+}][SO_4^{2-}] = S^2 = 1.1 \times 10^{-10}$
$\qquad S = 1.0488 \times 10^{-5} = \mathbf{1.0 \times 10^{-5} \text{ } M}$
The molar solubility of $BaSO_4$ in pure water is $1.0 \times 10^{-5} \text{ } M$.
b) Set up another reaction table with initial $[SO_4^{2-}] = 0.10 \text{ } M$ (from the Na_2SO_4).

Concentration (M)	$BaSO_4(s) \rightleftharpoons$	$Ba^{2+}(aq)$ +	$SO_4^{2-}(aq)$
Initial	—	0	0.10
Change	—	+ S	+ S
Equilibrium	—	S	0.10 + S

$K_{sp} = [Ba^{2+}][SO_4^{2-}] = S(0.10 + S) = 1.1 \times 10^{-10}$

Assume that $0.10 + S$ is approximately equal to 0.10, which appears to be a good assumption based on the fact that $0.10 > 1 \times 10^{-10}$, K_{sp}.

$$K_{sp} = S(0.10) = 1.1 \times 10^{-10}$$
$$S = 1.1 \times 10^{-9} \ M$$

Molar solubility of $BaSO_4$ in $0.10 \ M \ Na_2SO_4$ is **$1.1 \times 10^{-9} \ M$.**

Check: The solubility of $BaSO_4$ decreases when sulfate ions are already present in the solution. The calculated decrease is from $10^{-5} \ M$ to $10^{-9} \ M$, for a 10,000-fold decrease. This decrease is expected to be large because of the high concentration of sulfate ions.

19.9   Plan: First, write the solubility reaction for the salt. Then, check the ions produced when the salt dissolves to see if they will react with acid. Three cases are possible:
1) If OH^- is produced, then addition of acid will neutralize the hydroxide ions and shift the solubility equilibrium toward the products. This causes more salt to dissolve. Write the solubility and neutralization reactions.
2) If the anion from the salt is the conjugate base of a weak acid, it will react with the added acid in a neutralization reaction. Solubility of the salt increases as the anion is neutralized. Write the solubility and neutralization reactions.
3) If the anion from the salt is the conjugate base of a strong acid, it does not react with a strong acid. The solubility of the salt is unchanged by the addition of acid. Write the solubility reaction.
Solution:
a) Calcium fluoride, CaF_2
Solubility reaction: $CaF_2(s) \leftrightarrows Ca^{2+}(aq) + 2 \ F^-(aq)$
Fluoride ion is the conjugate base of HF, a weak acid. Thus, it will react with H_3O^+ from the strong acid, HNO_3.
Neutralization reaction: $F^-(aq) + H_3O^+(aq) \rightarrow HF(aq) + H_2O(l)$
The neutralization reaction decreases the concentration of fluoride ions, which causes the solubility equilibrium to shift to the right and more CaF_2 dissolves. The solubility of CaF_2 increases with the addition of HNO_3.
b) Zinc sulfide, ZnS
Solubility reaction: $ZnS(s) + H_2O(l) \leftrightarrows Zn^{2+}(aq) + HS^-(aq) + OH^-(aq)$
Two anions are formed because the sulfide ion from ZnS reacts almost completely with water to form HS^- and OH^-.
The hydroxide ion reacts with the added acid:
Neutralization reaction: $OH^-(aq) + H_3O^+(aq) \rightarrow 2 \ H_2O(l)$
In addition, the hydrogen sulfide ion, the conjugate base of the weak acid H_2S, reacts with the added acid:
Neutralization reaction: $HS^-(aq) + H_3O^+(aq) \rightarrow H_2S(aq) + H_2O(l)$
Both neutralization reactions decrease the concentration of products in the solubility equilibrium, which causes a shift to the right, and more ZnS dissolves. The addition of HNO_3 will increase the solubility of ZnS.
c) Silver iodide, AgI.
Solubility reaction: $AgI(s) \leftrightarrows Ag^+(aq) + I^-(aq)$
The iodide ion is the conjugate base of a strong acid, HI. So, I^- will not react with added acid. The solubility of AgI will not change with added HNO_3.

19.10   Plan: First, write the solubility equilibrium equation and ion-product expression. Use the given concentrations of calcium and phosphate ions to calculate Q_{sp}. Compare Q_{sp} to K_{sp}.
If $K_{sp} < Q_{sp}$, precipitation occurs. If $K_{sp} \geq Q_{sp}$ then, precipitation will not occur.
Solution:
Write the solubility equation:
$Ca_3(PO_4)_2(s) \leftrightarrows 3 \ Ca^{2+}(aq) + 2 \ PO_4^{3-}(aq)$
and ion-product expression: $Q_{sp} = [Ca^{2+}]^3[PO_4^{3-}]^2$
$$[Ca^{2+}] = [PO_4^{3-}] = 1.0 \times 10^{-9} \ M$$
$Q_{sp} = [Ca^{2+}]^3[PO_4^{3-}]^2 = (1.0 \times 10^{-9})^3(1.0 \times 10^{-9})^2 = 1.0 \times 10^{-45}$

Compare K_{sp} and Q_{sp}. $K_{sp} = 1.2 \times 10^{-29} > 1.0 \times 10^{-45} = Q_{sp}$
Precipitation will not occur because concentrations are below the level of a saturated solution as shown by the value of Q_{sp}.
Check: The concentrations of calcium and phosphate ions are quite low, but K_{sp} is also small so it is difficult to predict whether precipitation will occur by just looking at the numbers. One way to check is with an estimate of the concentrations of ions in a saturated solution.
$K_{sp} = [Ca^{2+}]^3[PO_4^{3-}]^2 = (3\,S)^3(2\,S)^2 = 108\,S^5 = 1 \times 10^{-29}$
$\qquad S = 6.213 \times 10^{-7}\,M$ (unrounded)
In a saturated solution
$[Ca^{2+}] = 3(6.213 \times 10^{-7}\,M) = 1.86 \times 10^{-6}\,M$ and $[PO_4^{3-}] = 2(6.213 \times 10^{-7}) = 1.24 \times 10^{-6}\,M$
Both of these concentrations are much greater than the actual concentrations of $1 \times 10^{-9}\,M$ for both ions, so the solution is unsaturated and no $Ca_3(PO_4)_2$ will precipitate.

19.11   Plan: First, write the solubility equilibrium equation and ion-product expression. For b) use the given amounts of nickel (II) and hydroxide ions to calculate Q_{sp}. Compare Q_{sp} to K_{sp}. For c) check the ions produced when the salt dissolves to see if they will react with acid. Three cases are possible:
1) If OH^- is produced, then addition of acid will neutralize the hydroxide ions and shift the solubility equilibrium toward the products. This causes more salt to dissolve.
2) If the anion from the salt is the conjugate base of a weak acid, it will react with the added acid in a neutralization reaction. Solubility of the salt increases as the anion is neutralized.
3) If the anion from the salt is the conjugate base of a strong acid, it does not react with a strong acid. The solubility of the salt is unchanged by the addition of acid.
Solution:
a) Write the solubility equation:
$Ni(OH)_2(s) \leftrightarrows Ni^{2+}(aq) + 2\,OH^-(aq)$
Scene 3 has the same relative number of ions as in the formula of $Ni(OH)_2$. The Ni^{2+} and OH^- ions are in a 1:2 ratio in Scene 3.
b) Write the ion-product expression: $Q_{sp} = [Ni^{2+}][OH^-]^2$
Calculate K_{sp} using Scene 3: $K_{sp} = [Ni^{2+}][OH^-]^2 = [2][4]^2 = 32$
Calculate Q_{sp} using Scene 1: $Q_{sp} = [Ni^{2+}][OH^-]^2 = [3][4]^2 = 48$
Calculate Q_{sp} using Scene 2: $Q_{sp} = [Ni^{2+}][OH^-]^2 = [4][2]^2 = 16$
Q_{sp} exceeds K_{sp} in Scene 1 (48 > 32) so additional solid will form in **Scene 1.**
c) Hydroxide ion is one of the products of the solubility equilibrium reaction. The hydroxide ion reacts with added **acid**:
$\qquad$ Neutralization reaction: $OH^-(aq) + H_3O^+(aq) \rightarrow 2\,H_2O(l)$
The neutralization reaction decreases the concentration of $OH^-(aq)$ in the solubility equilibrium, which causes a shift to the right, and more $Ni(OH)_2(s)$ dissolves.

19.12   Plan: Compare the K_{sp} values for the two salts. Since $CaSO_4$ is more soluble, calculate what concentration of sulfate ions that would need to be added to decrease the concentration of calcium ions to 99.99% of 0.025 M.
Solution:
The solubility equilibrium for $CaSO_4$ is
$CaSO_4(s) \leftrightarrows Ca^{2+}(aq) + SO_4^{2-}(aq)$
$\qquad$ and $K_{sp} = [Ca^{2+}][SO_4^{2-}] = 2.4 \times 10^{-5}$
$$\left[SO_4^{2-}\right] = \frac{K_{sp}}{[Ca^{2+}]}$$
The concentration of calcium ions will be 99.99% of 0.025 M. Calculate $[SO_4^{2-}]$:
$$[SO_4^{2-}] = \left(2.4 \times 10^{-5}\right)\left(\frac{100\%}{99.99\%}\right)\left(\frac{1}{0.025\ M}\right) = 9.60096 \times 10^{-4} = \mathbf{9.6 \times 10^{-4}\,M}$$

Check: An estimate of $[SO_4^{2-}]$ can be calculated by dividing 2.4×10^{-5} by 0.025 to give $1 \times 10^{-4}\,M$. This is approximately equal to the calculated result of $9.6 \times 10^{-4}\,M$.

19.13   Plan: Write the complex-ion formation equilibrium reaction. Calculate the initial concentrations of $Fe(H_2O)_6^{3+}$ and CN^-. The approach to complex ion equilibria problems is slightly different than for solubility equilibria because formation constants are generally large while solubility product constants are generally very small. The best mathematical approach is to assume that the equilibrium reaction goes to completion and then calculate back to find the equilibrium concentrations of reactants. So, assume that all of the $Fe(H_2O)_6^{3+}$ reacts to form $Fe(CN)_6^{3-}$ and calculate the concentration of $Fe(CN)_6^{3-}$ formed from the given concentrations of $Fe(H_2O)_6^{3+}$ and CN^- and the concentration of the excess reactant. Then use the complex ion formation equilibrium to find the equilibrium concentration of $Fe(H_2O)_6^{3+}$.

Solution:

Equilibrium reaction: $Fe(H_2O)_6^{3+}(aq) + 6\ CN^-(aq) \leftrightarrows Fe(CN)_6^{3-}(aq) + 6\ H_2O(l)$

Initial concentrations from a simple dilution calculation:

$$M_f = M_i V_i / V_f$$

$[Fe(H_2O)_6^{3+}] = (3.1 \times 10^{-2}\ M)\ (25.5\ mL)\ /\ (25.5 + 35.0)\ mL = 0.013066115\ M$ (unrounded)

$[CN^-] = (1.5\ M)\ (35.0\ mL)\ /\ (25.5 + 35.0)\ mL = 0.867768595\ M$ (unrounded)

Set up a reaction table:

Concentration (M)	$Fe(H_2O)_6^{3+}(aq)$	+	$6\ CN^-(aq)$	$\leftrightarrows$	$Fe(CN)_6^{3-}(aq) + 6\ H_2O(l)$	
Initial	0.013066115		0.867768595		0	—
Change	− 0.013066115 + x		− 6 (0.013066115)		+ 0.013066115	—
Equilibrium	x		0.789371905		0.013066115	—

$$K_f = 4.0 \times 10^{43} = \frac{\left[Fe(CN)_6^{3-}\right]}{\left[Fe(H_2O)_6^{3+}\right]\left[CN^-\right]^6} = \frac{[0.0130661]}{[x][0.78937190]^6}$$

$$x = 1.3501949 \times 10^{-45} = \mathbf{1.4 \times 10^{-45}\ M}$$

The concentration of $Fe(H_2O)_6^{3+}$ at equilibrium is $1.4 \times 10^{-45}\ M$. This concentration is so low that it is impossible to calculate it using the initial concentrations minus a variable x. The variable would have to be so close to the initial concentration that the initial concentration of x cannot be calculated to enough significant figures (43 in this case) to get a difference in concentrations of $1 \times 10^{-45}\ M$. Thus, the approach above is the best to calculate the very low equilibrium concentration of $Fe(H_2O)_6^{3+}$.

Check: The equilibrium concentration of $Fe(H_2O)_6^{3+}$ should be low because K_f is very large.

19.14   Plan: Write equations for the solubility equilibrium and formation of the silver-ammonia complex ion. Add the two equations to get the overall reaction. Set up a reaction table for the overall reaction with the given value for initial $[NH_3]$. Write equilibrium expressions from the overall balanced reaction and calculate $K_{overall}$ from K_f and K_{sp} values. Insert the equilibrium concentration values from the reaction table into the equilibrium expression and calculate solubility.

Solution: Equilibria:

$AgBr(s) \leftrightarrows Ag^+(aq) + Br^-(aq)$	$K_{sp} = 5.0 \times 10^{-13}$
$Ag^+(aq) + 2\ NH_3(aq) \leftrightarrows Ag(NH_3)_2^+(aq)$	$K_f = 1.7 \times 10^7$
$AgBr(s) + 2\ NH_3(aq) \leftrightarrows Ag(NH_3)_2^+(aq) + Br^-(aq)$	$K_{overall} = K_{sp}K_f$

Set up reaction table:

Concentration (M)	$AgBr(s)$	+	$2\ NH_3(aq)$	$\leftrightarrows$	$Ag(NH_3)_2^+(aq)$ +	$Br^-(aq)$
Initial	—		1.0		0	0
Change	—		− 2 S		+ S	+ S
Equilibrium	—		1.0 − 2 S		S	S

$$K_{overall} = \frac{\left[Ag(NH_3)_2^+\right]\left[Br^-\right]}{\left[NH_3\right]^2} = K_{sp}K_f = (5.0 \times 10^{-13})\ (1.7 \times 10^7) = 8.5 \times 10^{-6}$$

Calculate the solubility of AgBr:

$$K_{overall} = \frac{[S][S]}{[1.0 - 2S]^2} = 8.5 \times 10^{-6}$$

Assume that $1.0 - 2S$ is approximately equal to 1.0, which appears to be a good assumption based on the fact that $1.0 \gg 8.5 \times 10^{-6}, K_{overall}$.

$$\frac{[S][S]}{[1.0]^2} = 8.5 \times 10^{-6}$$

$S = 2.9154759 \times 10^{-3} = 2.9 \times 10^{-3} \ M$

The solubility of AgBr in ammonia is **less** than its solubility in hypo (sodium thiosulfate).

Check: Since the formation constant for $Ag(NH_3)_2^+$ is less than the formation constant of $Ag(S_2O_3)_2^{3-}$, the addition of ammonia will increase the solubility of AgBr less than the addition of thiosulfate ion increases its solubility.

END–OF–CHAPTER PROBLEMS

19.2   The weak acid component neutralizes added base and the weak base component neutralizes added acid so that the pH of the buffer solution remains relatively constant. The components of a buffer do not neutralize one another when they are a conjugate acid/base pair.

19.7   The buffer component ratio refers to the ratio of concentrations of the acid and base that make up the buffer. When this ratio is equal to 1, the buffer resists changes in pH with added acid to the same extent that it resists changes in pH with added base. The buffer range extends equally in both the acidic and basic direction. When the ratio shifts with higher [base] than [acid], the buffer is more effective at neutralizing added acid than base so the range extends further in the acidic than basic direction. The opposite is true for a buffer where [acid] > [base]. Buffers with a ratio equal to 1 have the greatest buffer range. The more the buffer component ratio deviates from 1, the smaller the buffer range.

19.9   a) The buffer component ratio and pH **increase** with added base. The OH^- reacts with HA to decrease its concentration and increase [NaA]. The ratio [NaA] / [HA] thus increases. The pH of the buffer will be more basic because the concentration of base, A^-, has increased and the concentration of acid, HA, decreased.
b) Buffer component ratio and pH **decrease** with added acid. The H_3O^+ reacts with A^- to decrease its concentration and increase [HA]. The ratio [NaA] / [HA] thus decreases. The pH of the buffer will be more acidic because the concentration of base, A^-, has decreased and the concentration of acid, HA, increased.
c) Buffer component ratio and pH **increase** with the added sodium salt. The additional NaA increases the concentration of both NaA and HA, but the relative increase in [NaA] is greater. Thus, the ratio increases and the solution becomes more basic. Whenever base is added to a buffer, the pH always increases, but only slightly if the amount of base is not too large.
d) Buffer component ratio and pH **decrease**. The concentration of HA increases more than the concentration of NaA, so the ratio is less and the solution is more acidic.

19.11   The buffer components are propanoic acid and propanoate ion. The sodium ions are spectator ions and are ignored because they are not involved in the buffer. The reaction table that describes this buffer is:

Concentration (M)	$CH_3CH_2COOH(aq) + H_2O(l) \rightleftharpoons$		$CH_3CH_2COO^-(aq) +$	$H_3O^+(aq)$
Initial	0.15	—	0.35	0
Change	− x	—	+ x	+ x
Equilibrium	0.15 − x	—	0.35 + x	x

Assume that x is negligible with respect to both 0.15 and 0.35 since both concentrations are much larger than K_a.

$$K_a = \frac{\left[H_3O^+\right]\left[CH_3CH_2COO^-\right]}{\left[CH_3CH_2COOH\right]} = \frac{[x][0.35 + x]}{[0.15 - x]} = \frac{[x][0.35]}{[0.15]} = 1.3 \times 10^{-5}$$

$$[H_3O^+] = K_a \frac{\left[CH_3CH_2COOH\right]}{\left[CH_3CH_2COO^-\right]} = \left(1.3 \times 10^{-5}\right)\left(\frac{0.15}{0.35}\right) = 5.57143 \times 10^{-6} = \mathbf{5.6 \times 10^{-6} \ M}$$

Check assumption: percent error = $(5.6 \times 10^{-6} / 0.15)100\% = 0.0037\%$. The assumption is valid.
        pH = −log $[H_3O^+]$ = −log (5.57143×10^{-6}) = 5.2540 = **5.25**.

Another solution path to find pH is using the Henderson–Hasselbalch equation:

$$pH = pK_a + \log\left(\frac{[base]}{[acid]}\right) \qquad pK_a = -\log(1.3 \times 10^{-5}) = 4.886$$

$$pH = 4.886 + \log\left(\frac{[CH_3CH_2COO^-]}{[CH_3CH_2COOH]}\right) = 4.886 + \log\left(\frac{[0.35]}{[0.15]}\right)$$

$$pH = 5.25398 = 5.25$$

19.13 The buffer components are nitrous acid, HNO_2, and nitrite ion, NO_2^-. The potassium ions are ignored because they are not involved in the buffer. Set up the problem with a reaction table.

Concentration (M)	$HNO_2(aq)$ +	$H_2O(l)$ $\leftrightarrows$	$NO_2^-(aq)$ +	$H_3O^+(aq)$
Initial	0.55	—	0.75	0
Change	− x	—	+ x	+ x
Equilibrium	0.55 − x	—	0.75 + x	x

Assume that x is negligible with respect to both 0.55 and 0.75 because both concentrations are much larger than K_a.

$$K_a = \frac{\left[H_3O^+\right]\left[NO_2^-\right]}{\left[HNO_2\right]} = \frac{[x][0.75 + x]}{[0.55 - x]} = \frac{[x][0.75]}{[0.55]} = 7.1 \times 10^{-4}$$

$$[H_3O^+] = K_a \frac{\left[HNO_2\right]}{\left[NO_2^-\right]} = \left(7.1 \times 10^{-4}\right)\frac{[0.55]}{[0.75]} = 5.2066667x \ 10^{-4} = \mathbf{5.2 \times 10^{-4} \ M}$$

Check assumption: percent error = $(5.2066667 \times 10^{-4} / 0.55)100\% = 0.095\%$. The assumption is valid.
pH = $-[H_3O^+]$ = $-\log (5.2066667 \times 10^{-4}) = 3.28344 = \mathbf{3.28}$
Verify the pH using the Henderson-Hasselbalch equation.

$$pH = pK_a + \log\left(\frac{[base]}{[acid]}\right) \qquad pK_a = -\log(7.1 \times 10^{-4}) = 3.149$$

$$pH = 3.149 + \log\left(\frac{[NO_2^-]}{[HNO_2]}\right) = 3.149 + \log\left(\frac{[0.75]}{[0.55]}\right)$$

$$pH = 3.2837 = 3.28$$

19.15 The buffer components are formic acid, HCOOH, and formate ion, $HCOO^-$. The sodium ions are ignored because they are not involved in the buffer. Calculate K_a from pK_a and write a reaction table for the dissociation of formic acid.

$K_a = 10^{-pK_a} = 10^{-3.74} = 1.8197 \times 10^{-4}$ (unrounded)

Concentration (M)	HCOOH(aq) +	$H_2O(l)$ $\leftrightarrows$	$HCOO^-(aq)$ +	$H_3O^+(aq)$
Initial	0.45	—	0.63	0
Change	− x	—	+ x	+ x
Equilibrium	0.45 − x	—	0.63 + x	x

Assume that x is negligible because both concentrations are much larger than K_a.

$$K_a = \frac{\left[H_3O^+\right]\left[HCOO^-\right]}{\left[HCOOH\right]} = \frac{[x][0.63 + x]}{[0.45 - x]} = \frac{[x][0.63]}{[0.45]} = 1.8197 \times 10^{-4} \text{ (unrounded)}$$

$$[H_3O^+] = K_a \frac{\left[HCOOH\right]}{\left[HCOO^-\right]} = \left(1.8197 \times 10^{-4}\right)\frac{[0.45]}{[0.63]} = 1.29979 \times 10^{-4} = 1.3 \times 10^{-4} \ M$$

Check assumption: percent error = $(1.29979 \times 10^{-4} / 0.45)100\% = 0.029\%$. The assumption is valid.
pH = $-[H_3O^+]$ = $-\log (1.29979 \times 10^{-4}) = 3.886127 = \mathbf{3.89}$

Verify the pH using the Henderson-Hasselbalch equation.

$$pH = pK_a + \log\left(\frac{[base]}{[acid]}\right)$$

$$pH = 3.74 + \log\left(\frac{[HCOO^-]}{[HCOOH]}\right) = 3.74 + \log\left(\frac{[0.63]}{[0.45]}\right)$$

$$pH = 3.8861 = 3.89$$

19.17   The buffer components are phenol, C_6H_5OH, and phenolate ion, $C_6H_5O^-$. The sodium ions are ignored because they are not involved in the buffer. Calculate K_a from pK_a and set up the problem with a reaction table.
$K_a = 10^{-pKa} = 10^{-10.00} = 1.0 \times 10^{-10}$

Concentration (M)	$C_6H_5OH(aq)$ +	$H_2O(l)$	⇆	$C_6H_5O^-(aq)$ +	$H_3O^+(aq)$
Initial	1.2	—		1.3	0
Change	– x	—		+ x	+ x
Equilibrium	1.2 – x	—		1.3 + x	x

Assume that x is negligible with respect to both 1.0 and 1.2 because both concentrations are much larger than K_a.

$$K_a = \frac{\left[H_3O^+\right]\left[C_6H_5O^-\right]}{\left[C_6H_5OH\right]} = \frac{[x][1.3 + x]}{[1.2 - x]} = \frac{[x][1.3]}{[1.2]} = 1.0 \times 10^{-10}$$

$$[H_3O^+] = K_a\frac{\left[C_6H_5OH\right]}{\left[C_6H_5O^-\right]} = \left(1.0 \times 10^{-10}\right)\left(\frac{1.2}{1.3}\right) = 9.23077 \times 10^{-11}\ M$$

Check assumption: percent error = $(9.23077 \times 10^{-11} / 1.2)100\% = 7.7 \times 10^{-9}\%$. The assumption is valid.
$pH = -\log (9.23077 \times 10^{-11}) = 10.03476 = \mathbf{10.03}$.
Verify the pH using the Henderson-Hasselbalch equation:

$$pH = pK_a + \log\left(\frac{[base]}{[acid]}\right)$$

$$pH = 10.00 + \log\left(\frac{[C_6H_5O^-]}{[C_6H_5OH]}\right) = 10.00 + \log\left(\frac{[1.3]}{[1.2]}\right)$$

$$\mathbf{pH = 10.03}$$

19.19   The buffer components are phenol, NH_3, and ammonium ion, NH_4^+. The chloride ions are ignored because they are not involved in the buffer. Calculate K_b from pK_b and set up the problem with a reaction table.
$K_b = 10^{-pKb} = 10^{-4.75} = 1.7782794 \times 10^{-5}$ (unrounded)

Concentration (M)	$NH_3(aq)$ +	$H_2O(l)$	⇆	$NH_4+(aq)$ +	$OH^-(aq)$
Initial	0.25	—		0.15	0
Change	– x	—		+ x	+ x
Equilibrium	0.25 – x	—		0.15 + x	x

Assume that x is negligible with respect to both 0.25 and 0.15 because both concentrations are much larger than K_b.
$K_b = 10^{-pKb} = 10^{-4.75} = 1.7782794 \times 10^{-5}$ (unrounded)

$$K_b = \frac{\left[NH_4^+\right]\left[OH^-\right]}{\left[NH_3\right]} = \frac{[0.15 + x]\left[OH^-\right]}{[0.25 - x]} = \frac{[0.15]\left[OH^-\right]}{[0.25]} = 1.7782794 \times 10^{-5} \text{ (unrounded)}$$

$$[OH^-] = K_b\frac{\left[NH_3\right]}{\left[NH_4^+\right]} = \left(1.7782794 \times 10^{-5}\right)\left(\frac{0.25}{0.15}\right) = 2.963799 \times 10^{-5}\ M \text{ (unrounded)}$$

Check assumption: percent error = $(2.963799 \times 10^{-5} / 0.25)100\% = 0.012\%$. The assumption is valid.
$pOH = -\log [OH^-] = -\log (2.963799 \times 10^{-5}) = 4.52815$ (unrounded)
$14.00 = pH + pOH$
$pH = 14.00 - pOH = 14.00 - 4.52815 = 9.4718 = \mathbf{9.47}$

Verify the pH using the Henderson-Hasselbalch equation. To do this, you must find the pK_a of the acid NH_4^+:
$14 = pK_a + pK_b$
$pK_a = 14 - pK_b = 14 - 4.75 = 9.25$

$$pH = pK_a + \log\left(\frac{[\text{base}]}{[\text{acid}]}\right)$$

$$pH = 9.25 + \log\left(\frac{[NH_3]}{[NH_4^+]}\right) = 9.25 + \log\left(\frac{[0.25]}{[0.15]}\right)$$

pH = 9.47

19.21 a) The buffer components are HCO_3^- from the salt $KHCO_3$ and CO_3^{2-} from the salt K_2CO_3. Choose the K_a value that corresponds to the equilibrium with these two components. K_{a1} refers to carbonic acid, H_2CO_3 losing one proton to produce HCO_3^-. This is not the correct K_a because H_2CO_3 is not involved in the buffer. K_{a2} is the correct K_a to choose because it is the equilibrium constant for the loss of the second proton to produce CO_3^{2-} from HCO_3^-.

b) Set up the reaction table and use K_{a2} to calculate pH.

Concentration (*M*)	$HCO_3^-{}_{(aq)}$ +	$H_2O(l)$ $\leftrightarrows$	$CO_3^{2-}(aq)$ +	$H_3O^+(aq)$
Initial	0.22	—	0.37	0
Change	$-x$	—	$+x$	$+x$
Equilibrium	$0.22 - x$	—	$0.37 + x$	x

Assume that x is negligible with respect to both 0.22 and 0.37 because both concentrations are much larger than K_a.

$$K_a = \frac{\left[H_3O^+\right]\left[CO_3^{2-}\right]}{\left[HCO_3^-\right]} = \frac{[x][0.37 + x]}{[0.22 - x]} = \frac{[x][0.37]}{[0.22]} = 4.7 \times 10^{-11}$$

$$[H_3O^+] = K_a \frac{\left[HCO_3^-\right]}{\left[CO_3^{2-}\right]} = \left(4.7 \times 10^{-11}\right)\left(\frac{0.22}{0.37}\right) = 2.79459 \times 10^{-11} \, M \text{ (unrounded)}$$

Check assumption: percent error = $(2.79459 \times 10^{-11} / 0.22)100\% = 1.3 \times 10^{-8}\%$. The assumption is valid.
$pH = -[H_3O^+] = -\log (2.79459 \times 10^{-11}) = 10.5537 = \textbf{10.55}$.
Verify the pH using the Henderson-Hasselbalch equation.

$$pH = pK_a + \log\left(\frac{[\text{base}]}{[\text{acid}]}\right) \qquad pK_a = -\log(4.7 \times 10^{-11}) = 10.328$$

$$pH = 10.328 + \log\left(\frac{[CO_3^{2-}]}{[HCO_3^-]}\right) = 10.328 + \log\left(\frac{[0.37]}{[0.22]}\right)$$

pH = 10.55

19.23 Given the pH and K_a of an acid, the buffer-component ratio can be calculated from the Henderson-Hasselbalch equation.
$pK_a = -\log K_a = -\log (1.3 \times 10^{-5}) = 4.8860566$ (unrounded)

$$pH = pK_a + \log\left(\frac{[\text{base}]}{[\text{acid}]}\right)$$

$$5.44 = 4.8860566 + \log\left(\frac{[Pr^-]}{[HPr]}\right)$$

$$0.5539434 = \log\left(\frac{[Pr^-]}{[HPr]}\right) \qquad \text{Raise each side to } 10^x.$$

$$\frac{[Pr^-]}{[HPr]} = 3.5805 = \textbf{3.6}$$

19.25 Given the pH and K_a of an acid, the buffer-component ratio can be calculated from the Henderson-Hasselbalch equation.

$$pK_a = -\log K_a = -\log(2.3 \times 10^{-9}) = 8.63827$$

$$pH = pK_a + \log\left(\frac{[base]}{[acid]}\right)$$

$$7.95 = 8.63827 + \log\left(\frac{[BrO^-]}{[HBrO]}\right)$$

$$-0.68827 = \log\left(\frac{[BrO^-]}{[HBrO]}\right) \qquad \text{Raise each side to } 10^x$$

$$\frac{[BrO^-]}{[HBrO]} = 0.204989 = \mathbf{0.20}$$

19.27 Determine the pK_a of the acid from the concentrations of the conjugate acid and base, and the pH of the solution. This requires the Henderson-Hasselbalch equation.

$$pH = pK_a + \log\left(\frac{[base]}{[acid]}\right)$$

$$3.35 = pK_a + \log\left(\frac{[A^-]}{[HA]}\right) = pK_a + \log\left(\frac{[0.1500]}{[0.2000]}\right)$$

$$3.35 = pK_a - 0.1249387$$
$$pK_a = 3.474939 = 3.47$$

Determine the moles of conjugate acid (HA) and conjugate base (A^-) using (M)(V) = moles

Moles HA = (0.5000 L) (0.2000 mol HA / L) = 0.1000 mol HA
Moles A^- = (0.5000 L) (0.1500 mol A^- / L) = 0.07500 mol A^-

The reaction is:

	HA(aq)	+	NaOH(aq)	$\rightarrow$	Na$^+$(aq)	+ A^-(aq)	+	H$_2$O(l)
Initial:	0.1000 mol		0.0015 mol			0.07500 mol		
Change:	− 0.0015 mol		−0.0015 mol			+ 0.0015 mol		
Final:	0.0985 mol		0 mol			0.0765 mol		

NaOH is the limiting reagent. The addition of 0.0015 mol NaOH produces an additional 0.0015 mol A^- and consumes 0.0015 mol of HA. Then

$$[A^-] = \frac{0.0765 \text{ mol } A^-}{0.5000 \text{ L}} = 0.153 \; M \; A^-$$

$$[HA] = \frac{0.0985 \text{ mol HA}}{0.5000 \text{ L}} = 0.197 \; M \text{ HA}$$

$$pH = pK_a + \log\left(\frac{[A^-]}{[HA]}\right)$$

$$pH = 3.474938737 + \log\left(\frac{[0.153]}{[0.197]}\right) = 3.365163942 = \mathbf{3.37}$$

19.29 Determine the pK_a of the acid from the concentrations of the conjugate acid and base and the pH of the solution. This requires the Henderson-Hasselbalch equation.

$$pH = pK_a + \log\left(\frac{[base]}{[acid]}\right)$$

$$8.77 = pK_a + \log\left(\frac{[Y^-]}{[HY]}\right) = pK_a + \log\left(\frac{[0.220]}{[0.110]}\right)$$

$$8.77 = pK_a + 0.301029996$$
$$pK_a = 8.46897 = 8.47$$

Determine the moles of conjugate acid (HY) and conjugate base (Y$^-$) using (M)(V) = moles.

Moles HY = (0.350 L) (0.110 mol HY / L) = 0.0385 mol HY

Moles Y$^-$ = (0.350 L) (0.220 mol Y$^-$ / L) = 0.077 mol Y$^-$)

The reaction is:

$$2\ HY(aq)\ +\ Ba(OH)_2(aq)\ \rightarrow\ Ba^{2+}(aq)\ +\ 2\ Y^-\ (aq) + 2\ H_2O(l)$$

Initial:	0.0385 mol	0.0015 mol	0.077 mol
Change:	– 0.0030 mol	–0.0015 mol	+ 0.0030 mol
Final:	0.0355 mol	0 mol	0.0800 mol

Ba(OH)$_2$ is the limiting reagent. The addition of 0.0015 mol Ba(OH)$_2$ will produce 2 x 0.0015 mol Y$^-$ and consume 2 x 0.0015 mol of HY.

Then

$$[Y^-] = \frac{0.0800\ \text{mol Y}^-}{0.350\ \text{L}} = 0.228571\ M\ Y^-$$

$$[HY] = \frac{0.0355\ \text{mol HY}}{0.350\ \text{L}} = 0.101429\ M\ HY$$

$$pH = pK_a + \log\left(\frac{[Y^-]}{[HY]}\right)$$

$$pH = 8.46897 + \log\left(\frac{[0.228571]}{[0.101429]}\right) = 8.82183 = \textbf{8.82}$$

19.31   a) The hydrochloric acid will react with the sodium acetate, NaC$_2$H$_3$O$_2$, to form acetic acid, HC$_2$H$_3$O$_2$:

$$HCl + NaC_2H_3O_2 \rightarrow HC_2H_3O_2 + NaCl$$

Calculate the number of moles of HCl and NaC$_2$H$_3$O$_2$. All of the HCl will be consumed to form HC$_2$H$_3$O$_2$, and the number of moles of C$_2$H$_3$O$_2$$^-$ will decrease.

$$\text{Initial moles HCl} = \left(\frac{0.452\ \text{mol HCl}}{L}\right)\left(\frac{10^{-3}\ L}{1\ mL}\right)(204\ mL) = 0.092208\ \text{mol HCl (unrounded)}$$

$$\text{Initial moles NaC}_2\text{H}_3\text{O}_2 = \left(\frac{0.400\ \text{mol NaC}_2\text{H}_3\text{O}_2}{L}\right)(0.500\ L) = 0.200\ \text{mol NaC}_2\text{H}_3\text{O}_2$$

	HCl	+	NaC$_2$H$_3$O$_2$	$\rightarrow$	HC$_2$H$_3$O$_2$	+	NaCl
Initial:	0.092208 mol		0.200 mol		0 mol		
Change:	–0.092208 mol		–0.092208 mol		+ 0.092208 mol		
Final:	0 mol		0.107792 mol		0.092208 mol		

Total volume = 0.500 L + (204 mL) (10^{-3} L / 1 mL) = 0.704 L

$$[HC_2H_3O_2] = \frac{0.092208\ \text{mol}}{0.704\ \text{L}} = 0.1309773\ M\ \text{(unrounded)}$$

$$[C_2H_3O_2^-] = \frac{0.107792\ \text{mol}}{0.704\ \text{L}} = 0.1531136\ M\ \text{(unrounded)}$$

$$pK_a = -\log K_a = -\log (1.8 \times 10^{-5}) = 4.744727495$$

$$pH = pK_a + \log\left(\frac{[C_2H_3O_2^-]}{[HC_2H_3O_2]}\right)$$

$$pH = 4.744727495 + \log\left(\frac{[0.1531136]}{[0.1309773]}\right) = 4.812545 = \textbf{4.81}$$

b) The addition of base would increase the pH, so the new pH is (4.81 + 0.15) = 4.96.
The new $[C_2H_3O_2^-]$ / $[HC_2H_3O_2]$ ratio is calculated using the Henderson-Hasselbalch equation.

$$pH = pK_a + \log\left(\frac{[C_2H_3O_2^-]}{[HC_2H_3O_2]}\right)$$

$$4.96 = 4.744727495 + \log\left(\frac{[C_2H_3O_2^-]}{[HC_2H_3O_2]}\right)$$

$$0.215272505 = \log\left(\frac{[C_2H_3O_2^-]}{[HC_2H_3O_2]}\right)$$

$$\frac{[C_2H_3O_2^-]}{[HC_2H_3O_2]} = 1.64162 \text{ (unrounded)}$$

From part (a), we know that $[HC_2H_3O_2] + [C_2H_3O_2^-] = (0.1309773\ M + 0.1531136\ M) = 0.2840909\ M$. Although the *ratio* of $[C_2H_3O_2^-]$ to $[HC_2H_3O_2]$ can change when acid or base is added, the *absolute amount* does not change unless acetic acid or an acetate salt is added.
Given that $[C_2H_3O_2^-]$ / $[HC_2H_3O_2] = 1.64162$ and $[HC_2H_3O_2] + [C_2H_3O_2^-] = 0.2840909\ M$, solve for $[C_2H_3O_2^-]$ and substitute into the second equation.

$[C_2H_3O_2^-] = 1.64162\ [HC_2H_3O_2]$ and $[HC_2H_3O_2] + 1.64162\ [HC_2H_3O_2] = 0.2840909\ M$
$[HC_2H_3O_2] = 0.1075441\ M$ and $[C_2H_3O_2^-] = 0.176547\ M$.
Moles of $C_2H_3O_2^-$ needed = $(0.176547\ \text{mol}\ C_2H_3O_2^- / L)(0.500\ L) = 0.0882735$ mol (unrounded)
Moles of $C_2H_3O_2^-$ initially = $(0.1531136\ \text{mol}\ C_2H_3O_2^- / L)(0.500\ L) = 0.0765568$ mol (unrounded)
This would require the addition of (0.0882735 mol – 0.0765568 mol)
= 0.0117167 mol $C_2H_3O_2^-$ (unrounded)
The KOH added reacts with $HC_2H_3O_2$ to produce additional $C_2H_3O_2^-$:

$$HC_2H_3O_2 + KOH \rightarrow C_2H_3O_2^- + K^+ + H_2O(l)$$

To produce 0.0117167 mol $C_2H_3O_2^-$ would require the addition of 0.0117167 mol KOH.

$$\text{Mass KOH} = (0.0117167\ \text{mol KOH})\left(\frac{56.11\ \text{g KOH}}{1\ \text{mol KOH}}\right) = 0.657424 = \textbf{0.66 g KOH}$$

19.33   Select conjugate pairs with K_a values close to the desired $[H_3O^+]$.
a) For pH $\approx$ 4.5, $[H_3O^+] = 10^{-4.5} = 3.2$ x 10^{-5} M. Some good selections are the $HOOC(CH_2)_4COOH$/ $HOOC(CH_2)_4COOH^-$ conjugate pair with K_a equal to 3.8 x 10^{-5} or $C_6H_5CH_2COOH/C_6H_5CH_2COO^-$ conjugate pair with K_a equal to 4.9 x 10^{-5}. From the base list, the $C_6H_5NH_2$ / $C_6H_5NH_3^+$ conjugate pair comes close with $K_a = 1.0$ x 10^{-14} / 4.0 x $10^{-10} = 2.5$ x 10^{-5}.
b) For pH $\approx$ 7.0, $[H_3O^+] = 10^{-7.0} = 1.0$ x 10^{-7} M. Two choices are the $H_2PO_4^-$ / HPO_4^{2-} conjugate pair with K_a of 6.3 x 10^{-8} and the $H_2AsO_4^-$ / $HAsO_4^{2-}$ conjugate pair with K_a of 1.1 x 10^{-7}.

19.35   Select conjugate pairs with pK_a values close to the desired pH. Convert pH to $[H_3O^+]$ for easy comparison to K_a values. Determine an appropriate base by $[OH^-] = K_w$ / $[H_3O^+]$.
a) For pH $\approx$ 3.5 ($[H_3O^+] = 10^{-pH} = 10^{-3.5} = 3.2$ x 10^{-4}), the best selection is the $HOCH_2CH(OH)COOH$/ $HOCH_2CH(OH)COOH^-$ conjugate pair with a $K_a = 2.9$ x 10^{-4}. The $CH_3COOC_6H_4COOH$ / $CH_3COOC_6H_4COO^-$ pair, with $K_a = 3.6$ x 10^{-4}, is also a good choice. The $[OH^-] = 1.0$ x 10^{-14} / 3.2 x $10^{-4} = 3.1$ x 10^{-11}, results in no reasonable K_b values from the appendix.
b) For pH $\approx$ 5.5 ($[H_3O^+] = 10^{-pH} = 3$ x 10^{-6}), no K_{a1} gives an acceptable pair; the K_{a2} values for adipic acid, malonic acid, and succinic acid are reasonable. The $[OH^-] = 1.0$ x 10^{-14} / 3 x $10^{-6} = 3$ x 10^{-9}, the K_b selection is C_5H_5N / $C_5H_5NH^+$.

19.38   The value of the K_a from the appendix: $K_a = 6.3$ x 10^{-8} (We are using K_{a2} since we dealing with the equilibrium in which the second hydrogen ion is being lost).
Determine the pK_a using $pK_a = -\log 6.3$ x $10^{-8} = 7.200659451$ (unrounded)

Use the Henderson-Hasselbalch equation:

$$pH = pKa + \log\left(\frac{[HPO_4^{2-}]}{[H_2PO_4^-]}\right)$$

$$7.40 = 7.200659451 + \log\left(\frac{[HPO_4^{2-}]}{[H_2PO_4^-]}\right)$$

$$0.19934055 = \log\left(\frac{[HPO_4^{2-}]}{[H_2PO_4^-]}\right)$$

$$\frac{[HPO_4^{2-}]}{[H_2PO_4^-]} = 1.582486 = \mathbf{1.6}$$

19.40   To see a distinct color in a mixture of two colors, you need one color to be about 10 times the intensity of the other. For this to take place, the concentration ratio [HIn] / [In⁻] needs to be greater than 10:1 or less than 1:10. This will occur when pH = pK_a – 1 or pH = pK_a + 1, respectively, giving a transition range of about two units.

19.42   The equivalence point in a titration is the point at which the number of moles of OH⁻ equals the number of moles of H_3O^+ (be sure to account for stoichiometric ratios, e.g.,1 mol of $Ca(OH)_2$ produces 2 moles of OH⁻). The endpoint is the point at which the added indicator changes color. If an appropriate indicator is selected, the endpoint is close to the equivalence point, but not normally the same. Using an indicator that changes color at a pH after the equivalence point means the equivalence point is reached first. However, if an indicator is selected that changes color at a pH before the equivalence point, then the endpoint is reached first.

19.44   a) The initial pH is lowest for flask solution of the strong acid, followed by the weak acid and then the weak base. In other words, *strong acid*-strong base < *weak acid*-strong base < strong acid-*weak base*.
b) At the equivalence point, the moles of H_3O^+ equal the moles of OH⁻, regardless of the type of titration. However, the strong acid-strong base equivalence point occurs at pH = 7.00 because the resulting cation-anion combination does not react with water. An example is the reaction $NaOH + HCl \rightarrow H_2O + NaCl$. Neither Na^+ nor Cl⁻ ions dissociate in water.
The weak acid-strong base equivalence point occurs at pH > 7, because the anion of the weak acid is weakly basic, whereas the cation of the strong base does not react with water. An example is the reaction HCOOH + $NaOH \rightarrow HCOO^- + H_2O + Na^+$. The conjugate base, HCOO⁻, reacts with water according to this reaction: $HCOO^- + H_2O \rightarrow HCOOH + OH^-$.
The strong acid-weak base equivalence point occurs at pH < 7, because the anion of the strong acid does not react with water, whereas the cation of the weak base is weakly acidic. An example is the reaction $HCl + NH_3 \rightarrow NH_4^+ + Cl^-$. The conjugate acid, NH_4^+, dissociates slightly in water: $NH_4^+ + H_2O \rightarrow NH_3 + H_3O^+$.
In rank order of pH of equivalence point, strong acid -*weak base* < *strong acid*-strong base < *weak acid* -strong base.

19.46   At the very center of the buffer region of a weak acid-strong base titration, the concentration of the weak acid and its conjugate base are equal, which means that at this point the pH of the solution equals the pK_a of the weak acid.

19.48   Indicators have a pH range that is approximated by $pK_a \pm 1$. The pK_a of cresol red is –log (3.5 x 10⁻⁹) = 8.5, so the indicator changes color over an approximate range of **7.5 to 9.5**.

19.50   Choose an indicator that changes color at a pH close to the pH of the equivalence point.
a) The equivalence point for a strong acid-strong base titration occurs at pH = 7.0. **Bromthymol blue** is an indicator that changes color around pH 7.
b)The equivalence point for a weak acid-strong base is above pH 7. Estimate the pH at equivalence point from equilibrium calculations.
At the equivalence point, all of the HCOOH and NaOH have been consumed; the solution is 0.050 M HCOO⁻. (The volume doubles because equal volumes of base and acid are required to reach the equivalence point. When the volume doubles, the concentration is halved.) The weak base HCOO– undergoes a base reaction:

Concentration, M $COOH^-(aq) + H_2O(l) \rightleftharpoons \quad HCOOH(aq) + OH^-(aq)$

Initial:	0.050 M	—	0	0
Change:	$-x$		$+x$	$+x$
Equilibrium:	$0.050 - x$		x	x

The K_a for HCOOH is 1.8×10^{-4}, so $K_b = 1.0 \times 10^{-14} / 1.8 \times 10^{-4} = 5.5556 \times 10^{-11}$ (unrounded)

$$K_b = \frac{[HCOOH][OH^-]}{[HCOO^-]} = \frac{[x][x]}{[0.050-x]} = \frac{[x][x]}{[0.050]} = 5.5556 \times 10^{-11}$$

$[OH^-] = x = 1.666673 \times 10^{-6} \, M$

$pOH = -\log (1.666673 \times 10^{-6}) = 5.7781496$ (unrounded)

$pH = 14.00 - pOH = 14.00 - 5.7781496 = 8.2218504 = 8.22$

Choose **thymol blue** or **phenolphthalein**.

19.52   a) The equivalence point for a weak base - strong acid is below pH 7. Estimate the pH at equivalence point from equilibrium calculations.

At the equivalence point, the solution is 0.25 M $(CH_3)_2NH_2^+$. (The volume doubles because equal volumes of base and acid are required to reach the equivalence point. When the volume doubles, the concentration is halved.)

$K_a = K_w / K_b = (1.0 \times 10^{-14}) / (5.9 \times 10^{-4}) = 1.6949152 \times 10^{-11}$ (unrounded)

Concentration, M $(CH_3)_2NH_2^+ (aq) + H_2O(l) \rightleftharpoons \quad (CH_3)_2NH (aq) + H_3O^+(aq)$

Initial:	0.25 M	—	0	0
Change:	$-x$		$+x$	$+x$
Equilibrium:	$0.25 - x$		x	x

$$K_a = \frac{[H_3O^+][(CH_3)_2NH]}{[(CH_3)_2NH_2^+]} = \frac{x^2}{0.25-x} = \frac{x^2}{0.25} = 1.6949152 \times 10^{-11}$$

$x = [H_3O^+] = 2.0584674 \times 10^{-6} \, M$ (unrounded)

$pH = -\log [H_3O^+] = -\log (2.0584674 \times 10^{-6}) = 5.686456 = 5.69$

Methyl red is an indicator that changes color around pH 5.7.

b) This is a strong acid - strong base titration; thus, the equivalence point is at pH = 7.00. **Bromthymol blue** is an indicator that changes color around pH 7.

19.54   The reaction occurring in the titration is the neutralization of H_3O^+ (from HCl) by OH^- (from NaOH):

$HCl(aq) + NaOH(aq) \rightarrow H_2O(l) + NaCl(aq)$ or, omitting spectator ions:

$H_3O^+(aq) + OH^-(aq) \rightarrow 2 H_2O(l)$

For the titration of a strong acid with a strong base, the pH before the equivalence point depends on the excess concentration of acid and the pH after the equivalence point depends on the excess concentration of base. At the equivalence point, there is not an excess of either acid or base so the pH is 7.0. The equivalence point occurs when 40.00 mL of base has been added. Use (M)(V) to determine the number of moles.

The initial number of moles of HCl = (0.1000 mol HCl / L) (10^{-3} L / 1 mL) (40.00 mL) = 4.000×10^{-3} mol HCl

a) At 0 mL of base added, the concentration of hydronium ion equals the original concentration of HCl.

     $pH = -\log (0.1000 \, M) = $ **1.0000**

b) Determine the moles of NaOH added:

     Moles of NaOH = (0.1000 mol NaOH / L) (10^{-3} L / 1 mL) (25.00 mL) = 2.500×10^{-3} mol NaOH

	$HCl(aq)$	+	$NaOH(aq)$	$\rightarrow$	$H_2O(l)$	+	$NaCl(aq)$
Initial:	4.000×10^{-3} mol		2.500×10^{-3} mol		—		0
Change:	-2.500×10^{-3} mol		-2.500×10^{-3} mol		—		$+2.500 \times 10^{-3}$ mol
Final:	1.500×10^{-3} mol		0				2.500×10^{-3} mol

The volume of the solution at this point is [(40.00 + 25.00) mL] (10^{-3} L / 1 mL) = 0.06500 L

The molarity of the excess HCl is (1.500×10^{-3} mol HCl) / (0.06500 L) = 0.02307 M (unrounded)

     $pH = -\log (0.02307) = $ **1.6368**

(Note that the NaCl product is a neutral salt that does not affect the pH).

c) Determine the moles of NaOH added:

Moles of NaOH = (0.1000 mol NaOH / L) (10^{-3} L / 1 mL) (39.00 mL) = 3.900 x 10^{-3} mol NaOH

	HCl(*aq*)	+	NaOH(*aq*)	→	H$_2$O(*l*)	+	NaCl(*aq*)
Initial:	4.000 x 10^{-3} mol		4.900 x 10^{-3} mol		–		0
Change:	– 3.900 x 10^{-3} mol		–3.900 x 10^{-3} mol		–		+ 3.900 x 10^{-3} mol
Final:	1.000 x 10^{-4} mol		0				3.900 x 10^{-3} mol

The volume of the solution at this point is [(40.00 + 39.00) mL] (10^{-3} L / 1 mL) = 0.07900 L
The molarity of the excess HCl is (1.00 x 10^{-4} mol HCl) / (0.07900 L) = 0.0012658 *M* (unrounded)

pH = –log (0.0012658) = **2.898**

d) Determine the moles of NaOH added:

Moles of NaOH = (0.1000 mol NaOH / L) (10^{-3} L / 1 mL) (39.90 mL) = 3.990 x 10^{-3} mol NaOH

	HCl(*aq*)	+	NaOH(*aq*)	→	H$_2$O(*l*)	+	NaCl(*aq*)
Initial:	4.000 x 10^{-3} mol		3.990 x 10^{-3} mol		–		0
Change:	– 3.990 x 10^{-3} mol		–3.990 x 10^{-3} mol		–		+ 3.990 x 10^{-3} mol
Final:	1.000 x 10^{-5} mol		0				3.900 x 10^{-3} mol

The volume of the solution at this point is [(40.00 + 39.90) mL] (10^{-3} L / 1 mL) = 0.07990 L
The molarity of the excess HCl is (1.0 x 10^{-5} mol HCl) / (0.07990 L) = 0.000125156 *M* (unrounded)

pH = –log (0.000125156) = **3.903**

e) Determine the moles of NaOH added:

Moles of NaOH = (0.1000 mol NaOH / L) (10^{-3} L / 1 mL) (40.00 mL) = 4.000 x 10^{-3} mol NaOH.

	HCl(*aq*)	+	NaOH(*aq*)	→	H$_2$O(*l*)	+	NaCl(*aq*)
Initial:	4.000 x 10^{-3} mol		4.000 x 10^{-3} mol		–		0
Change:	– 4.000 x 10^{-3} mol		–4.000 x 10^{-3} mol		–		+ 4.000 x 10^{-3} mol
Final:	0		0				4.000 x 10^{-3} mol

The NaOH will react with an equal amount of the acid and 0.0 mol HCl will remain. This is the equivalence point of a strong acid-strong base titration, thus, the pH is **7.00**. Only the neutral salt NaCl is in solution at the equivalence point.

f) The NaOH is now in excess. It will be necessary to calculate the excess base after reacting with the HCl. The excess strong base will give the pOH, which can be converted to the pH.

Determine the moles of NaOH added:

Moles of NaOH = (0.1000 mol NaOH / L) (10^{-3} L / 1 mL) (40.10 mL) = 4.010 x 10^{-3} mol NaOH
The HCl will react with an equal amount of the base, and 1.0 x 10^{-5} mol NaOH will remain.

	HCl(*aq*)	+	NaOH(*aq*)	→	H$_2$O(*l*)	+	NaCl(*aq*)
Initial:	4.000 x 10^{-3} mol		4.010 x 10^{-3} mol		–		0
Change:	– 4.000 x 10^{-3} mol		–4.000 x 10^{-3} mol		–		+ 4.000 x 10^{-3} mol
Final:	0		1.000 x 10^{-5} mol				4.000 x 10^{-3} mol

The volume of the solution at this point is [(40.00 + 40.10) mL] (10^{-3} L / 1 mL) = 0.08010 L
The molarity of the excess NaOH is (1.0 x 10^{-5} mol NaOH) / (0.08010 L) = 0.00012484 *M* (unrounded)

pOH = –log (0.00012484) = 3.9036 (unrounded)

pH = 14.00 – pOH = 14.00 – 3.9036 = 10.09637 = **10.10**

g) Determine the moles of NaOH added:

Moles of NaOH = (0.1000 mol NaOH / L) (10^{-3} L / 1 mL) (50.00 mL) = 5.000 x 10^{-3} mol NaOH
The HCl will react with an equal amount of the base, and 1.000 x 10^{-3} mol NaOH will remain.

	HCl(*aq*)	+	NaOH(*aq*)	→	H$_2$O(*l*)	+	NaCl(*aq*)
Initial:	4.000 x 10^{-3} mol		5.000 x 10^{-3} mol		–		0
Change:	– 4.000 x 10^{-3} mol		–4.000 x 10^{-3} mol		–		+4.000 x 10^{-3} mol
Final:	0		1.000 x 10^{-3} mol				4.000 x 10^{-3} mol

The volume of the solution at this point is [(40.00 + 50.00) mL] (10^{-3} L / 1 mL) = 0.09000 L
The molarity of the excess NaOH is (1.000 x 10^{-3} mol NaOH) / (0.09000 L) = 0.011111 *M* (unrounded)

pOH = –log (0.011111) = 1.95424 (unrounded)

pH = 14.00 – pOH = 14.00 – 1.95424 = 12.04576 = **12.05**

19.56   This is a titration between a weak acid and a strong base. The pH before addition of the base is dependent on the K_a of the acid (labeled HBut). Prior to reaching the equivalence point, the added base reacts with the acid to form butanoate ion (labeled But$^-$). The equivalence point occurs when 20.00 mL of base is added to the acid because at this point, moles acid = moles base. Addition of base beyond the equivalence point is simply the addition of excess OH$^-$.

The initial number of moles of HBut = (M)(V) =

$$(0.1000 \text{ mol HBut / L}) (10^{-3} \text{ L / 1 mL}) (20.00 \text{ mL}) = 2.000 \times 10^{-3} \text{ mol HBut}$$

a) At 0 mL of base added, the concentration of [H$_3$O$^+$] is dependent on the dissociation of butanoic acid:

	HBut	+	H$_2$O	$\leftrightharpoons$	H$_3$O$^+$	+	But$^-$
Initial:	0.100 M						0
Change:	$-x$						$+x$
Equilibrium:	0.100 $-$ x						x

$$K_a = \frac{\left[H_3O^+\right]\left[But^-\right]}{\left[HBut\right]} = \frac{x^2}{0.1000 - x} = \frac{x^2}{0.1000} = 1.54 \times 10^{-5}$$

$$x = [H_3O^+] = 1.2409673 \times 10^{-3} \ M \text{ (unrounded)}$$

$$pH = -\log [H_3O^+] = -\log (1.2409673 \times 10^{-3}) = 2.9062 = \mathbf{2.91}$$

b) Determine the moles of NaOH added:

Moles of NaOH = $(0.1000 \text{ mol NaOH / L}) (10^{-3} \text{ L / 1 mL}) (10.00 \text{ mL}) = 1.000 \times 10^{-3} \text{ mol NaOH}$

The NaOH will react with an equal amount of the acid, and 1.000×10^{-3} mol HBut will remain. An equal number of moles of But$^-$ will form.

	HBut(aq)	+	NaOH(aq)	$\rightarrow$	H$_2$O(l)	+	But$^-$(aq) + Na$^+$(aq)
Initial:	2.000×10^{-3} mol		1.000×10^{-3} mol		$-$		0 $-$
Change:	$- 1.000 \times 10^{-3}$ mol		-1.000×10^{-3} mol		$-$		$+1.000 \times 10^{-3}$ mol $-$
Final:	1.000×10^{-3} mol		0				1.000×10^{-3} mol

The volume of the solution at this point is [(20.00 + 10.00) mL] $(10^{-3}$ L / 1 mL) = 0.03000 L

The molarity of the excess HBut is $(1.000 \times 10^{-3}$ mol HBut) / (0.03000 L) = 0.03333 M (unrounded)

The molarity of the But$^-$ formed is $(1.000 \times 10^{-3}$ mol But$^-$) / (0.03000 L) = 0.03333 M (unrounded)

Using a reaction table for the equilibrium reaction of HBut:

	HBut	+	H$_2$O	$\leftrightharpoons$	H$_3$O$^+$	+	But$^-$
Initial:	0.03333 M						0.03333 M
Change:	$-x$						$+x$
Equilibrium:	0.03333 $-$ x						0.03333 + x

$$K_a = \frac{\left[H_3O^+\right]\left[But^-\right]}{\left[HBut\right]} = \frac{x(0.0333 + x)}{0.03333 - x} = \frac{x(0.03333)}{0.03333} = 1.54 \times 10^{-5}$$

$$x = [H_3O^+] = 1.54 \times 10^{-5} \ M \text{ (unrounded)}$$

$$pH = -\log [H_3O^+] = -\log (1.54 \times 10^{-5}) = 4.812479 = \mathbf{4.81}$$

c) Determine the moles of NaOH added:

Moles of NaOH = $(0.1000 \text{ mol NaOH / L}) (10^{-3} \text{ L / 1 mL}) (15.00 \text{ mL}) = 1.500 \times 10^{-3} \text{ mol NaOH}$

The NaOH will react with an equal amount of the acid, and 5.00×10^{-4} mol HBut will remain, and 1.500×10^{-3} moles of But$^-$ will form.

	HBut(aq)	+	NaOH(aq)	$\rightarrow$	H$_2$O(l)	+	But$^-$(aq) + Na$^+$(aq)
Initial:	2.000×10^{-3} mol		1.500×10^{-3} mol		$-$		0 $-$
Change:	-1.500×10^{-3} mol		-1.500×10^{-3} mol		$-$		$+ 1.500 \times 10^{-3}$ mol $-$
Final:	5.000×10^{-4} mol		0				1.500×10^{-3} mol

The volume of the solution at this point is [(20.00 + 15.00) mL] $(10^{-3}$ L / 1 mL) = 0.03500 L

The molarity of the excess HBut is $(5.00 \times 10^{-4}$ mol HBut) / (0.03500 L) = 0.0142857 M (unrounded)

The molarity of the But$^-$ formed is $(1.500 \times 10^{-3}$ mol But$^-$) / (0.03500 L) = 0.0428571 M (unrounded)

Using a reaction table for the equilibrium reaction of HBut:

	HBut	+	H$_2$O	$\leftrightharpoons$	H$_3$O$^+$	+	But$^-$
Initial:	0.0142857 M						0.0428571 M
Change:	$-x$						$+x$
Equilibrium:	0.0142857 $-$ x						0.0428571 + x

$$K_a = \frac{\left[H_3O^+\right]\left[But^-\right]}{\left[HBut\right]} = \frac{x\left(0.0428571 + x\right)}{0.0142857 - x} = \frac{x\left(0.0428571\right)}{0.0142857} = 1.54 \times 10^{-5}$$

$$x = [H_3O^+] = 5.1333 \times 10^{-6} \, M \text{ (unrounded)}$$
$$pH = -\log [H_3O^+] = -\log (5.1333 \times 10^{-6}) = 5.2896 = \mathbf{5.29}$$

d) Determine the moles of NaOH added:

Moles of NaOH = (0.1000 mol NaOH / L) (10^{-3} L / 1 mL) (19.00 mL) = 1.900 x 10^{-3} mol NaOH
The NaOH will react with an equal amount of the acid, and 1.00 x 10^{-4} mol HBut will remain, and 1.900 x 10^{-3} moles of But$^-$ will form.

	HBut(aq)	+	NaOH(aq)	→	H$_2$O(l)	+	But$^-$(aq)	+	Na$^+$(aq)
Initial:	2.000 x 10^{-3} mol		1.900 x 10^{-3} mol		–		0		–
Change:	– 1.900 x 10^{-3} mol		–1.900 x 10^{-3} mol		–		+1.900 x 10^{-3} mol		–
Final:	1.000 x 10^{-4} mol		0				1.900 x 10^{-3} mol		

The volume of the solution at this point is [(20.00 + 19.00) mL] (10^{-3} L / 1 mL) = 0.03900 L
The molarity of the excess HBut is (1.00 x 10^{-4} mol HBut) / (0.03900 L) = 0.0025641 M (unrounded)
The molarity of the But$^-$ formed is (1.900 x 10^{-3} mol But$^-$) / (0.03900 L) = 0.0487179 M (unrounded)
Using a reaction table for the equilibrium reaction of HBut:

	HBut	+	H$_2$O	⇄	H$_3$O$^+$	+	But$^-$
Initial:	0.0025641 M						0.0487179 M
Change:	–x						+x
Equilibrium:	0.0025641 – x						0.0487179 + x

$$K_a = \frac{\left[H_3O^+\right]\left[But^-\right]}{\left[HBut\right]} = \frac{x\left(0.0487179 + x\right)}{0.0025641 - x} = \frac{x\left(0.0487179\right)}{0.0025641} = 1.54 \times 10^{-5}$$

$$x = [H_3O^+] = 8.1052631 \times 10^{-7} \, M \text{ (unrounded)}$$
$$pH = -\log [H_3O^+] = -\log (8.1052631 \times 10^{-7}) = 6.09123 = \mathbf{6.09}$$

e) Determine the moles of NaOH added:

Moles of NaOH = (0.1000 mol NaOH / L) (10^{-3} L / 1 mL) (19.95 mL) = 1.995 x 10^{-3} mol NaOH
The NaOH will react with an equal amount of the acid, and 5 x 10^{-6} mol HBut will remain, and 1.995 x 10^{-3} moles of But$^-$ will form.

	HBut(aq)	+	NaOH(aq)	→	H$_2$O(l)	+	But$^-$(aq)	+	Na$^+$(aq)
Initial:	2.000 x 10^{-3} mol		1.995 x 10^{-3} mol		–		0		–
Change:	–1.995 x 10^{-3} mol		–1.995 x 10^{-3} mol		–		+1.995 x 10^{-3} mol		–
Final:	5.000 x 10^{-6} mol		0				1.995 x 10^{-3} mol		

The volume of the solution at this point is [(20.00 + 19.95) mL] (10^{-3} L / 1 mL) = 0.03995 L
The molarity of the excess HBut is (5 x 10^{-6} mol HBut) / (0.03995 L) = 0.000125156 M (unrounded)
The molarity of the But$^-$ formed is (1.995 x 10^{-3} mol But$^-$) / (0.03995 L) = 0.0499374 M (unrounded)
Using a reaction table for the equilibrium reaction of HBut:

	HBut	+	H$_2$O	⇄	H$_3$O$^+$	+	But$^-$
Initial:	0.000125156 M						0.0499374 M
Change:	–x						+x
Equilibrium:	0.000125156 – x						0.0499374 + x

$$K_a = \frac{\left[H_3O^+\right]\left[But^-\right]}{\left[HBut\right]} = \frac{x\left(0.0499374 + x\right)}{0.000125156 - x} = \frac{x\left(0.0499374\right)}{0.000125156} = 1.54 \times 10^{-5}$$

$$x = [H_3O^+] = 3.859637 \times 10^{-8} \, M \text{ (unrounded)}$$
$$pH = -\log [H_3O^+] = -\log (3.859637 \times 10^{-8}) = 7.41345 = \mathbf{7.41}$$

f) Determine the moles of NaOH added:

Moles of NaOH = (0.1000 mol NaOH / L) (10^{-3} L / 1 mL) (20.00 mL) = 2.000 x 10^{-3} mol NaOH
The NaOH will react with an equal amount of the acid, and 0 mol HBut will remain, and 2.000 x 10^{-3} moles of But$^-$ will form. This is the equivalence point.

$$\begin{array}{ccccccc} & \text{HBut}(aq) & + & \text{NaOH}(aq) & \rightarrow & \text{H}_2\text{O}(l) & + & \text{But}^-(aq) & + & \text{Na}^+(aq) \end{array}$$

	HBut(aq)	+	NaOH(aq)	→	H₂O(l)	+	But⁻(aq) + Na⁺(aq)
Initial:	2.000×10^{-3} mol		2.000×10^{-3} mol		–		0 —
Change:	-2.000×10^{-3} mol		-2.000×10^{-3} mol		–		$+2.000 \times 10^{-3}$ mol —
Final:	0		0				2.000×10^{-3} mol

The K_b of But$^-$ is now important.

The volume of the solution at this point is $[(20.00 + 20.00)$ mL$] (10^{-3}$ L $/ 1$ mL$) = 0.04000$ L

The molarity of the But$^-$ formed is $(2.000 \times 10^{-3}$ mol But$^-) / (0.04000$ L$) = 0.05000$ M (unrounded)

$\qquad K_b = K_w / K_a = (1.0 \times 10^{-14}) / (1.54 \times 10^{-5}) = 6.4935 \times 10^{-10}$ (unrounded)

Using a reaction table for the equilibrium reaction of But$-$:

	But⁻	+	H₂O	⇆	HBut	+	OH⁻
Initial:	0.05000 M				0		0
Change:	−x				+x		+x
Equilibrium:	0.05000 − x				x		x

$$K_b = \frac{\left[\text{HBut}\right]\left[\text{OH}^-\right]}{\left[\text{But}^-\right]} = \frac{[x][x]}{[0.05000 - x]} = \frac{[x][x]}{[0.05000]} = 6.4935 \times 10^{-10}$$

$[\text{OH}^-] = x = 5.6980259 \times 10^{-6}$ M

pOH $= -\log (5.6980259 \times 10^{-6}) = 5.244275575$ (unrounded)

pH $= 14.00 - $ pOH $= 14.00 - 5.244275575 = 8.755724425 = \textbf{8.76}$

g) After the equivalence point, the excess strong base is the primary factor influencing the pH.

Determine the moles of NaOH added:

Moles of NaOH $= (0.1000$ mol NaOH $/$ L$) (10^{-3}$ L $/ 1$ mL$) (20.05$ mL$) = 2.005 \times 10^{-3}$ mol NaOH

The NaOH will react with an equal amount of the acid, 0 mol HBut will remain, and 5×10^{-6} moles of NaOH will be in excess. There will be 2.000×10^{-3} mol of But$^-$ produced, but this weak base will not affect the pH compared to the excess strong base, NaOH.

	HBut(aq)	+	NaOH(aq)	→	H₂O(l)	+	But⁻(aq) + Na⁺(aq)
Initial:	2.000×10^{-3} mol		2.005×10^{-3} mol		–		0 —
Change:	-2.000×10^{-3} mol		-2.000×10^{-3} mol		–		$+2.000 \times 10^{-3}$ mol —
Final:	0		5.000×10^{-6} mol				2.000×10^{-3} mol

The volume of the solution at this point is $[(20.00 + 20.05)$ mL$] (10^{-3}$ L $/ 1$ mL$) = 0.04005$ L

The molarity of the excess OH$^-$ is $(5 \times 10^{-6}$ mol OH$^-) / (0.04005$ L$) = 1.2484 \times 10^{-4}$ M (unrounded)

$\qquad$ pOH $= -\log (1.2484 \times 10^{-4}) = 3.9036$ (unrounded)

$\qquad$ pH $= 14.00 - $ pOH $= 14.00 - 3.9036 = 10.0964 = \textbf{10.10}$

h) Determine the moles of NaOH added:

Moles of NaOH $= (0.1000$ mol NaOH $/$ L$) (10^{-3}$ L $/ 1$ mL$) (25.00$ mL$) = 2.500 \times 10^{-3}$ mol NaOH

The NaOH will react with an equal amount of the acid, 0 mol HBut will remain, and 5.00×10^{-4} moles of NaOH will be in excess.

	HBut(aq)	+	NaOH(aq)	→	H₂O(l)	+	But⁻(aq) + Na⁺(aq)
Initial:	2.000×10^{-3} mol		2.500×10^{-3} mol		–		0 —
Change:	-2.000×10^{-3} mol		-2.000×10^{-3} mol		–		$+2.000 \times 10^{-3}$ mol —
Final:	0		5.000×10^{-4} mol				2.000×10^{-3} mol

The volume of the solution at this point is $[(20.00 + 25.00)$ mL$] (10^{-3}$ L $/ 1$ mL$) = 0.04500$ L

The molarity of the excess OH$^-$ is $(5.00 \times 10^{-4}$ mol OH$^-) / (0.04500$ L$) = 1.1111 \times 10^{-2}$ M (unrounded)

$\qquad$ pOH $= -\log (1.1111 \times 10^{-2}) = 1.9542$ (unrounded)

$\qquad$ pH $= 14.00 - $ pOH $= 14.00 - 1.9542 = 12.0458 = \textbf{12.05}$

19.58   a) The balanced chemical equation is:

$\qquad$ NaOH(aq) + CH$_3$COOH(aq) → Na$^+(aq)$ + CH$_3$COO$^-(aq)$ + H$_2$O(l)

The sodium ions on the product side are written as separate species because they have no effect on the pH of the solution. Calculate the volume of NaOH needed:

Volume $=$

$$\left(\frac{0.0520 \text{ mol CH}_3\text{COOH}}{\text{L}}\right)\left(\frac{10^{-3} \text{ L}}{1 \text{ mL}}\right)(42.2 \text{ mL})\left(\frac{1 \text{ mol NaOH}}{1 \text{ mol CH}_3\text{COOH}}\right)\left(\frac{\text{L}}{0.0372 \text{ mol NaOH}}\right)\left(\frac{1 \text{ mL}}{10^{-3} \text{ L}}\right)$$

$= 58.989247 = \textbf{59.0 mL NaOH}$

Determine the moles of CH_3COOH present:

$$\text{Moles} = \left(\frac{0.0520 \text{ mol } CH_3COOH}{L}\right)\left(\frac{10^{-3} \text{ L}}{1 \text{ mL}}\right)(42.2 \text{ mL})$$

$$= 0.0021944 \text{ mol } CH_3COOH \text{ (unrounded)}$$

At the equivalence point, 0.0021944 mol NaOH will be added so the moles acid = moles base.

The NaOH will react with an equal amount of the acid, 0 mol CH_3COOH will remain, and 0.0021944 moles of CH_3COO^- will be formed.

	$CH_3COOH(aq)$	$+$	$NaOH(aq)$	$\rightarrow$	$H_2O(l)$	$+$	$CH_3COO^-(aq)$	$+$	$Na^+(aq)$
Initial:	0.0021944 mol		0.0021944 mol		–		0		–
Change:	– 0.0021944 mol		–0.0021944 mol		–		+0.0021944 mol		–
Final:	0		0				0.0021944 mol		

Determine the liters of solution present at the equivalence point:

Volume = [(42.0 + 58.989247) mL] $(10^{-3}$ L / 1 mL) = 0.100989 L (unrounded)

Concentration of CH_3COO^- at equivalence point:

Molarity = (0.0021944 mol CH_3COO^-) / (0.100989 L) = 0.021729 M (unrounded)

Calculate K_b for CH_3COO^-: $\qquad K_a$ $CH_3COOH = 1.8$ x 10^{-5}

$K_b = K_w / K_a = (1.0$ x $10^{-14}) / (1.8$ x $10^{-5}) = 5.556$ x 10^{-10} (unrounded)

Using a reaction table for the equilibrium reaction of CH_3COO^-:

	CH_3COO^-	$+$	H_2O	$\leftrightarrows$	CH_3COOH	$+$	OH^-
Initial:	0.021729 M				0		0
Change:	–x				+x		+x
Equilibrium:	0.021729 – x				x		x

Determine the hydroxide ion concentration from the K_b, and then determine the pH from the pOH.

$$K_b = \frac{\left[CH_3COOH\right]\left[OH^-\right]}{\left[CH_3COO^-\right]} = \frac{[x][x]}{[0.021729 - x]} = \frac{[x][x]}{[0.021729]} = 5.556 \text{ x } 10^{-10}$$

$[OH^-] = x = 3.4745693$ x 10^{-6} M (unrounded)

pOH = $-\log (3.4745693$ x $10^{-6}) = 5.459099012$ (unrounded)

pH = 14.00 – pOH = 14.00 – 5.459099012 = 8.54090 = **8.54**

b) The balanced chemical equations are:

$NaOH(aq) + H_2SO_3(aq) \rightarrow Na^+(aq) + HSO_3^- (aq) + H_2O(l)$

$NaOH(aq) + HSO_3^- (aq) \rightarrow Na^+(aq) + SO_3^{2-} (aq) + H_2O(l)$

The sodium ions on the product side are written as separate species because they have no effect on the pH of the solution. Calculate the volume of NaOH needed:

Volume =

$$\left(\frac{0.0850 \text{ mol } H_2SO_3}{L}\right)\left(\frac{10^{-3} \text{ L}}{1 \text{ mL}}\right)(28.9 \text{ mL})\left(\frac{1 \text{ mol NaOH}}{1 \text{ mol } H_2SO_3}\right)\left(\frac{L}{0.0372 \text{ mol NaOH}}\right)\left(\frac{1 \text{ mL}}{10^{-3} \text{ L}}\right)$$

$= 66.034943 = $ **66.0 mL NaOH**

It will require an equal volume to reach the second equivalence point for a total of 2 x 66.034943 = **132.1 mL**.

Determine the moles of HSO_3^- produced:

$$\text{Moles} = \left(\frac{0.0850 \text{ mol } H_2SO_3}{L}\right)\left(\frac{10^{-3} \text{ L}}{1 \text{ mL}}\right)(28.9 \text{ mL})\left(\frac{1 \text{ mol } HSO_3^{2-}}{1 \text{ mol } H_2SO_3}\right)$$

$= 0.0024565 \text{ mol } HSO_3^-$

An equal number of moles of SO_3^{2-} will be present at the second equivalence point.

Determine the liters of solution present at the first equivalence point:

Volume = [(28.9 + 66.034943) mL] $(10^{-3}$ L / 1 mL) = 0.094934943 L

Determine the liters of solution present at the second equivalence point:

Volume = [(28.9 + 66.034943 + 66.034943) mL] $(10^{-3}$ L / 1 mL) = 0.160969886 L

Concentration of HSO_3^- at equivalence point:

Molarity = (0.0024565 moles HSO_3^-) / (0.094934943 L) = 0.0258756 M

Concentration of SO_3^{2-} at equivalence point:

Molarity = (0.0024565 moles SO_3^{2-}) / (0.160969886 L) = 0.0152606 M

Calculate K_b for HSO_3^-: $K_a\ H_2SO_3 = 1.4 \times 10^{-2}$
$K_b = K_w / K_a = (1.0 \times 10^{-14}) / (1.4 \times 10^{-2}) = 7.142857 \times 10^{-13}$
Calculate K_b for SO_3^{2-}: $K_a\ HSO_3^- = 6.5 \times 10^{-8}$
$K_b = K_w / K_a = (1.0 \times 10^{-14}) / (6.5 \times 10^{-8}) = 1.53846 \times 10^{-7}$
For the first equivalence point:
Using a reaction table for the equilibrium reaction of HSO_3^-:

	HSO_3^-	+	H_2O	$\rightleftharpoons$	H_2SO_3	+	OH^-
Initial:	0.0258756 M				0		0
Change:	$-x$				$+x$		$+x$
Equilibrium:	$0.0258756 - x$				x		x

Determine the hydroxide ion concentration from the K_b, and then determine the pH from the pOH.

$$K_b = \frac{\left[H_2SO_3\right]\left[OH^-\right]}{\left[HSO_3^-\right]} = \frac{[x][x]}{[0.0258756 - x]} = \frac{[x][x]}{[0.0258756]} = 7.142857 \times 10^{-13}$$

$[OH^-] = x = 1.359506 \times 10^{-7}\ M$ (unrounded)
$pOH = -\log(1.359506 \times 10^{-7}) = 6.8666188$ (unrounded)
$pH = 14.00 - pOH = 14.00 - 6.8666188 = 7.13338 = \mathbf{7.13}$
For the second equivalence point:
Using a reaction table for the equilibrium reaction of SO_3^{2-}:

	SO_3^{2-}	+	H_2O	$\rightleftharpoons$	HSO_3^-	+	OH^-
Initial:	0.0152606 M				0		0
Change:	$-x$				$+x$		$+x$
Equilibrium:	$0.0152606 - x$				x		x

Determine the hydroxide ion concentration from the K_b, and then determine the pH from the pOH.

$$K_b = \frac{\left[HSO_3^-\right]\left[OH^-\right]}{\left[SO_3^{2-}\right]} = \frac{[x][x]}{[0.0152606 - x]} = \frac{[x][x]}{[0.0152606]} = 1.53846 \times 10^{-7}$$

$[OH^-] = x = 4.84539 \times 10^{-5}\ M$ (unrounded)
$pOH = -\log(4.84539 \times 10^{-5}) = 4.31467$ (unrounded)
$pH = 14.00 - pOH = 14.00 - 4.31467 = 9.68533 = \mathbf{9.69}$

19.60 a) The balanced chemical equation is:
$HCl(aq) + NH_3(aq) \rightarrow NH_4^+(aq) + Cl^-(aq)$
The chloride ions on the product side are written as separate species because they have no effect on the pH of the solution. Calculate the volume of HCl needed:
Volume =

$$\left(\frac{0.234\ mol\ NH_3}{L}\right)\left(\frac{10^{-3}\ L}{1\ mL}\right)(65.5\ mL)\left(\frac{1\ mol\ HCl}{1\ mol\ NH_3}\right)\left(\frac{L}{0.125\ mol\ HCl}\right)\left(\frac{1\ mL}{10^{-3}\ L}\right)$$

$= 122.616 = \mathbf{123\ mL\ HCl}$
Determine the moles of NH_3 present:

$$Moles = \left(\frac{0.234\ mol\ NH_3}{L}\right)\left(\frac{10^{-3}\ L}{1\ mL}\right)(65.5\ mL)$$

$= 0.015327\ mol\ NH_3$ (unrounded)
At the equivalence point, 0.05327 mol HCl will be added so the moles acid = moles base.
The HCl will react with an equal amount of the base, 0 mol NH_3 will remain, and 0.015327 moles of NH_4^+ will be formed.

	$HCl(aq)$	+	$NH_3(aq)$	$\rightarrow$	$NH_4^+(aq)$	+	$Cl^-(aq)$
Initial:	0.015327 mol		0.015327 mol		0		–
Change:	$-$ 0.015327 mol		-0.015327 mol		$+0.015327$ mol		–
Final:	0		0		0.015327 mol		

Determine the liters of solution present at the equivalence point:
Volume = $[(65.5 + 122.616)\ mL]\ (10^{-3}\ L\ /\ 1\ mL) = 0.188116\ L$

Concentration of NH_4^+ at equivalence point:

Molarity = (0.015327 mol NH_4^+) / (0.188116 L) = 0.081476 M

Calculate K_a for NH_4^+: K_b NH_3 = 1.76 x 10^{-5}

$K_a = K_w / K_b$ = (1.0 x 10^{-14}) / (1.76 x 10^{-5}) = 5.6818 x 10^{-10} (unrounded)

Using a reaction table for the equilibrium reaction of NH_4^+:

	NH_4^+	+	H_2O	$\rightleftarrows$	NH_3	+	H_3O^+
Initial:	0.081476 M				0		0
Change:	−x				+x		+x
Equilibrium:	0.081476 − x				x		x

Determine the hydrogen ion concentration from the K_a, and then determine the pH.

$$K_a = \frac{\left[H_3O^+\right]\left[NH_3\right]}{\left[NH_4^+\right]} = \frac{[x][x]}{[0.081476 - x]} = \frac{[x][x]}{[0.081476]} = 5.6818 \text{ x } 10^{-10}$$

x = [H_3O^+] = 6.803911 x 10^{-6} M (unrounded)

pH = −log [H_3O^+] = −log (6.803911 x 10^{-6}) = 5.1672 = **5.17**

b) The balanced chemical equation is:

$HCl(aq) + CH_3NH_2(aq) \rightarrow CH_3NH_3^+(aq) + Cl^-(aq)$

The chloride ions on the product side are written as separate species because they have no effect on the pH of the solution. Calculate the volume of HCl needed:

Volume =

$$\left(\frac{1.11 \text{ mol } CH_3NH_2}{L}\right)\left(\frac{10^{-3} \text{ L}}{1 \text{ mL}}\right)(21.8 \text{ mL})\left(\frac{1 \text{ mol HCl}}{1 \text{ mol } CH_3NH_2}\right)\left(\frac{L}{0.125 \text{ mol HCl}}\right)\left(\frac{1 \text{ mL}}{10^{-3} \text{ L}}\right)$$

= 193.584 = **194 mL HCl**

Determine the moles of CH_3NH_2 present:

$$\text{Moles} = \left(\frac{1.11 \text{ mol } CH_3NH_2}{L}\right)\left(\frac{10^{-3} \text{ L}}{1 \text{ mL}}\right)(21.8 \text{ mL})$$

= 0.024198 mol CH_3NH_2 (unrounded)

At the equivalence point, 0.024198 mol HCl will be added so the moles acid = moles base.

The HCl will react with an equal amount of the base, 0 mol CH_3NH_2 will remain, and 0.024198 moles of $CH_3NH_3^+$ will be formed.

	$HCl(aq)$	+	$CH_3NH_2(aq)$	$\rightarrow$	$CH_3NH_3^+(aq)$	+	$Cl^-(aq)$
Initial:	0.024198 mol		0.024198 mol		0		−
Change:	− 0.024198 mol		−0.024198 mol		+0.024198 mol		−
Final:	0		0		0.024198 mol		

Determine the liters of solution present at the equivalence point:

Volume = [(21.8 + 193.584) mL] (10^{-3} L / 1 mL) = 0.215384 L (unrounded)

Concentration of $CH_3NH_3^+$ at equivalence point:

Molarity = (0.024198 mol $CH_3NH_3^+$) / (0.215384 L) = 0.1123482 M

Calculate K_a for $CH_3NH_3^+$: K_b CH_3NH_2 = 4.4 x 10^{-4}

$K_a = K_w / K_b$ = (1.0 x 10^{-14}) / (4.4 x 10^{-4}) = 2.2727 x 10^{-11} (unrounded)

Using a reaction table for the equilibrium reaction of $CH_3NH_3^+$:

	$CH_3NH_3^+$	+	H_2O	$\rightleftarrows$	CH_3NH_2	+	H_3O^+
Initial:	0.1123482 M				0		0
Change:	−x				+x		+x
Equilibrium:	0.1123482 − x				x		x

Determine the hydrogen ion concentration from the K_a, and then determine the pH.

$$K_a = \frac{\left[H_3O^+\right]\left[CH_3NH_2\right]}{\left[CH_3NH_3^+\right]} = \frac{[x][x]}{[0.1123482 - x]} = \frac{[x][x]}{[0.1123482]} = 2.2727 \text{ x } 10^{-11}$$

x = [H_3O^+] = 1.5979 x 10^{-6} M (unrounded)

pH = −log [H_3O^+] = −log (1.5979 x 10^{-6}) = 5.7964 = **5.80**

19.63 Fluoride ion in BaF_2 is the conjugate base of the weak acid HF. The base hydrolysis reaction of fluoride ion
$$F^-(aq) + H_2O(l) \leftrightarrows HF(aq) + OH^-(aq)$$
therefore is influenced by the pH of the solution. As the pH increases, $[OH^-]$ increases and the equilibrium shifts to the left to decrease $[OH^-]$ and increase the $[F^-]$. As the pH decreases, $[OH^-]$ decreases and the equilibrium shifts to the right to increase $[OH^-]$ and decrease $[F^-]$. The changes in $[F^-]$ influence the solubility of BaF_2. Chloride ion is the conjugate base of a strong acid so it does not react with water. Thus, its concentration is not influenced by pH, and solubility of $BaCl_2$ does not change with pH.

19.65 Consider the reaction $AB(s) \leftrightarrows A^+(aq) + B^-(aq)$, where $Q_{sp} = [A^+][B^-]$. If $Q_{sp} > K_{sp}$, then there are more ions dissolved than expected at equilibrium, and the equilibrium shifts to the left and the compound AB precipitates. The excess ions precipitate as solid from the solution.

19.66 a) $Ag_2CO_3(s) \leftrightarrows 2\,Ag^+(aq) + CO_3^{2-}(aq)$
        Ion-product expression: $K_{sp} = [Ag^+]^2[CO_3^{2-}]$
   b) $BaF_2(s) \leftrightarrows Ba^{2+}(aq) + 2\,F^-(aq)$
        Ion-product expression: $K_{sp} = [Ba^{2+}][F^-]^2$
   c) $CuS(s) + H_2O(l) \leftrightarrows Cu^{2+}(aq) + HS^-(aq) + OH^-(aq)$
        Ion-product expression: $K_{sp} = [Cu^{2+}][HS^-][OH^-]$

19.68 a) $CaCrO_4(s) \leftrightarrows Ca^{2+}(aq) + CrO_4^{2-}(aq)$
        Ion-product expression: $K_{sp} = [Ca^{2+}][CrO_4^{2-}]$
   b) $AgCN(s) \leftrightarrows Ag^+(aq) + CN^-(aq)$
        Ion-product expression: $K_{sp} = [Ag^+][CN^-]$
   c) $NiS(s) + H_2O(l) \leftrightarrows Ni^{2+}(aq) + HS^-(aq) + OH^-(aq)$
        Ion-product expression: $K_{sp} = [Ni^{2+}][HS^-][OH^-]$

19.70 Write a reaction table, where S is the molar solubility of Ag_2CO_3:

Concentration (M)	$Ag_2CO_3(s)$ $\leftrightarrows$	$2\,Ag^+(aq)$ +	$CO_3^{2-}(aq)$
Initial	—	0	0
Change	—	+2 S	+ S
Equilibrium	—	2 S	S

$S = [Ag_2CO_3] = 0.032\ M$ so $[Ag^+] = 2\,S = 0.064\ M$ and $[CO_3^{2-}] = S = 0.032\ M$
$K_{sp} = [Ag^+]^2[CO_3^{2-}] = (0.064)^2(0.032) = 1.31072 \times 10^{-4} = \mathbf{1.3 \times 10^{-4}}$

19.72 The equation and ion-product expression for silver dichromate, $Ag_2Cr_2O_7$, is:
$$Ag_2Cr_2O_7(s) \leftrightarrows 2\,Ag^+(aq) + Cr_2O_7^{2-}(aq) \qquad K_{sp} = [Ag^+]^2[Cr_2O_7^{2-}]$$
The solubility of $Ag_2Cr_2O_7$, converted from g / 100 mL to M is:

Molar solubility $= S = \left(\dfrac{8.3 \times 10^{-3}\ \text{g } Ag_2Cr_2O_7}{100\ \text{mL}} \right)\left(\dfrac{1\ \text{mL}}{10^{-3}\ \text{L}} \right)\left(\dfrac{1\ \text{mol } Ag_2Cr_2O_7}{431.8\ \text{g } Ag_2Cr_2O_7} \right) = 0.00019221861\ M$ (unrounded)

Since 1 mole of $Ag_2Cr_2O_7$ dissociates to form 2 moles of Ag^+, the concentration of Ag^+ is $2\,S = 2(0.00019221861\ M) = 0.00038443723\ M$ (unrounded). The concentration of $Cr_2O_7^{2-}$ is $S = 0.00019221861\ M$ because 1 mole of $Ag_2Cr_2O_7$ dissociates to form 1 mole of $Cr_2O_7^{2-}$.
$K_{sp} = [Ag^+]^2[Cr_2O_7^{2-}] = (2\,S)^2(S) = (0.00038443723)^2(0.00019221861) = 2.8408 \times 10^{-11} = \mathbf{2.8 \times 10^{-11}}$.

19.74 The equation and ion-product expression for $SrCO_3$ is:
$$SrCO_3(s) \leftrightarrows Sr^{2+}(aq) + CO_3^{2-}(aq) \qquad K_{sp} = [Sr^{2+}][CO_3^{2-}]$$
a)   The solubility, S, in pure water equals $[Sr^{2+}]$ and $[CO_3^{2-}]$
Write a reaction table, where S is the molar solubility of $SrCO_3$:

Concentration (M)	$SrCO_3(s)$ $\leftrightarrows$	$Sr^{2+}(aq)$ +	$CO_3^{2-}(aq)$
Initial	—	0	0
Change	—	+ S	+ S
Equilibrium	—	S	S

$$K_{sp} = 5.4 \text{ x } 10^{-10} = [Sr^{2+}][CO_3^{2-}] = [S][S] = S^2$$
$$S = 2.32379 \text{ x } 10^{-5} = \mathbf{2.3 \text{ x } 10^{-5}} \textit{ M}$$

b) In 0.13 M $Sr(NO_3)_2$, the initial concentration of Sr^{2+} is 0.13 M.
Equilibrium $[Sr^{2+}]$ = 0.13 + S and equilibrium $[CO_3^{2-}]$ = S where S is the solubility of $SrCO_3$.

Concentration (M)	$SrCO_3(s)$ $\leftrightarrows$	$Sr^{2+}(aq)$	+	$CO_3^{2-}(aq)$
Initial	—	0.13		0
Change	—	+ S		+ S
Equilibrium	—	0.13 + S		S

$$K_{sp} = 5.4 \text{ x } 10^{-10} = [Sr^{2+}][CO_3^{2-}] = (0.13 + S)S$$
This calculation may be simplified by assuming S is small and setting 0.13 + S = 0.13.
$$K_{sp} = 5.4 \text{ x } 10^{-10} = (0.13)S$$
$$S = 4.1538 \text{ x } 10^{-9} = \mathbf{4.2 \text{ x } 10^{-9}} \textit{ M}$$

19.76   The equilibrium is: $Ca(IO_3)_2(s)$ $\leftrightarrows$ $Ca^{2+}(aq)$ + 2 $IO_3^-(aq)$. From the Appendix, $K_{sp}(Ca(IO_3)_2)$ = 7.1 x 10^{-7}.
a) Write a reaction table that reflects an initial concentration of Ca^{2+} = 0.060 M. In this case, Ca^{2+} is the common ion.

Concentration (M)	$Ca(IO_3)_2(s)$ $\leftrightarrows$	$Ca^{2+}(aq)$	+	2 $IO_3^-(aq)$
Initial	—	0.060		0
Change	—	+ S		+ 2 S
Equilibrium	—	0.060 + S		2 S

Assume that 0.060 + S $\approx$ 0.060 because the amount of compound that dissolves will be negligible in comparison to 0.060 M.
$$K_{sp} = [Ca^{2+}][IO_3^-]^2 = (0.060) (2 \text{ S})^2 = 7.1 \text{ x } 10^{-7}$$
$$S = 1.71998 \text{ x } 10^{-3} = 1.7 \text{ x } 10^{-3} \textit{ M}$$
Check assumption: (1.71998 x 10^{-3} M) / (0.060 M) x 100% = 2.9% < 5%, so the assumption is good.
S represents both the molar solubility of Ca^{2+} and $Ca(IO_3)_2$, so the molar solubility of $Ca(IO_3)_2$ is **1.7 x 10^{-3} M**.
b) Write a reaction table that reflects an initial concentration of IO_3^- = 0.060 M. IO_3^- is the common ion.

Concentration (M)	$Ca(IO_3)_2(s)$ $\leftrightarrows$	$Ca^{2+}(aq)$	+	2 $IO_3^-(aq)$
Initial	—	0		0.060
Change	—	+ S		+ 2 S
Equilibrium	—	S		0.060 + 2 S

The equilibrium concentration of Ca^{2+} is S, and the IO_3^- concentration is 0.060 + 2 S.
Assume that 0.060 + 2 S $\approx$ 0.060
$$K_{sp} = [Ca^{2+}][IO_3^-]^2 = (S) (0.060)^2 = 7.1 \text{ x } 10^{-7}$$
$$S = 1.97222 \text{ x } 10^{-4} = 2.0 \text{ x } 10^{-4} \textit{ M}$$
Check assumption: (1.97222 x 10^{-4} M) / (0.060 M) x 100% = 0.3% < 5%, so the assumption is good.
S represents both the molar solubility of Ca^{2+} and $Ca(IO_3)_2$, so the molar solubility of $Ca(IO_3)_2$ is **2.0 x 10^{-4} M**.

19.78   The larger the K_{sp}, the larger the molar solubility if the number of ions are equal.
a) **$Mg(OH)_2$** with K_{sp} = 6.3 x 10^{-10} has higher molar solubility than $Ni(OH)_2$ with K_{sp} = 6 x 10^{-16}.
b) **PbS** with K_{sp} = 3 x 10^{-25} has higher molar solubility than CuS with K_{sp} = 8 x 10^{-34}.
c) **Ag_2SO_4** with K_{sp} = 1.5 x 10^{-5} has higher molar solubility than MgF_2 with K_{sp} = 7.4 x 10^{-9}.

19.80   The larger the K_{sp}, the more water-soluble the compound if the number of ions are equal.
a) **$CaSO_4$** with K_{sp} = 2.4 x 10^{-5} is more water-soluble than $BaSO_4$ with K_{sp} = 1.1 x 10^{-10}.
b) **$Mg_3(PO_4)_2$** with K_{sp} = 5.2 x 10^{-24} is more water soluble than $Ca_3(PO_4)_2$ with K_{sp} = 1.2 x 10^{-29}.
c) **$PbSO_4$** with K_{sp} = 1.6 x 10^{-8} is more water soluble than AgCl with K_{sp} = 1.8 x 10^{-10}.

19.82   a) $AgCl(s)$ $\leftrightarrows$ $Ag^+(aq)$ + $Cl^-(aq)$
The chloride ion is the anion of a strong acid, so it does not react with H_3O^+. The solubility is not affected by pH.
b) $SrCO_3(s)$ $\leftrightarrows$ $Sr^{2+}(aq)$ + $CO_3^{2-}(aq)$
The strontium ion is the cation of a strong base, so pH will not affect its solubility.
The carbonate ion is the conjugate base of a weak acid and will act as a base:
$$CO_3^{2-}(aq) + H_2O(l) \leftrightarrows HCO_3^-(aq) + OH^-(aq)$$
$$\text{and } HCO_3^-(aq) + H_2O(l) \leftrightarrows H_2CO_3(aq) + OH^-(aq)$$

The H_2CO_3 will decompose to $CO_2(g)$ and $H_2O(l)$. The gas will escape and further shift the equilibrium. Changes in pH will change the $[CO_3^{2-}]$, so the solubility of $SrCO_3$ will increase with decreasing pH. **Solubility increases with addition of H_3O^+ (decreasing pH).** A decrease in pH will decrease $[OH^-]$, causing the base equilibrium to shift to the right which decreases $[CO_3^{2-}]$, causing the solubility equilibrium to shift to the right, dissolving more solid.

19.84   a) $Fe(OH)_2(s) \leftrightarrows Fe^{2+}(aq) + 2\ OH^-(aq)$
The hydroxide ion reacts with added H_3O^+:
$$OH^-(aq) + H_3O^+(aq) \rightarrow 2\ H_2O(l)$$
The added H_3O^+ consumes the OH^-, driving the equilibrium toward the right to dissolve more $Fe(OH)_2$. **Solubility increases with addition of H_3O^+ (decreasing pH).**
b) $CuS(s) + H_2O(l) \leftrightarrows Cu^{2+}(aq) + HS^-(aq) + OH^-(aq)$
Both HS^- and OH^- are anions of weak acids, so both ions react with added H_3O^+. **Solubility increases with addition of H_3O^+ (decreasing pH).**

19.86   The equilibrium is: $Cu(OH)_2(s) \leftrightarrows Cu^{2+}(aq) + 2\ OH^-(aq)$. The ion-product expression for is $K_{sp} = [Cu^{2+}][OH^-]^2$ and, from the Appendix, K_{sp} equals 2.2×10^{-20}.
To decide if a precipitate will form, calculate Q_{sp} with the given quantities and compare it to K_{sp}.

$$[Cu^{2+}] = \left(\frac{1.0 \times 10^{-3}\ mol\ Cu(NO_3)_2}{L} \right)\left(\frac{1\ mol\ Cu^{2+}}{1\ mol\ Cu(NO_3)_2} \right) = 1.0 \times 10^{-3}\ M\ Cu^{2+}$$

$$[OH^-] = \left(\frac{0.075\ g\ KOH}{1.0\ L} \right)\left(\frac{1\ mol\ KOH}{56.11\ g\ KOH} \right)\left(\frac{1\ mol\ OH^-}{1\ mol\ KOH} \right) = 1.33666 \times 10^{-3}\ M\ OH^-\ \text{(unrounded)}$$

$Q_{sp} = [Cu^{2+}][OH^-]^2 = (1.0 \times 10^{-3})(1.33666 \times 10^{-3})^2 = 1.7866599 \times 10^{-9}$ (unrounded)
Q_{sp} is greater than K_{sp} ($1.8 \times 10^{-9} > 2.2 \times 10^{-20}$), so **$Cu(OH)_2$ will precipitate**.

19.88   The equilibrium is: $Ba(IO_3)_2(s) \leftrightarrows Ba^{2+}(aq) + 2\ IO_3^-(aq)$. The ion-product expression for is $K_{sp} = [Ba^{2+}][IO_3^-]^2$ and, from the Appendix, K_{sp} equals 1.5×10^{-9}.
To decide if a precipitate will form, calculate Q_{sp} with the given quantities and compare it to K_{sp}.

$$[Ba^{2+}] = \left(\frac{7.5\ mg\ BaCl_2}{500.\ mL} \right)\left(\frac{10^{-3}\ g}{1\ mg} \right)\left(\frac{1\ mL}{10^{-3}\ L} \right)\left(\frac{1\ mol\ BaCl_2}{208.2\ g\ BaCl_2} \right)\left(\frac{1\ mol\ Ba^{2+}}{1\ mol\ BaCl_2} \right)$$
$$= 7.204611 \times 10^{-5}\ M\ Ba^{2+}\ \text{(unrounded)}$$

$$[IO_3^-] = \left(\frac{0.023\ mol\ NaIO_3}{L} \right)\left(\frac{1\ mol\ IO_3^-}{1\ mol\ NaIO_3} \right) = 0.023\ M\ IO_3^-$$

$Q_{sp} = [Ba^{2+}][IO_3^-]^2 = (7.204611 \times 10^{-5})(0.023)^2 = 3.81124 \times 10^{-8}$ (unrounded)
Since $Q_{sp} > K_{sp}$ ($3.8 \times 10^{-8} > 1.5 \times 10^{-9}$), **$Ba(IO_3)_2$ will precipitate**.

19.91   a) **$Fe(OH)_3$** will precipitate first because its K_{sp} (1.6×10^{-39}) is smaller than the K_{sp} for $Cd(OH)_2$ at 7.2×10^{-15}.
The precipitation reactions are:
$$Fe^{3+}(aq) + 3\ OH^-(aq) \rightarrow Fe(OH)_3(s) \qquad\qquad K_{sp} = [Fe^{3+}][OH^-]^3$$
$$Cd^{2+}(aq) + 2\ OH^-(aq) \rightarrow Cd(OH)_2(s) \qquad\qquad K_{sp} = [Cd^{2+}][OH^-]^2$$
The concentrations of Fe^{3+} and Cd^{2+} in the mixed solution are found from $M_{conc}V_{conc} = M_{dil}V_{dil}$.
$[Fe^{3+}] = [(0.50\ M)(50.0\ mL)] / [(50.0 + 125)\ mL] = 0.142857\ M\ Fe^{3+}$ (unrounded)
$[Cd^{2+}] = [(0.25\ M)(125\ mL)] / [(50.0 + 125)\ mL] = 0.178571\ M\ Cd^{2+}$ (unrounded)
The hydroxide ion concentration required to precipitate the metal ions comes from the metal ion concentrations and the K_{sp}.

$$[OH^-]_{Fe} = \sqrt[3]{\frac{K_{sp}}{[Fe^{3+}]}} = \sqrt[3]{\frac{1.6 \times 10^{-39}}{[0.142857]}} = 2.237 \times 10^{-13} = 2.2 \times 10^{-13}\ M$$

$$[OH^-]_{Cd} = \sqrt{\frac{K_{sp}}{[Cd^{2+}]}} = \sqrt{\frac{7.2 \times 10^{-15}}{[0.178571]}} = 2.0079864 \times 10^{-7} = 2.0 \times 10^{-7}\ M$$

A lower hydroxide ion concentration is required to precipitate the Fe^{3+}.

b) The two ions are separated by adding just enough NaOH to precipitate the iron(III) hydroxide, but precipitating no more than 0.01% of the cadmium. The Fe^{3+} is found in the solid precipitate while the Cd^{2+} remains in the solution.

c) A hydroxide concentration between the values calculated in part (a) will work. The best separation would be when $Q_{sp} = K_{sp}$ for $Cd(OH)_2$. This occurs when $[OH^-] = \mathbf{2.0 \times 10^{-7}\ M}$.

19.94   In the context of this equilibrium only, the increased solubility with added OH^- appears to be a violation of Le Châtelier's Principle. Adding OH^- should cause the equilibrium to shift towards the left, decreasing the solubility of PbS. Before accepting this conclusion, other possible equilibria must be considered. Lead is a metal ion and hydroxide ion is a ligand, so it is possible that a complex ion forms between the lead ion and hydroxide ion:
$$Pb^{2+}(aq) + n\ OH^-(aq) \leftrightarrows Pb(OH)_n^{2-n}(aq)$$
This decreases the concentration of Pb^{2+}, shifting the solubility equilibrium to the right to dissolve more PbS.

19.95   In many cases, a hydrated metal complex (e.g., $Hg(H_2O)_4^{2+}$) will exchange ligands when placed in a solution of another ligand (e.g., CN^-),
$$Hg(H_2O)_4^{2+}(aq) + 4\ CN^-(aq) \leftrightarrows Hg(CN)_4^{2-}(aq) + 4\ H_2O(l)$$
Note that both sides of the equation have the same "overall" charge of –2. The mercury complex changes from +2 to –2 because water is a neutral *molecular* ligand, whereas cyanide is an *ionic* ligand.

19.97   The two water ligands are replaced by two thiosulfate ion ligands. The +1 charge from the silver ion plus –4 charge from the two thiosulfate ions gives a net charge on the complex ion of –3.
$$Ag(H_2O)_2^+(aq) + 2\ S_2O_3^{2-}(aq) \leftrightarrows Ag(S_2O_3)_2^{3-}(aq) + 2\ H_2O(l)$$

19.99 $Ag^+(aq) + 2\ S_2O_3^{2-}(aq) \leftrightarrows Ag(S_2O_3)_2^{3-}(aq)$
The initial concentrations may be determined from $M_{con}V_{con} = M_{dil}V_{dil}$
$[Ag^+] = (0.044\ M)\ (25.0\ mL)\ /\ ((25.0 + 25.0)\ mL) = 0.022\ M\ Ag^+$
$[S_2O_3^{2-}] = (0.57\ M)\ (25.0\ mL)\ /\ ((25.0 + 25.0)\ mL) = 0.285\ M\ S_2O_3^{2-}$ (unrounded)
The reaction gives:

Concentration (M)	$Ag^+(aq)$	+	$2\ S_2O_3^{2-}(aq)$	→	$Ag(S_2O_3)_2^{3-}(aq)$	
Initial	0.022		0.285		0	
Change	–0.022		–2 (0.022)		+ 0.022	1:2:1 mole ratio
Equilibrium	0		0.241		0.022	

To reach equilibrium:

Concentration (M)	$Ag^+(aq)$	+	$2\ S_2O_3^{2-}(aq)$	$\leftrightarrows$	$Ag(S_2O_3)_2^{3-}(aq)$
Initial	0		0.241		0.022
Change	+ x		+ 2 x		– x
Equilibrium	x		0.241 + 2 x		0.022 – x

K_f is large, so $[Ag(S_2O_3)_2^{3-}] \approx 0.022\ M$ and $[S_2O_3^{2-}]_{equil} \approx 0.241\ M$ (unrounded)

$$K_f = \frac{\left[Ag(S_2O_3)_2^{3-}\right]}{\left[Ag^+\right]\left[S_2O_3^{2-}\right]^2} = \frac{[0.022]}{[x][0.241]^2} = 4.7 \times 10^{13}$$

$x = [Ag^+] = 8.0591778 \times 10^{-15} = \mathbf{8.1 \times 10^{-15}\ M}$

19.101   Write the ion-product equilibrium reaction and the complex-ion equilibrium reaction. Sum the two reactions to yield an overall reaction; multiply the two constants to obtain $K_{overall}$. Write a reaction table where $S = [Cr(OH)_3]_{dissolved} = [Cr(OH)_4^-]$.

Solubility-Product:	$Cr(OH)_3(s) \leftrightarrows Cr^{3+}(aq) + 3OH^-(aq)$	$K_{sp} = 6.3 \times 10^{-31}$
Complex-Ion	$Cr^{3+}(aq) + 4OH^-(aq) \leftrightarrows Cr(OH)_4^-(aq)$	$K_f = 8.0 \times 10^{29}$
Overall:	$Cr(OH)_3(s) + OH^-(aq) \leftrightarrows Cr(OH)_4^-(aq)$	$K = K_{sp}K_f = 0.504$ (unrounded)

At pH 13.0, the pOH is 1.0 and $[OH^-] = 10^{-1.0} = 0.1\ M$.
Reaction table:

Concentration (M)	$Cr(OH)_3(s)\ +$	$OH^-(aq) \leftrightarrows$	$Cr(OH)_4^-(aq)$
Initial	——	0.1	0
Change	——	– S	+ S
Equilibrium	——	0.1 – S	S

Assume that $0.1 - S \approx 0.1$.

$$K_{overall} = \frac{\left[Cr(OH)_4^-\right]}{\left[OH^-\right]} = \frac{[S]}{[0.1]} = 0.504$$

$S = [Cr(OH)_4^-] = 0.0504 = \mathbf{0.05\ M}$

19.103   The complex formation equilibrium is
$$Zn^{2+}(aq) + 4\ CN^-(aq) \leftrightarrows Zn(CN)_4^{2-}(aq) \qquad K_f = 4.2 \times 10^{19}$$
First, calculate the initial moles of Zn^{2+} and CN^-, then set up reaction table assuming that the reaction first goes to completion, and then calculate back to find the reactant concentrations.

$$\text{Moles } Zn^{2+} = (0.84\ g\ ZnCl_2)\left(\frac{1\ mol\ ZnCl_2}{136.31\ g\ ZnCl_2}\right)\left(\frac{1\ mol\ Zn^{2+}}{1\ mol\ ZnCl_2}\right) = 0.0061624\ mol\ Zn^{2+} \text{ (unrounded)}$$

$$\text{Moles } CN^- = \left(\frac{0.150\ mol\ NaCN}{L}\right)\left(\frac{10^{-3}\ L}{1\ mL}\right)(245\ mL)\left(\frac{1\ mol\ CN^-}{1\ mol\ NaCN}\right) = 0.03675\ mol\ CN^- \text{ (unrounded)}$$

The Zn^{2+} is limiting because there are significantly fewer moles of this ion, thus, $[Zn^{2+}] = 0$, and the moles of CN^- remaining are: $[0.03675 - 4(0.0061624)]$

$$[CN^-] = \frac{\left[0.03675 - 4(0.0061624)\right]mol\ CN^-}{(245\ mL)}\left(\frac{1\ mL}{10^{-3}\ L}\right) = 0.0493894\ M\ CN^-$$

The Zn^{2+} will produce an equal number of moles of the complex with the concentration:

$$[Zn(CN)_4^{2-}] = \left(\frac{0.0061624\ mol\ Zn^{2+}}{245\ mL}\right)\left(\frac{1\ mL}{10^{-3}\ L}\right)\left(\frac{1\ mol\ Zn(CN)_4^{2-}}{1\ mol\ Zn^{2+}}\right) = 0.025153\ M\ Zn(CN)_4^{2-}$$

Concentration (M)	$Zn^{2+}(aq)\ +$	$4\ CN^-(aq)\ \leftrightarrows$	$Zn(CN)_4^{2-}(aq)$
Initial	0	0.0493894	0.025153
Change	+ x	+ 4 x	– x
Equilibrium	x	0.0493894 + 4 x	0.025153 – x

Assume the –x and the +4x do not significantly change the associated concentrations.

$$K_f = \frac{\left[Zn(CN)_4^{2-}\right]}{\left[Zn^{2+}\right]\left[CN^-\right]^4} = \frac{[0.025153 - x]}{[x][0.0493894 + 4x]^4} = \frac{[0.025153]}{[x][0493894]^4} = 4.2 \times 10^{19}$$

$x = 1.006481 \times 10^{-16} = 1.0 \times 10^{-16}$
$[Zn^{2+}] = \mathbf{1.0 \times 10^{-16}\ M\ Zn^{2+}}$
$[Zn(CN)_4^{2-}] = 0.025153 - x = 0.025153 = \mathbf{0.025\ M\ Zn(CN)_4^{2-}}$
$[CN^-] = 0.0493894 + 4\ x = 0.0493894 = \mathbf{0.049\ M\ CN^-}$

19.105   Calculate the pH of the benzoic acid/benzoate buffer, using the Henderson-Hasselbalch equation. The K_a for benzoic acid is 6.3×10^{-5} (from the Appendix). The pK_a is $-log\ (6.3 \times 10^{-5}) = 4.201$. The reaction of benzoic acid with sodium hydroxide is:
$C_6H_5COOH(aq) + NaOH(aq) \rightarrow Na^+(aq) + C_6H_5COO^-(aq) + H_2O(l)$

$$\text{Moles of } C_6H_5COOH = \left(\frac{0.200\ mol\ C_6H_5COOH}{L}\right)\left(\frac{10^{-3}\ L}{1\ mL}\right)(475\ mL) = 0.0950\ mol\ C_6H_5COOH$$

$$\text{Moles NaOH} = \left(\frac{2.00 \text{ mol NaOH}}{L} \right) \left(\frac{10^{-3} \text{ L}}{1 \text{ mL}} \right) (25 \text{ mL}) = 0.050 \text{ mol NaOH}$$

NaOH is the limiting reagent:
The reaction table gives:

	$C_6H_5COOH(aq)$ +	$NaOH(aq)$ →	$Na^+(aq)$ +	$C_6H_5COO^-(aq)$ +	$H_2O(l)$
Initial	0.0950 mol	0.050 mol	—	0	—
Reacting	− 0.050 mol	− 0.050 mol		+ 0.050 mol	
Final	0.045 mol	0 mol		0.050 mol	

The concentrations after the reactions are:

$$[C_6H_5COOH] = \left(\frac{0.045 \text{ mol } C_6H_5COOH}{(475 + 25)\text{mL}} \right) \left(\frac{1 \text{ mL}}{10^{-3} \text{ L}} \right) = 0.090 \text{ } M \text{ } C_6H_5COOH$$

$$[C_6H_5COO^-] = \left(\frac{0.050 \text{ mol } C_6H_5COO^-}{(475 + 25)\text{mL}} \right) \left(\frac{1 \text{ mL}}{10^{-3} \text{ L}} \right) = 0.10 \text{ } M \text{ } C_6H_5COO^-$$

Calculating the pH from the Henderson-Hasselbalch equation:

$$pH = pKa + \log \left(\frac{[C_6H_5COO^-]}{[C_6H_5COOH]} \right) = 4.201 + \log \left(\frac{[0.10]}{[0.090]} \right) = 4.24676 = 4.2$$

Calculations on formic acid (HCOOH) also use the Henderson-Hasselbalch equation. The K_a for formic acid is (see the Appendix) 1.8×10^{-4} and the $pK_a = -\log (1.8 \times 10^{-4}) = 3.7447$.
The formate to formic acid ratio may now be determined:

$$pH = pKa + \log \left(\frac{[HCOO^-]}{[HCOOH]} \right)$$

$$4.24676 = 3.7447 + \log \left(\frac{[HCOO^-]}{[HCOOH]} \right)$$

$$0.50206 = \log \left(\frac{[HCOO^-]}{[HCOOH]} \right)$$

$$\left(\frac{[HCOO^-]}{[HCOOH]} \right) = 3.177313$$

$$[HCOO^-] = 3.177313 \text{ } [HCOOH]$$

Since the conjugate acid and the conjugate base are in the same volume, the mole ratio and the molarity ratios are identical.

Moles $HCOO^- = 3.177313$ mol HCOOH
The total volume of the solution is (500. mL) $(10^{-3}$ L / 1 mL) = 0.500 L
Let V_a = volume of acid solution added, and V_b = volume of base added. Thus:
$V_a + V_b = 0.500$ L
The reaction between the formic acid and the sodium hydroxide is:
$HCOOH(aq) + NaOH(aq) \rightarrow HCOONa(aq) + H_2O(l)$
The moles of NaOH added equals the moles of HCOOH reacted and the moles of HCOONa formed.
Moles NaOH = (2.00 mol NaOH / L) $(V_b) = 2.00 \text{ } V_b$ mol
Total moles HCOOH = (0.200 mol HCOOH / L) $(V_a) = 0.200 \text{ } V_a$ mol
The stoichiometric ratios in this reaction are all 1:1.
Moles HCOOH remaining after the reaction = $(0.200 \text{ } V_a - 2.00 \text{ } V_b)$ mol
Moles $HCOO^-$ = moles HCOONa = moles NaOH = 2.00 V_b
Using these moles and the mole ratio determined for the buffer gives:
Moles $HCOO^- = 3.177313$ mol HCOOH
$2.00 \text{ } V_b$ mol = 3.177313 $(0.200 \text{ } V_a - 2.00 \text{ } V_b)$ mol
$2.00 \text{ } V_b = 0.6354626 \text{ } V_a - 6.354626 \text{ } V_b$
$8.354626 \text{ } V_b = 0.6354626 \text{ } V_a$

The volume relationship given above gives $V_a = (0.500 - V_b)$ L

$8.354626\ V_b = 0.6354626\ (0.500 - V_b)$

$8.354626\ V_b = 0.3177313 - 0.6354626\ V_b$

$8.9900886\ V_b = 0.3177313$

$V_b = 0.0353424 = \textbf{0.035 L NaOH}$

$V_a = 0.500 - 0.0353424 = 0.4646576 = \textbf{0.465 L HCOOH}$

Limitations due to the significant figures lead to a solution with only an approximately correct pH.

19.107 A formate buffer contains formate ($HCOO^-$) as the base and formic acid (HCOOH) as the acid. From the Appendix, the K_a for formic acid is 1.8×10^{-4} and the $pK_a = -\log (1.8 \times 10^{-4}) = 3.7447$ (unrounded).

a)   The Henderson-Hasselbalch equation gives the component ratio, $[HCOO^-] / [HCOOH]$:

$$pH = pKa + \log\left(\frac{[HCOO^-]}{[HCOOH]}\right)$$

$$3.74 = 3.7447 + \log\left(\frac{[HCOO^-]}{[HCOOH]}\right)$$

$$-0.0047 = \log\left(\frac{[HCOO^-]}{[HCOOH]}\right)$$

$$\left(\frac{[HCOO^-]}{[HCOOH]}\right) = 0.989236 = \textbf{0.99}$$

b) To prepare solutions, set up equations for concentrations of formate and formic acid with x equal to the volume, in L, of 1.0 M HCOOH added. The equations are based on the neutralization reaction between HCOOH and NaOH that produces $HCOO^-$.

$$HCOOH(aq) + NaOH(aq) \rightarrow HCOO^-(aq) + Na^+(aq) + H_2O(l)$$

$$[HCOO^-] = (1.0\text{M NaOH})\left(\frac{(0.700-x)\text{L NaOH}}{0.700\text{ L solution}}\right)\left(\frac{1\text{ mol }HCOO^-}{1\text{ mol NaOH}}\right)$$

$$[HCOOH] = (1.0\text{ M HCOOH})\left(\frac{x\text{ L HCOOH}}{0.700\text{ L solution}}\right) - (1.0\text{ M NaOH})\left(\frac{(0.700-x)\text{L NaOH}}{0.700\text{ L solution}}\right)\left(\frac{1\text{ mol }HCOO^-}{1\text{ mol NaOH}}\right)$$

The component ratio equals 0.99 (from part a). Simplify the above equations and plug into ratio:

$$\frac{[HCOO^-]}{[HCOOH]} = \frac{\left[\left(\frac{0.700-x}{0.700}\right)\text{M }HCOO^-\right]}{\left[\left(\frac{x-(0.700-x)}{0.700}\right)\text{M HCOOH}\right]} = \frac{0.700-x}{2x-0.700} = 0.989236$$

Solving for x:

$x = 0.46751 = 0.468$ L

Mixing **0.468 L of 1.0 M HCOOH** and $0.700 - 0.468 = \textbf{0.232 L of 1.0 } \textbf{\textit{M}} \textbf{ NaOH}$ gives a buffer of pH 3.74.

c) The final concentration of HCOOH from the equation in part b:

$$[HCOOH] = (1.0\text{ M HCOOH})\left(\frac{0.468\text{ L HCOOH}}{0.700\text{ L solution}}\right) - (1.0\text{ M NaOH})\left(\frac{0.232\text{ L NaOH}}{0.700\text{ L solution}}\right)\left(\frac{1\text{ mole }HCOO^-}{1\text{ mole NaOH}}\right)$$

$$= 0.33714 = \textbf{0.34 } \textbf{\textit{M}} \textbf{ HCOOH}$$

19.110 The minimum urate ion concentration necessary to cause a deposit of sodium urate is determined by the K_{sp} for the salt. Convert solubility in g/100. mL to molar solubility and calculate K_{sp}. Substituting $[Na^+]$ and K_{sp} into the ion-product expression allows one to find $[Ur^-]$.

Molar solubility of NaUr:

$$[NaUr] = \left(\frac{0.085\text{ g NaUr}}{100.\text{ mL}}\right)\left(\frac{1\text{ mL}}{10^{-3}\text{ L}}\right)\left(\frac{1\text{ mol NaUr}}{190.10\text{ mol NaUr}}\right) = 4.47133 \times 10^{-3}\ M\text{ NaUr (unrounded)}$$

$4.47133 \times 10^{-3} \, M \, NaUr = [Na^+] = [Ur^-]$

$K_{sp} = [Na^+][Ur^-] = (4.47133 \times 10^{-3})(4.47133 \times 10^{-3}) = 1.999279 \times 10^{-5} \, M$ (unrounded)

When $[Na^+] = 0.15 \, M$:

$K_{sp} = 1.999279 \times 10^{-5} \, M = [0.15][Ur^-]$

$[Ur^-] = 1.33285 \times 10^{-4}$ (unrounded)

The minimum urate ion concentration that will cause precipitation of sodium urate is **$1.3 \times 10^{-4} \, M$**.

19.114　a) The solubility equilibrium for KCl is: $KCl(s) \leftrightarrows K^+(aq) + Cl^-(aq)$

The solubility of KCl is 3.7 M.

$K_{sp} = [K^+][Cl^-] = (3.7)(3.7) = 13.69 = \mathbf{14}$

b) Determine the total concentration of chloride ion in each beaker after the HCl has been added. This requires the moles originally present and the moles added.

Original moles from the KCl:

$$\text{Moles } K^+ = \text{Moles } Cl^- = \left(\frac{3.7 \text{ mol KCl}}{1 \text{ L}}\right)\left(\frac{10^{-3} \text{ L}}{1 \text{ mL}}\right)(100. \text{ mL})\left(\frac{1 \text{ mol } Cl^- \text{ ion}}{1 \text{ mol KCl}}\right) = 0.37 \text{ mol } Cl^-$$

Original moles from the 6.0 M HCl in the first beaker:

$$\text{Moles } Cl^- = \left(\frac{6.0 \text{ mol HCl}}{1 \text{ L}}\right)\left(\frac{10^{-3} \text{ L}}{1 \text{ mL}}\right)(100. \text{ mL})\left(\frac{1 \text{ mol } Cl^-}{1 \text{ mol HCl}}\right) = 0.60 \text{ mol } Cl^-$$

This results in $(0.37 + 0.60) \text{ mol} = 0.97 \text{ mol } Cl^-$

Original moles from the 12 M HCl in the second beaker:

$$\text{Moles } Cl^- = \left(\frac{12 \text{ mol HCl}}{1 \text{ L}}\right)\left(\frac{10^{-3} \text{ L}}{1 \text{ mL}}\right)(100. \text{ mL})\left(\frac{1 \text{ mol } Cl^-}{1 \text{ mol HCl}}\right) = 1.2 \text{ mol } Cl^-$$

This results in $(0.37 + 1.2) \text{ mol} = 1.57 \text{ mol } Cl^-$ (unrounded)

Volume of mixed solutions = $(100. \text{ mL} + 100. \text{ mL})(10^{-3} \text{ L} / 1 \text{ mL}) = 0.200 \text{ L}$

After the mixing:

$[K^+] = (0.37 \text{ mol } K^+) / (0.200 \text{ L}) = 1.85 \, M \, K^+$ (unrounded)

From 6.0 M HCl in the first beaker:

$[Cl^-] = (0.97 \text{ mol } Cl^-) / (0.200 \text{ L}) = 4.85 \, M \, Cl^-$ (unrounded)

From 12 M HCl in the second beaker:

$[Cl^-] = (1.57 \text{ mol } Cl^-) / (0.200 \text{ L}) = 7.85 \, M \, Cl^-$ (unrounded)

Determine a Q_{sp} value to see if K_{sp} is exceeded. If $Q < K_{sp}$, nothing will precipitate.

From 6.0 M HCl in the first beaker:

$Q_{sp} = [K^+][Cl^-] = (1.85)(4.85) = 8.9725 = 9.0 < 14$, so no KCl will precipitate.

From 12 M HCl in the second beaker:

$Q_{sp} = [K^+][Cl^-] = (1.85)(7.85) = 14.5225 = 15 > 14$, so KCl will precipitate.

The mass of KCl that will precipitate when 12 M HCl is added:

Equal amounts of K and Cl will precipitate. Let x be the molarity change.

$K_{sp} = [K^+][Cl^-] = (1.85 - x)(7.85 - x) = 13.69$

$x = 0.088697657 = 0.09$　　　　This is the change in the molarity of each of the ions.

$$\text{Mass KCl} = \left(\frac{0.088697657 \text{ mol } K^+}{L}\right)(0.200 \text{ L})\left(\frac{1 \text{ mol KCl}}{1 \text{ mol } K^+}\right)\left(\frac{74.55 \text{ g KCl}}{1 \text{ mol KCl}}\right) = 1.32248 = \mathbf{1 \text{ g KCl}}$$

19.117　a) Use the Henderson-Hasselbalch equation. $K_{a1} = 4.5 \times 10^{-7}$ (From Appendix C)

$pK_a = -\log K_a = -\log(4.5 \times 10^{-7}) = 6.34679$ (unrounded)

$$pH = pK_a + \log\left(\frac{[HCO_3^-]}{[H_2CO_3]}\right)$$

$$7.40 = 6.34679 + \log\left(\frac{[HCO_3^-]}{[H_2CO_3]}\right)$$

$$1.05321 = \log\left(\frac{[HCO_3^-]}{[H_2CO_3]}\right)$$

$$\frac{[HCO_3^-]}{[H_2CO_3]} = 11.3034235 \text{ (unrounded)}$$

$$\frac{[H_2CO_3]}{[HCO_3^-]} = 0.0884688 = \mathbf{0.088}$$

b) Use the Henderson-Hasselbalch equation.

$$pH = pK_a + \log\left(\frac{[HCO_3^-]}{[H_2CO_3]}\right)$$

$$7.20 = 6.34679 + \log\left(\frac{[HCO_3^-]}{[H_2CO_3]}\right)$$

$$0.85321 = \log\left(\frac{[HCO_3^-]}{[H_2CO_3]}\right)$$

$$\frac{[HCO_3^-]}{[H_2CO_3]} = 7.131978 \text{ (unrounded)}$$

$$\frac{[H_2CO_3]}{[HCO_3^-]} = 0.14021 = \mathbf{0.14}$$

19.118   The buffer components will be TRIS, $(HOCH_2)_3CNH_2$, and its conjugate acid TRISH$^+$, $(HOCH_2)_3CNH_3^+$. The conjugate acid is formed from the reaction between TRIS and HCl. Since HCl is the limiting reactant in this problem, the concentration of conjugate acid will equal the starting concentration of HCl, 0.095 M. The concentration of TRIS is the initial concentration minus the amount reacted.

$$\text{Mole TRIS} = \left(43.0 \text{ g TRIS}\right)\left(\frac{1 \text{ mol TRIS}}{121.14 \text{ g TRIS}}\right) = 0.354961 \text{ mol (unrounded)}$$

$$\text{Mole HCl added} = \left(\frac{0.095 \text{ mol HCl}}{L}\right)(1.00 \text{ L}) = 0.095 \text{ mol HCl} = \text{mol TRISH}^+$$

	$(HOCH_2)_3CNH_2$ (aq)	+ HCl(aq) $\leftrightarrows$	$(HOCH_2)_3CNH_3^+$ (aq)	+ Cl$^-$(aq)
Initial	0.354961 mol	0.095 mol	0	0
Reacting	−0.095 mol	− 0.095 mol	+ 0.095 mol	
Final	0.259961 mol	0 mol	0.095 mol	

Since there is 1.00 L of solution, the moles of TRIS and TRISH$^+$ equals their molarities.

$$pOH = pK_b + \log\frac{[acid]}{[base]} = 5.91 + \log\frac{[0.095]}{[0.259961]} = 5.472815 \text{ (unrounded)}$$

Therefore, pH of the buffer is 14.00 - 5.472815 = 8.527185 = **8.53**

19.120   Zinc sulfide, ZnS, is much less soluble than manganese sulfide, MnS. Convert $ZnCl_2$ and $MnCl_2$ to ZnS and MnS by saturating the solution with H_2S; $[H_2S]_{sat'd} = 0.10$ M. Adjust the pH so that the greatest amount of ZnS will precipitate and not exceed the solubility of MnS as determined by K_{sp}(MnS).

K_{sp}(MnS) = $[Mn^{2+}][HS^-][OH^-] = 3 \times 10^{-11}$

$[Mn^{2+}] = [MnCl_2] = 0.020$ M

$[HS^-]$ is calculated using the K_{a1} expression:

Concentration: $H_2S(aq) + H_2O(l) \leftrightarrows H_3O^+(aq) + HS^-(aq)$

	$H_2S(aq)$ + $H_2O(l)$ $\leftrightarrows$	$H_3O^+(aq)$	+ $HS^-(aq)$
Initial	0.10 mol	0	0
Reacting	−x	+ x	+ x
Final	0.10 − x	x	x

374

$$K_{a1} = \frac{\left[H_3O^+\right]\left[HS^-\right]}{\left[H_2S\right]} = \frac{\left[H_3O^+\right]\left[HS^-\right]}{\left[0.10-x\right]} = 9 \text{ x } 10^{-8} \quad \text{Assume } 0.10-x = 0.10$$

$[H_3O^+][HS^-] = 9 \text{ x } 10^{-9}$

$[HS^-] = 9 \text{ x } 10^{-9} / [H_3O^+]$

Substituting $[Mn^{2+}]$ and $[HS^-]$ into the K_{sp}(MnS) above gives:

K_{sp}(MnS) $= [Mn^{2+}][HS^-][OH^-] = 3 \text{ x } 10^{-11}$

K_{sp}(MnS) $= [Mn^{2+}] (9 \text{ x } 10^{-9} / [H_3O^+])[OH^-] = 3 \text{ x } 10^{-11}$

Substituting $K_w / [H_3O^+]$ for $[OH^-]$:

$$K_{sp}(\text{MnS}) = \left[Mn^{2+}\right]\left(\frac{9 \text{ x } 10^{-9}}{\left[H_3O^+\right]}\right)\left(\frac{K_w}{\left[H_3O^+\right]}\right) = 3 \text{ x } 10^{-11}$$

$$3 \text{ x } 10^{-11} \text{ x } [H_3O^+]^2 = \left[Mn^{2+}\right]\left(9 \text{ x } 10^{-9}\right)\left(K_w\right)$$

$$[H_3O^+] = \sqrt{\frac{\left[Mn^{2+}\right]\left[9 \text{ x } 10^{-9}\right]K_w}{3 \text{ x } 10^{-11}}} = \sqrt{\frac{(0.020)\left(9 \text{ x } 10^{-9}\right)\left(1.0 \text{ x } 10^{-14}\right)}{3 \text{ x } 10^{-11}}} = 2.4494897 \text{ x } 10^{-7} \text{ (unrounded)}$$

pH $= -\log [H_3O^+] = -\log (2.4494897 \text{ x } 10^{-7}) = 6.610924 = \textbf{6.6}$

Maintain the pH below 6.6 to separate the ions as their sulfides.

19.124 An indicator changes color when the buffer component ratio of the two forms of the indicator changes from a value greater than 1 to a value less than 1. The pH at which the ratio equals 1 is equal to pK_a. The midpoint in the pH range of the indicator is a good estimate of the pK_a of the indicator.

$\qquad$ pK_a = (3.4 + 4.8) / 2 = **4.1** $\qquad K_a = 10^{-4.1} = 7.943 \text{ x } 10^{-5} = \textbf{8 x } 10^{-5}$

19.126 a) A spreadsheet will help you to quickly calculate ΔpH / ΔV and average volume for each data point. At the equivalence point, the pH changes drastically when only a small amount of base is added, therefore, ΔpH / ΔV is at a maximum at the equivalence point.

Example calculation: For the first 2 lines of data: ΔpH = 1.22 – 1.00 = 0.22; ΔV = 10.00 – 0.00 = 10.00

$$\frac{\Delta\text{pH}}{\Delta\text{V}} = \frac{0.22}{10.00} = 0.022 \qquad\qquad V_{average}\text{(mL)} = (0.00 + 10.00)/2 = 5.00$$

V(mL)	pH	$\dfrac{\Delta\text{pH}}{\Delta\text{V}}$	$V_{average}$(mL)
0.00	1.00		
10.00	1.22	0.022	5.00
20.00	1.48	0.026	15.00
30.00	1.85	0.037	25.00
35.00	2.18	0.066	32.50
39.00	2.89	0.18	37.00
39.50	3.20	0.62	39.25
39.75	3.50	1.2	39.63
39.90	3.90	2.67	39.83
39.95	4.20	6	39.93
39.99	4.90	18	39.97
40.00	7.00	200	40.00
40.01	9.40	200	40.01
40.05	9.80	10	40.03
40.10	10.40	10	40.08
40.25	10.50	0.67	40.18
40.50	10.79	1.2	40.38
41.00	11.09	0.60	40.75
45.00	11.76	0.17	43.00
50.00	12.05	0.058	47.50

60.00	12.30	0.025	55.00
70.00	12.43	0.013	65.00
80.00	12.52	0.009	75.00

b)

Maximum slope (equivalence point) is at V = 40.00 mL

19.128   The equation that describes the behavior of a weak base in water is:
$$B(aq) + H_2O(l) \rightleftharpoons BH^+(aq) + OH^-(aq)$$

$$K_b = \frac{\left[BH^+\right]\left[OH^-\right]}{[B]}$$

$$-\log K_b = -\log \frac{\left[BH^+\right]\left[OH^-\right]}{[B]}$$

$$-\log K_b = -\log \frac{\left[BH^+\right]}{[B]} - \log [OH^-]$$

$$pK_b = -\log \frac{\left[BH^+\right]}{[B]} + pOH$$

$$pOH = pK_b + \log \frac{\left[BH^+\right]}{[B]}$$

19.129   Use HLac to indicate lactic acid and Lac⁻ to indicate the lactate ion. The Henderson-Hasselbalch equation gives the pH of the buffer. Determine the final concentrations of the buffer components from $M_{conc}V_{conc} = M_{dil}V_{dil}$. Determine the pK_a of the acid from the K_a.

   $pK_a = -\log K_a = -\log 1.38 \times 10^{-4} = 3.86012$ (unrounded)

Determine the molarity of the diluted buffer component as $M_{dil} = M_{conc}V_{conc} / V_{dil}$

   [HLac] = [(0.85 M) (225 mL)] / [(225 + 435) mL] = 0.28977 M HLac (unrounded)

   [Lac⁻] = [(0.68 M) (435 mL)] / [(225 + 435) mL] = 0.44818 M Lac⁻ (unrounded)

$$pH = pK_a + \log\left(\frac{[Lac^-]}{[HLac]}\right)$$

$$pH = 3.86012 + \log\left(\frac{[0.44818]}{[0.28977]}\right) = 4.049519 = \mathbf{4.05}$$

19.136   To determine which species are present from a buffer system of a polyprotic acid, check the pK_a values for the one that is closest to the pH of the buffer. The two components involved in the equilibrium associated with this K_a are the principle species in the buffer. For carbonic acid, pK_{a1} [$-\log(8 \times 10^{-7}) = 6.1$] is closest to the pH of 7.4, so H_2CO_3 and HCO_3^- are the species present. For phosphoric acid, pK_{a2} [$-\log(2.3 \times 10^{-7}) = 6.6$] is closest to the pH, so $H_2PO_4^-$ and HPO_4^{2-} are the principle species present.

$$H_2PO_4^-(aq) \ + \ H_2O(l) \ \leftrightarrows \ HPO_4^{2-}(aq) \ + \ H_3O^+(aq)$$

$$[H_3O^+] = 10^{-7.4} = 3.98107 \times 10^{-8} \ M \ \text{(unrounded)}$$

$$K_{a2} = \frac{\left[H_3O^+\right]\left[HPO_4^{2-}\right]}{\left[H_2PO_4^-\right]}$$

$$\frac{\left[HPO_4^{2-}\right]}{\left[H_2PO_4^-\right]} = \frac{K_{a2}}{\left[H_3O^+\right]} = \frac{2.3 \times 10^{-7}}{3.98107 \times 10^{-8}} = 5.777341 = \textbf{5.8}$$

19.139 Determine the minimum pH needed to cause the initial precipitation of NiS.

$$NiS(s) + H_2O(l) \leftrightarrows Ni^{2+}(aq) + HS^-(aq) + OH^-(aq)$$
$$K_{sp}(NiS) = 3 \times 10^{-16} = [Ni^{2+}][HS^-][OH^-]$$

$[Ni^{2+}] = 0.15 \ M$, so $[HS^-]$ and $[OH^-]$ must be found.
From $[H_2S] = 0.050 \ M$ and K_{a1} in the Appendix:

$$H_2S \ (aq) + H_2O(l) \leftrightarrows HS^- \ (aq) + H_3O^+ \ (aq)$$

$$K_{a1} = \frac{\left[H_3O^+\right]\left[HS^-\right]}{\left[H_2S\right]} = \frac{\left[H_3O^+\right]\left[HS^-\right]}{\left[0.050 - x\right]} = \frac{\left[H_3O^+\right]\left[HS^-\right]}{\left[0.050\right]} = 9 \times 10^{-8}$$

$[HS^-][H_3O^+] = 4.5 \times 10^{-9}$ or $[HS^-] = 4.5 \times 10^{-9} / [H_3O^+]$

$[OH^-] = K_w / [H_3O^+]$

$K_{sp}(NiS) = 3 \times 10^{-16} = [Ni^{2+}][HS^-][OH^-]$

$K_{sp}(NiS) = 3 \times 10^{-16} = (0.15)(4.5 \times 10^{-9} / [H_3O^+])(K_w / [H_3O^+])$

$3 \times 10^{-16} = (0.15)(4.5 \times 10^{-9})(1.0 \times 10^{-14}) / [H_3O^+]^2)$

$[H_3O^+]^2 = [(0.15)(4.5 \times 10^{-9})(1.0 \times 10^{-14})] / (3 \times 10^{-16}) = 2.25 \times 10^{-8}$ (unrounded)

$[H_3O^+] = 1.5 \times 10^{-4} \ M$ (unrounded)

$pH = -\log [H_3O^+] = -\log (1.5 \times 10^{-4} \ M) = 3.8239 = \textbf{3.8}$

19.141 a) To find the concentration of HCl after neutralizing the quinidine, calculate the concentration of quinidine and the amount of HCl required to neutralize it, remembering that the mole ratio for the neutralization is 2 mol HCl / 1 mol quinidine.

$$\text{Moles quinidine} = \left(33.85 \ \text{mg quinidine}\right)\left(\frac{10^{-3} \ g}{1 \ mg}\right)\left(\frac{1 \ \text{mol quinidine}}{324.41 \ \text{g quinidine}}\right)$$

$$= 1.0434326 \times 10^{-4} \ \text{mol quinidine (unrounded)}$$

$$\text{Moles HCl excess} = \left(6.55 \ mL\right)\left(\frac{10^{-3} \ L}{1 \ mL}\right)\left(\frac{0.150 \ \text{mol HCl}}{L}\right)$$

$$- \left(1.0434326 \times 10^{-4} \ \text{mol quinidine}\right)\left(\frac{2 \ \text{mol HCl}}{1 \ \text{mol quinidine}}\right) = 7.7381348 \times 10^{-4} \ \text{mol HCl}$$

(unrounded)

$$\text{Volume needed} = \left(7.7381348 \times 10^{-4} \ \text{mol HCl}\right)\left(\frac{1 \ \text{mol NaOH}}{1 \ \text{mol HCl}}\right)\left(\frac{1 \ L}{0.0133 \ \text{mol NaOH}}\right)\left(\frac{1 \ mL}{10^{-3} \ L}\right)$$

$$= 58.18146 = \textbf{58.2 mL NaOH solution}$$

b) Use the moles of quinidine and the concentration of the NaOH to determine the milliliters.

$$\text{Volume} = \left(1.043436 \times 10^{-4} \ \text{mol quinidine}\right)\left(\frac{1 \ \text{mol NaOH}}{1 \ \text{mol quinidine}}\right)\left(\frac{1 \ L}{0.0133 \ \text{mol NaOH}}\right)\left(\frac{1 \ mL}{10^{-3} \ L}\right)$$

$$= 7.84538 = \textbf{7.84 mL NaOH solution}$$

c) When quinidine (QNN) is first acidified, it has the general form QNH^+NH^+. At the first equivalence point, one of the acidified nitrogen atoms has completely reacted, leaving a singly protonated form, $QNNH^+$. This form of quinidine can react with water as either an acid or a base, so both must be considered. If the concentration of quinidine at the first equivalence point is greater than K_{b1}, then the $[OH^-]$ at the first equivalence point can be estimated as:

$$[OH^-] = \sqrt{K_{b1}K_{b2}} = \sqrt{(4.0 \times 10^{-6})(1.0 \times 10^{-10})} = 2.0 \times 10^{-8}\ M$$

$$[H_3O^+] = K_w / [OH^-] = (1.0 \times 10^{-14}) / (2.0 \times 10^{-8}) = 5.0 \times 10^{-7}\ M$$

$$pH = -\log [H_3O^+] = -\log (5.0 \times 10^{-7}\ M) = 6.3010 = \mathbf{6.30}$$

19.143 The Henderson-Hasselbalch equation demonstrates that the pH changes when the ratio of acid to base in the buffer changes (pK_a is constant at a given temperature):

$$pH = pK_a + \log\left(\frac{[A^-]}{[HA]}\right)$$

The pH of the A^- / HA buffer cannot be calculated because the identity of "A" and, thus, the value of pK_a are unknown. However, the change in pH can be described:

$$\Delta pH = \log\left(\frac{[A^-]}{[HA]}\right)_{final} - \log\left(\frac{[A^-]}{[HA]}\right)_{initial}$$

Since both [HA] and $[A^-]$ = 0.10 M, $\log\left(\dfrac{[A^-]}{[HA]}\right)_{initial}$ = 0 because [HA] = $[A^-]$, and log(1) = 0

So the change in pH is equal to the concentration ratio of base to acid after the addition of H_3O^+.
Consider the buffer prior to addition to the medium.

H_3O^+ (aq) +	A^-(aq)	→	HA(aq)
0.0010 mol	0.10 mol		0.10 mol
–0.0010 mol	–0.0010 mol		+0.0010 mol
0	0.099 mol		0.101 mol

When 0.0010 mol H_3O^+ is added to 1 L of the undiluted buffer, the $[A^-]$ / [HA] ratio changes from 0.10 / 0.10 to (0.099) / (0.101). The change in pH is:

$$\Delta pH = \log (0.099 / 0.101) = -0.008686 \text{ (unrounded)}$$

If the undiluted buffer changes 0.009 pH units with addition of 0.0010 mol H_3O^+, how much can the buffer be diluted and still not change by 0.05 pH units ($\Delta pH < 0.05$)?
Let x = fraction by which the buffer can be diluted. Assume 0.0010 mol H_3O^+ is added to 1 L.

$$\log \frac{[base]}{[acid]} = \log\left(\frac{(0.10x - 0.0010)}{(0.10x + 0.0010)}\right) = -0.05$$

$$\left(\frac{(0.10x - 0.0010)}{(0.10x + 0.0010)}\right) = 10^{-0.05} = 0.89125 \text{ (unrounded)}$$

$$0.10x - 0.0010 = 0.89125 (0.10x + 0.0010)$$
$$x = 0.173908 = 0.17$$

The buffer concentration can be decreased by a factor of 0.17, or **170 mL** of buffer can be diluted to 1 L of medium. At least this amount should be used to adequately buffer the pH change.

19.146 To find the volume of rain, first convert the inches to yards and find the volume in yd^3. Then convert units to cm^3 and on to L.

$$(10.0\ acres)\left(\frac{4.840 \times 10^3\ yd^2}{1\ acre}\right)\left(\frac{36\ in}{1\ yd}\right)^2 (1.00\ in)\left(\frac{2.54\ cm}{1\ in}\right)^3\left(\frac{1\ mL}{1\ cm^3}\right)\left(\frac{10^{-3}\ L}{1\ mL}\right) = 1.0279015 \times 10^6\ L \text{ (unrounded)}$$

a) At pH = 4.20, $[H_3O^+] = 10^{-4.20} = 6.3095734 \times 10^{-5}\ M$ (unrounded)
Mol H_3O^+ = $(6.3095734 \times 10^{-5}\ M)(1.0279015 \times 10^6\ L) = 64.8562 = \mathbf{65\ mol}$

b) Volume = $(10.0 \text{ acres})\left(\dfrac{4.840 \times 10^3 \text{ yd}^2}{1 \text{ acre}}\right)\left(\dfrac{36 \text{ in}}{1 \text{ yd}}\right)^2 (10.0 \text{ ft})\left(\dfrac{12 \text{ in}}{1 \text{ ft}}\right)\left(\dfrac{2.54 \text{ cm}}{1 \text{ in}}\right)^3\left(\dfrac{1 \text{ mL}}{1 \text{ cm}^3}\right)\left(\dfrac{10^{-3} \text{ L}}{1 \text{ mL}}\right)$

$\qquad$ = 1.23348 x 10^8 L (unrounded)

$\qquad$ Total volume of lake after rain = 1.23348 x 10^8 L + 1.0279015 x 10^6 L = 1.243759 x 10^8 L (unrounded)

$\qquad$ $[H_3O^+]$ = 64.8562 mol H_3O^+ / 1.243759 x 10^8 L = 5.214531 x 10^{-7} M

$\qquad$ pH = $-\log(5.214531 \times 10^{-7})$ = 6.2827847 = **6.28**

c) Each mol of H_3O^+ requires one mole of HCO_3^- for neutralization.

$\qquad$ Mass = $(64.8562 \text{ mol } H_3O^+)\left(\dfrac{1 \text{ mol } HCO_3^-}{1 \text{ mol } H_3O^+}\right)\left(\dfrac{61.02 \text{ g } HCO_3^-}{1 \text{ mol } HCO_3^-}\right)$ = 3.97575 x 10^3 = **4.0 x 10^3 g HCO_3^-**

19.148 $\quad$ Carbon dioxide dissolves in water to produce H_3O^+ ions:

$\qquad$ $CO_2(g) \leftrightarrows CO_2(aq)$

$\qquad$ $CO_2(aq) + H_2O(l) \leftrightarrows H_2CO_3(aq)$

$\qquad$ $H_2CO_3(aq) \leftrightarrows H_3O^+(aq) + HCO_3^-(aq)$

The molar concentration of CO_2, $[CO_2]$, depends on how much $CO_2(g)$ from the atmosphere can dissolve in pure water. At 25°C and 1 atm CO_2, 88 mL of CO_2 can dissolve in 100 mL of H_2O. The number of moles of CO_2 in 88 mL of CO_2 is:

$\qquad$ Moles CO_2 = PV / RT = $\dfrac{(1 \text{ atm})(88 \text{ mL})}{\left(0.0821\dfrac{L \bullet atm}{mol \bullet K}\right)\left((273 + 25) K\right)}\left(\dfrac{10^{-3} \text{ L}}{1 \text{ mL}}\right)\left(\dfrac{0.033\%}{100\%}\right)$

$\qquad$ = 1.186963 x 10^{-6} mol CO_2 (unrounded)

Since air is not pure CO_2, account for the volume fraction of air (0.033 L / 100 L) when determining the moles.

$[CO_2]$ = (1.186963 x 10^{-6} mol CO_2) / [(100 mL) (10^{-3} L / 1 mL)] = 1.186963 x 10^{-5} M CO_2 (unrounded)

$\qquad$ $K_{a1} = \dfrac{\left[H_3O^+\right]\left[HCO_3^-\right]}{\left[H_2CO_3\right]} = \dfrac{\left[H_3O^+\right]\left[HCO_3^-\right]}{\left[CO_2\right]}$ = 4.5 x 10^{-7}

$\qquad$ Let x = $[H_3O^+]$ = $[HCO_3^-]$

$\qquad$ $\dfrac{[x][x]}{\left[1.186963 \times 10^{-5} - x\right]}$ = 4.5 x 10^{-7}

$\qquad\qquad$ x^2 + 4.5 x 10^{-7} x $-$ 5.341336 x 10^{-12} = 0

$\qquad\qquad$ x = 2.0970596 x 10^{-6} M = $[H_3O^+]$ (unrounded)

$\qquad$ pH = $-\log(2.0970596 \times 10^{-6})$ = 5.678389 = **5.68**

19.150 $\quad$ Initial concentrations of Pb^{2+} and $Ca(EDTA)^{2-}$ before reaction based on mixing 100. mL of 0.10 M $Na_2Ca(EDTA)$ with 1.5 L blood:

$\qquad$ $[Pb^{2+}]$ = $\left(\dfrac{120 \text{ µg } Pb^{2+}}{100 \text{ mL}}\right)\left(\dfrac{1 \text{ mL}}{10^{-3} \text{ L}}\right)\left(\dfrac{1.5 \text{ L blood}}{1.6 \text{ L mixture}}\right)\left(\dfrac{10^{-6} \text{ g}}{1 \text{ µg}}\right)\left(\dfrac{1 \text{ mol } Pb^{2+}}{207.2 \text{ g } Pb^{2+}}\right)$ = 5.4295366 x 10^{-6} M Pb^{2+} (unrounded)

$M_{con}V_{con} = M_{dil}V_{dil}$

$[Ca(EDTA)^{2-}]$ = $M_{con}V_{con}$ / V_{dil} = [(0.10 M) (100 mL) (10^{-3} L / 1 mL)] / (1.6 L) = 6.25 x 10^{-3} M (unrounded)

Set up a reaction table assuming the reaction goes to completion:

Concentration (*M*)	$[Ca(EDTA)]^{2-}(aq)$ +	$Pb^{2+}(aq)$	$\leftrightarrows$	$[Pb(EDTA)]^{2-}(aq)$ +	$Ca^{2+}(aq)$
Initial	6.25 x 10^{-3}	5.4295366 x 10^{-6}		0	0
React	$-$ 5.4295366 x 10^{-6}	$-$5.4295366 x 10^{-6}		+ 5.4295366 x 10^{-6}	+ 5.4295366 x 10^{-6}
	6.24457 x 10^{-3}	0		5.4295366 x 10^{-6}	5.4295366 x 10^{-6}

Now set up a reaction table for the equilibrium process:

Concentration (*M*)	$[Ca(EDTA)]^{2-}(aq)$ +	$Pb^{2+}(aq)$	$\leftrightarrows$	$[Pb(EDTA)]^{2-}(aq)$ +	$Ca^{2+}(aq)$
Initial	6.24457 x 10^{-3}	0		5.4295366 x 10^{-6}	5.4295366 x 10^{-6}
Change	+ x	+ x		$-$x	$-$x
Equilibrium	6.24457 x 10^{-3} + $\quad$ x	x		5.4295366 x 10^{-6} $-$ x	5.4295366 x 10^{-6} $-$ x

$$K_c = 2.5 \times 10^7 = \frac{\left[Pb(EDTA)^{2-}\right]\left[Ca^{2+}\right]}{\left[Ca(EDTA)^{2-}\right]\left[Pb^{2+}\right]} = \frac{\left[5.4295366 \times 10^{-6}\right]\left[5.4295366 \times 10^{-6}\right]}{\left[6.24457 \times 10^{-3}\right]\left[x\right]}$$

$x = [Pb^{2+}] = 1.8883521 \times 10^{-16} M$ (unrounded)

Convert concentration from M to µg in 100 mL:

$$\left(\frac{1.8883521 \times 10^{-16} \text{ mol } Pb^{2+}}{L}\right)\left(\frac{10^{-3} \text{ L}}{1 \text{ mL}}\right)(100 \text{ mL})\left(\frac{207.2 \text{ g } Pb^{2+}}{1 \text{ mol } Pb^{2+}}\right)\left(\frac{1 \text{ µg}}{10^{-6} \text{ g}}\right) = 3.9126655 \times 10^{-9} \text{ µg } Pb^{2+} \text{ (unrounded)}$$

The final concentration is **3.9×10^{-9} µg / 100 mL**.

19.152   The molarity of a saturated NaCl solution must be found.

$$M \text{ NaCl} = \left(\frac{317 \text{ g NaCl}}{L}\right)\left(\frac{1 \text{ mol NaCl}}{58.44 \text{ g NaCl}}\right) = 5.42436687 \, M \text{ NaCl (unrounded)}$$

Determine the K_{sp} from the molarity just calculated.

NaCl(s) ⇄ Na$^+$(aq) + Cl$^-$(aq).

$K_{sp} = [Na^+][Cl^-] = S^2 = (5.42436687)^2 = 29.42375594 = 29.4$

$$\text{Moles of Cl}^- \text{ initially} = \left(\frac{5.4236687 \text{ mol NaCl}}{L}\right)(0.100 \text{ L})\left(\frac{1 \text{ mol Cl}^-}{1 \text{ mol NaCl}}\right) = 0.54236687 \text{ mol Cl}^- \text{ (unrounded)}$$

This is the same as the moles of Na$^+$ in the solution.

$$\text{Moles of Cl}^- \text{ added} = \left(\frac{8.65 \text{ mol HCl}}{L}\right)\left(\frac{10^{-3} \text{ L}}{1 \text{ mL}}\right)(28.5 \text{ mL})\left(\frac{1 \text{ mol Cl}^-}{1 \text{ mol HCl}}\right) = 0.246525 \text{ mol Cl}^- \text{ (unrounded)}$$

0.100 L of saturated solution contains 0.542 mol each Na$^+$ and Cl$^-$, to which you are adding 0.246525 mol of additional Cl$^-$ from HCl.

Volume of mixed solutions = 0.100 L + (28.5 mL) (10^{-3} L / 1 mL) = 0.1285 L (unrounded)

Molarity of Cl$^-$ in mixture = [(0.54236687 + 0.246525) mol Cl$^-$] / (0.1285 L) = 6.13924 M Cl$^-$ (unrounded)

Molarity of Na$^+$ in mixture = (0.54236687 mol Na$^+$) / (0.1285 L) = 4.220754 M Na$^+$ (unrounded)

Determine a Q value and compare this value to the K_{sp} to determine if precipitation will occur.

$Q_{sp} = [Na^+][Cl^-] = (4.220754)(6.13924) = 25.91222 = 25.9$

Since $Q_{sp} < K_{sp}$, no NaCl will precipitate.

19.153   a) For the solution to be a buffer, both HA and A$^-$ must be present in the solution. This situation occurs in **A** and **D**.

b)   Box A

The amounts of HA and A$^-$ are equal.

$$pH = pK_a + \log\left(\frac{[A^-]}{[HA]}\right) \qquad\qquad \left(\frac{[A^-]}{[HA]}\right) = 1 \text{ when the amounts of HA and A}^- \text{ are equal}$$

$pH = pK_a + \log 1$

$pH = pK_a = -\log (4.5 \times 10^{-5}) = 4.346787 = \textbf{4.35}$

Box B

Only A$^-$ is present at a concentration of 0.10 M.

The K_b for A$^-$ is needed.

$K_b = K_w / K_a = 1.0 \times 10^{-14} / 4.5 \times 10^{-5} = 2.222 \times 10^{-10}$ (unrounded)

A$^-$(aq) + H$_2$O(l) ⇄ OH$^-$(aq) + HA(aq)

Initial:	0.10 M	0	0
Change:	−x	−x	−x
Equilibrium:	0.10 − x	x	x

$$K_b = \frac{[HA][OH^-]}{[A^-]} = 2.222 \times 10^{-10}$$

$$K_b = \frac{[x][x]}{[0.10 - x]} = 2.222 \times 10^{-10} \qquad \text{Assume that x is small compared to 0.10}$$

$$K_b = 2.222 \times 10^{-10} = \frac{(x)(x)}{(0.10)}$$

x = 4.7138095 x 10^{-6} M OH⁻ (unrounded)

Check assumption: (4.7138095 x 10^{-6} / 0.10) x 100% = 0.005% error, so the assumption is valid.

$[H_3O]^+ = K_w / [OH^-] = (1.0 \times 10^{-14}) / (4.7138095 \times 10^{-6})$
= 2.1214264 x 10^{-9} M H_3O^+ (unrounded)

pH = –log [H_3O^+] = –log (2.1214264 x 10^{-9}) = 8.67337 = **8.67**

Box C

This is a 0.10 M HA solution. The hydrogen ion, and hence the pH, can be determined from the K_a.

Concentration	HA(aq)	+	H₂O(l) ⇌	H₃O⁺(aq) + A⁻(aq)	
Initial	0.10		—	0	0
Change	– x			+ x	+ x
Equilibrium	0.10 – x			x	x

(The H_3O^+ contribution from water has been neglected.)

$$K_a = 4.5 \times 10^{-5} = \frac{\left[H_3O^+\right]\left[A^-\right]}{[HA]}$$

$$K_a = 4.5 \times 10^{-5} = \frac{(x)(x)}{(0.10 - x)} \qquad \text{Assume that x is small compared to 0.10.}$$

$$K_a = 4.5 \times 10^{-5} = \frac{(x)(x)}{(0.10)}$$

[H_3O^+] = x = 2.12132 x 10^{-3} (unrounded)

Check assumption: (2.12132 x 10^{-3} / 0.10) x 100% = 2% error, so the assumption is valid.

pH = –log [H_3O^+] = –log (2.12132 x 10^{-3}) = 2.67339 = **2.67**

Box D

This is a buffer with a ratio of [A⁻] / [HA] = 5 / 3.
Use the Henderson-Hasselbalch equation for this buffer.

$$pH = pK_a + \log\left(\frac{[A^-]}{[HA]}\right)$$

$$pH = -\log(4.5 \times 10^{-5}) + \log\left[\frac{5}{3}\right] = 4.568636 = \textbf{4.57}$$

c) The initial stage in the titration would only have HA present. The amount of HA will decrease, and the amount of A⁻ will increase until only A⁻ remains. The sequence will be: **C, A, D, and B**.

d) At the equivalence point, all the HA will have reacted with the added base. This occurs in scene **B**.

CHAPTER 20 THERMODYNAMICS: ENTROPY, FREE ENERGY, AND THE DIRECTION OF CHEMICAL REACTIONS

FOLLOW–UP PROBLEMS

20.1   a) **PCl$_5$(g)**. For substances with the same type of atoms and in the same physical state, entropy increases with increasing number of atoms per molecule because more types of molecular motion are available.
b) **BaCl$_2$(s)**. Entropy increases with increasing atomic size. The Ba^{2+} ion and Cl^- ion are larger than the Ca^{2+} ion and F^- ion, respectively.
c) **Br$_2$(g)**. Entropy increases from solid → liquid → gas.

20.2   Plan: Predict the sign of ΔS^o_{sys} by comparing the randomness of the products with the randomness of the reactants. Calculate ΔS^o_{sys} using Appendix B values.

Solution:
a) $2\,NaOH(s) + CO_2(g) \rightarrow Na_2CO_3(s) + H_2O(l)$
The ΔS^o_{sys} is predicted to decrease ($\Delta S^o_{sys} < 0$) because the more random, gaseous reactant is transformed into a more ordered, liquid product.

$\Delta S^o_{sys} = \Sigma(m\,\Delta S^o_{products}) - \Sigma(n\,\Delta S^o_{reac\,tan\,ts})$

$\Delta S^o_{sys} = [(1\;mol\;Na_2CO_3)\,(S^{\circ}_{Na_2CO_3(s)}) + 1\;mol\;H_2O)\,(S^{\circ}_{H_2O(l)})] - [(2\;mol\;NaOH)\,(S^{\circ}_{NaOH(s)}) + 1\;mol\;CO_2)\,(S^{\circ}_{CO_2(g)})]$

$\Delta S^o_{sys} = [(1\;mol\;Na_2CO_3)\,(139\;J/mol \cdot K) + (1\;mol\;H_2O)\,(69.940\;J/mol \cdot K)]$

$\qquad\qquad - [(2\;mol\;NaOH)\,(64.454\;J/mol \cdot K) + (1\;mol\;CO_2)\,(213.7\;J/mol \cdot K)] = -133.668 = \mathbf{-134\;J/K}$

$\Delta S^o_{sys} < 0$ as predicted.

b) $2\,Fe(s) + 3\,H_2O(g) \rightarrow Fe_2O_3(s) + 3\,H_2(g)$
The change in gaseous moles is zero, so the sign of ΔS^o_{sys} is difficult to predict. Iron(III) oxide has greater entropy than Fe because it is more complex, but this is offset by the greater molecular complexity of H_2O versus H_2.

$\Delta S^o_{sys} = [(1\;mol\;Fe_2O_3)\,(S^{\circ}_{Fe_2O_3}) + (3\;mol\;H_2)\,(S^{\circ}_{H_2})] - [(2\;mol\;Fe)\,(S^{\circ}_{Fe}) + (3\;mol\;H_2O)\,(S^{\circ}_{H_2O})]$

$\qquad \Delta S^o_{sys} = [(1\;mol\;Fe_2O_3)\,(87.400\;J/mol \cdot K) + (3\;mol\;H_2)\,(130.6\;J/mol \cdot K)]$

$\qquad\qquad - [(2\;mol\;Fe)\,(27.3\;J/mol \cdot K) + (3\;mol\;H_2O)\,(188.72\;J/mol \cdot K)] = -141.56 = \mathbf{-141.6\;J/K}$

The negative ΔS^o_{sys} for equation (b) reflects that greater entropy of H_2O versus H_2 does outweigh the greater entropy of Fe_2O_3 versus Fe.

20.3   Plan: Write the balanced equation for the reaction and calculate the ΔS^o_{sys} using Appendix B. Determine the ΔS^o_{surr} by first finding ΔH^o_{rxn}. Add ΔS^o_{surr} to ΔS^o_{sys} to verify that ΔS_{univ} is positive.

Solution:
$2\,FeO(s) + 1/2\,O_2(g) \rightarrow Fe_2O_3(s)$

$\qquad\qquad \Delta S^o_{sys} = [(1\;mol\;Fe_2O_3)\,(S^{\circ}Fe_2O_3)] - [(2\;mol\;FeO)\,(S^{\circ}FeO) + (1/2\;mol\;O_2)(S^{\circ}O_2)]$

$\qquad\qquad \Delta S^o_{sys} = [(1\;mol\;Fe_2O_3)\,(87.400\;J/mol \cdot K)] - [(2\;mol\;FeO)\,(60.75\;J/mol \cdot K)$

$\qquad\qquad\qquad + (1/2\;mol\;O_2)\,(205.0\;J/mol \cdot K)]$

$\qquad\qquad \Delta S^o_{sys} = -136.6\;J/K$ (entropy change is expected to be negative because gaseous reactant is converted to

$\qquad\qquad\qquad$ solid product).

$$\Delta H^\circ_{sys} = [(1 \text{ mol Fe}_2\text{O}_3) \,(\Delta H^o_f \text{ Fe}_2\text{O}_3)] - [(2 \text{ mol FeO}) \,(\Delta H^o_f \text{ FeO}) + (1/2 \text{ mol O}_2)(\,\Delta H^o_f \text{ O}_2)]$$

$$\Delta H^\circ_{sys} = [(1 \text{ mol Fe}_2\text{O}_3) \,(-825.5 \text{ kJ/mol})] - [(2 \text{ mol FeO}) \,(-272.0 \text{ kJ/mol}) + (1/2 \text{ mol O}_2) \,(0)]$$

$$\Delta H^\circ_{sys} = -281.5 \text{ kJ}$$

$$\Delta S^o_{surr} = -\frac{\Delta H^o_{sys}}{T} = -\frac{-281.5 \text{ kJ}}{298 \text{ K}} = 0.94463 \text{ kJ/K}(10^3 \text{ J/1 kJ}) = 944.63 \text{ J/K} \quad \text{(unrounded)}$$

$$\Delta S^o_{univ} = \Delta S^o_{sys} + \Delta S^o_{surr} = (-136.6 \text{ J/K}) + (944.63 \text{ J/K}) = 808.03 = \textbf{808 J/K.}$$

Because ΔS^o_{univ} is positive, the reaction is spontaneous at 298 K.

Check: This process is also known as rusting. Common sense tells us that rusting occurs spontaneously. Although the entropy change of the system is negative, the increase in entropy of the surroundings is large enough to offset ΔS^o_{sys}.

20.4   Plan: Calculate the ΔH^o_{rxn} using ΔH^o_f values from Appendix B. Calculate ΔS^o_{rxn} from tabulated S° values and then use the relationship $\Delta G^o_{rxn} = \Delta H^o_{rxn} - T\Delta S^o_{rxn}$.

Solution:

$$\Delta H^\circ_{rxn} = [(2 \text{ mol NO}_2) \,(\Delta H^o_f \,(\text{NO}_2))] - [(2 \text{ mol NO}) \,(\Delta H^o_f \,(\text{NO})) + (1 \text{ mol O}_2) \,(\Delta H^o_f \,(\text{O}_2))]$$

$$\Delta H^\circ_{rxn} = [(2 \text{ mol NO}_2) \,(33.2 \text{ kJ/mol})] - [(2 \text{ mol NO}) \,(90.29 \text{ kJ/mol}) + (1 \text{ mol O}_2) \,(0)]$$

$$\Delta H^\circ_{rxn} = -114.18 = -114.2 \text{ kJ}$$

$$\Delta S^\circ_{rxn} = [(2 \text{ mol NO}_2) \,(S^o_f \,(\text{NO}_2))] - [(2 \text{ mol NO}) \,(S^o_f \,(\text{NO})) + (1 \text{ mol O}_2) \,(S^o_f \,(\text{O}_2))]$$

$$\Delta S^\circ_{rxn} = [(2 \text{ mol NO}_2) \,(239.9 \text{ J/mol·K})] - [(2 \text{ mol NO}) \,(210.65 \text{ J/mol·K}) + (1 \text{ mol O}_2) \,(205.0 \text{ J/mol·K})]$$

$$\Delta S^\circ_{rxn} = -146.5 \text{ J/K}$$

$$\Delta G^\circ_{rxn} = \Delta H^\circ_{rxn} - T\Delta S^\circ_{rxn} = -114.2 \text{ kJ} - [(298 \text{ K}) \,(-146.5 \text{ J/K}) \,(1 \text{ kJ}/10^3 \text{ J})] = -70.543 = \textbf{-70.5 kJ}$$

20.5   Plan: Use ΔG^o_f values from Appendix B to calculate ΔG°_{rxn}. $\Delta G^\circ_{sys} = \Sigma(\text{m} \Delta G^o_{f(products)}) - \Sigma(\text{n} \Delta G^o_{f(reac\tan ts)})$

Solution:

a) $\Delta G^\circ_{rxn} = [(2 \text{ mol NO}_2) \,(\Delta G^o_f \,(\text{NO}_2))] - [(2 \text{ mol NO}) \,(\Delta G^o_f \,(\text{NO})) + (1 \text{ mol O}_2) \,(\Delta G^o_f \,(\text{O}_2))]$

$$\Delta G^\circ_{rxn} = [(2 \text{ mol NO}_2) \,(51 \text{ kJ/mol})] - [(2 \text{ mol NO}) \,(86.60 \text{ kJ/mol}) + (1 \text{ mol O}_2) \,(0))]$$

$$\Delta G^\circ_{rxn} = -71.2 = \textbf{-71 kJ} \text{ (result agrees with answer in 20.4)}$$

b) $\Delta G^\circ_{rxn} = [(2 \text{ mol CO}) \,(\Delta G^o_f \,(\text{CO})] - [(2 \text{ mol C}) \,(\Delta G^o_f \,(\text{C})) + (1 \text{ mol O}_2) \,(\Delta G^o_f \,(\text{O}_2))]$

$$\Delta G^\circ_{rxn} = [(2 \text{ mol CO}) \,(-137.2 \text{ kJ/mol})] - [(2 \text{ mol C}) \,(0) + (1 \text{ mol O}_2) \,(0)]$$

$$\Delta G^\circ_{rxn} = \textbf{-274.4 kJ}$$

20.6   Plan: Predict the sign of ΔS^o_{sys} by comparing the randomness of the products with the randomness of the reactants. Use the relationship $\Delta G^o_{rxn} = \Delta H^o_{rxn} - T\Delta S^o_{rxn}$ to answer b)

Solution:

a) The reaction is $X_2Y_2(g) \rightarrow X_2(g) + Y_2(g)$. Since there are more moles of gaseous product than there are of gaseous reactant, entropy increases and $\Delta S > 0$.

b) The reaction is only spontaneous above 325°C or in other words, at high temperatures. In the relationship $\Delta G^\circ = \Delta H^\circ - T\Delta S^\circ$, when $\Delta S > 0$ so that $-T\Delta S^\circ$ is < 0, ΔG° will only be negative at high T if $\Delta H^\circ > 0$.

20.7   Plan: Examine the equation $\Delta G^\circ = \Delta H^\circ - T\Delta S^\circ$ and determine which combination of enthalpy and entropy will describe the given reaction.

Solution: Two choices can already be eliminated:

1) When $\Delta H > 0$ (endothermic reaction) and $\Delta S < 0$ (entropy decreases), the reaction is always nonspontaneous, regardless of temperature, so this combination does not describe the reaction.

2) When $\Delta H < 0$ (exothermic reaction) and $\Delta S > 0$ (entropy increases), the reaction is always spontaneous, regardless of temperature, so this combination does not describe the reaction.

Two combinations remain: 3) $\Delta H° > 0$ and $\Delta S° > 0$, or 4) $\Delta H° < 0$ and $\Delta S° < 0$. If the reaction becomes spontaneous at $-40°C$, this means that $\Delta G°$ becomes negative at lower temperatures. Case 3) becomes spontaneous at higher temperatures, when the $-T\Delta S°$ term is larger than the positive enthalpy term. By process of elimination, Case 4) describes the reaction.

Check: At a lower temperature, the negative $\Delta H°$ becomes larger than the positive $(-T\Delta S°)$ value, so $\Delta G°$ becomes negative.

20.8   Plan: Write the equilibrium expression for the reaction and calculate Q_c for each scene. A reaction that is proceeding to the right will have $\Delta G° < 0$ and a reaction that is proceeding to the left will have $\Delta G° > 0$. A reaction at equilibrium has $\Delta G° = 0$.
Solution:
a) $X_2(g) + 2\,Y_2(g) \leftrightarrows 2\,XY_2(g)$

$$Q_c = \frac{[XY_2]^2}{[X_2][Y_2]^2}$$

Mixture 1: $Q_c = \dfrac{[XY_2]^2}{[X_2][Y_2]^2} = \dfrac{[5]^2}{[2][1]^2} = 12.5$

Mixture 2: $Q_c = \dfrac{[XY_2]^2}{[X_2][Y_2]^2} = \dfrac{[4]^2}{[2][2]^2} = 2$

Mixture 3: $Q_c = \dfrac{[XY_2]^2}{[X_2][Y_2]^2} = \dfrac{[2]^2}{[4][2]^2} = 0.25$

Mixture 2 is at equilibrium since $Q_c = K_c$

b) $Q_c > K_c$ for Mixture 1 and the reaction is proceeding left to reach equilibrium; thus $\Delta G° > 0$. Mixture 2 is at equilibrium and $\Delta G° = 0$. Mixture 3 proceeds to the right to reach equilibrium since $Q_c < K_c$ and $\Delta G° < 0$. The ranking for most negative to most positive is **3 < 2 < 1.**

c) Any reaction mixture moves spontaneously towards equilibrium so both changes have a negative $\Delta G°$.

20.9   Plan: Write a balanced equation for the dissociation of hypobromous acid in water. The free energy of the reaction at standard state (part (a)) is calculated using $\Delta G° = -RT \ln K$. The free energy of the reaction under non-standard state conditions is calculated using $\Delta G = \Delta G° + RT \ln Q$.
Solution: $HBrO(aq) + H_2O(l) \leftrightarrows BrO^-(aq) + H_3O^+(aq)$
a) $\Delta G° = -RT \ln K = -(8.314 \text{ J/mol•K}) (298) \ln (2.3 \times 10^{-9}) = 4.927979 \times 10^4 = \textbf{4.9} \times \textbf{10}^4 \textbf{ J/mol}$

b) $Q = \dfrac{\left[H_3O^+\right]\left[BrO^-\right]}{\left[HBrO\right]} = \dfrac{\left[6.0 \times 10^{-4}\right]\left[0.10\right]}{\left[0.20\right]}$

$$\Delta G = \Delta G° + RT \ln Q = (4.927979 \times 10^4 \text{ J/mol}) + (8.314 \text{ J/mol•K}) (298 \text{ K}) \ln \frac{\left[6.0 \times 10^{-4}\right]\left[0.10\right]}{\left[0.20\right]}$$

$$= 2.9182399 \times 10^4 = \textbf{2.9} \times \textbf{10}^4 \textbf{ J/mol}$$

Check: The value of K_a is very small, so it makes sense that $\Delta G°$ is a positive number. The natural log of a negative exponent gives a negative number $(\ln 3.0 \times 10^{-4})$, so the value of ΔG decreases with concentrations lower than the standard state 1 M values.

END–OF–CHAPTER PROBLEMS

20.2   A spontaneous process occurs by itself (possibly requiring an initial input of energy) whereas a nonspontaneous process requires a continuous supply of energy to make it happen. It is possible to cause a nonspontaneous process to occur, but the process stops once the energy source is removed. A reaction that is found to be nonspontaneous under one set of conditions may be spontaneous under a different set of conditions (different temperature, different concentrations).

20.5   Vaporization is the change of a liquid substance to a gas so $\Delta S_{\text{vaporization}} = S_{\text{gas}} - S_{\text{liquid}}$. Fusion is the change of a solid substance into a liquid so $\Delta S_{\text{fusion}} = S_{\text{liquid}} - S_{\text{solid}}$. Vaporization involves a greater change in volume than fusion. Thus, the transition from liquid to gas involves a greater entropy change than the transition from solid to liquid.

20.6   In an exothermic process, the *system* releases heat to its *surroundings*. The entropy of the surroundings increases because the temperature of the surroundings increases ($\Delta S_{\text{surr}} > 0$). In an endothermic process, the system absorbs heat from the surroundings and the surroundings become cooler. Thus, the entropy of the surroundings decreases ($\Delta S_{\text{surr}} < 0$). A chemical cold pack for injuries is an example of a spontaneous, endothermic chemical reaction as is the melting of ice cream at room temperature.

20.8   a) **Spontaneous,** evaporation occurs because a few of the liquid molecules have enough energy to break away from the intermolecular forces of the other liquid molecules and move spontaneously into the gas phase.
        b) **Spontaneous,** a lion spontaneously chases an antelope without added force. This assumes that the lion has not just eaten.
        c) **Spontaneous,** an unstable substance decays spontaneously to a more stable substance.

20.10   a) **Spontaneous,** with a small amount of energy input, methane will continue to burn without additional energy (the reaction itself provides the necessary energy) until it is used up.
        b) **Spontaneous,** the dissolved sugar molecules have more states they can occupy than the crystalline sugar, so the reaction proceeds in the direction of dissolution.
        c) **Not spontaneous,** a cooked egg will not become raw again, no matter how long it sits or how many times it is mixed.

20.12   a) ΔS_{sys} **positive,** melting is the change in state from solid to liquid. The solid state of a particular substance always has lower entropy than the same substance in the liquid state. Entropy increases during melting.
        b) ΔS_{sys} **negative,** the entropy of most salt solutions is greater than the entropy of the solvent and solute separately, so entropy decreases as a salt precipitates.
        c) ΔS_{sys} **negative,** dew forms by the condensation of water vapor to liquid. Entropy of a substance in the gaseous state is greater than its entropy in the liquid state. Entropy decreases during condensation.

20.14   a) ΔS_{sys} **positive,** the process described is liquid alcohol becoming gaseous alcohol. The gas molecules have greater entropy than the liquid molecules.
        b) ΔS_{sys} **positive,** the process described is a change from solid to gas, an increase in possible energy states for the system.
        c) ΔS_{sys} **positive,** the perfume molecules have more possible locations in the larger volume of the room than inside the bottle. A system that has more possible arrangements has greater entropy.

20.16   a) ΔS_{sys} **negative,** reaction involves a gaseous reactant and no gaseous products, so entropy decreases. The number of particles also decreases, indicating a decrease in entropy.
        b) ΔS_{sys} **negative,** gaseous reactants form solid product and number of particles decreases, so entropy decreases.
        c) ΔS_{sys} **positive,** when a solid salt dissolves in water, entropy generally increases.

20.18   a) ΔS_{sys} **positive,** the reaction produces gaseous CO_2 molecules that have greater entropy than the physical states of the reactants.
        b) ΔS_{sys} **negative,** the reaction produces a net decrease in the number of gaseous molecules, so the system's entropy decreases.
        c) ΔS_{sys} **positive,** the reaction produces a gas from a solid.

20.20   a) ΔS_{sys} **positive,** decreasing the pressure increases the volume available to the gas molecules so entropy of the system increases.
        b) ΔS_{sys} **negative,** gaseous nitrogen molecules have greater entropy (more possible states) than dissolved nitrogen molecules.
        c) ΔS_{sys} **positive,** dissolved oxygen molecules have lower entropy than gaseous oxygen molecules.

20.22   a) **Butane** has the greater molar entropy because it has two additional C–H bonds that can vibrate and has greater rotational freedom around its bond. The presence of the double bond in 2–butene restricts rotation.
b) **Xe(g)** has the greater molar entropy because entropy increases with atomic size.
c) **CH$_4$(g)** has the greater molar entropy because gases in general have greater entropy than liquids.

20.24   a) Ethanol, **C$_2$H$_5$OH(l)**, is a more complex molecule than methanol, CH$_3$OH, and has the greater molar entropy.
b) A salt dissolves, giving an increase in the number of possible states for the ions increases. Thus, **KClO$_3$(aq)** has the greater molar entropy.
c) **K(s)** has greater molar entropy because K(s) has greater mass than Na(s).

20.26   a) **Diamond < graphite < charcoal.** Diamond has an ordered, 3–dimensional crystalline shape, followed by graphite with an ordered 2–dimensional structure, followed by the amorphous (disordered) structure of charcoal.
b) **Ice < liquid water < water vapor.** Entropy increases as a substance changes from solid to liquid to gas.
c) **O atoms < O$_2$ < O$_3$.** Entropy increases with molecular complexity because there are more modes of movement (e.g., bond vibration) available to the complex molecules.

20.28   a) **ClO$_4^-$(aq) > ClO$_3^-$(aq) > ClO$_2^-$(aq).** The decreasing order of molar entropy follows the order of decreasing molecular complexity.
b) **NO$_2$(g) > NO(g) > N$_2$(g).** N$_2$ has lower molar entropy than NO because N$_2$ consists of two of the same atoms while NO consists of two different atoms. NO$_2$ has greater molar entropy than NO because NO$_2$ consists of three atoms while NO consists of only two.
c) **Fe$_3$O$_4$(s) > Fe$_2$O$_3$(s) > Al$_2$O$_3$(s).** Fe$_3$O$_4$ has greater molar entropy than Fe$_2$O$_3$ because Fe$_3$O$_4$ is more complex and more massive. Fe$_2$O$_3$ and Al$_2$O$_3$ contain the same number of atoms but Fe$_2$O$_3$ has greater molar entropy because iron atoms are more massive than aluminum atoms.

20.31   A system at equilibrium does not spontaneously produce more products or more reactants. For either reaction direction, the entropy change of the system is exactly offset by the entropy change of the surroundings. Therefore, for system at equilibrium, $\Delta S_{univ} = \Delta S_{sys} + \Delta S_{surr} = 0$. However, for a system moving to equilibrium, $\Delta S_{univ} > 0$, because the Second Law states that for any spontaneous process, the entropy of the universe increases.

20.32   Since entropy is a state function, the entropy changes can be found by summing the entropies of the products and subtracting the sum of the entropies of the reactants.
For the given reaction $\Delta S^{\circ}_{rxn} = 2\, S^{\circ}_{HClO(g)} - (S^{\circ}_{H_2O(g)} + S^{\circ}_{Cl_2O(g)})$. Rearranging this expression to solve for $S^{\circ}_{Cl_2O(g)}$
gives $S^{\circ}_{Cl_2O(g)} = 2\, S^{\circ}_{HClO(g)} - S^{\circ}_{H_2O(g)} - \Delta S^{\circ}_{rxn}$

20.33   a) Prediction: ΔS° **negative** because number of moles of (Δn) gas decreases.
$\Delta S^{\circ} = [(1\ \text{mol}\ N_2O(g))\,(S^{\circ}(N_2O)) + (1\ \text{mol}\ NO_2(g))\,(S^{\circ}(NO_2))]$
$\qquad\qquad - [(3\ \text{mol}\ NO(g))\,(S^{\circ}(NO))]$
$\Delta S^{\circ} = [(1\ \text{mol}\ N_2O(g))\,(219.7\ \text{J/mol·K}) + (1\ \text{mol}\ NO_2(g))\,(239.9\ \text{J/mol·K})]$
$\qquad\qquad - [(3\ \text{mol}\ NO(g))\,(210.65\ \text{J/mol·K})]$
$\qquad \Delta S^{\circ} = -172.35 = \textbf{–172.4 J/K}$
b) Prediction: Sign difficult to predict because $\Delta n = 0$, but **possibly ΔS° positive** because water vapor has greater complexity than H$_2$ gas.
$\Delta S^{\circ} = [(2\ \text{mol}\ Fe(s))\,(S^{\circ}(Fe)) + (3\ \text{mol}\ H_2O(g))\,(S^{\circ}(H_2O))]$
$\qquad\qquad - [(3\ \text{mol}\ H_2(g))\,(S^{\circ}(H_2)) + (1\ \text{mol}\ Fe_2O_3(s))\,(S^{\circ}(Fe_2O_3))]$
$\Delta S^{\circ} = [(2\ \text{mol}\ Fe(s))\,(27.3\ \text{J/mol·K}) + (3\ \text{mol}\ H_2O(g))\,(188.72\ \text{J/mol·K})]$
$\qquad\qquad - [(3\ \text{mol}\ H_2(g))\,(130.6\ \text{J/mol·K}) + (1\ \text{mol}\ Fe_2O_3(s))\,(87.400\ \text{J/mol·K})]$
$\qquad \Delta S^{\circ} = 141.56 = \textbf{141.6 J/K}$
c) Prediction: ΔS°_{sys} **negative** because a gaseous reactant forms a solid product and also because the number of moles of gas (Δn) decreases.
$\Delta S^{\circ} = [(1\ \text{mol}\ P_4O_{10}(s))\,(S^{\circ}(P_4O_{10}))] - [(1\ \text{mol}\ P_4(s))\,(S^{\circ}(P_4)) + (5\ \text{mol}\ O_2(g))\,(S^{\circ}(O_2))]$
$\Delta S^{\circ} = [(1\ \text{mol}\ P_4O_{10}(s))\,(229\ \text{J/mol·K})] - [(1\ \text{mol}\ P_4(s))\,(41.1\ \text{J/mol·K}) + (5\ \text{mol}\ O_2(g))\,(205.0\ \text{J/mol·K})]$
$\qquad \Delta S^{\circ} = -837.1 = \textbf{–837 J/K}$

20.35   The balanced combustion reaction is
$$2\ C_2H_6(g) + 7\ O_2(g)\ \rightarrow\ 4\ CO_2(g) + 6\ H_2O(g)$$
$$\Delta S^{\circ}_{rxn} = [4\ mol\ (S^{\circ}(CO_2)) + 6\ mol\ (S^{\circ}(H_2O))] - [2\ mol\ (S^{\circ}(C_2H_6) + 7\ mol\ (S^{\circ}(CO_2)]$$
$$= [4\ mol\ (213.7\ J/mol{\cdot}K) + 6\ mol\ (188.72\ J/mol{\cdot}K)]$$
$$- [2\ mol\ (229.5\ J/mol{\cdot}K) + 7\ mol\ (205.0\ J/mol{\cdot}K)] = 93.12 = \mathbf{93.1\ J/K}$$
The entropy value is not per mole of C_2H_6 but per 2 moles. Divide the calculated value by 2 to obtain entropy per mole of C_2H_6.
Yes, the positive sign of ΔS is expected because there is a net increase in the number of gas molecules from 9 moles as reactants to 10 moles as products.

20.37   The balanced chemical equation for the described reaction is:
$$2\ NO(g) + 5\ H_2(g) \rightarrow 2\ NH_3(g) + 2\ H_2O(g)$$
Because the number of moles of gas decreases, i.e., $\Delta n = 4 - 7 = -3$, the entropy is expected to decrease.
$$\Delta S^{\circ} = [(2\ mol\ NH_3)\ (193\ J/mol{\cdot}K) + (2\ mol\ H_2O)\ (188.72\ J/mol{\cdot}K)]$$
$$- [(2\ mol\ NO)\ (210.65\ J/mol{\cdot}K) + (5\ mol\ H_2)\ (130.6\ J/mol{\cdot}K)]$$
$$\Delta S^{\circ} = -310.86 = \mathbf{-311\ J/K}$$
Yes, the calculated entropy matches the predicted decrease.

20.39   The reaction for forming Cu_2O from copper metal and oxygen gas is
$$2\ Cu(s) + 1/2\ O_2(g) \rightarrow Cu_2O(s)$$
$$\Delta S^{\circ}_{rxn} = [1\ mol\ (S^{\circ}(Cu_2O))] - [2\ mol\ (S^{\circ}(Cu)) + 1/2\ mol\ (S^{\circ}(O_2))]$$
$$= [1\ mol\ (93.1\ J/mol{\cdot}K)] - [2\ mol\ (33.1\ J/mol{\cdot}K) + 1/2\ mol\ (205.0\ J/mol{\cdot}K)]$$
$$= \mathbf{-75.6\ J/K}$$

20.41   One mole of methanol is formed from its elements in their standard states according to the following equation:
$$C(g) + 2\ H_2(g) + 1/2\ O_2(g) \rightarrow CH_3OH(l)$$
$$\Delta S^{\circ} = [1\ mol\ (S^{\circ}(CH_3OH))] - [1\ mol\ S^{\circ}(C(graphite)) + 2\ mol\ (S^{\circ}(H_2)) + 1/2\ mol\ (S^{\circ}(O_2))]$$
$$\Delta S^{\circ} = [(1\ mol\ CH_3OH)\ (127\ J/mol{\cdot}K)]$$
$$- [(1\ mol\ C)\ (5.686\ J/mol{\cdot}K) + (2\ mol\ H_2)\ (130.6\ J/mol{\cdot}K) + (1/2\ mol\ O_2)\ (205.0\ J/mol{\cdot}K)]$$
$$\Delta S^{\circ} = -242.386 = \mathbf{-242\ J/K}$$

20.44   Complete combustion of a hydrocarbon includes oxygen as a reactant and carbon dioxide and water as the products.
$$C_2H_2(g)\ +\ 5/2\ O_2(g)\ \rightarrow\ 2\ CO_2(g)\ +\ H_2O(g)$$
$$\Delta S^{\circ}_{rxn} = [2\ mol\ (S^{\circ}(CO_2)) + 1\ mol\ (S^{\circ}(H_2O))] - [1\ mol\ (S^{\circ}(C_2H_2)) + 5/2\ mol\ (S^{\circ}(O_2))]$$
$$= [2\ mol\ (213.7\ J/mol{\cdot}K) + 1\ mol\ (188.72\ J/mol{\cdot}K)]$$
$$- [1\ mol\ (200.85\ J/mol{\cdot}K) + 5/2\ mol\ (205.0\ J/mol{\cdot}K)] = -97.23 = \mathbf{-97.2\ J/K}$$

20.46   A spontaneous process has $\Delta S_{univ} > 0$. Since the Kelvin temperature is always positive, ΔG_{sys} must be negative ($\Delta G_{sys} < 0$) for a spontaneous process.

20.48 $\mathbf{\Delta H^{\circ}_{rxn}}$ **positive and** $\mathbf{\Delta S^{\circ}_{sys}}$ **positive.** The reaction is endothermic ($\Delta H^{\circ}_{rxn} > 0$) and requires a lot of heat from its surroundings to be spontaneous. The removal of heat from the surroundings results in $\Delta S^{\circ}_{surr} < 0$. The only way an endothermic reaction can proceed spontaneously is if $\Delta S^{\circ}_{sys} \gg 0$, effectively offsetting the decrease in surroundings entropy. In summary, the values of ΔH°_{rxn} and ΔS°_{sys} are both positive for this reaction. Melting is an example.

20.49   For a given substance, the entropy changes greatly from one phase to another, e.g., from liquid to gas. However, the entropy changes little within a phase. As long as the substance does not change phase, the value of ΔS° is relatively unaffected by temperature.

20.50 The ΔG^o_{rxn} can be calculated from the individual ΔG^o_f's of the reactants and products found in Appendix B.

$\Delta G^o_{rxn} = \Sigma[m \Delta G^o_f \text{ (products)}] - \Sigma[n \Delta G^o_f \text{ (reactants)}]$

a) $\Delta G^o_{rxn} = [(2 \text{ mol MgO}) (\Delta G^o_f \text{ MgO})] - [(2 \text{ mol Mg}) (\Delta G^o_f \text{ Mg}) + (1 \text{ mol O}_2) (\Delta G^o_f \text{ O}_2)]$

Both Mg(s) and $O_2(g)$ are the standard state forms of their respective elements, so their ΔG^o_f's are zero.

$\Delta G^o_{rxn} = [(2 \text{ mol MgO}) (-569.0 \text{ kJ/mol})] - [(2 \text{ mol Mg}) (0) + (1 \text{ mol O}_2) (0)] = \mathbf{-1138.0 \text{ kJ}}$

b) $\Delta G^o_{rxn} = [(2 \text{ mol CO}_2) (\Delta G^o_f \text{ CO}_2) + (4 \text{ mol H}_2\text{O}) (\Delta G^o_f \text{ H}_2\text{O})]$

$- [(2 \text{ mol CH}_3\text{OH}) (\Delta G^o_f \text{ CH}_3\text{OH}) + (3 \text{ mol O}_2) (\Delta G^o_f \text{ O}_2)]$

$\Delta G^o_{rxn} = [(2 \text{ mol CO}_2) (-394.4 \text{ kJ/mol}) + (4 \text{ mol H}_2\text{O}) (-228.60 \text{ kJ/mol})]$

$- [(2 \text{ mol CH}_3\text{OH}) (-161.9 \text{ kJ/mol}) + (3 \text{ mol O}_2) (0)]$

$\Delta G^o_{rxn} = \mathbf{-1379.4 \text{ kJ}}$

c) $\Delta G^o_{rxn} = [(1 \text{ mol BaCO}_3) (\Delta G^o_f \text{ BaCO}_3)] - [(1 \text{ mol BaO}) (\Delta G^o_f \text{ BaO}) + (1 \text{ mol CO}_2) (\Delta G^o_f \text{ CO}_2)]$

$\Delta G^o_{rxn} = [(1 \text{ mol BaCO}_3) (-1139 \text{ kJ/mol})] - [(1 \text{ mol BaO}) (-520.4 \text{ kJ/mol}) + (1 \text{ mol CO}_2) (-394.4 \text{ kJ/mol})]$

$\Delta G^o_{rxn} = -224.2 = \mathbf{-224 \text{ kJ}}$

20.52 The ΔH^o_{rxn} can be calculated from the individual ΔH^o_f values of the reactants and products found in Appendix B.

$\Delta H^o_{rxn} = \Sigma[m \Delta H^o_f \text{ (products)}] - \Sigma[n \Delta H^o_f \text{ (reactants)}]$

The S^o_{rxn} can be calculated from the individual S^o values of the reactants and products found in Appendix B.

$S^o_{rxn} = \Sigma[m S^o \text{ (products)}] - \Sigma[n S^o \text{ (reactants)}]$

ΔG^o_{rxn} can be calculated using $\Delta G^o_{rxn} = \Delta H^o_{rxn} - T \Delta S^o_{rxn}$

a) $\Delta H^o_{rxn} = [2 \text{ mol } (\Delta H^o_f \text{ (MgO)})] - [2 \text{ mol } (\Delta H^o_f \text{ (Mg)}) + 1 \text{ mol } (\Delta H^o_f \text{ (O}_2))]$

$= [2 \text{ mol } (-601.2 \text{ kJ/mol})] - [2 \text{ mol } (0) + 1 \text{ mol } (0)]$

$= -1202.4 \text{ kJ}$

$\Delta S^o_{rxn} = [2 \text{ mol } (S^o\text{(MgO)})] - [2 \text{ mol } (S^o\text{(Mg)}) + 1 \text{ mol } (S^o\text{(O}_2))]$

$= [2 \text{ mol } (26.9 \text{ J/mol} \cdot \text{K})] - [2 \text{ mol } (32.69 \text{ J/mol} \cdot \text{K}) + 1 \text{ mol } (205.0 \text{ J/mol} \cdot \text{K})]$

$= -216.58 \text{ J/K}$ (unrounded)

$\Delta G^o_{rxn} = \Delta H^o_{rxn} - T \Delta S^o_{rxn} = -1202.4 \text{ kJ} - [(298 \text{ K}) (-216.58 \text{ J/K}) (1 \text{ kJ} / 10^3 \text{ J})] = -1137.859 = \mathbf{-1138 \text{ kJ}}$

b) $\Delta H^o_{rxn} = [2 \text{ mol } (\Delta H^o_f \text{ (CO}_2)) + 4 \text{ mol } (\Delta H^o_f \text{ (H}_2\text{O}(g)))] - [2 \text{ mol } (\Delta H^o_f \text{ (CH}_3\text{OH)}) + 3 \text{ mol } (\Delta H^o_f \text{ (O}_2))]$

$= [2 \text{ mol } (-393.5 \text{ kJ/mol}) + 4 \text{ mol } (-241.826 \text{ kJ/mol})] - [2 \text{ mol } (-201.2 \text{ kJ/mol}) + 3 \text{ mol } (0)]$

$= -1351.904 \text{ kJ}$ (unrounded)

$\Delta S^o_{rxn} = [2 \text{ mol } (S^o\text{(CO}_2)) + 4 \text{ mol } (S^o\text{(H}_2\text{O}(g)))] - [2 \text{ mol } (S^o\text{(CH}_3\text{OH)}) + 3 \text{ mol } (S^o\text{(O}_2))]$

$= [2 \text{ mol } (213.7 \text{ J/mol} \cdot \text{K}) + 4 \text{ mol } (188.72 \text{ J/mol} \cdot \text{K})]$

$- [2 \text{ mol } (238 \text{ J/mol} \cdot \text{K}) + 3 \text{ mol } (205.0 \text{ J/mol} \cdot \text{K})] = 91.28 \text{ J/K}$ (unrounded)

$\Delta G^o_{rxn} = \Delta H^o_{rxn} - T \Delta S^o_{rxn} = -1351.904 \text{ kJ} - [(298 \text{ K}) (91.28 \text{ J/K}) (1 \text{ kJ} / 10^3 \text{ J})]$

$= -1379.105 = \mathbf{-1379 \text{ kJ}}$

c) $\Delta H^o_{rxn} = [1 \text{ mol } (\Delta H^o_f \text{ (BaCO}_3(s))] - [1 \text{ mol } (\Delta H^o_f \text{ (BaO)}) + 1 \text{ mol } (\Delta H^o_f \text{ (CO}_2))]$

$= [1 \text{ mol } (-1219 \text{ kJ/mol})] - [1 \text{ mol } (-548.1 \text{ kJ/mol}) + 1 \text{ mol } (-393.5 \text{ kJ/mol})]$

$= -277.4 \text{ kJ}$ (unrounded)

$\Delta S^o_{rxn} = [1 \text{ mol } (S^o\text{(BaCO}_3(s))] - [1 \text{ mol } (S^o\text{(BaO)}) + 1 \text{ mol } (S^o\text{(CO}_2))]$

$= [1 \text{ mol } (112 \text{ J/mol} \cdot \text{K})] - [1 \text{ mol } (72.07 \text{ J/mol} \cdot \text{K}) + 1 \text{ mol } (213.7 \text{ J/mol} \cdot \text{K})]$

$= -173.77 \text{ J/K}$ (unrounded)

$\Delta G^o_{rxn} = \Delta H^o_{rxn} - T \Delta S^o_{rxn} = -277.4 \text{ kJ} - [(298 \text{ K}) (-173.77 \text{ J/K}) (1 \text{ kJ} / 10^3 \text{ J})]$

$= -225.6265 = \mathbf{-226 \text{ kJ}}$

20.54   a) Entropy decreases ($\Delta S°$ **negative**) because the number of moles of gas decreases from reactants (1 ½ mol) to products (1 mole). The oxidation (combustion) of CO requires initial energy input to start the reaction, but then releases energy (exothermic, $\Delta H°$ **negative**) which is typical of all combustion reactions.

b) Method 1: Calculate $\Delta G°_{rxn}$ from $\Delta G°_f$'s of products and reactants.

$$\Delta G°_{rxn} = \Sigma[m \, \Delta G°_f \text{ (products)}] - \Sigma[n \, \Delta G°_f \text{ (reactants)}]$$

$$\Delta G°_{rxn} = [(1 \text{ mol } CO_2) \, (\Delta G°_f (CO_2))] - [(1 \text{ mol } CO) \, (\Delta G°_f (CO)) + (1/2 \text{ mol}) \, (\Delta G°_f (O_2))]$$

$$\Delta G°_{rxn} = [(1 \text{ mol } CO_2) \, (-394.4 \text{ kJ/mol})] - [(1 \text{ mol } CO) \, (-137.2 \text{ kJ/mol}) + (1/2 \text{ mol } O_2) \, (0)] = \textbf{–257.2 kJ}$$

Method 2: Calculate $\Delta G°_{rxn}$ from $\Delta H°$ and $\Delta S°$ at 298 K (the degree superscript indicates a reaction at standard state, given in the Appendix at 25°C).

$$\Delta H°_{rxn} = \Sigma[m \, \Delta H°_f \text{ (products)}] - \Sigma[n \, \Delta H°_f \text{ (reactants)}]$$

$$\Delta H°_{rxn} = [(1 \text{ mol } CO_2) \, (\Delta H°_f (CO_2))] - [(1 \text{ mol } CO) \, (\Delta H°_f (CO)) + (1/2 \text{ mol}) \, (\Delta H°_f (O_2))]$$

$$\Delta H°_{rxn} = [(1 \text{ mol } CO_2) \, (-393.5 \text{ kJ/mol})] - [(1 \text{ mol } CO) \, (-110.5 \text{ kJ/mol}) + (1/2 \text{ mol } O_2) \, (0)] = -283.0 \text{ kJ}$$

$$\Delta S°_{rxn} = \Sigma[m \, S° \text{ (products)}] - \Sigma[n \, S° \text{ (reactants)}]$$

$$\Delta S°_{rxn} = [(1 \text{ mol } CO_2) \, (\Delta S°_f (CO_2))] - [(1 \text{ mol } CO) \, (\Delta S°_f (CO)) + (1/2 \text{ mol}) \, (\Delta S°_f (O_2))]$$

$$\Delta S°_{rxn} = [(1 \text{mol } CO_2) \, (213.7 \text{ J/mol•K})] - [(1 \text{mol } CO) \, (197.5 \text{ J/mol•K}) + (1/2 \text{ mol } O_2) \, (205.0 \text{ J/mol•K})]$$

$$= -86.3 \text{ J/K}$$

$$\Delta G°_{rxn} = \Delta H°_{rxn} - T \, \Delta S°_{rxn} = (-283.0 \text{ kJ}) - [(298 \text{ K}) \, (-86.3 \text{ J/K}) \, (1 \text{ kJ} / 10^3 \text{ J})] = -257.2826 = \textbf{–257.3 kJ}$$

20.56   Reaction is $Xe(g) + 3 F_2(g) \rightarrow XeF_6(g)$

a) The change in entropy can be found from the relationship $\Delta G°_{rxn} = \Delta H°_{rxn} - T \, \Delta S°_{rxn}$.

$$\Delta S° = (\Delta H° - \Delta G°) / T$$
$$= [(-402 \text{ kJ/mol}) - (-280. \text{ kJ/mol})] / 298 \text{ K} = -0.40939597 = \textbf{–0.409 kJ/mol•K}$$

b) The change in entropy at 500. K is calculated from $\Delta G°_{rxn} = \Delta H°_{rxn} - T \, \Delta S°_{rxn}$ with T at 500. K.

$$\Delta G° = \Delta H° - T\Delta S° = (-402 \text{ kJ/mol}) - [(500. \text{ K}) \, (-0.40939597 \text{ kJ/mol•K})] = -197.302 = \textbf{–197 kJ/mol}$$

20.58   a) $\Delta H°_{rxn} = [(1 \text{ mol } CO) \, (\Delta H°_f (CO)) + (2 \text{ mol } H_2) \, (\Delta H°_f (H_2))] - [(1 \text{ mol } CH_3OH) \, (\Delta H°_f (CH_3OH))]$

$$\Delta H°_{rxn} = [(1 \text{ mol } CO) \, (-110.5 \text{ kJ/mol}) + (2 \text{ mol } H_2) \, (0)] - [(1 \text{ mol } CH_3OH) \, (-201.2 \text{ kJ/mol})]$$

$$\Delta H°_{rxn} = \textbf{90.7 kJ}$$

$$\Delta S°_{rxn} = [(1 \text{ mol } CO) \, (S°_f (CO)) + (2 \text{ mol } H_2) \, (S°_f (H_2))] - [(1 \text{ mol } CH_3OH) \, (S°_f (CH_3OH))]$$

$$\Delta S°_{rxn} = [(1 \text{ mol } CO) \, (197.5 \text{ J/mol•K}) + (2 \text{ mol } H_2) \, (130.6 \text{ J/mol•K}]$$
$$- [(1 \text{ mol } CH_3OH) \, (238 \text{ J/mol•K})]$$

$$\Delta S°_{rxn} = 220.7 = \textbf{221 J/K}$$

b) $T_1 = 28 + 273 = 301 \text{ K}$ $\Delta G° = 90.7 \text{ kJ} - [(301 \text{ K}) \, (220.7 \text{ J/K}) \, (1 \text{ kJ} / 10^3 \text{ J})] = 24.2693 = \textbf{24.3 kJ}$
$T_2 = 128 + 273 = 401 \text{ K}$ $\Delta G° = 90.7 \text{ kJ} - [(401 \text{ K}) \, (220.7 \text{ J/K}) \, (1 \text{ kJ} / 10^3 \text{ J})] = 2.1993 = \textbf{2.2 kJ}$
$T_3 = 228 + 273 = 501 \text{ K}$ $\Delta G° = 90.7 \text{ kJ} - [(501 \text{ K}) \, (220.7 \text{ J/K}) \, (1 \text{ kJ} / 10^3 \text{ J})] = -19.8707 = \textbf{–19.9 kJ}$

c) For the substances in their standard states, the reaction is nonspontaneous at 28°C, near equilibrium at 128°C and spontaneous at 228°C. Reactions with positive values of $\Delta H°_{rxn}$ and $\Delta S°_{rxn}$ become spontaneous at high temperatures.

20.60   At the normal boiling point, defined as the temperature at which the vapor pressure of the liquid equals 1 atm, the phase change from liquid to gas is at equilibrium. For a system at equilibrium, the change in Gibbs free energy is zero. Since the gas is at 1 atm and the liquid assumed to be pure, the system is at standard state and $\Delta G° = 0$. The temperature at which this occurs can be found from $\Delta G°_{rxn} = 0 = \Delta H°_{rxn} - T \, \Delta S°_{rxn}$.

$$\Delta H°_{rxn} = T \, \Delta S°_{rxn}$$
$$T_{bpt} = \Delta H° / \Delta S°$$

$$\Delta H^{\circ} = [1 \text{ mol } (\Delta H_f^{\circ} (Br_2(g)))] - [1 \text{ mol } (\Delta H_f^{\circ} (Br_2(l)))]$$

$$\Delta H^{\circ} = [1 \text{ mol}(30.91 \text{ kJ/mol})] - [1 \text{ mol } (\Delta H_f^{\circ} (0)] = \mathbf{30.91 \text{ kJ}}$$

$$\Delta S^{\circ} = [1 \text{ mol } (S^{\circ}(Br_2(g)))] - [1 \text{ mol } (S^{\circ}(Br_2(l)))]$$

$$\Delta S^{\circ} = [1 \text{ mol } (S^{\circ}(245.38 \text{ J/K} \bullet \text{mol})] - [1 \text{ mol } (S^{\circ}(152.23 \text{ J/K} \bullet \text{mol})] = \mathbf{93.15 \text{ J/K} = 0.09315 \text{ J/K}}$$

$$T_{bpt} = \frac{\Delta H^{o}}{\Delta S^{o}} = \frac{30.91 \text{ kJ}}{0.09315 \text{ kJ/K}} = 331.830 = \mathbf{331.8 \text{ K}}$$

20.62   a) The reaction for this process is $H_2(g) + 1/2 \ O_2(g) \rightarrow H_2O(g)$. The coefficients are written this way (instead of $2 \ H_2(g) + O_2(g) \rightarrow 2 \ H_2O(g)$) because the problem specifies thermodynamic values "per (1) mol H_2," not per 2 mol H_2.

$$\Delta H_{rxn}^{o} = [(1 \text{ mol } H_2O) \ (\Delta H_f^{o} (H_2O))] - [(1 \text{ mol } H_2) \ (\Delta H_f^{o} (H_2)) + (1/2 \text{ mol } O_2) \ (\Delta H_f^{o} (O_2))]$$

$$\Delta H_{rxn}^{o} = [(1 \text{ mol } H_2O) \ (-241.826 \text{ kJ/mol})] - [(1 \text{ mol } H_2) \ (0) + (1/2 \text{ mol } O_2) \ (0)]$$

$$\Delta H_{rxn}^{o} = \mathbf{-241.826 \text{ kJ}}$$

$$\Delta S_{rxn}^{o} = [(1 \text{ mol } H_2O) \ (S_f^{o} (H_2O))] - [(1 \text{ mol } H_2) \ (S_f^{o} (H_2)) + (1/2 \text{ mol } O_2) \ (S_f^{o} (O_2))]$$

$$\Delta S_{rxn}^{o} = [(1 \text{ mol } H_2O) \ (188.72 \text{ J/mol} \bullet \text{K})]$$
$$- [(1 \text{ mol } H_2) \ (130.6 \text{ J/mol} \bullet \text{K}) + (1/2 \text{ mol } O_2) \ (205.0 \text{ J/mol} \bullet \text{K})]$$

$$\Delta S_{rxn}^{o} = -44.38 = \mathbf{-44.4 \text{ J/K}}$$

$$\Delta G_{rxn}^{o} = [(1 \text{ mol } H_2O) \ (\Delta G_f^{o} (H_2O))] - [(1 \text{ mol } H_2) \ (\Delta G_f^{o} (H_2)) + (1/2 \text{ mol } O_2) \ (\Delta G_f^{o} (O_2))]$$

$$\Delta G_{rxn}^{o} = [(1 \text{ mol } H_2O) \ (-228.60 \text{ kJ/mol})] - [(1 \text{ mol } H_2) \ (0) + (1/2 \text{ mol } O_2) \ (0)]$$

$$\Delta G_{rxn}^{o} = \mathbf{-228.60 \text{ kJ}}$$

b) Because $\Delta H < 0$ and $\Delta S < 0$, the reaction will become nonspontaneous at higher temperatures because the positive ($-T\Delta S$) term becomes larger than the negative ΔH term.

c) The reaction becomes spontaneous below the temperature where $\Delta G_{rxn}^{o} = 0$

$$\Delta G_{rxn}^{o} = 0 = \Delta H_{rxn}^{o} - T \ \Delta S_{rxn}^{o}$$

$$\Delta H_{rxn}^{o} = T \ \Delta S_{rxn}^{o}$$

$$T = \Delta H^{\circ} / \Delta S^{\circ}$$

$$T = \frac{\Delta H^{o}}{\Delta S^{o}} = \frac{-241.826 \text{ kJ}}{-44.38 \text{ J/K}} \left(\frac{10^3 \text{ J}}{1 \text{ kJ}} \right) = 5448.986 = \mathbf{5.45 \times 10^3 \text{ K}}$$

20.64   a) An equilibrium constant that is much less than 1 indicates that very little product is made to reach equilibrium. The reaction, thus, is not spontaneous in the forward direction and ΔG° is a relatively large positive value.
b) A large negative ΔG° indicates that the reaction is quite spontaneous and goes almost to completion. At equilibrium, much more product is present than reactant so $K > 1$. Q depends on initial conditions, not equilibrium conditions, so its value cannot be predicted from ΔG°.

20.67   The standard free energy change, ΔG°, occurs when all components of the system are in their standard states (do not confuse this with ΔG_f^{o}, the standard free energy of formation). Standard state is defined as 1 atm for gases, 1 M for solutes, and pure solids and liquids. Standard state does not specify a temperature because standard state can occur at any temperature. $\Delta G^{\circ} = \Delta G$ when all concentrations equal 1 M and all partial pressures equal 1 atm. This occurs because the value of $Q = 1$ and $\ln Q = 0$ in the equation $\Delta G = \Delta G^{\circ} + RT \ln Q$.

20.68   For each reaction, first find $\Delta G°$, then calculate K from $\Delta G° = -RT \ln K$. Calculate ΔG^o_{rxn} using the ΔG^o_f values from Appendix B.

a) $\Delta G° = [1 \text{ mol } (\Delta G^o_f (NO_2(g)))] - [1 \text{ mol } (\Delta G^o_f (NO(g))) + \frac{1}{2} \text{ mol } (\Delta G^o_f (O_2(g)))]$

$= [1 \text{ mol } (51 \text{ kJ})] - [1 \text{ mol } (86.60 \text{ kJ}) + \frac{1}{2} \text{ mol } (0 \text{ kJ})] = -35.6 \text{ kJ (unrounded)}$

$$\ln K = \frac{\Delta G^o}{-RT} = \left(\frac{-35.6 \text{ kJ/mol}}{-(8.314 \text{ J/mol} \cdot K)(298 \text{ K})} \right) \left(\frac{10^3 \text{ J}}{1 \text{ kJ}} \right) = 14.3689 \text{ (unrounded)}$$

$K = e^{14.3689} = 1.7391377 \times 10^6 = \mathbf{1.7 \times 10^6}$

b) $\Delta G° = [1 \text{ mol } (\Delta G^o_f (H_2(g))) + 1 \text{ mol } (\Delta G^o_f (Cl_2(g)))] - [2 \text{ mol } (\Delta G^o_f (HCl(g)))]$

$= [1 \text{ mol } (0) + 1 \text{ mol } (0)kJ] - [2 \text{ mol } (-95.30 \text{ kJ/mol})] = 190.60 \text{ kJ}$

$$\ln K = \frac{\Delta G^o}{-RT} = \left(\frac{190.60 \text{ kJ/mol}}{-(8.314 \text{ J/mol} \cdot K)(298 \text{ K})} \right) \left(\frac{10^3 \text{ J}}{1 \text{ kJ}} \right) = -76.930 \text{ (unrounded)}$$

$K = e^{-76.930} = 3.88799 \times 10^{-34} = \mathbf{3.89 \times 10^{-34}}$

c) $\Delta G° = [2 \text{ mol } (\Delta G^o_f (CO(g)))] - [2 \text{ mol } (\Delta G^o_f (C(graphite))) + 1 \text{ mol } (\Delta G^o_f (O_2(g)))]$

$= [2 \text{ mol } (-137.2 \text{ kJ/mol})] - [2 \text{ mol } (0) + 1 \text{ mol } (0) \text{ kJ}] = 274.4 \text{ kJ}$

$$\ln K = \frac{\Delta G^o}{-RT} = \left(\frac{-274.4 \text{ kJ/mol}}{-(8.314 \text{ J/mol} \cdot K)(298 \text{ K})} \right) \left(\frac{10^3 \text{ J}}{1 \text{ kJ}} \right) = 110.75359 \text{ (unrounded)}$$

$K = e^{110.75359} = 1.2579778 \times 10^{48} = \mathbf{1.26 \times 10^{48}}$

Note: You may get a different answer depending on how you rounded in earlier calculations.

20.70   The equilibrium constant, K, is related to $\Delta G°$ through the equation $\Delta G° = -RT \ln K$. Calculate ΔG^o_{rxn} using the ΔG^o_f values from Appendix B.

a) $\Delta G^o_{rxn} = [(2 \text{ mol}) (\Delta G^o_f (H_2O)) + (2 \text{ mol}) (\Delta G^o_f (SO_2))] - [(2 \text{ mol}) (\Delta G^o_f (H_2S)) + (3 \text{ mol}) (\Delta G^o_f (O_2))]$

$\Delta G^o_{rxn} = [(2 \text{ mol}) (-228.60 \text{ kJ/mol}) + (2 \text{ mol}) (-300.2 \text{ kJ/mol})] - [(2 \text{ mol}) (-33 \text{ kJ/mol}) + (3 \text{ mol}) (0)]$

$\Delta G^o_{rxn} = -991.6 \text{ kJ (unrounded)}$

$$\ln K = \frac{\Delta G^o}{-RT} = \left(\frac{-991.6 \text{ kJ/mol}}{-(8.314 \text{ J/mol} \cdot K)(298 \text{ K})} \right) \left(\frac{10^3 \text{ J}}{1 \text{ kJ}} \right) = 400.2305 \text{ (unrounded)}$$

$K = e^{400.2305} = 6.571696 \times 10^{173} = \mathbf{6.57 \times 10^{173}}$

Comment: Depending on how you round ΔG^o_{rxn}, the value for K can vary by a factor of 2 or 3 because the inverse natural log varies greatly with small changes in ΔG^o_{rxn}. Your calculator might register "error" when trying to calculate e^{400} because it cannot calculate exponents greater than 99. In this case, divide 400.2305 by 2 $(= 200.115)$, and calculate $e^{200.115}e^{200.115} = (8.1066 \times 10^{86})^2 = (8.1066)^2 \times 10^{86 \times 2} = 65.71696 \times 10^{172}$ $= 6.57 \times 10^{173}$, which equals the first answer with rounding errors.

b) $\Delta G^o_{rxn} = [(1 \text{ mol}) (\Delta G^o_f (H_2O)) + (1 \text{ mol}) (\Delta G^o_f (SO_3))] - [(1 \text{ mol}) (\Delta G^o_f (H_2SO_4))]$

$\Delta G^o_{rxn} = [(1 \text{ mol}) (-237.192 \text{ kJ/mol}) + (1 \text{ mol}) (-371 \text{ kJ/mol})] - [(1 \text{ mol}) (-690.059 \text{ kJ/mol})]$

$\Delta G^o_{rxn} = 81.867 \text{ kJ (unrounded)}$

$$\ln K = \frac{\Delta G^o}{-RT} = \left(\frac{81.867 \text{ kJ/mol}}{-(8.314 \text{ J/mol} \cdot K)(298 \text{ K})} \right) \left(\frac{10^3 \text{ J}}{1 \text{ kJ}} \right) = -33.0432 \text{ (unrounded)}$$

$K = e^{-33.0432} = 4.4619 \times 10^{-15} = \mathbf{4.46 \times 10^{-15}}$

c) $\Delta G^o_{rxn} = [(1 \text{ mol}) (\Delta G^o_f (NaCN)) + (1 \text{ mol}) (\Delta G^o_f (H_2O))] - [(1 \text{ mol}) (\Delta G^o_f (HCN)) + (1 \text{ mol}) (\Delta G^o_f (NaOH))]$

NaCN(aq) and NaOH(aq) are not listed in Appendix B.
Converting the equation to net ionic form will simplify the problem:

$HCN(aq) + OH^-(aq) \leftrightarrows CN^-(aq) + H_2O(l)$

$\Delta G^o_{rxn} = [(1 \text{ mol}) (\Delta G^o_f (CN^-)) + (1 \text{ mol}) (\Delta G^o_f (H_2O))] - [(1 \text{ mol}) (\Delta G^o_f (HCN)) + (1 \text{ mol}) (\Delta G^o_f (OH^-))]$

$$\Delta G^{\circ}_{rxn} = [(1 \text{ mol}) (166 \text{ kJ/mol}) + (1 \text{ mol}) (-237.192 \text{ kJ/mol})]$$
$$- [(1 \text{ mol}) (112 \text{ kJ/mol}) + (1 \text{ mol}) (-157.30 \text{ kJ/mol})]$$
$$\Delta G^{\circ}_{rxn} = -25.892 \text{ kJ (unrounded)}$$

$$\ln K = \frac{\Delta G^{o}}{-RT} = \left(\frac{-25.892 \text{ kJ/mol}}{-(8.314 \text{ J/mol} \cdot \text{K})(298 \text{ K})} \right) \left(\frac{10^3 \text{ J}}{1 \text{ kJ}} \right) = 10.45055 \text{ (unrounded)}$$

$$K = e^{10.45055} = 3.4563 \times 10^4 = \mathbf{3.46 \times 10^4}$$

20.72 The solubility reaction for Ag_2S is
$$Ag_2S(s) + H_2O(l) \leftrightarrows 2\,Ag^+(aq) + HS^-(aq) + OH^-(aq)$$
$$\Delta G^{\circ} = [2 \text{ mol } (\Delta G^{o}_{f}\,Ag^+(aq)) + 1 \text{ mol } (\Delta G^{o}_{f}\,HS^-(aq)) + 1 \text{ mol } (\Delta G^{o}_{f}\,OH^-(aq)]$$
$$- [1 \text{ mol } (\Delta G^{o}_{f}\,Ag_2S(s)) + 1 \text{ mol } (\Delta G^{o}_{f}\,H_2O(l))]$$
$$= [(2 \text{ mol}) (77.111 \text{ kJ/mol}) + (1 \text{ mol}) (12.6 \text{ kJ/mol}) + (1 \text{ mol}) (-157.30 \text{ kJ/mol})]$$
$$- [(1 \text{ mol}) (-40.3 \text{ kJ/mol}) + (1 \text{ mol}) (-237.192 \text{ kJ/mol})]$$
$$= 287.014 \text{ kJ (unrounded)}$$

$$\ln K = \frac{\Delta G^{o}}{-RT} = \left(\frac{287.014 \text{ kJ/mol}}{-(8.314 \text{ J/mol} \cdot \text{K})(298 \text{ K})} \right) \left(\frac{10^3 \text{ J}}{1 \text{ kJ}} \right) = -115.8448675 \text{ (unrounded)}$$

$$K = e^{-115.8448675} = 4.8889241 \times 10^{-51} = \mathbf{4.89 \times 10^{-51}}$$

20.74 Calculate ΔG°_{rxn}, recognizing that $I_2(s)$, not $I_2(g)$, is the standard state for iodine.
Solve for K_p using the equation $\Delta G^{\circ} = -RT \ln K$.
$$\Delta G^{\circ}_{rxn} = [(2 \text{ mol}) (\Delta G^{o}_{f}\,(ICl))] - [(1 \text{ mol}) (\Delta G^{o}_{f}\,(I_2)) + (1 \text{ mol}) (\Delta G^{o}_{f}\,(Cl_2))]$$
$$\Delta G^{\circ}_{rxn} = [(2 \text{ mol}) (-6.075 \text{ kJ/mol})] - [(1 \text{ mol}) (19.38 \text{ kJ/mol}) + (1 \text{ mol}) (0)]$$
$$\Delta G^{\circ}_{rxn} = -31.53 \text{ kJ}$$

$$\ln K_p = \frac{\Delta G^{o}}{-RT} = \left(\frac{-31.53 \text{ kJ/mol}}{-(8.314 \text{ J/mol} \cdot \text{K})(298 \text{ K})} \right) \left(\frac{10^3 \text{ J}}{1 \text{ kJ}} \right) = 12.726169 \text{ (unrounded)}$$

$$K_p = e^{12.726169} = 3.3643794 \times 10^5 = \mathbf{3.36 \times 10^5}$$

20.76 $\Delta G^{\circ} = -RT \ln K = -(8.314 \text{ J/mol} \cdot \text{K}) (298 \text{ K}) \ln (1.7 \times 10^{-5}) = 2.72094 \times 10^4 \text{ J/mol} = \mathbf{2.7 \times 10^4 \text{ J/mol}}$
The large positive ΔG° indicates that it would not be possible to prepare a solution with the concentrations of lead and chloride ions at the standard state concentration of $1\,M$. A Q calculation using $1\,M$ solutions will confirm this: $PbCl_2(s) \leftrightarrows Pb^{2+}(aq) + Cl^-(aq)$
$$Q = [Pb^{2+}][Cl^-]^2$$
$$= (1\,M) (1\,M)^2$$
$$= 1$$
Since $Q > K_{sp}$, it is impossible to prepare a standard state solution of $Pb^{2+}(aq)$ and $Cl^-(aq)$.

20.78 a) The equilibrium constant, K, is related to ΔG° through the equation $\Delta G^{\circ} = -RT \ln K$.
$\Delta G^{\circ} = -RT \ln K = -(8.314 \text{ J/mol} \cdot \text{K}) (298 \text{ K}) \ln (9.1 \times 10^{-6}) = 2.875776 \times 10^4 = \mathbf{2.9 \times 10^4 \text{ J/mol}}$
b) Since ΔG°_{rxn} is positive, the reaction direction as written is nonspontaneous. The reverse direction, formation of reactants, is spontaneous, so the reaction proceeds to the left.
c) Calculate the value for Q and then use to find ΔG.
$$Q = \frac{\left[Fe^{2+}\right]^2 \left[Hg^{2+}\right]^2}{\left[Fe^{3+}\right]^2 \left[Hg_2^{2+}\right]} = \frac{[0.010]^2 [0.025]^2}{[0.20]^2 [0.010]} = 1.5625 \times 10^{-4} \text{ (unrounded)}$$

$\Delta G = \Delta G^{\circ} + RT \ln Q = 2.875776 \times 10^4 \text{ J/mol} + (8.314 \text{ J/mol} \cdot \text{K}) (298) \ln (1.5625 \times 10^{-4})$
$$= 7.044187 \times 10^3 = \mathbf{7.0 \times 10^3 \text{ J/mol}}$$
Because $\Delta G_{298} > 0$ and $Q > K$, the reaction proceeds to the left to reach equilibrium.

20.80   Formation of O_3 from O_2 : $3\ O_2(g) \rightleftarrows 2\ O_3(g)$ or per mole of ozone: $3/2\ O_2(g) \rightleftarrows O_3(g)$

a)   To decide when production of ozone is favored, both the signs of ΔH_f° and ΔS_f° for ozone are needed. From Appendix B, the values of ΔH_f° and S° can be used:

$\Delta H_f^\circ = [1\ \text{mol}\ O_3\ (\Delta H_f^\circ\ O_3)] - [3/2\ \text{mol}\ (\Delta H_f^\circ\ O_2)]$

$\Delta H_f^\circ = [1\ \text{mol}(143\ \text{kJ/mol}] - [3/2\ \text{mol}(0)] = 143\ \text{kJ/mol}$

$\Delta S_f^\circ = [1\ \text{mol}\ O_3\ (S^\circ\ O_3)] - [3/2\ \text{mol}\ O_2\ (S^\circ O_2)]$

$\Delta S_f^\circ = [1\ \text{mol}\ (238.82\ \text{J/mol•K}] - [3/2\ \text{mol}\ (205.0\ \text{J/mol•K})] = -68.68\ \text{J/mol•K}$ (unrounded).

The positive sign for ΔH_f° and the negative sign for ΔS_f° indicates the formation of ozone is favored at **no temperature**. The reaction is nonspontaneous at all temperatures.

b) At 298 K, ΔG° can be most easily calculated from ΔG_f° values for the reaction $3/2\ O_2(g) \rightleftarrows O_3(g)$.

From ΔG_f° : $\Delta G^\circ = [1\ \text{mol}\ O_3\ (\Delta G_f^\circ\ O_3)] - [3/2\ \text{mol}\ (\Delta G_f^\circ\ O_2)]$

$\Delta G^\circ = [\ 1\ \text{mol}\ (163\ \text{kJ/mol}\ O_3)] - [3/2\ \text{mol}\ (0)] = \textbf{163 kJ}$ for the formation of one mole of O_3.

c) Calculate the value for Q and then use to find ΔG.

$$Q = \frac{[O_3]}{[O_2]^{3/2}} = \frac{[5 \times 10^{-7}\ \text{atm}]}{[0.21\ \text{atm}]^{3/2}} = 5.195664 \times 10^{-6} \quad \text{(unrounded)}$$

$\Delta G = \Delta G^\circ + RT \ln Q = 163\ \text{kJ/mol} + (8.314\ \text{J/mol•K})\ (298)\ (1\ \text{kJ}\ /\ 10^3\ \text{J})\ \ln\ (5.195664 \times 10^{-6})$
    $= 132.85368 = \textbf{1} \times \textbf{10}^2\ \textbf{kJ/mol}$

20.83

	ΔS_{rxn}	ΔH_{rxn}	ΔG_{rxn}	Comment
(a)	+	–	–	**Spontaneous**
(b)	(+)	0	–	Spontaneous
(c)	–	+	(+)	Not spontaneous
(d)	0	(–)	–	Spontaneous
(e)	(–)	0	+	**Not spontaneous**
(f)	+	+	(–)	$T\Delta S > \Delta H$

a) The reaction is always spontaneous when $\Delta G_{rxn} < 0$, so there is no need to look at the other values other than to check the answer.
b) Because $\Delta G_{rxn} = \Delta H - T\Delta S = -T\Delta S$, ΔS must be positive for ΔG_{rxn} to be negative.
c) The reaction is always nonspontaneous when $\Delta G_{rxn} > 0$, so there is no need to look at the other values other than to check the answer.
d) Because $\Delta G_{rxn} = \Delta H - T\Delta S = \Delta H$, ΔH must be negative for ΔG_{rxn} to be negative.
e) Because $\Delta G_{rxn} = \Delta H - T\Delta S = -T\Delta S$, ΔS must be negative for ΔG_{rxn} to be positive.
f) Because $T\Delta S > \Delta H$, the subtraction of a larger positive term causes ΔG_{rxn} to be negative.

20.87   a) For the reaction, $K = \dfrac{[Hb \cdot CO][O_2]}{[Hb \cdot O_2][CO]}$ since the problem states that $[O_2] = [CO]$; the K expression simplifies to:

$$K = \frac{[Hb \cdot CO]}{[Hb \cdot O_2]}$$

The equilibrium constant, K, is related to $\Delta G°$ through the equation $\Delta G° = -RT \ln K$.

$$\ln K = \frac{\Delta G°}{RT} = \left(\frac{-14 \text{ kJ/mol}}{-(8.314 \text{ J/mol} \cdot \text{K})((273 + 37)\text{K})} \right) \left(\frac{10^3 \text{ J}}{1 \text{ kJ}} \right) = 5.431956979 \text{ (unrounded)}$$

$$K = e^{5.4319569798} = 228.596 = \mathbf{2.3 \times 10^2} = \frac{[\text{Hb} \cdot \text{CO}]}{[\text{Hb} \cdot \text{O}_2]}$$

b) By increasing the concentration of oxygen, the equilibrium can be shifted in the direction of Hb•O$_2$. Administer oxygen-rich air to counteract the CO poisoning.

20.90 Sum the two reactions to yield an overall reaction. Because the process occurs at 85°C (358 K), the $\Delta G°_{rxn}$ cannot be calculated simply from the $\Delta G°_f$'s, which are tabulated at 298 K. Instead, calculate $\Delta H°_{rxn}$ and $\Delta S°_{rxn}$ and use the equation $\Delta G = \Delta H - T\Delta S$.

$$UO_2(s) + 4\ HF(g) \rightarrow \cancel{UF_4(s)} + 2\ H_2O(g)$$
$$\underline{\cancel{UF_4(s)} + F_2(g) \rightarrow UF_6(s)}$$
$$UO_2(s) + 4\ HF(g) + F_2(g) \rightarrow UF_6(g) + 2\ H_2O(g) \text{ (overall process)}$$

$$\Delta H°_{rxn} = [(1 \text{ mol}) (\Delta H°_f (UF_6)) + (2 \text{ mol}) (\Delta H°_f (H_2O))]$$
$$- [(1 \text{ mol}) (\Delta H°_f (UO_2)) + (4 \text{ mol}) (\Delta H°_f (HF)) + (1 \text{ mol}) (\Delta H°_f (F_2))]$$

$$\Delta H°_{rxn} = [(1 \text{ mol}) (-2197 \text{ kJ/mol}) + (2 \text{ mol}) (-241.826 \text{ kJ/mol})]$$
$$- [(1 \text{ mol}) (-1085 \text{ kJ/mol}) + (4 \text{ mol}) (-273 \text{ kJ/mol}) + (1 \text{ mol}) (0)]$$

$$\Delta H°_{rxn} = -503.652 \text{ kJ (unrounded)}$$

$$\Delta S°_{rxn} = [(1 \text{ mol}) (S°_f (UF_6)) + (2 \text{ mol}) (S°_f (H_2O))] - [(1 \text{ mol}) (S°_f (UO_2)) + (4 \text{ mol}) (S°_f (HF)) + (1 \text{ mol}) (S°_f (F_2))]$$

$$\Delta S°_{rxn} = [(1 \text{ mol}) (225 \text{ J/mol}\cdot\text{K}) + (2 \text{ mol}) (188.72 \text{ J/mol}\cdot\text{K})]$$
$$- [(1 \text{ mol}) (77.0 \text{ J/mol}\cdot\text{K}) + (4 \text{ mol}) (173.67 \text{ J/mol}\cdot\text{K}) + (1 \text{ mol}) (202.7 \text{ J/mol}\cdot\text{K})]$$

$$\Delta S°_{rxn} = -371.94 \text{ J/K (unrounded)}$$

$$\Delta G°_{358} = \Delta H°_{rxn} - T\Delta S°_{rxn} = (-503.652 \text{ kJ}) - \{((273 + 85)\text{K}) (-371.94 \text{ J/K}) (\text{kJ} / 10^3 \text{ J})\} = -370.497 = \mathbf{-370.\ kJ}$$

20.92 a) $2\ N_2O_5(g) + 6\ F_2(g) \rightarrow 4\ NF_3(g) + 5\ O_2(g)$
b) Use the values from Appendix B to determine the value of $\Delta G°$.

$$\Delta G°_{rxn} = [(4 \text{ mol}) (\Delta G°_f (NF_3)) + (5 \text{ mol}) (\Delta G°_f (O_2))] - [(2 \text{ mol}) (\Delta G°_f (N_2O_5)) + (6 \text{ mol}) (\Delta G°_f (F_2))]$$

$$\Delta G°_{rxn} = [(4 \text{ mol}) (-83.3 \text{ kJ/mol})) + (5 \text{ mol}) (0 \text{ kJ/mol})] - [(2 \text{ mol}) (118 \text{ kJ/mol}) + (6 \text{ mol}) (0 \text{ kJ/mol})]$$

$$\Delta G°_{rxn} = -569.2 = \mathbf{-569\ kJ}$$

c) Calculate the value for Q and then use to find ΔG.

$$Q = \frac{[NF_3]^4 [O_2]^5}{[N_2O_5]^2 [F_2]^6} = \frac{[0.25 \text{ atm}]^4 [0.50 \text{ atm}]^5}{[0.20 \text{ atm}]^2 [0.20 \text{ atm}]^6} = 47.6837 \text{ (unrounded)}$$

$$\Delta G = \Delta G° + RT \ln Q = -569.2 \text{ kJ/mol} + (1 \text{ kJ} / 10^3 \text{ J}) (8.314 \text{ J/mol}\cdot\text{K}) (298) \ln (47.6837)$$
$$= -559.625 = \mathbf{-5.60 \times 10^2\ kJ/mol}$$

20.95 The reaction for the hydrogenation of ethene is
$$C_2H_4(g) + H_2(g) \rightarrow C_2H_6(g)$$
Calculate enthalpy, entropy, and free energy from values in Appendix B, which are given at 25°C.

$$\Delta H°_{rxn} = [1 \text{ mol} (\Delta H°_f (C_2H_6))] - [1 \text{ mol} (\Delta H°_f (C_2H_4)) + 1 \text{ mol} (\Delta H°_f (H_2))]$$
$$= [(1 \text{ mol})(-84.667 \text{ kJ})] - [(1 \text{ mol})(52.47 \text{ kJ}) + (1 \text{ mol}) (0 \text{ kJ})] = -137.137 = \mathbf{-137.14\ kJ}$$

$$\Delta S°_{rxn} = [1 \text{ mol} (S°(C_2H_6))] - [1 \text{ mol} (S°(C_2H_4)) + 1 \text{ mol} (S°(H_2))]$$
$$= 1(1 \text{ mol})(229.5 \text{ J/K})] - [(1 \text{ mol})(219.22 \text{ J/K}) + (1 \text{ mol}) (130.6 \text{ J/K})] = -120.32 = \mathbf{-120.3\ J/K}$$

$$\Delta G°_{rxn} = [(1 \text{ mol}) (\Delta G°_f (C_2H_6))] - [1 \text{ mol} (\Delta G°_f (C_2H_4)) + 1 \text{ mol} (\Delta G°_f (H_2))]$$
$$= (1 \text{ mol})(-32.89 \text{ kJ})] - [(1 \text{ mol})(68.36 \text{ kJ} + (1 \text{ mol})(0 \text{ kJ})] = \mathbf{-101.25\ kJ}$$

20.100   Determine the molar mass of glucose and tristearin. The molar mass and the mole ratios given in the problem combine, through Avogadro's number, to determine the molecules per gram. The mole ratios are assumed exact.

Glucose molar mass = 6 mol C (12.01 g/mol C) + 12 mol H (1.008 g/mol H) + 6 mol O (16.00 g/mol O)
   = 180.16 g/mol

Tristearin molar mass = 57 mol C (12.01 g/mol C) + 116 mol H (1.008 g/mol H) + 6 mol O (16.00 g/mol O)
   = 897.50 g/mol

a) $(1 \text{ g glucose}) \left(\dfrac{1 \text{ mol glucose}}{180.16 \text{ g glucose}} \right) \left[\dfrac{36 \text{ mol ATP}}{1 \text{ mol glucose}} \right] \left(\dfrac{6.022 \times 10^{23} \text{ molecules ATP}}{1 \text{ mol ATP}} \right)$

$= 1.20333 \times 10^{23} = \mathbf{1.203 \times 10^{23} \text{ molecules ATP/g glucose}}$

b) $(1 \text{ g tristearin}) \left(\dfrac{1 \text{ mol tristearin}}{897.50 \text{ g tristearin}} \right) \left[\dfrac{458 \text{ mol ATP}}{1 \text{ mol tristearin}} \right] \left(\dfrac{6.022 \times 10^{23} \text{ molecules ATP}}{1 \text{ mol ATP}} \right)$

$= 3.073065 \times 10^{23} = \mathbf{3.073 \times 10^{23} \text{ molecules ATP/g tristearin}}$

20.106   The given equilibrium is: G6P $\leftrightarrows$ F6P $K = 0.510$ at 298 K

a) Use the relationship: $\Delta G^{\circ} = -RT \ln K$
   $= -(8.314 \text{ J/mol} \cdot \text{K}) (298 \text{ K}) \ln (0.510) = 1.6682596 \times 10^3 = \mathbf{1.67 \times 10^3 \text{ J/mol}}$

b) Use the relationship: $\Delta G = \Delta G^{\circ} + RT \ln Q$
   $Q = [\text{F6P}] / [\text{G6P}] = 10.0$
   $\Delta G = 1.6682596 \times 10^3 \text{ J/mol} + (8.314 \text{ J/mol} \cdot \text{K}) (298 \text{ K}) \ln 10.0$
   $\Delta G = 7.3730799 \times 10^3 = \mathbf{7.37 \times 10^3 \text{ J/mol}}$

c) Repeat the calculation in part (b) with $Q = 0.100$
   $\Delta G = 1.6682596 \times 10^3 \text{ J/mol} + (8.314 \text{ J/mol} \cdot \text{K}) (298 \text{ K}) \ln 0.100$
   $\Delta G = -4.0365607 \times 10^3 = \mathbf{-4.04 \times 10^3 \text{ J/mol}}$

d) Use the relationship: $\Delta G = \Delta G^{\circ} + RT \ln Q$
   $(-2.50 \text{ kJ/mol}) (10^3 \text{ J} / 1 \text{ kJ}) = 1.6682596 \times 10^3 \text{ J/mol} + (8.314 \text{ J/mol} \cdot \text{K}) (298 \text{ K}) \ln Q$
   $(-4.1682596 \times 10^3 \text{ J/mol}) = (8.314 \text{ J/mol} \cdot \text{K}) (298 \text{ K}) \ln Q$
   $\ln Q = (-4.1682596 \times 10^3 \text{ J/mol}) / [(8.314 \text{ J/mol} \cdot \text{K}) (298 \text{ K})] = -1.68239696$ (unrounded)
   $Q = 0.18592778 = \mathbf{0.19}$

20.110   a) $\ln K = \dfrac{\Delta G^{\circ}}{RT}$

Since $\ln K_p = 0$ when $K_p = 1.00$, $\Delta G = 0$; combine this with $\Delta G = \Delta H - T\Delta S$. First, calculate ΔH and ΔS, using values in the Appendix.

$\Delta H^{\circ} = [(2 \text{ mol NH}_3) (\Delta H^{\circ}_f \text{ NH}_3)] - [(1 \text{ mol N}_2) (\Delta H^{\circ}_f \text{ N}_2) + (3 \text{ mol H}_2) (\Delta H^{\circ}_f \text{ H}_2)]$
   $= [(2 \text{ mol NH}_3) (-45.9 \text{ kJ/mol})] - [(1 \text{ mol N}_2) (0) + (3 \text{ mol H}_2) (0)] = -91.8 \text{ kJ}$

$\Delta S^{\circ} = [(2 \text{ mol NH}_3) (S^{\circ} \text{ NH}_3)] - [(1 \text{ mol N}_2) (S^{\circ} \text{ N}_2) + (3 \text{ mol H}_2) (S^{\circ} \text{ H}_2)]$
   $= [(2 \text{ mol NH}_3) (193 \text{ J/mol} \cdot \text{K})] - [(1 \text{ mol N}_2) (191.50 \text{ J/mol} \cdot \text{K}) + (3 \text{ mol H}_2) (130.6 \text{ J/mol} \cdot \text{K})]$
   $= -197.3 \text{ J/K}$ (unrounded)

   $\Delta G = 0$ (at equilibrium)
   $\Delta G = 0 = \Delta H - T\Delta S$
   $\Delta H = T\Delta S$

   $T = \dfrac{\Delta H^{\circ}}{\Delta S^{\circ}} = \dfrac{-91.8 \text{ kJ}}{-197.3 \text{ J/K}} \left(\dfrac{10^3 \text{ J}}{1 \text{ kJ}} \right) = 465.281 = \mathbf{465 \text{ K}}$

b) Use the relationships: $\Delta G = \Delta H - T\Delta S$ and $\Delta G = -RT \ln K$ with T = (273 + 400.) K = 673 K

$$\Delta G = \Delta H - T\Delta S = (-91.8 \text{ kJ}) (10^3 \text{ J} / 1 \text{ kJ}) - (673 \text{ K}) (-197.3 \text{ J/K})$$

$$\Delta G = 4.09829 \times 10^4 \text{ J (unrounded)}$$

$$\Delta G° = -RT \ln K$$

$$\ln K = \Delta G° / -RT = \left(\frac{4.09829 \times 10^4 \text{ J/mol}}{-(8.314 \text{ J/mol} \cdot \text{K})(673 \text{ K})} \right) = -7.32449 \text{ (unrounded)}$$

$$K = e^{-7.32449} = 6.591934 \times 10^{-4} = \mathbf{6.59 \times 10^{-4}}$$

c) The reaction rate is higher at the higher temperature. The time required (kinetics) overshadows the lower yield (thermodynamics).

CHAPTER 21 ELECTROCHEMISTRY: CHEMICAL CHANGE AND ELECTRICAL WORK

FOLLOW–UP PROBLEMS

21.1   <u>Plan:</u> Follow the steps for balancing a redox reaction in acidic solution:
           1. Divide into half-reactions
           2. For each half-reaction balance
                   a) Atoms other than O and H,
                   b) O atoms with H_2O,
                   c) H atoms with H^+ and
                   d) Charge with e^-.
           3. Multiply each half-reaction by an integer that will make the number of electrons lost equal to the number of electrons gained.
           4. Add the half-reactions and cancel substances appearing as both reactants and products.
           Then, add another step for basic solution:
           5. Add hydroxide ions to neutralize H^+. Cancel water.
        <u>Solution:</u>
           1. Divide into half-reactions: group the reactants and products with similar atoms.
                   $MnO_4^-(aq) \rightarrow MnO_4^{2-}(aq)$
                   $I^-(aq) \rightarrow IO_3^-(aq)$
           2. For each half-reaction balance
                   a) Atoms other than O and H
                   Mn and I are balanced so no changes needed
                   b) O atoms with H_2O,
                   $MnO_4^-(aq) \rightarrow MnO_4^{2-}(aq)$ O already balanced
                   $I^-(aq) + 3\ H_2O(l) \rightarrow IO_3^-(aq)$ Add 3 H_2O to balance oxygen
                   c) H atoms with H^+
                   $MnO_4^-(aq) \rightarrow MnO_4^{2-}(aq)$ H already balanced
                   $I^-(aq) + 3\ H_2O(l) \rightarrow IO_3^-(aq) + 6\ H^+(aq)$ Add 6H+ to balance hydrogen
                   d) Charge with e^-
                   $MnO_4^-(aq) + e^- \rightarrow MnO_4^{2-}(aq)$ Total charge of reactants is –1 and of products is –2,
                                                    so add 1 e^- to reactants to balance charge:
                   $I^-(aq) + 3\ H_2O(l) \rightarrow IO_3^-(aq) + 6\ H^+(aq) + 6\ e^-$ Total charge is –1 for reactants and +5 for
                                                    products, so add 6 e^- as product:
           3. Multiply each half-reaction by an integer that will make the number of electrons lost equal to the number of electrons gained. One electron is gained and 6 are lost so reduction must be multiplied by 6 for the number of electrons to be equal.
                   $6\ \{MnO_4^-(aq) + e^- \rightarrow MnO_4^{2-}(aq)\}$
                   $I^-(aq) + 3\ H_2O(l) \rightarrow IO_3^-(aq) + 6\ H^+(aq) + 6\ e^-$
           4. Add half-reactions and cancel substances appearing as both reactants and products.
                   $6\ MnO_4^-(aq) + \cancel{6\ e^-} \rightarrow 6\ MnO_4^{2-}(aq)$
                   $\underline{I^-(aq) + 3\ H_2O(l) \rightarrow IO_3^-(aq) + 6\ H^+(aq) + \cancel{6\ e^-}}$
        Overall: $6\ MnO_4^-(aq) + I^-(aq) + 3\ H_2O(l) \rightarrow 6\ MnO_4^{2-}(aq) + IO_3^-(aq) + 6\ H^+(aq)$
        5. Add hydroxide ions to neutralize H^+. Cancel water. The 6 H^+ are neutralized by adding 6 OH^-. The same number of hydroxide ions must be added to the reactants to keep the balance of O and H atoms on both sides of the reaction.
        $6\ MnO_4^-(aq) + I^-(aq) + 3\ H_2O(l) + 6\ OH^-(aq) \rightarrow 6\ MnO_4^{2-}(aq) + IO_3^-(aq) + 6\ H^+(aq) + 6\ OH^-(aq)$
        The neutralization reaction produces water: $6\ \{H^+ + OH^- \rightarrow H_2O\}$.

$6 \text{MnO}_4^-(aq) + \text{I}^-(aq) + 3 \text{H}_2\text{O}(l) + 6 \text{OH}^-(aq) \rightarrow 6 \text{MnO}_4^{2-}(aq) + \text{IO}_3^-(aq) + 6 \text{H}_2\text{O}(l)$

Cancel water:

$6 \text{MnO}_4^-(aq) + \text{I}^-(aq) + \cancel{3 \text{H}_2\text{O}(l)} + 6 \text{OH}^-(aq) \rightarrow 6 \text{MnO}_4^{2-}(aq) + \text{IO}_3^-(aq) + \cancel{6}\text{-H}_2\text{O}(l)$

Balanced reaction is

$6 \text{MnO}_4^-(aq) + \text{I}^-(aq) + 6 \text{OH}^-(aq) \rightarrow 6 \text{MnO}_4^{2-}(aq) + \text{IO}_3^-(aq) + 3\text{H}_2\text{O}(l)$

Balanced reaction including spectator ions is

$6 \text{KMnO}_4(aq) + \text{KI}(aq) + 6 \text{KOH}(aq) \rightarrow 6 \text{K}_2\text{MnO}_4(aq) + \text{KIO}_3(aq) + 3 \text{H}_2\text{O}(l)$

Check: Check balance of atoms and charge:

Reactants:	Products:
6 Mn atoms	6 Mn atoms
1 I atom	1 I atom
30 O atoms	30 O atoms
6 H atoms	6 atoms
−13 charge	−13 charge

21.2   Plan: Given the solution and electrode compositions, the two half cells involve the transfer of electrons 1) between chromium in $\text{Cr}_2\text{O}_7^{2-}$ and Cr^{3+} and 2) between Sn and Sn^{2+}. The negative electrode is the anode so the tin half-cell is where oxidation occurs. The graphite electrode with the chromium ion/chromate solution is where reduction occurs.

In the cell diagram, show the electrodes and the solutes involved in the half-reactions. Include the salt bridge and wire connection between electrodes. Set up the two half-reactions and balance. (Note that the $\text{Cr}^{3+}/\text{Cr}_2\text{O}_7^{2-}$ half-cell is in acidic solution.) Write the cell notation placing the anode half-cell first, then the salt bridge, then the cathode half-cell.

Solution: Cell diagram:

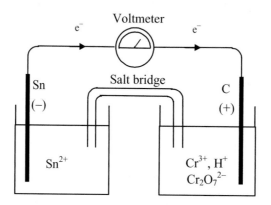

Balanced equations:

Anode is Sn/Sn^{2+} half-cell. Oxidation of Sn produces Sn^{2+}:

$\text{Sn}(s) \rightarrow \text{Sn}^{2+}(aq)$

All that needs to be balanced is charge:

$\text{Sn}(s) \rightarrow \text{Sn}^{2+}(aq) + 2 \text{e}^-$

Cathode is the $\text{Cr}^{3+}/\text{Cr}_2\text{O}_7^{2-}$ half-cell. Check the oxidation number of chromium in each substance to determine which is reduced. Cr^{3+} oxidation number is +3 and chromium in $\text{Cr}_2\text{O}_7^{2-}$ has oxidation number +6. Going from +6 to +3 involves gain of electrons so $\text{Cr}_2\text{O}_7^{2-}$ is reduced.

$\text{Cr}_2\text{O}_7^{2-}(aq) \rightarrow \text{Cr}^{3+}(aq)$

Balance Cr:

$\text{Cr}_2\text{O}_7^{2-}(aq) \rightarrow 2 \text{Cr}^{3+}(aq)$

Balance O:

$\text{Cr}_2\text{O}_7^{2-}(aq) \rightarrow 2 \text{Cr}^{3+}(aq) + 7 \text{H}_2\text{O}(l)$

Balance H:

$\text{Cr}_2\text{O}_7^{2-}(aq) + 14 \text{H}^+(aq) \rightarrow 2 \text{Cr}^{3+}(aq) + 7 \text{H}_2\text{O}(l)$

Balance charge:

$\text{Cr}_2\text{O}_7^{2-}(aq) + 14 \text{H}^+(aq) + 6 \text{e}^- \rightarrow 2 \text{Cr}^{3+}(aq) + 7 \text{H}_2\text{O}(l)$

Add two half-reactions multiplying the tin half-reaction by 3 to equalize the number of electrons transferred.

$$3\{Sn(s) \rightarrow Sn^{2+}(aq) + 2\ e^-\}$$
$$Cr_2O_7^{2-}(aq) + 14\ H^+(aq) + 6\ e^- \rightarrow 2\ Cr^{3+}(aq) + 7\ H_2O(l)$$

$$3\ Sn(s) + Cr_2O_7^{2-}(aq) + 14\ H^+(aq) \rightarrow 3\ Sn^{2+}(aq) + 2\ Cr^{3+}(aq) + 7\ H_2O(l)$$

Cell notation:

$$Sn(s) \mid Sn^{2+}(aq) \parallel H^+(aq), Cr_2O_7^{2-}(aq), Cr^{3+}(aq) \mid C(graphite)$$

21.3   Plan: Divide the reaction into half-reactions showing that Br_2 is reduced and V^{3+} is oxidized. Use the equation $E^o_{cell} = E^o_{cathode} - E^o_{anode}$ to solve for E^o_{anode} .

Solution:

Half-reactions:

Reduction (cathode): $Br_2(aq) + 2\ e^- \rightarrow 2\ Br^-(aq)$ $E^o_{cathode} = 1.07$ V from Appendix D

Oxidation (anode): $2\ V^{3+}(aq) + 2\ H_2O(l) \rightarrow 2\ VO^{2+}(aq) + 4\ H^+(aq) + 2\ e^-$

Overall: $Br_2(aq) + 2\ V^{3+}(aq) + 2\ H_2O(l) \rightarrow 2\ VO^{2+}(aq) + 4\ H^+(aq) + 2\ Br^-(aq)$ $E^o_{cell} = 1.39$ V from problem

$$E^o_{cell} = E^o_{cathode} - E^o_{anode}$$
$$E^o_{anode} = E^o_{cathode} - E^o_{cell} = 1.07\ V - 1.39\ V = \mathbf{-0.32\ V}$$

21.4   Plan: To determine if the reaction is spontaneous, divide into half-reactions and calculate E^o_{cell} . If E^o_{cell} is negative, the reaction is not spontaneous, so reverse the reaction to obtain the spontaneous reaction. Reducing strength increases with decreasing $E°$.

Solution: Divide into half-reactions and balance:

Reduction: $Fe^{2+}(aq) + 2\ e^- \rightarrow Fe(s)$ $E^o_{cathode} = -0.44$ V

Oxidation: $2\ \{Fe^{2+}(aq) \rightarrow 2\ Fe^{3+}(aq) + e^-\}$ $E^o_{anode} = 0.77$ V

The first half-reaction is reduction, so it is the cathode half-cell. The second half-reaction is oxidation, so it is the anode half-cell. Find the half-reactions in Appendix D.

$$E^o_{cell} = E^o_{cathode} - E^o_{anode} = -0.44\ V - 0.77\ V = -1.21\ V$$

The reaction is not spontaneous as written, so reverse the reaction:

$$Fe(s) + 2\ Fe^{3+}(aq) \rightarrow 3\ Fe^{2+}(aq)$$

E^o_{cell} is now +1.21 V, so the reversed reaction is spontaneous under standard state conditions.

When a substance acts as a reducing agent, it is oxidized. Both Fe and Fe^{2+} can be oxidized, so they can act as reducing agents. Since Fe^{3+} cannot lose more electrons, it cannot act as a reducing agent. The stronger reducing agent between Fe and Fe^{2+} is the one with the smaller standard reduction potential. $E°$ for Fe is -0.44, which is less than $E°$ for Fe^{2+}, $+0.77$. Therefore, Fe is a stronger reducing agent than Fe^{2+}. Ranking all three in order of decreasing reducing strength gives $\mathbf{Fe > Fe^{2+} > Fe^{3+}}$.

21.5   Plan: Reaction is $Cd(s) + Cu^{2+}(aq) \rightarrow Cd^{2+}(aq) + Cu(s)$.

Given $\Delta G°$, both K and E^o_{cell} can be calculated using the relationships

$\Delta G° = -RT \ln K$ and $\Delta G° = -nF\ E^o_{cell}$.

Solution:

$$\ln K = -\frac{\Delta G°}{RT} = -\left(\frac{-143\ kJ}{(8.314\ J/mol \cdot K)((273+25)K)}\right)\left(\frac{10^3\ J}{1\ kJ}\right) = 57.71779791\ (unrounded)$$

$$K = 1.1655237 \times 10^{25} = \mathbf{1.17 \times 10^{25}}$$

$\Delta G° = -nF\ E^o_{cell}$ $n = 2$ mol e^- for this reaction

$$E^o_{cell} = -\frac{\Delta G°}{nF} = -\left(\frac{-143\ kJ}{(2\ mol)(96485\ C/mol)}\right)\left(\frac{10^3\ J}{1\ kJ}\right)\left(\frac{C}{J/V}\right) = 0.7410478 = \mathbf{0.741\ V}$$

Note that an alternative way to calculate E^o_{cell} is as equal to $(RT/nF) \ln K$.

Check: Reverse the calculation of K:

$\Delta G° = -RT \ln K = -[(8.314 \text{ J/mol·K}) (298 \text{ K}) \ln (1.1655237 \times 10^{25})] (1 \text{ kJ} / 10^3 \text{ J}) = 143 \text{ kJ}$

The potential can easily be checked against the value calculated from Appendix D.

$E°_{cell} = E°_{cathode} - E°_{anode} = (0.34 \text{ V}) - (-0.40 \text{ V}) = 0.74 \text{ V}$

The potential value agrees with this literature value.

21.6   Plan: The problem is asking for the concentration of iron ions when $E_{cell} = E°_{cell} + 0.25 \text{ V}$.

Use the Nernst equation, $E_{cell} = E°_{cell} - \dfrac{RT}{nF} \ln Q$ to find $[Fe^{2+}]$.

Solution: Determining the cell reaction and $E°_{cell}$:

Oxidation:	$Fe(s) \rightarrow Fe^{2+}(aq) + 2e^-$	$E° = -0.44 \text{ V}$
Reduction:	$Cu^{2+}(aq) + 2e^- \rightarrow Cu(s)$	$E° = 0.34 \text{ V}$

Overall $Fe(s) + Cu^{2+}(aq) \rightarrow Fe^{2+}(aq) + Cu(s)$

$E°_{cell} = E°_{cathode} - E°_{anode} = 0.34 \text{ V} - (-0.44 \text{ V}) = 0.78 \text{ V}$

For the reaction $Q = [Fe^{2+}]/[Cu^{2+}]$, so the Nernst equation is

$E_{cell} = E°_{cell} - \dfrac{0.0592}{n} \log \dfrac{[Fe^{2+}]}{[Cu^{2+}]}$

Substituting in values from the problem:

$E_{cell} = E°_{cell} + 0.25 \text{ V} = 0.78 \text{ V} + 0.25 \text{ V} = 1.03 \text{ V}$

$1.03 \text{ V} = 0.78 \text{ V} - \dfrac{0.0592}{n} \log \dfrac{[Fe^{2+}]}{[Cu^{2+}]}$

$0.25 \text{ V} = -\dfrac{0.0592}{2} \log \dfrac{[Fe^{2+}]}{[0.30]}$

$-8.44595 = \log \dfrac{[Fe^{2+}]}{[0.30]}$

$3.58141 \times 10^{-9} = \dfrac{[Fe^{2+}]}{[0.30]}$

$[Fe^{2+}] = 1.074423 \times 10^{-9} = \mathbf{1.1 \times 10^{-9} \ \mathit{M}}$

Check: When $[Fe^{2+}] = [Cu^{2+}]$, the log of $[Fe^{2+}]/[Cu^{2+}]$ equals 0 and $E_{cell} = E°_{cell}$. A change to $[Fe^{2+}] = (0.1)[Cu^{2+}]$ gives log $([Fe^{2+}]/[Cu^{2+}]) = -1$, and the potential changes by $-(-1)(0.03 \text{ V}) = 0.03 \text{ V}$. Use this to estimate the potential change that should occur if $[Fe^{2+}] = 1 \times 10^{-9} \ M$ and $[Cu^{2+}] = 0.1 \ M$. Estimating the ratio $[Fe^{2+}]/[Cu^{2+}]$ $= 1 \times 10^{-9}/0.1 = 1 \times 10^{-8}$ gives log of the ratio equal to -8 to give a potential change of $-(-8)(0.06 \text{ V}/2) = 0.24 \text{ V}$, which is almost the given difference in potential.

21.7   Plan: Half-cell B contains a higher concentration of gold ions, so the ions will be reduced to decrease the concentration while in half-cell A, with a lower $[Au^{3+}]$, gold metal will be oxidized to increase the concentration of gold ions. In the overall cell reaction, the lower $[Au^{3+}]$ appears as a product and the higher $[Au^{3+}]$ appears as a reactant. This means that $Q = [Au^{3+}]_{lower}/[Au^{3+}]_{higher}$. Use the Nernst equation with $E°_{cell} = 0$ to find E_{cell}. Oxidation of gold metal occurs at the anode, half-cell A, which is negative.
Solution:

$E_{cell} = E°_{cell} - \dfrac{RT}{nF} \ln Q = E°_{cell} - \dfrac{0.0592}{n} \log \dfrac{[Au^{3+}_{lower}]}{[Au^{3+}_{higher}]}$ $n = 3 \text{ mol } e^-$ for the reduction of Au^{3+} to Au

$E_{cell} = 0 - \dfrac{0.0592}{3} \log \dfrac{[7.0 \times 10^{-4}]}{[2.5 \times 10^{-2}]} = 0.0306427 = \mathbf{0.031 \text{ V}}$

Check: The potential of the cell can be estimated by realizing that each tenfold difference in concentration between the two half cells gives a 20 mV change in potential. The difference is more than tenfold but less than 100 times, so potential should fall between 20 and 40 mV.

21.8   Plan: In the electrolysis, the more easily reduced metal ion will be reduced at the cathode. Compare K^+ and Al^{3+} as to their location on the periodic table to find which has the higher ionization energy (IE increases up and across the table). Higher ionization energy of the metal means more easily reduced. At the anode, the more easily oxidized nonmetal ion will be oxidized. For oxidation, compare the electronegativity of the two nonmetals. The ion from the less electronegative element will be more easily oxidized. Note that the standard reduction potentials cannot be used in the case of molten salts because the $E°$ values are based on aqueous solutions of ions, not liquid salts.
        Solution: Al is above and to the right of K, so Al^{3+} is more easily reduced than K^+. The reaction at the cathode is
$$Al^{3+}(l) + 3\ e^- \rightarrow Al(s)$$
F is the most electronegative element, so Br^- must be more easily oxidized than F^-. The reaction at the anode is
$$2\ Br^-\ (l) \rightarrow Br_2(g) + 2\ e^-$$
To add the two half-reaction, the number of electrons must be equal:
$$2\ \{Al^{3+}(l) + 3\ e^- \rightarrow Al(s)\}$$
$$\underline{3\ \{2\ Br^-\ (l)\ \rightarrow\ Br_2(g) + 2\ e^-\}}$$
Overall: $2\ Al^{3+}(l) + 6\ Br^-\ (l) \rightarrow 2\ Al(s) + 3\ Br_2(g)$
This is the overall reaction.

21.9   Plan: In aqueous $AuBr_3$, the species present are $Au^{3+}(aq)$, $Br^-(aq)$, H_2O, and very small amounts of H^+ and OH^-. The possible half-reactions are reduction of either Au^{3+} or H_2O and oxidation of either Br^- or H_2O. Whichever reduction and oxidation half-reactions are more spontaneous will take place, with consideration of the overvoltage.
        Solution: The two possible reductions are
$$Au^{3+}(aq) + 3\ e^- \rightarrow Au(s) \qquad\qquad E° = 1.50\ V$$
$$2\ H_2O(l) + 2\ e^- \rightarrow H_2(g) + 2\ OH^-(aq) \qquad E = -1\ V\ \text{(with overvoltage)}$$
The reduction of gold ions occurs because it has a higher reduction potential (more spontaneous) than reduction of water.
        The two possible oxidations are
$$2\ Br^-(aq) \rightarrow Br_2(l) + 2\ e^- \qquad\qquad E° = 1.07\ V$$
$$2\ H_2O(l) \rightarrow O_2(g) + 4\ H^+(aq) + 4\ e^- \qquad E = 1.4\ V\ \text{(with overvoltage)}$$
The oxidation with the less positive reduction potential is the more spontaneous oxidation so bromide ions are oxidized at the anode.
        The two half-reactions that are predicted are
        Cathode: $Au^{3+}(aq) + 3\ e^- \rightarrow Au(s)$ $E° = 1.50\ V$
        Anode: $2\ Br^-(aq) \rightarrow Br_2(l) + 2\ e^-$ $E° = 1.07\ V$

21.10   Plan: Current is charge per time, so to find the time, divide the charge by the current. To find the charge transferred, first write the balanced half-reaction, then calculate the charge from the grams of copper and moles of electrons transferred per molar mass of copper.
        Solution:
$$Cu^{2+}(aq) + 2\ e^- \rightarrow Cu(s),\ \text{so 2 mol }e^-\text{ per mole of Cu}$$

$$(1.50\ \text{g Cu})\left(\frac{1\ \text{mol Cu}}{63.55\ \text{g Cu}}\right)\left(\frac{2\ \text{mol }e^-}{1\ \text{mol Cu}}\right)\left(\frac{96485\ \text{C}}{1\ \text{mol }e^-}\right)\left(\frac{A}{C/s}\right)\left(\frac{1}{4.75\ A}\right)\left(\frac{1\ \text{min}}{60\ \text{s}}\right) = 15.9816 = \textbf{16.0 min}$$

END–OF–CHAPTER PROBLEMS

21.1   Oxidation is the loss of electrons (resulting in a higher oxidation number), while reduction is the gain of electrons (resulting in a lower oxidation number). In an oxidation-reduction reaction, electrons transfer from the oxidized substance to the reduced substance. The oxidation number of the reactant being oxidized increases while the oxidation number of the reactant being reduced decreases.

21.3 **No**, one half-reaction cannot take place independently of the other because there is always a transfer of electrons from one substance to another. If one substance loses electrons (oxidation half-reaction), another substance must gain those electrons (reduction half-reaction).

21.6 To remove protons from an equation, add an equal number of hydroxide ions to both sides to neutralize the H^+ and produce water: $H^+(aq) + OH^-(aq) \rightarrow H_2O(l)$.

21.8 Spontaneous reactions, $\Delta G_{sys} < 0$, take place in voltaic cells, which are also called galvanic cells. Non-spontaneous reactions take place in electrolytic cells and result in an increase in the free energy of the cell ($\Delta G_{sys} > 0$).

21.10 Assign oxidation numbers:

$$\overset{+1}{16}\ H^+(aq) + 2\ \overset{\overset{-8}{+7\ -2}}{MnO_4^-}(aq) + 10\ \overset{-1}{Cl^-}(aq) \rightarrow 2\ \overset{+2}{Mn^{2+}}(aq) + 5\ \overset{0}{Cl_2}(g) + 8\ \overset{\overset{+2}{+1\ -2}}{H_2O}(l)$$

a) To decide which reactant is oxidized, look at oxidation numbers. **Cl⁻** is oxidized because its oxidation number increases from –1 in Cl⁻ to 0 in Cl₂.
b) **MnO₄⁻** is reduced because the oxidation number of Mn decreases from +7 in MnO₄⁻ to +2 in Mn²⁺.
c) The oxidizing agent is the substance that causes the oxidation by accepting electrons. The oxidizing agent is the substance reduced in the reaction, so **MnO₄⁻** is the oxidizing agent.
d) **Cl⁻** is the reducing agent because it loses the electrons that are gained in the reduction.
e) **From Cl⁻**, which is losing electrons, **to MnO₄⁻**, which is gaining electrons.
f) $8\ H_2SO_4(aq) + 2\ KMnO_4(aq) + 10\ KCl(aq) \rightarrow 2\ MnSO_4(aq) + 5\ Cl_2(g) + 8\ H_2O(l) + 6\ K_2SO_4(aq)$

21.12 a) Divide into half-reactions:

$ClO_3^-(aq) \rightarrow Cl^-(aq)$
$I^-(aq) \rightarrow I_2(s)$

Balance elements other than O and H

$ClO_3^-(aq) \rightarrow Cl^-(aq)$	chlorine is balanced
$2\ I^-(aq) \rightarrow I_2(s)$	iodine now balanced

Balance O by adding H₂O

$ClO_3^-(aq) \rightarrow Cl^-(aq) + 3\ H_2O(l)$	add 3 waters to add 3 O atoms to product
$2\ I^-(aq) \rightarrow I_2(s)$	no change

Balance H by adding H⁺

$ClO_3^-(aq) + 6\ H^+(aq) \rightarrow Cl^-(aq) + 3\ H_2O(l)$	add 6 H⁺ to reactants
$2\ I^-(aq) \rightarrow I_2(s)$	no change

Balance charge by adding e⁻

$ClO_3^-(aq) + 6\ H^+(aq) + 6\ e^- \rightarrow Cl^-(aq) + 3\ H_2O(l)$	add 6 e⁻ to reactants for a –1 charge on each side
$2\ I^-(aq) \rightarrow I_2(s) + 2\ e^-$	add 2 e⁻ to products for a –2 charge on each side

Multiply each half-reaction by an integer to equalize the number of electrons

$ClO_3^-(aq) + 6\ H^+(aq) + 6\ e^- \rightarrow Cl^-(aq) + 3\ H_2O(l)$	multiply by 1 to give 6 e⁻
$3\{2\ I^-(aq) \rightarrow I_2(s) + 2\ e^-\}$ or	multiply by 3 to give 6 e⁻
$6\ I^-(aq) \rightarrow 3\ I_2(s) + 6\ e^-$	

Add half-reactions to give balanced equation in acidic solution.

$$ClO_3^-(aq) + 6\ H^+(aq) + 6\ I^-(aq) \rightarrow Cl^-(aq) + 3\ H_2O(l) + 3\ I_2(s)$$

Check balancing:

Reactants:		Products:	
	1 Cl		1 Cl
	3 O		3 O
	6 H		6 H
	6 I		6 I
	–1 charge		–1 charge

Oxidizing agent is ClO₃⁻ and reducing agent is I⁻.

b) Divide into half-reactions:

$$MnO_4^-(aq) \rightarrow MnO_2(s)$$
$$SO_3^{2-}(aq) \rightarrow SO_4^{2-}(aq)$$

Balance elements other than O and H

$MnO_4^-(aq) \rightarrow MnO_2(s)$	Mn is balanced
$SO_3^{2-}(aq) \rightarrow SO_4^{2-}(aq)$	S is balanced

Balance O by adding H_2O

$MnO_4^-(aq) \rightarrow MnO_2(s) + 2\ H_2O(l)$	add 2 H_2O to products
$SO_3^{2-}(aq) + H_2O(l) \rightarrow SO_4^{2-}(aq)$	add 1 H_2O to reactants

Balance H by adding H^+

$MnO_4^-(aq) + 4\ H^+(aq) \rightarrow MnO_2(s) + 2\ H_2O(l)$	add 4 H^+ to reactants
$SO_3^{2-}(aq) + H_2O(l) \rightarrow SO_4^{2-}(aq) + 2\ H^+(aq)$	add 2 H^+ to products

Balance charge by adding e^-

$MnO_4^-(aq) + 4\ H^+(aq) + 3\ e^- \rightarrow MnO_2(s) + 2\ H_2O(l)$ add 3 e^- to reactants for a 0 charge on each side
$SO_3^{2-}(aq) + H_2O(l) \rightarrow SO_4^{2-}(aq) + 2\ H^+(aq) + 2\ e^-$ add 2 e^- to products for a –2 charge on each side

Multiply each half-reaction by an integer to equalize the number of electrons

$2\{MnO_4^-(aq) + 4\ H^+(aq) + 3\ e^- \rightarrow MnO_2(s) + 2\ H_2O(l)\}$ or   multiply by 2 to give 6 e^-
$2\ MnO_4^-(aq) + 8\ H^+(aq) + 6\ e^- \rightarrow 2\ MnO_2(s) + 4\ H_2O(l)$
$3\{SO_3^{2-}(aq) + H_2O(l) \rightarrow SO_4^{2-}(aq) + 2\ H^+(aq) + 2\ e^-\}$ or   multiply by 3 to give 6 e^-
$3\ SO_3^{2-}(aq) + 3\ H_2O(l) \rightarrow 3\ SO_4^{2-}(aq) + 6\ H^+(aq) + 6\ e^-$

Add half-reactions and cancel substances that appear as both reactants and products

$2\ MnO_4^-(aq) + \cancel{8}H^+(aq) + 3\ SO_3^{2-}(aq) + \cancel{3\ H_2O(l)} \rightarrow 2\ MnO_2(s) + \cancel{4}H_2O(l) + 3\ SO_4^{2-}(aq) + \cancel{6\ H^+(aq)}$

The balanced equation in acidic solution is:

$2\ MnO_4^-(aq) + 2\ H^+(aq) + 3\ SO_3^{2-}(aq) \rightarrow 2\ MnO_2(s) + H_2O(l) + 3\ SO_4^{2-}(aq)$

To change to basic solution, add OH^- to both sides of equation to neutralize H^+.

$2\ MnO_4^-(aq) + 2\ H^+(aq) + 2\ OH^-(aq) + 3\ SO_3^{2-}(aq) \rightarrow 2\ MnO_2(s) + H_2O(l) + 3\ SO_4^{2-}(aq) + 2\ OH^-(aq)$
$2\ MnO_4^-(aq) + \cancel{2}\ H_2O(l) + 3\ SO_3^{2-}(aq) \rightarrow 2\ MnO_2(s) + \cancel{H_2O(l)} + 3\ SO_4^{2-}(aq) + 2\ OH^-(aq)$

Balanced equation in basic solution:

$2\ MnO_4^-(aq) + H_2O(l) + 3\ SO_3^{2-}(aq) \rightarrow 2\ MnO_2(s) + 3\ SO_4^{2-}(aq) + 2\ OH^-(aq)$

Check balancing:

Reactants:		Products:	
	2 Mn		2 Mn
	18 O		18 O
	2 H		2 H
	3 S		3 S
	–8 charge		–8 charge

Oxidizing agent is MnO_4^- and reducing agent is SO_3^{2-}.

c) Divide into half-reactions:

$$MnO_4^-(aq) \rightarrow Mn^{2+}(aq)$$
$$H_2O_2(aq) \rightarrow O_2(g)$$

Balance elements other than O and H

$MnO_4^-(aq) \rightarrow Mn^{2+}(aq)$	Mn is balanced
$H_2O_2(aq) \rightarrow O_2(g)$	No other elements to balance

Balance O by adding H_2O

$MnO_4^-(aq) \rightarrow Mn^{2+}(aq) + 4\ H_2O(l)$	add 4 H_2O to products
$H_2O_2(aq) \rightarrow O_2(g)$	O is balanced

Balance H by adding H^+

$MnO_4^-(aq) + 8\ H^+(aq) \rightarrow Mn^{2+}(aq) + 4\ H_2O(l)$	add 8 H^+ to reactants
$H_2O_2(aq) \rightarrow O_2(g) + 2\ H^+(aq)$	add 2 H^+ to products

Balance charge by adding e^-

$MnO_4^-(aq) + 8\ H^+(aq) + 5\ e^- \rightarrow Mn^{2+}(aq) + 4\ H_2O(l)$ add 5 e^- to reactants for +2 on each side
$H_2O_2(aq) \rightarrow O_2(g) + 2\ H^+(aq) + 2\ e^-$ add 2 e^- to products for 0 charge on each side

Multiply each half-reaction by an integer to equalize the number of electrons

$2\{MnO_4^-(aq) + 8\ H^+(aq) + 5\ e^- \rightarrow Mn^{2+}(aq) + 4\ H_2O(l)\}$ or multiply by 2 to give 10 e⁻
$2\ MnO_4^-(aq) + 16\ H^+(aq) + 10\ e^- \rightarrow 2\ Mn^{2+}(aq) + 8\ H_2O(l)$
$5\{H_2O_2(aq) \rightarrow O_2(g) + 2\ H^+(aq) + 2\ e^-\}$ or multiply by 5 to give 10 e⁻
$5\ H_2O_2(aq) \rightarrow 5\ O_2(g) + 10\ H^+(aq) + 10\ e^-$

Add half-reactions and cancel substances that appear as both reactants and products

$2\ MnO_4^-(aq) + \cancel{16}\ H^+(aq) + 5\ H_2O_2(aq) \rightarrow 2\ Mn^{2+}(aq) + 8\ H_2O(l) + 5\ O_2(g) + \cancel{10\ H^+(aq)}$

The balanced equation in acidic solution

$2\ MnO_4^-(aq) + 6\ H^+(aq) + 5\ H_2O_2(aq) \rightarrow 2\ Mn^{2+}(aq) + 8\ H_2O(l) + 5\ O_2(g)$

Check balancing:

Reactants:		Products:	
	2 Mn		2 Mn
	18 O		18 O
	16 H		16 H
	+4 charge		+4 charge

Oxidizing agent is MnO₄⁻ and reducing agent is H₂O₂.

21.14 a) Balance the reduction half-reaction:

$Cr_2O_7^{2-}(aq) \rightarrow 2\ Cr^{3+}(aq)$ balance Cr
$Cr_2O_7^{2-}(aq) \rightarrow 2\ Cr^{3+}(aq) + 7\ H_2O(l)$ balance O by addin H_2O
$Cr_2O_7^{2-}(aq) + 14\ H^+(aq) \rightarrow 2\ Cr^{3+}(aq) + 7\ H_2O(l)$ balance H by adding H^+
$Cr_2O_7^{2-}(aq) + 14\ H^+(aq) + 6\ e^- \rightarrow 2\ Cr^{3+}(aq) + 7\ H_2O(l)$ balance charge by adding 6 e⁻

Balance the oxidation half-reaction:

$Zn(s) \rightarrow Zn^{2+}(aq) + 2\ e^-$ balance charge by adding 2 e⁻

Add the two half-reactions multiplying the oxidation half-reaction by 3 to equalize the electrons.

$Cr_2O_7^{2-}(aq) + 14\ H^+(aq) + 6\ e^- \rightarrow 2\ Cr^{3+}(aq) + 7\ H_2O(l)$
$\underline{3\ Zn(s) \rightarrow 3\ Zn^{2+}(aq) + 6\ e^-}$
$Cr_2O_7^{2-}(aq) + 14\ H^+(aq) + 3\ Zn(s) \rightarrow 2\ Cr^{3+}(aq) + 7\ H_2O(l) + 3\ Zn^{2+}(aq)$

Oxidizing agent is Cr₂O₇²⁻ and reducing agent is Zn.

b) Balance the reduction half-reaction:

$MnO_4^-(aq) \rightarrow MnO_2(s) + 2\ H_2O(l)$ balance O by adding H_2O
$MnO_4^-(aq) + 4\ H^+(aq) \rightarrow MnO_2(s) + 2\ H_2O(l)$ balance H by adding H^+
$MnO_4^-(aq) + 4\ H^+(aq) + 3\ e^- \rightarrow MnO_2(s) + 2\ H_2O(l)$ balance charge by adding 3e⁻

Balance the oxidation half-reaction

$Fe(OH)_2(s) + H_2O(l) \rightarrow Fe(OH)_3(s)$ balance O by adding H_2O
$Fe(OH)_2(s) + H_2O(l) \rightarrow Fe(OH)_3(s) + H^+(aq)$ balance H by adding H^+
$Fe(OH)_2(s) + H_2O(l) \rightarrow Fe(OH)_3(s) + H^+(aq) + e^-$ balance charge by adding 1 e⁻

Add half-reactions after multiplying oxidation half-reaction by 3.

$MnO_4^-(aq) + 4\ H^+(aq) + 3\ e^- \rightarrow MnO_2(s) + 2\ H_2O(l)$
$\underline{3\ Fe(OH)_2(s) + 3\ H_2O(l) \rightarrow 3\ Fe(OH)_3(s) + 3\ H^+(aq) + 3\ e^-}$
$MnO_4^-(aq) + \cancel{4}\ H^+(aq) + 3\ Fe(OH)_2(s) + \cancel{3}\ H_2O(l) \rightarrow MnO_2(s) + \cancel{2\ H_2O(l)} + 3\ Fe(OH)_3(s) + \cancel{3}\ H^+(aq)$
$MnO_4^-(aq) + H^+(aq) + 3\ Fe(OH)_2(s) + H_2O(l) \rightarrow MnO_2(s) + 3\ Fe(OH)_3(s)$

Add OH⁻ to both sides to neutralize the H⁺ and convert $H^+ + OH^- \rightarrow H_2O$

$MnO_4^-(aq) + H^+(aq) + OH^-(aq) + 3\ Fe(OH)_2(s) + H_2O(l) \rightarrow MnO_2(s) + 3\ Fe(OH)_3(s) + OH^-(aq)$
$MnO_4^-(aq) + 3\ Fe(OH)_2(s) + 2\ H_2O(l) \rightarrow MnO_2(s) + 3\ Fe(OH)_3(s) + OH^-(aq)$

Oxidizing agent is MnO₄⁻ and reducing agent is Fe(OH)₂.

c) Balance the reduction half-reaction:

$2\ NO_3^-(aq) \rightarrow N_2(g)$ balance N
$2\ NO_3^-(aq) \rightarrow N_2(g) + 6\ H_2O(l)$ balance O by adding H_2O
$2\ NO_3^-(aq) + 12\ H^+(aq) \rightarrow N_2(g) + 6\ H_2O(l)$ balance H by adding H^+
$2\ NO_3^-(aq) + 12\ H^+(aq) + 10\ e^- \rightarrow N_2(g) + 6\ H_2O(l)$ balance charge by adding 10 e⁻

Balance the oxidation half-reaction:

$Zn(s) \rightarrow Zn^{2+}(aq) + 2\ e^-$ balance charge by adding 2 e⁻

Add the half-reactions after multiplying the reduction half-reaction by 1 and the oxidation half-reaction by 5.

$$2 \text{ NO}_3^-(aq) + 12 \text{ H}^+(aq) + 10 \text{ e}^- \rightarrow \text{N}_2(g) + 6 \text{ H}_2\text{O}(l)$$

$$\underline{5 \text{ Zn}(s) \rightarrow 5 \text{ Zn}^{2+}(aq) + 10 \text{ e}^-}$$

$$2 \text{ NO}_3^-(aq) + 12 \text{ H}^+(aq) + 5 \text{ Zn}(s) \rightarrow \text{N}_2(g) + 6 \text{ H}_2\text{O}(l) + 5 \text{ Zn}^{2+}(aq)$$

Oxidizing agent is NO_3^- and reducing agent is Zn.

21.16   a) Balance the reduction half-reaction:

$\text{NO}_3^-(aq) \rightarrow \text{NO}(g) + 2 \text{ H}_2\text{O}(l)$	balance O by adding H_2O
$\text{NO}_3^-(aq) + 4 \text{ H}^+(aq) \rightarrow \text{NO}(g) + 2 \text{ H}_2\text{O}(l)$	balance H by adding H^+
$\text{NO}_3^-(aq) + 4 \text{ H}^+(aq) + 3 \text{ e}^- \rightarrow \text{NO}(g) + 2 \text{ H}_2\text{O}(l)$	balance charge by adding 3 e^-

Balance oxidation half-reaction:

$4 \text{ Sb}(s) \rightarrow \text{Sb}_4\text{O}_6(s)$	balance Sb
$4 \text{ Sb}(s) + 6 \text{ H}_2\text{O}(l) \rightarrow \text{Sb}_4\text{O}_6(s)$	balance O by adding H_2O
$4 \text{ Sb}(s) + 6 \text{ H}_2\text{O}(l) \rightarrow \text{Sb}_4\text{O}_6(s) + 12 \text{ H}^+(aq)$	balance H by adding H^+
$4 \text{ Sb}(s) + 6 \text{ H}_2\text{O}(l) \rightarrow \text{Sb}_4\text{O}_6(s) + 12 \text{ H}^+(aq) + 12 \text{ e}^-$	balance charge by adding 12 e^-

Multiply each half-reaction by an integer to equalize the number of electrons

$4\{ \text{NO}_3^-(aq) + 4 \text{ H}^+(aq) + 3 \text{ e}^- \rightarrow \text{NO}(g) + 2 \text{ H}_2\text{O}(l)\}$	Multiply by 4 to give 12 e^-
$1\{4 \text{ Sb}(s) + 6 \text{ H}_2\text{O}(l) \rightarrow \text{Sb}_4\text{O}_6(s) + 12 \text{ H}^+(aq) + 12 \text{ e}^-\}$	Multiply by 1 to give 12 e^-

This gives:

$$4 \text{ NO}_3^-(aq) + 16 \text{ H}^+(aq) + 12 \text{ e}^- \rightarrow 4 \text{ NO}(g) + 8 \text{ H}_2\text{O}(l)$$

$$4 \text{ Sb}(s) + 6 \text{ H}_2\text{O}(l) \rightarrow \text{Sb}_4\text{O}_6(s) + 12 \text{ H}^+(aq) + 12 \text{ e}^-$$

Add half-reactions. Cancel common reactants and products.

$$4 \text{ NO}_3^-(aq) + \cancel{16} \text{ H}^+(aq) + 4 \text{ Sb}(s) + \cancel{6 \text{ H}_2\text{O}}(l) \rightarrow 4 \text{ NO}(g) + \cancel{8} \text{ H}_2\text{O}(l) + \text{Sb}_4\text{O}_6(s) + \cancel{12 \text{ H}^+(aq)}$$

Balanced equation in acidic solution:

$$4 \text{ NO}_3^-(aq) + 4 \text{ H}^+(aq) + 4 \text{ Sb}(s) \rightarrow 4 \text{ NO}(g) + 2 \text{ H}_2\text{O}(l) + \text{Sb}_4\text{O}_6(s)$$

Oxidizing agent is NO_3^- and reducing agent is Sb.

b) Balance reduction half-reaction:

$\text{BiO}_3^-(aq) \rightarrow \text{Bi}^{3+}(aq) + 3 \text{ H}_2\text{O}(l)$	balance O by adding H_2O
$\text{BiO}_3^-(aq) + 6 \text{ H}^+(aq) \rightarrow \text{Bi}^{3+}(aq) + 3 \text{ H}_2\text{O}(l)$	balance H by adding H^+
$\text{BiO}_3^-(aq) + 6 \text{ H}^+(aq) + 2 \text{ e}^- \rightarrow \text{Bi}^{3+}(aq) + 3 \text{ H}_2\text{O}(l)$	balance charge to give +3 on each side

Balance oxidation half-reaction:

$\text{Mn}^{2+}(aq) + 4 \text{ H}_2\text{O}(l) \rightarrow \text{MnO}_4^-(aq)$	balance O by adding H_2O
$\text{Mn}^{2+}(aq) + 4 \text{ H}_2\text{O}(l) \rightarrow \text{MnO}_4^-(aq) + 8 \text{ H}^+(aq)$	balance H by adding H^+
$\text{Mn}^{2+}(aq) + 4 \text{ H}_2\text{O}(l) \rightarrow \text{MnO}_4^-(aq) + 8 \text{ H}^+(aq) + 5 \text{ e}^-$	balance charge to give +2 on each side

Multiply each half-reaction by an integer to equalize the number of electrons

$5\{\text{BiO}_3^-(aq) + 6 \text{ H}^+(aq) + 2 \text{ e}^- \rightarrow \text{Bi}^{3+}(aq) + 3 \text{ H}_2\text{O}(l)\}$	Multiply by 5 to give 10 e^-
$2\{\text{Mn}^{2+}(aq) + 4 \text{ H}_2\text{O}(l) \rightarrow \text{MnO}_4^-(aq) + 8 \text{ H}^+(aq) + 5 \text{ e}^-\}$	Multiply by 2 to give 10 e^-

This gives:

$$5 \text{ BiO}_3^-(aq) + 30 \text{ H}^+(aq) + 10 \text{ e}^- \rightarrow 5 \text{ Bi}^{3+}(aq) + 15 \text{ H}_2\text{O}(l)$$

$$2 \text{ Mn}^{2+}(aq) + 8 \text{ H}_2\text{O}(l) \rightarrow 2 \text{ MnO}_4^-(aq) + 16 \text{ H}^+(aq) + 10 \text{ e}^-$$

Add half-reactions. Cancel H_2O and H^+ in reactants and products.

$$5 \text{ BiO}_3^-(aq) + \cancel{30} \text{ H}^+(aq) + 2 \text{ Mn}^{2+}(aq) + \cancel{8 \text{ H}_2\text{O}}(l) \rightarrow 5 \text{ Bi}^{3+}(aq) + \cancel{15} \text{ H}_2\text{O}(l) + 2 \text{ MnO}_4^-(aq) + \cancel{16 \text{ H}^+(aq)}$$

Balanced reaction in acidic solution:

$$5 \text{ BiO}_3^-(aq) + 14 \text{ H}^+(aq) + 2 \text{ Mn}^{2+}(aq) \rightarrow 5 \text{ Bi}^{3+}(aq) + 7 \text{ H}_2\text{O}(l) + 2 \text{ MnO}_4^-(aq)$$

BiO_3^- is the oxidizing agent and Mn^{2+} is the reducing agent.

c) Balance the reduction half-reaction:

$\text{Pb(OH)}_3^-(aq) \rightarrow \text{Pb}(s) + 3 \text{ H}_2\text{O}(l)$	balance O by adding H_2O
$\text{Pb(OH)}_3^-(aq) + 3 \text{ H}^+(aq) \rightarrow \text{Pb}(s) + 3 \text{ H}_2\text{O}(l)$	balance H by adding H^+
$\text{Pb(OH)}_3^-(aq) + 3 \text{ H}^+(aq) + 2 \text{ e}^- \rightarrow \text{Pb}(s) + 3 \text{ H}_2\text{O}(l)$	balance charge to give 0 on each side

Balance the oxidation half-reaction

$\text{Fe(OH)}_2(s) + \text{H}_2\text{O}(l) \rightarrow \text{Fe(OH)}_3(s)$	balance O by adding H_2O
$\text{Fe(OH)}_2(s) + \text{H}_2\text{O}(l) \rightarrow \text{Fe(OH)}_3(s) + \text{H}^+(aq)$	balance H by adding H^+
$\text{Fe(OH)}_2(s) + \text{H}_2\text{O}(l) \rightarrow \text{Fe(OH)}_3(s) + \text{H}^+(aq) + \text{e}^-$	balance charge to give 0 on each side

Multiply each half-reaction by an integer to equalize the number of electrons

$1\{Pb(OH)_3^-(aq) + 3\ H^+(aq) + 2\ e^- \rightarrow Pb(s) + 3\ H_2O(l)\}$ Multiply by 1 to give 2 e⁻

$2\{Fe(OH)_2(s) + H_2O(l) \rightarrow Fe(OH)_3(s) + H^+(aq) + e^-\}$ Multiply be 2 to give 2 e⁻

This gives:

$Pb(OH)_3^-(aq) + 3\ H^+(aq) + 2\ e^- \rightarrow Pb(s) + 3\ H_2O(l)$

$2\ Fe(OH)_2(s) + 2\ H_2O(l) \rightarrow 2\ Fe(OH)_3(s) + 2\ H^+(aq) + 2\ e^-$

Add the two half-reactions. Cancel H_2O and H^+.

$Pb(OH)_3^-(aq) + \cancel{3}\ H^+(aq) + 2\ Fe(OH)_2(s) + \cancel{2\ H_2O}(l) \rightarrow Pb(s) + \cancel{3}\ H_2O(l) + 2\ Fe(OH)_3(s) + \cancel{2\ H^+}(aq)$

$Pb(OH)_3^-(aq) + H^+(aq) + 2\ Fe(OH)_2(s) \rightarrow Pb(s) + H_2O(l) + 2\ Fe(OH)_3(s)$

Add one OH⁻ to both sides to neutralize H^+.

$Pb(OH)_3^-(aq) + \cancel{H^+(aq) + OH}^-(aq) + 2\ Fe(OH)_2(s) \rightarrow Pb(s) + H_2O(l) + 2\ Fe(OH)_3(s) + OH^-(aq)$

$Pb(OH)_3^-(aq) + \cancel{H_2O(l)} + 2\ Fe(OH)_2(s) \rightarrow Pb(s) + \cancel{H_2O(l)} + 2\ Fe(OH)_3(s) + OH^-(aq)$

Balanced reaction in basic solution:

$Pb(OH)_3^-(aq) + 2\ Fe(OH)_2(s) \rightarrow Pb(s) + 2\ Fe(OH)_3(s) + OH^-(aq)$

$Pb(OH)_3^-$ is the oxidizing agent and $Fe(OH)_2$ is the reducing agent.

21.18   a) Balance reduction half-reaction:

$MnO_4^-(aq) \rightarrow Mn^{2+}(aq) + 4\ H_2O(l)$ balance O by adding H_2O

$MnO_4^-(aq) + 8\ H^+(aq) \rightarrow Mn^{2+}(aq) + 4\ H_2O(l)$ balance H by adding H^+

$MnO_4^-(aq) + 8\ H^+(aq) + 5\ e^- \rightarrow Mn^{2+}(aq) + 4\ H_2O(l)$ balance charge by adding 5 e⁻

Balance oxidation half-reaction:

$As_4O_6(s) \rightarrow 4\ AsO_4^{3-}(aq)$ balance As

$As_4O_6(s) + 10\ H_2O(l) \rightarrow 4\ AsO_4^{3-}(aq)$ balance O by adding H_2O

$As_4O_6(s) + 10\ H_2O(l) \rightarrow 4\ AsO_4^{3-}(aq) + 20\ H^+(aq)$ balance H by adding H^+

$As_4O_6(s) + 10\ H_2O(l) \rightarrow 4\ AsO_4^{3-}(aq) + 20\ H^+(aq) + 8\ e^-$ balance charge by adding 8 e⁻

Multiply reduction half-reaction by 8 and oxidation half-reaction by 5 to transfer 40 e⁻ in overall reaction. Add the half-reactions and cancel H_2O and H^+.

$8\ MnO_4^-(aq) + 64\ H^+(aq) + 40\ e^- \rightarrow 8\ Mn^{2+}(aq) + 32\ H_2O(l)$

$\underline{5\ As_4O_6(s) + 50\ H_2O(l) \rightarrow 20\ AsO_4^{3-}(aq) + 100\ H^+(aq) + 40\ e^-}$

$5\ As_4O_6(s) + 8\ MnO_4^-(aq) + \cancel{64\ H^+(aq)} + \cancel{50}\ H_2O(l) \rightarrow 20\ AsO_4^{3-}(aq) + 8\ Mn^{2+}(aq) + \cancel{32\ H_2O(l)} + \cancel{100}\ H^+(aq)$

Balanced reaction in acidic solution:

$5\ As_4O_6(s) + 8\ MnO_4^-(aq) + 18\ H_2O(l) \rightarrow 20\ AsO_4^{3-}(aq) + 8\ Mn^{2+}(aq) + 36\ H^+(aq)$

Oxidizing agent is MnO_4^- and reducing agent is As_4O_6.

b) The reaction gives only one reactant, P_4. Since both products contain phosphorus, divide the half-reactions so each include P_4 as the reactant.

Balance reduction half-reaction:

$P_4(s) \rightarrow 4\ PH_3(g)$ balance P

$P_4(s) + 12\ H^+(aq) \rightarrow 4\ PH_3(g)$ balance H by adding H^+

$P_4(s) + 12\ H^+(aq) + 12\ e^- \rightarrow 4\ PH_3(g)$ balance charge by adding 12 e⁻

Balance oxidation half-reaction:

$P_4(s) \rightarrow 4\ HPO_3^{2-}(aq)$ balance P

$P_4(s) + 12\ H_2O(l) \rightarrow 4\ HPO_3^{2-}(aq)$ balance O by adding H_2O

$P_4(s) + 12\ H_2O(l) \rightarrow 4\ HPO_3^{2-}(aq) + 20\ H^+(aq)$ balance H by adding H^+

$P_4(s) + 12\ H_2O(l) \rightarrow 4\ HPO_3^{2-}(aq) + 20\ H^+(aq) + 12\ e^-$ balance charge by adding 12 e⁻

Add two half-reactions and cancel H^+.

$P_4(s) + 12\ H^+(aq) + 12\ e^- \rightarrow 4\ PH_3(g)$

$\underline{P_4(s) + 12\ H_2O(l) \rightarrow 4\ HPO_3^{2-}(aq) + 20\ H^+(aq) + 12\ e^-}$

$2\ P_4(s) + \cancel{12\ H^+(aq)} + 12\ H_2O(l) \rightarrow 4\ HPO_3^{2-}(aq) + 4\ PH_3(g) + \cancel{20}\ H^+(aq)$

Balanced reaction in acidic solution:

$2\ P_4(s) + 12\ H_2O(l) \rightarrow 4\ HPO_3^{2-}(aq) + 4\ PH_3(g) + 8\ H^+(aq)$ or

$P_4(s) + 6\ H_2O(l) \rightarrow 2\ HPO_3^{2-}(aq) + 2\ PH_3(g) + 4\ H^+(aq)$

P_4 is both the oxidizing agent and reducing agent.

c) Balance the reduction half-reaction:

$$MnO_4^-(aq) \rightarrow MnO_2(s) + 2\ H_2O(l) \qquad \text{balance O by adding } H_2O$$
$$MnO_4^-(aq) + 4\ H^+(aq) \rightarrow MnO_2(s) + 2\ H_2O(l) \qquad \text{balance H by adding } H^+$$
$$MnO_4^-(aq) + 4\ H^+(aq) + 3\ e^- \rightarrow MnO_2(s) + 2\ H_2O(l) \qquad \text{balance charge by adding 3 } e^-$$

Balance oxidation half-reaction:

$$CN^-(aq) + H_2O(l) \rightarrow CNO^-(aq) \qquad \text{balance O by adding } H_2O$$
$$CN^-(aq) + H_2O(l) \rightarrow CNO^-(aq) + 2\ H^+(aq) \qquad \text{balance H by adding } H^+$$
$$CN^-(aq) + H_2O(l) \rightarrow CNO^-(aq) + 2\ H^+(aq) + 2\ e^- \qquad \text{balance charge by adding 2 } e^-$$

Multiply the oxidation half-reaction by 3 and reduction half-reaction by 2 to transfer 6 e^- in overall reaction. Add two half-reactions. Cancel the H_2O and H^+.

$$2\ MnO_4^-(aq) + 8\ H^+(aq) + 6\ e^- \rightarrow 2\ MnO_2(s) + 4\ H_2O(l)$$
$$\underline{3\ CN^-(aq) + 3\ H_2O(l) \rightarrow 3\ CNO^-(aq) + 6\ H^+(aq) + 6\ e^-}$$
$$2\ MnO_4^-(aq) + 3\ CN^-(aq) + \cancel{8}\ H^+(aq) + \cancel{3\ H_2O}(l) \rightarrow 2\ MnO_2(s) + 3\ CNO^-(aq) + \cancel{6\ H^+(aq)} + \cancel{4}H_2O(l)$$
$$2\ MnO_4^-(aq) + 3\ CN^-(aq) + 2\ H^+(aq) \rightarrow 2\ MnO_2(s) + 3\ CNO^-(aq) + H_2O(l)$$

Add 2 OH^- to both sides to neutralize H^+ and form H_2O.

$$2\ MnO_4^-(aq) + 3\ CN^-(aq) + 2\ H^+(aq) + 2\ OH^-(aq) \rightarrow 2\ MnO_2(s) + 3\ CNO^-(aq) + H_2O(l) + 2\ OH^-(aq)$$
$$2\ MnO_4^-(aq) + 3\ CN^-(aq) + 2\ \cancel{H_2O(l)} \rightarrow 2\ MnO_2(s) + 3\ CNO^-(aq) + \cancel{H_2O(l)} + 2\ OH^-(aq)$$

Balanced reaction in basic solution:

$$2\ MnO_4^-(aq) + 3\ CN^-(aq) + H_2O(l) \rightarrow 2\ MnO_2(s) + 3\ CNO^-(aq) + 2\ OH^-(aq)$$

Oxidizing agent is MnO_4^- and reducing agent is CN^-.

21.21 a) Balance reduction half-reaction:

$$NO_3^-(aq) \rightarrow NO_2(g) + H_2O(l) \qquad \text{balance O by adding } H_2O$$
$$NO_3^-(aq) + 2\ H^+(aq) \rightarrow NO_2(g) + H_2O(l) \qquad \text{balance H by adding } H^+$$
$$NO_3^-(aq) + 2\ H^+(aq) + e^- \rightarrow NO_2(g) + H_2O(l) \qquad \text{balance charge to give 0 on each side}$$

Balance oxidation half-reaction:

$$Au(s) + 4\ Cl^-(aq) \rightarrow AuCl_4^-(aq) \qquad \text{balance Cl}$$
$$Au(s) + 4\ Cl^-(aq) \rightarrow AuCl_4^-(aq) + 3\ e^- \qquad \text{balance charge to } -4 \text{ on each side}$$

Multiply each half-reaction by an integer to equalize the number of electrons

$$3\{NO_3^-(aq) + 2\ H^+(aq) + e^- \rightarrow NO_2(g) + H_2O(l)\} \qquad \text{Multiply by 3 to give 3 } e^-$$
$$1\{Au(s) + 4\ Cl^-(aq) \rightarrow AuCl_4^-(aq) + 3\ e^-\} \qquad \text{Multiply be 1 to give 3 } e^-$$

This gives:

$$3\ NO_3^-(aq) + 6\ H^+(aq) + 3\ e^- \rightarrow 3\ NO_2(g) + 3\ H_2O(l)$$
$$Au(s) + 4\ Cl^-(aq) \rightarrow AuCl_4^-(aq) + 3\ e^-$$

Add half-reactions.

$$Au(s) + 3\ NO_3^-(aq) + 4\ Cl^-(aq) + 6\ H^+(aq) \rightarrow AuCl_4^-(aq) + 3\ NO_2(g) + 3\ H_2O(l)$$

b) Oxidizing agent is **NO_3^-** and reducing agent is **Au**.

c) The HCl provides chloride ions that combine with the unstable gold ion to form the stable ion, $AuCl_4^-$.

21.22 a) **A** is the anode because by convention the anode is shown on the left.

b) **E** is the cathode because by convention the cathode is shown on the right.

c) **C** is the salt bridge providing electrical connection between the two solutions.

d) **A** is the anode, so oxidation takes place there. Oxidation is the loss of electrons, meaning that electrons are leaving the anode.

e) **E** is assigned a positive charge because it is the cathode.

f) **E** gains mass because the reduction of the metal ion produces the solid metal which plates out on E.

21.25 An active electrode is a reactant or product in the cell reaction, whereas an inactive electrode is neither a reactant nor a product. An inactive electrode is present only to conduct electricity when the half-cell reaction does not include a metal. Platinum and graphite are commonly used as inactive electrodes.

21.26 a) The metal **A** is being oxidized to form the metal cation. To form positive ions, an atom must always lose electrons, so this half-reaction is always an oxidation.

b) The metal ion **B** is gaining electrons to form the metal **B**, so it is displaced.

c) The anode is the electrode at which oxidation takes place, so metal **A** is used as the anode.
d) Acid oxidizes metal **B** and metal **B** oxidizes metal **A**, so acid will oxidize metal **A** and **bubbles will form** when metal **A** is placed in acid. The same answer results if strength of reducing agents is considered. The fact that metal **A** is a better reducing agent than metal **B** indicates that if metal **B** reduces acid, then metal **A** will also reduce acid.

21.27   a) If the zinc electrode is negative, oxidation takes place at the zinc electrode:
            $Zn(s) \rightarrow Zn^{2+}(aq) + 2\ e^-$
Reduction half-reaction: $Sn^{2+}(aq) + 2\ e^- \rightarrow Sn(s)$
Overall reaction: $Zn(s) + Sn^{2+}(aq) \rightarrow Zn^{2+}(aq) + Sn(s)$
b)

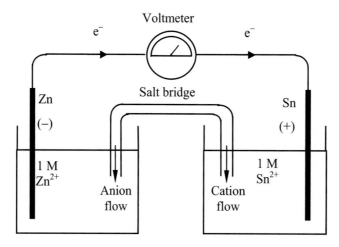

21.29   a) Electrons flow from the anode to the cathode, so **from the iron half-cell to the nickel half-cell**, left to right in the figure. By convention, the anode appears on the left and the cathode on the right.
b) Oxidation occurs at the anode, which is the electrode in the **iron** half-cell.
c) Electrons enter the reduction half-cell, the **nickel** half-cell in this example.
d) Electrons are consumed in the reduction half-reaction. Reduction takes place at the cathode, **nickel** electrode.
e) The anode is assigned a negative charge, so the **iron** electrode is negatively charged.
f) Metal is oxidized in the oxidation half-cell, so the **iron** electrode will decrease in mass.
g) The solution must contain nickel ions, so any nickel salt can be added. **1 M NiSO$_4$** is one choice.
h) KNO$_3$ is commonly used in salt bridges, the ions being **K$^+$ and NO$_3^-$**. Other salts are also acceptable answers.
i) **Neither**, because an inactive electrode could not replace either electrode since both the oxidation and the reduction half-reactions include the metal as either a reactant or a product.
j) Anions will move toward the half-cell in which positive ions are being produced. The oxidation half-cell produces Fe^{2+}, so salt bridge anions move **from right** (nickel half-cell) **to left** (iron half-cell).
k)   Oxidation half-reaction: $Fe(s) \rightarrow Fe^{2+}(aq) + 2\ e^-$
        Reduction half-reaction: $Ni^{2+}(aq) + 2\ e^- \rightarrow Ni(s)$
        Overall cell reaction: $Fe(s) + Ni^{2+}(aq) \rightarrow Fe^{2+}(aq) + Ni(s)$

21.31   a) The cathode is assigned a positive charge, so the iron electrode is the cathode.

Reduction half-reaction: $Fe^{2+}(aq) + 2\ e^- \rightarrow Fe(s)$

Oxidation half-reaction: $Mn(s) \rightarrow Mn^{2+}(aq) + 2\ e^-$

Overall cell reaction: $Fe^{2+}(aq) + Mn(s) \rightarrow Fe(s) + Mn^{2+}(aq)$

b)

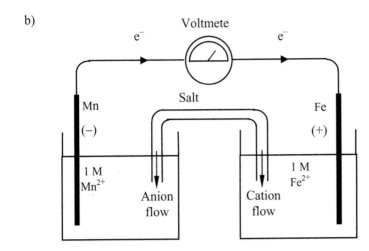

21.33   In cell notation, the oxidation components of the anode compartment are written on the left of the salt bridge and the reduction components of the cathode compartment are written to the right of the salt bridge. A double vertical line separates the anode from the cathode and represents the salt bridge. A single vertical line separates species of different phases.

Anode || Cathode

a) Al is oxidized, so it is the anode and appears first in the cell notation. There is a single vertical line separating the solid metals from their solutions.

$Al(s)|Al^{3+}(aq)||Cr^{3+}(aq)|Cr(s)$

b) Cu^{2+} is reduced, so Cu is the cathode and appears last in the cell notation. The oxidation of SO_2 does not include a metal, so an inactive electrode must be present. Hydrogen ion must be included in the oxidation half-cell.

$Pt|SO_2(g)|SO_4^{2-}(aq), H^+(aq)||Cu^{2+}(aq)|Cu(s)$

21.36   A negative E^o_{cell} indicates that the cell reaction is not spontaneous, $\Delta G^\circ > 0$. The reverse reaction is spontaneous with $E^o_{cell} > 0$.

21.37   Similar to other state functions, the sign of E° changes when a reaction is reversed. Unlike ΔG°, ΔH° and S°, E° is an intensive property, the ratio of energy to charge. When the coefficients in a reaction are multiplied by a factor, the values of ΔG°, ΔH° and S° are multiplied by the same factor. However, E° does not change because both the energy and charge are multiplied by the factor and their ratio remains unchanged.

21.38   a) Divide the balanced equation into reduction and oxidation half-reactions and add electrons. Add water and hydroxide ion to the half-reaction that includes oxygen.

Oxidation: $Se^{2-}(aq) \rightarrow Se(s) + 2\ e^-$

Reduction: $2\ SO_3^{2-}(aq) + 3\ H_2O(l) + 4\ e^- \rightarrow S_2O_3^{2-}(aq) + 6\ OH^-(aq)$

b) $E^o_{cell} = E^o_{cathode} - E^o_{anode}$

$E^o_{anode} = E^o_{cathode} - E^o_{cell} = -0.57\ V - 0.35\ V = \mathbf{-0.92\ V}$

21.40   The greater (more positive) the reduction potential, the greater the strength as an oxidizing agent.
   a)From Appendix D:
   $Fe^{3+}(aq) + e^- \rightarrow Fe^{2+}(aq)$ $E° = 0.77$ V
   $Br_2(l) + 2 e^- \rightarrow 2Br^-(aq)$ $E° = 1.07$ V
   $Cu^{2+}(aq) + e^- \rightarrow Cu(s)$ $E° = 0.34$ V
   When placed in order of decreasing strength as oxidizing agents: $\mathbf{Br_2 > Fe^{3+} > Cu^{2+}}$.
   b)   From Appendix D:
   $Ca^{2+}(aq) + 2e– \rightarrow Ca(s)$ $E° = –2.87$ V
   $Cr_2O_7^{2-}(aq) + 14H^+(aq)\ 6e^- \rightarrow 2Cr^{3+}(aq) + 7H_2O(l)$ $E° = 1.33$ V
   $Ag^+(aq) + e^- \rightarrow Ag(s)$ $E° = 0.80$ V
   When placed in order of increasing strength as oxidizing agents: $\mathbf{Ca^{2+} < Ag^+ < Cr_2O_7^{2-}}$.

21.42 $E^o_{cell} = E^o_{cathode} - E^o_{anode}$ E^o values are found in Appendix D. Spontaneous reactions have $E^o_{cell} > 0$.
   a)   Oxidation: $Co(s) \rightarrow Co^{2+}(aq) + 2 e^-$ $E° = –0.28$ V
          Reduction: $2 H^+(aq) + 2 e^- \rightarrow H_2(g)$ $E° = 0.00$ V
          Overall reaction: $Co(s) + 2 H^+(aq) \rightarrow Co^{2+}(aq) + H_2(g)$

                                                      $E^o_{cell} = 0.00$ V $-$ $(–0.28$ V$) = \mathbf{0.28}$ V

          Reaction is **spontaneous** under standard state conditions because E^o_{cell} is positive.

   b)   Oxidation: $2\{Mn^{2+}(aq) + 4 H_2O(l) \rightarrow MnO_4^-(aq) + 8 H^+(aq) + 5 e^-\}$ $E° = +1.51$ V
          Reduction: $5\{Br_2(l) + 2 e^- \rightarrow 2 Br^-(aq)\}$ $E° = +1.07$ V
          Overall: $2 Mn^{2+}(aq) + 5 Br_2(l) + 8 H_2O(l) \rightarrow 2 MnO_4^-(aq) + 10 Br^-(aq) + 16 H^+(aq)$
                                                      $E^o_{cell} = 1.07$ V $- 1.51$ V $= \mathbf{–0.44}$ V

          Reaction is **not spontaneous** under standard state conditions with $E^o_{cell} < 0$.

   c)   Oxidation: $Hg_2^{2+}(aq) \rightarrow 2 Hg^{2+}(aq) + 2 e^-$ $E° = +0.92$ V
          Reduction: $Hg_2^{2+}(aq) + 2 e^- \rightarrow 2 Hg(l)$ $E° = +0.85$ V
          Overall: $2 Hg_2^{2+}(aq) \rightarrow 2 Hg^{2+}(aq) + 2 Hg(l)$
                               or $Hg_2^{2+}(aq) \rightarrow Hg^{2+}(aq) + Hg(l)$ $E^o_{cell} = 0.85$ V$- 0.92$ V $= \mathbf{–0.07}$ V

          Negative E^o_{cell} indicates reaction is **not spontaneous** under standard state conditions.

21.44 $E^o_{cell} = E^o_{cathode} - E^o_{anode}$ E^o values are found in Appendix D. Spontaneous reactions have $E^o_{cell} > 0$.
   a)   Oxidation: $2\{Ag(s) \rightarrow Ag^+(aq) + e^-\}$ $E° = 0.80$ V
          Reduction: $Cu^{2+}(aq) + 2 e^- \rightarrow Cu(s)$ $E° = +0.34$ V
          Overall: $2 Ag(s) + Cu^{2+}(aq) \rightarrow 2 Ag^+(aq) + Cu(s)$

                                                      $E^o_{cell} = +0.34$ V $- 0.80$ V $= \mathbf{–0.46}$ V

          The reaction is **not spontaneous**.
   b)   Oxidation: $3\{Cd(s) \rightarrow Cd^{2+}(aq) + 2 e^-\}$ $E° = –0.40$ V
          Reduction: $Cr_2O_7^{2-}(aq) + 14 H^+(aq) + 6 e^- \rightarrow 2 Cr^{3+}(aq) + 7 H_2O(l)$ $E° = +1.33$ V
          Overall: $Cr_2O_7^{2-}(aq) + 3 Cd(s) + 14 H^+(aq) \rightarrow 2 Cr^{3+}(aq) + 3 Cd^{2+}(aq) + 7 H_2O(l)$
                                                      $E^o_{cell} = +1.33$ V$- (–0.40$ V$) = \mathbf{+1.73}$

          The reaction is **spontaneous**.
   c)   Oxidation: $Pb(s) \rightarrow Pb^{2+}(aq) + 2 e^-$ $E° = –0.13$ V
          Reduction: $Ni^{2+}(aq) + 2 e^- \rightarrow Ni(s)$ $E° = –0.25$ V
          Overall: $Pb(s) + Ni^{2+}(aq) \rightarrow Pb^{2+}(aq) + Ni(s)$
                                                      $E^o_{cell} = –0.25$ V$- (–0.13$ V$) = \mathbf{–0.12}$ V

          The reaction is **not spontaneous**.

21.46 Spontaneous reactions have $E^{\circ}_{cell} > 0$. All three reactions are written as reductions. When two half-reactions are paired, one half-reaction must be reversed and written as an oxidation. Reverse the half-reaction that will result in a positive value of E°_{cell}.

Adding (1) and (2) to give a spontaneous reaction involves converting (1) to oxidation:
Oxidation: $2\{Al(s) \rightarrow Al^{3+}(aq) + 3\,e^-\}$ $E^{\circ} = -1.66$ V
Reduction: $3\{N_2O_4(g) + 2\,e^- \rightarrow 2\,NO_2^-(aq)\}$ $E^{\circ} = 0.867$ V
$3\,N_2O_4(g) + 2\,Al(s) \rightarrow 6\,NO_2^-(aq) + 2\,Al^{3+}(aq)$

$$E^{\circ}_{cell} = 0.867 \text{ V} - (-1.66 \text{ V}) = \textbf{2.53 V}$$

Oxidizing agents: $N_2O_4 > Al^{3+}$; reducing agents: $Al > NO_2^-$
Adding (1) and (3) to give a spontaneous reaction involves converting (1) to oxidation:
Oxidation: $2\{Al(s) \rightarrow Al^{3+}(aq) + 3\,e^-\}$ $E^{\circ} = -1.66$ V
Reduction: $3\{SO_4^{2-}(aq) + H_2O(l) + 2\,e^- \rightarrow SO_3^{2-}(aq) + 2\,OH^-(aq)\}$ $E^{\circ} = 0.93$ V
$2\,Al(s) + 3\,SO_4^{2-}(aq) + 3\,H_2O(l) \rightarrow 2\,Al^{3+}(aq) + 3\,SO_3^{2-}(aq) + 6\,OH^-(aq)$

$$E^{\circ}_{cell} = 0.93 \text{ V} - (-1.66 \text{ V}) = \textbf{2.59 V}$$

Oxidizing agents: $SO_4^{2-} > Al^{3+}$; reducing agents: $Al > SO_3^{2-}$
Adding (2) and (3) to give a spontaneous reaction involves converting 2 to oxidation:
Oxidation: $2\,NO_2^-(aq) \rightarrow N_2O_4(g) + 2\,e^-$ $E^{\circ} = 0.867$ V
Reduction: $SO_4^{2-}(aq) + H_2O(l) + 2\,e^- \rightarrow SO_3^{2-}(aq) + 2\,OH^-(aq)$ $E^{\circ} = 0.93$ V
$SO_4^{2-}(aq) + 2\,NO_2^-(aq) + H_2O(l) \rightarrow SO_3^{2-}(aq) + N_2O_4(g) + 2\,OH^-(aq)$

$$E^{\circ}_{cell} = 0.93 \text{ V} - 0.867 \text{ V} = \textbf{0.06 V}$$

Oxidizing agents: $SO_4^{2-} > N_2O_4$; reducing agents: $NO_2^- > SO_3^{2-}$
Rank oxidizing agents (substance being reduced) in order of increasing strength:
     $Al^{3+} < N_2O_4 < SO_4^{2-}$
Rank reducing agents (substance being oxidized) in order of increasing strength:
     $SO_3^{2-} < NO_2^- < Al$

21.48 Spontaneous reactions have $E^{\circ}_{cell} > 0$. All three reactions are written as reductions. When two half-reactions are paired, one half-reaction must be reversed and written as an oxidation. Reverse the half-reaction that will result in a positive value of E°_{cell}.

Adding (1) and (2) to give a spontaneous reaction involves converting (2) to oxidation:
Oxidation: $Pt(s) \rightarrow Pt^{2+}(aq) + 2\,e^-$ $E^{\circ} = 1.20$ V
Reduction: $2\,HClO(aq) + 2\,H^+(aq) + 2\,e^- \rightarrow Cl_2(g) + 2\,H_2O(l)$ $E^{\circ} = 1.63$ V
       $2\,HClO(aq) + Pt(s) + 2\,H^+(aq) \rightarrow Cl_2(g) + Pt^{2+}(aq) + 2\,H_2O(l)$

$$E^{\circ}_{cell} = 1.63 \text{ V} - 1.20 \text{ V} = \textbf{0.43 V}$$

Oxidizing agents: $HClO > Pt^{2+}$; reducing agents: $Pt > Cl_2$
Adding (1) and (3) to give a spontaneous reaction involves converting (3) to oxidation:
Oxidation: $Pb(s) + SO_4^{2-}(aq) \rightarrow PbSO_4(s) + 2\,e^-$ $E^{\circ} = -0.31$ V
Reduction: $2\,HClO(aq) + 2\,H^+(aq) + 2\,e^- \rightarrow Cl_2(g) + 2\,H_2O(l)$ $E^{\circ} = 1.63$ V
       $2\,HClO(aq) + Pb(s) + SO_4^{2-}(aq) + 2\,H^+(aq) \rightarrow Cl_2(g) + PbSO_4(s) + 2\,H_2O(l)$

$$E^{\circ}_{cell} = 1.63 \text{ V} - (-0.31 \text{ V}) = \textbf{1.94 V}$$

Oxidizing agents: $HClO > PbSO_4$; reducing agents: $Pb > Cl_2$
Adding (2) and (3) to give a spontaneous reaction involves converting (3) to oxidation:
Oxidation: $Pb(s) + SO_4^{2-}(aq) \rightarrow PbSO_4(s) + 2\,e^-$ $E^{\circ} = -0.31$ V
Reduction: $Pt^{2+}(aq) + 2\,e^- \rightarrow Pt(s)$ $E^{\circ} = 1.20$ V
       $Pt^{2+}(aq) + Pb(s) + SO_4^{2-}(aq) \rightarrow Pt(s) + PbSO_4(s)$

$$E^{\circ}_{cell} = 1.20 \text{ V} - (-0.31 \text{ V}) = \textbf{1.51 V}$$

Oxidizing agents: $Pt^{2+} > PbSO_4$; reducing agents: $Pb > Pt$
Order of increasing strength as oxidizing agent: **$PbSO_4 < Pt^{2+} < HClO$**
Order of increasing strength as reducing agent: **$Cl_2 < Pt < (Pb + SO_4^{2-})$**

21.50 Metal A + Metal B salt → solid colored product on metal A

    Conclusion: Product is solid metal B. B is undergoing reduction and plating out on A. A is better reducing agent than B.

Metal B + acid → gas bubbles

    Conclusion: Product is H_2 gas produced as result of reduction of H^+. B is better reducing agent than acid.

Metal A + Metal C salt → no reaction

    Conclusion: C is not undergoing reduction. C must be a better reducing agent than A.

Since C is a better reducing agent than A, which is a better reducing agent than B and B reduces acid, then **C would also reduce acid to form H_2 bubbles**.

The order of strength of reducing agents is: **C > A > B**.

21.53 At the negative (anode) electrode, oxidation occurs so the overall cell reaction is

$A(s) + B^+(aq) \rightarrow A^+(aq) + B(s)$ with $Q = [A^+] / [B^+]$.

a) The reaction proceeds to the right because with $E_{cell} > 0$ (voltaic cell), the spontaneous reaction occurs. As the cell operates, **$[A^+]$ increases and $[B^+]$ decreases**.

b) **E_{cell} decreases** because the cell reaction takes place to approach equilibrium, $E_{cell} = 0$.

c) E_{cell} and E^o_{cell} are related by the Nernst equation: $E_{cell} = E^o_{cell} - (RT / nF)\ln([A^+] / [B^+])$.

$E_{cell} = E^o_{cell}$ when $(RT / nF)\ln([A^+] / [B^+]) = 0$. This occurs when $\ln([A^+] / [B^+]) = 0$. Recall that $e^0 = 1$, so **$[A^+]$ must equal $[B^+]$** for E_{cell} to equal E^o_{cell}.

d) **Yes**, it is possible for E_{cell} to be less than E^o_{cell} when **$[A^+] > [B^+]$**.

21.55 In a concentration cell, the overall reaction takes place to decrease the concentration of the more concentrated electrolyte. The more concentrated electrolyte is reduced, so it is in the **cathode** compartment.

21.56 The equilibrium constant can be found by using $\ln K = \dfrac{nFE^\circ}{RT}$ or $\log K = \dfrac{nE^\circ}{0.0592}$. Use E° values from Appendix D to calculate E^o_{cell} and then calculate K.

a) Oxidation: $Ni(s) \rightarrow Ni^{2+}(aq) + 2\,e^-$ $E^\circ = -0.25$ V

    Reduction: $2\{Ag^+(aq) + 1\,e^- \rightarrow Ag(s)\}$ $E^\circ = 0.80$ V

        $Ni(s) + 2\,Ag^+(aq) \rightarrow Ni^{2+}(aq) + 2\,Ag(s)$

    $E^o_{cell} = E^o_{cathode} - E^o_{anode} = 0.80$ V $- (-0.25$ V$) = 1.05$ V; 2 electrons are transferred.

$$\log K = \frac{nE^\circ}{0.0592} = \frac{2(1.05\text{ V})}{0.0592\text{ V}} = 35.47297 \text{ (unrounded)}$$

    $K = 10^{35.47297}$

    $K = 2.97146 \times 10^{35} = \mathbf{3 \times 10^{35}}$

b) Oxidation: $3\{Fe(s) \rightarrow Fe^{2+}(aq) + 2\,e^-\}$ $E^\circ = -0.44$ V

    Reduction: $2\{Cr^{3+}(aq) + 3\,e^- \rightarrow Cr(s)\}$ $E^\circ = -0.74$ V

    $E^o_{cell} = E^o_{cathode} - E^o_{anode} = -0.74$ V $- (-0.44$ V$) = -0.30$ V; 6 electrons are transferred.

$$\log K = \frac{nE^\circ}{0.0592} = \frac{6(-0.30\text{ V})}{0.0592\text{ V}} = -30.4054 \text{ (unrounded)}$$

    $K = 10^{-30.405}$

    $K = 3.9318 \times 10^{-31} = \mathbf{4 \times 10^{-31}}$

21.58 The equilibrium constant can be found by using $\ln K = \dfrac{nFE^\circ}{RT}$ or $\log K = \dfrac{nE^\circ}{0.0592}$. Use E° values from Appendix D to calculate E^o_{cell} and then calculate K.

a) Oxidation: $2\{Ag(s) \rightarrow Ag^+(aq) + 1\,e^-\}$ $E^\circ = 0.80$ V

    Reduction: $Mn^{2+}(aq) + 2\,e^- \rightarrow Mn(s)$ $E^\circ = -1.18$ V

        $Ni(s) + 2\,Ag^+(aq) \rightarrow Ni^{2+}(aq) + 2\,Ag(s)$

    $E^o_{cell} = E^o_{cathode} - E^o_{anode} = -1.18$ V $- (0.80$ V$) = -1.98$V; 2 electrons are transferred.

$$\log K = \frac{nE^\circ}{0.0592} = \frac{2(-1.98 \text{ V})}{0.0592 \text{ V}} = -66.89189 \text{ (unrounded)}$$

$$K = 10^{-66.89189}$$

$$K = 1.2826554 \times 10^{-67} = \mathbf{1 \times 10^{-67}}$$

b) Oxidation: $2 \text{ Br}^-(aq) \rightarrow \text{Br}_2(l) + 2 \text{ e}^-$ $E^\circ = 1.07 \text{ V}$

     Reduction: $\text{Cl}_2(g) + 2 \text{ e}^- \rightarrow 2 \text{ Cl}^-(aq)$ $E^\circ = 1.36 \text{ V}$

         $2 \text{ Br}^-(aq) + \text{Cl}_2(g) \rightarrow \text{Br}_2(l) + 2 \text{ Cl}^-(aq)$

$E^\circ_{cell} = 1.36 \text{ V} - 1.07 \text{ V} = 0.29 \text{ V}$;   2 electrons transferred.

$$\log K = \frac{nE^\circ}{0.0592} = \frac{2(0.29 \text{ V})}{0.0592 \text{ V}} = 9.797297 \text{ (unrounded)}$$

$$K = 10^{9.797297}$$

$$K = 6.2704253 \times 10^9 = \mathbf{6 \times 10^9}$$

21.60   Substitute J / C for V.

     a) $\Delta G^\circ = -nFE^\circ = -(2 \text{ mol e}^-)(96485 \text{ C / mol e}^-)(1.05 \text{ J / C}) = -2.026185 \times 10^5 = \mathbf{-2.03 \times 10^5 \text{ J}}$

     b) $\Delta G^\circ = -nFE^\circ = -(6 \text{ mol e}^-)(96485 \text{ C / mol e}^-)(-0.30 \text{ J / C}) = 1.73673 \times 10^5 = \mathbf{1.73 \times 10^5 \text{ J}}$

21.62   Substitute J / C for V.

     a) $\Delta G^\circ = -nFE^\circ = -(2 \text{ mol e}^-)(96485 \text{ C / mol e}^-)(-1.98 \text{ J / C}) = 3.820806 \times 10^5 = \mathbf{3.82 \times 10^5 \text{ J}}$

     b) $\Delta G^\circ = -nFE^\circ = -(2 \text{ mol e}^-)(96485 \text{ C / mol e}^-)(0.29 \text{ J / C}) = -5.59613 \times 10^4 = \mathbf{-5.6 \times 10^4 \text{ J}}$

21.64   Find ΔG° from the fact that $\Delta G^\circ = -RT \ln K$. Then use ΔG° value to find E°_{cell} from $\Delta G^\circ = -nFE^\circ$.

     $T = (273 + 25)\text{K} = 298 \text{ K}$

     $\Delta G^\circ = -RT \ln K = -(8.314 \text{ J/mol•K}) (298 \text{ K}) \ln (5.0 \times 10^4) = -2.68067797 \times 10^4 = \mathbf{-2.7 \times 10^4 \text{ J}}$

$$E^\circ = -\frac{\Delta G}{nF} = -\frac{-2.68067797 \times 10^4 \text{ J}}{(1 \text{ mol e}^-)(96485 \text{ C/mol e}^-)}\left(\frac{1 \text{ V}}{1 \text{ J/C}}\right) = 0.27783365 = \mathbf{0.28 \text{ V}}$$

21.66   Use $\log K = \frac{nE^\circ_{cell}}{0.0592}$ and $\Delta G^\circ = -nFE^\circ$. $T = (273 + 25)\text{K} = 298 \text{ K}$

$$E^\circ_{cell} = \frac{0.0592}{n} \log K = \frac{0.0592}{2} \log 65 = 0.0536622 = \mathbf{0.054 \text{ V}}$$

     $\Delta G^\circ = -RT \ln K = -(8.314 \text{ J/mol•K}) (298 \text{ K}) \ln (65) = -1.034234 \times 10^4 = \mathbf{-1.0 \times 10^4 \text{ J}}$

21.68   Since this is a voltaic cell, a spontaneous reaction is occurring. For a spontaneous reaction between H_2/H^+ and Cu/Cu^{2+}, Cu^{2+} must be reduced and H_2 must be oxidized:

     Oxidation: $\text{H}_2(g) \rightarrow \text{Cu}(s) + 2 \text{ H}^+(aq) + 2 \text{ e}^-$ $E^\circ = 0.00 \text{ V}$

     Reduction: $\text{Cu}^{2+}(aq) + 2 \text{ e}^- \rightarrow \text{Cu}(s)$ $E^\circ = 0.34 \text{ V}$

        $\text{Cu}^{2+}(aq) + \text{H}_2(g) \rightarrow \text{Cu}(s) + 2 \text{ H}^+(aq)$

   $E^\circ_{cell} = E^\circ_{cathode} - E^\circ_{anode}$

          $= 0.34 \text{ V} - 0.00 \text{ V}$

          $= 0.34 \text{ V}$

$$E_{cell} = E^\circ_{cell} - \frac{0.0592}{n} \log Q$$

$$E_{cell} = E^\circ_{cell} - \frac{0.0592}{n} \log \frac{\left[\text{H}^+\right]^2}{\left[\text{Cu}^{2+}\right]\left[\text{H}_2\right]}$$

        For a standard hydrogen electrode $[\text{H}^+] = 1.0 \text{ } M$ and $[\text{H}_2] = 1.0 \text{ atm}$

$$0.22 \text{ V} = 0.34 \text{ V} - \frac{0.0592}{2} \log \frac{1.0}{\left[Cu^{2+}\right]1.0}$$

$$0.22 \text{ V} - 0.34 \text{ V} = -\frac{0.0592}{2} \log \frac{1.0}{\left[Cu^{2+}\right]1.0}$$

$$-0.12 \text{ V} = -\frac{0.0592}{2} \log \frac{1.0}{\left[Cu^{2+}\right]1.0}$$

$$4.054054 = \log \frac{1.0}{\left[Cu^{2+}\right]1.0} \qquad \text{Raise each side to } 10^x.$$

$$1.132541 \times 10^4 = \frac{1}{[Cu^{2+}]}$$

$[Cu^{2+}] = 8.8296999 \times 10^{-5} = \textbf{8.8} \times \textbf{10}^{-5} \textbf{\textit{M}}$

21.70   The spontaneous reaction (voltaic cell) involves the oxidation of Co and the reduction of Ni^{2+}.

Oxidation: $Co(s) \rightarrow Co^{2+}(aq) + 2 e^-$ $E^\circ = -0.28 \text{ V}$

Reduction: $Ni^{2+}(aq) + 2 e^- \rightarrow Ni(s)$ $E^\circ = -0.25 \text{ V}$

$$Ni^{2+}(aq) + Co(s) \rightarrow Ni(s) + Co^{2+}(aq)$$

$$E^\circ_{cell} = E^\circ_{cathode} - E^\circ_{anode} = -0.25 \text{ V} - (-0.28 \text{ V}) = 0.03 \text{ V}$$

a) Use the Nernst equation: $E_{cell} = E^\circ_{cell} - \dfrac{0.0592}{n} \log \dfrac{\left[Co^{2+}\right]}{\left[Ni^{2+}\right]}$ $n = 2 \text{ e}^-$

$$E_{cell} = 0.03 \text{ V} - \frac{0.0592}{2} \log \frac{[0.20]}{[0.80]}$$

$$= 0.047820975 \text{ V} = \textbf{0.05 V}$$

b) From part (a), notice that an increase in $[Co^{2+}]$ leads to a decrease in cell potential. Therefore, the concentration of cobalt ion must increase further to bring the potential down to 0.03 V. Thus, the new concentrations will be

$[Co^{2+}] = 0.20 \text{ } M + x$ and $[Ni^{2+}] = 0.80 \text{ } M - x$ (There is a 1:1 mole ratio)

$$0.03 \text{ V} = 0.03 \text{ V} - \frac{0.0592}{2} \log \frac{[0.20 + x]}{[0.80 - x]}$$

$$0 = -\frac{0.0592}{2} \log \frac{[0.20 + x]}{[0.80 - x]}$$

$$0 = \log \frac{[0.20 + x]}{[0.80 - x]} \qquad \text{Raise each side to } 10^x.$$

$$1 = \frac{[0.20 + x]}{[0.80 - x]}$$

$0.20 + x = 0.80 - x$

$x = 0.30 \text{ M}$

$[Ni^{2+}] = 0.80 - 0.30 = \textbf{0.50 \textit{M}}$

c) At equilibrium $E_{cell} = 0.00$, to decrease the cell potential to 0.00, $[Co^{2+}]$ increases and $[Ni^{2+}]$ decreases.

$$0.00 \text{ V} = 0.03 \text{ V} - \frac{0.0592}{2} \log \frac{[0.20 + x]}{[0.80 - x]}$$

$$-0.03 \text{ V} = -0.0296 \log \frac{[0.20 + x]}{[0.80 - x]}$$

$$1.0135135 = \log \frac{[0.20 + x]}{[0.80 - x]} \qquad \text{Raise each side to } 10^x.$$

$$10.316052 = \frac{[0.20 + x]}{[0.80 - x]}$$

x = 0.7116332 (unrounded)
$[Co^{2+}]$ = 0.20 + 0.7116332 = 0.9116332 = **0.91 M**
$[Ni^{2+}]$ = 0.80 – 0.7116332 = 0.08837 = **0.09 M**

21.72   The overall cell reaction proceeds to increase the 0.10 M H^+ concentration and decrease the 2.0 M H^+ concentration. Therefore, half-cell **A is the anode** because it has the lower concentration.
Oxidation: $H_2(g{:}0.95 \, atm) \rightarrow 2H^+(aq{:} \, 0.10 \, M) + 2 \, e^-$ $E^° = 0.00$ V
Reduction: $2H^+(aq{:} \, 2.0 \, M) + 2 \, e^- \rightarrow H_2(g{:} \, 0.60 \, atm)$ $E^° = 0.00$ V
    $2H^+(aq{:} \, 2.0 \, M) + H_2(g{:}0.95 \, atm) \rightarrow 2H^+(aq{:} \, 0.10 \, M) + H_2(g{:} \, 0.60 \, atm)$
      $E^o_{cell} = 0.00$ V   n = 2 e^-

Q for the cell equals $\dfrac{\left[H^+\right]^2_{anode} \; P_{H(cathode)}}{\left[H^+\right]^2_{cathode} \; P_{H(anode)}} = \dfrac{(0.10)^2 (0.60)}{(2.0)^2 (0.95)} = 0.00157895$ (unrounded)

$E_{cell} = 0.00 \text{ V} - \dfrac{0.0592}{2} \log (0.00157895) = 0.0829283 = \mathbf{0.083 \text{ V}}$

21.74   Electrons flow from the anode, where oxidation occurs, to the cathode, where reduction occurs. The electrons always flow from the anode to the cathode, no matter what type of cell.

21.76   A D-sized battery is much larger than an AAA-sized battery, so the D-sized battery contains a greater amount of the cell components. The potential, however, is an intensive property and does not depend on the amount of the cell components. (Note that amount is different from concentration.) The total amount of charge a battery can produce does depend on the amount of cell components, so the D-sized battery produces more charge than the AAA-sized battery.

21.78   The Teflon spacers keep the two metals separated so the copper cannot conduct electrons that would promote the corrosion of the iron skeleton. Oxidation of the iron by oxygen causes rust to form and the metal to corrode.

21.81   Sacrificial anodes are metals with $E^°$ less than that for iron, –0.44 V, so they are more easily oxidized than iron.
a) $E^°$(aluminum) = –1.66. Yes, except aluminum resists corrosion because once a coating of its oxide covers it, no more aluminum corrodes. Therefore, it would not be a good choice.
b) $E^°$(magnesium) = –2.37 V. Yes, magnesium is appropriate to act as a sacrificial anode.
c) $E^°$(sodium) = –2.71 V. Yes, except sodium reacts with water, so it would not be a good choice.
d) $E^°$(lead) = –0.13 V. No, lead is not appropriate to act as a sacrificial anode because its $E^°$ value is too high.
e) $E^°$(nickel) = –0.25 V. No, nickel is inappropriate as a sacrificial anode because its $E^°$ value is too high.
f) $E^°$(zinc) = –0.76 V. Yes, zinc is appropriate to act as a sacrificial anode.
g) $E^°$(chromium) = –0.74 V. Yes, chromium is appropriate to act as a sacrificial anode.

21.83 $3 \, Cd^{2+}(aq) + 2 \, Cr(s) \rightarrow 3 \, Cd(s) + 2 \, Cr^{3+}(aq)$
      $E^o_{cell} = -0.40 \text{ V} - (-0.74 \text{ V}) = 0.34$ V
To reverse the reaction requires 0.34 V with the cell in its standard state. A 1.5 V supplies more than enough potential, so the cadmium metal oxidizes to Cd^{2+} and chromium plates out.

21.85   The oxidation number of nitrogen in the nitrate ion, NO_3^-, is +5 and cannot be oxidized further since nitrogen has only five electrons in its outer level. In the nitrite ion, NO_2^-, on the other hand, the oxidation number of nitrogen is +3, so it can be oxidized to the +5 state.

21.87 a) At the anode, bromide ions are oxidized to form bromine (**Br$_2$**). $2 \, \text{Br}^- \, (l) \; \rightarrow \; \text{Br}_2(l) \, + \, 2 \, \text{e}^-$
       b) At the cathode, sodium ions are reduced to form sodium metal (**Na**). $\text{Na}^+(l) \, + \, \text{e}^- \rightarrow \text{Na}(s)$

21.89 Either iodide ions or fluoride ions can be oxidized at the anode. The ion that more easily loses an electron will form. Since I is less electronegative than F, I^- will more easily lose its electron and be oxidized at the anode. The product at the **anode is I$_2$** gas. The iodine is a gas because the temperature is high to melt the salts.
Either potassium or magnesium ions can be reduced at the cathode. Magnesium has greater ionization energy than potassium because magnesium is located up and to the right of potassium on the periodic table. The greater ionization energy means that magnesium ions will more readily add an electron (be reduced) than potassium ions. The product at the cathode is **magnesium** (liquid).

21.91 **Bromine** gas forms at the anode because the electronegativity of bromine is less than that of chlorine. **Calcium** metal forms at the cathode because its ionization energy is greater than that of sodium.

21.93 Possible reductions:

$\text{Cu}^{2+}(aq) \, + \, 2 \, \text{e}^- \rightarrow \; \text{Cu}(s)$	$E° = +0.34 \text{ V}$
$\text{Ba}^{2+}(aq) \, + \, 2 \, \text{e}^- \rightarrow \; \text{Ba}(s)$	$E° = -2.90 \text{ V}$
$\text{Al}^{3+}(aq) \, + \, 3 \, \text{e}^- \rightarrow \; \text{Al}(s)$	$E° = -1.66 \text{ V}$
$2 \, \text{H}_2\text{O}(l) + 2 \, \text{e}^- \rightarrow \text{H}_2(g) + 2 \, \text{OH}^-(aq)$	$E = -1 \text{ V with overvoltage}$

Copper can be prepared by electrolysis is its aqueous salt since its reduction half-cell potential is more positive than the potential for the reduction of water. The reduction of copper is more spontaneous than the reduction of water. Since the reduction potentials of Ba^{2+} and Al^{3+} are more negative and therefore less spontaneous than the reduction of water, these ions cannot be reduced in the presence of water since the water is reduced instead.
Possible oxidations:

$2\text{Br}^-(aq) \rightarrow \; \text{Br}_2(l) + 2 \, \text{e}^-$	$E° = +1.07 \text{ V}$
$2 \, \text{H}_2\text{O}(l) \rightarrow \text{O}_2(g) + 4 \, \text{H}^+(aq) + 4 \, \text{e}^-$	$E = 1.4 \text{ V with overvoltage}$

Bromine can be prepared by electrolysis of its aqueous salt because its reduction half-cell potential is more negative than the potential for the oxidation of water with overvoltage. The more negative reduction potential for Br^- indicates that its oxidation is more spontaneous than the oxidation of water.

21.95 Possible reductions:

$\text{Li}^+(aq) \, + \, 1 \, \text{e}^- \rightarrow \; \text{Li}(s)$	$E° = -3.05 \text{ V}$
$\text{Zn}^{2+}(aq) \, + \, 2 \, \text{e}^- \rightarrow \; \text{Zn}(s)$	$E° = -0.76 \text{ V}$
$\text{Ag}^+(aq) \, + \, \text{e}^- \rightarrow \; \text{Ag}(s)$	$E° = 0.80 \text{ V}$
$2 \, \text{H}_2\text{O}(l) + 2 \, \text{e}^- \rightarrow \text{H}_2(g) + 2 \, \text{OH}^-(aq)$	$E = -1 \text{ V with overvoltage}$

Zinc and silver can be prepared by electrolysis of their aqueous salt solutions since their reduction half-cell potentials are more positive than the potential for the reduction of water. The reduction of zinc and silver is more spontaneous than the reduction of water. Since the reduction potential of Li^+ is more negative and therefore less spontaneous than the reduction of water, this ionsa cannot be reduced in the presence of water since the water is reduced instead.
Possible oxidations:

$2\text{I}^-(aq) \rightarrow \; \text{I}_2(l) + 2 \, \text{e}^-$	$E° = +0.53 \text{ V}$
$2 \, \text{H}_2\text{O}(l) \rightarrow \text{O}_2(g) + 4 \, \text{H}^+(aq) + 4 \, \text{e}^-$	$E = 1.4 \text{ V with overvoltage}$

Iodine can be prepared by electrolysis of its aqueous salt because its reduction half-cell potential is more negative than the potential for the oxidation of water with overvoltage. The more negative reduction potential for I^- indicates that its oxidation is more spontaneous than the oxidation of water.

21.97 a) Possible oxidations:

$2 \, \text{H}_2\text{O}(l) \rightarrow \text{O}_2(g) + 4 \, \text{H}^+(aq) + 4 \, \text{e}^-$	$E = 1.4 \text{ V with overvoltage}$
$2 \, \text{F}^- \rightarrow \text{F}_2(g) + 2 \, \text{e}^-$	$E° = 2.87 \text{ V}$

Since the reduction potential of water is more negative than the reduction potential for F^-, the oxidation of water is more spontaneous than that of F^-. The oxidation of water produces oxygen gas (**O$_2$**), and hydronium ions (**H$_3$O$^+$**) at the anode.

Possible reductions:

$$2 H_2O(l) + 2 e^- \rightarrow H_2(g) + 2 OH^-(aq) \qquad E = -1 \text{ V with overvoltage}$$
$$Li^+(aq) + e^- \rightarrow Li(s) \qquad E° = -3.05 \text{ V}$$

Since the reduction potential of water is more positive than that of Li^+, the reduction of water is more spontaneous than the reduction of Li^+. The reduction of water produces $\mathbf{H_2}$ gas and $\mathbf{OH^-}$ at the cathode.

b) Possible oxidations:

$$2 H_2O(l) \rightarrow O_2(g) + 4 H^+(aq) + 4 e^- \qquad E = 1.4 \text{ V with overvoltage}$$

The oxidation of water produces oxygen gas ($\mathbf{O_2}$), and hydronium ions ($\mathbf{H_3O^+}$) at the anode. The SO_4^{2-} ion cannot oxidize as S is already in its highest oxidation state in SO_4^{2-}.

Possible reductions:

$$2 H_2O(l) + 2 e^- \rightarrow H_2(g) + 2 OH^-(aq) \qquad E = -1 \text{ V with overvoltage}$$
$$Sn^{2+}(aq) + 2 e^- \rightarrow Sn(s) \qquad E° = -0.14 \text{ V}$$
$$SO_4^{2-}(aq) + 4 H^+(aq) + 2 e^- \rightarrow SO_2(g) + 2 H_2O(l) \qquad E = -0.63 \text{ V (approximate)}$$

The potential for sulfate reduction is estimated from the Nernst equation using standard state concentrations and pressures for all reactants and products except H^+, which in pure water is 1×10^{-7} M.

$$E = 0.20 \text{ V} - (0.0592 / 2) \log [1 / (1 \times 10^{-7})^4] = -0.6288 = -0.63 \text{ V}$$

The most easily reduced ion is Sn^{2+} with the most positive reduction potential, so **tin metal** forms at the cathode.

21.99   a) Possible oxidations:

$$2 H_2O(l) \rightarrow O_2(g) + 4 H^+(aq) + 4 e^- \qquad E = 1.4 \text{ V with overvoltage}$$

The oxidation of water produces oxygen gas ($\mathbf{O_2}$), and hydronium ions ($\mathbf{H_3O^+}$) at the anode. NO_3^- cannot oxidize since N is in its highest oxidation state in NO_3^-.

Possible reductions:

$$2 H_2O(l) + 2 e^- \rightarrow H_2(g) + 2 OH^-(aq) \qquad E = -1 \text{ V with overvoltage}$$
$$Cr^{3+}(aq) + 3 e^- \rightarrow Cr(s) \qquad E° = -0.74 \text{ V}$$
$$NO_3^-(aq) + 4 H^+(aq) + 3 e^- \rightarrow NO(g) + 2 H_2O(l) \qquad E = +0.13 \text{ V (approximate)}$$

The potential for nitrate reduction is estimated from the Nernst equation using standard state concentrations and pressures for all reactants and products except H^+, which in pure water is 1×10^{-7} M.

$$E = 0.96 \text{ V} - (0.0592 / 2) \log [1 / (1 \times 10^{-7})^4] = 0.1312 = 0.13 \text{ V}$$

The most easily reduced ion is NO_3^-, with the most positive reduction potential so **NO gas** is formed at the cathode.

b) Possible oxidations:

$$2 H_2O(l) \rightarrow O_2(g) + 4 H^+(aq) + 4 e^- \qquad E = 1.4 \text{ V with overvoltage}$$
$$2 Cl^-(aq) \rightarrow Cl_2(g) + 2 e^- \qquad E° = 1.36 \text{ V}$$

The oxidation of chloride ions to produce **chlorine gas** occurs at the anode. Cl^- has a more negative reduction potential showing that it is more easily oxidized than water.

Possible reductions:

$$2 H_2O(l) + 2 e^- \rightarrow H_2(g) + 2 OH^-(aq) \qquad E = -1 \text{ V with overvoltage}$$
$$Mn^{2+}(aq) + 2 e^- \rightarrow Mn(s) \qquad E° = -1.18 \text{ V}$$

It is easier to reduce water than to reduce manganese ions, so **hydrogen gas and hydroxide ions** form at the cathode. The reduction potential of Mn^{2+} is more negative than that of water showing that its reduction is less spontaneous than that of water.

21.101 $Mg^{2+} + 2 e^- \rightarrow Mg$

a) $(45.6 \text{ g Mg}) \left(\dfrac{1 \text{ mol Mg}}{24.31 \text{ g Mg}} \right) \left(\dfrac{2 \text{ mol } e^-}{1 \text{ mol Mg}} \right) = 3.75154257 = \mathbf{3.75 \text{ mol } e^-}$

b) $(3.75154257 \text{ mol } e^-) \left(\dfrac{96485 \text{ C}}{\text{mol } e^-} \right) = 3.619676 \times 10^5 = \mathbf{3.62 \times 10^5 \text{ coulombs}}$

c) $\left(\dfrac{3.619676 \times 10^5 \text{ C}}{3.50 \text{ h}} \right) \left(\dfrac{1 \text{ h}}{3600 \text{ s}} \right) \left(\dfrac{A}{C/s} \right) = 28.727586 = \mathbf{28.7 \text{ A}}$

21.103 $Ra^{2+} + 2e^- \rightarrow Ra$

In the reduction of radium ions, Ra^{2+}, to radium metal, the transfer of two electrons occurs.

$$(235 \text{ C})\left(\frac{1 \text{ mol } e^-}{96485 \text{ C}}\right)\left(\frac{1 \text{ mol Ra}}{2 \text{ mol } e^-}\right)\left(\frac{226 \text{ g Ra}}{1 \text{ mol Ra}}\right) = 0.275224 = \mathbf{0.275 \text{ g Ra}}$$

21.105 $Zn^{2+} + 2e^- \rightarrow Zn$

$$\text{Time} = (65.5 \text{ g Zn})\left(\frac{1 \text{ mol Zn}}{65.41 \text{ g Zn}}\right)\left(\frac{2 \text{ mol } e^-}{1 \text{ mol Zn}}\right)\left(\frac{96485 \text{ C}}{1 \text{ mol } e^-}\right)\left(\frac{1}{21.0 \text{ A}}\right)\left(\frac{1 \text{ A}}{C/s}\right) = 9.20169 \times 10^3 = \mathbf{9.20 \times 10^3 \text{ seconds}}$$

21.107 a) The sodium sulfate ionizes to produce Na^+ and SO_4^{2-} ions which make the water conductive; therefore the current will flow through the water to complete the circuit, increasing the rate of electrolysis. Pure water, which contains very low (10^{-7} M) concentrations of H^+ and OH^-, conducts electricity very poorly.
b) The reduction of H_2O has a more positive half-potential (–1 V) than the reduction of Na^+ (–2.71 V); the more spontaneous reduction of water will occur instead of the less spontaneous reduction of sodium ion. The oxidation of H_2O is the only oxidation possible because SO_4^{2-} cannot be oxidized under these conditions. In other words, it is easier to reduce H_2O than Na^+ and easier to oxidize H_2O than SO_4^{2-}.

21.109 $Zn^{2+} + 2e^- \rightarrow Zn$

$$\text{Mass} = (0.855 \text{ A})\left(\frac{C/s}{A}\right)\left(\frac{3600 \text{ s}}{1 \text{ h}}\right)\left(\frac{24 \text{ h}}{1 \text{ day}}\right)(2.50 \text{ day})\left(\frac{1 \text{ mol } e^-}{96485 \text{ C}}\right)\left(\frac{1 \text{ mol Zn}}{2 \text{ mol } e^-}\right)\left(\frac{65.41 \text{ g Zn}}{1 \text{ mol Zn}}\right)$$
$$= 62.599998 = \mathbf{62.6 \text{ g Zn}}$$

21.111 The half-reaction is: $2 H^+(aq) + 2 e^- \rightarrow H_2(g)$
a) First, find the moles of hydrogen gas.

$$n = \frac{PV}{RT} = \frac{(12.0 \text{ atm})(3.5 \times 10^6 \text{ L})}{\left(0.0821 \frac{L \cdot atm}{mol \cdot K}\right)((273 + 25)K)} = 1.716682 \times 10^6 \text{ mol } H_2 \text{ (unrounded)}$$

Then, find the coulombs knowing that there are two electrons transferred per mol of H_2.

$$(1.716682 \times 10^6 \text{ mole } H_2)\left(\frac{2 \text{ mol } e^-}{1 \text{ mol } H_2}\right)\left(\frac{96485 \text{ C}}{1 \text{ mol } e^-}\right) = 3.3126814 \times 10^{11} = \mathbf{3.3 \times 10^{11} \text{ coulombs}}$$

a) Remember that 1 V equals 1 J / C.

$$\left(\frac{1.44 \text{ J}}{C}\right)(3.31268 \times 10^{11} \text{ C}) = 4.77026 \times 10^{11} = \mathbf{4.8 \times 10^{11} \text{ J}}$$

c) $(4.77026 \times 10^{11} \text{ J})\left(\frac{1 \text{ kJ}}{10^3 \text{ J}}\right)\left(\frac{1 \text{ kg}}{4.0 \times 10^4 \text{ kJ}}\right) = 1.19257 \times 10^4 = \mathbf{1.2 \times 10^4 \text{ kg}}$

21.114 From the current 65.0% of the moles of product will be copper and 35.0% zinc. Assume a current of exactly 100 coulombs. The amount of current used to generate copper would be (65.0% / 100%) (100 C) = 65.0 C, and the amount of current used to generate zinc would be (35.0% / 100%) (100 C) = 35.0 C.
The half-reactions are: $Cu^{2+}(aq) + 2 e^- \rightarrow Cu(s)$ and $Zn^{2+}(aq) + 2 e^- \rightarrow Zn(s)$.

$$\text{Mass copper} = (65.0 \text{ C})\left(\frac{1 \text{ mol } e^-}{96485 \text{ C}}\right)\left(\frac{1 \text{ mol Cu}}{2 \text{ mol } e^-}\right)\left(\frac{63.55 \text{ g Cu}}{1 \text{ mol Cu}}\right) = 0.021406177 \text{ g Cu (unrounded)}$$

$$\text{Mass zinc} = (35.0 \text{ C})\left(\frac{1 \text{ mol } e^-}{96485 \text{ C}}\right)\left(\frac{1 \text{ mol Zn}}{2 \text{ mol } e^-}\right)\left(\frac{65.41 \text{ g Zn}}{1 \text{ mol Zn}}\right) = 0.01186376 \text{ g Zn (unrounded)}$$

$$\text{Mass \% copper} = \left(\frac{0.021406177 \text{ g Cu}}{\left(0.021406177 + 0.01186376 \right) \text{g Sample}} \right) \times 100\% = 64.340900 = \textbf{64.3\% Cu}$$

21.116 The reaction is: $Au^{3+}(aq) + 3 \text{ e}^- \rightarrow Au(s)$
a) Find the volume of gold needed to plate the earring and then use density to find the mass and moles of gold needed. The volume of the gold is the volume of a cylinder (see the inside back cover of the text).
$V = \pi r^2 h$

$$V = \pi \left(\frac{4.00 \text{ cm}}{2} \right)^2 (0.25 \text{ mm}) \left(\frac{10^{-3} \text{ m}}{1 \text{ mm}} \right) \left(\frac{1 \text{ cm}}{10^{-2} \text{ m}} \right) = 0.314159265 \text{ cm}^3$$

$$\left(0.314159265 \text{ cm}^3 \right) \left(\frac{19.3 \text{ g Au}}{1 \text{ cm}^3} \right) \left(\frac{1 \text{ mol Au}}{197.0 \text{ g Au}} \right) = 0.03077803 \text{ mol Au (unrounded)}$$

$$\text{Time} = \left(0.03077803 \text{ mol Au} \right) \left(\frac{3 \text{ mol e}^-}{1 \text{ mol Au}} \right) \left(\frac{96485 \text{ C}}{1 \text{ mol e}^-} \right) \left(\frac{A}{C/s} \right) \left(\frac{1}{0.013 \text{ A}} \right) \left(\frac{1 \text{ h}}{3600 \text{ s}} \right) \left(\frac{1 \text{ day}}{24 \text{ h}} \right)$$

$$= 7.931675 = \textbf{8 days}$$
b) The time required doubles once for the second earring of the pair and doubles again for the second side, thus it will take four times as long as one side of one earring.
$\qquad$ Time = (4) (7.931675 days) = 31.7267 = **32 days**
c) Start by multiplying the moles of gold from part (a) by four to get the moles for the earrings. Convert this moles to grams, then to troy ounces, and finally to dollars.

$$\text{Cost} = (4)\left(0.03077803 \text{ mol Au} \right) \left(\frac{197.0 \text{ g Au}}{1 \text{ mol Au}} \right) \left(\frac{1 \text{ Troy Ounce}}{31.10 \text{ g}} \right) \left(\frac{\$320}{\text{Troy Ounce}} \right) = 249.549 = \textbf{\$250}$$

21.119 The half-reactions and the cell reaction are:
$\qquad Zn(s) + 2\,\cancel{OH^-}\,\cancel{(aq)} \rightarrow ZnO(s) + \cancel{H_2O(l)} + \cancel{2\,e^-}$
$\qquad \underline{Ag_2O(s) + \cancel{H_2O(l)} + \cancel{2\,e^-} \rightarrow 2\,Ag(s) + 2\,\cancel{OH^-}\,\cancel{(aq)}}$
$\qquad Zn(s) + Ag_2O(s) \rightarrow ZnO(s) + 2\,Ag(s)$
The key is the moles of zinc. From the moles of zinc, the moles of electrons and the moles of Ag_2O may be found.

$$\text{Moles Zn} = \left(0.75 \text{ g Zn} \right) \left(\frac{80\%}{100\%} \right) \left(\frac{1 \text{ mol Zn}}{65.41 \text{ g Zn}} \right) = 0.00917291 \text{ mol Zn (unrounded)}$$

The 80% is assumed to have two significant figures.

a) Time = $\left(0.00917291 \text{ mol Zn} \right) \left(\frac{2 \text{ mol e}^-}{1 \text{ mol Zn}} \right) \left(\frac{96485 \text{ C}}{1 \text{ mol e}^-} \right) \left(\frac{A}{C/s} \right) \left(\frac{1 \text{ } \mu A}{10^{-6} \text{ A}} \right) \left(\frac{1}{0.85 \text{ } \mu A} \right) \left(\frac{1 \text{ h}}{3600 \text{ s}} \right) \left(\frac{1 \text{ day}}{24 \text{ h}} \right)$

$$= 2.410262 \times 10^4 = \textbf{2.4 x 10}^4 \textbf{ days}$$
b) Mass Ag = $\left(0.00917291 \text{ mol Zn} \right) \left(\frac{1 \text{ mol Ag}_2O}{1 \text{ mol Zn}} \right) \left(\frac{100\%}{95\%} \right) \left(\frac{2 \text{ mol Ag}}{1 \text{ mol Ag}_2O} \right) \left(\frac{107.9 \text{ g Ag}}{1 \text{ mol Ag}} \right)$

$$= 2.0836989 = \textbf{2.1 g Ag}$$
c) Cost = $\left(2.0836989 \text{ g Ag} \right) \left(\frac{95\%}{100\%} \right) \left(\frac{1 \text{ troy oz}}{31.10 \text{ g Ag}} \right) \left(\frac{\$13.00}{\text{troy oz}} \right) \left(\frac{1}{2.410262 \times 10^4 \text{ days}} \right)$

$$= 3.433027 \times 10^{-5} = \textbf{\$ 3.4 x 10}^{-5}\textbf{/day}$$

21.122 a) Cell with SHE and Pb / Pb^{2+}:
$\qquad$ Oxidation: $Pb(s) \rightarrow Pb^{2+}(aq) + 2e^- \qquad\qquad E^\circ = -0.13 \text{ V}$
$\qquad$ Reduction: $2\,H^+(aq) + 2e^- \rightarrow H_2(g) \qquad\qquad E^\circ = 0.0 \text{ V}$
$\qquad E^\circ_{cell} = E^\circ_{cathode} - E^\circ_{anode} = 0.0 \text{ V} - (-0.13 \text{ V}) = \textbf{0.13 V}$

Cell with SHE and Cu / Cu^{2+}:

Oxidation: $H_2(g) \rightarrow 2\ H^+(aq) + 2e^-$ $E° = 0.0$ V

Reduction: $Cu^{2+}(aq) + 2e^- \rightarrow Cu(s)$ $E° = 0.34$ V

$E^o_{cell} = E^o_{cathode} - E^o_{anode} = 0.34\ V - 0.00\ V = \mathbf{0.34\ V}$

b) The anode (negative electrode) in cell with SHE and Pb / Pb^{2+} is **Pb**.

The anode in cell with SHE and Cu / Cu^{2+} is **platinum** in the SHE.

c) The precipitation of PbS decreases $[Pb^{2+}]$. Use Nernst equation to see how this affects potential. Cell reaction is

$$Pb(s) + 2\ H^+(aq) \rightarrow Pb^{2+}(aq) + H_2(g)$$

$$E_{cell} = E^o_{cell} - \frac{0.0592}{n} \log \frac{\left[[Pb^{2+}]P_{H_2} \right]}{\left[[H^+]^2 \right]} \qquad n = 2\ e^-$$

$$E_{cell} = E^o_{cell} - \frac{0.0592}{2} \log \frac{\left[[Pb^{2+}]P_{H_2} \right]}{\left[[H^+]^2 \right]}$$

Decreasing the concentration of lead ions gives a negative value for the term

$$\frac{0.0592}{2} \log \frac{\left[[Pb^{2+}]P_{H_2} \right]}{\left[[H^+]^2 \right]}.$$

When this negative value is subtracted from E^o_{cell}, cell potential **increases**.

d) The $[H^+] = 1.0$ M and the $H_2 = 1$ atm in the SHE.

Cell reaction: $Cu^{2+}(aq) + H_2(g) \rightarrow Cu(s) + 2\ H^+(aq)$

$$E_{cell} = E^o_{cell} - \frac{0.0592}{2} \log \frac{\left[[H^+]^2 \right]}{\left[[Cu^{2+}]P_{H_2} \right]}$$

$$E_{cell} = 0.34\ V - \frac{0.0592}{2} \log \frac{\left[[1\ ]^2 \right]}{\left[[1 \times 10^{-16}]1\ atm \right]}$$

$$E_{cell} = -0.1336 = \mathbf{-0.13\ V}$$

21.125 The three steps equivalent to the overall reaction $M^+(aq) + e^- \rightarrow M(s)$ are

1) $M^+(aq) \rightarrow M^+(g)$ Energy is $-\Delta H_{hydration}$

2) $M^+(g) + e^- \rightarrow M(g)$ Energy is $-IE$ or $-\Delta H_{ionization}$

3) $M(g) \rightarrow M(s)$ Energy is $-\Delta H_{atomization}$

The energy for step 3 is similar for all three elements, so the difference in the energy for the overall reaction depends on the values for $-\Delta H_{hydration}$ and $-IE$. The lithium ion has a much more negative hydration energy than Na^+ and K^+ because it is a smaller ion with large charge density that holds the water molecules more tightly. The amount of energy required to remove the waters surrounding the lithium ion offsets the lower ionization energy to make the overall energy for the reduction of lithium larger than expected.

21.127 The key factor is that the table deals with electrode potentials in aqueous solution. The very high and low standard electrode potentials involve extremely reactive substances, such as F_2 (a powerful oxidant), Li (a powerful reductant). These substances react directly with water, rather than according to the desired half-reactions. An alternative (essentially equivalent) explanation is that any aqueous cell with a voltage of more than 1.23 V has the ability to electrolyze water into hydrogen and oxygen. When two electrodes with 6 V across them are placed in water, electrolysis of water will occur.

21.129　a) Aluminum half-reaction: $Al^{3+}(aq) + 3\,e^- \rightarrow Al(s)$, so n = 3. Remember that 1 A = 1 C / s.

$$\text{Time} = \left(1000 \text{ kg Al}\right)\left(\frac{10^3 \text{ g}}{1 \text{ kg}}\right)\left(\frac{1 \text{ mol Al}}{26.98 \text{ g Al}}\right)\left(\frac{3 \text{ mol e}^-}{1 \text{ mol Al}}\right)\left(\frac{96485 \text{ C}}{1 \text{ mol e}^-}\right)\left(\frac{A}{C/s}\right)\left(\frac{1}{100{,}000 \text{ A}}\right)$$

$$= 1.0728502 \times 10^5 = \mathbf{1.073 \times 10^5 \text{ s}}$$

The molar mass of aluminum limits the significant figures.

b) Multiply the time by the current and voltage, remembering that 1 A = 1 C / s (thus, 100,000 A is 100,000 C / s) and 1 V = 1 J / C (thus, 5.0 V = 5.0 J / C). Change units of J to kW•h.

$$\left(1.0728502 \times 10^5 \text{ s}\right)\left(\frac{100{,}000 \text{ C}}{s}\right)\left(\frac{5.0 \text{ J}}{C}\right)\left(\frac{1 \text{ kJ}}{10^3 \text{ J}}\right)\left(\frac{1 \text{ kW} \cdot \text{h}}{3.6 \times 10^3 \text{ kJ}}\right) = 1.4900698 \times 10^4 = \mathbf{1.5 \times 10^4 \text{ kW•h}}$$

c) From part (b), the 1.5×10^4 kW•h calculated is per 1000 kg of aluminum. Use the ratio of kW•h to mass to find kW•h / lb and then use efficiency and cost per kW•h to find cost per pound.

$$\text{Cost} = \left(\frac{1.4900698 \times 10^4 \text{ kW} \cdot \text{h}}{1000 \text{ kg Al}}\right)\left(\frac{1 \text{ kg}}{2.205 \text{ lb}}\right)\left(\frac{0.90 \text{ cents}}{1 \text{ kW} \cdot \text{h}}\right)\left(\frac{100\%}{90.\%}\right) = 6.757686 = \mathbf{6.8\text{¢ / lb Al}}$$

21.131　Statement: metal D + hot water → reaction　Conclusion: D reduces water to produce $H_2(g)$.

Statement: D + E salt → no reaction　　　Conclusion: D does not reduce E salt, so E reduces D salt. E is better reducing agent than D.

Statement: D + F salt → reaction　　　Conclusion: D reduces F salt. D is better reducing agent than F.

If E metal and F salt are mixed, the salt F would be reduced producing F metal because E has the greatest reducing strength of the three metals (E is stronger than D and D is stronger than F). The ranking of increasing reducing strength is **F < D < E.**

21.134　Substitute J / C for V.

a) Cell I: Oxidation number (O.N.) of H changes from 0 to +1, so 1 electron is lost from each of 4 hydrogens for a total of 4 electrons. O.N. of oxygen changes from 0 to –2, indicating that 2 electrons are gained by each of the two oxygens for a total of 4 electrons. There is a transfer of **four electrons** in the reaction. The potential given in the problem allows the calculation of ΔG:

$$\Delta G^\circ = -nFE^\circ = -(4 \text{ mol e}^-)(96485 \text{ C / mol e}^-)(1.23 \text{ J / C}) = -4.747062 \times 10^5 = \mathbf{-4.75 \times 10^5 \text{ J}}$$

Cell II: In $Pb(s) \rightarrow PbSO_4$, O.N. of Pb changes from 0 to +2 and in $PbO_2 \rightarrow PbSO_4$, O.N. of PB changes from +4 to +2. There is a transfer of **two electrons** in the reaction.

$$\Delta G^\circ = -nFE^\circ = -(2 \text{ mol e}^-)(96485 \text{ C / mol e}^-)(2.04 \text{ J / C}) = -3.936588 \times 10^5 = \mathbf{-3.94 \times 10^5 \text{ J}}$$

Cell III: O.N. of each of two Na atoms changes from 0 to +1 and O.N. of Fe changes from +2 to 0. There is a transfer of **two electrons** in the reaction.

$$\Delta G^\circ = -nFE^\circ = -(2 \text{ mol e}^-)(96485 \text{ C / mol e}^-)(2.35 \text{ J / C}) = -4.534795 \times 10^5 = \mathbf{-4.53 \times 10^5 \text{ J}}$$

b)　　Cell I:　Mass of reactants = $\left(2 \text{ mol H}_2\right)\left(\dfrac{2.016 \text{ g H}_2}{1 \text{ mol H}_2}\right) + \left(1 \text{ mol O}_2\right)\left(\dfrac{32.00 \text{ g O}_2}{1 \text{ mol O}_2}\right) = 36.032 \text{ g (unrounded)}$

$$\frac{w_{max}}{\text{Reactant Mass}} = \left(\frac{-4.747062 \times 10^5 \text{ J}}{36.032 \text{ g}}\right)\left(\frac{1 \text{ kJ}}{10^3 \text{ J}}\right) = -13.17457 = \mathbf{-13.2 \text{ kJ/g}}$$

Cell II:　Mass of reactants =

$$\left(1 \text{ mol Pb}\right)\left(\frac{207.2 \text{ g Pb}}{1 \text{ mol Pb}}\right) + \left(1 \text{ mol PbO}_2\right)\left(\frac{239.2 \text{ g PbO}_2}{1 \text{ mol PbO}_2}\right) + \left(2 \text{ mol H}_2SO_4\right)\left(\frac{98.09 \text{ g H}_2SO_4}{1 \text{ mol H}_2SO_4}\right)$$

$$= 642.58 \text{ g (unrounded)}$$

$$\frac{w_{max}}{\text{Reactant Mass}} = \left(\frac{-3.936588 \times 10^5 \text{ J}}{642.58 \text{ g}}\right)\left(\frac{1 \text{ kJ}}{10^3 \text{ J}}\right) = -0.612622 = \mathbf{-0.613 \text{ kJ/g}}$$

Cell III:　Mass of reactants = $\left(2 \text{ mol Na}\right)\left(\dfrac{22.99 \text{ g Na}}{1 \text{ mol Na}}\right) + \left(1 \text{ mol FeCl}_2\right)\left(\dfrac{126.75 \text{ g FeCl}_2}{1 \text{ mol FeCl}_2}\right)$

$$= 172.73 \text{ g (unrounded)}$$

$$\frac{W_{max}}{\text{Reactant Mass}} = \left(\frac{-4.534795 \times 10^5 \text{ J}}{172.73 \text{ g}}\right)\left(\frac{1 \text{ kJ}}{10^3 \text{ J}}\right) = -2.625366 = \textbf{-2.62 kJ/g}$$

Cell I has the highest ratio (most energy released per gram) because the reactants have very low mass while Cell II has the lowest ratio because the reactants are very massive.

21.138 The cell reaction is: $Cu^{2+}(aq) + Sn(s) \rightarrow Cu(s) + Sn^{2+}(aq)$

a) Use the current and time to calculate total charge. Remember that the unit 1 A is 1C / s, so the time must be converted to seconds. From the total charge, the number of electrons transferred to form copper is calculated by dividing total charge by the charge per mole of electrons. Each mole of copper deposited requires 2 moles of electrons, so divide moles of electrons by 2 to get moles of copper. Then convert to grams of copper.

$$\text{Mass Cu} = (0.17 \text{ A})\left(\frac{C/s}{A}\right)\left(\frac{3600 \text{ s}}{1 \text{ h}}\right)(48.0 \text{ h})\left(\frac{1 \text{ mol e}^-}{96485 \text{ C}}\right)\left(\frac{1 \text{ mol Cu}}{2 \text{ mol e}^-}\right)\left(\frac{63.55 \text{ g Cu}}{1 \text{ mol Cu}}\right) = 9.674275 = \textbf{9.7 g Cu}$$

b) The initial concentration of Cu^{2+} is 1.00 M (standard condition) and initial volume is 345 mL. Use this to calculate the initial moles of copper ions, then subtract the number of moles of copper ions converted to copper metal.

$$M \, Cu^{2+} = \frac{\left[\left(1.00 \frac{\text{mol Cu}^{2+}}{L}\right)(345 \text{ mL})\left(\frac{10^{-3} \text{ L}}{1 \text{ mL}}\right)\right] - \left[(9.674275 \text{ g Cu})\left(\frac{1 \text{ mol Cu}}{63.55 \text{ g Cu}}\right)\right]}{(345 \text{ mL})\left(\frac{10^{-3} \text{ L}}{1 \text{ mL}}\right)}$$

$$= 0.558751 = \textbf{0.56 } \textit{M} \textbf{ Cu}^{2+}$$

21.140 Examine each reaction to determine which reactant is the oxidizing agent by which reactant gains electrons in the reaction.

From reaction between $U^{3+} + Cr^{3+} \rightarrow Cr^{2+} + U^{4+}$, find that Cr^{3+} oxidizes U^{3+}.
From reaction between $Fe + Sn^{2+} \rightarrow Sn + Fe^{2+}$, find that Sn^{2+} oxidizes Fe.
From the fact that there is no reaction that occurs between Fe and U^{4+}, find that Fe^{2+} oxidizes U^{3+}.
From reaction between $Cr^{3+} + Fe \rightarrow Cr^{2+} + Fe^{2+}$, find that Cr^{3+} oxidizes Fe.
From reaction between $Cr^{2+} + Sn^{2+} \rightarrow Sn + Cr^{3+}$, find that Sn^{2+} oxidizes Cr^{2+}.

Notice that nothing oxidizes Sn, so Sn^{2+} must be the strongest oxidizing agent. Both Cr^{3+} and Fe^{2+} oxidize U^{3+}, so U^{4+} must be the weakest oxidizing agent. Cr^{3+} oxidizes iron so Cr^{3+} is a stronger oxidizing agent than Fe^{2+}.

The half–reactions in order from strongest to weakest oxidizing agent:
$Sn^{2+}(aq) + 2 \text{ e}^- \rightarrow Sn(s)$
$Cr^{3+}(aq) + \text{e}^- \rightarrow Cr^{2+}(aq)$
$Fe^{2+}(aq) + 2 \text{ e}^- \rightarrow Fe(s)$
$U^{4+}(aq) + \text{e}^- \rightarrow U^{3+}(aq)$

21.144 a) The calomel half-cell is the anode and the silver half-cell is the cathode. The overall reaction is
$$2 \, Ag^+(aq) + 2 \, Hg(l) + 2 \, Cl^-(aq) \rightarrow 2 \, Ag(s) + Hg_2Cl_2(s)$$

$E^o_{cell} = E^o_{cathode} - E^o_{anode}$

$E^o_{cell} = 0.80 \text{ V} - 0.24 \text{ V} = 0.56 \text{ V}$ with n = 2.

Use the Nernst equation to find $[Ag^+]$ when $E_{cell} = 0.060$ V.

$$E = E^o - \frac{0.0592 \text{ V}}{n} \log Q$$

$$E = E^o - \frac{0.0592 \text{ V}}{2} \log \frac{1}{\left[Ag^+\right]^2\left[Cl^-\right]^2}$$

$$0.060 \text{ V} = 0.56 \text{ V} - \frac{0.0592 V}{2} \log \frac{1}{\left[Ag^+\right]^2\left[Cl^-\right]^2}$$

The problem suggests assuming that $[Cl^-]$ is constant. Assume it is 1.00 M.

$$(0.060 \text{ V} - 0.56 \text{ V})(-2 / 0.0592 \text{ V}) = \log \frac{1}{\left[Ag^+\right]^2 [1.00]^2}$$

$$16.89189 = \log \frac{1}{\left[Ag^+\right]^2 [1.00]^2}$$

$\log [Ag^+]^2 = -16.89189$ (unrounded)   {math note: $\log (1 / x) = -\log (x)$}
$[Ag^+]^2 = 10^{-16.89189}$
$[Ag^+]^2 = 1.2826554 \times 10^{-17}$
$[Ag^+] = 3.5814179 \times 10^{-9} = \mathbf{3.6 \times 10^{-9}\ \textit{M}}$

Two significant figures come from subtracting 0.56 from 0.060 that gives 0.50 with only two significant figures.

b) Again use the Nernst equation and assume $[Cl^-] = 1.00$ M.

$$E = E^\circ - \frac{0.0592 \text{ V}}{2} \log \frac{1}{\left[Ag^+\right]^2 \left[Cl^-\right]^2}$$

$$0.53 \text{ V} = 0.56 \text{ V} - \frac{0.0592 \text{ V}}{2} \log \frac{1}{\left[Ag^+\right]^2 \left[Cl^-\right]^2}$$

$$(0.53 \text{ V} - 0.56 \text{ V})(-2 / 0.0592 \text{ V}) = \log \frac{1}{\left[Ag^+\right]^2 [1.00]^2}$$

$$1.0135135 = \log \frac{1}{\left[Ag^+\right]^2 [1.00]^2}$$

$\log [Ag^+]^2 = -1.0135135$ (unrounded)   {math note: $\log (1 / x) = -\log (x)$}
$[Ag^+]^2 = 10^{-1.0135135}$
$[Ag^+]^2 = 0.0969363$
$[Ag^+] = 0.300346 = \mathbf{0.3\ \textit{M}}$

One significant figure comes from subtracting 0.53 from 0.57 that gives 0.03 with only one significant figure.

21.146 a) The reaction is $Ag^+(aq) \rightarrow Ag(s) + 1e^-$

Nonstandard cell: $E_{waste} = E^o_{cell} - \left(\dfrac{0.0592}{1\ e^-}\right) \log [Ag^+]_{waste}$

Standard cell: $E_{standard} = E^o_{cell} - \left(\dfrac{0.0592}{1\ e^-}\right) \log [Ag^+]_{standard}$

b) To find $[Ag^+]_{waste}$: $E^o_{cell} = E_{standard} + \left(\dfrac{0.0592}{1\ e^-}\right) \log [Ag^+]_{standard} = E_{waste} + \left(\dfrac{0.0592}{1\ e^-}\right) \log [Ag^+]_{waste}$

$$E_{waste} - E_{standard} = -\left(\frac{0.0592}{1\ e^-}\right) \log \frac{[Ag^+]_{waste}}{[Ag^+]_{standard}}$$

$$E_{standard} - E_{waste} = (0.0592 \text{ V})(\log [Ag^+]_{waste} - \log [Ag^+]_{standard})$$

$$\frac{E_{standard} - E_{waste}}{0.0592 \text{ V}} = (\log [Ag^+]_{waste} - \log [Ag^+]_{standard})$$

$$\log [Ag^+]_{waste} = \frac{E_{standard} - E_{waste}}{0.0592} + \log [Ag^+]_{standard})$$

$$[Ag^+]_{waste} = [\text{antilog} \left(\frac{E_{standard} - E_{waste}}{0.0592 \text{ V}}\right)]([Ag^+]_{standard})$$

c) Convert M to ng / L for both $[Ag^+]_{waste}$ and $[Ag^+]_{standard}$:

$$E_{waste} - E_{standard} = (-0.0592\ V\ /\ 1)\log\frac{[Ag^+]_{waste}}{[Ag^+]_{standard}}$$

If both silver ion concentrations are in the same units, in this case ng / L, the "conversions" cancel and the equation derived in part (b) applies if the standard concentration is in ng / L.

$$C\ (Ag^+)_{waste} = \left[antilog\left(\frac{E_{standard} - E_{waste}}{0.0592\ V}\right)\right]\left(C\left(Ag^+\right)_{standard}\right)$$

d) Plug the values into the answer for part (c).

$$[Ag^+]_{waste} = \left[antilog\left(\frac{-0.003}{0.0592\ V}\right)\right](1000.\ ng/L)\ = 889.8654 = \textbf{900 ng / L}$$

e) Temperature is included in the RT / nF term, which equals 0.0592 V / n at 25°C. To account for different temperatures, insert the RT / nF term in place of 0.0592 V / n.

$$E_{standard} + \left(\frac{2.303RT}{nF}\right)\log[Ag^+]_{standard} = E_{waste} + \left(\frac{2.303RT}{nF}\right)\log[Ag^+]_{waste}$$

$$E_{standard} - E_{waste} = \left(\frac{2.303R}{nF}\right)(T_{waste}\log[Ag^+]_{waste} - T_{standard}\log[Ag^+]_{standard})$$

$$(E_{standard} - E_{waste})\left(\frac{nF}{2.303R}\right) = T_{waste}\log[Ag^+]_{waste} - T_{standard}\log[Ag^+]_{standard}$$

$$(E_{standard} - E_{waste})\left(\frac{nF}{2.303R}\right) + T_{standard}\log[Ag^+]_{standard} = T_{waste}\log[Ag^+]_{waste}$$

$$\log[Ag^+]_{waste} = \left(\frac{\left(E_{standard} - E_{waste}\right)(nF\ /\ 2.303\ R) + T_{standard}\log\left[Ag^+\right]_{standard}}{T_{waste}}\right)$$

$$[Ag^+]_{waste} = antilog\left(\frac{\left(E_{standard} - E_{waste}\right)(nF\ /\ 2.303\ R) + T_{standard}\log\left[Ag^+\right]_{standard}}{T_{waste}}\right)$$

21.148 a) Determine the total charge the cell can produce.

$$Capacity = (300.\ mA \cdot h)\left(\frac{10^{-3}\ A}{1\ mA}\right)\left(\frac{3600\ s}{1\ h}\right)\left(\frac{1\ C}{1\ A \cdot s}\right) = \textbf{1.08 x 10}^3\ \textbf{C}$$

b) The half-reactions are:

$Cd^0 \rightarrow Cd^{2+} + 2\ e^-$ and $2\ (NiO(OH) + H_2O(l) + e^- \rightarrow Ni(OH)_2 + OH^-)$

2 mol e^- flow per mole of reaction.

Assume 100% conversion of reactants.

$$Mass\ Cd = (1080\ C)\left(\frac{1\ mol\ e^-}{96485\ C}\right)\left(\frac{1\ mol\ Cd}{2\ mol\ e^-}\right)\left(\frac{112.4\ g\ Cd}{1\ mol\ Cd}\right) = 0.62907 = \textbf{0.629 g Cd}$$

$$Mass\ NiO(OH) = (1080\ C)\left(\frac{1\ mol\ e^-}{96485\ C}\right)\left(\frac{1\ mol\ NiO(OH)}{1\ mol\ e^-}\right)\left(\frac{91.70\ g\ NiO(OH)}{1\ mol\ NiO(OH)}\right) = 1.026439 = \textbf{1.03 g NiO(OH)}$$

$$Mass\ H_2O = (1080\ C)\left(\frac{1\ mol\ e^-}{96485\ C}\right)\left(\frac{1\ mol\ H_2O}{1\ mol\ e^-}\right)\left(\frac{18.02\ g\ H_2O}{1\ mol\ H_2O}\right)$$

$$= 0.20170596 = \textbf{0.202 g H}_2\textbf{O}$$

Total mass of reactants = 0.62907 g Cd + 1.026439 g NiO(OH) + 0.20170596 g H_2O

= 1.85721 = **1.86 g reactants**

c) Mass % reactants $= \left(\dfrac{1.85721\ g}{18.3\ g}\right) \times 100\% = 10.14872 = \mathbf{10.1\%}$

21.150 Place the elements in order of increasing (more positive) $E°$.
Reducing agent strength: Li > Ba > Na > Al > Mn > Zn > Cr > Fe > Ni > Sn > Pb > Cu > Ag > Hg > Au
Metals with potentials lower than that of water (–0.83 V) can displace hydrogen from water by reducing the hydrogen in water. These can displace H_2 from water: Li, Ba, Na, Al, and Mn
Metals with potentials lower than that of hydrogen (0.00 V) can displace hydrogen from acids by reducing the H^+ in acid. These can displace H_2 from acid: Li, Ba, Na, Al, Mn, Zn, Cr, Fe, Ni, Sn, and Pb
Metals with potentials above that of hydrogen (0.00 V) cannot displace (reduce) hydrogen.
These cannot displace H_2: Cu, Ag, Hg, and Au

21.153 a) The reference half-reaction is: $Cu^{2+}(aq) + 2\ e^- \rightarrow Cu(s)$ $E° = 0.34\ V$
Before the addition of the ammonia, $E_{cell} = 0$. The addition of ammonia lowers the concentration of copper ions through the formation of the complex $Cu(NH_3)_4{}^{2+}$. The original copper ion concentration is $[Cu^{2+}]_{original}$, and the copper ion concentration in the solution containing ammonia is $[Cu^{2+}]_{ammonia}$.
The Nernst equation is used to determine the copper ion concentration in the cell containing ammonia.

$$E = E° - \dfrac{0.0592\ V}{n}\ \log Q$$

$$0.129\ V = 0.00\ V - \dfrac{0.0592\ V}{2}\ \log \dfrac{\left[Cu^{2+}\right]_{ammonia}}{\left[Cu^{2+}\right]_{original}}$$

$$0.129\ V = -\dfrac{0.0592\ V}{2}\ \log \dfrac{\left[Cu^{2+}\right]_{ammonia}}{\left[0.0100\right]_{original}}$$

$$(0.129\ V)\ (-2\ /\ 0.0592) = \log \dfrac{\left[Cu^{2+}\right]_{ammonia}}{\left[0.0100\right]_{original}}$$

$$-4.358108108 = \log \dfrac{\left[Cu^{2+}\right]_{ammonia}}{\left[0.0100\right]_{original}}$$

$$4.3842154 \times 10^{-5} = \dfrac{\left[Cu^{2+}\right]_{ammonia}}{\left[0.0100\right]_{original}}$$

$[Cu^{2+}]_{ammonia} = 4.3842154 \times 10^{-7}\ M$ (unrounded)
This is the concentration of the copper ion that is not in the complex. The concentration of the complex and of the uncomplexed ammonia must be determined before K_f may be calculated.
The original number of moles of copper and the original number of moles of ammonia are found from the original volumes and molarities:

$$\text{Original moles of copper} = \left(\dfrac{0.0100\ mol\ Cu(NO_3)_2}{L}\right)\left(\dfrac{1\ mol\ Cu^{2+}}{1\ mol\ Cu(NO_3)_2}\right)\left(\dfrac{10^{-3}\ L}{1\ mL}\right)(90.0\ mL)$$

$$= 9.00 \times 10^{-4}\ mol\ Cu^{2+}$$

$$\text{Original moles of ammonia} = \left(\dfrac{0.500\ mol\ NH_3}{L}\right)\left(\dfrac{10^{-3}\ L}{1\ mL}\right)(10.0\ mL) = 5.00 \times 10^{-3}\ mol\ NH_3$$

Determine the moles of copper still remaining uncomplexed.

$$\text{Remaining moles of copper} = \left(\dfrac{4.3842154 \times 10^{-7}\ mol\ Cu^{2+}}{L}\right)\left(\dfrac{10^{-3}\ L}{1\ mL}\right)(100.0\ mL)$$

$$= 4.3842154 \times 10^{-8}\ mol\ Cu$$

The difference between the original moles of copper and the copper ion remaining in solution is the copper in the complex (= moles of complex). The molarity of the complex may now be found.

$$\text{Moles copper in complex} = (9.00 \times 10^{-4} - 4.3842154 \times 10^{-8})\ \text{mol}\ Cu^{2+}$$
$$= 8.9995615 \times 10^{-4}\ \text{mol}\ Cu^{2+}\ \text{(unrounded)}$$

$$\text{Molarity of complex} = \left(\frac{8.9995615 \times 10^{-4}\ \text{mol}\ Cu^{2+}}{100.0\ \text{mL}}\right)\left(\frac{1\ \text{mol}\ Cu(NH_3)_4^{2+}}{1\ \text{mol}\ Cu^{2+}}\right)\left(\frac{1\ \text{mL}}{10^{-3}\ \text{L}}\right)$$

$$= 8.9995615 \times 10^{-3}\ M\ Cu(NH_3)_4^{2+}\ \text{(unrounded)}$$

The concentration of the remaining ammonia is found as follows:

$$\text{Molarity of ammonia} = \left(\frac{\left(5.00 \times 10^{-3}\ \text{mol}\ NH_3\right) - \left(8.9995615 \times 10^{-4}\ \text{mol}\ Cu^{2+}\right)\left(\dfrac{4\ \text{mol}\ NH_3}{1\ \text{mol}\ Cu^{2+}}\right)}{100.0\ \text{mL}}\right)\left(\frac{1\ \text{mL}}{10^{-3}\ \text{L}}\right)$$

$$= 0.014001753\ M\ \text{ammonia (unrounded)}$$

The K_f equilibrium is:

$$Cu^{2+}(aq) + 4\ NH_3(aq) \rightleftharpoons Cu(NH_3)_4^{2+}(aq)$$

$$K_f = \frac{\left[Cu(NH_3)_4^{2+}\right]}{\left[Cu^{2+}\right]\left[NH_3\right]^4} = \frac{\left[8.9995615 \times 10^{-3}\right]}{\left[4.3842154 \times 10^{-7}\right]\left[0.014001753\right]^4} = 5.34072 \times 10^{11} = \mathbf{5.3 \times 10^{-11}}$$

b) The K_f will be used to determine the new concentration of free copper ions.
    Moles uncomplexed ammonia before the addition of new ammonia =
        $(0.014001753\ \text{mol}\ NH_3\ /\ L)\ (10^{-3}\ L\ /\ 1\ mL)\ (100.0\ mL) = 0.001400175\ \text{mol}\ NH_3$
    Moles ammonia added = 5.00×10^{-3} mol NH_3 (same as original moles of ammonia)
From the stoichiometry:

	$Cu^{2+}(aq)$	+	$4\ NH_3(aq)$	$\rightarrow$	$Cu(NH_3)_4^{2+}(aq)$
Initial moles:	4.3842154×10^{-8} mol		0.001400175 mol		8.9995615×10^{-4} mol
Added moles:			5.00×10^{-3} mol		
Cu^{2+} is limiting:	$-(4.3842154 \times 10^{-8}\ \text{mol})$		$-4(4.3842154 \times 10^{-8}\ \text{mol})$		$+(4.3842154 \times 10^{-8}\ \text{mol})$
After the reaction:	0		0.006400 mol		9.00000×10^{-4} mol

Determine concentrations before equilibrium:
    $[Cu^{2+}] = 0$
    $[NH_3] = (0.006400\ \text{mol}\ NH_3\ /\ 110.0\ mL)\ (1\ mL\ /\ 10^{-3}\ L) = 0.0581818\ M\ NH_3$
    $[Cu(NH_3)_4^{2+}] = (9.00000 \times 10^{-4}\ \text{mol}\ Cu(NH_3)_4^{2+}\ /\ 110.0\ mL)\ (1\ mL\ /\ 10^{-3}\ L)$
        $= 0.008181818\ M\ Cu(NH_3)_4^{2+}$

Now allow the system to come to equilibrium:

	$Cu^{2+}(aq)$	+	$4\ NH_3(aq)$	$\rightleftharpoons$	$Cu(NH_3)_4^{2+}(aq)$
Initial molarity:	0		0.0581818		0.008181818
Change:	$+x$		$+4x$		$-x$
Equilibrium:	x		$0.0581818 + 4x$		$0.008181818 - x$

$$K_f = \frac{\left[Cu(NH_3)_4^{2+}\right]}{\left[Cu^{2+}\right]\left[NH_3\right]^4} = \frac{\left[0.008181818 - x\right]}{[x]\left[0.0581818 + 4x\right]^4} = 5.34072 \times 10^{11}$$

Assume $-x$ and $+4x$ are negligible when compared to their associated numbers:

$$K_f = \frac{\left[0.008181818\right]}{[x]\left[0.0581818\right]^4} = 5.34072 \times 10^{11}$$

$$x = [Cu^{2+}] = 1.3369 \times 10^{-9}\ M\ Cu^{2+}$$

Use the Nernst equation to determine the new cell potential:

$$E = 0.00 \text{ V} - \frac{0.0592 \text{ V}}{2} \log \frac{\left[Cu^{2+}\right]_{ammonia}}{\left[Cu^{2+}\right]_{original}}$$

$$E = -\frac{0.0592 \text{ V}}{2} \log \frac{\left[1.3369 \times 10^{-9}\right]}{\left[0.0100\right]}$$

$$E = 0.203467 = \mathbf{0.20 \text{ V}}$$

c) The first step will be to do a stoichiometry calculation of the reaction between copper ions and hydroxide ions.

$$\text{Moles OH}^- = \left(\frac{0.500 \text{ mol NaOH}}{L}\right)\left(\frac{1 \text{ mol OH}^-}{1 \text{ mol NaOH}}\right)\left(\frac{10^{-3} \text{ L}}{1 \text{ mL}}\right)(10.0 \text{ mL}) = 5.00 \times 10^{-3} \text{ mol OH}^-$$

The initial moles of copper ions were determined earlier: 9.00×10^{-4} mol Cu^{2+}
The reaction:

	$Cu^{2+}(aq)$	+	$2 \text{ OH}^-(aq)$	$\rightarrow$	$Cu(OH)_2(s)$
Initial moles:	9.00×10^{-4} mol		5.00×10^{-3} mol		
Cu^{2+} is limiting:	$-(9.00 \times 10^{-4}$ mol$)$		$-2(9.00 \times 10^{-4}$ mol$)$		
After the reaction:	0		0.0032 mol		

Determine concentrations before equilibrium:

$[Cu^{2+}] = 0$
$[NH_3] = (0.0032 \text{ mol OH}^- / 100.0 \text{ mL}) (1 \text{ mL} / 10^{-3} \text{ L}) = 0.032 \ M \ OH^-$

Now allow the system to come to equilibrium:

	$Cu(OH)_2(s)$	$\leftrightarrows$	$Cu^{2+}(aq)$	+	$2 \text{ OH}^-(aq)$
Initial molarity:			0.0		0.032
Change:			$+x$		$+2x$
Equilibrium:			x		$0.032 + 2x$

$K_{sp} = [Cu^{2+}][OH^-]^2 = 2.2 \times 10^{-20}$
$K_{sp} = [x][0.032 + 2x]^2 = 2.2 \times 10^{-20}$
   Assume 2x is negligible compared to 0.032 M.
$K_{sp} = [x][0.032]^2 = 2.2 \times 10^{-20}$
$x = [Cu^{2+}] = 2.1487375 \times 10^{-17} = 2.1 \times 10^{-17} \ M$

Use the Nernst equation to determine the new cell potential:

$$E = 0.00 \text{ V} - \frac{0.0592 \text{ V}}{2} \log \frac{\left[Cu^{2+}\right]_{hydroxide}}{\left[Cu^{2+}\right]_{original}}$$

$$E = -\frac{0.0592 \text{ V}}{2} \log \frac{\left[2.1487375 \times 10^{-17}\right]}{\left[0.0100\right]}$$

$$E = 0.434169 = \mathbf{0.43 \text{ V}}$$

d) Use the Nernst equation to determine the copper ion concentration in the half-cell containing the hydroxide ion.

$$E = 0.00 \text{ V} - \frac{0.0592 \text{ V}}{2} \log \frac{\left[Cu^{2+}\right]_{hydroxide}}{\left[Cu^{2+}\right]_{original}}$$

$$0.340 = -\frac{0.0592 \text{ V}}{2} \log \frac{\left[Cu^{2+}\right]_{hydroxide}}{\left[0.0100\right]}$$

$$(0.340 \text{ V}) (-2 / 0.0592) = \log \frac{\left[Cu^{2+}\right]_{hydroxide}}{\left[0.0100\right]}$$

$$-11.486486 = \log \frac{\left[Cu^{2+}\right]_{hydroxide}}{[0.0100]}$$

$$3.2622256 \times 10^{-12} = \frac{\left[Cu^{2+}\right]_{hydroxide}}{[0.0100]}$$

$[Cu^{2+}]_{hydroxide} = 3.2622256 \times 10^{-14}\ M$ (unrounded)

Now use the K_{sp} relationship:

$K_{sp} = [Cu^{2+}][OH^-]^2 = 2.2 \times 10^{-20}$
$K_{sp} = [3.2622256 \times 10^{-14}][OH^-]^2 = 2.2 \times 10^{-20}$
$[OH^-]^2 = 6.743862 \times 10^{-7}$
$[OH^-] = 8.2121 \times 10^{-4} = 8.2 \times 10^{-4}\ M\ OH^- = \mathbf{8.2 \times 10^{-4}\ \textit{M}\ NaOH}$

21.156 a) The chemical equation for the combustion of octane is:

$2\ C_8H_{18}(l) + 25\ O_2(g) \rightarrow 16\ CO_2(g) + 18\ H_2O(g)$

The heat of reaction may be determined from heats of formation in the Appendix or in the problem.

$\Delta H° = \Sigma n \Delta H_f° - \Sigma m \Delta H_f°$

$\Delta H° = [(16\ mol)\ (\Delta H_f°(CO_2(g)) + (18\ mol)\ (\Delta H_f°(H_2O(g))]$
$\quad\quad - [(2\ mol)\ (\Delta H_f°(C_8H_{18}(l)) + (25\ mol)\ (\Delta H_f°(O_2(g)))]$

$\Delta H° = [(16\ mol)\ (-393.5\ kJ/mol) + (18\ mol)\ (-241.826\ kJ/mol)]$
$\quad\quad - [(2\ mol)\ (-250.1\ kJ/mol) + (25\ mol)\ (0\ kJ/mol)]$

$\Delta H° = [-10648.868\ kJ] - [-500.2\ kJ] = -10148.668 = -10148.7\ kJ$

The energy from 1.00 gallon of gasoline is:

$$(1.00\ gal)\left(\frac{4\ qt}{1\ gal}\right)\left(\frac{1\ L}{1.057\ qt}\right)\left(\frac{1\ mL}{10^{-3}\ L}\right)\left(\frac{0.7028\ g}{mL}\right)\left(\frac{1\ mol\ C_8H_{18}}{114.22\ g\ C_8H_{18}}\right)\left(\frac{-10148.7\ kJ}{2\ mol\ C_8H_{18}}\right)$$

$$= -1.1815165 \times 10^5 = \mathbf{-1.18 \times 10^5\ kJ}$$

b) The energy from the combustion of hydrogen must be found using the balanced chemical equation and the values from the Appendix.

$H_2(g) + 1/2\ O_2(g) \rightarrow H_2O(g)$

With the reaction written this way, the heat of reaction is simply the heat of formation of water vapor, and no additional calculations are necessary.

$\Delta H° = -241.826\ kJ$

The moles of hydrogen needed to produce the energy from part (a) are:

$$\text{Moles } H_2 = \left(-1.1815165 \times 10^5\ kJ\right)\left(\frac{1\ mol\ H_2}{-241.826\ kJ}\right) = 488.5812755\ mol\ \text{(unrounded)}$$

Finally, use the ideal gas equation to determine the volume.

$$V = nRT / P = \frac{\left(488.5812755\ mol\ H_2\right)\left(0.0821\ \frac{L \bullet atm}{mol \bullet K}\right)\left((273 + 25)K\right)}{\left(1.00\ atm\right)} = 1.195353 \times 10^4 = \mathbf{1.20 \times 10^4\ L}$$

c) This part of the problem requires the half-reaction for the electrolysis of water to produce hydrogen gas.

$2\ H_2O(l) + 2\ e^- \rightarrow H_2(g) + 2\ OH^-(aq)$

Use 1 A = 1 C / s

$$\text{Time} = \left(488.5812755\ mol\ H_2\right)\left(\frac{2\ mol\ e^-}{1\ mol\ H_2}\right)\left(\frac{96485\ C}{1\ mol\ e^-}\right)\left(\frac{s}{1.00 \times 10^3\ C}\right) = 9.428153 \times 10^4 = \mathbf{9.43 \times 10^4\ seconds}$$

d) This is a straight factor-label problem.

$$\text{Power} = \left(\frac{1.00 \times 10^3\ C}{s}\right)(6.00\ V)\left(\frac{1\ J}{1\ C \bullet V}\right)\left(9.428153 \times 10^4\ s\right)\left(\frac{1\ kW \bullet h}{3.6 \times 10^6\ J}\right) = 157.1359 = \mathbf{157\ kW\bullet h}$$

e) The process is only 88.0% efficient, additional electricity is necessary to produce sufficient hydrogen. This is the purpose of the (100% / 88.0%) factor.

$$\text{Cost} = (157.13588 \text{ kW}\cdot\text{h}) \left(\frac{0.950 \text{ cents}}{1 \text{ kW}\cdot\text{h}} \right) \left(\frac{100\%}{88\%} \right) = 169.635 = \mathbf{170. \ cents}$$

21.158   The half-reactions are (from the Appendix):

Oxidaton:	$H_2(g) \rightarrow 2 \ H^+(aq) + 2 \ e^-$	$E = 0.00 \text{ V}$
Reduction:	$\underline{2(Ag^+(aq) + 1 \ e^- \rightarrow Ag(s))}$	$E = 0.80 \text{ V}$
Overall:	$2 \ Ag^+(aq) + H_2(g) \rightarrow 2 \ Ag(s) + 2 \ H^+(aq)$	$E_{cell} = 0.80 \text{ V} - 0.0 \text{ V} = 0.80 \text{ V}$

The hydrogen ion concentration can now be found from the Nernst equation.

$$E = E^\circ - \frac{0.0592 \text{ V}}{n} \log Q$$

$$0.915 \text{ V} = 0.80 \text{ V} - \frac{0.0592 \text{ V}}{2} \log \frac{\left[H^+ \right]^2}{\left[Ag^+ \right]^2 P_{H_2}}$$

$$0.915 \text{ V} - 0.80 \text{ V} = - \frac{0.0592 \text{ V}}{2} \log \frac{\left[H^+ \right]^2}{\left[0.100 \right]^2 (1.00)}$$

$$(0.915 \text{ V} - 0.80 \text{ V})(-2 / 0.0592 \text{ V}) = \log \frac{\left[H^+ \right]^2}{\left[0.100 \right]^2 (1.00)}$$

$$-3.885135 = \log \frac{\left[H^+ \right]^2}{\left[0.0100 \right]}$$

$$1.30276 \times 10^{-4} = \frac{\left[H^+ \right]^2}{\left[0.0100 \right]}$$

$$[H^+] = 1.1413851 \times 10^{-3} \ M \text{ (unrounded)}$$

$$\text{pH} = -\log [H^+] = -\log (1.1413851 \times 10^{-3}) = 2.94256779 = \mathbf{2.94}$$

CHAPTER 22 THE ELEMENTS IN NATURE AND INDUSTRY

FOLLOW-UP PROBLEMS

There are no follow-up problems in this chapter.

END-OF-CHAPTER PROBLEMS

22.2   Iron forms iron(III) oxide, Fe_2O_3 (commonly known as hematite). Calcium forms calcium carbonate, $CaCO_3$, (commonly known as limestone). Sodium is commonly found in sodium chloride (halite), $NaCl$. Zinc is commonly found in zinc sulfide (sphalerite), ZnS.

22.3   a) Differentiation refers to the processes involved in the formation of the earth into regions (core, mantle, and crust) of differing composition. Substances separated according to their densities, with the more dense material in the core and less dense in the crust.
b) The four most abundant elements are oxygen, silicon, aluminum, and iron in order of decreasing abundance.
c) Oxygen is the most abundant element in the crust and mantle but is not found in the core. Silicon, aluminum, calcium, sodium, potassium, and magnesium are also present in the crust and mantle but not in the core.

22.7   Plants produced O_2, slowly increasing the oxygen concentration in the atmosphere and creating an oxidative environment for metals. Evidence of Fe(II) deposits pre-dating Fe(III) deposits suggests this hypothesis is true. The decay of plant material and its incorporation into the crust increased the concentration of carbon in the crust and created large fossil fuel deposits.

22.9   Fixation refers to the process of converting a substance in the atmosphere into a form more readily usable by organisms. Examples are the fixation of carbon by plants in the form of carbon dioxide and of nitrogen by bacteria in the form of nitrogen gas. Fixation of carbon dioxide gas by plants converts the CO_2 into carbohydrates during photosynthesis. Fixation of nitrogen gas by nitrogen-fixing bacteria involves the conversion of N_2 to ammonia and ammonium ions.

22.12   Atmospheric nitrogen is utilized by three fixation pathways: atmospheric, industrial, and biological. Atmospheric fixation requires high-temperature reactions (e.g., lightning) to convert N_2 into NO and other oxidized species. Industrial fixation involves mainly the formation of ammonia, NH_3, from N_2 and H_2. Biological fixation occurs in nitrogen-fixing bacteria that live in the roots of legumes. "Human activity" is referred to as "industrial fixation." Human activity is a significant factor, contributing about 17% of the nitrogen removed.

22.14   a) No gaseous phosphorus compounds are involved in the phosphorus cycle, so the atmosphere is not included in the phosphorus cycle.
b) Two roles of organisms in phosphorus cycle: 1) Plants excrete acid from their roots to convert PO_4^{3-} ions into more soluble $H_2PO_4^-$ ions, which the plant can absorb. 2) Through excretion and decay after death, organisms return soluble phosphate compounds to the cycle.

22.17   a) First determine the amount of F present in 100. kg of fluorapatite, using the molar mass of $Ca_5(PO_4)_3F$ (504.31 g/mol); take 15% of this amount. Convert mol F to mol SiF_4, and use the ideal gas law to determine volume of SiF_4 gas.

$$\text{Moles F} = \left(100.\ \text{kg } Ca_5(PO_4)_3F\right)\left(\frac{10^3\ \text{g } Ca_5(PO_4)_3F}{1\ \text{kg } Ca_5(PO_4)_3F}\right)\left(\frac{1\ \text{mol } Ca_5(PO_4)_3F}{504.31\ \text{g } Ca_5(PO_4)_3F}\right)\left(\frac{1\ \text{mol F}}{1\ \text{mol } Ca_5(PO_4)_3F}\right)\left(\frac{15\%}{100\%}\right)$$

$$= 29.74361\ \text{mol F (unrounded)}$$

$$\text{Moles } SiF_4 = \left(29.74361\ \text{mol F}\right)\left(\frac{1\ \text{mol } SiF_4}{4\ \text{mol F}}\right) = 7.43590\ \text{mol } SiF_4 \text{ (unrounded)}$$

$$V = nRT / P = \dfrac{\left(7.43590 \text{ mol SiF}_4\right)\left(0.0821 \dfrac{\text{L} \cdot \text{atm}}{\text{mol} \cdot \text{K}}\right)\left((273 + 1450.)\text{K}\right)}{\left(1.00 \text{ atm}\right)}$$

$$= 1051.86977 = \mathbf{1.1 \times 10^3 \text{ L}}$$

b) Assume that all of the fluorine in sodium hexafluorosilicate is available as F^- when redissolved in drinking water. First, calculate the moles of Na_2SiF_6 that form from 7.436 mol of SiF_4 according to the reaction stoichiometry. Then, calculate the mass of F^- available from the Na_2SiF_6 that is produced in the reaction.

$$\text{Mole Na}_2\text{SiF}_6 = \left(7.43590 \text{ mol SiF}_4\right)\left(\dfrac{1 \text{ mol Na}_2\text{SiF}_6}{2 \text{ mol SiF}_4}\right) = 3.71795 \text{ mol Na}_2\text{SiF}_6$$

$$\text{Mass F}^- = \left(3.71795 \text{ mol Na}_2\text{SiF}_6\right)\left(\dfrac{6 \text{ mol F}^-}{1 \text{ mol Na}_2\text{SiF}_6}\right)\left(\dfrac{19.00 \text{ g F}^-}{1 \text{ mol F}^-}\right) = 423.8463 = 4.2 \times 10^2 \text{ g F}^-$$

The definition of ppm ("parts per million") states that 1 ppm $F^- = (1 \text{ g } F^- / (10^6 \text{ g } H_2O)$. If 1 g F^- will fluoridate 10^6 g H_2O to a level of 1 ppm, how many grams (converted to mL using density, then converted from mL to L to m^3) of water can 424 g F^- fluoridate to the 1 ppm level? Necessary conversion factors: 1 m^3 = 1000 L, density of water = 1.00 g/mL .

$$\text{Volume} = \left(423.8463 \text{ g F}^-\right)\left(\dfrac{10^6 \text{ g H}_2\text{O}}{1 \text{ g F}^-}\right)\left(\dfrac{\text{mL H}_2\text{O}}{1.00 \text{ g H}_2\text{O}}\right)\left(\dfrac{1 \text{ cm}^3}{1 \text{ mL}}\right)\left(\dfrac{10^{-2} \text{ m}}{1 \text{ cm}}\right)^3$$

$$= 423.8463 = \mathbf{4.2 \times 10^2 \text{ m}^3}$$

22.18   a) The iron ions form an insoluble salt, $Fe_3(PO_4)_2$, that decreases the yield of phosphorus. This salt is of limited value.
b) Each mole of $Ca_3(PO_4)_2$ contains 2 moles of P atoms and produces, at 100% yield, 0.5 mol of P_4. Use conversion factor 1 metric ton (T) = 1 x 10^3 kg.

$$\text{Mass Ca}_3(\text{PO}_4)_2 = \left(50. \text{ T}\right)\left(\dfrac{98\%}{100\%}\right)\left(\dfrac{10^3 \text{ kg}}{1 \text{ T}}\right)\left(\dfrac{10^3 \text{ g}}{1 \text{ kg}}\right)\left(\dfrac{1 \text{ mol Ca}_3(\text{PO}_4)_2}{310.18 \text{ g Ca}_3(\text{PO}_4)_2}\right) \text{ x}$$

$$\left(\dfrac{1 \text{ mol P}_4}{2 \text{ mol Ca}_3(\text{PO}_4)_2}\right)\left(\dfrac{123.88 \text{ g P}_4}{1 \text{ mol P}_4}\right)\left(\dfrac{1 \text{ kg}}{10^3 \text{ g}}\right)\left(\dfrac{1 \text{ T}}{10^3 \text{ kg}}\right)\left(\dfrac{90.\%}{100\%}\right)$$

$$= 8.80635 = \mathbf{8.8 \text{ metric tons P}_4}$$

22.20   a) Roasting involves heating the mineral in air (O_2) at high temperatures to convert the mineral to the oxide.
b) Smelting is the reduction of the metal oxide to the free metal using heat and a reducing agent such as coke.
c) Flotation is a separation process in which the ore is removed from the gangue by exploiting the different abilities of the two to interact with detergent. The gangue sinks to the bottom and the lighter ore-detergent mix is skimmed off the top.
d) Refining is the final step in the purification process to yield the pure metal with no impurities.

22.25   a) Slag is a waste product of iron metallurgy formed by the reaction
$$\text{CaO}(s) + \text{SiO}_2(s) \rightarrow \text{CaSiO}_3(l)$$
In other words, slag is a by-product of steel-making and contains the impurity SiO_2.
b) Pig iron, used to make cast iron products, is the impure product of iron metallurgy (containing 3 – 4% C) that is purified to steel.
c) Steel refers to the products of iron metallurgy, specifically alloys of iron containing small amounts of other elements including 1 – 1.5% carbon.
d) Basic-oxygen process refers to the process used to purify pig iron to form steel. The pig iron is melted and oxygen gas under high pressure is passed through the liquid metal. The oxygen oxidizes impurities to their oxides, which then react with calcium oxide to form a liquid that is decanted. Molten steel is left after the basic-oxygen process.

22.27 Iron and nickel are more easily oxidized than copper, so they are separated from the copper in the roasting step and conversion to slag. In the electrorefining process, all three metals are oxidized into solution, but only Cu^{2+} ions are reduced at the cathode to form $Cu(s)$.

22.30 Le Châtelier's principle says that the system shifts toward formation of K as the gaseous metal leaves the cell.

22.31 a) Aqueous salt solutions are mixtures of ions and water. When two half-reactions are possible at an electrode, the one with the more positive electrode potential occurs (Section 21.7). In this case, the two half-reactions are

$$M^+ + e^- \rightarrow M^0 \qquad\qquad E°_{red} = -3.05 \text{ V}, -2.93 \text{ V}, \text{ and } -2.71 \text{ V}$$
$$\text{for Li}^+, K^+, \text{and Na}^+, \text{respectively}$$
$$2\,H_2O + 2\,e^- \rightarrow H_2 + 2\,OH^- \qquad E_{red} = -0.42 \text{ V, with an overvoltage of about } -1 \text{ V}$$

In all of these cases, it is energetically more favorable to reduce H_2O to H_2 than to reduce M^+ to M.
b) The question asks if Ca could chemically reduce RbX, i.e., convert Rb^+ to Rb^0. In order for this to occur, Ca^0 loses electrons ($Ca^0 \rightarrow Ca^{2+} + 2\,e^-$) and each Rb^+ gains an electron ($2\,Rb^+ + 2\,e^- \rightarrow 2\,Rb^0$). The reaction is written as follows:

$$2\,RbX + Ca \rightarrow CaX_2 + 2\,Rb$$

where $\Delta H = IE_1(Ca) + IE_2(Ca) - 2\,IE_1(Rb) = 590 + 1145 - 2(403) = +929 \text{ kJ/mol}$.
Recall that Ca^0 acts as a *reducing agent* for the Rb^+ ion because it *oxidizes*. The energy required to remove an electron is the ionization energy. It requires more energy to ionize calcium's electrons, so it seems unlikely that Ca^0 could reduce Rb^+. Based on IE's and a positive ΔH for the forward reaction, it seems more reasonable that Rb^0 would reduce Ca^{2+}.
c) If the reaction is carried out at a temperature greater than 688°C (the boiling point of rubidium), the product mixture will contain gaseous Rb. This can be removed from the reaction vessel, causing a shift in equilibrium to form more Rb product. If the reaction is carried out between 688°C and 1484°C (b.p. for Ca), then Ca remains in the molten phase and remains separated from gaseous Rb.
d) The reaction of calcium with molten CsX is written as follows:

$$2\,CsX + Ca \rightarrow CaX_2 + 2\,Cs$$

where $\Delta H = IE_1(Ca) + IE_2(Ca) - 2\,IE_1(Cs) = 590 + 1145 - 2(376) = +983 \text{ kJ/mol}$. This reaction is more unfavorable than for Rb, but Cs has a lower boiling point of 671°C. If the reaction is carried out between 671°C and 1484°C, then calcium can be used to separate gaseous Cs from molten CsX.

22.32 a) For each mole of Na metal produced, 0.5 mole of Cl_2 is produced. Calculate amount of chlorine gas from stoichiometry, then use ideal gas law to find volume of chlorine gas.

$$2\,NaCl(l) \rightarrow 2\,Na(l) + Cl_2(g)$$

$$\text{Moles } Cl_2 = \left(31.0 \text{ kg Na}\right)\left(\frac{10^3 \text{ g}}{1 \text{ kg}}\right)\left(\frac{1 \text{ mol Na}}{22.99 \text{ g Na}}\right)\left(\frac{1 \text{ mol } Cl_2}{2 \text{ mol Na}}\right) = 674.2062 \text{ mol } Cl_2 \text{ (unrounded)}$$

$$V = nRT / V = \frac{\left(674.2062 \text{ mol } Cl_2\right)\left(0.0821 \dfrac{L \cdot atm}{mol \cdot K}\right)\left((273 + 540.)K\right)}{(1.0 \text{ atm})}$$

$$= 4.50014 \times 10^4 = \mathbf{4.5 \times 10^4 \ L}$$

b) Two moles of electrons are passed through the cell for each mole of Cl_2 produced.

$$\text{Coulombs} = \left(674.2062 \text{ mol } Cl_2\right)\left(\frac{2 \text{ mol } e^-}{1 \text{ mol } Cl_2}\right)\left(\frac{96485 \text{ C}}{1 \text{ mol } e^-}\right)$$

$$= 1.30101570 \times 10^8 = \mathbf{1.30 \times 10^8 \ Coulombs}$$

c) Current is charge per time with the amp unit equal to C / s.

$$\left(1.30101570 \times 10^8 \text{ C}\right)\left(\frac{1 \text{ s}}{77 \text{ C}}\right) = 1.689631 \times 10^6 = \mathbf{1.69 \times 10^6 \ seconds}$$

22.35 a) Mg^{2+} is more difficult to reduce than H_2O, so $H_2(g)$ would be produced instead of Mg metal. $Cl_2(g)$ forms at the anode due to overvoltage.
b) The ΔH_f° for $MgCl_2(s)$ is -641.6 kJ/mol: $Mg(s) + Cl_2(g) \rightarrow MgCl_2(s) + heat$
High temperature favors the reverse reaction (which is endothermic), so high temperatures favor the formation of magnesium metal and chlorine gas.

22.37 a) Sulfur dioxide is the reducing agent and is oxidized to the +6 state (as sulfate ion, SO_4^{2-})
b) The sulfate ion formed reacts as a base in the presence of acid to form the hydrogen sulfate ion.

$$SO_4^{2-}(aq) + H^+(aq) \rightarrow HSO_4^- \; aq)$$

c) Skeleton redox equation: $\quad SO_2(g) + H_2SeO_3(aq) \rightarrow Se(s) + HSO_4^-(aq)$

Reduction half-reaction: $\quad H_2SeO_3(aq) \rightarrow Se(s)$

balance O and H $\quad H_2SeO_3(aq) + 4\,H^+(aq) \rightarrow Se(s) + 3\,H_2O(l)$

balance charge $\quad H_2SeO_3(aq) + 4\,H^+(aq) + 4\,e^- \rightarrow Se(s) + 3\,H_2O(l)$

Oxidation half-reaction: $\quad SO_2(g) \rightarrow HSO_4^- \; aq)$

balance O and H $\quad SO_2(g) + 2\,H_2O(l) \rightarrow HSO_4^-(aq) + 3\,H^+(aq)$

balance charge $\quad SO_2(g) + 2\,H_2O(l) \rightarrow HSO_4^-(aq) + 3\,H^+(aq) + 2\,e^-$

Multiply oxidation half-reaction by 2 and add the two half-reactions.

$$H_2SeO_3(aq) + 2\,SO_2(g) + H_2O(l) \rightarrow Se(s) + 2\,HSO_4^-(aq) + 2\,H^+(aq)$$

22.42 a) The oxidation state of copper in Cu_2S is +1 (sulfur is –2).
The oxidation state of copper in Cu_2O is +1 (oxygen is –2).
The oxidation state of copper in Cu is 0 (ox state is always 0 in the elemental form).
b) The reducing agent is the species that is *oxidized*. Sulfur changes oxidation state from –2 in Cu_2S to +4 in SO_2. Therefore, Cu_2S is the reducing agent, leaving Cu_2O as the oxidizing agent.

22.44 a) Use the surface area and thickness of the film to calculate its volume. Multiply the volume by the density to find the mass of the aluminum oxide. Convert mass to moles of aluminum oxide. The oxidation of aluminum involves the loss of 3 electrons for each aluminum atom or 6 electrons for each Al_2O_3 compound formed. From this, find the number of electrons produced and multiply by Faraday's constant to find coulombs.

$$\text{Moles } Al_2O_3 = \left(2.5 \text{ m}^2\right)\left(23. \times 10^{-6}\text{ m}\right)\left(\frac{1 \text{ cm}}{10^{-2}\text{ m}}\right)^3\left(\frac{3.97 \text{ g } Al_2O_3}{\text{cm}^3}\right)\left(\frac{1 \text{ mol } Al_2O_3}{101.96 \text{ g } Al_2O_3}\right)$$

$$= 2.2388682 \text{ mol } Al_2O_3$$

$$\text{Coulombs} = \left(2.2388682 \text{ mol } Al_2O_3\right)\left(\frac{6 \text{ mol } e^-}{1 \text{ mol } Al_2O_3}\right)\left(\frac{96485 \text{ C}}{1 \text{ mol } e^-}\right) = 1.296103 \times 10^6 = \mathbf{1.3 \times 10^6 \text{ C}}$$

b) Current in amps can be found by dividing charge in coulombs by the time in seconds.
$[(1.296103 \times 10^6 \text{ C}) / [15 \text{ min } (60 \text{ s } / \text{ min})] \times (1 \text{ A } / (\text{C } / \text{ s})) = 1200 = \mathbf{1.2 \times 10^3 \text{ A}}$

22.47 Free energy, ΔG, is the measure of the ability of a reaction to proceed spontaneously. The direct reduction of ZnS follows the reaction:

$2\,ZnS(s) + C(\text{graphite}) \rightarrow 2\,Zn(s) + CS_2(g)$

$\Delta G^\circ_{rxn} = [2\,\Delta H_f^\circ(Zn) + 1\Delta H_f^\circ(CS_2)] - [2\Delta H_f^\circ(ZnS) + 1\Delta H_f^\circ(C)]$

$\Delta G^\circ_{rxn} = [(2 \text{ mol})(0) + (1 \text{ mol})(66.9 \text{ kJ/mol})] - [(2 \text{ mol})(-198 \text{ kJ/mol}) + (1 \text{ mol})(0)]$

$\Delta G^\circ_{rxn} = 462.9 = +463 \text{ kJ}$

Since ΔG°_{rxn} is positive, this reaction is not spontaneous at standard-state conditions. The stepwise reduction involves the conversion of ZnS to ZnO, followed by the final reduction step:

$2\,ZnO(s) + C(s) \rightarrow 2\,Zn(s) + CO_2(g)$

$\Delta G^\circ_{rxn} = [2\,\Delta H_f^\circ(Zn) + 1\Delta H_f^\circ(CO_2)] - [2\Delta H_f^\circ(ZnO) + 1\Delta H_f^\circ(C)]$

$\Delta G^\circ_{rxn} = [(2 \text{ mol})(0) + (1 \text{ mol})(-394.4 \text{ kJ/mol})] - [(2 \text{ mol})(-318.2 \text{ kJ/mol}) + (1 \text{ mol})(0)]$

$\Delta G^\circ_{rxn} = +242.0 \text{ kJ}$

This reaction is also not spontaneous, but the oxide reaction is less unfavorable.

22.48 The formation of sulfur trioxide is very slow at ordinary temperatures. Increasing the temperature can speed up the reaction, but the reaction is exothermic, so increasing the temperature decreases the yield. Recall that yield, or the extent to which a reaction proceeds, is controlled by the thermodynamics of the reaction. Adding a catalyst increases the rate of the formation reaction but does not impact the thermodynamics, so a lower temperature can be used to enhance the yield. Catalysts are used in such a reaction to allow control of both rate and yield of the reaction.

22.51 a) The chlor-alkali process yields Cl_2, H_2, and NaOH.
b) The mercury-cell process yields higher purity NaOH, but produces Hg-polluted waters that are discharged into the environment.

22.52 a) The balanced equation for the oxidation of SO_2 to SO_3: $2 SO_2(g) + O_2(g) \rightarrow 2 SO_3(g)$
$\Delta G° = [(2 \text{ mol } SO_3) (\Delta G_f^o \text{ of } SO_3)] - [(2 \text{ mol } SO_2) (\Delta G_f^o \text{ of } SO_2) + (1 \text{ mol } O_2) (\Delta G_f^o \text{ of } O_2)]$

$\Delta G° = [(2 \text{ mol } SO_3) (-371 \text{ kJ / mol } SO_3)] - [(2 \text{ mol } SO_2) (-300.2 \text{ kJ / mol } SO_2) + (1 \text{ mol } O_2) (0)]$
$= -141.6 = -142 \text{ kJ}$
The reaction is spontaneous at 25°C.
b) The rate of the reaction is very slow at 25°C, so the reaction does not produce significant amounts of SO_3 at room temperature.
c) Calculate standard state free energy at 500.°C (773 K), using values for $\Delta H°$ and $\Delta S°$ at 25°C with the assumption that these values are constant with temperature.
$\Delta H° = [(2 \text{ mol } SO_3) (\Delta H_f^o SO_3)] - [(2 \text{ mol } SO_2) (\Delta H_f^o SO_2) + (1 \text{ mol } O_2) (\Delta H_f^o O_2)]$

$\Delta H° = [(2 \text{ mol } SO_3) (-396 \text{ kJ / mol } SO_3)] - [(2 \text{ mol } SO_2) (-296.8 \text{ kJ / mol } SO_2) + (1 \text{ mol } O_2) (0)]$
$= -198.4 = -198 \text{ kJ}$
$\Delta S° = [(2 \text{ mol } SO_3) (S° \text{ of } SO_3)] - [(2 \text{ mol } SO_2) (S° \text{ of } SO_2) + (1 \text{ mol } O_2) (S° \text{ of } O_2)]$
$\Delta S° = [(2 \text{ mol } SO_3) (256.66 \text{ J/mol·K})] - [(2 \text{ mol } SO_2) (248.1 \text{ J/mol·K}) + (1 \text{ mol } O_2) (205.0 \text{ J/mol·K})]$
$= -187.88 = -187.9 \text{ J/K}$
$\Delta G_{500} = \Delta H° - T\Delta S° = -198.4 \text{ kJ} - ((273 + 500.) \text{ K}) (-187.88 \text{ J/K}) (1 \text{ kJ} / 10^3 \text{ J}) = -53.16876 = \mathbf{-53 \text{ kJ}}$
The reaction is spontaneous at 500°C since free energy is negative.
d) The equilibrium constant at 500°C is smaller than the equilibrium constant at 25°C because the free energy at 500°C indicates the reaction does not go as far to completion as it does at 25°C.
Equilibrium constants can be calculated from $\Delta G° = -RT \ln K$.

At 25°C $\ln K = -\dfrac{\Delta G°}{RT} = -\dfrac{(-142 \text{ kJ})\left(\dfrac{10^3 \text{ J}}{1 \text{ kJ}}\right)}{(8.314 \text{ J/mol·K}) ((273 + 25)K)} = 57.31417694$

$K_{25} = e^{57.31417694} = 7.784501 \times 10^{24} = 7.8 \times 10^{24}$

At 500°C $\ln K = -\dfrac{\Delta G°}{RT} = -\dfrac{(-53 \text{ kJ})\left(\dfrac{10^3 \text{ J}}{1 \text{ kJ}}\right)}{(8.314 \text{ J/mol·K}) ((273 + 500)K)} = 8.246817$

$K_{500} = e^{8.2468} = 3.815 \times 10^3 = 3.8 \times 10^3$
e) Temperature below which the reaction is spontaneous at standard state can be calculated by setting $\Delta G°$ equal to zero and using the enthalpy and entropy values at 25°C to calculate the temperature.

$T = \dfrac{\Delta H°}{\Delta S°} = \dfrac{-198 \text{ kJ}}{-187.9 \text{ J/K}\left(\dfrac{1 \text{ kJ}}{10^3 \text{ J}}\right)} = 1.05375 \times 10^3 = \mathbf{1.05 \times 10^3 \text{ K}}$

22.53 This problem deals with the stoichiometry of electrolysis. The balanced oxidation half reaction for the chlor-alkali process is given in the chapter:
$$2 Cl^-(aq) \rightarrow Cl_2(g) + 2 e^-$$
Use the Faraday constant, F, (1 $F = 9.6485 \times 10^4$ C / mol e$^-$) and the fact that 1 mol of Cl_2 produces 2 mol e$^-$ or 2 F, to convert coulombs to moles of Cl_2.

$$\text{Mass Cl}_2 = \left(3 \times 10^4 \, A\right)\left(\frac{C/s}{A}\right)\left(\frac{3600 \, s}{1 \, h}\right)(8 \, h)\left(\frac{1 \, \text{mol e}^-}{96485 \, C}\right)\left(\frac{1 \, \text{mol Cl}_2}{2 \, \text{mol e}^-}\right)\left(\frac{70.90 \, \text{g Cl}_2}{1 \, \text{mol Cl}_2}\right)\left(\frac{1 \, \text{kg}}{10^3 \, \text{g}}\right)\left(\frac{2.205 \, \text{lb}}{1 \, \text{kg}}\right)$$

$$= 699.9689 = \textbf{7} \times \textbf{10}^\textbf{2} \textbf{ pounds Cl}_\textbf{2}$$

22.57   a) Because P_4O_{10} is a drying agent, water is incorporated in the formation of phosphoric acid, H_3PO_4.

$$P_4O_{10}(s) + 6 \, H_2O(l) \rightarrow 4 \, H_3PO_4(l)$$

b) First calculate the concentration (M) of the resulting phosphoric acid solution.

$$\text{Molarity} = \left(\frac{8.5 \, \text{g P}_4\text{O}_{10}}{0.750 \, \text{L}}\right)\left(\frac{1 \, \text{mol P}_4\text{O}_{10}}{283.88 \, \text{g P}_4\text{O}_{10}}\right)\left(\frac{4 \, \text{mol H}_3\text{PO}_4}{1 \, \text{mol P}_4\text{O}_{10}}\right) = 0.1596919 \, M \, H_3PO_4 \text{ (unrounded)}$$

Phosphoric acid is a weak acid and only partly dissociates to form H_3O^+ in water, based on its K_a. Since $K_{a1} \gg K_{a2} \gg K_{a3}$, assume that the H_3O^+ contributed by the second and third dissociation is negligible in comparison to the H_3O^+ contributed by the first dissociation.

$$H_3PO_4(l) + H_2O(l) \rightarrow H_3O^+(aq) + H_2PO_4^-(aq)$$

$$K_a = 7.2 \times 10^{-3} = \frac{\left[H_3O^+\right]\left[H_2PO_4^-\right]}{\left[H_3PO_4\right]}$$

$$K_a = 7.2 \times 10^{-3} = \frac{[x][x]}{(0.1596919 - x)}$$

The problem will need to be solved as a quadratic.

$$x^2 = K_a(0.1596919 - x) = (7.2 \times 10^{-3})(0.1596919 - x) = 1.14978 \times 10^{-3} - 7.2 \times 10^{-3} \, x$$
$$x^2 + 7.2 \times 10^{-3} \, x - 1.14978 \times 10^{-3} = 0$$
$$a = 1 \qquad b = 7.2 \times 10^{-3} \qquad c = -1.14978 \times 10^{-3}$$

$$x = \frac{-7.2 \times 10^{-3} \pm \sqrt{\left(7.2 \times 10^{-3}\right)^2 - 4(1)\left(-1.14978 \times 10^{-3}\right)}}{2(1)}$$

$$x = 3.049897 \times 10^{-2} \, M \, H_3O^+ \text{ (unrounded)}$$
$$pH = -\log [H^+] = -\log (3.049897 \times 10^{-2}) = 1.51571 = \textbf{1.52}$$

22.59   The reaction taking place in the blast furnace to produce iron is

$$Fe_2O_3(s) + 3 \, CO(g) \rightarrow 2 \, Fe(s) + 3 \, CO_2(g)$$

(t = ton = 2000 lbs, and not T = metric ton = 1000 kg)
a) Three moles of carbon dioxide are produced for every 2 moles of Fe.

$$\text{Mass} = \left(8400.\text{t Fe}\right)\left(\frac{2000 \, \text{lb}}{1 \, \text{t}}\right)\left(\frac{1 \, \text{kg}}{2.205 \, \text{lb}}\right)\left(\frac{10^3 \, \text{g}}{1 \, \text{kg}}\right)\left(\frac{1 \, \text{mol Fe}}{55.85 \, \text{g Fe}}\right)\left(\frac{3 \, \text{mol CO}_2}{2 \, \text{mol Fe}}\right)\left(\frac{44.01 \, \text{g CO}_2}{1 \, \text{mol CO}_2}\right)$$

$$= 9.005755 \times 10^9 = \textbf{9.006} \times \textbf{10}^\textbf{9} \textbf{ g CO}_\textbf{2}$$

b) Combustion reaction for octane:

$$2 \, C_8H_{18}(l) + 25 \, O_2 \rightarrow 16 \, CO_2(g) + 18 \, H_2O(l)$$

$$\left(1.0 \times 10^6 \, \text{autos}\right)\left(\frac{5.0 \, \text{gal}}{1 \, \text{auto}}\right)\left(\frac{4 \, \text{qt}}{1 \, \text{gal}}\right)\left(\frac{1 \, \text{L}}{1.057 \, \text{qt}}\right)\left(\frac{1 \, \text{mL}}{10^{-3} \, \text{L}}\right)\left(\frac{0.74 \, \text{g}}{\text{mL}}\right)\left(\frac{1 \, \text{mol C}_8\text{H}_{18}}{114.22 \, \text{g C}_8\text{H}_{18}}\right)\left(\frac{16 \, \text{mol CO}_2}{2 \, \text{mol C}_8\text{H}_{18}}\right)\left(\frac{44.01 \, \text{g CO}_2}{1 \, \text{mol CO}_2}\right)$$

$$= 4.31604 \times 10^{10} = \textbf{4.3} \times \textbf{10}^\textbf{10} \textbf{ g CO}_\textbf{2}$$

The CO_2 production by automobiles is much greater than that from steel production.

22.61   a) Step 2 of "Isolation and Uses of Magnesium" describes this reaction. Sufficient OH^- must be added to precipitate $Mg(OH)_2$. The necessary amount of OH^- can be determined from the K_{sp} of $Mg(OH)_2$.

$$Mg(OH)_2(s) \rightarrow Mg^{2+}(aq) + 2 \, OH^-(aq)$$
$$K_{sp} = 6.3 \times 10^{-10} = [Mg^{2+}][OH^-]^2$$
$$[OH^-] = [(6.3 \times 10^{-10}) / (0.052)]^{1/2} = 1.100699 \times 10^{-4} = 1.1 \times 10^{-4} \, M$$
Thus, if $[OH^-] > 1.1 \times 10^{-4} \, M$ (i.e., if pH > 10.04), $Mg(OH)_2$ will precipitate.

b) The amount of OH^- that a saturated solution of $Ca(OH)_2$ provides is dependent on its K_{sp}.

$$Ca(OH)_2(s) \rightarrow Ca^{2+}(aq) + 2OH^-(aq)$$

$K_{sp} = 6.5 \times 10^{-6} = [Ca^{2+}][OH^-]^2 = (x)(2x)^2 = 4x^3$

$x = 0.011756673$; $[OH^-] = 2x = 0.023513346\ M$ (unrounded)

Substitute this amount into the K_{sp} expression for $Mg(OH)_2$ to determine how much Mg^{2+} reacts with this amount of OH^-.

$K_{sp}\ (Mg(OH)_2) = 6.3 \times 10^{-10} = [Mg^{2+}][OH^-]^2$

$[Mg^{2+}] = (6.3 \times 10^{-10})\ /\ (0.023513346)^2 = 1.1394929 \times 10^{-6}\ M$ (unrounded)

This concentration is the amount of the original $0.052\ M\ Mg^{2+}$ that was not precipitated. The percent precipitated is the difference between the remaining $[Mg^{2+}]$ and the initial $[Mg^{2+}]$ divided by the initial concentration.

Fraction Mg^{2+} remaining = $(1.1394929 \times 10^{-6}\ M)\ /\ (0.052\ M) = 2.19129 \times 10^{-5}$ (unrounded)

Mg^{2+} precipitated = $1 - 2.19129 \times 10^{-5} = 0.9999781 = 1$ (to the limit of the significant figures, all the magnesium has precipitated.)

22.63   It will be necessary to calculate K_p from ΔG. The equation $\Delta G = \Delta H - T\Delta S$ will be used to calculate ΔG at the two temperatures. Then use $\Delta G = -RT \ln K$ to determine K. Subscripts indicate the temperature, and (1) or (2) indicate the reaction.

For the first reaction (1):

ΔH°_{rxn} = [4 mol NO (90.29 kJ/mol) + 6 mol H_2O (–241.826 kJ/mol)] – [4 mol NH_3 (–45.9 kJ/mol)
    + 5 mol O_2 (0 kJ/mol)] = –906.196 kJ (unrounded)

ΔS°_{rxn} = [4 mol NO (210.65 J/K•mol) + 6 mol H_2O (188.72 J/K•mol)] – [4 mol NH_3 (193 J/K•mol)
    + 5 mol O_2 (205.0 J/K•mol)]}(1 kJ / 10^3 J) = 0.17792 kJ/K (unrounded)

$\Delta G = \Delta H - T\Delta S$

ΔG_{25} = –906.196 kJ – [(273 + 25)K](0.17792 kJ/K) = –959.216 kJ (unrounded)

ΔG_{900} = –906.196 kJ – [(273 + 900.)K](0.17792 kJ/K) = –1114.896 kJ (unrounded)

For the second reaction (2):

ΔH°_{rxn} = [2 mol N_2 (0 kJ/mol) + 6 mol H_2O (–241.826 kJ/mol)]

    – [4 mol NH_3(–45.9 kJ/mol) + 3 mol O_2 (0 kJ/mol)]

    = –1267.356 kJ (unrounded)

ΔS°_{rxn} = [2 mol N_2 (191.5 J/K•mol) + 6 mol H_2O (188.72 J/K•mol)] – [4 mol NH_3 (193 J/K•mol)
    + 3 mol O_2 (205.0 J/K•mol)]}(1 kJ / 10^3 J) = 0.12832 J/K (unrounded)

$\Delta G = \Delta H - T\Delta S$

ΔG_{25} = –1267.356 kJ – [(273 + 25)K](0.12832 kJ/K) = –1305.595 kJ (unrounded)

ΔG_{900} = –1267.356 kJ – [(273 + 900.)K](0.12832 kJ/K) = –1417.875 kJ (unrounded)

a) $\ln [K_{25}(1)] = -\dfrac{\Delta G}{RT} = -\dfrac{(-959.216\ kJ/mol)}{(8.314\ J/mol \cdot K)((273 + 25)K)}\left(\dfrac{10^3\ J}{1\ kJ}\right) = 387.159687$ (unrounded)

$K_{25}(1) = 1.38 \times 10^{168} = \mathbf{1 \times 10^{168}}$

$\ln [K_{25}(2)] = -\dfrac{\Delta G}{RT} = -\dfrac{(-1305.595\ kJ/mol)}{(8.314\ J/mol \cdot K)((273 + 25)K)}\left(\dfrac{10^3\ J}{1\ kJ}\right) = 526.9655$ (unrounded)

$K_{25}(2) = 7.2146 \times 10^{228} = \mathbf{7 \times 10^{228}}$

b) $\ln [K_{900}(1)] = -\dfrac{\Delta G}{RT} = -\dfrac{(-1114.896\ kJ/mol)}{(8.314\ J/mol \cdot K)((273 + 900)K)}\left(\dfrac{10^3\ J}{1\ kJ}\right) = 114.3210817$ (unrounded)

$K_{900}(1) = 4.4567 \times 10^{49} = \mathbf{4.6 \times 10^{49}}$

$\ln [K_{900}(2)] = -\dfrac{\Delta G}{RT} = -\dfrac{(-1417.875\ kJ/mol)}{(8.314\ J/mol \cdot K)((273 + 900)K)}\left(\dfrac{10^3\ J}{1\ kJ}\right) = 145.388$ (unrounded)

$K_{900}(2) = 1.38422 \times 10^{63} = \mathbf{1.4 \times 10^{63}}$

c) Cost (\$) Pt $= \left(1.01 \times 10^7 \text{ t HNO}_3\right)\left(\dfrac{175 \text{ mg Pt}}{1 \text{ t HNO}_3}\right)\left(\dfrac{10^{-3} \text{ g}}{1 \text{ mg}}\right)\left(\dfrac{1 \text{ kg}}{10^3 \text{ g}}\right)\left(\dfrac{32.15 \text{ Troy Oz}}{1 \text{ kg}}\right)\left(\dfrac{\$1260}{1 \text{ Troy Oz}}\right)$

$\qquad = 7.159966 \times 10^7 = \textbf{\$ 7.2} \times \textbf{10}^\textbf{7}$

d) $\left(1.01 \times 10^7 \text{ t HNO}_3\right)\left(\dfrac{175 \text{ mg Pt}}{1 \text{ t HNO}_3}\right)\left(\dfrac{72\%}{100\%}\right)\left(\dfrac{10^{-3} \text{ g}}{1 \text{ mg}}\right)\left(\dfrac{1 \text{ kg}}{10^3 \text{ g}}\right)\left(\dfrac{32.15 \text{ troy oz}}{1 \text{ kg}}\right)\left(\dfrac{\$1260}{1 \text{ troy oz}}\right)$

$\qquad = \$5.155175 \times 10^7 = \textbf{\$5.16} \times \textbf{10}^\textbf{7}$

22.67 (1) $2 \text{ H}_2\text{O}(l) + 2 \text{ FeS}_2(s) + 7 \text{ O}_2(g) \rightarrow 2 \text{ Fe}^{2+}(aq) + 4 \text{ SO}_4^{2-}(aq) + 4 \text{ H}^+(aq)$ Increase acidity
       (2) $4 \text{ H}^+(aq) + 4 \text{ Fe}^{2+}(aq) + \text{O}_2(g) \rightarrow 4 \text{ Fe}^{3+}(aq) + 2 \text{ H}_2\text{O}(l)$
       (3) $\text{Fe}^{3+}(aq) + 3 \text{ H}_2\text{O}(l) \rightarrow \text{Fe(OH)}_3(s) + 3 \text{ H}^+(aq)$ Increase acidity
       (4) $8 \text{ H}_2\text{O}(l) + \text{FeS}_2(s) + 14 \text{ Fe}^{3+}(aq) \rightarrow 15 \text{ Fe}^{2+}(aq) + 2 \text{ SO}_4^{2-}(aq) + 16 \text{ H}^+(aq)$ Increase acidity

22.69 The carbon fits into interstitial positions. The cell size will increase slightly to accommodate the carbons atoms. The increase in size is assumed to be negligible. The mass of the carbon added depends upon the carbon in the unit cell.
Ferrite:

$$\left(\dfrac{7.86 \text{ g}}{\text{cm}^3}\right) + \left(\dfrac{7.86 \text{ g}}{\text{cm}^3}\right)\left(\dfrac{0.0218\%}{100\%}\right) = 7.86171 = \textbf{7.86 g/cm}^\textbf{3}$$

Austenite:

$$\left(\dfrac{7.40 \text{ g}}{\text{cm}^3}\right) + \left(\dfrac{7.40 \text{ g}}{\text{cm}^3}\right)\left(\dfrac{2.08\%}{100\%}\right) = 7.55392 = \textbf{7.55 g/cm}^\textbf{3}$$

22.71 a) Decide what is reduced and what is oxidized of the two ions present in molten NaOH. Then, write half-reactions and balance them.
       At the cathode, sodium ions are reduced: $\text{Na}^+(l) + \text{e}^- \rightarrow \text{Na}(l)$
       At the anode, hydroxide ions are oxidized: $4 \text{ OH}^- (l) \rightarrow \text{O}_2(g) + 2 \text{ H}_2\text{O}(g) + 4 \text{ e}^-$
       b) The overall cell reaction is $4 \text{ Na}^+(l) + 4 \text{ OH}^-(l) \rightarrow 4 \text{ Na}(l) + \text{O}_2(g) + 2 \text{ H}_2\text{O}(g)$ and the reaction between sodium metal and water is $2 \text{ Na}(l) + 2 \text{ H}_2\text{O}(g) \rightarrow 2 \text{ NaOH}(l) + \text{H}_2(g)$.
       For each mole of water produced, 2 moles of sodium are produced, but for each mole of water that reacts only 1 mole of sodium reacts. So, the maximum amount of sodium that reacts with water is half of the sodium produced. In the balanced cell reaction, 4 moles of electrons are transferred for 4 moles of sodium. Since 1/2 of the Na reacts with H_2O, the maximum efficiency is 1/2 mol Na/mol e⁻, or **50%**.

22.74 a) The reaction for the fixation of carbon dioxide includes CO_2 and H_2O as reactants and $(\text{CH}_2\text{O})_n$ and O_2 as products. The balanced equation is:

$$n \text{ CO}_2(g) + n \text{ H}_2\text{O}(l) \rightarrow (\text{CH}_2\text{O})_n(s) + n \text{ O}_2(g)$$

b) The moles of carbon dioxide fixed equal the moles of O_2 produced.

$$\text{Mole O}_2 / \text{day} = \left(\dfrac{48 \text{ g CO}_2}{\text{day}}\right)\left(\dfrac{1 \text{ mol CO}_2}{44.01 \text{ g CO}_2}\right)\left(\dfrac{1 \text{ mol O}_2}{1 \text{ mol CO}_2}\right) = 1.090661 \text{ mol O}_2 / \text{day}$$

To find the volume of oxygen, use the ideal gas equation. Temperature must be converted to K.

$$T = \dfrac{5}{9}\left[T(^\circ\text{F}) - 32\right] + 273.15 = \dfrac{5}{9}\left[T(78.^\circ) - 32\right] + 273.15 = 298.7056 \text{ K}$$

$$V = nRT / P = \dfrac{\left(1.090661 \text{ mol O}_2\right)\left(0.0821 \text{ L} \cdot \text{atm/mol} \cdot \text{K}\right)\left(298.7056 \text{ K}\right)}{\left(1.0 \text{ atm}\right)} = 26.74708 = \textbf{27 L}$$

c) The moles of air containing 1.0090661 mol of CO_2 is

$$\text{Mole } CO_2 \text{ / day} = \left(\frac{48 \text{ g } CO_2}{\text{day}}\right)\left(\frac{1 \text{ mol } CO_2}{44.01 \text{ g } CO_2}\right)\left(\frac{100 \text{ mol \% air}}{0.035 \text{ mol \% } CO_2}\right) = 3.11617 \times 10^3 \text{ mol air / day}$$

$$V = nRT / P = \frac{\left(3.11617 \times 10^3 \text{ mol air}\right)\left(0.0821 \text{ L} \cdot \text{atm/mol} \cdot \text{K}\right)\left(298.7056 \text{ K}\right)}{(1.0 \text{ atm})} = 7.6420 \times 10^4 = \mathbf{7.6 \times 10^4 \text{ L}}$$

22.75 This problem requires a series of conversion steps:

Concentration =

$$\left(\frac{210. \text{ kg}(NH_4)_2SO_4}{1000. \text{ m}^3}\right)\left(\frac{10^3 \text{ g}}{1 \text{ kg}}\right)\left(\frac{1 \text{ mol } (NH_4)_2SO_4}{132.15 \text{ g } (NH_4)_2SO_4}\right)\left(\frac{2 \text{ mol } NO_3^-}{1 \text{ mol } (NH_4)_2SO_4}\right)\left(\frac{37\%}{100\%}\right)\left(\frac{62.01 \text{ g } NO_3^-}{1 \text{ mol } NO_3^-}\right)\left(\frac{1 \text{ mg}}{10^{-3} \text{ g}}\right)\left(\frac{10^{-3} \text{ m}^3}{1 \text{ L}}\right)$$

$$= 72.9198 = \mathbf{73 \text{ mg/L}}$$

This assumes the plants in the field absorb none of the fertilizer.

22.77 The balanced chemical equation for this reaction is:

$$6 \text{ HF}(g) + Al(OH)_3(s) + 3 \text{ NaOH}(aq) \rightarrow Na_3AlF_6(aq) + 6 \text{ H}_2O(l)$$

Determine which one of the three reactants is the limiting reactant.

1) The amount of $Al(OH)_3$ requires conversion of kg to g, and then determination of moles of cryolite formed.

$$\text{mol } Na_3AlF_6 = \left(365 \text{ kg } Al(OH)_3\right)\left(\frac{10^3 \text{ g}}{1 \text{ kg}}\right)\left(\frac{1 \text{ mol } Al(OH)_3}{78.00 \text{ g } Al(OH)_3}\right)\left(\frac{1 \text{ mol } Na_3AlF_6}{1 \text{ mol } Al(OH)_3}\right)$$

$$= 4.6795 \times 10^3 \text{ mol } Na_3AlF_6 \text{ (unrounded)}$$

2) The amount of NaOH requires conversion from amount of solution to g, then determination of moles of cryolite formed.

$$\text{mol } Na_3AlF_6 = \left(1.20 \text{ m}^3\right)\left(\frac{1 \text{ L}}{10^{-3} \text{ m}^3}\right)\left(\frac{1 \text{ mL}}{10^{-3} \text{ L}}\right)\left(\frac{1.53 \text{ g}}{\text{mL}}\right)\left(\frac{50.0\% \text{ NaOH}}{100\%}\right)\left(\frac{1 \text{ mol NaOH}}{40.00 \text{ g NaOH}}\right)\left(\frac{1 \text{ mol } Na_3AlF_6}{3 \text{ mol NaOH}}\right)$$

$$= 7.650 \times 10^3 \text{ mol } Na_3AlF_6 \text{ (unrounded)}$$

3) The amount of HF requires conversion from volume to moles using the ideal gas law, and then determination of moles of cryolite formed.

$n = PV/RT$

$$\text{mol } Na_3AlF_6 = \left[\frac{(305 \text{ kPa})(265 \text{ m}^3)}{(0.0821 \text{ L/mol} \cdot \text{K})((273.2 + 91.5)\text{K})}\left(\frac{10^3 \text{ Pa}}{1 \text{ kPa}}\right)\left(\frac{1 \text{ atm}}{1.01325 \times 10^5 \text{ Pa}}\right)\left(\frac{1 \text{ L}}{10^{-3} \text{ m}^3}\right)\right]\left(\frac{1 \text{ mol } Na_3AlF_6}{6 \text{ mol HF}}\right)$$

$$= 4.4402 \times 10^3 \text{ mol } Na_3AlF_6 \text{ (unrounded)}$$

Because the 265 m^3 of HF produces the smallest amount of Na_3AlF_6, it is the limiting reactant. Convert moles Na_3AlF_6 to grams Na_3AlF_6, using the molar mass, convert to kg, and multiply by 95.6% to determine the final yield.

$$\text{mass cryolite} = \left(4.4402 \times 10^3 \text{ mol } Na_3AlF_6\right)\left(\frac{209.95 \text{ g } Na_3AlF_6}{1 \text{ mol } Na_3AlF_6}\right)\left(\frac{1 \text{ kg}}{10^3 \text{ g}}\right)\left(\frac{95.6\%}{100\%}\right) = 891.195 = \mathbf{891 \text{ kg } Na_3AlF_6}$$

22.78 a) The rate of effusion is inversely proportional to the square root of the molar mass of the molecule.

$$\frac{\text{rate}_H}{\text{rate}_D} = \sqrt{\frac{M_D}{M_H}} = \sqrt{\frac{4.00 \text{ g/mol}}{2.016 \text{ g/mol}}} = 1.40859 \text{ (unrounded)}$$

The time for D_2 to effuse is 1.41 times greater than that for H_2.

$\text{Time}_{D_2} = (1.40859)(16.5 \text{ min}) = 23.2417 = \mathbf{23.2 \text{ min}}$

b) Set x equal to the number of effusion steps. The ratio of mol H_2 to mol D_2 is 99:1.

Set up the equation to solve for x: $99 / 1 = (1.40859)^x$.

When solving for an exponent, take the log of both sides.

$\log(99) = \log(1.40859)^x$

Remember that $\log(a^b) = b \log(a)$

log(99) = x log(1.40859)

x = 13.4129 (unrounded)

To separate H_2 and D_2 to 99% purity requires **14 effusion steps**.

22.80 a) Use the stoichiometric relationships found in the balanced chemical equation to find mass of Al_2O_3. Assume that 1 metric ton Al is an exact number.

$$2\ Al_2O_3\ (\text{in } Na_3AlF_6) + 3\ C(gr) \rightarrow 4\ Al(l) + 3\ CO_2(g)$$

$$\text{mass } Al_2O_3 = \left(1\ t\ Al\right)\left(\frac{10^3\ kg}{1\ t}\right)\left(\frac{10^3 g}{1\ kg}\right)\left(\frac{1\ mol\ Al}{26.98\ g\ Al}\right)\left(\frac{2\ mol\ Al_2O_3}{4\ mol\ Al}\right)\left(\frac{101.96\ g\ Al_2O_3}{1\ mol\ Al_2O_3}\right)\left(\frac{1\ kg}{10^3\ g}\right)\left(\frac{1\ t}{10^3\ kg}\right)$$

= 1.8895478 = **1.890 metric tons Al_2O_3**

Therefore, **1.890 tons of Al_2O_3** are consumed in the production of 1 ton of pure Al.

b) Use a ratio of 3 mol C: 4 mol Al to find mass of graphite consumed.

$$\text{mass } C = \left(1\ t\ Al\right)\left(\frac{10^3\ kg}{1\ t}\right)\left(\frac{10^3 g}{1\ kg}\right)\left(\frac{1\ mol\ Al}{26.98\ g\ Al}\right)\left(\frac{3\ mol\ C}{4\ mol\ Al}\right)\left(\frac{12.01\ g\ C}{1\ mol\ C}\right)\left(\frac{1\ kg}{10^3\ g}\right)\left(\frac{1\ t}{10^3\ kg}\right)$$

= 0.3338584 = **0.3339 tons C**

Therefore, **0.3339 tons of C** are consumed in the production of 1 ton of pure Al, assuming 100% efficiency.

c) The percent yield with respect to Al_2O_3 is **100%** because the actual plant requirement of 1.89 tons Al_2O_3 equals the theoretical amount calculated in part (a).

d) The amount of graphite used in reality to produce 1 ton of Al is greater than the amount calculated in (b). In other words, a 100% efficient reaction takes only 0.3339 tons of graphite to produce a ton of Al, whereas real production requires more graphite and is less than 100% efficient. Calculate the efficiency using a simple ratio:

(0.45 t) (x) = (0.3338584 t) (100%)

x = 74.19076 = **74%**

e) For every 4 moles of Al produced, 3 moles of CO_2 are produced.

$$\text{moles } C = \left(1\ t\ Al\right)\left(\frac{10^3\ kg}{1\ t}\right)\left(\frac{10^3\ g}{1\ kg}\right)\left(\frac{1\ mol\ Al}{26.98\ g\ Al}\right)\left(\frac{3\ mol\ CO_2}{4\ mol\ Al}\right) = 2.7798 \times 10^4\ mol\ CO_2\ \text{(unrounded)}$$

The problem states that 1 atm is exact. Use the ideal gas law to calculate volume, given moles, temperature and pressure.

$$V = nRT / P = \left(\frac{\left(2.7798 \times 10^4\ mol\ CO_2\right)\left(0.08206\ L \cdot atm/mol \cdot K\right)\left((273 + 960.)K\right)}{1\ atm}\right)\left(\frac{10^{-3}\ m^3}{1\ L}\right)$$

= 2.812601×10^3 = **$2.813 \times 10^3\ m^3$**

22.85 Acid rain increases the leaching of PO_4^{3-} into the groundwater, due to the protonation of PO_4^{3-} to form HPO_4^{2-} and $H_2PO_4^{-}$. As shown in calculations a) and b), solubility increases from $6.4 \times 10^{-7}\ M$ (in pure water) to $1.1 \times 10^{-2}\ M$ (in acidic rainwater).

a) Solubility of a salt can be calculated from its K_{sp}. From the table in Appendix C, K_{sp} for $Ca_3(PO_4)_2$ is 1.2×10^{-29}.

	$Ca_3(PO_4)_2(s) \leftrightarrows$	$Ca^{2+}(aq)$	+	$2\ PO_4^{3-}(aq)$
Initial	—	0		0
Change	−x	+3x		+2x
Equilibrium	—	3x		2x

$K_{sp} = [Ca^{2+}]^3[PO_4^{3-}]^2 = (3\ x)^3(2\ x)^2 = 108\ x^5 = 1.2 \times 10^{-29}$.

x = 6.4439×10^{-7} = **$6.4 \times 10^{-7}\ M\ Ca_3(PO_4)_2$**

b) Phosphate is derived from a weak acid, so the pH of the solution impacts exactly where the acid-base equilibrium lies. Phosphate can gain one H^+ to form HPO_4^{2-}. Gaining another H^+ gives $H_2PO_4^{-}$ and a last H^+ added gives H_3PO_4. The K_a values for phosphoric acid are $K_{a1} = 7.2 \times 10^{-3}$, $K_{a2} = 6.3 \times 10^{-8}$, $K_{a3} = 4.2 \times 10^{-13}$. To find the ratios of the various forms of phosphate, use the equilibrium expressions and the concentration of H^+:

$[H_3O^+] = 10^{-4.5} = 3.162 \times 10^{-5}\ M$ (unrounded).

$$H_3PO_4(aq) + H_2O(l) \leftrightarrows \quad H_2PO_4^-(aq) + H_3O^+(aq)$$

$$K_{a1} = \frac{\left[H_2PO_4^-\right]\left[H_3O^+\right]}{\left[H_3PO_4\right]}$$

$$\frac{\left[H_2PO_4^-\right]}{\left[H_3PO_4\right]} = \frac{K_{a1}}{\left[H_3O^+\right]} = \frac{7.2 \times 10^{-3}}{3.162 \times 10^{-5}} = 227.70 \text{ (unrounded)}$$

$$H_2PO_4^-(aq) + H_2O(l) \leftrightarrows \quad HPO_4^{2-}(aq) + H_3O^+(aq)$$

$$K_{a2} = \frac{\left[HPO_4^{2-}\right]\left[H_3O^+\right]}{\left[H_2PO_4^-\right]}$$

$$\frac{\left[HPO_4^{2-}\right]}{\left[H_2PO_4^-\right]} = \frac{K_{a2}}{\left[H_3O^+\right]} = \frac{6.3 \times 10^{-8}}{3.162 \times 10^{-5}} = 1.9924 \times 10^{-3} \text{ (unrounded)}$$

$$HPO_4^{2-}(aq) + H_2O(l) \leftrightarrows \quad PO_4^{3-}(aq) + H_3O^+(aq)$$

$$K_{a3} = \frac{\left[PO_4^{3-}\right]\left[H_3O^+\right]}{\left[HPO_4^{2-}\right]}$$

$$\frac{\left[PO_4^{3-}\right]}{\left[HPO_4^{2-}\right]} = \frac{K_{a3}}{\left[H_3O^+\right]} = \frac{4.2 \times 10^{-13}}{3.162 \times 10^{-5}} = 1.32827 \times 10^{-8} \text{ (unrounded)}$$

From the ratios, the concentration of $H_2PO_4^-$ is at least 100 times more than the concentration of any other species, so assume that dihydrogen phosphate is the dominant species and find the value for $[PO_4^{3-}]$.

$$\frac{\left[PO_4^{3-}\right]}{\left[HPO_4^{2-}\right]} = 1.32827 \times 10^{-8} \quad \text{and} \quad \frac{\left[HPO_4^{2-}\right]}{\left[H_2PO_4^-\right]} = 1.9924 \times 10^{-3}$$

$$[PO_4^{3-}] = (1.32827 \times 10^{-8})[HPO_4^{2-}] \quad \text{and} \quad [HPO_4^{2-}] = (1.9924 \times 10^{-3})[H_2PO_4^-]$$
$$[PO_4^{3-}] = (1.32827 \times 10^{-8})[HPO_4^{2-}] = (1.32827 \times 10^{-8})(1.9924 \times 10^{-3})[H_2PO_4^-].$$
$$[PO_4^{3-}] = (2.6464 \times 10^{-11})(\ )[H_2PO_4^-])$$

Substituting this into the K_{sp} expression gives
$$K_{sp} = [Ca^{2+}]^3[PO_4^{3-}]^2 =$$
$$K_{sp} = [Ca^{2+}]^3\{(2.6464 \times 10^{-11})[H_2PO_4^-]\}^2$$

Rearranging gives
$$K_{sp} / (2.6464 \times 10^{-11})^2 = [Ca^{2+}]^3[H_2PO_4^-]^2$$

The concentration of calcium ions is still represented as 3 x and the concentration of dihydrogen phosphate ion as 2 x, since each $H_2PO_4^-$ comes from one PO_4^{3-}.
$$K_{sp} / (2.6464 \times 10^{-11})^2 = (3\ x)^3(2\ x)^2$$
$$(1.2 \times 10^{-29}) / (2.6464 \times 10^{-11})^2 = 108\ x^5$$
$$x = 1.0967 \times 10^{-2} = \mathbf{1.1 \times 10^{-2}\ M}$$

22.86   a) The mole % of oxygen missing may be estimated from the discrepancy between the actual and ideal formula.

$$\text{mol \% O missing} = \left(\frac{(2.000 - 1.98)\text{ mol O}}{2.000 \text{ mol O}}\right) \times 100\% = \mathbf{1.00\%}$$

b) The molar mass is determined like a normal molar mass except that the quantity of oxygen is not an integer value.

molar mass = 1 mol Pb (207.2 g/mol) + 1.98 mol O (16.00 g/mol) = 238.88 = **238.9 g/mol**

22.88   It may be advantageous to review Chapter 12.

$$\text{Density Silver} = \frac{\left(\dfrac{4 \text{ Ag atoms}}{\text{unit cell}}\right)\left(\dfrac{1 \text{ mol Ag}}{6.022 \times 10^{23} \text{ Ag atoms}}\right)\left(\dfrac{107.9 \text{ g Ag}}{1 \text{ mol Ag}}\right)}{\left(\dfrac{(408.6 \text{ pm})^3}{\text{unit cell}}\right)\left(\dfrac{10^{-12} \text{ m}}{1 \text{ pm}}\right)^3\left(\dfrac{1 \text{ cm}}{10^{-2} \text{ m}}\right)^3} = 10.506198 = \mathbf{10.51 \text{ g/cm}^3}$$

The calculation will now be repeated replacing the atomic mass of silver with the "atomic mass" of sterling silver. The "atomic mass" of sterling silver is simply a weighted average of the masses of the atoms present (Ag and Cu):

$$\text{Atomic mass of sterling silver} = \left(\frac{107.9 \text{ g}}{\text{mol}}\right)\left(\frac{(100.0 - 7.5)\%}{100\%}\right) + \left(\frac{63.55 \text{ g}}{\text{mol}}\right)\left(\frac{7.5\%}{100\%}\right) = 104.57375 \text{ g/mol (unrounded)}$$

$$\text{Density Sterling Silver} = \frac{\left(\dfrac{4 \text{ Ag atoms}}{\text{unit cell}}\right)\left(\dfrac{1 \text{ mol Ag}}{6.022 \times 10^{23} \text{ Ag atoms}}\right)\left(\dfrac{104.57375 \text{ g"Ag"}}{1 \text{ mol Ag}}\right)}{\left(\dfrac{(408.6 \text{ pm})^3}{\text{unit cell}}\right)\left(\dfrac{10^{-12} \text{ m}}{1 \text{ pm}}\right)^3\left(\dfrac{1 \text{ cm}}{10^{-2} \text{ m}}\right)^3} = 10.182322 = \mathbf{10.2 \text{ g/cm}^3}$$

CHAPTER 23 THE TRANSITION ELEMENTS AND THEIR COORDINATION COMPOUNDS

FOLLOW–UP PROBLEMS

23.1 Plan: Locate the element on the periodic table and use its position and atomic number to write a partial electron configuration. Add or subtract electrons to obtain the configuration for the ion. Partial electron configurations do not include the noble gas configuration and any filled f sublevels, but do include outer level electrons (n-level) and the n–1 level d orbitals. Remember that ns electrons are removed before n–1 electrons.
Solution:
a) Ag is in the fifth row of the periodic table, so $n = 5$, and it is in group 1B(11). The partial electron configuration of the element is $5s^1 4d^{10}$. For the ion, Ag^+, remove one electron from the $5s$ orbital. The partial electron configuration of Ag^+ is $\mathbf{4d^{10}}$.
b) Cd is in the fifth row and group 2B(12), so the partial electron configuration of the element is $5s^2 4d^{10}$. Remove two electrons from the $5s$ orbital for the ion. The partial electron configuration of Cd^{2+} is $\mathbf{4d^{10}}$.
c) Ir is in the sixth row and group 8B(9). The partial electron configuration of the element is $6s^2 5d^7$. Remove two electrons from the $6s$ orbital and one from $5d$ for ion Ir^{3+}. The partial electron configuration of Ir^{3+} is $\mathbf{5d^6}$.
Check: Total the electrons to make sure they agree with configuration.
a) Ag^+ should have $47 – 1 = 46$ e⁻. Configuration has 36 e⁻ (from Kr configuration) plus the 10 electrons in $4d^{10}$. This also totals 46 e⁻.
b) Cd^{2+} should have $48 – 2 = 46$ e⁻. Configuration is the same as Ag^+, so 46 e⁻ as well.
c) Ir^{3+} should have $77 – 3 = 74$ e⁻. Total in configuration is 54 e⁻ (noble gas configuration) plus 14 e⁻ (in f-sublevel) plus 6 e⁻ (in the partial configuration) which equals 74 e⁻.

23.2 Plan: Find Er on the periodic table and write the electron configuration for Er^{3+}. Place the outer level electrons in an electron diagram, making sure to follow the Aufbau principle and Hund's rule. Count the number of unpaired electrons.
Solution:
Er is a lanthanide element in row 6 and with atomic number 68. The electron configuration of the element is $[Xe]6s^2 5d^1 4f^{11}$. Remove three electrons: 2 from the $6s$ and 1 from $5d$ for the Er^{3+} ion, $\mathbf{[Xe]4f^{11}}$.

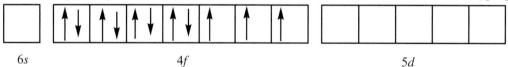

6s   4f   5d

The number of unpaired electrons is **three**, all in the $4f$ sublevel.
Check: Total number of electrons for Er^{3+} is $68 – 3 = 65$ e⁻. The configuration has 54 ([Xe] configuration) plus 11 in $4f$ to equal 65 e⁻.

23.3 Plan: Name the compound by following rules outlined in section 23.4. Use Table 23.8 for the names of ligands. Write the formula from the name by breaking down the name into pieces that follow naming rules.
Solution:
a) The compound consists of the cation $[Cr(H_2O)_5Br]^{2-}$ and anion Cl^-. The metal ion in the cation is chromium and its charge is found from the charges on the ligands and the total charge on the ion.

   total charge = (charge on Cr) + 5(charge on H_2O) + (charge on Br^-)
   -2 = (charge on Cr) + 5(0) + (–1)
   charge on Cr = +3

The name of the cation will end with chromium(III) to indicate the oxidation state of the chromium ion since there is more than one possible oxidation state for chromium. The water ligand is named aqua. There are five water ligands, so pentaaqua describes the $(H_2O)_5$ ligands. The bromide ligand is named bromo. The ligands are named in alphabetical order, so aqua comes before bromo. The name of the cation is pentaaquabromochromium(III). Add the chloride for the anion to complete the name: **pentaaquabromochromium(III) chloride**.

b) The compound consists of a cation, barium ion, and the anion hexacyanocobaltate(III). The formula of the anion consists of six (from hexa) cyanide (from cyano) ligands and cobalt (from cobaltate) in the +3 oxidation state (from (III)). Putting the formula together gives $[Co(CN)_6]^{n+}$. To find the charge on the complex ion, calculate from charges on the ligands and the metal ion:

$$\text{total charge} = 6(\text{charge on } CN^-) + (\text{charge on cobalt ion})$$
$$= 6(-1) + (+3)$$
$$= -3$$

The formula of complex ion is $[Co(CN)_6]^{3-}$. Combining this with the cation, Ba^{2+}, gives the formula for the compound: $\textbf{Ba}_3\textbf{[Co(CN)}_6\textbf{]}_2$. The three barium ions give a +6 charge and two anions give a –6 charge, so the net result is a neutral salt.

23.4 Plan: The given complex ion has a coordination number of 6 (*en* is bidentate), so it will have an octahedral arrangement of ligands. Stereoisomers can be either geometric or optical. For geometric isomers, check if the ligands can be arranged either next to (*cis*) or opposite (*trans*) each other. Then, check if the mirror image of any of the geometric isomers is not superimposable. If mirror image is not superimposable, the structure is an optical isomer.

Solution:
The complex ion contains two NH_3 ligands and two chloride ligands, both of which can be arranged in either the *cis* or *trans* geometry. The possible combinations are 1) *cis*-NH_3 and *cis*-Cl, 2) *trans*-NH_3 and *cis*-Cl, and 3) *cis*-NH_3 and *trans*-Cl. Both NH_3 and Cl *trans* is not possible because the two bonds to ethylenediamine can be arranged only in this *cis*-position, which leaves only one set of *trans* positions.

cis-NH₃ *cis*-Cl   *trans*-NH₃ *cis*-Cl   *cis*-NH₃ *trans*-Cl

The mirror images of the second two structures are superimposable since two of the ligands, either ammonia or chloride ion, are arranged in the *trans* position. When both types of ligands are in the *cis* arrangement, the mirror image is not a superimposable optical isomer:

cis-NH₃ *cis*-Cl   *cis*-NH₃ *cis*-Cl

There are four stereoisomers of $[Co(NH_3)_2(en)Cl_2]^+$.

23.5 Plan: Compare the two ions for oxidation state of the metal ion and the relative ability of ligands to split the *d*-orbital energies.

Solution:
The oxidation number of vanadium in both ions is the same, so compare the two ligands. Ammonia is a stronger field ligand than water, so the complex ion $\textbf{[V(NH}_3\textbf{)}_6\textbf{]}^{3+}$ absorbs visible light of higher energy than $[V(H_2O)_6]^{3+}$ absorbs.

23.6   Plan: Determine the charge on the manganese ion in the complex ion and the number of electrons in its d-orbitals. Since it is an octahedral ligand, the d-orbitals split into three lower energy orbitals and two higher energy orbitals. Check the spectrochemical series for whether CN^- is a strong or weak field ligand. If it is a strong field ligand, fill the three lower energy orbitals before placing any electrons in the higher energy d-orbitals. If it is a weak field ligand, place one electron in each of the five d orbitals, low energy before high energy, before pairing any electrons. After filling orbitals, count the number of unpaired electrons. The complex ion is low-spin if the ligand is a strong field ligand and high-spin if the ligand is weak field.
Solution:
Find the charge on manganese:
        total charge = (charge on Mn) + 6(charge on CN^-)
charge on Mn = $-3 - [6(-1)] = +3$
The electron configuration of Mn is $[Ar]4s^23d^5$; the electron configuration of Mn^{3+} is $[Ar]3d^4$.
The ligand is CN^-, which is a strong field ligand, so the four d electrons will fill the lower energy d-orbitals before any are placed in the higher energy d-orbitals:

Higher energy d-orbitals

Lower energy d-orbitals

Two electrons are unpaired in $[Mn(CN)_6]^{3-}$. The complex is **low spin** since the cyanide ligand is a strong field ligand. The splitting energy is greater than the electron pairing energy.

END–OF–CHAPTER PROBLEMS

23.2   a) All transition elements in Period 5 will have a "base" configuration of $[Kr]5s^2$, and will differ in the number of d electrons (x) that the configuration contains. Therefore, the general electron configuration is
$\mathbf{1s^22s^22p^63s^23p^64s^23d^{10}4p^65s^24d^x}$.
b) A general electron configuration for Period 6 transition elements includes f sublevel electrons, which are lower in energy than the d sublevel. The configuration is $\mathbf{1s^22s^22p^63s^23p^64s^23d^{10}4p^65s^24d^{10}5p^66s^24f^{14}5d^x}$.

23.4   The maximum number of unpaired d electrons is **five** since there are five d orbitals. An example of an atom with five unpaired d electrons is Mn with electron configuration $[Ar]4s^23d^5$. An ion with five unpaired electrons is Mn^{2+} with electron configuration $[Ar]3d^5$.

23.6   a) One would expect that the elements would increase in size as they increase in mass from Period 5 to 6. Because there are 14 inner transition elements in Period 6, the effective nuclear charge increases significantly. As effective charge increases, the atomic size decreases or "contracts." This effect is significant enough that Zr^{4+} and Hf^{4+} are almost the same size but differ greatly in atomic mass.
b) The size increases from Period 4 to 5, but stays fairly constant from Period 5 to 6.
c) Atomic mass increases significantly from Period 5 to 6, but atomic radius (and thus volume) hardly increases, so Period 6 elements are very dense.

23.9   a) A paramagnetic substance is attracted to a magnetic field, while a diamagnetic substance is slightly repelled by one.
b) Ions of transition elements often have unfilled d-orbitals whose unpaired electrons make the ions paramagnetic. Ions of main group elements usually have a noble gas configuration with no partially filled levels. When orbitals are filled, electrons are paired and the ion is diamagnetic.

c) The d-orbitals in the transition element ions are not filled, which allows an electron from a lower energy d-orbital to move to a higher energy d-orbital. The energy required for this transition is relatively small and falls in the visible wavelength range. All orbitals are filled in a main-group element ion, so enough energy would have to be added to move an electron to a higher energy level, not just another orbital within the same energy level. This amount of energy is relatively large and outside the visible range of wavelengths.

23.10   a) V: $1s^2 2s^2 2p^6 3s^2 3p^6 4s^2 3d^3$ Check: $2 + 2 + 6 + 2 + 6 + 2 + 3 = 23\ e^-$
        b) Y: $1s^2 2s^2 2p^6 3s^2 3p^6 4s^2 3d^{10} 4p^6 5s^2 4d^1$ Check: $2 + 2 + 6 + 2 + 6 + 2 + 10 + 6 + 2 + 1 = 39\ e^-$
        c) Hg: $[Xe]6s^2 4f^{14} 5d^{10}$ Check: $54 + 2 + 14 + 10 = 80\ e^-$

23.12   a) Osmium is in row 6 and group 8B(8). Electron configuration of Os is $[Xe]6s^2 4f^{14} 5d^6$.
        b) Cobalt is in row 4 and group 8B(9). Electron configuration of Co is $[Ar]4s^2 3d^7$.
        c) Silver is in row 5 and group 1B(11). Electron configuration of Ag is $[Kr]5s^1 4d^{10}$. Note that the filled d orbital is the preferred arrangement, so the configuration is not $5s^2 4d^9$.

23.14   Transition metals lose their s orbital electrons first in forming cations.
        a)   The two $4s$ electrons and one $3d$ electron are removed to form Sc^{3+}:
        Sc: $[Ar]4s^2 3d^1$; Sc^{3+}: $[Ar]$ or $1s^2 2s^2 2p^6 3s^2 3p^6$. There are **no unpaired electrons**.
        b)   The single $4s$ electron and one $3d$ electron are removed to form Cu^{2+}:
        Cu: $[Ar]4s^1 3d^{10}$; Cu^{2+}: $[Ar]3d^9$. There is **one unpaired electron**.
        c)   The two $4s$ electrons and one $3d$ electron are removed to form Fe^{3+}:
        Fe: $[Ar]4s^2 3d^6$; Fe^{3+}: $[Ar]3d^5$. There are **five unpaired electrons** since each of the five d electrons occupies its own orbital.
        d)   The two $5s$ electrons and one $4d$ electron are removed to form Nb^{3+}:
        Nb: $[Kr]5s^2 4d^3$; Nb^{3+}: $[Kr]4d^2$. There are **two unpaired electrons**.

23.16   For groups 3B(3) to 7B(7), the highest oxidation state is equal to the group number. The highest oxidation state Occurs when both ns electrons and all $(n–1)d$ electrons have been removed.
        a)   Tantalum, Ta, is in group 5B(5), so the highest oxidation state is **+5**. The electron configuration of Ta is $[Xe]\ 6s^2 4f^{14} 5d^3$ with a total of 5 electrons in the $6s$ and $5d$ orbitals.
        b)   Zirconium, Zr, is in group 4B(4), so the highest oxidation state is **+4**. The electron configuration of Zr is $[Kr]\ 5s^2 4d^2$ with a total of 4 electrons in the $5s$ and $4d$ orbitals.
        c)   Manganese, Mn, is in group 7B(7), so the highest oxidation state is **+7**. The electron configuration of Mn is $[Ar]\ 4s^2 3d^5$ with a total of 7 electrons in the $4s$ and $3d$ orbitals.

23.18   The elements in Group 6B(6) exhibit an oxidation state of +6. These elements have a total of 6 electrons in the outermost s orbital and d orbitals: $ns^1(n–1)d^5$. These elements include **Cr, Mo, and W**. Sg (Seaborgium) is also in Group 6B(6), but its lifetime is so short that chemical properties, like oxidation states within compounds, are impossible to measure.

23.20   Transition elements in their lower oxidation states act more like metals. The oxidation state of chromium in CrF_2 is +2 and in CrF_6 is +6 (use –1 oxidation state of fluorine to find oxidation state of Cr). CrF_2 exhibits greater metallic behavior than CrF_6 because the chromium is in a lower oxidation state in CrF_2 than in CrF_6.

23.22   While atomic size increases slightly down a group of transition elements, the nuclear charge increases much more, so the first ionization energy generally increases. The reduction potential for Mo is lower, so **it is more difficult to oxidize Mo** than Cr. In addition, the ionization energy of Mo is higher than that of Cr, so it is more difficult to remove electrons from, i.e., oxidize, Mo.

23.24   Oxides of transition metals become less basic (or more acidic) as oxidation state increases. The oxidation state of chromium in CrO_3 is +6 and in CrO is +2, based on the –2 oxidation state of oxygen. The oxide of the higher oxidation state, **CrO_3**, produces a more acidic solution.

23.28   a) The f-block contains **7** orbitals, so if one electron occupied each orbital, a maximum of **7** electrons would be unpaired.
        b) The maximum number of unpaired electrons corresponds to a half-filled f subshell.

23.30 a) Lanthanum is a transition element in row 6 with atomic number 57. La: $[Xe]6s^2 5d^1$
b) Cerium is in the lanthanide series with atomic number 58. Ce: $[Xe]6s^2 4f^1 5d^1$, so Ce^{3+}: $[Xe]4f^1$
c) Einsteinium is in the actinide series with atomic number 99. Es: $[Rn]7s^2 5f^{11}$
d) Uranium is in the actinide series with atomic number 92. U: $[Rn]7s^2 5f^3 6d^1$. Removing four electrons gives U^{4+} with configuration $[Rn]5f^2$.

23.32 a) Europium is in the lanthanide series with atomic number 63. The configuration of Eu is $[Xe]6s^2 4f^7$. The stability of the half-filled f sublevel explains why the configuration is not $[Xe]6s^2 4f^6 5d^1$. The two $6s$ electrons are removed to form the Eu^{2+} ion, followed by electron removal in the f-block to form the other two ions:
   Eu^{2+}: $[Xe]4f^7$
   Eu^{3+}: $[Xe]4f^6$
   Eu^{4+}: $[Xe]4f^5$
   The stability of the half-filled f sublevel makes Eu^{2+} most stable.
b) Terbium is in the lanthanide series with atomic number 65. The configuration of Tb is $[Xe]6s^2 4f^9$. The two $6s$ electrons are removed to form the Tb^{2+} ion, followed by electron removal in the f-block to form the other two ions:
   Tb^{2+}: $[Xe]4f^9$
   Tb^{3+}: $[Xe]4f^8$
   Tb^{4+}: $[Xe]4f^7$
   Tb would demonstrate a +4 oxidation state because it has the half-filled sublevel.

23.34 The lanthanide element **gadolinium, Gd**, has electron configuration $[Xe]6s^2 4f^7 5d^1$ with **eight** unpaired electrons. The ion Gd^{3+} has **seven** unpaired electrons: $[Xe]4f^7$.

23.36 Valence-state electronegativity refers to the effective electronegativity of an element in a given oxidation state. As manganese is bonded to more oxygen atoms in its different oxides, its oxidation number becomes more positive. In solution a water molecule is attracted to the increasingly positive manganese and one of its protons becomes easier to lose. Acidity of a metal oxide increases with the metal's increasing oxidation state.

23.40 Mercury demonstrates an unusual physical property in that it exists in the liquid state at room temperature. There are two reasons for the weaker forces between mercury atoms. First, in its crystal structure each mercury atom is surrounded by 6 nearest neighbors instead of the usual 12. This decreases the forces holding mercury atoms, so they can move more freely than other metal atoms. Second, the bonding between metal atoms is metallic bonding, which involves only the $6s$ electrons in mercury compared to the s and d electrons in other metals. Mercury occurs in the +1 oxidation state, unusual for its group, because two mercury +1 ions form a dimer, Hg_2^{2+} by sharing their unpaired $6s$ electrons in a covalent bond.

23.42 A reaction is favorable if the $E°$ for the reaction, as written, is positive. Since $Cr^{2+}(aq)$ is *prepared*, it must appear on the right side of the equation. $Cr(s)$ and $Cr^{3+}(aq)$ must appear as reactants. Multiply the first equation by 2 so that the electron transfer is balanced and reverse the third equation.
   Reduction: $2(Cr^{3+}(aq) + e^- \rightarrow Cr^{2+})$ $E° = -0.41$ V (changing coefficients does not change $E°$)
   Oxidation: $Cr(s) \rightarrow Cr^{2+}(aq) + 2 e^-$ $E° = +0.91$ V
   Overall: $2 Cr^{3+}(aq) + Cr(s) \rightarrow 3 Cr^{2+}$ $E° = +0.50$ V
   Since $E° > 0$, the reaction is spontaneous. So, **yes**, Cr^{2+} can be prepared from Cr and Cr^{3+}.

23.44 Check reduction standard potentials in Appendix D.
   The standard potential for the reduction of MnO_4^- to Mn^{2+} is +1.51 V and the standard potential for the oxidation of Mn^{2+} to MnO_2 is –1.23 V. The reaction between MnO_4^- and Mn^{2+} to produce MnO_2 is +0.28 V, which indicates a spontaneous reaction:
   $5(Mn^{2+}(aq) + 2 H_2O(l) \rightarrow MnO_2(s) + 4 H^+(aq) + 2 e^-)$ $E° = -(1.23)$ V
   $2(5 e^- + MnO_4^-(aq) + 8 H^+(aq) \rightarrow Mn^{2+}(aq) + 4 H_2O(l))$ $E° = +1.51$ V
   $5 Mn^{2+}(aq) + 10 H_2O(l) + 2 MnO_4^-(aq) + 16 H^+(aq) \rightarrow 5 MnO_2(s) + 20 H^+(aq) + 2 Mn^{2+}(aq) + 8 H_2O(l)$
   Overall: $3 Mn^{2+}(aq) + 2 MnO_4^-(aq) + 2 H_2O(l) \rightarrow 5 MnO_2(s) + 4 H^+(aq)$ $E° = +0.28$ V
   Since $E° > 0$, the reaction is spontaneous and produces brown MnO_2.

23.47   The coordination number indicates the number of ligand atoms bonded to the central metal ion. The oxidation number represents the number of electrons lost to form the ion. The coordination number is unrelated to the oxidation number.

23.49   Coordination number of two indicates **linear** geometry.
Coordination number of four indicates either **tetrahedral** or **square planar** geometry.
Coordination number of six indicates **octahedral** geometry.

23.52   The *-ate* ending signifies that the complex ion has a negative charge.

23.55   a) The oxidation state of nickel is found from the total charge on the ion (+2 because two Cl^- charges equals –2) and the charge on ligands:
        charge on nickel = +2 – 6(0 charge on water) = +2
Name nickel as nickel(II) to indicate oxidation state. Ligands are six (hexa-) waters (aqua). Put together with chloride anions to give **hexaaquanickel(II) chloride**.
b) The cation is $[Cr(en)_3]^{n+}$ and the anion is ClO_4^-, the perchlorate ion (see Chapter 2 for naming polyatomic ions). The charge on the cation is +3 to make a neutral salt in combination with 3 perchlorate ions. The ligand is ethylenediamine, which has 0 charge. The charge of the cation equals the charge on chromium ion, so chromium(III) is included in the name. The three ethylenediamine ligands, abbreviated en, are indicated by the prefix tris because the name of the ligand includes a numerical indicator, di-. The complete name is **tris(ethylenediamine)chromium(III) perchlorate**.
c) The cation is K^+ and the anion is $[Mn(CN)_6]^{4-}$. The charge of 4– is deduced from the four potassium ions in the formula. The oxidation state of Mn is $-4 - \{6(-1)\} = +2$. The name of CN^- ligand is cyano and six ligands are represented by the prefix hexa. The name of manganese anion is manganate(II). The -ate suffix on the complex ion is used to indicate that it is an anion. The full name of compound is **potassium hexacyanomanganate(II)**.

23.57   The charge of the central metal atom was determined in 22.24 because the Roman numeral indicating oxidation state is part of the name. The coordination number, or number of ligand atoms bonded to the metal ion, is found by examining the bonded entities inside the square brackets to determine if they are unidentate, bidentate, or polydentate.
a) The Roman numeral "II" indicates a **+2** oxidation state. There are 6 water molecules bonded to Ni and each ligand is unidentate, so the coordination number is **6**.
b) The Roman numeral "III" indicates a **+3** oxidation state. There are 3 ethylenediamine molecules bonded to Cr, but each ethylenediamine molecule contains two donor N atoms (bidentate). Therefore, the coordination number is **6**.
c) The Roman numeral "II" indicates a **+2** oxidation state. There are 6 unidentate cyano molecules bonded to Mn, so the coordination number is **6**.

23.59   a) The cation is K^+, potassium. The anion is $[Ag(CN)_2]^-$ with the name dicyanoargentate (I) ion for the two cyanide ligands and the name of silver in an anion, argentate(I). The Roman numeral (I) indicates the oxidation number on Ag. O.N. for $Ag = -1 - \{2(-1)\} = +1$ since the complex ion has a charge of –1 and the cyanide ligands are also –1. The complete name is **potassium dicyanoargentate(I)**.
b) The cation is Na^+, sodium. Since there are two +1 sodium ions, the anion is $[CdCl_4]^{2-}$ with a charge of 2–. The anion is the tetrachlorocadmate(II) ion. With four –1 chloride ligands, the oxidation state of cadmium is +2 and the name of cadmium in an anion is cadmate. The complete name is **sodium tetrachlorocadmate(II)**.
c) The cation is $[Co(NH_3)_4(H_2O)Br]^{2+}$. The 2+ charge is deduced from the 2Br– ions. The cation has the name tetraamineaquabromocobalt(III) ion, with four ammonia ligands (tetraammine), one water ligand (aqua) and one bromide ligand (bromo). The oxidation state of cobalt is +3: $2 - \{4(0) + 1(0) + 1(-1)\}$. The oxidation state is indicated by (III), following cobalt in the name. The anion is Br^-, bromide. The complete name is **tetraammineaquabromocobalt(III) bromide**.

23.61   a) The counter ion is K^+, so the complex ion is $[Ag(CN)_2]^-$. Each cyano ligand has a –1 charge, so silver has a **+1** charge: $+1 + 2(-1) = -1$. Each cyano ligand is unidentate, so the coordination number is **2**.
b) The counter ion is Na^+, so the complex ion is $[CdCl_4]^{2-}$. Each chloride ligand has a –1 charge, so Cd has a **+2** charge: $+2 + 4(-1) = -2$. Each chloride ligand is unidentate, so the coordination number is **4**.

c) The counter ion is Br⁻, so the complex ion is $[Co(NH_3)_4(H_2O)Br]^{2+}$. Both the ammine and aqua ligands are neutral. The bromide ligand has a –1 charge, so Co has a **+3** charge: $+3 + 4(0) + 0 + (-1) = +2$. Each ligand is unidentate, so the coordination number is **6**.

23.63   a) The cation is tetramminezinc ion. The tetraammine indicate four NH_3 ligands. Zinc has an oxidation state of +2, so the charge on the cation is +2. The anion is SO_4^{2-}. Only one sulfate is needed to make a neutral salt. The formula of the compound is **[Zn(NH₃)₄]SO₄**.
b) The cation is pentaamminechlorochromium(III) ion. The ligands are 5 NH_3 from pentaammine, and one chloride from chloro. The chromium ion has a charge of +3, so the complex ion has a charge equal to +3 from chromium, plus 0 from ammonia, plus –1 from chloride for a total of +2. The anion is chloride, Cl⁻. Two chloride ions are needed to make a neutral salt. The formula of compound is **[Cr(NH₃)₅Cl]Cl₂**.
c) The anion is bis(thiosulfato)argentate(I). Argentate(I) indicates silver in the +1 oxidation state, and bis(thiosulfato) indicates 2 thiosulfate ligands, $S_2O_3^{2-}$. The total charge on the anion is +1 plus 2(–2) to equal –3. The cation is sodium, Na⁺. Three sodium ions are needed to make a neutral salt. The formula of compound is **Na₃[Ag(S₂O₃)₂]**.

23.65   Coordination compounds act like electrolytes, i.e., they dissolve in water to yield charged species. However, the complex ion itself does not dissociate. The "number of individual ions per formula unit" refers to the number of ions that would form per coordination compound upon dissolution in water.
a) The counter ion is SO_4^{2-}, so the complex ion is $[Zn(NH_3)_4]^{2+}$. Each ammine ligand is unidentate, so the coordination number is **4**. Each molecule dissolves in water to form one SO_4^{2-} ion and one $[Zn(NH_3)_4]^{2+}$ ion, so **2 ions** form per formula unit.
b) The counter ion is Cl⁻, so the complex ion is $[Cr(NH_3)_5Cl]^{2+}$. Each ligand is unidentate, so the coordination number is **6**. Each molecule dissolves in water to form two Cl⁻ ions and one $[Cr(NH_3)_5Cl]^{2+}$ ion, so **3 ions** form per formula unit.
c) The counter ion is Na⁺, so the complex ion is $[Ag(S_2O_3)_2]^{3-}$. Assuming that the thiosulfate ligand is unidentate, the coordination number is **2**. Each molecule dissolves in water to form 3 Na⁺ ions and one $[Ag(S_2O_3)_2]^{3-}$ ion, so **4 ions** form per formula unit.

23.67   a) The cation is hexaaquachromium(III) with the formula $[Cr(H_2O)_6]^{3+}$. The total charge of the ion equals the charge on chromium because water is a neutral ligand. Six water ligands are indicated by a *hexa* prefix to aqua. The anion is SO_4^{2-}. To make the neutral salt requires 3 sulfate ions for 2 cations. The compound formula is
[Cr(H₂O)₆]₂(SO₄)₃.
b) The anion is tetrabromoferrate(III) with the formula $[FeBr_4]^-$. The total charge on the ion equals +3 charge on iron plus 4 x –1 charge on each bromide ligand for –1 overall. The cation is barium, Ba^{2+}. Two anions are needed for each barium ion to make a neutral salt. The compound formula is **Ba[FeBr₄]₂**.
c) The anion is bis(ethylenediamine)platinum(II) ion. Charge on platinum is +2 and this equals the total charge on the complex ion because ethylenediamine is a neutral ligand. The *bis* preceding ethylenediamine indicates two ethylenediamine ligands, which are represented by the abbreviation "en." The cation is $[Pt(en)_2]^{2+}$. The anion is CO_3^{2-}. One carbonate combines with one cation to produce a neutral salt. The compound formula is **[Pt(en)₂]CO₃**.

23.69   a) The counter ion is SO_4^{2-}, so the complex ion is $[Cr(H_2O)_6]^{3+}$. Each aqua ligand is unidentate, so the coordination number is **6**. Each molecule dissolves in water to form three SO_4^{2-} ions and two $[Cr(H_2O)_6]^{3+}$ ions, so **5 ions** form per formula unit.
b) The counter ion is Ba^{2+}, so the complex ion is $[FeBr_4]^-$. Each bromo ligand is unidentate, so the coordination number is **4**. Each molecule dissolves in water to form one Ba^{2+} ion and 2 $[FeBr_4]^-$ ions, so **3 ions** form per formula unit.
c) The counter ion is CO_3^{2-}, so the complex ion is $[Pt(en)_2]^{2+}$. Each ethylenediamine ligand is bidentate, so the coordination number is **4**. Each molecule dissolves in water to form one CO_3^{2-} ion and one $[Pt(en)_2]^{2+}$ ion, so **2 ions** forms per formula unit.

23.71   Ligands that form linkage isomers have two different possible donor atoms.
        a) The nitrite ion **forms linkage isomers** because it can bind to the metal ion through the lone pair on the N
        atom or any lone pair on either O atom. Resonance Lewis structures are

$$\left[\ddot{O}=\ddot{N}-\ddot{O}\!:\right]^{-} \longleftrightarrow \left[:\ddot{O}-\ddot{N}=\ddot{O}\right]^{-}$$

        b) Sulfur dioxide molecules **form linkage isomers** because the lone pair on the S atom or any lone pair on either
        O atom can bind the central metal ion.

$$\ddot{O}=\ddot{S}=\ddot{O}$$

        c) Nitrate ions have an N atom with no lone pair and three O atoms, all with lone pairs than can bond to the
        metal ion. But all of the O atoms are equivalent so these ions **do not form linkage isomers**.

$$\left[:\ddot{O}-N=\ddot{O} \;\; :\ddot{O}:\right]^{-} \longleftrightarrow \left[:\ddot{O}-N-\ddot{O}: \;\; \ddot{O}\right]^{-} \longleftrightarrow \left[\ddot{O}=N-\ddot{O}: \;\; :\ddot{O}:\right]^{-}$$

23.73   a) Platinum ion, Pt^{2+}, is a d^{8} ion so the ligand arrangement is square planar. A *cis* and *trans* geometric isomer
        exist for this complex ion:

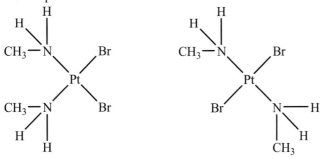

            cis isomer   *trans* isomer

        No optical isomers exist because the mirror images of both compounds are superimposable on the original
        molecules. In general, a square planar molecule is superimposable on its mirror image.
        b) A *cis* and *trans* geometric isomer exist for this complex ion. No optical isomers exist because the mirror
        images of both compounds are superimposable on the original molecules.

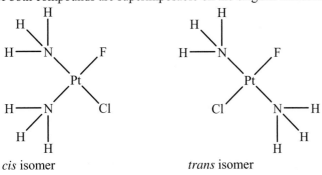

            cis isomer   *trans* isomer

c) Three geometric isomers exist for this molecule, although they are not named *cis* or *trans* because all the ligands are different. A different naming system is used to indicate the relation of one ligand to another.

23.75　Types of isomers for coordination compounds are coordination isomers with different arrangements of ligands and counterions, linkage isomers with different donor atoms from the same ligand bound to the metal ion, geometric isomers with differences in ligand arrangement relative to other ligands, and optical isomers with mirror images that are not superimposable.

a) Platinum ion, Pt^{2+}, is a d^8 ion, so the ligand arrangement is square planar. The ligands are 2 Cl^- and 2 Br^-, so the arrangement can be either both ligands *trans* or both ligands *cis* to form geometric isomers.

b) The complex ion can form linkage isomers with the NO_2 ligand. Either the N or an O may be the donor.

c) In the octahedral arrangement, the two iodide ligands can be either *trans* to each other, 180° apart, or *cis* to each other, 90° apart.

23.77　The traditional formula does not correctly indicate that at least some of the chloride ions serve as counter ions. The NH_3 molecules serve as ligands, so *n* could equal 6 to satisfy the coordination number requirement. This complex has the formula $[Cr(NH_3)_6]Cl_3$. This is a correct formula because the name "chromium(III)" means that the chromium has a +3 charge, and this balances with the –3 charge provided by the three chloride counter ions. However, when this compound dissociates in water, it produces 4 ions ($[Cr(NH_3)_6]^{3+}$ and 3 Cl^-), whereas NaCl only produces two ions. Other possible compounds are *n*=5, $CrCl_3 \cdot 5NH_3$, with formula $[Cr(NH_3)_5Cl]Cl_2$; *n*=4, $CrCl_3 \cdot 4NH_3$, with formula $[Cr(NH_3)_4Cl_2]Cl$; *n*=3, $CrCl_3 \cdot 3NH_3$, with formula $[Cr(NH_3)_3Cl_3]$. The compound **$[Cr(NH_3)_4Cl_2]Cl$** has a coordination number equal to 6 and produces two ions when dissociated in water, so it has an electrical conductivity similar to an equimolar solution of NaCl.

23.79 First find the charge on the palladium ion, then arrange ligands and counter ions to form the complex.

a) Charge on palladium = $-$ [(+1 on K^+) + (0 on NH_3) + 3(–1 on Cl^-)] = +2

Palladium(II) forms four-coordinate complexes. The four ligands in the formula are one NH_3 and three Cl^- ions. The formula of the complex ion is $[Pd(NH_3)Cl_3]^-$. Combined with the potassium cation, the compound formula is **$K[Pd(NH_3)Cl_3]$.**

b) Charge on palladium = $-$ [2(–1 on Cl^-) + 2(0 on NH_3)] = +2

Palladium(II) forms four-coordinate complexes. The four ligands are 2 chloride ions and 2 ammonia molecules. The formula is **$[PdCl_2(NH_3)_2]$**.

c) Charge on palladium = $-$ [2(+1 on K^+) + 6(–1 on Cl^-)] = +4

Palladium(IV) forms six-coordinate complexes. The six ligands are the 6 chloride ions. The formula is **$K_2[PdCl_6]$.**

d) Charge on palladium = $-$ [4(0 on NH_3) + 4(–1 on Cl^-)] = +4

Palladium(IV) forms six-coordinate complexes. The ammonia molecules have to be ligands. The other two ligand bonds are formed with two of the chloride ions. The remaining two chloride ions are the counter ions. The formula is **$[Pd(NH_3)_4Cl_2]Cl_2$**.

23.81 a) Four empty orbitals of equal energy are "created" to receive the donated electron pairs from four ligands. The four orbitals are hybridized from an s, two p, and one d orbital from the previous n level to form 4 $\textbf{\textit{dsp}}^2$ orbitals.

b) One s and three p orbitals become four $\textbf{\textit{sp}}^3$ hybrid orbitals.

23.84 Absorption of **orange** or **yellow** light gives a blue solution.

23.85 a) The crystal field splitting energy is the energy difference between the two sets of d-orbitals that result from the bonding of ligands to a central transition metal atom.

b) In an octahedral field of ligands, the ligands approach along the x, y, and z axes. The $d_{x^2-y^2}$ and d_{z^2} orbitals are located along the x, y, and z axes, so ligand interaction is higher in energy than the other orbital-ligand interactions. The other orbital-ligand interactions are lower in energy because the d_{xy}, d_{yz}, and d_{xz} orbitals are located between the x, y, and z axes.

c) In a tetrahedral field of ligands, the ligands do not approach along the x, y, and z axes. The ligand interaction is greater for the d_{xy}, d_{yz}, and d_{xz} orbitals and lesser for the $d_{x^2-y^2}$ and d_{z^2} orbitals. The crystal field splitting is reversed, and the d_{xy}, d_{yz}, and d_{xz} orbitals are higher in energy than the $d_{x^2-y^2}$ and d_{z^2} orbitals.

23.88 If Δ is greater than $E_{pairing}$, electrons will preferentially pair spins in the lower energy d-orbitals before adding as unpaired electrons to the higher energy d-orbitals. If Δ is less than $E_{pairing}$, electrons will preferentially add as unpaired electrons to the higher d-orbitals before pairing spins in the lower energy d-orbitals. The first case gives a complex that is low-spin and less paramagnetic than the high-spin complex formed in the latter case.

23.90 To determine the number of d electrons in a central metal ion, first write the electron configuration for the metal atom. Examine the formula of the complex to determine the charge on the central metal ion, and then write the ion's configuration.

a) Electron configuration of Ti: $[Ar]4s^2 3d^2$

Charge on Ti: Each chloride ligand has a –1 charge, so Ti has a +4 charge {+4 + 6(–1)} = 2– ion. Both of the $4s$ electrons and both $3d$ electrons are removed.

Electron configuration of Ti^{4+}: $[Ar]$

Ti^{4+} has **no d electrons**.

b) Electron configuration of Au: $[Xe]6s^1 4f^{14} 5d^{10}$

Charge on Au: The complex ion has a –1 charge ($[AuCl_4]^-$) since K has a +1 charge. Each chloride ligand has a –1 charge, so Au has a +3 charge {+3 + 4(–1)} = 1– ion. The $6s$ electron and two d electrons are removed.

Electron configuration of Au^{3+}: $[Xe]4f^{14} 5d^8$

Au^{3+} has **8 d electrons**.

c) Electron configuration of Rh: $[Kr]5s^2 4d^7$

Charge on Rh: Each chloride ligand has a –1 charge, so Rh has a +3 charge {+3 + 6(–1)} = 3– ion. The $5s$ electrons and one $4d$ electron are removed.

Electron configuration of Rh^{3+}: $[Kr]4d^6$

Rh^{3+} has **6 d electrons**.

23.92   a) $[(+2 \text{ on } Ca^{2+}) + 6(-1 \text{ on } F^-) + Ir = 0]$
Charge on iridium $= -[(+2 \text{ on } Ca^{2+}) + 6(-1 \text{ on } F^-)] = +4$
Configuration of Ir is $[Xe]6s^2 4f^{14} 5d^7$.
        Configuration of Ir^{4+} is $[Xe]4f^{14} 5d^5$. **5** d electrons in Ir^{4+}.
b) $[Hg + 4(-1 \text{ on } I^-)] = -2$
 Charge on mercury $= -[4(-1 \text{ on } I^-)] - 2 = +2$
        Configuration of Hg is $[Xe]6s^2 4f^{14} 5d^{10}$.
        Configuration of Hg^{2+} is $[Xe]4f^{14} 5d^{10}$. **10** d electrons in Hg^{2+}.
c) $[Co + (-4 \text{ on } EDTA)] = -2$
Charge on cobalt $= -[-4 \text{ on } EDTA] - 2 = +2$
        Configuration of Co is $[Ar]4s^2 3d^7$.
        Configuration of Co^{2+} is $[Ar]3d^7$. **7** d electrons in Co^{2+}.

23.94

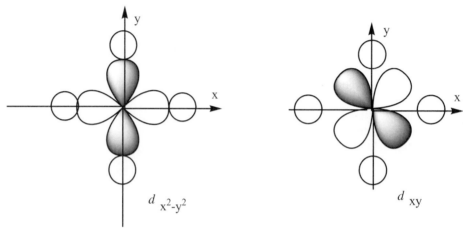

$d_{x^2-y^2}$ d_{xy}

The unequal shading of the lobes of the d orbitals is a consequence of the quantum mechanical derivation of these orbitals, and does not affect the current discussion.
In an octahedral field of ligands, the ligands approach along the x, y, and z axes. The $d_{x^2-y^2}$ orbital is located

along the x and y axes, so ligand interaction is greater. The d_{xy} orbital is offset from the x and y axes by 90°, so ligand interaction is less. The greater interaction of the $d_{x^2-y^2}$ orbital results in its higher energy.

23.96   a) Ti: $[Ar]4s^2 3d^2$. The electron configuration of Ti^{3+} is $[Ar]3d^1$. With only one electron in the d-orbitals, the titanium(III) ion **cannot form** high and low-spin complexes – all complexes will contain one unpaired electron and have the same spin.
b) Co: $[ar]4s^2 3d^7$. The electron configuration of Co^{2+} is $[Ar]3d^7$ and will form high and low-spin complexes with 7 electrons in the d orbital.
c) Fe: $[Ar]4s^2 3d^6$. The electron configuration of Fe^{2+} is $[Ar]3d^6$ and will form high and low-spin complexes with 6 electrons in the d orbital.
d) Cu: $[Ar]4s^1 3d^{10}$. The electron configuration of Cu^{2+} is $[Ar]3d^9$, so in complexes with both strong and weak field ligands, one electron will be unpaired and the spin in both types of complexes is identical. Cu^{2+} **cannot form** high and low-spin complexes.

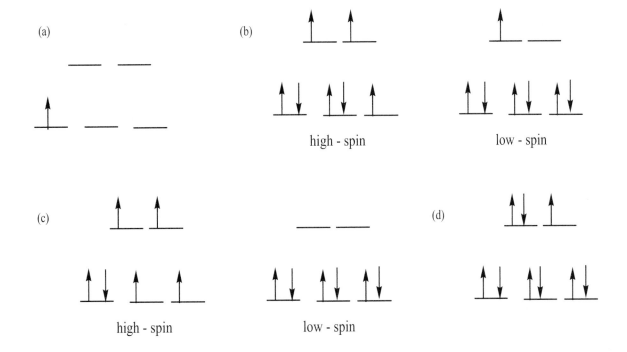

(a)

(b) high - spin   low - spin

(c) high - spin   low - spin

(d)

23.98   To draw the orbital-energy splitting diagram, first determine the number of d electrons in the transition metal ion. Examine the formula of the complex ion to determine the electron configuration of the metal ion, remembering that the ns electrons are lost first. Determine the coordination number from the number of ligands, recognizing that 6 ligands result in an octahedral arrangement and 4 ligands result in a tetrahedral or square planar arrangement. Weak-field ligands give the maximum number of unpaired electrons (high-spin) while strong-field ligands lead to electron pairing (low-spin).

a) Electron configuration of Cr: $[Ar]4s^13d^5$

Charge on Cr: The aqua ligands are neutral, so the charge on Cr is +3.

Electron configuration of Cr^{3+}: $[Ar]3d^3$

Six ligands indicate an octahedral arrangement. Using Hund's rule, fill the lower energy t_{2g} orbitals first, filling empty orbitals before pairing electrons within an orbital.

b) Electron configuration of Cu: $[Ar]4s^13d^{10}$

Charge on Cu: The aqua ligands are neutral, so Cu has a +2 charge.

Electron configuration of Cu^{2+}: $[Ar]3d^9$

Four ligands and a d^9 configuration indicate a square planar geometry (only filled d sublevel ions exhibit tetrahedral geometry). Use Hund's rule to fill in the 9 d electrons. Therefore, the correct orbital-energy splitting diagram shows one unpaired electron.

c) Electron configuration of Fe: $[Ar]4s^23d^6$

Charge on Fe: Each fluoride ligand has a –1 charge, so Fe has a +3 charge to make the overall complex charge equal to –3.

Electron configuration of Fe^{3+}: $[Ar]3d^5$

Six ligands indicate an octahedral arrangement. Use Hund's rule to fill the orbitals.

F^- is a weak field ligand, so the splitting energy, Δ, is not large enough to overcome the resistance to electron pairing. The electrons remain unpaired, and the complex is called *high-spin*.

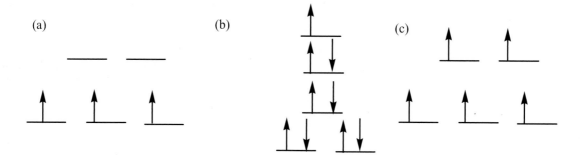

(a) (b) (c)

23.100 To draw the orbital-energy splitting diagram, first determine the number of d electrons in the transition metal ion. Examine the formula of the complex ion to determine the electron configuration of the metal ion, remembering that the ns electrons are lost first. Determine the coordination number from the number of ligands, recognizing that 6 ligands result in an octahedral arrangement and 4 ligands result in a tetrahedral or square planar arrangement. Weak-field ligands give the maximum number of unpaired electrons (high-spin) while strong-field ligands lead to electron pairing (low-spin).
a) Electron configuration of Mo: $[Kr]5s^1 3d^5$
Charge on Mo: Each chloride ligand has a –1 charge, so Mo has a +3 charge to make the overall complex charge equal to –3.
Electron configuration of Mo^{3+}: $[Kr]4d^3$
Six ligands indicate an octahedral arrangement. Using Hund's rule, fill the lower energy t_{2g} orbitals first, filling empty orbitals before pairing electrons within an orbital.
b) Electron configuration of Ni: $[Ar]4s^2 3d^8$
Charge on Ni: The aqua ligands are neutral, so the charge on Ni is +2.
Electron configuration of Ni^{2+}: $[Ar]3d^8$
Six ligands indicate an octahedral arrangement. Use Hund's rule to fill the orbitals.
H_2O is a weak field ligand, so the splitting energy, Δ, is not large enough to overcome the resistance to electron pairing. One electron occupies each of the 5 d orbitals before pairing in the t_{2g} orbitals, and the complex is called *high-spin*.
c) Electron configuration of Ni: $[Ar]4s^2 3d^8$
Charge on Ni: Each cyanide ligand has a –1 charge, so Ni has a +2 charge to make the overall complex charge equal to –2.
Electron configuration of Ni^{2+}: $[Ar]3d^8$
The coordination number is 4 and the complex is square planar.
The complex is low spin because CN^- is a strong field ligand. Electrons pair in one set of orbitals before occupying orbitals of higher energy.

(a) (b) (c)

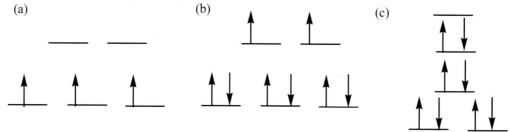

23.102 The spectrochemical series describes the spectrum of splitting energy, Δ. NO_2^- is a stronger ligand than NH_3, which is stronger than H_2O. NO_2^- produces the largest Δ, followed by NH_3 and then H_2O with the smallest Δ value. The energy of light absorbed increases as Δ increases since more energy is required to excite an electron from a lower energy orbital to a higher energy orbital when Δ is very large.
$$[Cr(H_2O)_6]^{3+} < [Cr(NH_3)_6]^{3+} < [Cr(NO_2)_6]^{3-}$$

23.104 A violet complex absorbs yellow-green light. The light absorbed by a complex with a weaker ligand would be at a lower energy and longer wavelength. Light of lower energy than yellow-green light is yellow, orange, or red light. The color observed would be **blue** or **green**.

23.107 The aqua ligand is weaker than the ammine ligand. The weaker ligand results in a lower splitting energy and absorbs a lower energy of visible light. The green hexaaqua complex appears green because it absorbs red light (opposite side of the color wheel). The hexaammine complex appears violet because it absorbs yellow light, which is higher in energy (shorter λ) than red light.

23.111 The electron configuration of Hg is $[Xe]6s^2 4f^{14} 5d^{10}$ and that of Hg^+ is $[Xe]6s^1 4f^{14} 5d^{10}$. The electron configuration of Cu is $[Ar]4s^1 3d^{10}$ and that of Cu^+ is $[Ar]3d^{10}$. In the mercury(I) ion, there is one electron in the $6s$ orbital that can form a covalent bond with the electron in the $6s$ orbital of another Hg^+ ion. In the copper(I) ion, there are no electrons in the s orbital to bond with another copper(I) ion.

23.113 a) The coordination number of cobalt is **6**. The two Cl^- ligands are unidentate and the two ethylenediamine ligands are bidentate, so a total of 6 ligand atoms are connected to the central metal ion.
b) The counter ion is Cl^-, so the complex ion is $[Co(en)_2Cl_2]^+$. Each chloride ligand has a -1 charge and each en ligand is neutral, so cobalt has a **+3** charge: $+3 + 2(0) + 2(-1) = +1$.
c) One mole of complex dissolves in water to yield one mole of $[Co(en)_2Cl_2]^+$ ions and one mole of Cl^- ions. Therefore, each formula unit yields **2** individual ions.
d) One mole of complex dissolves to form one mole of Cl^- ions, which reacts with the Ag^+ ion (from $AgNO_3$) to form **one mole of AgCl precipitate**.

23.123 Geometric (*cis-trans*) and linkage isomerism. The SCN^- ligand can bond to the metal through either the S atom or the N atom.

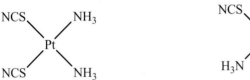

cis-diamminedithiocyanatoplatinum(II)   *trans*-diamminedithiocyanatoplatinum(II)

cis-diamminediisothiocyanatoplatinum(II)   *trans*-diamminediisothiocyanatoplatinum(II)

cis-diamminethiocyanatoisothiocyanatoplatinum(II)   *trans*-diamminethiocyanatoisothiocyanatoplatinum(II)

23.125 $[Co(NH_3)_4(H_2O)Cl]^{2+}$ tetraammineaquachlorocobalt(III) ion
2 geometric isomers

trans Cl and H$_2$O   *cis* Cl and NH$_3$

$[Cr(H_2O)_3Br_2Cl]$ triaquadibromochlorochromium(III)
3 geometric isomers

Br's *trans*   Br's *cis*   Br's *cis*
                      H$_2$O's facial   H$_2$O's meridional

(Unfortunately meridional and facial isomers are not covered in the text. Facial (fac) isomers have three adjacent corners of the octahedron occupied by similar groups. Meridional (mer) isomers have three similar groups around the outside of the complex.)

$[Cr(NH_3)_2(H_2O)_2Br_2]^+$ diamminediaquadibromochromium(III) ion
6 isomers (5 geometric)

All pairs are *trans*   Only NH$_3$'s are *trans*

Only H$_2$O's are *trans*   Only Br's are *trans*

All pairs are *cis*. These are optical isomers of each other.

456

23.128 a) The oxidation state of Mn in MnO_4^{2-} is **+6**, in MnO_4^- is **+7**, and in MnO_2 is **+4**.
   b) The two half-reactions are

   reduction: $MnO_4^{2-}(aq) \rightarrow MnO_2(s)$
   balancing gives: $MnO_4^{2-}(aq) + 4\,H^+(aq) + 2\,e^- \rightarrow MnO_2(s) + 2\,H_2O(l)$
   oxidation: $MnO_4^{2-}(aq) \rightarrow MnO_4^-(aq) + e^-$
   Multiply the oxidation half-reaction by 2 and add to get
   reduction: $MnO_4^{2-}(aq) + 4\,H^+(aq) + 2\,e^- \rightarrow MnO_2(s) + 2\,H_2O(l)$
   oxidation: $2\,MnO_4^{2-}(aq) \rightarrow 2\,MnO_4^-(aq) + 2e^-$
$$\mathbf{3\,MnO_4^{2-}(aq) + 4\,H^+(aq) \rightarrow 2\,MnO_4^-(aq) + MnO_2(s) + 2\,H_2O(l)}$$

23.135 a) A plane that includes both ammonia ligands and the zinc ion is a plane of symmetry. The complex **does not have optical isomers** because it has a plane of symmetry.
   b) The Pt^{2+} ion is d^8, so the complex is square planar. Any square planar complex has a plane of symmetry, so the complex **does not have optical isomers**.
   c) The *trans* octahedral complex has the two chloride ions opposite each other. A plane of symmetry can be passed through the two chlorides and platinum ion, so the *trans*-complex **does not have optical isomers**.
   d) **No optical isomers** for same reason as in part (c).
   e) The *cis*-isomer does not have a plane of symmetry, so it **does have optical isomers**.

23.136 Assume a 100-g sample (making the percentages of each element equal to the mass present in grams) and convert the mass of each element to moles using the element's molar mass. Divide each mole amount by the smallest mole amount to determine whole number ratios of the elements to determine the empirical formula.
   Note, the intermediate values are unrounded to minimize rounding errors.

$$\text{moles Pt} = (38.8 \text{ g Pt})\left(\frac{1 \text{ mol Pt}}{195.1 \text{ g Pt}}\right) = 0.198872 \text{ mol Pt}$$

$$\text{moles Cl} = (14.1 \text{ g Cl})\left(\frac{1 \text{ mol Cl}}{35.45 \text{ g Cl}}\right) = 0.397743 \text{ mol Cl}$$

$$\text{moles C} = (28.7 \text{ g C})\left(\frac{1 \text{ mol C}}{12.01 \text{ g C}}\right) = 2.389675 \text{ mol C}$$

$$\text{moles P} = (12.4 \text{ g P})\left(\frac{1 \text{ mol P}}{30.97 \text{ g P}}\right) = 0.400387 \text{ mol P}$$

$$\text{moles H} = (6.02 \text{ g H})\left(\frac{1 \text{ mol H}}{1.008 \text{ g H}}\right) = 5.972222 \text{ mol H}$$

Divide each of the moles just calculated by the smallest value (Pt) gives:
    Pt = 1 Cl = 2 C = 12 P = 2 H = 30
The empirical formula for the compound is $PtCl_2C_{12}P_2H_{30}$. Each triethylphosphine ligand, $P(C_2H_5)_3$ accounts for one P atom, 6 C atoms, and 15 H atoms. According to the empirical formula, there are two triethylphosphine ligands: 2 $P(C_2H_5)_3$ "equals" $C_{12}P_2H_{30}$. The compound must have two $P(C_2H_5)_3$ ligands and two chloro ligands per Pt ion. The formula is $[Pt[P(C_2H_5)_3]_2Cl_2]$. The central Pt ion has 4 ligands and is square planar, existing as either a *cis* or *trans* compound.

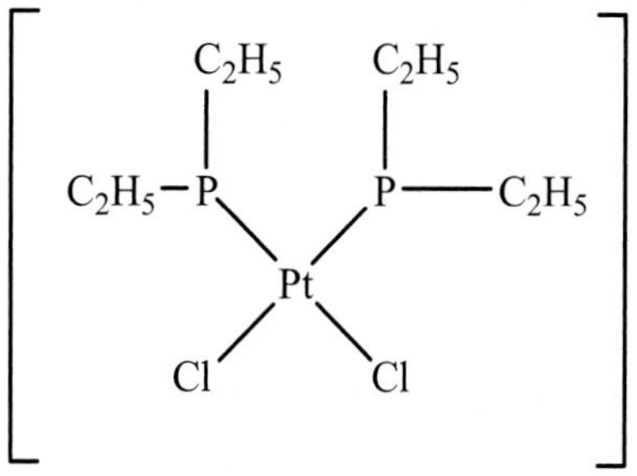

C₂H₅ ... these are chemical structure diagrams.

C_2H_5 C_2H_5

C_2H_5-P $P-C_2H_5$

Pt

Cl   Cl

C_2H_5

C_2H_5-P Cl

Pt

Cl $P-C_2H_5$

C_2H_5

cis-dichlorobis(triethylphosphine)platinum(II) 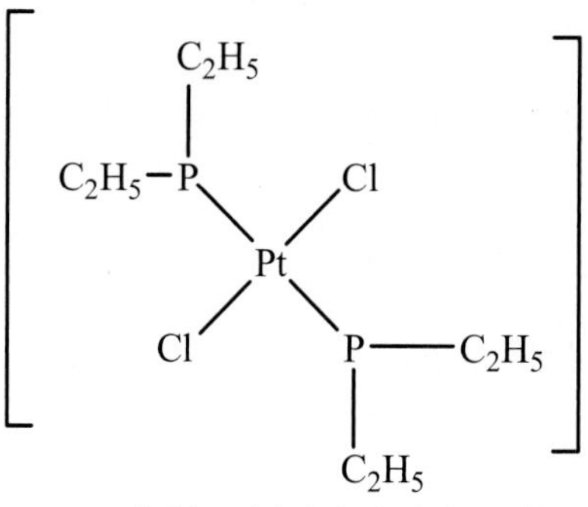   *trans*-dichlorobis(triethylphosphine)platinum(II)

23.140   a) The first reaction shows no change in the number of particles. In the second reaction, the number of reactant particles is greater than the number of product particles. A decrease in the number of particles means a decrease in entropy, while no change in number of particles indicates little change in entropy. Based on entropy change only, the first reaction is favored.

b) The ethylenediamine complex is more stable with respect to ligand exchange with water because the entropy change is unfavorable.

CHAPTER 24 NUCLEAR REACTIONS AND THEIR APPLICATIONS

FOLLOW–UP PROBLEMS

24.1   Plan: Write a skeleton equation that shows an unknown nuclide, $_{Z}^{A}X$, undergoing beta decay, $_{-1}^{0}\beta$, to form cesium–133, $_{55}^{133}Cs$. Conserve mass and atomic number by ensuring the superscripts and subscripts equal one another on both sides of the equation. Determine the identity of X by using the periodic table to identify the element with atomic number equal to Z.
Solution:
The unknown nuclide yields cesium–133 and a β particle:
$$_{Z}^{A}X \rightarrow {}_{55}^{133}Cs + {}_{-1}^{0}\beta$$
To conserve atomic number, Z must equal 54. Element is xenon.
To conserve mass number, A must equal 133.
The identity of $_{Z}^{A}X$ is $_{54}^{133}Xe$.
The balanced equation is: $_{54}^{133}Xe \rightarrow {}_{55}^{133}Cs + {}_{-1}^{0}\beta$
Check: A = 133 = 133 + 0 and Z = 54 = 55 + (–1), so mass is conserved.

24.2   Plan: Nuclear stability is found in nuclides with an N/Z ratio that falls within the band of stability. Nuclides with an even N and Z, especially those nuclides that have magic numbers, are exceptionally stable. Examine the two nuclides to see which of these criteria can explain the difference in stability.
Solution:
Phosphorus–31 has 16 neutrons and 15 protons, with an N/Z ratio of 1.07. Phosphorus–30 has 15 neutrons and 15 protons, with an N/Z ratio of 1.00. ^{31}P has an even N while ^{30}P has both an odd Z and an odd N. ^{31}P has a slightly higher N/Z ratio that is closer to the band of stability.

24.3   Plan: Examine the N/Z ratio and determine which mode of decay will yield a product nucleus that is closer to the band.
Solution:
a) Iron–61 has an N/Z ratio of (61 – 26)/26 = 1.35, which is too high for this region of the band. Iron–61 will undergo β decay.
$$_{26}^{61}Fe \rightarrow {}_{27}^{61}Co + {}_{-1}^{0}\beta \qquad \text{new } N/Z = (61 - 27)/27 = 1.26$$
b) Americium–241 has Z > 83, so it undergoes α decay.
$$_{95}^{241}Am \rightarrow {}_{93}^{237}Np + {}_{2}^{4}He$$

24.4   Plan: Use the half-life of ^{24}Na to find k. Substitute the value of k, initial activity (A_0), and time of decay (4 days) into the integrated first-order rate equation to solve for activity at a later time (A_t).
Solution:
$k = (\ln 2) / t_{1/2} = (\ln 2) / 15\ h = 0.0462098\ h^{-1}$ (unrounded)
$\ln A_0 - \ln A_t = kt$
$\ln A_t = -kt + \ln A_0 = -(0.0462098\ h^{-1})\ (4.0\ days)\ (24\ h/day) + \ln (2.5 \times 10^9) = 17.2034$ (unrounded)
$\ln A_t = 17.2034$
$A_t = 2.96034 \times 10^7 = \mathbf{3.0 \times 10^7\ d/s}$
Check: In 4 days, a little more than 6 half-lives (6.4) have passed. A quick calculation on the calculator shows that $(2.5 \times 10^9) / 2 = 1.25 \times 10^9$ (first half-life). Dividing by 2 again shows 6.25×10^8 (second half-life). Continued division by 2 shows that this value comes to 3.9×10^7 d/s after six divisions by 2. This number is slightly higher than the calculated answer because 6 half-lives have been calculated, not 6.4 half-lives.

24.5   Plan: The wood from the case came from a living organism, so A_0 equals 15.3 d/min•g. Substitute the current activity of the case (A_t), A_0 and the $t_{1/2}$ of carbon (5730 yr) into the first-order rate expression and solve for t.
Solution:

$$t = \frac{1}{k}\ln\frac{A_0}{A_t} = \frac{1}{\left(\ln 2\middle/t_{1/2}\right)}\ln\frac{A_0}{A_t} = \frac{1}{\left(\ln 2\middle/5730\ \text{yr}\right)}\ln\frac{(15.3\ \text{d}/\text{min•g})}{(9.41\ \text{d}/\text{min•g})} = 4018.2 = \textbf{4.02 x 10}^3\ \textbf{years}$$

Check: Since the activity of the bones is a little more than half of the modern activity, the age should be a little less than the half-life of carbon.

24.6   Plan: Uranium–235 has 92 protons and 143 neutrons in its nucleus. Calculate the mass defect (Δm) in one ^{235}U atom, convert to MeV and divide by 235 to obtain binding energy/nucleon.
Solution:
Δm = [(92 x mass H atom) + (143 x mass neutron)] – mass ^{235}U atom
Δm = [(92 x 1.007825 amu) + (143 x 1.008665)] – 235.043924 amu
Δm = 1.915071 amu
Binding Energy / nucleon = [(1.915071 amu) (931.5 MeV/amu)] / (235 nucleons)
        = 7.591015 = **7.591 MeV / nucleon**
The BE/nucleon of ^{12}C is 7.680 MeV/nucleon. The energy per nucleon holding the ^{235}U nucleus together is less than that for ^{12}C (7.591 < 7.680), so ^{235}U is less stable than ^{12}C.
Check: The answer is consistent with the expectation that uranium is less stable than carbon.

END–OF–CHAPTER PROBLEMS

24.1   a) Chemical reactions are accompanied by relatively small changes in energy while nuclear reactions are accompanied by relatively large changes in energy.
b) Increasing temperature increases the rate of a chemical reaction but has no effect on a nuclear reaction.
c) Both chemical and nuclear reaction rates increase with higher reactant concentrations.
d) If the reactant is limiting in a chemical reaction, then more reactant produces more product and the yield increases in a chemical reaction. The presence of more radioactive reagent results in more decay product, so a higher reactant concentration increases the yield in a nuclear reaction.

24.2   a) The percentage of sulfur atoms that are sulfur-32 is 95.02%, the same as the relative abundance of ^{32}S.
b) The atomic mass is larger than the isotopic mass of ^{32}S. Sulfur-32 is the lightest isotope, as stated in the problem, so the other 5% of sulfur atoms are heavier than 31.972070 amu. The average mass of all the sulfur atoms will therefore be greater than the mass of a sulfur-32 atom.

24.4   Radioactive decay that produces a different element requires a change in *atomic number* (Z, number of protons).
$$_Z^A X \qquad A = \text{mass number (protons + neutrons)}$$
Z = number of protons (positive charge)
X = symbol for the particle
$N = A - Z$ (number of neutrons)
a) Alpha decay produces an atom of a different element, i.e., a daughter with two less protons and two less neutrons.
$$_Z^A X \rightarrow {}_{Z-2}^{A-4}Y + {}_2^4 He \quad \text{2 fewer protons, 2 fewer neutrons}$$
b) Beta decay produces an atom of a different element, i.e., a daughter with one more proton and one less neutron. A neutron is converted to a proton and β particle in this type of decay.
$$_Z^A X \rightarrow {}_{Z+1}^{A}Y + {}_{-1}^0 \beta \quad \text{1 more proton, 1 less neutron}$$
c) Gamma decay does not produce an atom of a different element and Z and N remain unchanged.
$$_Z^A X^* \rightarrow {}_Z^A X + {}_0^0 \gamma \quad (_Z^A X^* = \text{energy rich state}), \text{ no change in number of protons or neutrons.}$$
d) Positron emission produces an atom of a different element, i.e., a daughter with one less proton and one more neutron. A proton is converted into a neutron and positron in this type of decay.
$$_Z^A X \rightarrow {}_{Z-1}^{A}Y + {}_{+1}^0 \beta \quad \text{1 less proton, 1 more neutron}$$

e) Electron capture produces an atom of a different element, i.e., a daughter with one less proton and one more neutron. The net result of electron capture is the same as positron emission, but the two processes are different.

$$^A_Z X + ^0_{-1}e \rightarrow ^A_{Z-1}Y \qquad \text{1 less proton, 1 more neutron}$$

A different element is produced in all cases except (c).

24.6   A neutron-rich nuclide decays to convert neutrons to protons while a neutron-poor nuclide decays to convert protons to neutrons. The conversion of neutrons to protons occurs by beta decay:

$$^1_0 n \rightarrow ^1_1 p + ^0_{-1}\beta$$

The conversion of protons to neutrons occurs by either positron decay:

$$^1_1 p \rightarrow ^1_0 n + ^0_1 \beta$$

or electron capture:

$$^1_1 p + ^0_{-1}e \rightarrow ^1_0 n$$

Neutron-rich nuclides, with a high N/Z, undergo β decay. Neutron-poor nuclides, with a low N/Z, undergo positron decay or electron capture.

24.8   In a balanced nuclear equation, the total of mass numbers and the total of charges on the left side and the right side must be equal.

a) $^{234}_{92}U \rightarrow ^4_2He + ^{230}_{90}Th$ Mass: $234 = 4 + 230$;   Charge: $92 = 2 + 90$

b) $^{232}_{93}Np + ^0_{-1}e \rightarrow ^{232}_{92}U$ Mass: $232 + 0 = 232$;   Charge: $93 + (-1) = 92$

c) $^{12}_7 N \rightarrow ^0_1\beta + ^{12}_6 C$ Mass: $12 = 0 + 12$;   Charge: $7 = 1 + 6$

24.10   a) The process converts a neutron to proton, so the mass number is the same, but the atomic number increases by one.

$$^{27}_{12}Mg \rightarrow ^0_{-1}\beta + ^{27}_{13}Al$$

b) Positron emission decreases atomic number by one, but not mass number.

$$^{23}_{12}Mg \rightarrow ^0_1\beta + ^{23}_{11}Na$$

c) The electron captured by the nucleus combines with a proton to form a neutron, so mass number is constant, but atomic number decreases by one.

$$^{103}_{46}Pd + ^0_{-1}e \rightarrow ^{103}_{45}Rh$$

24.12   a) In other words, an unknown nuclide decays to give Ti–48 and a positron.

$$^{48}_{23}V \rightarrow ^{48}_{22}Ti + ^0_1\beta$$

b) In other words, an unknown nuclide captures an electron to form Ag–107.

$$^{107}_{48}Cd + ^0_{-1}e \rightarrow ^{107}_{47}Ag$$

c) In other words, an unknown nuclide decays to give Po–206 and an alpha particle.

$$^{210}_{86}Rn \rightarrow ^{206}_{84}Po + ^4_2He$$

24.14   a) $^{186}_{78}Pt + ^0_{-1}e \rightarrow ^{186}_{77}Ir$

b) $^{225}_{89}Ac \rightarrow ^{221}_{87}Fr + ^4_2He$

c) $^{129}_{52}Te \rightarrow ^{129}_{53}I + ^0_{-1}\beta$

24.16   Look at the N/Z ratio, the ratio of the number of neutrons to the number of protons. If the N/Z ratio falls in the band of stability, the nuclide is predicted to be stable. For stable nuclides of elements with atomic number greater than 20, the ratio of number of neutrons to number of protons (N/Z) is greater than one. In addition, the ratio increases gradually as atomic number increases. Also check for exceptionally stable numbers of neutrons and /or protons – the "magic" number of 2, 8, 20, 28, 50, 82 and (N = 126). Also, even numbers of protons and or neutrons are related to stability whereas odd numbers are related to instability.

(a) $^{20}_{8}O$ appears stable because its Z (8) value is a magic number, but its N/Z ratio $(20 - 8)/8 = 1.50$ is too high and this nuclide is above the band of stability; $^{20}_{8}O$ is unstable.

(b) $^{59}_{27}Co$ might look unstable because its Z value is an odd number, but its N/Z ratio $(59 - 27)/27 = 1.19$ is in the band of stability, so $^{59}_{27}Co$ appears stable.

(c) $^{9}_{3}Li$ appears unstable because its N/Z ratio $(9 - 3)/3 = 2.00$ is too high and is above the band of stability.

24.18 For stable nuclides of elements with atomic number greater than 20, the ratio of number of neutrons to number of protons (N/Z) is greater than one. In addition, the ratio increases gradually as atomic number increases.
a) For the element iodine $Z = 53$. For iodine-127, $N = 127 - 53 = 74$. The N/Z ratio for ^{127}I is $74/53 = 1.4$. Of the examples of stable nuclides given in the book ^{107}Ag has the closest atomic number to iodine. The N/Z ratio for ^{107}Ag is 1.3. Thus, it is likely that iodine with 6 additional protons is stable with an N/Z ratio of 1.4.
b) Tin is element number 50 ($Z = 50$). From part (a), the stable nuclides of tin would have N/Z ratios approximately 1.3 to 1.4. The N/Z ratio for ^{106}Sn is $(106 - 50)/50 = 1.1$. The nuclide ^{106}Sn is unstable with an N/Z ratio that is too low.
c) For ^{68}As, $Z = 33$ and $N = 68 - 33 = 35$ and $N/Z = 1.1$. The ratio is within the range of stability, but the nuclide is most likely unstable because there is an odd number of both protons and neutrons.

24.20 a) $^{238}_{92}U$: Nuclides with $Z > 83$ decay through α decay.

b) The N/Z ratio for $^{48}_{24}Cr$ is $(48 - 24)/24 = 1.00$. This number is below the band of stability because N is too low and Z is too high. To become more stable, the nucleus decays by converting a proton to a neutron, which is positron decay. Alternatively, a nucleus can capture an electron and convert a proton into a neutron through electron capture.

c) The N/Z ratio for $^{50}_{25}Mn$ is $(50 - 25)/25 = 1.00$. This number is also below the band of stability, so the nuclide undergoes positron decay or electron capture.

24.22 a) For carbon-15, $N/Z = 9/6 = 1.5$, so the nuclide is neutron-rich. To decrease the number of neutrons and increase the number of protons, carbon-15 decays by beta decay.
b) The N/Z ratio for ^{120}Xe is $66/54 = 1.2$. Around atomic number 50, the ratio for stable nuclides is larger than 1.2, so ^{120}Xe is proton-rich. To decrease the number of protons and increase the number of neutrons, the xenon-120 nucleus either undergoes positron emission or electron capture.
c) Thorium-224 has an N/Z ratio of $134/90 = 1.5$. All nuclides of elements above atomic number 83 are unstable and decay to decrease the number of both protons and neutrons. Alpha decay by thorium-224 is the most likely mode of decay.

24.24 Stability results from a favorable N/Z ratio, even numbers of N and/or Z, and the occurrence of magic numbers. The N/Z ratio of $^{52}_{24}Cr$ is $(52 - 24)/24 = 1.17$, which is within the band of stability. The fact that Z is even does not account for the variation in stability because all isotopes of chromium have the same Z. However, $^{52}_{24}Cr$ has 28 neutrons, so N is both an even number and a magic number for this isotope only.

24.28 The equation for the nuclear reaction is
$$^{235}_{92}U \rightarrow\ ^{207}_{82}Pb + \underline{\ \ }\ ^{0}_{-1}\beta + \underline{\ \ }\ ^{4}_{2}He$$
To determine the coefficients, notice that the beta particles will not impact the mass number. Subtracting the mass number for lead from the mass number for uranium will give the total mass number for the alpha particles released, $235 - 207 = 28$. Each alpha particle is a helium nucleus with mass number 4. The number of helium atoms is determined by dividing the total mass number change by 4, $28/4 = 7$ helium atoms or 7 alpha particles. The equation is now
$$^{235}_{92}U \rightarrow\ ^{207}_{82}Pb + \underline{\ \ }\ ^{0}_{-1}\beta + 7\ ^{4}_{2}He$$

To find the number of beta particles released, examine the difference in number of protons (atomic number) between the reactant and products. Uranium, the reactant, has 92 protons. The atomic number in the products, lead atom and 7 helium nuclei, total 96. To balance the atomic numbers, four electrons (beta particles) must be emitted to give the total atomic number for the products as 96 − 4 = 92, the same as the reactant. In summary, 7 alpha particles and 4 beta particles are emitted in the decay of uranium-235 to lead-207.

$$^{235}_{92}\text{U} \rightarrow {}^{207}_{82}\text{Pb} + 4\ {}^{0}_{-1}\beta + 7\ {}^{4}_{2}\text{He}$$

24.31 No, it is not valid to conclude that $t_{1/2}$ equals 1 minute because the number of nuclei is so small (6 nuclei). Decay rate is an average rate and is only meaningful when the sample is macroscopic and contains a large number of nuclei, as in the second case. Because the second sample contains 6×10^{12} nuclei, the conclusion that $t_{1/2}$ = 1 minute is valid.

24.33 Specific activity of a radioactive sample is its decay rate per gram. Calculate the specific activity from the number of particles emitted per second (disintegrations per second = dps) and the mass of the sample.
Specific activity = decay rate per gram.
1 Ci = 3.70×10^{10} dps

$$\text{Specific Activity} = \left(\frac{1.56 \times 10^6 \text{ dps}}{1.65 \text{ mg}}\right)\left(\frac{1 \text{ mg}}{10^{-3} \text{ g}}\right)\left(\frac{1 \text{ Ci}}{3.70 \times 10^{10} \text{ dps}}\right) = 2.55528 \times 10^{-2} = \mathbf{2.56 \times 10^{-2} \text{ Ci/g}}$$

24.35 A becquerel is a disintegration per second (d/s). A disintegration can be any of the radioactive decay particles mentioned in the text (α, β, positron, etc.). Therefore, the emission of one α particle equals one disintegration.

$$\text{Specific Activity} = \left(\frac{\left(\frac{7.4 \times 10^4 \text{ d}}{\text{min}}\right)\left(\frac{1 \text{ min}}{60 \text{ s}}\right)}{8.58\ \mu\text{g}\left(\frac{10^{-6} \text{ g}}{1\ \mu\text{g}}\right)}\right)\left(\frac{1 \text{ bq}}{1 \text{ dps}}\right) = 1.43745 \times 10^8 = \mathbf{1.4 \times 10^8 \text{ Bq/g}}$$

24.37 Decay constant is the rate constant for the first order reaction:
Decay rate = $-(\Delta N / \Delta t) = k\text{N}$
$-(-1 \text{ atom} / 1 \text{ d}) = k (1 \times 10^{12} \text{ atom})$
$k = \mathbf{1 \times 10^{-12} \text{ d}^{-1}}$

24.39 The rate constant, k, relates the number of radioactive nuclei to their decay rate through the equation $A = k\text{N}$. The number of radioactive nuclei is calculated by converting moles to atoms using Avogadro's number. The decay rate is 1.39×10^5 d/yr or more simply, 1.39×10^5 yr^{-1} (the disintegrations are assumed).
Decay rate = $-(\Delta N / \Delta t) = k\text{N}$

$$-\frac{-1.39 \times 10^5 \text{ atom}}{1.00 \text{ yr}} = k\left(\frac{1.00 \times 10^{-12} \text{ mol}}{}\right)\left(\frac{6.022 \times 10^{23} \text{ atom}}{1 \text{ mol}}\right)$$

1.39×10^5 atom /yr = k (6.022×10^{11} atom)
$k = (1.39 \times 10^5$ atom /yr) / 6.022×10^{11} atom
$k = 2.30820 \times 10^{-7} = \mathbf{2.31 \times 10^{-7} \text{ yr}^{-1}}$

24.41 Radioactive decay is a first-order process, so the integrated rate law is $\ln[N]_t = \ln[A]_0 - kt$
To calculate the fraction of bismuth–212 remaining after 3.75×10^3 h, first find the value of k from the half-life, then calculate the fraction remaining with N_0 set to 1 (exactly).
$t_{1/2} = 1.01$ yr $t = 3.75 \times 10^3$ h
$k = (\ln 2) / (t_{1/2}) = (\ln 2) / (1.01 \text{ yr}) = 0.686284 \text{ yr}^{-1}$ (unrounded)
$\ln[N]_t = \ln[A]_0 - kt$

$$\ln[N]_t = \ln[2.00 \text{ mg}] - (0.686284 \text{ yr}^{-1})(3.75 \times 10^3 \text{ h})\left(\frac{1 \text{ d}}{24 \text{ h}}\right)\left(\frac{1 \text{ y}}{365 \text{ d}}\right)$$

$$\ln[N]_t = 0.399361$$
$$N_t = e^{0.399361}$$
$$N_t = 1.49087 = \textbf{1.49 mg}$$

24.43   Lead–206 is a stable daughter of ^{238}U. Since all of the ^{206}Pb came from ^{238}U, the starting amount of ^{238}U was $(270\ \mu mol + 110\ \mu mol) = 380\ \mu mol = N_0$. The amount of ^{238}U at time t (current) is $270\ \mu mol = N_t$. Find k from the first-order rate expression for half-life, and then substitute the values into the integrated rate law and solve for t.

$t_{1/2} = (\ln 2)\ /\ k$ $\qquad\qquad k = (\ln 2)\ /\ t_{1/2} = (\ln 2)\ /\ (4.5 \times 10^9\ yr) = 1.540327 \times 10^{-10}\ yr^{-1}$ (unrounded)

$$\ln\frac{N_0}{N_t} = kt$$

$$\ln\frac{380\ \mu mol}{270\ \mu mol} = (1.540327 \times 10^{-10}\ yr^{-1})t$$

$$0.341749293 = (1.540327 \times 10^{-10}\ yr^{-1})t$$

$$t = (0.341749293)\ /\ (1.540327 \times 10^{-10}\ yr^{-1}) = 2.21868 \times 10^9 = \textbf{2.2} \times \textbf{10}^\textbf{9}\ \textbf{yr}$$

24.45   Use the conversion factor, $1\ Ci = 3.70 \times 10^{10}$ disintegrations per second (dps).

$$Activity = \left(\frac{6 \times 10^{-11}\ mCi}{mL}\right)\left(\frac{10^{-3}\ Ci}{1\ mCi}\right)\left(\frac{3.70 \times 10^{10}\ dps}{1\ Ci}\right)\left(\frac{60\ s}{1\ min}\right)\left(\frac{1000\ mL}{1.057\ qt}\right)\left(\frac{1\ qt}{4\ cups}\right)\left(\frac{1\ cup}{8\ oz}\right)(8\ oz)$$

$$= 31.50426 = \textbf{30 dpm}$$

24.47   Both N_t and N_0 are given: the number of nuclei present currently, N_t, is found from the moles of ^{232}Th. Each fission track represents one nucleus that disintegrated, so the number of nuclei disintegrated is added to the number of nuclei currently present to determine the initial number of nuclei, N_0. The rate constant, k, is calculated from the half-life. All values are substituted into the first-order decay equation to find t.

$t_{1/2} = (\ln 2)\ /\ k$

$\qquad k = (\ln 2)\ /\ t_{1/2} = (\ln 2)\ /\ (1.4 \times 10^{10}\ yr) = 4.9510512 \times 10^{-11}\ yr^{-1}$ (unrounded)

$N_t = (3.1 \times 10^{-15}\ mol\ Th)\ (6.022 \times 10^{23}\ atoms\ Th\ /\ mol\ Th) = 1.86682 \times 10^9\ atoms\ Th$ (unrounded)

$$\ln\frac{N_0}{N_t} = kt$$

$$\ln\frac{\left(1.86682 \times 10^9 + 9.5 \times 10^4\right)atoms}{1.86682 \times 10^9\ atoms} = (4.9510512 \times 10^{-11}\ yr^{-1})t$$

$$5.08874 \times 10^{-5} = (4.9510512 \times 10^{-11}\ yr^{-1})t$$

$$t = (5.08874 \times 10^{-5})\ /\ (4.9510512 \times 10^{-11}\ yr^{-1}) = 1.027809 \times 10^6 = \textbf{1.0} \times \textbf{10}^\textbf{6}\ \textbf{yr}$$

24.50   Both gamma radiation and neutron beams have no charge, so neither is deflected by electric or magnetic fields. Neutron beams differ from gamma radiation in that a neutron has mass approximately equal to that of a proton. Researchers observed that a neutron beam could induce the emission of protons from a substance. Gamma rays do not cause such emissions.

24.52   Protons are repelled from the target nuclei due to the interaction of like (positive) charges. Higher energy is required to overcome the repulsion.

24.53   a) An alpha particle is a reactant with ^{10}B and a neutron is one product. The mass number for the reactants is $10 + 4 = 14$. So, the missing product must have a mass number of $14 - 1 = 13$. The total atomic number for the reactants is $5 + 2 = 7$, so the atomic number for the missing product is 7.

$$^{10}_{5}B + ^{4}_{2}He \rightarrow ^{1}_{0}n + ^{13}_{7}N$$

b) For the reactants, the mass number is $28 + 2 = 30$ and the atomic number is $14 + 1 = 15$. The given product has mass number 29 and atomic number 15, so the missing product particle has mass number 1 and atomic number 0. The particle is thus a neutron.

$$^{28}_{14}Si + ^{2}_{1}H \rightarrow ^{1}_{0}n + ^{29}_{15}P$$

c) The products are 2 neutrons and ^{244}Cf with a total mass number of $2 + 244 = 246$, and an atomic number of 98. The given reactant particle is an alpha particle with mass number 4 and atomic number 2. The missing reactant must have mass number of $246 - 4 = 242$ and atomic number $98 - 2 = 96$. Element 96 is Cm.

$$^{242}_{96}\text{Cm} + ^4_2\text{He} \rightarrow 2\,^1_0\text{n} + ^{244}_{98}\text{Cf}$$

24.58 Ionizing radiation is more dangerous to children because their rapidly dividing cells are more susceptible to radiation than an adult's slowly dividing cells.

24.60 a) The rad is the amount of radiation energy absorbed in J per body mass in kg.

$$\left(\frac{3.3 \times 10^{-7}\text{ J}}{135\text{ lb}}\right)\left(\frac{2.205\text{ lb}}{1\text{ kg}}\right)\left(\frac{1\text{ rad}}{1 \times 10^{-2}\text{ J/kg}}\right) = 5.39 \times 10^{-7} = \mathbf{5.4 \times 10^{-7}\text{ rad}}$$

b) Conversion factor is 1 rad = 0.01 Gy
$$(5.39 \times 10^{-7}\text{ rad})(0.01\text{ Gy} / 1\text{ rad}) = 5.39 \times 10^{-9} = \mathbf{5.4 \times 10^{-9}\text{ Gy}}$$

24.62 a) Convert the given information to units of J/kg. 1 rad = 0.01 J/kg = 0.01 Gy

$$\text{Dose} = \frac{\left(6.0 \times 10^5\,\beta\right)\left(8.74 \times 10^{-14}\text{ J/}\beta\right)}{70.\text{ kg}}\left(\frac{1\text{ rad}}{0.01\text{ J/}_{\text{kg}}}\right)\left(\frac{0.01\text{ Gy}}{1\text{ rad}}\right) = 7.4914 \times 10^{-10} = \mathbf{7.5 \times 10^{-10}\text{ Gy}}$$

b) Convert grays to rads and multiply rads by RBE to find rems. Convert rems to mrems.

$$\text{rem} = \text{rads} \times \text{RBE} = \left(7.4914 \times 10^{-10}\text{ Gy}\right)\left(\frac{1\text{ rad}}{0.01\text{ Gy}}\right)(1.0)\left(\frac{1\text{ mrem}}{10^{-3}\text{ rem}}\right) = 7.4914 \times 10^{-5} = \mathbf{7.5 \times 10^{-5}\text{ mrem}}$$

c) 1 rem = 0.01 Sv
$$\text{Sv} = (7.5 \times 10^{-5}\text{ mrem})(10^{-3}\text{ rem} / 1\text{ mrem})(0.01\text{ Sv} / \text{rem}) = 7.4914 \times 10^{-10} = \mathbf{7.5 \times 10^{-10}\text{ Sv.}}$$

24.65 Use the time and disintegrations per second (Bq) to find the number of ^{60}Co atoms that disintegrate, which equals the number of β particles emitted. The dose in rads is calculated as energy absorbed per body mass.

$$\text{Dose} = \left(\frac{475\text{ Bq}}{1.858\text{ g}}\right)\left(\frac{10^3\text{ g}}{1\text{ kg}}\right)\left(\frac{1\text{ dps}}{1\text{ Bq}}\right)\left(\frac{5.05 \times 10^{-14}\text{ J}}{1\text{ disint.}}\right)(24.0\text{ min})\left(\frac{60\text{ s}}{1\text{ min}}\right)\left(\frac{1\text{ rad}}{0.01\text{ J/}_{\text{kg}}}\right) = 1.8591 \times 10^{-3} = \mathbf{1.86 \times 10^{-3}\text{ rad}}$$

24.67 NAA does not destroy the sample while chemical analyses does. Neutrons bombard a non-radioactive sample, "activating" or energizing individual atoms within the sample to create radioisotopes. The radioisotopes decay back to their original state (thus, the sample is not destroyed) by emitting radiation that is different for each isotope.

24.73 Energy is released when a nuclide forms from nucleons. The nuclear binding energy is the amount of energy holding the nucleus together. Energy is absorbed to break the nucleons into nucleons and is released when nucleons "come together."

24.75 Apply the appropriate conversions from the chapter or the inside back cover.

a) Energy $= \left(0.01861\text{ MeV}\right)\left(\frac{10^6\text{ eV}}{1\text{ MeV}}\right) = \mathbf{1.861 \times 10^4\text{ eV}}$

b) Energy $= \left(0.01861\text{ MeV}\right)\left(\frac{10^6\text{ eV}}{1\text{ MeV}}\right)\left(\frac{1.602 \times 10^{-19}\text{ J}}{1\text{ eV}}\right) = 2.981322 \times 10^{-15} = \mathbf{2.981 \times 10^{-15}\text{ J}}$

24.77 Convert moles of ^{239}Pu to atoms of ^{239}Pu using Avogadro's number. Multiply the number of atoms by the energy per atom (nucleus) and convert the MeV to J using the conversion 1 eV = 1.602 x 10^{-19} J.

$$\text{Energy} = \left(1.5 \text{ mol}\,^{239}\text{Pu}\right)\left(\frac{6.022 \times 10^{23} \text{ atoms}}{\text{mol}}\right)\left(\frac{5.243 \text{ MeV}}{1 \text{ atom}}\right)\left(\frac{10^6 \text{ eV}}{1 \text{ MeV}}\right)\left(\frac{1.602 \times 10^{-19} \text{ J}}{1 \text{ eV}}\right)$$

$$= 7.587075 \times 10^{11} = \textbf{7.6} \times \textbf{10}^{\textbf{11}} \textbf{ J}$$

24.79 Use the conversion in the chapter (1 amu = 931.5 MeV)
The mass defect is calculated from the mass of 8 protons (^{1}H) and 8 neutrons versus the mass of the oxygen nuclide.
Mass of 8 ^{1}H atoms = 8 x 1.007825 = 8.062600 amu
Mass of 8 neutrons = 8 x 1.008665 = 8.069320 amu
        Total mass   16.131920 amu
Mass defect = Δm = 16.131920 – 15.994915 = 0.137005 amu / ^{16}O = 0.137005 g/mol ^{16}O

a) Binding energy = $\left(\dfrac{0.137005 \text{ amu}\,^{16}\text{O}}{16 \text{ nucleons}}\right)\left(\dfrac{931.5 \text{ MeV}}{1 \text{ amu}}\right)$ = 7.976259844 = **7.976 MeV/nucleon**

b) Binding energy = $\left(\dfrac{0.137005 \text{ amu}\,^{16}\text{O}}{1 \text{ atom}}\right)\left(\dfrac{931.5 \text{ MeV}}{1 \text{ amu}}\right)$ = 127.6201575 = **127.6 MeV/atom**

c) Use $\Delta E = \Delta mc^2$

$$\text{Binding energy} = \left(\frac{0.137005 \text{ g}\,^{16}\text{O}}{\text{mol}}\right)\left(\frac{1 \text{ kg}}{10^3 \text{ g}}\right)\left(2.99792 \times 10^8 \text{ m/s}\right)^2\left(\frac{1 \text{ J}}{\text{kg} \cdot \text{m}^2 / \text{s}^2}\right)\left(\frac{1 \text{ kJ}}{10^3 \text{ J}}\right)$$

$$= 1.2313357 \times 10^{10} = \textbf{1.23134} \times \textbf{10}^{\textbf{10}} \textbf{ kJ/mol}$$

24.81 Calculate Δm, convert the mass defect to MeV, and divide by 59 nucleons.
Mass of 27 ^{1}H atoms = 27 x 1.007825 = 27.211275 amu
Mass of 32 neutrons = 32 x 1.008665 = 32.27728 amu
        Total mass   59.488555 amu
Mass defect = Δm = 59.488555 – 58.933198 = 0.555357 amu/^{59}Co = 0.555357 g/mol ^{59}Co

a) Binding energy = $\left(\dfrac{0.555357 \text{ amu}\,^{59}\text{Co}}{59 \text{ nucleons}}\right)\left(\dfrac{931.5 \text{ MeV}}{1 \text{ amu}}\right)$ = 8.768051619 = **8.768 MeV/nucleon**

b) Binding energy = $\left(\dfrac{0.555357 \text{ amu}\,^{59}\text{Co}}{1 \text{ atom}}\right)\left(\dfrac{931.5 \text{ MeV}}{1 \text{ amu}}\right)$ = 517.3150 = **517.3 MeV/atom**

c) Use $\Delta E = \Delta mc^2$

$$\text{Binding energy} = \left(\frac{0.555357 \text{ g}\,^{59}\text{Co}}{\text{mol}}\right)\left(\frac{1 \text{ kg}}{10^3 \text{ g}}\right)\left(2.99792 \times 10^8 \text{ m/s}\right)^2\left(\frac{1 \text{ J}}{\text{kg} \cdot \text{m}^2 / \text{s}^2}\right)\left(\frac{1 \text{ kJ}}{10^3 \text{ J}}\right)$$

$$= 4.9912845 \times 10^{10} = \textbf{4.99128} \times \textbf{10}^{\textbf{10}} \textbf{ kJ/mol}$$

24.85 In both radioactive decay and fission, radioactive particles are emitted, but the process leading to the emission is different. Radioactive decay is a spontaneous process in which unstable nuclei emit radioactive particles and energy. Fission occurs as the result of high-energy bombardment of nuclei with small particles that cause the nuclides to break into smaller nuclides, radioactive particles, and energy.
In a chain reaction, all fission events are not the same. The collision between the small particle emitted in the fission and the large nucleus can lead to splitting of the large nuclei in a number of ways to produce several different products.

24.88 The water serves to slow the neutrons so that they are better able to cause a fission reaction. Heavy water ($_1^2H_2O$ or D_2O) is a better moderator because it does not absorb neutrons as well as light water ($_1^1H_2O$) does, so more neutrons are available to initiate the fission process. However, D_2O does not occur naturally in great abundance, so production of D_2O adds to the cost of a heavy water reactor. In addition, if heavy water does absorb a neutron, it becomes *tritiated*, i.e., it contains the isotope tritium, $_1^3H$, which is radioactive.

24.91 In the *s*-process, a nucleus captures a neutron, emitting a γ ray. Then the nucleus emits a beta particle to form another element. The stable isotopes of most heavy elements form by the *s*-process. The *r*-process forms less stable isotopes by multiple neutron captures, followed by multiple beta decays.

24.95 a) Use the values given in the problem to calculate the mass change (reactant – products) for:

$$_{96}^{243}Cm \rightarrow \ _{94}^{239}Pu + \ _2^4He$$

$$\Delta mass \ (kg) = [243.0614 \ amu - (4.0026 + 239.0522) \ amu]\left(\frac{1.66054 \times 10^{-24} \ g}{amu}\right)\left(\frac{1 \ kg}{10^3 \ g}\right)$$

$$= 1.095956 \times 10^{-29} = \mathbf{1.1 \times 10^{-29} \ kg}$$

b) $E = \Delta mc^2 = \left(1.095956 \times 10^{-29} \ kg\right)\left(2.99792 \times 10^8 \ m/s\right)^2 \left(\dfrac{1 \ J}{kg \bullet m^2\big/s^2}\right) = 9.84993 \times 10^{-13} = \mathbf{9.8 \times 10^{-13} \ J}$

c) E released $= (9.84993 \times 10^{-13} \ J/reaction) \ (6.022 \times 10^{23} \ reactions/mol) \ (1 \ kJ \ / \ 10^3 \ J)$
$= 5.9316 \times 10^8 = \mathbf{5.9 \times 10^8 \ kJ/mol}$
This is approximately 1 million times larger than a typical heat of reaction.

24.97 Determine k for ^{14}C using the half-life (5730 yr):
$k = (\ln 2) \ / \ t_{1/2} = (\ln 2) \ / \ (5730 \ yr) = 1.2096809 \times 10^{-4} \ yr^{-1}$ (unrounded)
Determine the mass of carbon in 4.58 grams of $CaCO_3$:

$$mass = \left(4.58 \ g \ CaCO_3\right)\left(\frac{1 \ mol \ CaCO_3}{100.09 \ g \ CaCO_3}\right)\left(\frac{1 \ mol \ C}{1 \ mol \ CaCO_3}\right)\left(\frac{12.01 \ g \ C}{1 \ mol \ C}\right) = 0.5495634 \ g \ C \ (unrounded)$$

The activity is: $(3.2 \ dpm) / (0.5495634 \ g \ C) = 5.8228 \ dpm \bullet g^{-1}$
Using the integrated rate law:

$$\ln\left(\frac{N_t}{N_0}\right) = -kt \qquad N_0 = 15.3 \ dpm \ g^{-1} \ (the \ ratio \ of \ ^{12}C{:}^{14}C \ in \ living \ organisms)$$

$$\ln\left(\frac{5.8228 \ dpm \ g^{-1}}{15.3 \ dpm \ g^{-1}}\right) = -(1.2096809 \times 10^{-4} \ yr^{-1})t$$

$t = 7986.17 = \mathbf{8.0 \times 10^3 \ yr}$

24.101 Determine how many grams of AgCl are dissolved in 1 mL of solution. The activity of the radioactive Ag^+ indicates how much AgCl dissolved, given a starting sample with a specific activity (175 nCi/g).

$$Concentration = \left(\frac{1.25 \times 10^{-2} \ Bq}{mL}\right)\left(\frac{1 \ dps}{1 \ Bq}\right)\left(\frac{1 \ Ci}{3.70 \times 10^{10} \ dps}\right)\left(\frac{1 \ nCi}{10^{-9} \ Ci}\right)\left(\frac{1 \ g \ AgCl}{175 \ nCi}\right)$$

$$= 1.93050 \times 10^{-6} \ g \ AgCl/mL \ (unrounded)$$

Convert g/mL to mol/L (molar solubility) using the molar mass of AgCl.

$$Molarity = \left(\frac{1.93050 \times 10^{-6} \ g \ AgCl}{mL}\right)\left(\frac{1 \ mol \ AgCl}{143.4 \ g \ AgCl}\right)\left(\frac{1 \ mL}{10^{-3} \ L}\right)$$

$$= 1.34623 \times 10^{-5} = \mathbf{1.35 \times 10^{-5} \ M \ AgCl}$$

24.103 Determine the value of k from the half-life. Then determine the fraction from the integrated rate law.

$k = (\ln 2) / t_{1/2} = (\ln 2) / (7.0 \times 10^8 \text{ yr}) = 9.90210 \times 10^{-10} \text{ yr}^{-1}$ (unrounded)

$\ln\left(\dfrac{N_t}{N_0}\right) = -kt = (9.90210 \times 10^{-10} \text{ yr}^{-1})(2.8 \times 10^9 \text{ yr}) = -2.772588$ (unrounded)

$\left(\dfrac{N_t}{N_0}\right) = 0.062500 = \mathbf{6.2 \times 10^{-2}}$

24.106 a) Find the rate constant, k, using any two data pairs (the greater the time between the data points, the greater the reliability of the calculation). Calculate $t_{1/2}$ using k.

$\ln\left(\dfrac{N_t}{N_0}\right) = -kt$

$\ln\left(\dfrac{495 \text{ photons/s}}{5000 \text{ photons/s}}\right) = -k(20 \text{ h})$

$-2.312635 = -k(20 \text{ h})$

$k = 0.11563 \text{ h}^{-1}$ (unrounded)

$t_{1/2} = (\ln 2) / k = (\ln 2) / (0.11563 \text{ h}^{-1}) = 5.9945 = \mathbf{5.99 \text{ h}}$ (Assuming the times are exact, and the emissions have three significant figures.)

b) The percentage of isotope *remaining* is the fraction remaining after 2.0 h (N_t where t = 2.0 h) divided by the initial amount (N_0), i.e., fraction remaining is N_t / N_0. Solve the first order rate expression for N_t / N_0, and then subtract from 100% to get fraction *lost*.

$\ln\left(\dfrac{N_t}{N_0}\right) = -kt$

$\ln\left(\dfrac{N_t}{N_0}\right) = -(0.11563 \text{ h}^{-1})(2.0 \text{ h}) = -0.23126$ (unrounded)

$\left(\dfrac{N_t}{N_0}\right) = 0.793533$ (unrounded) $\qquad$ $\left(\dfrac{N_t}{N_0}\right) \times 100\% = 79.3533\%$ (unrounded)

The fraction lost is $100\% - 79.3533\% = 20.6467\% = \mathbf{21\%}$ of the isotope is lost upon preparation.

24.108 a) fraction remaining after 10.0 yr $= \left(\dfrac{1}{2}\right)^{t/t_{1/2}} = \left(\dfrac{1}{2}\right)^{10.0/5730} = 0.998791 = \mathbf{0.999}$

b) fraction remaining after 10.0×10^3 yr $= \left(\dfrac{1}{2}\right)^{10.0 \times 10^3 / 5730} = 0.298292 = \mathbf{0.298}$

c) fraction remaining after 10.0×10^4 yr $= \left(\dfrac{1}{2}\right)^{10.0 \times 10^4 / 5730} = 5.5772795 \times 10^{-6} = \mathbf{5.58 \times 10^{-6}}$

d) Radiocarbon dating is more reliable for (b) because a significant quantity of ^{14}C has decayed and a significant quantity remains. Therefore, a change in the amount of ^{14}C would be noticeable. For the fraction in (a), very little ^{14}C has decayed and for (c) very little ^{14}C remains. In either case, it will be more difficult to measure the change so the error will be relatively large.

24.110 At 1 half-life, the fraction of sample is 0.500.

$(0.900)^n = 0.500$

$n \ln(0.900) = \ln(0.500)$

$n = (\ln 0.500) / (\ln 0.900) = 6.578813 = \mathbf{6.58 \text{ h}}$

24.114 The *production rate* of radon gas (volume/hour) is also the *decay rate* of ^{226}Ra. The decay rate, or activity, is proportional to the number of radioactive nuclei decaying, or the number of atoms in 1.000 g of ^{226}Ra, using the relationship A = kN. Calculate the number of atoms in the sample, and find k from the half-life. Convert the activity in units of nuclei/time (also disintegrations per unit time) to volume/time using the ideal gas law.

$$^{226}_{88}Ra \rightarrow {}^{4}_{2}He + {}^{222}_{86}Rn$$

$k = (\ln 2) / t_{1/2} = (\ln 2) / [(1599 \text{ yr}) (8766 \text{ h/yr})] = 4.9451051 \times 10^{-8} \text{ h}^{-1}$ (unrounded)

The mass of ^{226}Ra is 226.025402 amu/atom or 226.025402 g/mol.

$$N = (1.000 \text{ g Ra})\left(\frac{1 \text{ mol Ra}}{226.025402 \text{ g Ra}}\right)\left(\frac{6.022 \times 10^{23} \text{ Ra atoms}}{1 \text{ mol Ra}}\right) = 2.6643023 \times 10^{21} \text{ Ra atoms (unrounded)}$$

$A = kN = (4.9451051 \times 10^{-8} \text{ h}^{-1}) (2.6643023 \times 10^{21} \text{ Ra atoms}) = 1.3175254 \times 10^{14} \text{ Ra atoms/h}$ (unrounded)

This result means that 1.318×10^{14} ^{226}Ra nuclei are decaying into ^{222}Rn nuclei every hour. Convert atoms of ^{222}Rn into volume of gas using the ideal gas law.

moles = $(1.3175254 \times 10^{14} \text{ Ra atoms} / \text{h}) (1 \text{ atom Rn} / 1 \text{ atom Ra}) (1 \text{ mol Rn} / 6.022 \times 10^{23} \text{ Rn})$
$= 2.1878536 \times 10^{-10} \text{ mol Rn} / \text{h}$ (unrounded)

$$V = nRT / P = \frac{\left(2.1878536 \times 10^{-10} \text{ mol Rn/h}\right)\left(0.08206 \dfrac{\text{L} \cdot \text{atm}}{\text{mol} \cdot \text{K}}\right)(273.15 \text{ K})}{1 \text{ atm}}$$

$= 4.904006 \times 10^{-9} = \textbf{4.904} \times \textbf{10}^{-9} \textbf{ L/h}$

Therefore, radon gas is produced at a rate of 4.904×10^{-9} L/h. Note: Activity could have been calculated as decay in moles/time, removing Avogadro's number as a multiplication and division factor in the calculation.

24.117 Determine k from the half-life

$k = (\ln 2) / t_{1/2} = (\ln 2) / (29 \text{ yr}) = 0.023902 \text{ yr}^{-1}$ (unrounded)

$$\ln\left(\frac{N_t}{N_0}\right) = -kt$$

$$\ln\left(\frac{1.0 \times 10^4 \text{ particles}}{7.0 \times 10^4 \text{ particles}}\right) = -(0.023902 \text{ yr}^{-1})t$$

$-1.945910 = -(0.023902 \text{ yr}^{-1})t$

$t = 81.412022 = \textbf{81 yr}$

24.119 a) Convert pCi to Bq using the conversion factors 1 Ci = 3.70×10^{10} Bq and 1 pCi = 10^{-12} Ci.

$$\text{Activity} = \left(\frac{4.0 \text{ pCi}}{\text{L}}\right)\left(\frac{10^{-12} \text{ Ci}}{1 \text{ pCi}}\right)\left(\frac{3.70 \times 10^{10} \text{ Bq}}{1 \text{ Ci}}\right) = 0.148 = 0.15 \text{ Bq/L}$$

The safe level is **0.15 Bq/L**.

b) Use the first-order rate expression to find the activity at the later time (t = 9.5 days).

$k = (\ln 2) / 3.82 \text{ days} = 0.181452 \text{ days}^{-1}$ (unrounded)

$$\ln\left(\frac{N_t}{N_0}\right) = -kt$$

$$\ln\left(\frac{N_t}{41.5 \text{ pCi/L}}\right) = -(0.181452 \text{ days}^{-1}) (9.5 \text{ days}) = -1.723794 \text{ (unrounded)}$$

$$\left(\frac{N_t}{41.5 \text{ pCi/L}}\right) = 0.17838806 \text{ (unrounded)}$$

$N_t = 7.403104$ pCi/L

$$\text{Activity} = \left(\frac{7.403104 \text{ pCi}}{\text{L}}\right)\left(\frac{10^{-12} \text{ Ci}}{1 \text{ pCi}}\right)\left(\frac{3.70 \times 10^{10} \text{ Bq}}{1 \text{ Ci}}\right) = 0.2739148 = \textbf{0.27 Bq/L}$$

c) The desired activity is 0.15 Bq/L, however, the room air currently contains 0.27 Bq/L. Rearrange the first-order rate expression to solve for time.

$$\ln\left(\frac{N_t}{N_0}\right) = -kt$$

$$\ln\left(\frac{0.15\ \text{Bq/L}}{0.2739148\ \text{Bq/L}}\right) = -(0.181452\ \text{days}^{-1})t$$

$$-0.602182 = -(0.181452\ \text{days}^{-1})t$$

$$t = 3.318685 = \textbf{3.3 days}$$

It takes 3.3 days more to reach the recommended EPA level. A total of 12.8 (3.3 + 9.5) days is required to reach recommended levels when the room was initially measured at 41.5 pCi/L.

24.122   The energy per time is calculated from the disintegrations per time (1.0 mCi) and the energy per disintegration (5.59 MeV).

$$(1.0\ \text{mCi})\left(\frac{10^{-3}\ \text{Ci}}{1\ \text{mCi}}\right)\left(\frac{3.70\times10^{10}\ \text{dps}}{1\ \text{Ci}}\right)\left(\frac{5.59\ \text{MeV}}{1\ \text{disint.}}\right)\left(\frac{1.602\times10^{-13}\ \text{J}}{1\ \text{MeV}}\right) = 3.3134166\times10^{-5}\ \text{J/s (unrounded)}$$

Convert the mass of the child to kg:

$$(54\ \text{lb})\left(\frac{1\ \text{kg}}{2.205\ \text{lb}}\right) = 24.4897959\ \text{kg (unrounded)}$$

Time for 1.0 mrad to be absorbed:

$$\text{Time} = (1.0\ \text{mrad})\left(\frac{10^{-3}\ \text{rad}}{1\ \text{mrad}}\right)\left(\frac{0.01\ \text{J/kg}}{1\ \text{rad}}\right)\left(\frac{24.4897959\ \text{kg}}{3.3134166\times10^{-5}\ \text{J/s}}\right) = 7.3911007 = \textbf{7.4 s}$$

24.125   Because the 1941 wine has a little over twice as much tritium in it, just over one half-life has passed between the two wines. Therefore, the older wine was produced before 1929 (1941– 12.26) but not much earlier than that. To find the number of years back in time, use the first order rate expression, where $N_0 = 2.32\ N$, $N_t = N$ and t = years transpired between the manufacture date and 1941.

$$k = (\ln 2)\ /\ t_{1/2} = (\ln 2)\ /\ 12.26\ \text{yr} = 0.05653729\ \text{yr}^{-1}\ \text{(unrounded)}$$

$$\ln\left(\frac{N_t}{N_0}\right) = -kt$$

$$\ln\left(\frac{N}{2.32\ N}\right) = -(0.05653729\ \text{yr}^{-1})t$$

$$-0.841567 = -(0.05653729\ \text{yr}^{-1})t$$

$$t = 14.8851669 = 14.9\ \text{yr}$$

The wine was produced in (1941 – 15) = **1926**.

24.128   Determine the mass defect for the reaction:

$$\Delta m = \text{mass of reactants - mass of products}$$
$$= (14.003074 + 1.008665) - (14.003241 + 1.007825)$$
$$= 15.011739 - 15.011066 = 0.000673\ \text{amu}$$

$$\text{Energy} = (0.000673\ \text{amu})\left(\frac{931.5\ \text{MeV}}{1\ \text{amu}}\right)\left(\frac{10^6\ \text{eV}}{1\ \text{MeV}}\right) = 6.268995\times10^5 = \textbf{6.27}\times\textbf{10}^{\textbf{5}}\ \textbf{eV}$$

$$\text{Energy} = (6.268995\times10^5\ \text{eV})\left(\frac{1.602\times10^{-19}\ \text{J}}{1\ \text{eV}}\right)\left(\frac{1\ \text{kJ}}{10^3\ \text{J}}\right)\left(\frac{6.022\times10^{23}}{1\ \text{mol}}\right) = 6.0478524\times10^7 = \textbf{6.05}\times\textbf{10}^{\textbf{7}}\ \textbf{kJ/mol}$$

24.131 a) Kinetic energy $= 1/2\, mv^2 = (1/2)\, m\, (3\, RT\, /\, M_H) = (3/2)\, (RT\, /\, N_A)$

$$\text{Energy} = \left(\frac{3}{2}\right)\frac{(8.314\ \text{J/mol} \cdot \text{K})(1.00 \times 10^6\,\text{K})}{6.022 \times 10^{23}\ \text{atom/mol}} = 2.0709066 \times 10^{-17} = \mathbf{2.07 \times 10^{-17}\ J/atom}\ {}_1^1\text{H}$$

b) A kilogram of ^{1}H will annihilate a kilogram of anti–H; thus, two kilograms will be converted to energy:
Energy $= mc^2 = (2.00\ \text{kg})(2.99792 \times 10^8\ \text{m/s})^2(\text{J}/(\text{kg}\cdot\text{m}^2/\text{s}^2) = 1.7975048 \times 10^{17}\ \text{J}$ (unrounded)

$$\text{H atoms} = \left(\frac{1.7975048 \times 10^{17}\ \text{J}}{1\ \text{kg H}}\right)\left(\frac{1\ \text{kg}}{10^3\ \text{g}}\right)\left(\frac{1.0078\ \text{g H}}{1\ \text{mol H}}\right)\left(\frac{1\ \text{mol H}}{6.022 \times 10^{23}\ \text{atoms H}}\right)\left(\frac{\text{H atoms}}{2.0709066 \times 10^{-17}\ \text{J}}\right)$$

$$= 1.4525903 \times 10^7 = \mathbf{1.45 \times 10^7\ H\ atoms}$$

c) $4\,{}_1^1\text{H} \rightarrow {}_2^4\text{He} + 2\,{}_1^0\beta$ (Positrons have the same mass as electrons.)

$\Delta m = [4(1.007825\ \text{amu})] - [4.00260\ \text{amu} - 2(0.000549\ \text{amu})] = 0.027602\ \text{amu}\ /\ {}_2^4\text{He}$

$$\Delta m = \left(\frac{0.027602\ \text{g}}{\text{mol He}}\right)\left(\frac{1\ \text{kg}}{10^3\ \text{g}}\right)\left(\frac{1\ \text{mol He}}{4\ \text{mol H}}\right)\left(\frac{1\ \text{mol H}}{6.022 \times 10^{23}\ \text{H atoms}}\right)\left(1.4525903 \times 10^7\ \text{H atoms}\right)$$

$$= 1.6644967 \times 10^{-22}\ \text{kg (unrounded)}$$

Energy $= (\Delta m)c^2 = (1.6644967 \times 10^{-22}\ \text{kg})(2.99792 \times 10^8\ \text{m/s})^2(\text{J}/(\text{kg}\cdot\text{m}^2/\text{s}^2)$
$$= 1.4959705 \times 10^{-5} = \mathbf{1.4960 \times 10^{-5}\ J}$$

d) Calculate the energy generated in part (b):

$$\text{Energy} = \left(\frac{1.7975048 \times 10^{17}\ \text{J}}{1\ \text{kg H}}\right)\left(\frac{1\ \text{kg}}{10^3\ \text{g}}\right)\left(\frac{1.0078\ \text{g H}}{1\ \text{mol H}}\right)\left(\frac{1\ \text{mol H}}{6.022 \times 10^{23}\ \text{atoms H}}\right) = 3.0081789 \times 10^{-10}\ \text{J}$$

Energy increase $= (1.4959705 \times 10^{-5} - 3.0081789 \times 10^{-10})\ \text{J} = 1.4959404 \times 10^{-5} = \mathbf{1.4959 \times 10^{-5}\ J}$

e) $3\,{}_1^1\text{H} \rightarrow {}_2^3\text{He} + 1\,{}_1^0\beta$

$\Delta m = [3(1.007825\ \text{amu})] - [3.01603\ \text{amu} + 0.000549\ \text{amu}]$
$$= 0.006896\ \text{amu}\ /\ {}_2^3\text{He} = 0.006896\ \text{g/mol}\ {}_2^3\text{He}$$

$$\Delta m = \left(\frac{0.006896\ \text{g}}{\text{mol He}}\right)\left(\frac{1\ \text{kg}}{10^3\ \text{g}}\right)\left(\frac{1\ \text{mol He}}{3\ \text{mol H}}\right)\left(\frac{1\ \text{mol H}}{6.022 \times 10^{23}\ \text{H atoms}}\right)\left(1.4525903 \times 10^7\ \text{H atoms}\right)$$

$$= 5.5447042 \times 10^{-23}\ \text{kg (unrounded)}$$

Energy $= (\Delta m)c^2 = (5.5447042 \times 10^{-23}\ \text{kg})(2.99792 \times 10^8\ \text{m/s})^2(\text{J}/(\text{kg}\cdot\text{m}^2/\text{s}^2)$
$$= 4.9833164 \times 10^{-6} = \mathbf{4.983 \times 10^{-6}\ J}$$

No, the Chief Engineer should advise the Captain to keep the current technology.

24.134 The difference in the energies of the two α particles gives the energy of the γ-ray released to get from excited state I to the ground state. The energy of the 2% α particle is equal to the highest energy α particle minus the energy of the two γ-rays (the rays are from excited state II to excited state I, and from excited state I to the ground state).

a) $(4.816 - 4.773)\ \text{MeV} = 0.043\ \text{MeV} = $ energy of γ-ray. Use $E = hc\,/\,\lambda$ to determine the wavelength.

$\lambda = hc\,/\,E$

$$\lambda = \frac{(6.626 \times 10^{-34}\ \text{J} \cdot \text{s})(2.99792 \times 10^8\ \text{m/s})}{(0.043\ \text{MeV})}\left(\frac{1\ \text{MeV}}{1.602 \times 10^{-13}\ \text{J}}\right) = 2.8836 \times 10^{-11} = 2.9 \times 10^{-11}\ \text{m}$$

b) $(4.816 - 0.043 - 0.060)\ \text{MeV} = \mathbf{4.713\ MeV}$

24.138 Multiply each of the half lives by 20 (the number of half lives is considered to be exact).
a) ^{242}Cm   20 (163 days) $= \mathbf{3.26 \times 10^3\ days}$
b) ^{214}Po   20 (1.6 $\times 10^{-4}$ s) $= \mathbf{3.2 \times 10^{-3}\ s}$
c) ^{232}Th   20 (1.39 $\times 10^{10}$ yr) $= \mathbf{2.78 \times 10^{11}\ yr}$

24.140 a) When 1.00 kg of antimatter annihilates 1.00 kg of matter, the change in mass is
$\Delta m = 0 - 2.00\ \text{kg} = -2.00\ \text{kg}$. The energy released is calculated from $\Delta E = \Delta mc^2$.
$\Delta E = (-2.00\ \text{kg})(2.99792 \times 10^8\ \text{m/s})^2(\text{J}/(\text{kg}\cdot\text{m}^2/\text{s}^2)) = -1.7975048 \times 10^{17} = \mathbf{-1.80 \times 10^{17}\ J}$
The negative value indicates the energy is released.

b) Assuming that four hydrogen atoms fuse to form the two protons and two neutrons in one helium atom and release two positrons, the energy released can be calculated from the binding energy of helium-4.

$$4\,^1_1\text{H} \rightarrow \,^4_2\text{He} + 2\,^0_{+1}\beta$$

$$\Delta m = [4(1.007825 \text{ amu})] - [4.00260 \text{ amu} + 2(0.000549 \text{ amu})]$$
$$= 0.02760 \text{ amu per } ^4_2\text{He formed}$$

$$\text{total } \Delta m = \left(1.00 \times 10^5 \text{ H atoms}\right)\left(\frac{0.02760 \text{ amu}}{^4\text{He}}\right)\left(\frac{1\,^4\text{He}}{4 \text{ H atoms}}\right) = 6.90 \times 10^2 \text{ amu}$$

$$\text{Energy} = \left(6.90 \times 10^2 \text{ amu}\right)\left(\frac{931.5 \text{ MeV}}{1 \text{ amu}}\right)\left(\frac{1.602 \times 10^{-16} \text{ kJ}}{1 \text{ MeV}}\right)$$
$$= 1.02966 \times 10^{-10} \text{ kJ per anti-H collision (unrounded)}$$

$$\text{Anti-H atoms} = \left(1.00 \text{ kg}\right)\left(\frac{10^3 \text{ g}}{1 \text{ kg}}\right)\left(\frac{1 \text{ mol anti} - \text{H}}{1.008 \text{ g anti} - \text{H}}\right)\left(\frac{6.022 \times 10^{23} \text{ anti} - \text{H}}{1 \text{ mol anti} - \text{H}}\right)$$
$$= 5.9742 \times 10^{26} \text{ anti-H (unrounded)}$$

$$\text{Energy released} = \left(\frac{1.02966 \times 10^{-10} \text{ kJ}}{\text{anti} - \text{H}}\right)\left(5.9742 \times 10^{26} \text{ anti} - \text{H}\right) = 6.15139 \times 10^{16} = \textbf{6.15} \times \textbf{10}^{\textbf{16}} \textbf{ kJ}$$

c) From the above calculations, the procedure in part b) with excess hydrogen produces more energy per kilogram of antihydrogen.

24.141 Einstein's equation is $E = mc^2$, which is modified to $E = \Delta mc^2$ to reflect a mass defect. The speed of light, c, is 2.99792×10^8 m/s. The mass of exactly one amu is 1.66054×10^{-27} kg (inside back cover). When the quantities are multiplied together, the unit will be kg•m^2/s^2, which is also the unit of joules. Convert J to MeV using the conversion factor 1.602×10^{-13} J = 1 MeV.

$$\Delta E = (\Delta m)c^2$$

$$\Delta E = \left[\left(1 \text{ amu}\right)\left(\frac{1.66054 \times 10^{-27} \text{ kg}}{1 \text{ amu}}\right)\right]\left(2.99792 \times 10^8 \text{ m/s}\right)^2\left(\frac{\text{J}}{\text{kg} \bullet \text{m}^2/\text{s}^2}\right)\left(\frac{1 \text{ MeV}}{1.602 \times 10^{-13} \text{ J}}\right)$$

$$= 9.3159448 \times 10^2 = \textbf{9.316} \times \textbf{10}^{\textbf{2}} \textbf{ MeV}$$

24.145 The rate of formation of plutonium-239 depends on the rate of decay of neptunium-239 with a half-life of 2.35 days.

$$k = (\ln 2) / t_{1/2} = (\ln 2) / 2.35 \text{ days} = 0.294956 \text{ day}^{-1} \text{ (unrounded)}$$

$$\ln\left(\frac{N_t}{N_0}\right) = -kt$$

$$\ln\left(\frac{\left(1.00 \text{ kg}\right)\left(90.\%/100\%\right)}{1.00 \text{ kg}}\right) = -\left(0.294956 \text{ day}^{-1}\right)t$$

$$-0.105360515 = -\left(0.294956 \text{ day}^{-1}\right)t$$
$$t = 0.357207568 = \textbf{0.357 days}$$

Notes

Notes

Notes

Notes